High Altitude Medicinal Plants

Suman Chandra • Hemant Lata
Editors

High Altitude Medicinal Plants

Botany, Conservation and Cultivation

Springer

Editors
Suman Chandra
School of Pharmacy
The University of Mississippi
University, MS, USA

Hemant Lata
School of Pharmacy
The University of Mississippi
University, MS, USA

ISBN 978-3-031-99405-0 ISBN 978-3-031-99406-7 (eBook)
https://doi.org/10.1007/978-3-031-99406-7

Cover Image: Sinopodophyllum hexandrum (Berberidaceae). Photo credit: Dr. Yang Niu and Dr. Hang Sun (authors of Chapter 25)

This Springer imprint is published by the registered company Springer Nature Switzerland AG
The registered company address is: Gewerbestrasse 11, 6330 Cham, Switzerland

To our parents and our children,
Rishi and Riddhi

Foreword

In the pursuit of health and well-being, humans have always looked to nature for guidance and healing. For millennia, people living in mountainous regions have relied on the unique medicinal plants that thrive in these challenging environments. Some of these plants, often growing at the edges of the world's tallest mountains, have evolved remarkable resilience and therapeutic properties, making them indispensable to the indigenous peoples who have passed down their knowledge of their uses through generations. As science advances, we are increasingly recognizing the value of these plants—not just as folklore, but as potent sources of modern-day medicine.

This book, *High Altitude Medicinal Plants: Botany, Conservation and Cultivation*, is a vital resource that highlights the significance of these unique group of plants and the urgent need to protect them. The authors offer a thorough and insightful exploration of the scientific, ecological, and cultural dimensions of high-altitude medicinal plants. Through detailed chapters on botany, conservation challenges, and cultivation methods, this book provides a comprehensive understanding of these plants' immense potential and their vulnerability in the face of environmental change and human activity.

The importance of high-altitude plants cannot be overstated. Many of these species are not only critical to local health systems but are also sources of valuable compounds that may hold the key to developing future medicines. Yet, their survival is threatened by climate change, overharvesting, and the degradation of their fragile habitats. This book serves as a timely call to action, urging us to adopt responsible practices that ensure the continued availability of these plants for future generations.

In addition to presenting the scientific facts, the authors celebrate the deep cultural ties between high-altitude communities and these plants. Traditional knowledge and modern science come together in this volume, underscoring the importance of an integrated approach to conservation and cultivation. By combining indigenous wisdom with cutting-edge research, we are better equipped to address the pressing environmental challenges that threaten these plants.

This book is more than just a collection of botanical data; it is a valuable resource for anyone interested in the intersection of nature, medicine, and sustainability. It provides essential information for researchers, conservationists, herbalists, practitioners, academicians, and students alike, offering practical guidance for the cultivation and preservation of these incredible plants. Through this work, we can foster a deeper appreciation for the importance of high-altitude plants and ensure that their legacy is preserved for the benefit of both our health and the health of the planet.

I am confident that this book will inspire readers to engage more thoughtfully with the natural world, to recognize the incredible value of the plants from mountainous regions, and to take action in support of their conservation and sustainable use. It is a vital contribution to our growing understanding of the interconnectedness of life on Earth and our responsibility to safeguard it. I congratulate the editors and authors for their endeavors and wish the book all the success.

Pitamber Prasad Dhyani

Director (Ret.), GB Pant National Institute
of Himalayan Environment (An Autonomous
Institute of Ministry of Environment, Forest
& Climate Change, Government of India)
Almora, Uttarakhand, India
Former Vice-Chancellor of Shri Guru Ram Rai University
Dehradun, Uttarakhand, India
Former Vice-Chancellor of Shri Dev Suman Uttarakhand University
Tehri, Uttarakhand, India
Former Vice-Chancellor of VMSV Uttarakhand Technical University
Dehradun, India
Former Vice-Chancellor of Haridwar University
Roorkee, Uttarakhand, India

Preface

Medicinal plants have played a vital role in the development of modern drugs, serving as a rich source of bioactive compounds with therapeutic potential. For centuries, traditional medicine systems around the world have relied on plants to treat a wide range of ailments, and many of these natural remedies have led to the discovery of important pharmaceuticals. For example, compounds like quinine, morphine, and artemisinin were all derived from plants and have had a profound impact on medicine. In modern drug development, scientists isolate, analyze, and sometimes modify plant-derived compounds to create effective and safer medications. As research continues, medicinal plants remain a promising frontier for discovering novel drugs, especially in the face of rising antibiotic resistance and emerging diseases.

Plants found in high mountainous regions such as the Alps, Himalayas, and Rockies are highly valued for their unique bioactive compounds adapted to harsh environmental conditions. These plants are often used in traditional healing systems like Ayurveda, Tibetan, and Chinese medicine. However, they are becoming increasingly rare or endangered due to a combination of overharvesting, climate change, habitat degradation, and unsustainable collection practices. Rising demand for natural remedies has led to excessive harvesting without allowing time for regeneration, while warming temperatures and shifting weather patterns threaten their fragile alpine habitats. Conservation efforts and sustainable harvesting practices are urgently needed to protect these rare and valuable plant species for future generations.

Having spent many momentous years of our lives in the Himalayas, we, as editors, see this book not only as a scholarly work rather as a heartfelt tribute. The mountains have been our mentors, their rugged terrains reflecting both the challenges and rewards of our journey. Our bond with these high-altitude landscapes runs deep, shaped by years of exploration and research—ranging from the laboratories of the High Altitude Plant Physiology Research Center in Srinagar, Garhwal (550 m above mean sea level), to its Alpine Research Station at Tungnath (3600 m above mean sea level) at HNB Garhwal University, and the GB Pant National Institute of Himalayan Environment in Almora. During our PhD and postdoctoral studies, we ventured into remote areas to explore rare and endangered medicinal plants—not just as researchers but as humble students of nature. This book emerges from our shared respect for these herbs and mountains, and from the hope that their legacy will continue to flourish—inspiring conservation, sustainable use, and deep respect for the wisdom nestled in high mountain regions.

The purpose of this book, *High Altitude Medicinal Plants: Botany, Conservation and Cultivation,* is to serve in a single volume the comprehensive knowledge and experiences of renowned researchers and scientists in the field of high altitude medicinal plants research. Each chapter is independently written by experts in their field of endeavor. The book comprises 25 chapters, beginning with an introductory chapter titled "The Mountain Environment, Its Biodiversity and Home of Rare Plant Species" by Prof. (Dr.) Christine Körner. We had many insightful and detailed email exchanges with him regarding the best way to describe the vertical dimension—whether as elevation, altitude, or height—in ecological, climatological, and broader geophysical studies of the Earth. The book is broadly categorized in sections that briefly talk about the physiological adaptations of alpine plants to varying altitudes and climatic conditions; their ethnobotanical and traditional uses; phytochemistry and bioactive compounds with potential for phytopharmaceutical applications; conservation strategies, both short- and long-term, including micropropagation and cryopreservation; cultivation practices; issues related to uncontrolled wild harvesting, trade, and marketing; their role in mountain livelihoods and socioeconomic development; and future strategies for sustainable use.

Editing this book has been a genuine pleasure, largely thanks to the exceptional collaboration of the contributors, their adherence to deadlines, and the richness of the material they provided. We extend our sincere gratitude and heartfelt thanks to each author for their generous dedication of time and effort. We also wish to express our sincere gratitude to Prof. (Dr.) Mahmoud A. ElSohly at The University of Mississippi for his mentorship and to Dr. Christina Eckey, Dr. Eliana G. Acosta, and Ramya Ravichandran at Springer Nature for their patience and invaluable support.

University, MS, USA

Suman Chandra
Hemant Lata

Contents

About the Editors

Suman Chandra, PhD earned his MS degree in Physics, MPhil in Environmental Plant Physiology, and PhD in Plant Physiology from the High Altitude Plant Physiology Research Center at HNB Garhwal University, India. He worked at the High Altitude Plant Physiology Research Center and GB Pant National Institute of Himalayan Environment in India between 1992 and 2000. In 2001, he joined the University of Mississippi, where he currently holds the positions of Principal Scientist at the National Center for Natural Products Research and Research Professor of BioMolecular Sciences at the School of Pharmacy. With over 30 years of experience in medicinal plant research, his areas of interest include conservation, cultivation, and enhancement of medicinal plants, as well as studying the effects of climate change on the physiology and secondary metabolites of medicinally important plants. He has authored more than 100 publications, delivered over 100 presentations, and has 2 books to his credit.

Hemant Lata, PhD earned her PhD in Plant Physiology from the High Altitude Plant Physiology Research Center at HNB Garhwal University, India. She received her MS in Biotechnology from Guru Nanak Dev University, India, and BS in Botany (Hons.) from the University of Delhi, India. She completed her Post-Doctoral training at the Laboratory of Cellular and Molecular Cytogenetics in the Department of Botany at the University of Delhi. In 2000, she joined The University of Mississippi as a Post-Doctoral Research Associate and was promoted to a research faculty position in December 2003. Currently, she serves as a Principal Scientist at the National Center for Natural Products Research and Research Professor in

BioMolecular Sciences at the School of Pharmacy. She has developed and published micropropagation protocols for several high-value medicinal plants. With over 30 years of experience in plant biotechnology, her research focus on screening, micropropagation, conservation, and multiplication of medicinal plants. She has authored more than 100 publications and edited 2 books on medicinally important plants.

The Mountain Environment, Its Biodiversity, and Home of Rare Plant Species

1

Christian Körner

Abstract

In this chapter, I will summarize a number of features of mountain ecosystems that apply globally. After providing a brief statement on mountain definitions and some mountain statistics, I will recall the major climatic and atmospheric drivers and how their action is modified by geomorphology and type of vegetation. Mountains host a much higher diversity of plant species than expected from land area and the demanding life conditions. Many of the species are regional endemics, and land use, including plant collection for medical applications, is likely to threaten mountain biodiversity more than climatic change. Mountain biomes, the ones above the climatic treeline in particular, are less vulnerable to climatic change than most other biomes. It is the human population in mountains, their infrastructure, and well-being that is vulnerable to climatic change, especially when hydrological regime changes take place.

This introductory chapter is a combination of a revised reprint of a text I had provided to the *Nepalese Journal of Zoology* at the occasion of their annual assembly in Kathmandu in 2021 (Körner 2022), ideas I collected for a review on mountain biodiversity in Körner (2004), and a number of specific aspects of plant life at high elevation that apply to the topic of this volume. I hope these lines will provide a

This chapter includes text excerpts from the following publication: Körner C (2022) A global framework of mountain ecology. Nepal J Zool 6(S1): 42–44. The original publication is accessible at https://doi.org/10.3126/njz.v6iS1.xxxxx.

C. Körner (✉)
Department of Environmental Sciences, University of Basel, Basel, Switzerland
e-mail: ch.koerner@unibas.ch

S. Chandra, H. Lata (eds.), *High Altitude Medicinal Plants*,
https://doi.org/10.1007/978-3-031-99406-7_1

framework for the environmental and biodiversity conditions medicinal plants face in mountains. My focus will be on open land, where plants are directly exposed to the environmental conditions, in contrast to more sheltered life conditions within montane forests.

1.1 Mountain Definitions

Mountains are geomorphological features resulting from plate tectonics (including volcanism). They can neither be defined by elevation, given the many elevated table lands, nor by a certain climate, given the latitudinal and elevational climatic gradients. The unifying principle is topographic roughness often addressed as ruggedness. Which degree of ruggedness one considers mountainous is a matter of scale and common sense, and cannot, in a strict sense, be defined from first principle. A widely held approximation that meets most people's perception is a minimum vertical amplitude of somewhere between 150 and 300 m of elevation within an area of around 2 × 2 km, commonly obtained from nine adjacent grid points in a 30 arcsec geographical grid. One can attribute the ruggedness obtained from such a core area to variable sizes of surrounding land, causing the terrain globally considered to be mountainous to vary between 12.5% and 30% (Körner et al. 2021). With the narrower concept preferred here, using a ruggedness threshold of 200 m projected onto a 21 km^2 reference window at the equator, 16.5 Mio km^2 of land outside Antarctica is mountainous. This definition includes c. 2 km of non-rugged foreland and all valleys less than 4 km wide. Statistically, this land area is home to c. 0.5 billion people, three-quarters of which are living in urban areas close to mountains (Ehrlich et al. 2021) and one-quarter are actually living on mountains, mostly at lower, warmer elevations (Körner et al. 2017). So, Kathmandu, La Paz, or Innsbruck are clearly mountain cities. Some of the wider mountain definitions include, for instance, Hong Kong and other megacities, because they have some hills on their outskirts, and the reference area is so large, and thus inclusive for such mountain features. This is to be kept in mind when judging the various statistics published in the literature (Körner et al. 2021). 21.5% of that 16.5 Mio km^2 land is above the climatic treeline (by definition alpine) and 78.5% below (montane). The alpine fraction corresponds to 2.6% of the global land area outside Antarctica but hosts an over-proportional fraction of global biodiversity (Körner 2004). In terms of terminology, elevated land should be addressed by elevation, while altitude applies to the free atmosphere only (McVicar and Körner 2013), something non-native speakers like myself had long ignored.

1.2 Principles of Mountain Climate

There are only four climatic factors that change with altitude (in the free atmosphere) in all mountains: (1) barometric pressure (and with it, partial pressures of oxygen and CO_2) declines by about 10% per 1000 m. (2) Due to the expansion of the atmosphere, air temperature declines by c. 0.55 K per 100 m (the use of the Kelvin is encouraged for differences, irrespective of the °C scale for temperature, to avoid confusion between the two), but this "laps rate" of air temperature varies with season and differs between means and extremes of temperature (Kollas et al. 2014). (3) Clear sky turbidity diminishes, hence clear sky solar radiation increases with altitude, and (4) the fraction of UV-B radiation of any given total solar radiation increases with altitude (Körner 2007). All other changes reflect local peculiarities. While the atmosphere can hold less vapor, the cooler the air (saturation vapor pressure declines with altitude), the actually effective humidity (best expressed as saturation vapor pressure deficit, vpd) depends on local circulation, fog, and land surface temperature. Depending on cloudiness, solar radiation that reaches the land surface may increase or decrease with elevation. There are also no rules for precipitation, although many temperate mountains receive increasing precipitation with elevation, there may be a sharp decrease above the condensation layer in subtropical and tropical mountains. These patterns may change from front range to inner ranges, as is well exemplified by data for Pakistan (Körner 2007). While summits can be quite windy, the mountains of the world belong to the least windy environments because their roughness slows atmospheric circulation, and inner mountain valleys belong to the calmest regions globally. Isotherms, that is, lines of a given air temperature, rise in elevation from front to inner ranges, called the "mass elevation effect" because this effect is larger, the greater the mass of a mountain range is (for a summary, see Körner 2021a. The important message for ecologists is: do not use elevation as a surrogate for change in temperature or other atmospheric factors (except for the four listed above) but measure and apply the actual environmental conditions. These may deviate substantially from what weather stations report (Körner and Hiltbrunner 2018).

There is one common geomorphological feature that applies to all mountain ranges: the land area becomes smaller with elevation. While globally 5.1% of all land area is above 2000 m, 2.7% are above 3000 m, 1.7% above 4000 m, and only 0.5% above 5000 m (Körner 2007). This is an ecologically important point because biodiversity correlates with available land area. As one moves upslope, elevational belts occupy less space. As the climate warms, upslope-moving biota risk some crowding. To some extent, this effect becomes moderated by the microclimate effects of micro-topography, discussed in the next paragraph.

1.3 Climate–Topography Interaction and the Role of Plant Size for Micro-climate

The above climatic trends apply to the free atmosphere. The climate actually experienced by plants, animals, and microbes cannot be approximated by these trends, except for trees. By their upright architecture that evolved for outcompeting neighbors, trees are aerodynamically well coupled to the free atmosphere and experience, in a first approximation, a climate as it is measured by a nearby weather station. This is also the reason for treeline formation, often exhibiting patterns like the shore line of a hydrolelectic reservoir, reflecting the thermal layering of the atmosphere. The treeline is often fragmented by disturbances or trees may be absent from the treeline due to land use, though (Körner 2021b; Körner and Hoch 2023) Indeed, the global treeline phenomenon follows a common isotherm of air temperature (Paulsen and Körner 2014), with a seasonal mean near 6 °C the best approximation. As the vegetation above treeline becomes shorter and shorter, the thermal decoupling from the free atmosphere becomes more pronounced. Hence, by reducing their size, alpine plants trap solar heat (by slowing heat convection) and arrive at much warmer daytime temperature than treeline trees, not dissimilar from temperatures in plants growing at much lower elevation. Being "small by design," the high-elevation vegetation above treeline and all other small organisms associated with them, is "engineering" a micro-environment that is so different from the free atmosphere that weather station data become useless (Scherrer and Körner 2011). These effects become accelerated by topography, that is, exposure to the sun (slope and compass direction) and by shelter (microtopographic roughness). Topography also influences moisture and snow distribution; the latter is a key driver of species distribution at higher latitude mountains. Snow pack protects against freezing but shortens the season in so-called snowbed habitats. The message for alpine ecologists: obtain your own micro-environmental data. While modern, cheap, and reliable instrumentation makes this easier, the handling of such tools requires training in the basics of microclimatology. For instance, sensors must never be exposed to full sunshine, and nonprofessional shelters and screens can make things worse (Körner and Hiltbrunner 2018).

Above the treeline, topography becomes a dominant factor of life. Life conditions may change over a few meters of horizontal distance by more than they do for trees across 1000 m of elevation (Scherrer and Körner 2011; Körner 2021a). These mosaics of life conditions are not tied to elevation and associated isotherms of air temperature, but to micro-topography. A sheltered niche on an equator-facing slope at 5000 m elevation in the Himalayas may be warmer or as warm during the day as a treeless habitat on a poleward or wind-exposed location at 3000 m elevation. Scherrer and Körner (2011) illustrated that the alpine vegetation on a single mountain slope of 2 km width may host thermal habitats at close proximity that differ by 8 K seasonal mean plant temperature (corresponding to c. 1500 m elevation in the free atmosphere). This enormous thermal habitat diversity within a small area is one of the reasons why alpine organism can be expected to be far less vulnerable to climatic change than any other type of biome. Alpine organisms can escape unfavorable temperature conditions by moving over very short distances. For that reason, the alpine flora has been considered most robust against climatic warming, an effect that becomes enhanced by the predominant clonal nature of these plants (Körner and Hiltbrunner 2021). This does not hold for trees because by their stature they are aerodynamically coupled to ambient atmospheric conditions (Körner and Hoch 2023).

1.4 Experiments by Nature

These sharp micro-environmental gradients host organismic assemblages for thousands of years. Hence, these communities and the associated soils can be considered a long-term answer of nature to these conditions. No experiment in an experimental garden or in a growth chamber can ever arrive at such a steady-state situation. Thermal or moisture gradients, but also season length gradients in regions with regular snow cover, produce species range limits at the 0.1 m scale. Exploring such micro-range limits and relating these limits with microclimatic data helps formulating and testing hypothesis (a recent example by Von Bühren and Hiltbrunner 2022). Such "experiments" by nature can help unravel the controlling mechanisms of range limits, which can then be applied to larger scales. Long-term monitoring of such micro-patterns of species distribution can help visualize responses to ongoing global change. The only problem is the year-to-year stochasticity of weather conditions, calling for sufficiently long-time frames for initial data collection. Embedded in a permanent plot concept, such data collection can also serve as a reference for future re-visitations (Körner et al. 2022).

1.5 Secondary Compounds in Mountain Plants

Mountain plants are praised for their bright colors, often strong scent and for containing compounds that make them attractive as spices, for aromatic honey, tea, and alcoholic beverages or for medicinal purposes. The reasons for these traits are manifold, but also relate to strong solar radiation and/or its UV fraction (e.g., Ambach et al. 1993), periodic physiological stress, slow growth, and defense against herbivory or pathogens (e.g., Spitaler et al. 2006). Some of these traits have phylogenetic roots, meaning that low-elevation representatives of certain genera and families are also known for these characteristics. For instance, families from which species are often collected, such as Lamiaceae and Apiacea, contribute over-proportionally to the high-elevation flora of the mountains of Eurasia.

It is not always clear which factor leads to a more pronounced production of certain secondary compounds that are considered to be of medicinal value. In fact, as was exemplified for one of the most established medicinal plants of the Alps, *Arnica montana* (Fig. 1.1) some compounds respond to increasing UV, while others respond to the decline in temperature, along elevational gradients (Spitaler et al. 2006; Albert et al. 2009). In addition, soil quality (moisture and nitrogen availability in particular) may exert significant effects, as was shown for the two prominent medicinal plants in central Nepal, *Aconitum naviculare* and *Neopicrorhiza scrophulariiflora* by Shresta and Jah (2008).

1.6 A Need to Objectively Establish the Effectiveness of Horticultural Production Rather Than Collection

Many mountain plants, alpine ones in particular, are collected for their attractiveness, for uses in alcoholic beverages (Fig. 1.2) for cosmetic or medicinal purposes or, in the case of shrubs, even as fuel wood substitute, in the absence of trees. Because of that interest, some of these species became really threatened and may be driven to extinction. It depends on local governance whether such activities can be prohibited. The central issue here is, does the purpose justify the utilization in the first place?

Fig. 1.1 *Arnica montana* (Asteraceae) an established medicinal plant in the European Alps. This alpine species can also be grown in high-elevation medicinal herb gardens and alcoholic extracts are traditionally used as an externally applied anti-inflammatory drug

Fig. 1.2 Extraction of alpine plants for non-medicinal purposes. Roots of *Gentiana punctata* and *G. purpurea* excavated at 2500 m elevation in the Swiss Central Alps (Furka Pass region) for producing a slightly bitter "Schnaps," a traditional distillate, often also mixed into digestives

While people living in remote places in mountains, far away from modern medicinal support, and also for cost reasons, often rely on traditional medicinal plants (e.g., Pei et al. 2009; Petelka et al. 2020; Naveen et al. 2021). Some of these stand testing of effectiveness, but many do not, but simply are attributed with effects that relate to their attractiveness, color, shape, taste, sent, or simply belief (Fig. 1.3). Many of these herbal drugs are applied as a sort of placebo, with the recovery of the patients often commented as to occur in 7 days, rather than within a week without the drug. In addition, such herbal medicines contain a suite of secondary compounds, evolutionarily selected for their defense nature, thus deterring or harming herbivores. In other words, the desired effect is likely to be paired with unwanted side effects of such cocktails of secondary compounds, which are not always obvious in the short term. When plant species are collected in the wild, and thus, become potentially threatened, while there is no evidenced medicinal value, such land use is in particular conflict with biodiversity conservation targets. Because beneficial placebo effects are often real and had been shown to reach up to the benefit of established medicinal treatments (Howick et al. 2013; Locher et al. 2017), the discouragement of collecting such plants is rather difficult.

The preferred way of providing effective herbal drugs from mountains would always be the herbal garden, rather than risking the decline or extinction of wild populations (e.g., NMPB 2008). It is an important field of research to develop the adequate techniques for such horticultural drug production and sort out species by proven effectiveness (for a recent review, see Petelka et al. 2020; see several contributions in this volume). Both the high-elevation climatic induction of secondary compound production and the need for appropriate soil conditions appear to be the main challenges. Mild (low elevation) climate and nutrient-rich soils may increase yield, but often diminish the desired herb quality. There is an immense literature on such alpine or high-elevation herb gardening, with a focus either on simply providing a botanical attraction, or an emphasis loaded with a lot of often esoteric beliefs.

Fig. 1.3 A rare alpine species (*Saussurea medusa*, Asteraceae) that is collected in the SE Himalayan at elevations >4600 m for its peculiar appearance, and sold, although it has no whatsoever medicinal value.

The published literature on sound scientific alpine or high mountain medicinal herb growing is very limited (for examples of systematic trials, see Sher et al. 2010; NMBP 2008).

While screening for the actual phytochemistry of potential medicinal plants is very important (e.g., Tiwari et al. 2023), it bears the danger that the taxa that emerge as particularly powerful become popularized before their potential for horticultural production has been explored, thus encouraging collections of wild plants in their natural habitat.

1.7 Rare Plants Endangered by Extraction for Fuel

Finally, because of the devastation of high-elevation landscapes over millennia of land use, many montane areas have lost their natural tree cover (e.g., Noroozi and Körner 2018). For instance, most of the Tibetan plateau that receives >300 mm of precipitation (Paulsen and Körner 2014) is not naturally treeless, but it became custom to address this terrain as "alpine," although it is, by definition, montane, as long as the land is below the climatic tree limit (4500–4900 m depending on region). Similarly, much of what is addressed as paramos in the northern (tropical) Andes or altiplano in the more southern (subtropical) Andes is treeless because of either natural or man-made fires and land use. As a consequence, there is a shortage of firewood and people collect shrubs, or, even worse, endangered cushion plants for heating or even semi-industrial processing of minerals (Fig. 1.4).

1.8 Conclusions

With these comments on the high-elevation environment, its bioclimatic features, and the areal delineation, and with a few thoughts on the causes of high plant diversity at high elevation, I want to close this prelude with a warning. Whoever is concerned with environmental changes at high elevation is well advised not to overestimate climate change effects. These are much less important than many believe, simply because small plants are nested in their own microenvironment. Land use, either by intensivation or abandonment and collecting specific plant species, may impose more severe threats. The often-claimed high vulnerability of mountains to climatic change does in large *not* apply to the wild biota, but are threats to people living in mountains. So, it is the human mountain population that might be vulnerable for all sorts of reasons, most related to the hydrological changes, but not or hardly the high-elevation biota (Körner and Hiltbrunner 2021). This climate change-related vulnerability debate has been misleading in the past for understandable political reasons, but it does not apply to the majority of organisms, native to high mountains, the ones above the climatic treeline in particular.

Acknowledgments I thank the *Nepalese Journal of Zoology* for the permission to reprint the text parts on mountain definitions and mountain climatology.

Fig. 1.4 In the lack of firewood, the highly threatened, very long-lived high-Andean giant cushion plant (*Azorella compacta*, Apiaceae) is collected to fuel drying ovens for the mineral borax in the Bolivian Altiplano. The boron salt is used as a cleaning agent and is also attributed with rather dubious medicinal effects

References

Albert A, Sareedenchai V, Heller W, Seidlitz HK, Zidorn C (2009) Temperature is the key to altitudinal variation of phenolics in *Arnica montana* L. cv. Arbo. Oecologia 160:1–8

Ambach W, Blumthaler M, Schöpf T (1993) Increase of biologically effective ultraviolet radiation with altitude. J Wilderness Med 4:198–197

Ehrlich D, Melchiorri M, Capitani C (2021) Population trends and urbanisation in mountain ranges of the world. Land 10:255–255

Howick J, Friedemann C, Tsakok M, Watson R, Tsakok T, Thomas J, Perera R, Fleming S, Heneghan C (2013) Are treatments more effective than placebos? A systematic review and meta-analysis. PLoS One 8(5):e62599. https://doi.org/10.1371/journal.pone.0062599

Kollas C, Randin CF, Vitasse Y, Körner C (2014) How accurately can minimum temperatures at the cold limits of tree species be extrapolated from weather station data? Agric For Meteorol 184:257–266

Körner C (2004) Mountain biodiversity, its causes and function. Ambio Spec Rep 13:11–17

Körner C (2007) The use of "altitude" in ecological research. Trends Ecol Evol 22:569–574

Körner C (2021b) The cold range limit of trees. Trends Ecol Evol 36:979–989

Körner C (2021a) Alpine plant life. Springer, Cham

Körner C (2022) A global framework of mountain ecology. Nepal J Zool 6(S1):42–44. https://doi.org/10.3126/njz.v6iS1.xxxxx

Körner C, Hiltbrunner E (2018) The 90 ways to describe plant temperature. Perspect Plant Ecol Evol Syst 30:16–21

Körner C, Hiltbrunner E (2021) Why is alpine flora comparatively robust against climatic warming? Diversity 13:383. https://doi.org/10.3390/d13080383

Körner C, Jetz W, Paulsen J, Payne D, Rudmann-Maurer K, Spehn EM (2017) A global inventory of mountains for bio-geographical applications. Alp Bot 127:1–15

Körner C, Hoch G (2023) Not every high-latitude or high-elevation forest edge is a treeline. J Biogeopgraphy 50:838–845. https://doi.org/10.1111/jbi.14593

Körner C, Urbach D, Paulsen C (2021) Mountain definitions and their consequences. Alp Bot 131:213–217. https://doi.org/10.1007/s00035-021-00265-8

Körner C, Berninger UG, Daim A, Eberl T, Fernández Mendoza F, Füreder L, Grube M, Hainzer E, Kaiser R, Meyer E, Newesely C, Niedrist G, Niedrist GH, Petermann JS, Seeber J, Tappeiner U, Wickham S (2022) Long-term monitoring of high elevation terrestrial and aquatic ecosystems in the Alps—a five-year synthesis. Ecomont 14:48–69. https://doi.org/10.1553/eco.mont-14-2s48

Locher C, Koechlin H, Zion SR, Werner C, Pine DS, Kirsch I, Kessler RC, Kossowsky J (2017) Efficacy and safety of selective serotonin reuptake inhibitors, serotonin-norepinephrine reuptake inhibitors, and placebo for common psychiatric disorders among children and adolescents: a systematic review and meta-analysis. JAMA Psychiatry 1(74):1011–1020. https://doi.org/10.1001/jamapsychiatry.2017.2432

McVicar TR, Körner C (2013) On the use of elevation, altitude, and height in the ecological and climatological literature. Oecologia 171:335–337

Naveen C, Gajendra S, Shashank L, Ishwari R, Lalit MT (2021) Alpine medicinal and aromatic plants in the western Himalaya, India: an ecological review. Indian J Ecol 18:319–331

NMPB (2008) Agro-techniques of selected medicinal plants. National Medicinal Plant Board, India, TERI Press, New Delhi

Noroozi J, Körner C (2018) A bioclimatic characterization of high elevation habitats in the Alborz mountains of Iran. Alp Bot 128:1–11. https://doi.org/10.1007/s00035-018-0202-9

Paulsen J, Körner C (2014) A climate-based model to predict potential treeline position around the globe. Alp Bot 124:1–12

Pei S, Huai H, Yang L (2009) Medicinal plants and their conservation in China with reference to the Chinese Himalayan region. Asian Med 5:273–290

Petelka J, Plagg B, Säumel I et al (2020) Traditional medicinal plants in South Tyrol (northern Italy, southern Alps): biodiversity and use. J Ethnobiol Ethnomed 16:74. https://doi.org/10.1186/s13002-020-00419-8

Scherrer D, Körner C (2011) Topographically controlled thermal-habitat differentiation buffers alpine plant diversity against climate warming. J Biogeogr 38:406–416

Sher H, Alyemeni MN, Dullah F (2010) Cultivation and domestication study of high value medicinal plant species (its economic potential and linkages with commercialization). Afr J Agric Res 5(18):2462–2470

Shrestha BB, Jha PK (2008) Habiat range of two Alpine medicinal plants in a Trans-Himalayan dry valley, Central Nepal. J Mt Sci 6:66–77

Spitaler R, Schlorhaufer PD, Ellmerer EP, Merfort I, Bortenschlager S, Stuppner H, Zidorn C (2006) Altitudinal variation of secondary metabolite profiles in flowering heads of *Arnica montana* cv. Arbo. Phytochemistry 67:409–417

Tiwari D, Kewlani P, Gaira KS et al (2023) Predicting phytochemical diversity of medicinal and aromatic plants (MAPs) across eco-climatic zones and elevation in Uttarakhand using generalized additive model. Sci Rep 13:10888. https://doi.org/10.1038/s41598-023-37495-1

Von Bühren RS, Hiltbrunner E (2022) Low winter temperatures and diverging freezing resistance set the cold range limit of widespread alpine graminoids. J Biogeogr 49:1562–1575

Climatic Change and Physiological Adaptations in Alpine Medicinal Plants

2

Sudeep Chandra, Vijaylaxmi Trivedi, V. K. Purohit, M. C. Nautiyal, and Suman Chandra

Abstract

This chapter explores the impact of climate change on alpine medicinal plants, focusing on how these plants have evolved physiological and morphological adaptations to survive in harsh and changing environments. Alpine regions, known for their extreme weather conditions, are home to a variety of plants that have developed unique mechanisms to cope with cold temperatures, high radiation, and fluctuating weather patterns. As global temperatures rise, these plants face new challenges, including shifts in precipitation patterns, warmer winters, and altered growing seasons. This chapter examines the specific adaptive traits, such as changes in metabolism, growth patterns, morphology, and reproductive strategies, that enable these plants to thrive despite environmental stressors. By understanding these adaptations, the chapter highlights the potential for alpine medicinal plants to maintain their properties and show resilience or sensitivity in terms of growth and survival, even as their habitats are altered, offering valuable insights for their conservation and sustainable use in the face of climate change.

S. Chandra (✉) · V. Trivedi · V. K. Purohit · M. C. Nautiyal
High Altitude Plant Physiology Research Center, HNB Garhwal University, Srinagar Garhwal, Uttarakhand, India

S. Chandra
School of Pharmacy, The University of Mississippi, University, MS, USA

S. Chandra, H. Lata (eds.), *High Altitude Medicinal Plants*,
https://doi.org/10.1007/978-3-031-99406-7_2

2.1 Introduction

Approximately 15% of the world's land surface is covered by alpine environments, which are found in high-latitude and -altitude regions (Bliss 1971). These areas are characterized by a low-temperature climate with a short growing season, and average temperatures generally do not exceed 6–8 °C, marking the lower distribution limit of the alpine zone globally (Körner and Paulsen 2004). The term "alpine" refers to treeless areas above the treeline, characterized by low temperatures in high-elevation areas (Grabherr et al. 2003; Körner 1995, 2003; Nagy and Grabherr 2009). Alpine environments are marked by unique atmospheric conditions, such as low temperatures and reduced partial pressure of gases. In these regions, both total and partial gas pressures of the atmosphere decrease almost linearly with increasing altitude. For example, at 2600 m altitude, the total pressure is about 750 hPa, compared to 1013 hPa at sea level, a 26% reduction. At the upper limits of higher plant life (5500–6000 m in the Himalayas and Andes), plants and animals exist at approximately half the pressure found at sea level, with the partial pressures of CO_2 and oxygen similarly reduced because the atmospheric composition remains constant across these altitudes (Peacock 1998).

As altitude increases, air temperature drops due to several factors: the reduced density of air, the condensation of water vapor (which releases heat during the formation of fog or clouds), and the forced movement of air, such as thermal uplift on slopes or horizontal heat advection by wind. This decrease in atmospheric temperature leads to a reduction in the moisture content of saturated air, lowering the vapor pressure deficit at any given relative humidity (Körner 2021). Alpine areas, due to their high elevation, also receive high-intensity radiation with a strong UV index. The radiation climate in mountains shows some consistent features: solar radiation flux increases with altitude under both clear skies and overcast conditions, although high radiation extremes and the frequency of local cloud cover also increase with altitude (Dirmhirn and Eaton 1975). However, climate characteristics in alpine regions often reflect regional rather than global patterns, complicating functional explanations of alpine plant traits (Körner 2021). Solar radiation may not change consistently with elevation and can even decrease in some regions, such as the mountains of Scotland (Barry 1981). Similarly, precipitation patterns do not follow a general trend with elevation and can vary widely, such as in the western Himalayas, where precipitation first increases and then decreases with elevation (Lauscher 1976; Rundel 1994; Hemp 2002; Körner 2007).

Alpine plants play a crucial role in alpine ecosystems, biodiversity, and as indicators of climate change. These plants are characterized by unique traits such as short stature, cushioned growth, bright-colored flowers, and the accumulation of secondary metabolites. Alpine plant communities are structurally simple, primarily composed of long-lived perennial herbaceous plants with relatively low plant diversity. The flora includes flowering plants (Angiosperms), bryophytes, lichens, and some ferns, with most Angiosperms being herbaceous perennials or very low shrubs. Annual plants are rare in these environments. The main life forms of perennial herbs in alpine regions are cushion or polster plants, rosette plants, leafy-stemmed plants, and grass-form plants (Billings and Mooney 1968). These plants are well adapted to the harsh climatic conditions through modulations in their secondary metabolism and production of solutes to maintain homeostasis (Holmström et al. 1996).

Climate change is significantly impacting alpine zones by altering temperature patterns, precipitation regimes, and extreme weather events. In alpine regions, temperatures are rising at a rate faster than the global average. This warming trend is leading to earlier snowmelt and a longer growing season for plants. Warmer temperatures can promote plant growth in some species, but they also pose risks such as increased respiration rates that lead to higher carbon loss and potential overheating in plants adapted to cooler climates. The accelerated warming is also causing a shift in species distribution (Rana et al. 2022), with some plants migrating to higher altitudes to find suitable conditions, while cold-adapted species face habitat loss and increased competition from lower-altitude species migrating upward (Körner 2016; Gottfried et al. 2012).

Although alpine areas are defined by the absence of trees, taller shrub communities are expanding in some regions, and subalpine plants are moving upward, often at the expense of smaller shrubs. This encroachment of shrubs and trees is one of the major impacts of climate change on alpine environments, potentially reducing the extent of these unique ecosystems (Körner 2021). A more subtle change was noted in a 23-year study in the Adirondack Mountains, where there was a decrease in bryophytes and lichens and an increase in vascular plants, particularly in undisturbed areas (Robinson et al. 2010). Global warming is expected to alter abiotic conditions that influence individual plant performance, with alpine and arctic ecosystems being particularly vulnerable. Changes in soil moisture, increased nutrient availability, and direct warming effects may impact plant growth, reproduction, and resource allocation (Callaghan and Jonasson 1995; Guisan et al. 1998; Körner 1995, 1999).

Precipitation patterns in alpine zones are becoming more variable due to climate change, affecting water availability and the overall hydrology of these regions. Changes include increased frequency and intensity of precipitation events, which can lead to more extreme snowmelt episodes and flooding. Conversely, some areas may experience prolonged dry spells, increasing the risk of drought stress for alpine plants. These alterations in precipitation regimes can disrupt the delicate balance of alpine ecosystems, affecting soil moisture and nutrient availability, which are critical for plant growth and survival (Körner 2002; Scherrer and Körner 2010).

Extreme weather events, such as heatwaves and heavy storms, are becoming more common in alpine regions as a result of climate change. These events can cause immediate and severe damage to alpine flora, including physical damage from storms and heat stress during unusually hot periods. Increased frequency of frost events following early snowmelt can also harm early-blooming plants. The compounded stress from these extreme events can reduce plant reproductive success and resilience, making alpine ecosystems more vulnerable to long-term changes. As a result, the overall biodiversity and functioning of these unique environments are at risk, highlighting the urgent need for conservation efforts to mitigate the impacts of climate change (Inouye 2008).

2.2 Adaptations to Increasing Temperature and CO_2

Alpine plants have evolved remarkable adaptations to cope with the low temperatures and harsh conditions characteristic of high-altitude environments. One key adaptation is the development of a compact, cushion-like growth form, which helps reduce heat loss and protects against chilling winds (Körner 2003). These cushion plants create microenvironments that are significantly warmer than the surrounding air, sometimes by as much as 10 °C, thus providing a stable and favorable habitat for plant tissues and associated organisms (Körner and Larcher 1988). Another notable morphological adaptation is the presence of hairy or woolly surfaces on leaves and stems. These trichomes (hair-like structures) serve multiple functions: they reduce water loss by trapping a layer of humid air close to the plant surface, provide insulation against cold temperatures, and protect against intense ultraviolet (UV) radiation by reflecting excess light (Körner 1995). Some alpine plants, such as species in the genus *Saussurea*, have evolved dense woolly coverings that effectively shield their leaves from the extreme cold and UV radiation common at high altitudes (Streb et al. 1998).

Adaptation also involves the physiological changes in the plants' cellular structures, such as the accumulation of solutes like sugars and amino acids, which act as antifreeze compounds, lowering the freezing point of cell sap and protecting cellular structures from ice crystal formation (Holmström et al. 1996). Alpine plants also exhibit a high degree of plasticity in their photosynthetic processes. They can efficiently perform photosynthesis at lower temperatures, which is crucial for survival and growth during the short growing season when temperatures can still drop significantly at night (Körner 1995). Furthermore, these plants often have a slow but steady growth rate, allowing them to allocate resources efficiently over time and survive prolonged periods of adverse conditions (Billings and Mooney 1968). Another critical adaptation is the production of anthocyanins, pigments that not only give plants their characteristic bright colors but also protect against UV radiation and oxidative stress (Streb et al. 1998). These pigments absorb harmful UV rays and mitigate damage from reactive oxygen species produced under high light intensity conditions typical of alpine environments. By employing these multifaceted adaptations, alpine plants maintain their metabolic functions and reproductive capabilities despite the challenges posed by their extreme habitats, illustrating a remarkable evolutionary ingenuity to thrive in conditions that are often inhospitable to most other forms of life (Körner 2021).

Heat stress primarily leads to the disruption of photosynthetic machinery, resulting in reduced photosynthetic rates, less carbon uptake, and increased photorespiration. Higher temperatures increase the closing of stomata, which reduces the intake of CO_2 and hinders the efficiency of photosynthesis as in *Rhododendron hybridum* (Shen et al. 2017), *Clematis cadmia,* and *Clematis crassifolia* (Hu et al. 2021). According to Shen et al. (2017), *R. hybridum* showed a decreasing trend in chlorophyll and carotenoid synthesis as heat stress increased. As a result, medicinal plants undergo a decrease in the amount of biomass they accumulate, their growth is affected, and their yields are lowered when exposed to extended heat stress conditions. In addition, elevated temperatures worsen oxidative stress in medicinal plants, leading to the build-up of reactive oxygen species (ROS) and causing oxidative harm to cellular components. Reactive oxygen species (ROS), including as superoxide radicals, hydrogen peroxide, and hydroxyl radicals, are produced as secondary products of photosynthesis and respiration during heat stress conditions. The excessive build-up of reactive oxygen species (ROS) surpasses the body's ability to counteract them with antioxidant defense mechanisms, resulting in the oxidation of lipids, proteins, and DNA. Medicinal plants utilized a variety of enzymatic and nonenzymatic antioxidants, such as superoxide dismutase, catalase (CAT), glutathione, and ascorbate, to prevent oxidative stress. Earlier research revealed that, CAT, POD, and SOD activity decreased in *Rhododendron hybridum, Clematis cadmia,* and *Clematis crassifolia* as heat stress increased (Shen et al. 2017; Hu et al. 2021). These antioxidants help remove reactive oxygen species and minimize harm to cells. Additionally, secondary metabolites such as phenol, saponin, tannins, alkaloid, and flavonoid content increased with increasing heat stress in *Catharanthus roseus* and *Mentha piperita* (Alhaithloul et al. 2019). Moreover, elevated temperatures disrupt the hormonal equilibrium in medicinal plants, influencing the production and communication of crucial phytohormones that play a role in both stress responses and growth regulation. Abscisic acid is a crucial stress hormone, rapidly accumulates in response to temperature stress, inducing stomatal closure and facilitating osmotic adjustment to improve water retention. Moreover, ABA regulates the activity of stress-responsive genes and coordinates antioxidant defense mechanisms to reduce the negative effects caused by heat (Zandalinas et al. 2016). In contrast, higher temperatures hinder the production and communication of growth-stimulating hormones, such as auxins and gibberellins, which leads to a reduction in cell elongation, meristematic activity, and reproductive development (Jan et al. 2022). Furthermore, elevated temperatures induce changes in the production and buildup of secondary metabolites in medicinal plants, which in turn affect their pharmacological characteristics and therapeutic effectiveness. For example, antimicrobial, antifungal, and anticancer activity decreased as *Catharanthus roseus* and *Mentha piperita* exposed to heat stress (Alhaithloul et al. 2019). Heat stress also induces the synthesis of heat shock proteins (HSPs), which function as molecular chaperones to stabilize protein structures and protect cells against thermal damage (Khan and Shahwar 2020). Previous research (Hu et al. 2021) demonstrated that, as heat increased, upregulation of genes related to HSP also observed in medicinal plants such as *Clematis cadmia* (c194329_g3, c195983_g1, and c194434_g1) and *Clematis crassifolia* (C194329_G3, C188817_g1, and C194434_G1). Osmolytes play a significant role during heat stress. Earlier reports suggested that proline, glycine betaine, sugars, mannitol, and inositol decreased with increasing heat stress in *Catharanthus roseus* and *Mentha piperita* (Alhaithloul et al. 2019). On the other hand, electrolyte leakage increased with increasing heat stress in *Rhododendron hybridum* (Shen et al. 2017).

Climatic changes not only lead to heat stress but can cause erratic precipitation and snowfall patterns in high-altitude plants, consequently leading to other types of stresses, for example, chilling or frost injury. Low temperature or chilling stress primarily leads to the disruption of cellular membranes. As temperatures decrease, the fluidity of cell membranes is compromised, hindering the transport of essential nutrients and metabolites (Ghassemi et al. 2021). Additionally, chilling stress disrupts the photosynthetic machinery, reducing energy production and further impacting plant growth. Chilling stress also triggers the production of reactive oxygen species (ROS) within plant cells. ROS are highly reactive molecules that can damage cellular components like proteins, DNA, and membranes (Ghassemi et al. 2021). A previous report revealed that a significant amount of MDA and H_2O_2 content accumulated under cold stress in *Rheum spiciforme*. To combat this oxidative stress, plants activate antioxidant defense mechanisms, including enzymes like superoxide dismutase, ascorbate peroxidase, glutathione reductase, and guaiacol peroxidase as in *Rheum spiciforme* (Khan et al. 2023; Ghassemi et al. 2021).

However, prolonged chilling stress can overwhelm these defenses, leading to oxidative damage and impaired cellular function (Zobayed et al. 2005). Chilling stress can manifest in several ways at the whole-plant level. Seed germination and seedling emergence are often negatively affected with reduced root growth hindering water and nutrient uptake (Ghassemi et al. 2021). Chilling stress can also delay flowering and fruit development, impacting yields (Staniak et al. 2021). In response to chilling stress, medicinal plants undergo a series of metabolic adjustments. Plants may accumulate primary and secondary metabolites. For example, in *Rheum spiciforme* accumulation of anthraquinone such as aloe-emodin, rhein, emodin, and physcion, stilbenoid (piceatannol and resveratrol), along with flavonoid compounds such as rutin and quercetin increased under cold stress (Khan et al. 2023). Proline content increased in *Sibbaldia procumbens* and *Bistorta affinis* and *Sibbaldia procumbens* Proline, ABA, protein, and sugar content increased and IAA decreased with decreasing temperature (Rahman et al. 2019). Gaseous exchange parameters are also affected by cold stress in alpine plants. Earlier report demonstrated that chlorophyll content, photochemical, and nonphotochemical quenching and Fv/Fm decreased significantly under cold stress in *Chorispora bungeana*. However, there was not a significant difference observed in the PSII quantum yield between plants growing at 4 °C and −4 °C (Li et al. 2010).

Jeon et al. (2018), conducted a study to examine the impact of cold stress on the genetic material and chemical compounds in *Fagopyrum tataricum*. Their findings indicate that the levels of sugars and their by-products, particularly maltose, increased by over 20.6 times. Additionally, anthocyanins, proanthocyanidins, and certain organic acids showed a substantial rise in response to cold stress. Conversely, some amino acids as well as their by-products decreased in plants subjected to cold treatment compared to those grown at 25 °C. Nevertheless, it is proposed that low temperatures influence the process of sugar production, leading to higher levels of sugars. These increased sugars may then serve as signaling molecules, triggering the activation of genes responsible for the production of anthocyanin biosynthetic enzymes. Consequently, this promotes the accumulation of anthocyanin. Additionally, the production of stress hormones like abscisic acid, JA, and SA also increased. These hormones play a central role in regulating plant responses to stress, including stomatal closure (reducing water loss) and the activation of

Table 2.1 Major physiological changes observed in alpine plants to climate change and warming

Physiological changes/adaptations	Positive effects	Negative effects	Recent scientific references
Phenological shifts	Extended growing season. Earlier flowering and fruiting. Flowering anomalies e.g., no floral induction.	Potential mismatches with pollinators. Increased risk of frost damage to early blooms.	Thórhallsdóttir (1998), Bjorkman et al. (2017) and Chandra et al. (2020, 2022)
Photosynthesis and respiration	Increased photosynthetic rates in some species.	Increased respiration rates may lead to higher carbon loss. Potential for reduced water-use efficiency.	Körner (1999) and Dawes et al. (2015)
Water relations	Some species may develop deeper root systems.	Increased drought stress. Reduced soil moisture availability.	Comas et al. (2013) and Scherrer and Körner (2010)
Growth and biomass	Enhanced growth in some species due to warmer temperatures.	Reduced growth in species unable to cope with temperature changes.	Shi et al. (2015) and Frei et al. (2014)
Reproductive success	Higher seed production in some species.	Reduced reproductive success due to stress factors like drought and temperature extremes.	Curtis et al. (1994) and Inouye (2008)
Species distribution and community composition	Range expansion for some species. Increased diversity in some areas.	Habitat loss for cold-adapted species. Increased competition with invasive species.	Gottfried et al. (2012) and Pauli et al. (2012)
Physiological stress responses	Increased production of protective compounds (e.g., antioxidants).	Elevated stress hormones (e.g., abscisic acid). Cellular damage from heat and drought stress.	Alhaithloul et al. (2019) and Körner (2016)

stress-responsive genes. The higher concentration of JA leading to higher survival rates of *Arabis alpina* suggested that JA signaling is positively involved in the freezing tolerance of *Arabis alpina* (Kolaksazov et al. 2013; Table 2.1).

2.3 Photosynthesis and Light Adaptations

Alpine plants have evolved various strategies to optimize light capture in the harsh, low-temperature environments they inhabit. One key adaptation involves modifications in leaf structure and orientation. Alpine plants often develop small, thick leaves that are densely packed with chloroplasts, enhancing their ability to capture and utilize available light efficiently (Jiang et al. 2021). These leaves also frequently exhibit morphological adaptations, for example, layer of fine hairs that helps reflect excess light, thus preventing photoinhibition while maintaining an optimal light environment for photosynthesis (Körner 2003; Niinemets 2007).

Additionally, many alpine plants display heliotropic movements, whereby their leaves and flowers track the sun's movement across the sky to maximize light capture during the day. This sun-tracking behavior allows these plants to make the most of the limited sunlight available in their short growing season (Vesk and Westoby 2004). Research has shown that heliotropic movements can significantly enhance light interception and improve overall photosynthetic efficiency, which is crucial for survival in such extreme conditions (Baskin and Baskin 1998; Biskup et al. 2016).

At the cellular level, alpine plants have adapted their photosynthetic processes to function efficiently at low temperatures. This includes modifications to the photosystems and associated proteins to maintain high levels of activity despite the cold. For instance, studies have found that the enzyme Rubisco, essential for carbon fixation, exhibits higher activity at lower temperatures in alpine plants compared to lowland species (Huner et al. 1993; Sage and Kubien 2007). These biochemical adjustments ensure that photosynthesis can proceed effectively, even when temperatures are not optimal.

Moreover, alpine plants often form symbiotic relationships with mycorrhizal fungi, which enhance their ability to capture light and nutrients from the soil. These fungi extend the root surface area, facilitating better nutrient uptake, which supports more robust photosynthetic activity and overall plant health. This mutualistic relationship is especially important in nutrient-poor alpine soils, where the efficient use of available light and resources is critical for survival (Read and Haselwandter 1981; Heinemeyer and Fitter 2004).

Recent research also highlights the role of phenotypic plasticity in alpine plants' adaptation to fluctuating environmental conditions. Alpine plants can dynamically adjust their photosynthetic capacity in response to seasonal variations in light intensity and temperature. During the short growing season, they ramp up their photosynthetic machinery to maximize carbon gain, while in the off-season, they reduce metabolic activity to conserve energy (Germino and Smith 2000; Körner 2021). This flexibility allows them to exploit favorable conditions fully while minimizing energy expenditure during less favorable periods. These adaptations, ranging from leaf structure modifications and heliotropic movements to the production of protective pigments and symbiotic relationships with mycorrhizal fungi, underscore the remarkable resilience and versatility of these plants in some of the most challenging habitats on Earth.

2.4 Adaptation to Solar Radiation (UV Protection)

Mountainous regions, for example, alpines experience greater UV radiation levels due to the synchronized effects of shorter radiation path length, lesser aerosol load, and higher albedo of the snow surfaces (Blumthaler 2007). This enhanced UV incidence has a pronounced effect on the morpho-physiology of alpine plants with a number of adaptations employed by them to overcome the excessive UV damage stress.

Another crucial adaptation is the production of anthocyanins and other protective pigments, which serve to filter harmful ultraviolet (UV) radiation and reduce oxidative stress. These pigments absorb excess light energy and dissipate it as heat, thereby protecting the photosynthetic machinery from damage. The red and purple hues observed in many alpine plants are indicative of high anthocyanin content, which is directly correlated with increased UV protection and improved light capture efficiency in cold environments (Hughes and Smith 2007; Streb et al. 1998).

2.5 Adaptations to Optimize Water and Nutrient Uptake

Alpine areas are characterized as limited ecosystems because of limited availability of the factors affecting growth (e.g., light, water, nutrients, favorable temperature), as well as limited growing season itself. Alpine plants have evolved a range of adaptations to optimize water relations and enhance drought tolerance, critical for survival in their harsh, high-altitude environments. One key adaptation is the development of deep root systems that allow plants to access water stored in deeper soil layers, even during prolonged dry periods. Additionally, many alpine species possess small, thick leaves with a high density of trichomes (hair-like structures) or a waxy cuticle, which reduce water loss by minimizing transpiration. These leaves also often have a low surface area to volume ratio, further conserving water by reducing the amount of exposed surface from which water can evaporate (Slattery 2015; Soukup et al. 2021).

Another important adaptation is the ability to regulate stomatal function efficiently. Alpine plants tend to have stomata that can close rapidly in response to water stress, thereby conserving water while still allowing for gas exchange when conditions are favorable. Moreover, some alpine species exhibit a form of drought tolerance known as "xerophytism," where they can tolerate significant dehydration and recover when water becomes available. These physiological and morphological adaptations enable alpine plants to maintain their metabolic activities in environments where water is a limiting factor, ensuring their survival in extreme conditions (Pool 1923).

Additionally, some alpine plants exhibit a fibrous root system with a high root density, which increases the surface area for water absorption. This dense network of roots is particularly effective in capturing water from the soil during brief periods of moisture availability, such as after rainfall or snowmelt (Pohl et al. 2011). The roots may also have specialized structures, such as mycorrhizal associations, where fungi form symbiotic relationships with plant roots, enhancing water and nutrient uptake. These root modifications are essential for alpine plants to maintain water balance and survive in their challenging, high-altitude habitats (Becklin and Galen 2009).

2.6 Documented Shifts in Plant Distributions

The resurgence in alpine studies in a few decades showed evidence of the change in plant distribution pattern in alpine summits across the globe. In search of the hospitable environment, the alpine plants are documented on move, mostly upward to submit but sometime also downward. A major factor that pushed this shifting is rising temperature, in search of the cooler temperature plants are moving to upper summit that are comparatively cooler. Factors like changes in snow cover affect water availability to curb such situations the plants are moving toward the areas with more consistent snow cover. Not only the abiotic factors but the changing competition pattern also force the migration of the plants. Other factors, including changing seasonal patterns and change in precipitation, also promote the shifting of the plants in more favorable locations in alpines. Due to this upward shifting, the species richness is highly affected in alpine summit, and changes in plant species richness are caused by warming. The alpine summit's species richness increases as a result of warming because it offers the ideal refuge for lower-altitude plants that are climbing upward in search of a suitable environment. Though this was a very short-term study, a 4-year investigation by Hamid et al. (2020) revealed an increase in the species richness on alpine summits during 4 years in the Indian Himalaya. Another study, spanning 7 years in the Australian alpines, found that the mean species richness increased from 45 to 50 species (about 12%) throughout the course of the study, representing a pace of nearly one new species each year and 15 new species documented at a single summit. The species with the biggest growth in this case were shrub and graminoid (Venn et al. 2012). Another study conducted on four summits at Dovrefjell, Norway, over a 15-year period between 2001 and 2015 (Vanneste et al. 2019), with a focus primarily on cryptogams, revealed a decline in lichen richness and an increase in bryophyte richness, with no change in the richness of vascular plants.

According to Wang et al. (2019), the intrusion of herbaceous plants had changed tundra vegetation to a significant extent over the last three decades, between 1985 and 2015, especially within low and middle alpine tundra regions in Changbai Mountains, China. As per the study, the herbaceous species would continue migrating upward and expanding while their dwarf shrub counterparts would also continue shifting upward but shrinking over the next three decades under the RCP 4.5 and RCP 8.5 scenarios. Kelly and Goulden (2008) discovered that during the 30 years prior to the second survey, the climate in the Santa Rosa Mountains of Southern California warmed at the surface, precipitation variability increased, and snowfall dropped. Because of global warming, plants are moving poleward along latitudes and upward along altitudes in quest of isotherms (Zu et al. 2021).

In general, plants can either fly or battle their way to thrive in a changing climate. They adapt to a limited range of environmental circumstances in order to tolerate change in situ; but, if the degree of climate change is significant, this limited range of acceptable habitat will also be degraded (Jackson and Overpeck 2000; Broennimann et al. 2006). Since many alpine endemics already inhabit the highest altitude habitats within their range, upslope migration is not unheard of in that scenario. Given the dispersal hurdles, geographic isolation, and low dispersal capabilities of many endemics, escape via long-distance migration is unthinkable (Cody and Overton 1996; Dullinger et al. 2012; Thomas 2011). Lowland species are driven upward by warming temperatures, and there might not be any other species that have adapted to higher temperatures to make up for the loss (Colwell et al. 2008). In fact, because of global warming, lowland species from warmer microhabitats may move to cooler refuges at the same height (Bush et al. 2004). Summit trap phenomena, however, may affect alpine flora (Rana et al. 2017; Salick et al. 2019). Because they must contend with non-alpine plants that are also moving up vertically in mountains in search of favorable conditions, the upward shifting of alpine plants is not as tranquil as it might seem. According to Dainese et al. (2017), who examined records of 1334 plant species gathered over a 20-year period in the European Alps, the rate of upward shifting is greater for non-native alpine plants. Globally, anthropogenic pressures have led to an increase in non-native plant species in alpine habitats, which may have negative effects on native plant communities and ecosystem services. According to several lines of evidence, non-native species in mountains may react to climate change more quickly and drastically than native species (Pauchard et al. 2009).

Studies have indicated that alpine plants have not only moved their range upward along mountains, but also downward. Lenoir et al. (2010) entailed the plants that moved lower in search of an appropriate habitat in one such study. These changes, however, were more a result of the biotic interaction than the direct effect of the abiotic climate change and its components. Temperature changes in the habitat cause changes in species interactions and the emergence of novel species interactions, such as competition and facilitation (Hughes 2000; Archaux 2004). Plants that gain or lose from these novel forms of interaction occasionally choose to move downhill, which offers facilitation all of a sudden. When Bhatta et al. (2018) conducted a follow-up survey of Langtang National Park in central Nepal after 25 years, they observed one such scenario in which there had been a downward shift in the plant species in the alpines. They proposed that this shifting was caused by increasing disturbance brought on by both changes in land-use patterns and climatic warming. The significance of competition as the limiting factor on species distributions is diminished when it is absent or diminished as a result of changes in land-use patterns. Additionally, species may shift downslope as a direct result of habitat modification, with or without competitive release. This can happen as a result of other local changes in habitat suitability, human-induced disturbances or permanent habitat changes (like land-use changes, recreational activities, and avalanches), or natural disturbances like windthrows, fires, and avalanches. Hättenschwiler and Körner (1995), for instance, proposed that in the Swiss Central Alps the high concentration of nitrogen deposition and the end of forest cattle grazing may have caused denser and more abundant ground vegetation, which in turn enhanced the replacement of *Pinus sylvestris* seedling populations by *P. cembra* populations below the current lower margin of *P. cembra* adult trees. Research indicates that decreased snow cover and early melting may also contribute to plants moving downward because snow acts as an ideal insulator for the soil microclimate in alpine regions. Under such circumstances, the soil would freeze, making it easy for plant roots to be damaged by frequent frost (Inouye 2008).

In general, the changes in the distribution patterns of alpine plants highlight the intricate relationship between biotic interactions, climate change, and human activities. These plants' continuous movement serves as a reminder of how dynamic alpine ecosystems are and how important it is to maintain biodiversity in these areas through regular monitoring and adaptive conservation measures. It is crucial to comprehend and lessen the effects of climate change on these delicate ecosystems as these shifts present both opportunities and challenges for conservation efforts.

2.7 Phenological Shifts

The majority of plant phenological events are dependent on temperature and light; hence, a change in temperature can significantly alter a plant's phenology (Sherry et al. 2007), which can have important implications for alpine plants. Snowmelt time has an early impact on seed germination and other aspects of plant phenology. It is known that in alpine plants cold stratification throughout the winter months or too low temperatures to promote emergence after autumn dissemination causes spring germination to take precedence (Schwienbacher et al. 2011; Walck et al. 2011). In certain species, spring emergence shifts to autumn emergence due to climate warming; nevertheless, this shift is primarily caused by the dormant state of the seeds. In such circumstances, seeds that need winter stratification do not germinate; nevertheless, a shorter winter and less snow cover can permanently stop germination in such seeds. For species that are now suited to emerge in the spring, emerging in the autumn could have significant consequences (Mondoni et al. 2012). Orsenigo et al. (2021) used lab simulations to replicate the two heatwave episodes that transpired in spring 2003 and autumn 2011 to study the effect of the heat wave on seed germination. When there were no heat waves, seeds germination mostly happened in the spring after they had gone through the autumn and winter seasons. On the other hand, heat waves caused a notable rise in 30% of the species' spring germination and 50% of their autumn germination. Alpine plants respond to hear waves either by delaying their of emergence from spring to autumn or by simply increasing the percentage of emerging seedlings. But only seeds that are either non-dormant or conditionally dormant exhibit this transition (Baskin and Baskin 1998). Although they can influence the germination percentages, autumn heat waves rarely changed the spring germination in seeds in a deep dormant state. According to Orsenigo et al. (2021), it considerably reduces summer germination in *Centaurea, Cirsium, Hypericum, Rumex*, and *Solidago*, but significantly increases summer germination in species like *Allium, Gentiana kochiana, Brachypodium, Geum*, and *Homogyne*. Due to the plasticity of seed features, even the same species responded differently to warmer temperatures, as seen in *Wahlenbergia ceracea*, where seed traits had a significant impact on germination (Notarnicola et al. 2023). In several cases, heat accelerated seed germination, especially for larger seeds. Wang et al. (2018) found in another study that warming enhanced the phenology of seedling emergence in *Primula alpicola, Pedicularis fletcheri, Meconopsis integrifolia*, and *M. racemosa*. According to Galen and Stanton (1999), certain species may be able to mitigate the advancement of phenology brought on by warming by delaying seedling emergence through prolonged snow cover. According to several research (Bibi et al. 2018; Piao et al. 2019), warming speeds up the ecodormancy break, which advances spring phenology in plants, but slows down the endodormancy break, which delays spring phenology. By decreasing the availability of soil water, warming can also directly or indirectly promote fall phenology (Estiarte and Peñuelas 2015; Liu et al. 2016).

Irregularity in germination time can start a trail of irregular and altered phenology. Any alteration in phenology can sometimes prove dangerous to the whole ecosystem. A path of irregular and changed phenology can be initiated by irregularity in the germination time. Any change in phenology might occasionally prove hazardous to the ecology as a whole. Temporal asynchrony or occasionally decreased interspecies complementarity could be the outcome and may impact interspecific competition for resources or the efficacy of plant pollination. In some cases, longer growing seasons brought on by an earlier start to growth may initially enhance plant productivity at the expense of nutrient reserve depletion, as well as lower long-term productivity and reproductive effort (Chapin and Shaver 1996). Through altering processes like nitrogen cycling and net primary production, subsequent changes in community composition may have an impact on organisms at all trophic levels (Eviner and Chapin 2003). The most well-documented phenological event affecting alpine plants is flowering, which is primarily regulated by photoperiod; other growth phases are influenced by temperature and the date of snowmelt. As observed in *Saxifraga oppositifolia*, the date of snowmelt determined shoot growth, flowering time, and seed development, whereas photoperiod alone influenced the onset of the flowering primordia. Keller and Körner's experiment addressed the relationships between temperature, photoperiodism, and alpine plant phenology (2003). They claimed that whereas species were more responsive to temperature regimes at shorter day lengths, photoperiodism had a greater influence on flowering dynamics at longer day lengths. Furthermore, it was found that certain species were insensitive to both light and temperature, suggesting that they were influenced by other ecological factors such as the melting of snow.

The response of the alpine plant life cycle to various environmental and climatic variables has been studied through phenology on a regular basis (Ernakovich et al. 2014; Khorsand et al. 2015; Ding et al. 2016). Phenology is also a useful indicator for determining whether a species can adapt to the changing environment. Furthermore, since the timing of sensitive phenophases is critical to a species' survival, phenology is a useful criterion to test whether or not a species will be able to adjust with the progressive anticipation of snowmelt (Schmid et al. 2017). In a manipulative experiment, Smith (2011) demonstrated that temperature, N availability, and increasing snowpack (winter precipitation) affect when flowers bloom, how many flowers bloom, and when seedlings germinate in alpine tundra. They observed that flowering might occur sooner in response to warming or later in response to altered snowpack. Phenological reactions to the addition of N, on the other hand, were more surprising and did not match findings from lowland grassland systems (Cleland et al. 2006). Suzuki and Kudo (2000) observed earlier leaf and flower emergence in the deciduous shrubs *Vaccinium uliginosum*, *Arctous alpinus,* and evergreen shrub *Ledum palustre* in manipulated OTC study with extended growing season. Individual leaf lifespan and foliage period were observed to be longer in the OTCs than in the control plots in this study.

The date of flowering and the total number of flowers produced are the two most significant parameters to consider when analyzing the flowering phenology. In both arctic and alpine plants, warming has been shown to both accelerate and delay the initial flowering date (Lee et al. 2011; Prevéy et al. 2019). According to Dorji et al. (2020), warming-related shifts in plant flowering date have an impact on the number of flowers that are produced in alpine habitats. In two species of *Potentilla*, *P. fruticosa* and *P. saundersiana*, they discovered that the timing of first flowering was significantly advanced in response to warming; no other significant impacts were observed on the length or productivity of flowering. In addition to warming, other unpredictable weather conditions can also impact flowering in some plant species. For example, in the same study by Dorji et al. (2020), simulated snowstorms caused *P. saundersiana* to delay its first flowering date. Flower production increases when the date of the last flowering is extended due to such phenological shifts. If the last flowering date is also extended, the phenological shift will result in an increase in flower production. In the same study, both *Potentilla* species showed an increase in flower production as the last date of flowering was delayed. These findings suggest that advances in the timing of the first flowering alone will not always translate into longer flowering times and higher reproductive effort when climate change modifies the phenology of alpine plants. Rather, results highlighted how the last flowering date plays a crucial role in influencing plant reproductive success and effort. Additionally, it has been observed that distinct phenophases have varied temperature sensitivity thresholds (Meng et al. 2016). According to Jiang et al. (2016), fruit set time is comparatively more consistent in response to warming and cooling. The length of the flowering period is significant since it can give plants more chances to reproduce (Dieringer 1991; Schwartz 2003; Nagahama et al. 2018). In a study carried out in Norway with an ambient climate, the majority of the plant species observed advanced the dates of their first flowers after warmer and wetter winters. However, only a small number of species that began flowering earlier also prolong their flowering span (Post and Stenseth 1999). Warming has been demonstrated to affect flowering duration in different ways, including shortening it (Črepinšek et al. 2012) and lengthening it (Dunne et al. 2003; Millerrushing et al. 2007). The response's direction depends on species, location, and functional traits (Dunne et al. 2003; Morales et al. 2007; Bock et al. 2014; Tao et al. 2017; Prevéy et al. 2019). For example, it has been observed that early flowering species respond to warmer temperatures and earlier snowmelt by extending their flowering time; however, this response is not significantly correlated with soil moisture (Dunne et al. 2003).

Plant phenological responses to climate change may also be influenced by biotic interacts and increased in the study by Dorji et al. (2018) and Hopping et al. (2018). Given that alterations in plant species interactions can impact phenology, it is possible that the reduced competition from *Kobresia pygmaea* contributed to the earlier flowering of *Potentilla. fruticosa* and *P. saundersiana* (Diez et al. 2012; Wolf et al. 2017). Furthermore, growing *K. pygmaea* (Dorji et al. 2013, 2018) implies that *P. saundersiana* may have flowered later due to increased competition with *K. pygmaea* in snow addition plots (Liu et al. 2017; Panchen and Gorelick 2017). Not only are rising temperatures and CO_2 wreaking havoc on alpine ecosystems, but heavy rains, severe storms, and flooding are also contributing causes that are often overlooked. The effects of these components on alpine, tropical, temperate subtropical plants were demonstrated by Zdenka Křenová et al. (2022). Due to their synchronized behavior, temperate and alpine plants are particularly susceptible to such conditions, which could harm their reproductive systems or sabotage relationships among plants and pollinators (Memmott et al. 2007).

In summary, phenological events in alpine plants are greatly influenced by the intricate interactions between temperature, photoperiod, snowmelt timing, and extreme weather events. These interactions frequently result in changes to germination, flowering, and overall reproductive success, which can have a cascading effect on alpine ecosystems.

2.8 Alterations in Growth Rates and Biomass Production

The response of the plants was modified by the seasonality of many environmental parameters, such as temperature, radiation, precipitation, and day length (Piao et al. 2019). In higher latitudes and altitudes, plant growth often follows a particular pattern: it increases in the spring at the start of the growing season, reaches its maximum during the mid-growing season, and then decreases in the autumn at the end of the growing season (Billings and Mooney 1968; Körner 2003). However, due to climate changes this growth pattern is changing mainly due to alleviated low-temperature constraints on plant growth in cold regions (Park et al. 2019). The warming that has occurred over the past few decades has changed spring phenology (Shen et al. 2015), advanced the timing of maximum photosynthesis (Park et al. 2019), and delayed fall phenology (Barichivich et al. 2013; Liu et al. 2016), according to satellite data. Increasing temperature and CO_2 have direct effects on plant metabolic process, and lower temperature and lower CO_2 partial pressure have been typical features of the alpine habitats. However, due to the global warming, the scenario is changing, and growing season in alpines, which is typically shorter due to the lower temperature now becoming longer day by day. Various remote-sensing-based studies highlighted greening of alpines due to accelerated snowmelt. Even though the growth season has been extended, alpine plants do not always delay senescence in response to accelerated snowmelt. According to Möhl et al. (2022), in certain studies, early senescence was also noted. They demonstrated experimentally that climatic warming-induced early snowmelt causes early senescence in both above and belowground alpine plants. As a result, it is worth noting that alpine plant growth patterns and carbon uptake are heavily influenced by internal controls rather than growing season length, which may reflect an evolutionary adaptation to a short growing season. Changes in plant traits may not be apparent in one or two generations, but may become evident after a few generations. One such experiment, conducted by White et al. (2023), revealed that plant features in *Viscaria alpina* had altered over the last 20 years, with newer generations showing higher drought adaptation. These findings show that alpine species have adapted to the changing environment, but it takes a few generations to fully adjust. Warming has been shown to promote or decrease alpine plant growth; nonetheless, it can enhance growth by increasing air temperature, which is often lower than the ideal temperature necessary for plant growth (Gonsamo et al. 2018). However, the likelihood of spring frost damage increases with warming, which can reduce plant development throughout the warming-induced prolongation of the growing season (Liu et al. 2018).

Alpine shrubs function better at higher elevations as a result of warming, but less well at lower distributional limits due to warming-induced drought limitation. Dolezal et al. (2020) found that climate has a considerable influence on alpine forb recruitment and growth, demonstrating their ability to respond quickly to climatic fluctuation. In another study, Zhou et al. (2024) reported changes in the ecophysiological characteristics of *Elymus nutan* and *Potentilla anserina* due to experimental warming, including increased leaf light utilization efficiency, maximum photochemical efficiency, non-photochemical quenching, relative water content, and leaf area. However, these changes can be species specific, with an individual species being particularly variable in its adaptation to environmental conditions such as warming. As in the case of *E. nutans*, where Zhou et al. (2024) showed an increase in stomatal conductance, photosynthetic rate, transpiration rate, specific leaf volume, and total water content with increasing temperature, *P. anserina* exhibited the opposite features. As a result, the two species' adaptive methods in response to climate change will undoubtedly differ. Regardless of how warming initially benefits plants by breaking the low-temperature barrier in alpines, decreased soil moisture due to increased air and soil temperature eventually becomes a limiting factor, negatively affecting plant photosynthetic capacity (Allison et al. 2018). However, warming caused a shift in plant functional traits, including larger leaves, higher photosynthetic resource-use efficiency, thinner roots, and greater specific root length and nutrient concentrations, resulting in an increase in acquisitive characteristics (Wei et al. 2023).

Plant responses to climate change are not only species-specific, but also uneven throughout alpine growth forms. In one such study, Yao et al. (2023) observed a wide range of changes in the net photosynthetic rate (Pn) in sedge, grass, and shrubs under future temperature scenarios. Other traits such as leaf area and nutritional content, have also been reported to be highly susceptible to climatic variations (Buzzard et al. 2019; Bjorkman et al. 2020). Climate change has also altered the balance between resource availability and utilization in plants, particularly nitrogen. Plant photosynthetic capacity is positively connected with leaf nitrogen levels, which can either increase (Šímová et al. 2017) or decrease (Storkey and Macdonald 2022) when temperatures rise. The correlation is attributed to photosynthetic enzymes and chlorophyll in leaf tissues. Leaf characteristics eventually influenced plant net primary production (NPP); however, this NPP behaved differently aboveground and belowground to warming (Quan et al. 2020). A 4-year warming experiment revealed a significant rise in belowground NPP but had no effect on aboveground NPP. The one other metric that must be examined here is biomass, which is thought to be heavily influenced by warming and other climate change factors due to rapid growth, but this is not necessarily true. Wang et al. (2020) found the first in situ evidence of altering growth patterns in alpine plants using a 35-year dataset of seasonal biomass dynamics in a Tibetan alpine grassland. This demonstrated that, despite early phenology and faster growth in plants, their annual biomass production remained unaltered, but increased in spring due to warming-induced early commencement of plant growth. Global data acquired from 46 open-top chambers (OTC) located throughout the world's alpine mountain belts investigated the possible effects of warming on plant performance. The overall growth and reproductive production response of 91 plant species was significantly positive, with arctic tundra plants responding more positively than alpine tundra plants to warming. Shrubs were the most beneficial plant forms as they grew much larger and reproduced more effectively when temperatures increased. However, forbs and graminoids continue to underperform when compared to shrubs. The other component was increasing precipitation, to which tundra plants responded more positively than alpines.

2.9 Conclusion

The resilience of alpine plants is a direct result of their specialized adaptations to extreme environmental conditions, including low temperatures, intense UV radiation, and limited water availability. However, these ecosystems are increasingly at risk due to climate change, which is altering key abiotic factors such as temperature and precipitation patterns. Systematic monitoring of alpine plant communities is essential to track these changes and assess their impacts on plant physiology, species distribution, and ecosystem dynamics. This data is vital for developing conservation strategies aimed at preserving alpine biodiversity and ecosystem function. Coordinated efforts in research, conservation, and policy are necessary to mitigate the effects of climate change on these sensitive and ecologically significant environments.

Despite significant progress in understanding alpine plant adaptations, several research gaps remain. One key area is the need for more detailed studies on the genetic and physiological mechanisms underlying tolerance and climate change susceptibility in alpine plants. While the broad strategies of morpho-physiological adaptations are known, the specific genes and molecular pathways involved are not fully elucidated. Additionally, there is a lack of long-term data on how alpine plants respond to the combined stresses of warming temperatures, altered precipitation patterns, and increased frequency of extreme weather events.

Future research should focus on integrating genomic approaches with field experiments to identify the adaptive traits that confer resilience to changing environmental conditions. Moreover, there is a need for more extensive monitoring programs that cover a broader range of alpine ecosystems, particularly in understudied regions. This would provide a more comprehensive understanding of how different species and communities are responding to climate change. Cross-disciplinary collaborations that link ecological, physiological, and genetic studies with climate modelling will be critical in predicting future changes and informing conservation efforts. Additionally, research should prioritize the development of management practices that enhance the resilience of alpine ecosystems, such as promoting species with high adaptive potential or implementing measures to reduce human impacts.

References

Alhaithloul HA, Soliman MH, Ameta KL, El-Esawi MA, Elkelish A (2019) Changes in ecophysiology, osmolytes, and secondary metabolites of the medicinal plants of *Mentha piperita* and *Catharanthus roseus* subjected to drought and heat stress. Biomol Ther 10(1):43

Allison SD, Romero-Olivares AL, Lu Y, Taylor JW, Treseder KK (2018) Temperature sensitivities of extracellular enzyme Vmax and Km across thermal environments. Glob Chang Biol 24:2884–2897. https://doi.org/10.1111/gcb.14045

Archaux F (2004) Breeding upwards when climate is becoming warmer: no bird response in the French Alps. Ibis 146:138–144

Barichivich J, Briffa K, Myneni R, Osborn T, Melvin T, Ciais P et al (2013) Large-scale variations in the vegetation growing season and annual cycle of atmospheric CO2 at high northern latitudes from 1950 to 2011. Glob Chang Biol 19:3167–3183

Barry RG (1981) Mountain weather and climate. London: Methuen. 313 pp. £17.50 Prog Phys Geogr Earth Environ, 7(2), 301–302

Baskin CC, Baskin JM (1998) Seeds: ecology, biogeography, and evolution of dormancy and germination, 2nd edn. Academic, San Diego

Becklin KM, Galen C (2009) Intra-and interspecific variation in mycorrhizal associations across a heterogeneous habitat gradient in alpine plant communities. Arct Antarct Alp Res 41(2):183–190. https://doi.org/10.1657/1938-4246-41.2.183

Bhatta KP, Grytnes JA, Vetaas OR (2018) Downhill shift of alpine plant assemblages under contemporary climate and land-use changes. Ecosphere 9(1):e02084. https://doi.org/10.1002/ecs2.2084

Bibi S, Wang L, Li X, Zhou J, Chen D, Yao T (2018) Climatic and associated cryospheric, biospheric, and hydrological changes on the Tibetan Plateau: a review. Int J Climatol 38:e1–e17

Billings WD, Mooney HA (1968) The ecology of Arctic and Alpine plants. Biol Rev Camb Philos Soc 43(4):481–529. https://doi.org/10.1111/j.1469-185X.1968.tb00968.x

Biskup B et al (2016) Heliotropic leaf movement enhances light capture and photosynthesis in alpine environments. J Plant Res 129(6):851–860

Bjorkman AD et al (2017) Plant functional trait change across a warming tundra biome. Nature 562(7725):57–62

Bjorkman AD, García Criado M, Myers-Smith IH, Ravolainen V, Jónsdöttir IS, Westergaard KB et al (2020) Status and trends in Arctic vegetation: evidence from experimental warming and long-term monitoring. Ambio 49:678–692. https://doi.org/10.1007/s13280-019-01161-6
Bliss LC (1971) Arctic and alpine plant life cycles. Annu Rev Ecol Syst 2:405–438
Blumthaler M (2007) Factors, trends and scenarios of UV radiation in arctic-alpine environments. In: Arctic alpine ecosystems and people in a changing environment. Springer, Berlin/Heidelberg, pp 181–193
Bock A, Sparks TH, Estrella N, Jee N, Casebow A, Schunk C, Menzel A (2014) Changes in first flowering dates and flowering duration of 232 plant species on the island of Guernsey. Glob Chang Biol 20(11):3508–3519
Broennimann O, Thuiller W, Hughes G, Midgley GF, Alkemade JR, Guisan A (2006) Do geographic distribution, niche property and life form explain plants' vulnerability to global change? Glob Chang Biol 12(6):1079–1093. https://doi.org/10.1111/j.1365-2486.2006.01157.x
Bush MB, Silman MR, Urrego DH (2004) 48,000 years of climate and forest change in a biodiversity hotspot. Science 303:827–829. https://doi.org/10.1126/science.1090795
Buzzard V, Michaletz ST, Deng Y, He Z, Ning D, Shen L et al (2019) Continental scale structuring of forest and soil diversity via functional traits. Nat Ecol Evol 3:1298–1308. https://doi.org/10.1038/s41559-019-0954-7
Callaghan TV, Jonasson S (1995) Arctic terrestrial ecosystems and environmental change. Philos Trans R Soc Lond Ser A Phys Eng Sci 352(1699):259–276
Chandra S, Chandola V, Nautiyal MC, Purohit VK (2020) Elevated CO_2 causes earlier flowering in an alpine medicinal herb *Aconitum heterophyllum* Wall. Curr Sci 118(11):1650–1651
Chandra S, Singh A, Mathew JR et al (2022) Phenocam observed flowering anomaly of *Rhododendron arboreum* Sm. in Himalaya: a climate change impact perspective. Environ Monit Assess 194:877. https://doi.org/10.1007/s10661-022-10466-1
Chapin FS III, Shaver GR (1996) Physiological and growth responses of arctic plants to a field experiment simulating climatic change. Ecology 77(3):822–840
Cleland EE, Chiariello NR, Loarie SR, Mooney HA, Field CB (2006) Diverse responses of phenology to global changes in a grassland ecosystem. Proc Natl Acad Sci 103(37):13740–13744
Cody ML, McC J (1996) Short-term evolution of reduced dispersal in island plant populations. J Ecol 84:53–61. https://doi.org/10.2307/2261699
Colwell RK, Brehm G, Cardelús CL, Gilman AC, Longino JT (2008) Global warming, elevational range shifts, and lowland biotic attrition in the wet tropics. Science 322:258–261. https://doi.org/10.1126/science.1162547
Comas LH, Becker SR, Cruz VMV, Byrne PF, Dierig DA (2013) Root traits contributing to plant productivity under drought. Front Plant Sci 4:442. https://doi.org/10.3389/fpls.2013.00442
Črepinšek Z, Štampar F, Kajfež-Bogataj L, Solar A (2012) The response of *Corylus avellana* L. phenology to rising temperature in north-eastern Slovenia. Int J Biometeorol 56:681–694
Curtis PS, Snow AA, Miller AS (1994) Genotype-specific effects of elevated CO_2 on fecundity in wild radish (*Raphanus raphanistrum*). Oecologia 97:100–105
Dawes MA et al (2015) Species-specific tree growth responses to 9 years of CO_2 enrichment at the alpine treeline. J Ecol 103(3):906–917. https://doi.org/10.1111/j.1365-2745.2010.01764.x
Dainese M, Aikio S, Hulme PE, Bertolli A, Prosser F, Marini L (2017) Human disturbance and upward expansion of plants in a warming climate. Nature Climate Change 7(8):577–580. https://doi.org/10.1038/nclimate3337
Dieringer G (1991) Variation in individual flowering time and reproductive success of *Agalinis strictifolia* (Scrophulariaceae). Am J Bot 78(4):497–503
Diez JM, D'Antonio CM, Dukes JS, Grosholz ED, Olden JD, Sorte CJ, Miller LP (2012) Will extreme climatic events facilitate biological invasions? Front Ecol Environ 10(5):249–257
Ding MJ, Li LH, Nie Y, Chen Q, Zhang YL (2016) Spatio-temporal variation of spring phenology in Tibetan Plateau and its linkage to climate change from 1982 to 2012. J Mt Sci 13:83–94
Dirmhirn I, Eaton FD (1975) Some characteristics of the Albedo of snow. J Appl Meteorol Climatol 14:375–379

Dolezal J, Kurnotova M, Stastna P, Klimesova J (2020) Alpine plant growth and reproduction dynamics in a warmer world. New Phytol 228(4):1295–1305

Dorji T, Totland Ø, Moe SR, Hopping KA, Pan J, Klein JA (2013) Plant functional traits mediate reproductive phenology and success in response to experimental warming and snow addition in Tibet. Glob Chang Biol 19(2):459–472

Dorji T, Hopping KA, Wang S, Piao S, Tarchen T, Klein JA (2018) Grazing and spring snow counteract the effects of warming on an alpine plant community in Tibet through effects on the dominant species. Agric For Meteorol 263:188–197

Dorji T, Hopping KA, Meng F, Wang S, Jiang L, Klein JA (2020) Impacts of climate change on flowering phenology and production in alpine plants: the importance of end of flowering. Agric Ecosyst Environ 291:106795

Dullinger S, Willner W, Plutzar C, Englisch T, Schratt-Ehrendorfer L, Moser D, Niklfeld H (2012) Post-glacial migration lag restricts range filling of plants in the European Alps. Glob Ecol Biogeogr 21(8):829–840. https://doi.org/10.1111/j.1466-8238.2011.00732.x

Dunne JA, Harte J, Taylor KJ (2003) Subalpine meadow flowering phenology responses to climate change: integrating experimental and gradient methods. Ecol Monogr 73(1):69–86

Ernakovich JG, Hopping KA, Berdanier AB, Simpson RT, Kachergis EJ, Steltzer H, Wallenstein MD (2014) Predicted responses of arctic and alpine ecosystems to altered seasonality under climate change. Glob Chang Biol 20(10):3256–3269

Estiarte M, Peñuelas J (2015) Alteration of the phenology of leaf senescence and fall in winter deciduous species by climate change: effects on nutrient proficiency. Glob Chang Biol 21:1005–1017

Eviner VT, Chapin FS III (2003) Functional matrix: a conceptual framework for predicting multiple plant effects on ecosystem processes. Annu Rev Ecol Evol Syst 34(1):455–485

Frei ER, Ghazoul J, Matter P, Heggli M, Pluess AR (2014) Plant population differentiation and climate change: responses of grassland species along an elevational gradient. Glob Chang Biol 20(2):441–455

Galen C, Stanton ML (1999) Seedling establishment in alpine buttercups under experimental manipulations of growing-season length. Ecology 80:2033–2044

Germino MJ, Smith WK (2000) High resistance to low-temperature photoinhibition in two alpine, snowbank species. Physiol Plant 110(1):89–95. https://doi.org/10.1034/j.1399-3054.2000.110112.x

Ghassemi S, Delangiz N, Lajayer BA, Saghafi D, Maggi F (2021) Review and future prospects on the mechanisms related to cold stress resistance and tolerance in medicinal plants. Acta Ecol Sin 41(2):120–129

Gonsamo A, Chen JM, Ooi YW (2018) Peak season plant activity shift towards spring is reflected by increasing carbon uptake by extratropical ecosystems. Glob Chang Biol 24:2117–2128

Gottfried M, Pauli H, Futschik A et al (2012) Continent-wide response of mountain vegetation to climate change. Nat Clim Chang 2:111–115. https://doi.org/10.1038/nclimate1329

Grabherr G et al (2003) Climate effects on mountain plants. Nature 426(6967):376–377

Guisan A, Theurillat JP, Kienast F (1998) Predicting the potential distribution of plant species in an alpine environment. J Veg Sci 9(1):65–74

Hamid M, Khuroo AA, Malik AH, Ahmad R, Singh CP, Dolezal J, Haq SM (2020) Early evidence of shifts in alpine summit vegetation: a case study from Kashmir Himalaya. Front Plant Sci 11:421

Hättenschwiler S, Körner C (1995) Responses to recent climate warming of Pinus sylvestris and Pinus cembra within their montane transition zone in the Swiss Alps. J Veg Sci 6:357–368

Heinemeyer A, Fitter AH (2004) Impact of temperature on the arbuscular mycorrhizal (AM) symbiosis: growth responses of the host plant and its AM fungal partner. J Exp Bot 55(396):525–534. https://doi.org/10.1093/jxb/erh049

Hemp A (2002) Ecology of the pteridophytes on the southern slopes of Mt. Kilimanjaro—I. Altitudinal distribution. Plant Ecol 159:211–239. https://doi.org/10.1023/A:1015569125417

Holmström KO, Mäntylä E, Welin B, Mandal A, Palva ET, Tunnela OE, Londesborough J (1996) Drought tolerance in tobacco. Nature 379(6567):683–684

Hopping KA, Knapp AK, Dorji T, Klein JA (2018) Warming and land use change concurrently erode ecosystem services in Tibet. Glob Chang Biol 24(11):5534–5548
Hu Q, Qian R, Zhang Y, Zhang X, Ma X, Zheng J (2021) Physiological and gene expression changes of *Clematis crassifolia* and *Clematis cadmia* in response to heat stress. Front Plant Sci 12:624875
Hughes L (2000) Biological consequences of global warming: is the signal already apparent? Trends Ecol Evol 15:56–61
Hughes NM, Smith WK (2007) Attenuation of incident light in Galax urceolata (Diapensiaceae): concerted influence of adaxial and abaxial anthocyanic layers on photoprotection. Am J Bot 94(5):784–790
Huner NP, Öquist G, Hurry VM, Krol M, Falk S, Griffith M (1993) Photosynthesis, photoinhibition and low temperature acclimation in cold tolerant plants. Photosynth Res 37:19–39
Inouye DW (2008) Effects of climate change on phenology, frost damage, and floral abundance of montane wildflowers. Ecology 89(2):353–362. https://doi.org/10.1890/06-2128.1
Jackson ST, Overpeck JT (2000) Responses of plant populations and communities to environmental changes of the late Quaternary. Paleobiology 26(S4):194–220. https://doi.org/10.1017/S0094837300026932
Jan R, Khan M, Adnan M, Asaf S, Asif S, Kim KM, Murad W (2022) Exogenous phytohormones and fertilizers enhance *Jatropha curcas* L. growth through the regulation of physiological, morphological, and biochemical parameters. Plan Theory 11(24):3584
Jeon J, Kim JK, Wu Q, Park SU (2018) Effects of cold stress on transcripts and metabolites in tartary buckwheat (*Fagopyrum tataricum*). Environ Exp Bot 155:488–496
Jiang LL, Wang SP, Meng FD, Duan JC, Niu HS, Xu GP, Wang GJ (2016) Relatively stable response of fruiting stage to warming and cooling relative to other phenological events. Ecology 97(8):1961–1969
Jiang Y, Cheng X, Zi H, Huang X (2021) Physiological and ecological responses of an alpine plant *Picea likiangensis* at different altitudinal gradients. Pak J Bot 53(6):2233–2240
Keller F, Körner C (2003) The role of photoperiodism in alpine plant development. Arct Antarct Alp Res 35(3):361–368
Kelly AE, Goulden ML (2008) Rapid shifts in plant distribution with recent climate change. Proc Natl Acad Sci 105(33):11823–11826. https://doi.org/10.1073/pnas.0802891105
Khan Z, Shahwar D (2020) Role of heat shock proteins (HSPs) and heat stress tolerance in crop plants. In: Sustainable agriculture in the era of climate change. Springer, pp 211–234
Khan MI, Bashir N, Pandith SA, Patil SS, Pable AA, Shah MA, Shahzad A (2023) Low temperature stress modulates the biochemical, metabolic, and molecular behavior of the Trans-Himalayan medicinal herb *Rheum spiciforme* Royle (spiked rhubarb). Ind Crop Prod 193:116154
Khorsand Rosa R, Oberbauer SF, Starr G, Parker La Puma I, Pop E, Ahlquist L, Baldwin T (2015) Plant phenological responses to a long-term experimental extension of growing season and soil warming in the tussock tundra of Alaska. Glob Chang Biol 21(12):4520–4532
Kolaksazov M, Laporte F, Ananieva K, Dobrev P, Herzog M, Ananiev ED (2013) Effect of chilling and freezing stresses on jasmonate content in *Arabis alpina*. Bulgarian J Agr Sci 19(2):15–17
Körner C (1995) Alpine plant life: functional plant ecology of high mountain ecosystems. Springer
Körner C (2007) The use of 'altitude'in ecological research. Trends in ecology & evolution 22(11): 569–574. https://doi.org/10.1016/j.tree.2007.09.006
Körner C (1999) Alpine plant life: functional plant ecology of high mountain ecosystems. Springer
Körner C (2002) Impact of atmospheric changes on high mountain vegetation. In: Mountain environments in changing climates. Routledge, pp 124–132
Körner C (2003) Alpine plant life: functional plant ecology of high mountain ecosystems. Springer, Heidelberg
Körner C (2016) The use of 'altitude' in ecological research. Trends Ecol Evol 21(8):569–574
Körner C (2021) Alpine plant life: functional plant ecology of high mountain ecosystems. Springer
Körner C, Larcher W (1988) Plant life in cold climates. Symp Soc Exp Biol 42:25–57
Körner C, Paulsen J (2004) A worldwide study of high-altitude treeline temperatures. J Biogeogr 31(5):713–732

Křenová Z, Kindlmann P, Shelly JS, Sklenář P, Sivila S, Bílá K, Jaramillo R (2022) Are temperate alpine plants with distinct phenology more vulnerable to extraordinary climate events than their continuously flowering relatives in tropical mountains? Front Ecol Evol 9:804102

Lauscher F (1976) Weltweite Typen der Höhenabhängigkeit des Niederschlags (Worldwide types of altitude dependence of precipitation). Wetter Leben 28:80–90

Lee EK, Cibrian-Jaramillo A, Kolokotronis SO, Katari MS, Stamatakis A, Ott M, DeSalle R (2011) A functional phylogenomic view of the seed plants. PLoS Genet 7(12):e1002411

Lenoir J, Gégout JC, Guisan A, Vittoz P, Wohlgemuth T, Zimmermann NE, Dullinger S, Pauli H, Willner W, Svenning JC (2010) Going against the flow: potential mechanisms for unexpected downslope range shifts in a warming climate. Ecography 33(2): 295–303. https://doi.org/10.1111/j.1600-0587.2010.06279.x

Li YH, Sun ZL, Xu XL, Jin M, Liu YJ, Zhang H, An LZ (2010) Influence of low temperatures on photosystem II photochemistry and expression of the NADPH: protochlorophyllide oxidoreductase in the alpine, subnival perennial, *Chorispora bungeana*. Photosynthetica 48:457–468

Liu Q, Piao S, Janssens IA, Fu Y, Peng S, Lian X, Ciais P, Myneni RB, Peñuelas J, Wang T (2018) Extension of the growing season increases vegetation exposure to frost. Nature Communications 9(1): 426. https://doi.org/10.1038/s41467-017-02690-y

Liu Q, Fu YH, Zhu Z, Liu Y, Liu Z, Huang M et al (2016) Delayed autumn phenology in the Northern Hemisphere is related to change in both climate and spring phenology. Glob Chang Biol 22:3702–3711

Liu X, Lyu SM, Sun DX, Bradshaw CJA, Zhou SR (2017) Species decline under nitrogen fertilization increases community-level competence of fungal diseases. Proc R Soc B Biol Sci 284:20162621

Memmott J, Craze PG, Waser NM, Price MV (2007) Global warming and the disruption of plant-pollinator interactions. Ecol Lett 10:710–717

Meng F, Zhou Y, Wang S, Duan J, Zhang Z, Niu H, Inouye DW (2016) Temperature sensitivity thresholds to warming and cooling in phenophases of alpine plants. Clim Chang 139:579–590

Miller-Rushing AJ, Katsuki T, Primack RB, Ishii Y, Lee SD, Higuchi H (2007) Impact of global warming on a group of related species and their hybrids: cherry tree (Rosaceae) flowering at Mt. Takao, Japan. Am J Bot 94(9):1470–1478

Möhl P, von Büren RS, Hiltbrunner E (2022) Data from: growth of alpine grassland will start and stop earlier under climate warming. Figshare. https://doi.org/10.6084/m9.figshare.20440497

Mondoni A, Rossi G, Orsenigo S, Probert RJ (2012) Climate warming could shift the timing of seed germination in alpine plants. Ann Bot 110(1):155–164

Morales P, Hickler T, Rowell DP, Smith B, Sykes MT (2007) Changes in European ecosystem productivity and carbon balance driven by regional climate model output. Glob Chang Biol 13(1):108–122

Nagahama A, Kubota Y, Satake A (2018) Climate warming shortens flowering duration: a comprehensive assessment of plant phenological responses based on gene expression analyses and mathematical modeling. Ecol Res 33:1059–1068

Nagy L, Grabherr G (2009) The biology of alpine habitats. Oxford University Press

Niinemets U (2007) Photosynthesis and resource distribution through plant canopies. Plant Cell Environ 30(9):1052–1071. https://doi.org/10.1111/j.1365-3040.2007.01683.x

Notarnicola RF, Nicotra AB, Kruuk LE, Arnold PA (2023) Effects of warming temperatures on germination responses and trade-offs between seed traits in an alpine plant. J Ecol 111(1):62–76

Orsenigo S, Fenu G, Gargano D, Montagnani C, Abeli T, Alessandrini A et al (2021) Red list of threatened vascular plants in Italy. Plant Biosyst 155(2):310–335

Panchen ZA, Gorelick R (2017) Prediction of Arctic plant phenological sensitivity to climate change from historical records. Ecol Evol 7(5):1325–1338

Park T, Chen C, Macias-Fauria M, Tømmervik H, Choi S, Winkler A et al (2019) Changes in timing of seasonal peak photosynthetic activity in northern ecosystems. Glob Chang Biol 25:2382–2395

Pauchard A, Kueffer C, Dietz H, Daehler CC, Alexander J, Edwards PJ et al (2009) Ain't no mountain high enough: plant invasions reaching new elevations. Front Ecol Environ 7(9):479–486

Pauli H, Gottfried M, Dullinger S, Abdaladze O, Akhalkatsi M, Alonso JLB et al (2012) Recent plant diversity changes on Europe's mountain summits. Science 336(6079):353–355. https://doi.org/10.1126/science.1219033

Peacock AJ (1998) ABC of oxygen: oxygen at high altitude. BMJ 317(7165):1063–1063. https://doi.org/10.1136/bmj.317.7165.1063

Piao S, Liu Q, Chen A, Janssens IA, Fu Y, Dai J, Zhu X (2019) Plant phenology and global climate change: current progresses and challenges. Glob Chang Biol 25(6):1922–1940

Pohl M, Stroude R, Buttler A, Rixen C (2011) Functional traits and root morphology of alpine plants. Ann Bot 108(3):537–545. https://doi.org/10.1093/aob/mcr169

Pool RJ (1923) Xerophytism and comparative leaf anatomy in relation to transpiring power. Bot Gaz 76(3):221–240

Post E, Stenseth NC (1999) Climatic variability, plant phenology, and northern ungulates. Ecology 80(4):1322–1339

Prevéy JS, Rixen C, Rüger N, Høye TT, Bjorkman AD, Myers-Smith IH et al (2019) Warming shortens flowering seasons of tundra plant communities. Nat Ecol Evol 3(1):45–52

Quan Q, Zhang F, Meng C, Ma F, Zhou Q, Sun F, Niu S (2020) Shifting biomass allocation determines community water use efficiency under climate warming. Environmental Research Letters 15(9): 094041. https://doi.org/10.1088/1748-9326/aba472

Rahman IU, Hart R, Afzal A, Iqbal Z, Alqarawi AA, Abd-Allah EF, Calixto ES (2019) Ecophysiological plasticity and cold stress adaptation in Himalayan alpine herbs: *Bistorta affinis* and *Sibbaldia procumbens*. Plan Theory 8(10):378

Rana SK, Rana HK, Ghimire SK, Shrestha KK, Ranjitkar S (2017) Predicting the impact of climate change on the distribution of two threatened Himalayan medicinal plants of Liliaceae in Nepal. J Mt Sci 14:558–570. https://doi.org/10.1007/s11629-015-3822-1

Rana SK, Rana HK, Stöcklin J, Ranjitkar S, Sun H, Song B (2022) Global warming pushes the distribution range of the two alpine 'glasshouse' Rheum species north and upwards in the Eastern Himalayas and the Hengduan Mountains. Front Plant Sci 13:925296

Read DJ, Haselwandter K (1981) Observations on the mycorrhizal status of some alpine plant communities. New Phytol 88(3):341–352. https://doi.org/10.1111/j.1469-8137.1981.tb01729.x

Robinson SC, Ketchledge EH, Fitzgerald BT, Raynal DJ, Kimmerer RW (2010) A 23-year assessment of vegetation composition and change in the Adirondack alpine zone, New York State. Rhodora 112(952):355–377

Rundel PW (1994) Tropical alpine environments: plant form and function. Cambridge University Press

Sage RF, Kubien DS (2007) The temperature response of C3 and C4 photosynthesis. Plant Cell Environ 30(9):1086–1106. https://doi.org/10.1111/j.1365-3040.2007.01682.x

Salick J, Fang ZD, Hart R (2019) Rapid changes in eastern Himalayan alpine flora with climate change. Am J Bot 106:520–530. https://doi.org/10.1002/ajb2.1263

Scherrer D, Koerner C (2010) Infra-red thermometry of alpine landscapes challenges climatic warming projections. Glob Chang Biol 16(9):2602–2613

Schmid SF, Stöcklin J, Hamann E, Kesselring H (2017) High-elevation plants have reduced plasticity in flowering time in response to warming compared to low-elevation congeners. Basic Appl Ecol 21:1–12

Schwartz MD (ed) (2003) Phenology: an integrative environmental science, vol 132. Kluwer Academic Publishers

Schwienbacher E, Navarro-Cano JA, Neuner G, Erschbamer B (2011) Correspondence of seed traits with niche position in glacier foreland succession. Plant Ecol 213:371–382

Shen M, Piao S, Dorji T, Liu Q, Cong N, Chen X et al (2015) Plant phenological responses to climate change on the Tibetan Plateau: research status and challenges. Natl Sci Rev 2:454–467

Shen HF, Zhao B, Xu JJ, Liang W, Huang WM, Li HH (2017) Effects of heat stress on changes in physiology and anatomy in two cultivars of Rhododendron. S Afr J Bot 112:338–345

Sherry RA, Zhou X, Gu S, Arnone JA III, Schimel DS, Verburg PS, Luo Y (2007) Divergence of reproductive phenology under climate warming. Proc Natl Acad Sci 104(1):198–202

Shi P et al (2015) Contrasting growth responses of seedling and adult trees of two dominant species of subtropical forests to experimental warming. Sci Rep 5:9252
Šímová I, Rueda M, Hawkins BA (2017) Stress from cold and drought as drivers of functional trait spectra in North American angiosperm tree assemblages. Ecol Evol 7:7548–7559. https://doi.org/10.1002/ece3.3297
Slattery D (2015) Alpine vegetation—what grows where? In: Australian Alps. CSIRO Publishing, p 59. https://doi.org/10.1071/978148630172007.59.86.2015.7
Smith MD (2011) The ecological role of climate extremes: current understanding and future prospects. J Ecol 99(3):651–655
Soukup A, Pecková E, Ježková B, Sklenář P (2021) Structural adaptations in plants from the humid equatorial Andes indicate a trade-off between hydraulic transport efficiency and safety. Am J Bot 108(11):2127–2142. https://doi.org/10.1002/ajb2.1799
Suzuki S, Kudo G (2000) Responses of alpine shrubs to simulated environmental change during three years in the mid-latitude mountain northern Japan. Ecography 23(5): 553-564. https://doi.org/10.1111/j.1600-0587.2000.tb00174.x
Staniak M, Czopek K, Stępień-Warda A, Kocira A, Przybyś M (2021) Cold stress during flowering alters plant structure, yield and seed quality of different soybean genotypes. Agronomy 11(10):2059
Storkey J, Macdonald AJ (2022) The role of long-term experiments in validating trait-based approaches to achieving multifunctionality in grasslands. Front Agric Sci Eng 9:187–196. https://doi.org/10.15302/J-FASE-2021438
Streb P, Shang W, Feierabend J et al (1998) Divergent strategies of photoprotection in high-mountain plants. Planta 207:313–324. https://doi.org/10.1007/s004250050488
Tao Z, Shen L, Gu X, Wang Y, Yu H, He Y (2017) Embryonic epigenetic reprogramming by a pioneer transcription factor in plants. Nature 551(7678):124–128
Thomas CD (2011) Translocation of species, climate change, and the end of trying to recreate past ecological communities. Trends Ecol Evol 26:216–221. https://doi.org/10.1016/j.tree.2011.02.006
Thórhallsdóttir TE (1998) Flowering phenology in the central highland of Iceland and implications for climatic warming in the Arctic. Oecologia 114(1):43–49
Vanneste T, Valdés A, Verheyen K, Perring MP, Bernhardt-Römermann M, Andrieu E et al (2019) Functional trait variation of forest understorey plant communities across Europe. Basic Appl Ecol 34:1–14. https://doi.org/10.1016/j.baae.2018.09.004
Venn S, Pickering C, Green K (2012) Short-term variation in species richness across an altitudinal gradient of alpine summits. Biodivers Conserv 21:3157–3186
Vesk PA, Westoby M (2004) Funding the bud bank: a review of the costs of buds. Oikos 106(1):200–208
Walck JL, Hidayati SN, Dixon KW, Thompson KEN, Poschlod P (2011) Climate change and plant regeneration from seed. Glob Chang Biol 17(6):2145–2161
Wang G, Baskin CC, Baskin JM, Yang X, Liu G, Ye X et al (2018) Effects of climate warming and prolonged snow cover on phenology of the early life history stages of four alpine herbs on the southeastern Tibetan Plateau. Am J Bot 105(6):967–976
Wang L, Wang WJ, Wu Z, Du H, Zong S, Ma S (2019) Potential distribution shifts of plant species under climate change in Changbai Mountains, China. Forests 10(6):498
Wang H, Liu H, Cao G, Ma Z, Li Y, Zhang F et al (2020) Alpine grassland plants grow earlier and faster but biomass remains unchanged over 35 years of climate change. Ecol Lett 23(4):701–710. https://doi.org/10.1111/ele.13474
Wei B, Zhang DY, Wang GQ, Liu Y, Li QL, Zheng ZH et al (2023) Experimental warming altered plant functional traits and their coordination in a permafrost ecosystem. New Phytol 240:1802–1816. https://doi.org/10.1111/nph.19115
White FJ, Hay FR, Abeli T, Mondoni A (2023) Two decades of climate change alters seed longevity in an alpine herb: implications for ex situ seed conservation. Alp Bot 133(1):11–20
Wolf AA, Zavaleta ES, Selmants PC (2017) Flowering phenology shifts in response to biodiversity loss. Proc Natl Acad Sci 114(13):3463–3468

Yao HY, Li XY, Zhang CC, Wang P, Shi FZ, Deng YJ (2023) Tradeoff in the supply and demand for CO_2 dominates the divergence of net photosynthesis rates of functional plants in alpine ecosystems. Ecohydrology 16:e2487. https://doi.org/10.1002/eco.2487

Zandalinas SI, Balfagón D, Arbona V, Gómez-Cadenas A, Inupakutika MA, Mittler R (2016) ABA is required for the accumulation of APX1 and MBF1c during a combination of water deficit and heat stress. J Exp Bot 67(18):5381–5390

Zhou H, Li Z, Zhang B, Du F, Xue J (2024) Sources and chemical weathering implications of strontium and hydrochemistry in an inland alpine permafrost basin. Geochem Geophys Geosyst 25(6):e2024GC011432

Zobayed SMA, Afreen F, Kozai T (2005) Temperature stress can alter the photosynthetic efficiency and secondary metabolite concentrations in St. John's wort. Plant Physiol Biochem 43(10–11):977–984

Zu K, Wang Z, Zhu X et al (2021) Upward shift and elevational range contractions of subtropical mountain plants in response to climate change. Sci Total Environ 783:146896. https://doi.org/10.1016/j.scitotenv.2021.146896

3 Changes in Growth and Metabolism of Alpine Plants Along Altitudinal and Microhabitat Gradients

Vaishali Chandola, Pallavi Sati, Sudeep Chandra, V. K. Purohit, A. R. Nautiyal, and Suman Chandra

Abstract

Alpine ecosystems comprise unique plant communities with diverse floral characteristics and growth habits with adaptations to low temperatures and partial pressure of gases and short growing season, and therefore are particularly rich in secondary metabolites. The floral community of the alpines is shaped by a number of abiotic and biotic factors often working in interaction with each other. These factors are significant determinants of the growth, metabolism, and structure of a plant community. Changes in any of the factors owing mostly to the anthropogenic and/or natural reasons might alter the overall dynamics of the alpine community. The present review thus focuses and discusses the factors that influence alpine vegetation and the changes that have been studied in alpine communities along altitudinal and habitat gradient in relation to the biotic and abiotic components of the ecosystem.

V. Chandola (✉) · P. Sati (✉) · S. Chandra · V. K. Purohit · A. R. Nautiyal
High Altitude Plant Physiology Research Center, HNB Garhwal University, Srinagar Garhwal, Uttarakhand, India

S. Chandra
School of Pharmacy, The University of Mississippi, University, MS, USA

S. Chandra, H. Lata (eds.), *High Altitude Medicinal Plants*,
https://doi.org/10.1007/978-3-031-99406-7_3

Abbreviations

Asl	Above sea level
Pn	Photosynthetic rate
CO_2	Carbon dioxide
UV	Ultraviolet
SOD	Superoxide dismutase
CAT	Catalase
APX	Ascorbate peroxidase
POD	Peroxidase
DPPH	2,2-Diphenyl-1-picrylhydrazyl
DHAR	Dehydroascorbate reductase
GR	Glutathione reductase
MDHAR	Monodehydroascorbate reductase
RUBISCO	Ribulose-1,5-bisphosphate carboxylase/oxygenase
N	Nitrogen
P	Phosphorus
K	Potassium
MDA	Melondialdehyde
IAA	Indole-3-acetic acid
GA	Gibberellic acid
ABA	Abscisic acid
SLA	Specific leaf area

3.1 Introduction

Altitude and microhabitat are critical determinants of plant growth and metabolism, shaping physiological processes and ecological interactions in diverse ways. Altitude imposes various environmental stressors, including changes in temperature, precipitation, and radiation levels, which directly impact plant physiology. As altitude increases, temperature decreases and so do the partial pressure of gases like oxygen and carbon dioxide, leading to physiological adaptations in plants to cope with these stressors. For instance, alpine plants often exhibit modifications in leaf morphology, such as reduced leaf size and increased thickness, to minimize water loss and enhance photosynthetic efficiency (Körner 2003). Additionally, changes in atmospheric pressure and gas concentrations at higher altitudes can influence carbon assimilation rates and respiratory metabolism in plants (Volkov et al. 2020). Furthermore, microhabitat characteristics, such as soil composition, moisture levels, and topography, vary along altitudinal gradients, creating unique niches for plant communities. These microhabitat differences influence nutrient availability, root distribution, and microbial interactions, thereby shaping plant growth and metabolism (Jones et al. 2016).

Moreover, microhabitat heterogeneity within alpine ecosystems further modulates plant responses to altitude. Sheltered microsites, such as rock crevices and snow beds, can provide refuge from extreme environmental conditions and facilitate plant establishment and growth (Alexander et al. 2015). These microhabitats often exhibit higher soil moisture levels and nutrient concentrations compared to exposed areas, promoting favorable conditions for plant growth. Alpine plants exhibit plasticity in their physiological and morphological traits in response to microhabitat variability, allowing them to exploit available resources efficiently (Cavieres et al. 2016). Thus, understanding the interactive effects of altitude and microhabitat on plant growth and metabolism is crucial for elucidating plant adaptations to alpine environments and predicting ecosystem responses to climate change.

At altitude and microhabitat levels, both abiotic and biotic factors play crucial roles in shaping the growth and metabolism of plants in alpine ecosystems. Abiotic factors include temperature, precipitation, soil composition, and radiation levels. Temperature decreases with increasing altitude, leading to physiological adjustments in plants to cope with cold stress (Körner 2003). Precipitation patterns vary along altitudinal gradients, influencing water availability and drought stress, particularly in arid alpine regions (Dai et al. 2020). Soil characteristics, such as texture, nutrient content, and pH, change with altitude, affecting nutrient availability and root growth (Körner 2016). Furthermore, radiation levels increase at higher altitudes due to reduced atmospheric filtration, influencing photosynthetic rates and plant metabolism (Volkov et al. 2020). Biotic factors, including competition, herbivory, mutualistic interactions, and microbial communities, also influence plant growth and metabolism. Competition for resources intensifies at lower altitudes, where plant density is higher, leading to reduced growth rates and altered resource allocation patterns (Li et al. 2022). Herbivory pressure increases with altitude due to reduced predator abundance, impacting plant defense strategies and leaf morphology (Wang et al. 2023). Additionally, mutualistic interactions with mycorrhizal fungi and microbial communities influence nutrient uptake and metabolism in alpine plants, with species composition varying along altitudinal gradients (Johnson et al. 2015; Yang et al. 2021). Understanding the combined effects of these abiotic and biotic factors is essential for predicting plant responses to environmental changes in alpine ecosystems and informing conservation strategies.

Alpine plants have evolved various adaptations at the genetic, morphological, physiological, and biochemical levels to thrive in the extreme conditions of high-altitude environments. These adaptations enable them to cope with challenges such as low temperatures, high ultraviolet (UV) radiation, and nutrient-poor soils (García-Plazaola et al. 2015). Because of their adaptability, plants may live in a variety of environments along altitudinal gradients. Thus, by examining these adaptive traits and connecting them to different abiotic and biotic variables, it will be easier to comprehend the limitations placed on plant distribution by altitudinal gradients. The process of adapting to cold stress involves significant modifications in gene expression, leading to changes in the makeup of the proteome, metabolome, and transcriptome. Many alpine plants exhibit a compact growth form, minimizing exposure to harsh environmental conditions and reducing heat loss (Magaña Ugarte et al. 2019). Some species form cushion-like structures close to the ground, providing insulation and protection from wind and temperature extremes such as *Silene acaulis*,

Ligusticum porteri, *Saxifraga* spp., *Dryas octopetala*, *Azorella* spp., *Ranunculus uniflorus,* and *Oreopolus glacialis* (Musarella et al. 2020). Alpine plants often have physiological mechanisms to tolerate cold temperatures, such as the production of antifreeze proteins (Deswal and Sharma 2014) or the ability to supercool without ice formation in their tissues (Rathore et al. 2018). Efficient photosynthesis is crucial in low-temperature environments. Alpine plants may have adapted photosynthetic machinery to function optimally at low temperatures (Streit et al. 2014). Production of secondary metabolites, such as flavonoids and phenolics, is common in alpine plants (Nataraj et al. 2022). These compounds serve as antioxidants and herbivory and UV protectants (Shi et al. 2022). Enzymes involved in metabolic processes may have higher temperature optima, ensuring biochemical reactions proceed effectively in colder conditions (Bashyal 2023).

Alpine plants often flower early in the growing season to maximize the utilization of the short favorable period for reproduction (Shi et al. 2022). Many alpine plants rely on wind pollination as pollinators may be scarce at high altitudes (Holtmeier and Broll 2010). As plants from high-altitude areas are very sensitive to fluctuation of different variables at higher elevation, it is necessary to understand various changes in plants at genetic, morphological, physiological, and biochemical levels. In this chapter, we have discussed various changes associated with high-altitude plants along an altitudinal gradient to understand that what is the magnitude of the changes at different levels (genetic, morphological, physiological, anatomical, and biochemical) in different plant groups (tree, shrubs, and herbs). Additionally, how these modifications are useful to understand the plant adaptation with changing elevations.

3.2 Major Abiotic and Biotic Factors Working at Altitudinal and Microhabitat Gradient

3.2.1 Abiotic Factors

3.2.1.1 Temperature

Temperature decreases with increasing altitude, affecting plant physiological processes such as photosynthesis, respiration, and growth. Microclimatic variations within microhabitats further modulate temperature regimes, influencing plant distribution and phenology. Temperature influences plant growth, development, and metabolic processes, with cold stress affecting enzyme activities and carbon assimilation rates. Temperature gradients within microhabitats can create thermal niches for plant species, influencing their physiological responses (Körner 2003; Jones et al. 2016).

Warming temperatures generally lead to earlier snowmelt, longer growing seasons, and shifts in plant phenology, such as earlier flowering and leaf-out (Chandra et al. 2020). These changes can result in mismatches between plant life cycles and the availability of pollinators, potentially reducing reproductive success (Hülber et al. 2010). Additionally, higher temperatures can cause an upward shift in the distribution of alpine plants as they migrate to higher elevations in search of suitable climatic conditions, which may lead to increased competition and changes in community composition (Lenoir et al. 2008). Conversely, decline in temperature or increased frequency of extreme cold events can damage plant tissues, reduce growth rates, and limit the geographical range of species (Bjorkman et al. 2020). Cold-induced stress can particularly affect seedlings and juvenile plants, leading to higher mortality rates and reduced recruitment (Inouye 2000). Both warming and cooling trends can thus disrupt the delicate balance of alpine ecosystems, affecting biodiversity and ecosystem functioning (Körner 2003; Walther et al. 2005).

3.2.1.2 Precipitation

Precipitation patterns vary along altitudinal gradients, impacting water availability and drought stress in alpine plants. Microhabitat characteristics, such as soil moisture levels and topography, further influence water availability and plant–water relations. Precipitation patterns and soil moisture levels influence water availability for alpine plants, affecting stomatal conductance, transpiration, and nutrient uptake. Water stress can lead to reduced growth rates, altered carbon allocation, and changes in leaf morphology (Dai et al. 2020; Alexander et al. 2015).

Changes in precipitation, whether increases or decreases, significantly affect alpine plants by altering soil moisture, nutrient availability, and plant physiology. Increased precipitation can enhance soil moisture and promote plant growth, but excessive water can lead to waterlogging and root diseases, adversely affecting plant health (Pauli et al. 2012). Conversely, decreased precipitation and prolonged drought conditions can reduce soil moisture, leading to water stress, stunted growth, and higher mortality rates among alpine plants (Körner 2003). Drought conditions also exacerbate competition for limited water resources, impacting plant community structure and composition (Winkler et al. 2019). Furthermore, changes in precipitation patterns can disrupt the timing of plant life cycles, such as flowering and seed set, thereby affecting reproductive success and long-term survival (Galen et al. 1999). Both increases and decreases in precipitation thus pose challenges for alpine plants, highlighting the importance of understanding hydrological impacts on these sensitive ecosystems.

3.2.1.3 Soil Composition

Soil properties, including texture, nutrient content, and pH, change with altitude and microhabitat, influencing nutrient availability and root growth. Soil microbial communities also vary along gradients, affecting nutrient cycling and plant–microbe interactions. Soil nutrient availability and microbial activity influence nutrient uptake and metabolism in alpine plants. Nutrient limitations can constrain plant growth and productivity, particularly at higher altitudes and in nutrient-poor microhabitats (Bardgett et al. 2020; Körner 2016).

3.2.1.4 Radiation

Solar radiation levels increase at higher altitudes due to reduced atmospheric filtration, affecting photosynthetic rates, and plant metabolism. Microhabitat characteristics, such as slope aspect and vegetation cover, further influence radiation levels and plant responses. Solar radiation influences photosynthetic rates, carbon fixation, and secondary metabolism in alpine plants, with variations in radiation levels affecting plant distribution and community composition. Shade-adapted species may exhibit different metabolic responses compared to sun-adapted species, depending on microhabitat characteristics (Volkov et al. 2020; Scherrer and Körner 2011).

3.2.2 Biotic Factors

3.2.2.1 Competition

Competition for resources, such as water, nutrients, and light, intensifies at lower altitudes and within microhabitats characterized by favorable conditions. Studies by Li et al. (2022) demonstrated reduced growth rates and altered resource allocation patterns in alpine plants subjected to high competition levels. Competition for resources limits plant growth and metabolism, particularly at lower altitudes and in resource-rich microhabitats. Resource availability, including water and nutrients, shapes plant physiological processes and carbon allocation strategies (Körner 2016).

3.2.2.2 Herbivory

Herbivory pressure varies with altitude and microhabitat, influenced by factors such as predator abundance and plant defense strategies. Wang et al. (2023) highlighted the role of herbivory in shaping leaf morphology and photosynthetic efficiency in alpine ecosystems. Herbivory alters plant morphology, reduces leaf area, and affects carbon assimilation rates, leading to changes in plant growth and metabolic pathways (Fürstenberg-Hägg et al. 2020).

3.2.2.3 Symbiotic Interactions

Mutualistic associations with mycorrhizal fungi and microbial communities influence nutrient uptake and metabolism in alpine plants. Research by Johnson et al. (2015) showed that mycorrhizal symbioses enhance plant resilience to environmental stressors and contribute to ecosystem productivity. Mutualistic associations with mycorrhizal fungi enhance nutrient acquisition and modulate plant responses to environmental stress, influencing metabolic pathways and ecosystem functioning (Yang et al. 2021). An illustrative diagram of the local factors affecting vegetation growth in alpine regions is shown in Fig. 3.1.

3.3 Changes in Alpine Plants Along Altitudinal and Microhabitat Gradient

3.3.1 Changes in Gene Expression

3.3.1.1 Transcriptomic Changes

The process of adapting to cold stress involves significant modifications in gene expression, leading to changes in the makeup of transcriptome. Transcriptomic analysis in shrub species like *Rhododendron sanguineum* showed variation in plant

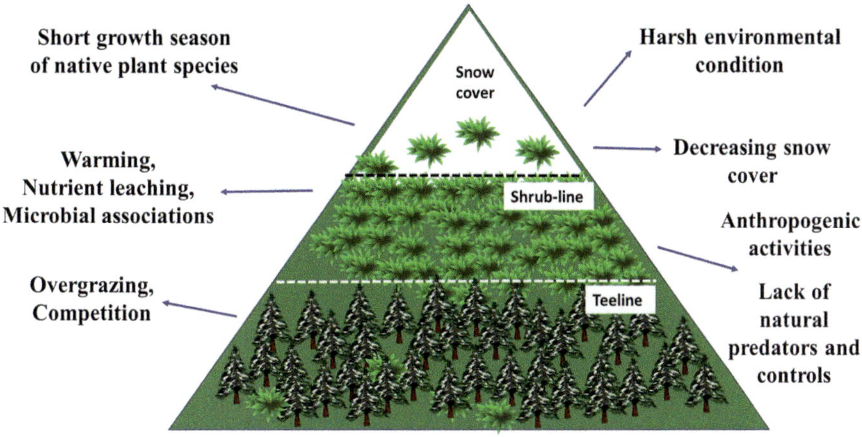

Fig. 3.1 Local factors affecting vegetation growth at alpine regions

species along an altitudinal gradient (Ye et al. 2023). The gene expression patterns in the flower and leaf of *Rhododendron sanguineum* were observed distinct to each organ. Regarding flowers, the identified genes (RsHMGR, RsTPS) were mostly associated with terpenoid metabolism. On the other hand, in leaves, the identified genes (RsCHS, RsF3'5'H) were primarily associated with anthocyanin biosynthesis. Terpenoids constitute the primary constituents of floral aroma, which is believed to serve as an attraction for insects to facilitate pollination. As a result of a decrease in pollinators at higher elevations, it seems that floral organs create a significantly smaller amount of fragrance in order to save energy. The presence of secondary metabolites, such as anthocyanins, in leaves plays a crucial role in determining the plants' alternative adaptive response to harsh environments. These metabolites help the plants cope with selection pressures caused by insect herbivory resulting from environmental changes, as well as competition for resources in biosynthetic pathways at high altitudes (Ye et al. 2023). Zhang et al. (2021a, b) conducted a transcriptome investigation on *Saussurea*, a herbaceous species found at higher elevations and identified over 100 genes to have been involved in DNA repair, membrane transport, reaction to UV-B, reproductive activities, and hypoxia, along with some metabolic functions related to nutrition consumption. These genes are probably important for facilitating *Saussurea* species an ability to adapt to the challenging conditions prevalent at high altitudes. Previous studies have also investigated alterations in the transcriptome of some tree species, for example, *Picea crassifolia* (Shi et al. 2020). The research found that the majority of genes linked to photosynthesis were suppressed, while several genes associated with the production of cuticle lipids and flavonoids were activated in alpine environments and these genes help in regulating plant adaptations and survival at higher elevations.

3.3.1.2 Proteomic Changes

The adaptation to cold stress necessitates substantial modifications in gene expression, resulting in variations to the proteome composition. Proteomic research uncovers a range of stress-responsive proteins, particularly antioxidants, that have accumulated to cope up with the combined stress effects in *Picrorhiza kurroa* along altitudinal gradient (Kumari and Kumar 2024). A total number of 58 protein sites had been determined in *Kobresia pygmaea* that showed considerable variations in expression levels among leaf samples collected from four distinct altitude sites (4600–5100 masl). All of the proteins increased with increasing elevation that helped plants to adapt at extreme environment at alpine area (Li et al. 2014). Another study detected and characterized 83 proteins that were expressed differently in *Lamiophlomis rotate* (Ma et al. 2015), and it was discovered that the levels as well as enzyme activity of proteins involved in the production of hydrogen sulfide significantly increased with increasing altitude. It was concluded that these modifications constitute a response to adaptation shown by plant species in high-altitude environments.

3.3.2 Metabolomic Changes

For the purpose of analyzing the dynamic of metabolites, metabolomics is an extremely useful tool. Production of plant secondary metabolites can enhance the ability of plants to cope up with adverse environmental conditions. Metabolomic analysis showed that a total number of 50 metabolites in the leaves of *Herpetospermum pedunculosum* showed positive correlation with increasing elevation. Metabolomic analysis revealed that phenylpropanoid biosynthesis, phenylalanine metabolism, tyrosine, tryptophan, and phenylalanine biosynthesis, related to the process of biosynthesis of flavonoids, were found to be upregulated in the leaf sample of *Draba oreades* Schrenk collected from higher elevations (Li et al. 2022) that may strengthen the ecological adaptability of plant species in higher-altitude regions. The results of the metabolomic as well as transcriptome analysis revealed that 31 flavonoids had a significant correlation with 227 differentially expressed genes that led to the formation of 412 linked pairings between flavonoids along with the genes that were differently expressed (283 positive and 129 negative). Possibly the processes believed to be responsible for flavonoids accumulation at various altitudes are due to changes in the transport and localization of flavonoids in *Cyclocarya paliurus*. Accumulation of flavonoids increased with increasing elevation. Upregulation of genes that are associated with the production of energy and proteins with increasing elevation makes the plant stronger in terms of adaptability in extreme environmental conditions (Du et al. 2021). An interactive illustration of growth and metabolic changes in alpine plants in response to environmental factors is shown in Fig. 3.2.

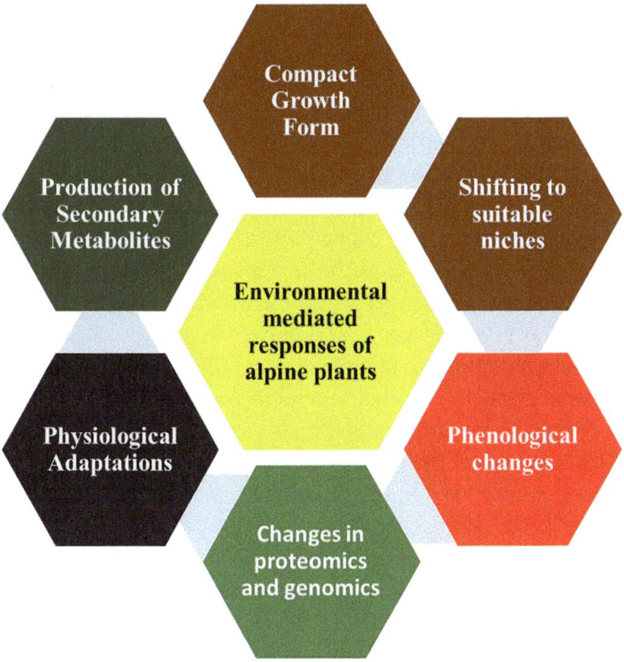

Fig. 3.2 Growth and metabolic changes in alpine plants to environmental factors

3.3.3 Physiological Changes

The influence of elevation on physiological parameters has become the subject of considerable debate, with data indicating that the observed net photosynthetic rate (Pn) at higher elevations is often equivalent to, or even higher, those at lower altitudes (Atkin et al. 2006; Wang et al. 2017). Nevertheless, it is important to acknowledge that the correlation between Pn and elevation may not exhibit a consistent trend across different vegetation categories, such as trees, shrubs, and herbs. Trees at higher elevation exhibited greater variations in Pn compared to herbs and shrubs (Streb and Cornic 2011; García-Plazaola et al. 2015; Shi et al. 2006; Zhai et al. 2024). In comparison to low-altitude locations, the high-altitude zones had lower temperatures and carbon dioxide concentrations, which led to a rise in Pn, 16% for herbaceous plants, 60% for shrubs, and 43% for trees (Zhai et al. 2024); however, results have been found to be species specific and contradictory in some cases. For example, tree species inhabiting higher altitude areas, namely *Polylepis tarapacana* and *Polylepis rugulosa,* showed maximum photosynthetic rate in higher elevation (García-Plazaola et al. 2015). On the other hand, in a herbaceous plant species Pn showed negative correlation with increasing elevation (Streb and Cornic 2011). However, in another study, photosynthetic capacity increased in shrub species *Buddleja davidii* (Shi et al. 2006) at higher elevation. Net photosynthesis is lower at higher elevations due to several factors. The primary reasons include lower temperatures, reduced atmospheric pressure, and limited availability of key resources. Cold temperatures slow down metabolic processes, affecting enzyme activity crucial for photosynthesis. Additionally, lower atmospheric pressure reduces the partial pressure of carbon dioxide, limiting its diffusion into plant stomata. This hinders the carbon fixation process in photosynthesis. Moreover, high elevations may lead to nutrient-poor soils and shorter growing seasons, further constraining plant growth. In combination, these factors contribute to the overall reduction in net photosynthesis at higher elevations (Streb and Cornic 2011; García-Plazaola et al. 2015; Shi et al. 2006; Zhai et al. 2024).

Stomatal conductance is essential for efficient carbon dioxide uptake during photosynthesis. Plants at high elevations may have adapted to maintain sufficient stomatal conductance to optimize photosynthesis under these conditions. Stomatal conduction showed negative correlation with increasing elevation in *Buddleja davidii* (Shi et al. 2006). Plants may adjust their stomatal conductance to balance the need for gas exchange with the risk of excessive water loss and potential freezing damage in colder conditions. On the other hand, high elevations are characterized by lower oxygen levels. Stomatal conductance can influence gas exchange, including oxygen uptake. Some high-altitude plants may have adaptations to enhance oxygen uptake efficiency. For example, Pn and stomatal conductance have been reported increased with increasing elevation in *Herpetospermum pedunculosum* while florescence and transpiration rate showed no significant change with increasing elevation (Zhao et al. 2019).

3.3.4 Biochemical Changes

3.3.4.1 Pigments

Plant photosynthetic capacity is directly impacted by photosynthetic pigments, which are sensitive to environmental conditions and high light intensity. Stress from water, strong sunshine, and high temperatures may reduce the concentration of chlorophyll by delaying the synthesis of chlorophyll, speeding up its breakdown, or damaging the structures of chloroplasts. Leaf chlorophyll content decreased in alpine species such as *Leymus secalinus* (Cui et al. 2018), *Rhododendron anthopogon* (Rathore et al. 2018), *Podophyllum hexandrum,* and *Aconitum balfourii* (Chandra and Lata 2022). In *Picea likiangensis,* chlorophyll content and carotenoid content showed humped shaped pattern along with altitudinal gradient (Jiang et al. 2021). Carotenoids are adaptable substances that are necessary for photosynthesis and the formation of complexes that capture light. They may also effectively eliminate accumulated reactive oxygen species (ROS) to shield chloroplasts from damage caused by photooxidation. Carotenoid content increased with increasing elevation in *Leymus secalinus* (Cui et al. 2018) and *Rhododendron anthopogon* (Rathore et al. 2018), which can protect plants from high light intensity in higher elevation. These could be linked to a decline in photoinhibition along with photosynthesis. While it is well known that high altitudes reduce the concentration of carotenoid, some prior research has shown the opposite trend (Ahmad et al. 2016).

Anthocyanins are pigments responsible for the red, purple, and blue colors in many fruits, flowers, and leaves of plants. At higher elevations, there is increased exposure to ultraviolet (UV) radiation due to thinner atmospheres (Sedej et al. 2020). Anthocyanins can act as a natural sunscreen, absorbing and dissipating UV light, thereby protecting plant tissues from damage caused by excessive radiation. Anthocyanins have antioxidant properties and can help protect plant cells from damage caused by reactive oxygen species, which are produced under stressful conditions (Zhao et al. 2019). Anthocyanin pigments can attract pollinators, such as bees and butterflies, which are crucial for plant reproduction (Landi et al. 2015). Anthocyanin content increased with increasing elevation in *Plantago lanceolata* (Stiles et al. 2007), *Vaccinium* spp. (Zoratti et al. 2015), and *Sambucus nigra* (Senica et al. 2017). Although an opposite trend was also observed in *Herpetospermum pedunculosum* (Zhao et al. 2019). At higher elevations where pollinator activity might be limited, having vibrant anthocyanin-colored flowers can increase the chances of successful pollination.

3.3.4.2 Primary Metabolites

Soluble protein plays a crucial role as osmotic regulators in plants under abiotic stress, which may be caused either directly or indirectly by several environmental variables. Protein content increased in *Pedicularis punctata*, *Plantago major* (Rahman et al. 2020), and *Picea likiangensis* (Jiang et al. 2021) with increasing elevation. The presence of soluble protein and soluble sugar can elevate the concentration of cellular fluid, providing stability to cell membranes and helpful in maintaining osmotic potential. Elevated altitudes may lead to an augmentation in the concentration of soluble proteins, hence enhancing the overall content of functional proteins and promoting efficient cellular metabolism. Previous studies have shown that different plant species accumulated total soluble sugar in response to cold stress. For instance, Bano and Fatima (2009) found that the Hunza Valley in Pakistan's higher-altitude alpine plants had the greatest concentration of total soluble sugar. Total soluble sugar accumulation along the elevational gradient is a great way to shield plant components from cold stress, such as cell membranes. Sugar content increased in *Pedicularis punctata* and *Plantago major* (Rahman et al. 2020), and *Picea likiangensis* (Jiang et al. 2021) with increasing elevation. The elevated levels of soluble sugars found in plants inhabiting alpine environments suggest a greater need for carbohydrate or soluble sugar transportation in order to sustain osmotic equilibrium.

3.3.4.3 Secondary Metabolites

Secondary metabolite increased with increasing altitude in various alpine species. Phenolic chemicals shelter plants from UV-B radiation, which may cause structural damage. Compared to low elevations, plants in alpine areas are likely to have higher phenolics due to more acute UV-B radiation damage. Total phenolic content showed positive correlation with increasing altitude in *Taxus wallichiana* (Adhikari et al. 2022); on the other hand, some species showed negative correlation with increasing elevation such as *Rhododendron anthopogon* (Rathore et al. 2018). It has been suggested that plants use rising phenolic concentrations as a defense mechanism against herbivore assaults (Chadwick et al. 2013). Tannin content increased with increasing elevation in *Taxus wallichiana* (Adhikari et al. 2022). Alkaloids such as aconitine in *Aconitum violaceum* (Rawat et al. 2014), aconitine, and pseudoaconitine in *Aconitum heterophyllum* and *Aconitum balfourii,* respectively (Bahuguna et al. 2013), increased with increasing elevation. Tetracyclic diterpenoid (Taxol) in *Taxus wallichiana* (Adhikari et al. 2022) also increased with increasing elevation. These secondary metabolites play a key role in defense mechanism in plants at high altitudes.

3.3.4.4 Antioxidant Activities

Abiotic stressors may disrupt the equilibrium of free-radical metabolism, leading to an excessive buildup of ROS and subsequent oxidative harm to plants. Plants have created an antioxidant defense mechanism, including of enzymes like SOD, CAT, AP, and POD, to combat oxidative damage under many unfavorable circumstances. SOD and POD activity increased in *Rhododendron anthopogon, Leymus secalinus,* and *Picea likiangensis* (Jiang et al. 2021) with increasing elevation (Rathore et al. 2018; Cui et al. 2018). SOD acts as a principal scavenger of superoxide anion radicals. It facilitates the conversion of superoxide anion radicals into hydrogen peroxide. The level of SOD activity is directly proportional to the extent of protection against oxidative stress-induced damage. POD has the ability to protect ROS during the early stages of senescence. However, it may also trigger the breakdown of chlorophyll and lead to lipid peroxidation during the late stages of senescence. Catalase activity also increased with increasing elevation in *Leymus secalinus* (Cui et al. 2018) and *Picea likiangensis* (Jiang et al. 2021). Ascorbate peroxidase also increased with increasing elevation in *Abies georgei* var. smithii (Guo et al. 2016). Nonenzymatic antioxidant activity (DPPH) increased with increasing elevation in *Taxus wallichiana* (Adhikari et al. 2022). APX, DHAR, GR, SOD, and MDHAR increased in leaves of *Herpetospermum pedunculosum* (Zhao et al. 2019) with increasing elevation. These modifications may facilitate the acclimatization of plants to higher altitudes (Cui et al. 2018). RUBISCO is the enzyme responsible for the process of carbon fixation in photosynthesis. It has a crucial impact on plant growth under various growth circumstances, such as high light intensity and temperatures (Amaral et al. 2024). RUBISCO and phosphoenolpyruvate carboxylase activity decreased with increasing altitude in *Aconitum naviculare* and *Aconitum spicatum* (Bashyal 2023). On the other hand, no significant changes were observed in RUBISCO activity along an altitudinal gradient in *Buddleja davidii* (Shi et al. 2006). The results of these investigations showed that severe weather conditions at higher elevations reduce photosynthetic activity, which has a direct impact on the growth of plants (Amaral et al. 2024).

3.3.4.5 Nutrient Content

Variations in climatic circumstances also lead to major differences in soil physical and chemical properties and vegetation, which are associated with reduced levels of macro- and micronutrients along with increasing elevation (Rathore et al. 2018).

Understanding the relationship between vegetation and soil characteristics along an altitudinal gradient might help us better understand plant communities in alpine environments. Previous research has shown that the development of alpine plant species is greatly influenced by the availability of nutrients. It has been discovered that nitrogen and carbon rise with elevation and have a positive correlation with precipitation and the yearly mean temperature (Zhang et al. 2021a, b). Consequently, it was discovered that there was an increasing tendency toward an altitudinal gradient in pH, soil temperature, C:N ratio, and electrical conductivity (Mangral et al. 2023). However, a reverse trend was recorded in *Rhododendron anthopogon* (Rathore et al. 2018). The total availability of nutrients in the soil is impacted by the decomposition of litter and changes in soil pH. Because it promotes circumstances that accelerate oxidation and immobilization, soil pH is a crucial component in controlling the availability of nutrients at higher elevation. Plants at higher altitudes encounter limited nutrient availability, which may adversely impact their growth and development. Leaf nitrogen and potassium content decreased with increasing altitude in *Buddleja davidii* (Shi et al. 2006), *Aconitum naviculare, Aconitum spicatum* (Bashyal 2023), and *Rheum emodi* (Thapliyal and Shukla 2022). *Aconitum carmichaeli* exhibited decreased growth in low mineral availability (Hou et al. 2006). In *Picea likiangensis,* nutrient contents like nitrogen and phosphorus content showed humped-shaped pattern along with altitudinal gradient (Jiang et al. 2021). Acidic pH and soil mineral deficiency at higher elevation inhibited the growth of *Rhododendron anthopogon* at higher elevation (Rathore et al. 2018).

3.3.4.6 Hormones

One of the stress hormones, abscisic acid is essential for controlling plant water status, tolerance to osmotic stress, and ability to respond to both biotic and abiotic challenges. Hormonal changes are also reported with changing elevation. According to Fahad and Bano (2012), plant leaf concentrations of ABA, IAA, and GA were significantly impacted by changes in altitude. Gibberellins are crucial for controlling a variety of activities that occur throughout the growth and development of plants at higher elevation (Gao et al. 2017). IAA increased with increasing elevation in *Pedicularis punctata* and *Plantago major*. Similarly, ABA also increased in *Herpetospermum pedunculosum* (Zhao et al. 2019) as altitude increased. A potential reason for the accumulation of the stress hormone (ABA) might be a reaction to external stressors, such as extreme cold (Rahman et al. 2020).

3.3.4.7 Membrane-Bound Contents

The most important osmotic regulator in abiotic stress situations is proline, which enables plants to withstand unfavorable circumstances. Prior observations through studies on proline accumulation with increasing altitude have been helpful to understand the stress tolerance via reduction of the freezing point to protect species like *R. anthopogon* (Mangral et al. 2023), and *Picea likiangensis* (Jiang et al. 2021) from freeze damage, underscoring the importance of proline accumulation in plant adaptability. However, reverse trends were also observed in some plants from higher elevation such as in *Pedicularis punctata*, *Plantago major* (Rahman et al. 2020), and *Rhododendron anthopogon* (Rathore et al. 2018). Lipid peroxidation changes, specifically measured as malonaldehyde equivalents, are used to assess the extent of oxidative damage. MDA is produced as a secondary product of processes involving ROS. *Rhododendron anthopogon* and *Leymus secalinus* showed a significant decrease in MDA accumulation with increasing elevation (Rathore et al. 2018; Cui et al. 2018). The decrease in MDA with increasing elevation indicates that plants have a strong capacity to prevent lipid peroxidation, which reflects their stronger resilience to stress. In addition, a low percentage of soil moisture at lower altitudes may lead to oxidative stress by increasing the generation of ROS, which in turn leads to higher rates of lipid peroxidation (Noctor et al. 2014).

3.3.5 Morphological Changes

3.3.5.1 Plant Height

Plant height increased in *Rheum emodi* (Thapliyal and Shukla 2022) and *Aconitum violaceum* (Rawat et al. 2014) as altitude increased. On the other hand, a contradictory trend was also observed in alpine species such as *Aconitum naviculare* and *Aconitum spicatum* (Bashyal 2023). In colder environments, trees might allocate more resources to belowground growth and structural support rather than vertical growth, resulting in shorter tree heights. Wind exerts mechanical stress on trees, especially on their aboveground parts, including branches and stems. Strong winds can cause flexing, swaying, and bending of tree stems, leading to mechanical damage or breakage. In response to wind stress, some trees may develop shorter, sturdier trunks or have a more flexible structure to better withstand wind forces. At higher elevation, ramets of *Vaccinium myrtillus* L. were younger, with shorter shoots and narrower growth rings (Boscutti et al. 2022). Other growth parameters were also reported to decrease with increasing altitude as in alpine tree species such as *Abies alba* and *Fagus sylvatica* (Maxime and Hendrik 2011). Elevation has a positive correlation with both the mean segment lengths and mean growth rates in the tree-ring series in *Juniper tibetica* (He et al. 2013). These changes are associated with plant adaptation at higher elevation where climatic conditions are harsh.

3.3.5.2 Leaf and Needle Orientation

The change in leaf orientation, often observed as a reduction in leaf size or a more compact growth form, in high-altitude plants is typically an adaptive response to the challenging environmental conditions prevalent at higher elevations. Wind can influence the orientation of leaves or needles on a tree. In response to wind, some trees may adjust the angle of their foliage to minimize wind resistance and reduce the risk of wind-induced damage. Leaf morphological characteristics, including length, specific leaf area, and width, show a decline as elevation increased in *Rhododendron companulatum* (Jamloki et al. 2023). This adaptation can help optimize photosynthesis by maintaining an efficient gas exchange. Compact leaf forms and reduced leaf size help minimize heat loss and maintain a more stable microclimate around the plant. This adaptation aids in coping with temperature extremes and conserving energy. Although some opposite trend is also reported in alpine plant species in *Buddleja davidii* (Shi et al. 2006) and *Rhododendron companulatum* (Jamloki et al. 2023), SLA increased in these species with increasing elevation.

3.3.5.3 Leaf Thickness

It increased with increasing altitude in herbaceous plant species such as *Aconitum heterophyllum*, *Aconitum balfourii* (Bahuguna et al. 2013), *Rheum tanguticum* (Shen et al. 2023), *Saxifraga hostii* (Sedej et al. 2020), *Aciphylla glacialis, Oreomyrrhis eriopoda, Wahlenbergia ceracea* (Geange et al. 2017), *Aconitum spicatum,* and *Aconitum naviculare* (Bashyal 2023). However, leaf thickness showed an increase at higher elevations in shrub species *Rhododendron companulatum* (Jamloki et al. 2023). Thicker leaves are often observed to withstand colder temperatures, increased ultraviolet radiation, and lower atmospheric pressure. The thicker cuticle and densely packed mesophyll cells in these leaves contribute to better water retention, minimizing dehydration risks in the harsh mountainous environments. Additionally, the increased thickness may serve as a protective mechanism against potential physical damage from wind or ice (Shen et al. 2023).

3.3.5.4 Plant Biomass

Prior research has also shown the impact of altitudinal gradient on plant biomass. According to the study by Liu et al. (2015), biomass of *Potentilla fruticose* decreased with increasing altitude, whereas *Caragana jubata* showed hump shaped pattern with increasing altitude. Similarly, biomass decreased with increasing altitude in *Aciphylla glacialis*, *Oreomyrrhis eriopoda,* and *Wahlenbergia ceracea* (Geange et al. 2017). Cold temperatures impede enzymatic activity, slowing down growth, while lower atmospheric pressure limits gas exchange. Scarce nutrients, coupled with a shorter growing season, hinder optimal development. These combined stress factors lead to reduced carbon assimilation, resulting in diminished biomass. Alpine plants, thus, undergo adaptive strategies emphasizing resource conservation and resilience in the face of challenging high-altitude conditions (Liu et al. 2015). Root biomass increased with increasing elevation in *Aconitum naviculare* and *Aconitum spicatum* (Bashyal 2023). According to the study, it was due to water scarcity at higher elevation.

3.3.5.5 Trichome

In order to protect themselves from a wide variety of biotic and abiotic challenges, plants have developed a wide variety of defense mechanisms. Through their role as mediators of relationships between the plant and its environment, the form and density of trichomes have an impact on a number of elements of plant eco-physiology. Trichome density increased in *Herpetospermum pedunculosum* (Zhao et al. 2019) with increasing elevation. Together with the stomata, cutin, and wax on the epidermis, trichomes were responsible for a variety of protective tasks. These functions included the synthesis, storage, and secretion of a great deal of essential compounds (Shen et al. 2023).

3.3.5.6 Anatomical Changes

Stomata are tiny openings found on the surfaces of leaves, which function as valves regulated by turgor pressure to regulate the gas exchange. The distribution and dimensions of stomata are crucial adaptation characteristics that exhibit significant variation in response to changing altitude. As altitude rises, there is a drop in air temperature, as well as in the partial pressure of CO_2 and O_2. Stomatal density has been seen to vary along these gradients, although the patterns vary with species. For instance, a study found that as altitude increased, the number of stomata per unit area on the leaves of six pteridophyte species also increased. This adaptation allows for enhanced absorption of oxygen and carbon dioxide. However, no such relationship was detected for five other species (Kessler et al. 2007). In contrast, stomatal density showed decreasing trend in some other alpine species such as *Lamiophlomis rotata* (Ma et al. 2015), *Arabis alpina* (Kammer et al. 2015), and *Herpetospermum pedunculosum* (Zhao et al. 2019). On the other hand, the leaf epidermis serves as a crucial protective layer, shielding plants from potential harm caused by excessive water loss through transpiration and intense exposure to sunlight at elevated elevations. The findings of the previous study demonstrated a positive correlation between altitude and the thickness of both the lower and upper epidermis in *Rhododendron chrysanthum* (Zhang et al. 2014), *Abies georgei, Pinus yunnanensis, Larix potaninii,* and *Quercus aquifolioides* (Liu et al. 2020). However, the palisade, along with spongy mesophyll, plays a crucial role in photosynthesis because these structures contain the chloroplasts, which are the primary sites for the photosynthetic processes. Plant species possessing a denser mesophyll layer may exhibit enhanced photosynthetic efficiency. The thicknesses of spongy mesophyll and palisade exhibited a positive correlation with increasing altitudes in *Polygonum paleaceum* (Wang et al. 2016), *Epilobium amurense, Pedicularis densispica,* and *Potentilla fulgens* (Liu et al. 2020). The primary vascular structure of the leaf is a crucial tissue since it supplies water to the leaf lamina. The vein thickness exhibited an increase in *Polygonum paleaceum* (Wang et al. 2016), *Epilobium amurense, Pedicularis densispica,* and *Potentilla fulgens* (Liu et al. 2020) as the altitudes increased.

3.3.5.7 Phenological Changes

Plants adapt essential physiological, physical characteristics, and reproductive to climatic variations. Phenology refers to the cyclic patterns of plant life, making it a highly responsive bio-indicator for climate change. Multiple studies have demonstrated that changes in altitude modify the sequence and length of phenological events in specific plant species (Wani et al. 2023). According to prior research, altitudinal gradients affected phenological changes in *Rhododendron arboretum* (Singh et al. 2015). The study revealed that flowering pattern was affected by temperature. Furthermore, findings have demonstrated that the reactions of phenological occurrences to climate change are distinctive to each species and differ depending on the blooming functional group, such as plants that flower early in spring or in mid-summer (Wani et al. 2023). Early flowering was observed in lower elevation while the plants grown at higher elevation showed late flowering.

In *Rhododendron grande* and *Rhododendron kenderickii*, maximum seedling mortality was seen at higher altitudes (Bharali et al. 2012). In another alpine shrub species *Juniperus communis,* growth showed negative correlation with increasing elevation; however, the number of seeds showed positive correlation with altitude (Ortiz et al. 2002; Gruwez et al. 2014). Cui et al. (2018) demonstrated that the distribution, growth rates, and reproduction of *Leymus secalinus* were negatively correlated with an increase in altitude on the Qinghai-Tibetan Plateau (Cui et al. 2018). Chandola et al. (2023) also reported a significant difference of 35 ± 5 days in the emergence of *Trillium govanianum*, an alpine herb, at an altitudinal gradient, with an earlier emergence at the lower elevations. However, at elevations of >3500 masl, the emergence was observed to be late with an earlier senescence.

3.3.5.8 Changes in Habitat

Unique biodiversity is supported by mountainous environments, which have diverse vegetation zones varying in altitude. Even with a little height difference, they show significant microclimate change, which impacts the distribution and composition of plant communities. With rising temperatures, it is predicted that plant species' distributional ranges would shift to higher elevations. Phenotypic plasticity allows for quick reproduction, which makes up for the short growth season in a number of alpine as well as subalpine species. Higher altitudes have been the site of more frequent observations of plants exhibiting greater phenotypic plasticity (Rathore et al. 2018; Jamloki et al. 2023; Mangral et al. 2023), enabling them to adjust to the shifting surroundings. The previous findings clearly demonstrated a favorable state of regeneration and indicated the upward shift of the *Rhododendron companulatum* (Jamloki et al. 2023).

3.4 Recent Advances and Future Directions

Recent research has provided insights into the interactive effects of abiotic and biotic factors on alpine plant growth and metabolism. Advances in remote sensing and modeling techniques have improved our understanding of spatial patterns and temporal dynamics in alpine ecosystems (Dobrowski 2011). Molecular studies have elucidated genetic mechanisms underlying plant defense responses and symbiotic interactions (Hao et al. 2023), while ecological research has highlighted the importance of keystone species in mediating plant environmental interactions (Yang and Wang 2024). Future research should focus on integrating multidisciplinary approaches to assess the combined impacts of abiotic and biotic factors across altitudinal and microhabitat gradients, with implications for ecosystem management and conservation.

3.5 Conclusion

The first terrestrial plants appeared about 470 million years ago during the Ordovician epoch, surviving the metamorphoses of the planet through the Devonian, Triassic, Jurassic, Cretaceous, and Tertiary periods. These organisms know how to survive and the key is in their genetics. So, understanding the overall physiology of plants in relation to the ecological drivers of habitat, altitude, biotic, and abiotic factors is crucial.

Plant metabolism alters with the habitat in which it thrives, and recent studies employing advanced technologies are delving deeply to understand the patterns, processes, and complexities of the plant kingdom. The discussion on various adaptive features of alpine plants in relation to habitat and altitudinal gradients provides scientific confirmations on the plasticity of plants. However, with the changing climatic conditions and predicted future scenario, it is of utmost importance to further strengthen the understanding of plant metabolic regulation to be able to sustain life.

Acknowledgments The authors are thankful to the former directors of HAPPRC, Prof. M.C. Nautiyal, Prof. P. Prasad, and colleagues at HAPPRC, for their support and encouragement.

References

Adhikari P, Joshi K, Singh M, Pandey A (2022) Influence of altitude on secondary metabolites, antioxidants, and antimicrobial activities of Himalayan yew (*Taxus wallichiana*). Plant Biosyst 156(1):187–195

Ahmad KS, Hameed M, Fatima S, Ashraf M, Ahmad F, Naseer M (2016) Morpho-anatomical and physiological adaptations to high altitude in some Aveneae grasses from Neelum Valley, Western Himalayan Kashmir. Acta Physiol Plant 38:93. https://doi.org/10.1007/s11738-016-2114-x

Alexander JM, Diez JM, Levine JM (2015) Novel competitors shape species' responses to climate change. Nature 525(7567):515–518

Amaral J, Lobo AK, Carmo-Silva E (2024) Regulation of Rubisco activity in crops. New Phytol 241(1):35–51

Atkin OK, Scheurwater I, Pons TL (2006) High thermal acclimation potential of both photosynthesis and respiration in two lowland Plantago species in contrast to an alpine congeneric. Glob Chang Biol 12(3):500–515

Bahuguna R, Prakash V, Bisht H (2013) Quantitative enhancement of active content and biomass of two aconitum species through suitable cultivation technology. Int J Conserv Sci 4(1):101–106

Bano A, Fatima M (2009) Salt tolerance in *Zea mays* (L.) following inoculation with Rhizobium and Pseudomonas. Biol Fertil Soils 45:405–413. https://doi.org/10.1007/s00374-008-0344-9

Bardgett RD, van der Putten WH, Bezemer TM, Bradford MA, Christensen S, Eppinga MB, Fukami T, Manning P, Martijn Bezemer T, Mckenzie SC, O'Connor M (2020) Below-ground biotic interactions: a systematic review and synthesis of studies examining plant–soil biota interactions and ecosystem processes. New Phytol 215(2):1302–1317

Bashyal S (2023) Plant functional traits of *Aconitum spicatum* and *Aconitum naviculare* along the elevation gradients in Central Nepal. Doctoral dissertation, Department of Botany

Bharali S, Paul A, Khan ML, Singha LB (2012) Survival and growth of seedlings of two Rhododendron tree species along an altitudinal gradient in a temperate broad leaved forest of Arunachal Pradesh, India. Int J Plant Res 2(1):39–46

Bjorkman AD, García Criado M, Myers-Smith IH et al (2020) Status and trends in Arctic vegetation: evidence from experimental warming and long-term monitoring. Ambio 49:678–692. https://doi.org/10.1007/s13280-019-01161-6

Boscutti F, Lami F, Pellegrini E, Buccheri M, Busato F, Martini F, ... & Marini L (2022) Urban sprawl facilitates invasions of exotic plants across multiple spatial scales. Biol Invasions 24(5):1497–1510

Cavieres LA, Hernandez-Fuentes C, Sierra-Almeida A (2016) Facilitation among plants as an insurance policy for diversity in alpine communities. Funct Ecol 30(1):52–59

Chandola V, Chandra S, Kumar RR et al (2023) Reproductive behavior of an endangered Himalayan medicinal herb *Trillium govanianum* Wall. ex D. Don. Vegetos 36:1437–1445. https://doi.org/10.1007/s42535-022-00548-3

Chadwick M, Trewin H, Gawthrop F, & Wagstaff C (2013) Sesquiterpenoids lactones: benefits to plants and people. Int J Mol Sci 14(6):12780–12805

Chandra S, Lata H (2022) Effect of altitude on biochemical and photosynthetic characteristics of *Aconitum balfourii* and *Podophyllum hexandrum*: high value endangered medicinal herbs from Himalayas. Int J Trop Agric 40(3–4):195–203

Chandra S, Chandola V, Nautiyal MC, Purohit VK (2020) Elevated CO_2 causes earlier flowering in an alpine medicinal herb *Aconitum heterophyllum* Wall. Curr Sci 118(11):1650–1651

Cui G, Li B, He W, Yin X, Liu S, Lian L, Zhang P (2018) Physiological analysis of the effect of altitudinal gradients on *Leymus secalinus* on the Qinghai-Tibetan Plateau. PLoS One 13(9):e0202881

Dai A, Luo D, Song M, Liu J (2020) Arctic amplification is caused by sea-ice loss under increasing CO_2. Nat Commun 11(1):1–12

Deswal R, Sharma B (2014) Antifreeze proteins in plants: an overview with an insight into the detection techniques including nanobiotechnology. J Proteins Proteomics 5(2):89–107

Dobrowski SZ (2011) A climatic basis for microrefugia: the influence of terrain on climate. Glob Chang Biol 17(2):1022–1035. https://doi.org/10.1111/j.1365-2486.2010.02263.x

Du Y, Fan L, Xu Z, Wen Z, Cai T, Feijo A, ... & Qiao H (2021) A multi-faceted comparative perspective on elevational beta-diversity: The patterns and their causes. Proc R Soc B 288(1949):20210343

Fürstenberg-Hägg J, Zagrobelny M, Bak S (2020) Plant defense against insect herbivores. Int J Mol Sci 21(3):983. https://doi.org/10.3390/ijms140510242

Fahad S, & Bano A (2012) Effect of salicylic acid on physiological and biochemical characterization of maize grown in saline area. Pak J Bot 44(4):1433–1438

Galen, C., Sherry, R. A., & Carroll, A. B. (1999). Are flowers physiological sinks or faucets? Costs and correlates of water use by flowers of Polemonium viscosum. Oecologia, 118(4), 461-470.

Gao X, Zhang Y, He Z, Fu X (2017) Gibberellins. In: Li JY, Li CY, Smith SM (eds) Hormone metabolism & signaling in plants, vol 4. Academic, pp 107–160

García-Plazaola JI, Rojas R, Christie DA, Coopman RE (2015) Photosynthetic responses of trees in high-elevation forests: comparing evergreen species along an elevation gradient in the Central Andes. AoB Plants 7:plv058

Geange SR, Briceño VF, Aitken NC, Ramirez-Valiente JA, Holloway-Phillips MM, Nicotra AB (2017) Phenotypic plasticity and water availability: responses of alpine herb species along an elevation gradient. Clim Chang Responses 4(1):1–12

Gruwez R, De Frenne P, De Schrijver A, Leroux O, Vangansbeke P, Verheyen K (2014) Negative effects of temperature and atmospheric depositions on the seed viability of common juniper (*Juniperus communis*). Ann Bot 113(3):489–500

Guo QQ, Li HE, Zhang WH (2016) Variations in leaf functional traits and physiological characteristics of *Abies georgei* var. smithii along the altitude gradient in the Southeastern Tibetan Plateau. J Mt Sci 13:1818–1828

Hao Z, Li X, Liu X, Zhang S (2023) Molecular mechanisms underlying plant defense responses to herbivory. Front Plant Sci 14:104

He M, Yang B, Bräuning A (2013) Tree growth–climate relationships of *Juniperus tibetica* along an altitudinal gradient on the southern Tibetan Plateau. Trees 27:429–439

Hou D, Shu G, Ren Z (2006) Effect of nitrogen, phosphorus, and potassium on several nitrogen metabolism enzymes of *Aconitum carmichaeli* Dexb. (Ranunculaceae). Southwest China J Agric Sci 19(5):857–862

Holtmeier KF, & Broll G (2010) Altitudinal and polar treelines in the northern hemisphere causes and response to climate change (Obere und polare Baumgrenze auf der nördlichen Hemisphäre Ursachen und Antwort auf den Klimawandel). Polarforsch 79(3):139–153.

Hülber K, Winkler M, Grabherr G (2010) Biotic interactions in the alpine landscape. In: Carnic JCGG, Virtanen R, Luoto M (eds) Biotic interactions and global change. Springer, pp 217–226

Inouye DW (2000) The ecological and evolutionary significance of frost in the context of climate change. Ecol Lett 3(5):457–463

Jamloki A, Singh A, Chandra S, Shukla V, Nautiyal MC, Malik ZA (2023) Population structure, regeneration potential and leaf morphological traits of *Rhododendron campanulatum* D. Don along an altitudinal gradient in Western Himalaya. Plant Biosyst 157(1):159–174

Jiang Y, Cheng X, Zi H, Huang X (2021) Physiological and ecological responses of an alpine plant *Picea likiangensis* at different altitudinal gradients. Pak J Bot 53(6):2233–2240

Johnson NC, Wilson GWT, Bowker MA, Wilson JA, Miller RM (2015) Resource limitation is a driver of local adaptation in mycorrhizal symbioses. Proc Natl Acad Sci 107(5):2093–2098

Jones HG, Linderholm HW, Schär C (2016) A new perspective for managing climate change risks for alpine river basins. J Hydrol 541:1116–1123

Kammer PM, Steiner JS, Schöb C (2015) *Arabis alpina* and *Arabidopsis thaliana* have different stomatal development strategies in response to high altitude pressure conditions. Alp Bot 125(2):1–12

Kessler M, Siorak Y, Wunderlich M, Wegner C (2007) Patterns of morphological leaf traits among pteridophytes along humidity and temperature gradients in the Bolivian Andes. Funct Plant Biol 34(11):963–971

Körner C (2003) Alpine plant life: functional plant ecology of high mountain ecosystems. Springer

Körner C (2016) Plant adaptation to cold climates. In: Thomas B, Murray B, Murphy D (eds) Encyclopedia of applied plant sciences. Academic, pp 173–178

Kumari M, Kumar R (2024) Functional trait correlation network and proteomic analysis reveal multifactorial adaptation mechanisms to a climatic gradient associated with high altitude in the Himalayan region. Plant Cell Environ 47(5):1556–1574. https://doi.org/10.1111/pce.14830

Landi M, Tattini M, Gould KS (2015) Multiple functional roles of anthocyanins in plant-environment interactions. Environ Exp Bot 119:4–17

Lenoir J, Gégout JC, Marquet PA, De Ruffray P, Brisse H (2008) A significant upward shift in plant species optimum elevation during the 20th century. Science 320(5884):1768–1771. https://doi.org/10.1126/science.1156831

Li X, Yang Y, Ma L, Sun X, Yang S, Kong X, Yang Y (2014) Comparative proteomics analyses of *Kobresia pygmaea* adaptation to environment along an elevational gradient on the central Tibetan Plateau. PLoS One 9(6):e98410

Li Y, Liu W, Wang T, Zhao W, Jiang X, Ma W (2022) Effects of competition on plant growth and physiological responses along an altitudinal gradient. Ecol Evol 12(1):e8534

Liu ZW, Chen RS, Song YX, Han CT (2015) Distribution and estimation of aboveground biomass of alpine shrubs along an altitudinal gradient in a small watershed of the Qilian Mountains, China. J Mt Sci 12:961–971. https://doi.org/10.1007/s11629-013-2854-7

Liu W, Zheng L, Qi D (2020) Variation in leaf traits at different altitudes reflects the adaptive strategy of plants to environmental changes. Ecol Evol 10(15):8166–8175. https://doi.org/10.1002/ece3.6519

Ma L, Yang L, Zhao J, Wei J, Kong X, Wang C, Zhang X, Yang Y, Hu X (2015) Comparative proteomic analysis reveals the role of hydrogen sulfide in the adaptation of the alpine plant *Lamiophlomis rotata* to altitude gradient in the northern Tibetan plateau. Planta 241(4):887–906

Magaña Ugarte R, Escudero A, Gavilán RG (2019) Metabolic and physiological responses of Mediterranean high-mountain and alpine plants to combined abiotic stresses. Physiol Plant 165(2):403–412

Mangral ZA, Islam SU, Tariq L, Kaur S, Ahmad R, Malik AH, Dar TUH (2023) Altitudinal gradient drives significant changes in soil physico-chemical and eco-physiological properties of *Rhododendron anthopogon*: a case study from Himalaya. Front For Glob Change 6:1181299

Maxime C, Hendrik D (2011) Effects of climate on diameter growth of co-occurring *Fagus sylvatica* and *Abies alba* along an altitudinal gradient. Trees 25(2):265–276

Musarella CM, Brullo S, Del Galdo GG (2020) Contribution to the orophilous cushion-like vegetation of central-southern and insular Greece. Plan Theory 9(12):1678

Nataraj N, Hussain M, Ibrahim M, Hausmann AE, Rao S, Kaur S, Olsson SB (2022) Effect of altitude on volatile organic and phenolic compounds of *Artemisia brevifolia* Wall ex Dc. from the Western Himalayas. Front Ecol Evol 10:864728

Noctor G, Mhamdi A, Foyer CH (2014) The roles of reactive oxygen metabolism in drought: not so cut and dried. Plant Physiol 164(4):1636–1648

Ortiz PL, Arista M, Talavera S (2002) Sex ratio and reproductive effort in the dioecious *Juniperus communis* subsp. alpina (Suter) Čelak. (Cupressaceae) along an altitudinal gradient. Ann Bot 89(2):205–211

Pauli H, Gottfried M, Dullinger S, Abdaladze O, Akhalkatsi M, Benito Alonso JL, Coldea G, Dick J, Erschbamer B, Fernández Calzado R, Ghosn D, Holten JI, Kanka R, Kazakis G, Kollár J, Larsson P, Moiseev P, Moiseev D, Molau U, Molero Mesa J, Nagy L, Pelino G, Puşcaş M, Rossi G, Stanisci A, Syverhuset AO, Theurillat JP, Tomaselli M, Unterluggauer P, Villar L, Vittoz P, Grabherr G (2012) Recent plant diversity changes on Europe's mountain summits. Science 336(6079):353–355. https://doi.org/10.1126/science.1219033

Rahman IU, Afzal A, Iqbal Z, Hart R, Abd-Allah EF, Alqarawi AA, Bussmann RW (2020) Response of plant physiological attributes to altitudinal gradient: plant adaptation to temperature variation in the Himalayan region. Sci Total Environ 706:135714

Rathore N, Thakur D, Chawla A (2018) Seasonal variations coupled with elevation gradient drives significant changes in eco-physiological and biogeochemical traits of a high altitude evergreen broadleaf shrub, *Rhododendron anthopogon*. Plant Physiol Biochem 132:708–719

Rawat JM, Rawat B, Mishra S, Negi R, Mishra SN, Chandra A, Nautiyal S (2014) Altitudinal and seasonal variation in bioactive compound aconitine in *Aconitum violaceum*, a threatened medicinal plant of Indian Himalayan region. Int J Adv Res 2:981–988

Scherrer D, Körner C (2011) Topographically controlled thermal-habitat differentiation buffers alpine plant diversity against climate warming. J Biogeogr 38(2):406–416

Sedej TT, Erznožnik T, Rovtar J (2020) Effect of UV radiation and altitude characteristics on the functional traits and leaf optical properties in *Saxifraga hostii* at the alpine and montane sites in the Slovenian Alps. Photochem Photobiol Sci 19:180–192

Senica M, Stampar F, Veberic R, Mikulic-Petkovsek M (2017) The higher the better? Differences in phenolics and cyanogenic glycosides in Sambucus nigra leaves, flowers and berries from different altitudes. J Sci Food Agric 97(8):2623–2632

Shen N, Xu W, Guo F, Cui Y, Chen Y, Su X (2023) Effect of fertilizer and harvesting stage on growth and bioactive compounds content of *Rheum tanguticum*. J Plant Nutr 46(18):4462–4476

Shi Z, Liu S, Liu X, Centritto M (2006) Altitudinal variation in photosynthetic capacity, diffusional conductance and δ13C of butterfly bush (*Buddleja davidii*) plants growing at high elevations. Physiol Plant 128(4):722–731

Shi Z, Deng X, Bai D, Lei J, Li M, Zeng L, Xiao W (2020) Transcriptomic analysis reveals the mechanism of *Picea crassifolia* survival for alpine treeline condition. Forests 11:156. https://doi.org/10.3390/f11020156

Shi S, Shi R, Li T, Zhou D (2022) UV-B radiation effects on the alpine plant *Kobresia humilis* in a Qinghai-Tibet Alpine Meadow. Plan Theory 11(22):3102

Singh N, Ram J, Tewari A, & Yadav RP (2015) Phenological events along the elevation gradient and effect of climate change on Rhododendron arboreum Sm. in Kumaun Himalaya. Curr Sci 106–110

Stiles EA, Cech NB, Dee SM, Lacey EP (2007) Temperature-sensitive anthocyanin production in flowers of *Plantago lanceolata*. Physiol Plant 129(4):756–765

Streb P, Cornic G (2011) Photosynthesis and antioxidative protection in alpine herbs. In: Plants in alpine regions: cell physiology of adaption and survival strategies. Springer Vienna, Vienna, pp 75–97

Streit K, Siegwolf RT, Hagedorn F, Schaub M, Buchmann N (2014) Lack of photosynthetic or stomatal regulation after 9 years of elevated [CO_2] and 4 years of soil warming in two conifer species at the alpine treeline. Plant Cell Environ 37(2):315–326

Thapliyal J, Shukla V (2022) The elevational effect on soil parameters and biochemical constituents of *Rheum emodi* Wall. Ex Meisn: an important medicinal plant of Garhwal Himalaya. Int J Emerg Technol Innov Res 9(11):62–74. http://www.jetir.org/papers/JETIR2211510.pdf

Volkov V, Wang B, Dominy J, Song W, She Y (2020) Thermal ecology of plants along elevation gradients: a global review and outlook. Plant Ecol 221(3):191–207

Walther GR, Beißner S, Burga CA (2005) Trends in the upward shift of alpine plants. J Veg Sci 16(5): 541–548. https://doi.org/10.1658/1100-9233(2005)16[541,TITUSO]2.0.CO;2

Wang YY, Qi DH, Liu WS, Liang WB (2016) Comparison on leaf phenotypic and anatomical structures of *Polygonum paleaceum* along altitudinal gradients at Yulong Mountains. Acta Botan Boreali-Occiden Sin 36:70–77

Wang Y, Liu J, Li T, Ma Y, Ding J, Lu Z (2023) Herbivory alters plant defense strategies and resource allocation patterns in alpine ecosystems. Ecology 104(3):e03456

Wani ZA, Negi VS, Bhat JA, Satish KV, Kumar A, Khan S et al (2023) Elevation, aspect, and habitat heterogeneity determine plant diversity and compositional patterns in the Kashmir Himalaya. Front For Glob Chang 6:1019277. https://doi.org/10.3389/ffgc.2023.1019277

Winkler DE, Lubetkin KC, Carrell AA, Jabis MD, Yang Y, Kueppers LM (2019) Chapter 12—responses of alpine plant communities to climate warming. In: Ecosystem consequences of soil warming. Academic, pp 297–346. https://doi.org/10.1016/B978-0-12-813493-1.00013-2

Yang Y, Wang G (2024) Keystone species mediate biotic interactions in alpine ecosystems. J Ecol 112(4):821–832

Yang Z, Zhang Y, Li X, Wu Y, Gao Y (2021) Differential responses of mycorrhizal symbionts to altitudinal gradients in alpine regions. Fungal Ecol 49:101018

Ye LJ, Möller M, Luo YH, Zou JY, Zheng W, Liu J, Gao LM (2023) Variation in gene expression along an elevation gradient of *Rhododendron sanguineum* var. *haemaleum* assessed in a comparative transcriptomic analysis. Front Plant Sci 14:1133065

Zhai B, Hu Z, Sun S, Tang Z, Wang G (2024) Characteristics of photosynthetic rates in different vegetation types at high-altitude in mountainous regions. Sci Total Environ 907:168071

Zhang CY, Wang Y, Li MS, Xu MZ, Liu XH (2014) Comparative analysis of anatomical structure of *Rhododendron chrysanthum* leaves at different altitudes in Changbai Mountain area. J Agric Sci Yanbian Univ 36:39–43

Zhang Y, Ai J, Sun Q, Li Z, Hou L, Song L et al (2021a) Soil organic carbon and total nitrogen stocks as affected by vegetation types and altitude across the mountainous regions in the Yunnan Province, South-Western China. Catena 196:104872. https://doi.org/10.1016/j.catena.2020.104872

Zhang X, Sun Y, Landis JB, Shen J, Zhang H, Kuang T, Sun W, Sun J, Tiamiyu BB, Deng T et al (2021b) Transcriptomes of Saussurea (Asteraceae) provide insights into high-altitude adaptation. Plan Theory 10:1715. https://doi.org/10.3390/plants10081715

Zhao Y, Xu F, Liu J, Guan F, Quan H, Meng F (2019) The adaptation strategies of *Herpetospermum pedunculosum* (Ser.) Baill at altitude gradient of the Tibetan plateau by physiological and metabolomic methods. BMC Genomics 20:1–15

Zoratti L, Jaakola L, Häggman H, Giongo L (2015) Anthocyanin profile in berries of wild and cultivated Vaccinium spp. along altitudinal gradients in the Alps. J Agric Food Chem 63(39):8641–8650

High-Altitude Medicinal Plants: Traditional Uses and Distribution Patterns

4

Chandra Prakash Kala

Abstract

The high altitudes of the Himalaya have always been the center of curiosity and the explorations for being the rich repositories of medicinal plants. The unique climatic, geographical, geological, and hydrological conditions across the Himalayan range, apart from anthropogenic factors, determine the distribution and diversity of medicinal plant species. Historically, these high-altitude medicinal plants have been one of the major sources of evolution of ancient therapeutic systems, including Ayurveda, Tibetan, and other nonformal traditional systems of medicine. Some of the medicinal plant species have restricted distribution range; hence, they are used locally, whereas others are used across the communities due to wide range of their distribution. This chapter aims to examine the patterns in the distribution of medicinal plant species across the natural habitats in the high-altitude areas of the Himalaya, apart from the traditional uses of these plants by the communities.

4.1 Introduction

The high-altitude areas have always been a subject of considerable interest among the ecologists and plant explorers due to the unique mountain ecosystem, exceptional floristic diversity, and unique patterns in plants' adaptation to extreme environmental conditions. Several pastoral communities use such areas as summer grazing grounds and others earn their livelihood by collecting valuable medicinal herbs as the high-altitude Himalayan and trans-Himalayan flora are well known for considerable medicinal properties, hence the Ayurvedic, Tibetan, and other

C. P. Kala (✉)
Indian Institute of Forest Management, Bhopal, Madhya Pradesh, India

S. Chandra, H. Lata (eds.), *High Altitude Medicinal Plants*,
https://doi.org/10.1007/978-3-031-99406-7_4

nonformal traditional systems of medicine have been using these plant species from time immemorial (Kala 1998, 2005a, b). The existence of a variety of traditional systems of medicines still today is the manifestation of the close relationship between the high-altitude plants and the human societies. The earlier secluded landscapes and the communities depending on such high-altitude areas are no longer disconnected from urban and global societies, leading to significant social and cultural transformations (Kala 2023). Unfortunately, a significant amount of traditional knowledge accumulated through centuries of experimentation by the high-altitude communities on the use of medicinal plants has been lost in recent times. The reasons for the loss of traditional knowledge are many, including the spread of modern marketing forces and weakening of customary norms (Kala 2003; Aswani et al. 2018).

Most of the indigenous communities, especially in the high-altitude areas, are not able to cope up and adapt to the changing scenario. Besides, the outside forces in the form of commercialization compel people to resort to the overexploitation of natural resources, resulting in severe environmental degradation and the shortage of precious natural resources, including lifesaving medicinal plant species (Kala 2005c). Traditional knowledge on medicinal plants can be used to provide better healthcare facilities and also to conserve the precious floristic diversity (Khakurel et al. 2022). In this backdrop, an attempt is made to review the existing information on the traditional uses of high-altitude medicinal plants. Besides, realizing the importance of availability and distribution of such an important healthcare commodity, the patterns in distribution of medicinal plants across the natural habitats in the high-altitude areas of the Himalaya were also explored.

4.2 Traditional Uses of High-Altitude Medicinal Plants

The high-altitude Himalayan region is still occupied by traditional local communities, and their livelihood is closely associated with locally available natural resources. Besides, different types of ailments are prevalent in human communities living in the Himalayan region and elsewhere, and these communities historically depend on the plant species growing in their surrounding areas for the treatment of various such ailments. Communities living in the high-altitude areas remained isolated for a long period due to poor accessibility and harsh climatic and geographical conditions; hence, such isolation has shaped their systems of therapy (Kala 2003, 2023). Moreover, a large number of medicinal plants growing in the high-altitude areas are used in Ayurveda, which is one of the prominent systems of therapy in India (Kala 2005b). On the other hand, in the trans-Himalayan region, the local communities had developed their own healthcare system, which is known as the Tibetan medical system. Since amchis are the practitioners of this system of medicine, it is also known as the Amchi Medical System. In due course, Bauddha Bhiksus (monks) introduced the knowledge of "Ayurveda" in the trans-Himalayan region, and propagated it among their disciples wherever they preached (Kala 2005a). Plants remain the major ingredients in both Ayurveda and Tibetan Medicine Systems, the prominent traditional therapeutic systems in the Indian Himalaya. Traditional uses and plant parts used of some prominent high-altitude medicinal plants in the Himalayas are described in Table 4.1.

Apart from the practitioners of Ayurveda and Amchi Medical Systems, the traditional herbal healers across the ethnic communities have accumulated a rich wealth of knowledge on medicinal plants. For instance, the traditional healers in Monpa community are called Bonpa, and they believe that high-altitude plants possess some specific properties; hence, they are more effective in curing diseases than the low-altitude plant species (Tsering and Tag 2015). Besides, medicinal plants have a strong acceptance in religious activities of the people inhabiting in the Himalayan region, who worship the plants in the form of various gods, goddesses, and local deities (Kala 2009). *Origanum vulgare, Saussurea obvallata, Juniperus communis,* and *Nardostachys grandiflora* are examples of the high-altitude plant species highly used for medicinal as well as religious purposes by the Hindus. Different plant parts such as flowers, fruits, roots, tubers, bark, and stems are used for making herbal medicines. The herbal healers combine number of plant species or the plant parts for the preparation of herbal formulation, which is finally prescribed for the treatment of ailments. Apart from higher plants, the lower plants, such as lichens, are also used by amchis (Kala 2005a). For example, certain lichens are mixed with other ingredients for curing chronic fever and poisoning (Nayaka et al. 2010; Devkota et al. 2017). Many plant species are used for common purposes. At the same time, different parts of the same plant are used for different purposes by different ethnic groups in the Himalaya. Apart from the human use, animal husbandry uses many plant species as its primary source of healthcare.

Some of the medicinal plants being inhabited across the Himalaya are used widely across the communities, whereas others have a narrow range of distribution and hence are used locally. *Picrorhiza kurrooa*, listed in the CITES, is an important herb that is extensively used in Ayurveda, and it grows across the high-altitude region of northwest, eastern, and trans-Himalayan areas. On the other hand, *Rheum nobile*, a well-known medicinal plant, grows in the eastern Himalaya such as Sikkim and has restricted local uses. It is eaten raw to relieve thirst and restlessness and also its rhizomes are used for the treatment of jaundice, dyspepsia, edema, and dysentery (Fei et al. 2017). *Eriophyton wallichii* is another medicinal plant species used as an antiseptic and grows in high-altitude areas of Sikkim (Tsarong 1994).

Saussurea tridactyla or snow lotus grows on the rocky habitats over 3600–5700 m in parts of Sikkim has brilliant white flowers appearing over dark green leaves. It is harvested in July and August and is used as a tonic for weakness, menstrual disorders, and arthritis. It has been overexploited and hence has become rare (www.flowersofindia.net). Like snow lotus, the population of many other important high-altitude medicinal plant species has gone down over the years and the increasing scarcity of such plants may lead to the loss of traditional knowledge on medicinal uses of such species. For example, *Aconitum heterophyllum, Aconitum violaceum, Paris polyphylla, Dactylorhiza hatageria, Podophyllum hexandrum, Picrorhiza kurrooa, Rheum webbianum,* and *Rheum moorcroftianum* have become rare and endangered (Kala 2005c); hence, the traditional knowledge associated with such

Table 4.1 Traditional uses and plant parts used of some prominent high-altitude medicinal plants in the Himalaya

Species	Plant part used	Uses
Acantholimon lycopodioides (Girard) Boiss.	Whole	Cardiac disorder
Aconitum balfourii Stapf	Root	Sciatica
Aconitum heterophyllum Wall. ex. Royle	Root	Tonic, diarrhea, aphrodisiac
Aconitum violaceum Jac. Ex Stapf	Root	Cough
Angelica glauca Edgew.	Root	Gastric complaints
Arnebia euchroma (Royle ex Benth.) Johnston	Root	Hair tonic
Astragalus candoleanus Royle ex Benth	Root	Skin diseases, blood purifier
Bergenia stracheyi (Hk. f. & Th.) Engl.	Root, flower	Blisters
Codonopsis ovata Benth.	Root	Bruises
Corydalis flabellata Edgew.	Whole plant	Fever
Dactylorhiza hatagirea (D.Don) Soo.	Tuber	Kidney complaints, aphrodisiac
Delphinium brunonianum Royle	Seed	Intestinal worm, insomnia
Delphinium cashmerianum Royle	Seed, shoot, flower	Colic, throat-ache
Delphinium viscosum Hk. & Th.	Root	Rheumatism
Ephedra gerardiana Wall. ex Stapf.	Stem, root	Asthma, fever, bronchitis
Euphorbia thomsoniana Boiss.	Whole plant	Skin diseases
Euphorbia tibetica Boiss.	Root	Vomiting
Ferula jaeschkeana Vatke	Root	Chest pain, toothache
Fritillaria roylei Hk.	Bulb	Asthma, tonic
Gentiana kurroo Royle	Flower	Cough, stomachache
Gentianella moorcroftiana (Wall. ex Griseb.) Airy Shaw	Whole plant	Headache, blood purifier
Gentianella tenella (Rottb.) Borner	Whole plant	Fever
Geranium himalayense Klot.	Root	Body ache
Geranium sibiricum L.	Leaf	Diarrhea
Geranium wallichianum Don ex Sw.	Root	Toothache
Geum elatum Wall. ex G. Don	Whole plant	Dysentery
Hyoscyamus niger L.	Fruit	Toothache
Hyssopus officinalis L.	Leaf	Urinary problem
Inula racemosa Hk. f.	Root, flower	Gastrointestinal, Rheumatism
Iris kashmiriana Baker	Bulb	Rheumatism
Juniperus communis L.	Fruit, leaf	Kidney disorder
Jurinea macrocephala (Royle) Aswal & Goel	Leaf	Antiseptic, fever
Lancea tibetica Hk. f. Th.	Seed	Cough, blood vomiting
Lagotis cashmiriana (Royle) Rupr.	Whole plant	Fever
Lloydia serotina (L.) Reichb.	Tuber, leaf	Blood purification, fever, eye infection

(continued)

Table 4.1 (continued)

Species	Plant part used	Uses
Lindelofia longifolia (Benth.) Baill	Leaf	Diarrhoea
Nardostachys grandiflora DC.	Root	Heart disorders, snake bite
Nepeta glutinosa Benth.	Whole plant	Fever
Malaxis muscifera (Lindl.) Ktze.	Tuber	Tonic
Parnassia pusilla Hk.f.	Root	Wounds
Physochlaina praealta (Decne.) Miers	Seed, flower	Toothache
Picrorhiza kurrooa Royle ex Benth.	Root, flower	Fever, blood purification
Podophyllum hexandrum Wall.	Fruit	Cancer
Polygonum affine Don	Root	Diarrhea
Potentilla fulgens Wall. ex Hk.f.	Root	Toothache
Rheum australe D. Don	Root	Internal pain
Rheum moorcroftianum Royle	Root	Internal injury
Rheum nobile Hook.f. & Thomson	Rhizome	Restlessness
Rheum webbianum Royle	Root	Internal injury
Rhododendron anthopogon D. Don	Leaf, flower	Internal pain, aphrodisiac, cold
Rhododendron lepidotum Wall. ex D. Don	Leaf, flower	Internal pain, aphrodisiac
Saussurea bracteata Decne.	Flower bud, leaf	Boils, headache, cough
Saussurea gnaphaloides (Royle ex DC.) Sch.Bip.	Whole	Kidney problem, cough and cold
Saussurea gossypiphora D. Don	Flower	Fever
Saussurea lappa (Decne.) Sch-Bip.	Root	Dysentery, ulcer, stomachache
Saussurea obvallata (DC.) Edgew.	Flower bud	Hydrocele, Reproductive disorders
Saussurea simpsoniana (Field. & Gard.) Lipsch.	Flower, root	Reproductive disorder, fever
Selinum vaginatum (Edgew.) Cl.	Root	Skin diseases
Swertia petiolata Don	Whole plant	Fever, headache
Tanacetum dolichophyllum Kitam.	Whole plant	Fever, body ache
Tanacetum nanam Cl.	Leaf	Intestinal worms
Thymus linearis Benth.	Whole	Tonic, child birth
Waldhemia glabra (Decne.) Regel	Whole	Septic, wounds
Waldheimia tomentosa (Decne.) Regel	Leaf, flower	Septic, wounds

species needs to be documented before it vanishes. A study reports that out of 20 important medicinal plant species, besides *Taraxacum officinale*, a common medicinal plant species, *Aconitum heterophyllum,* is the most frequently used medicinal plant species in the Kashmir valley of northwest Himalaya (Tali et al. 2019). In addition, *Ephedra gerardiana*, a high-altitude medicinal plant growing in Himachal Pradesh, Uttarakhand, and Jammu and Kashmir, is among the bestselling herbal medicines in the developed countries (Kaul 2010).

Saussurea costus, a critically endangered species and included in Schedule VI of the Wildlife (Protection) Act, 1972, is reported to grow naturally in small pockets of Kashmir valley in northwest Himalaya. The roots of this species are used as medicine. In the 1940s, local people in the Lahaul valley of Himachal Pradesh began to cultivate this important species, which is being cultivated as a part of the local agricultural practices (Goraya and Ved 2017). Though domestication of high-altitude medicinal plants is a challenging task, however, cultivation of such rare and endangered medicinal plant species may not only help conserve the concerned species but also preserve the traditional knowledge associated with such important species.

4.3 Distribution of High-Altitude Medicinal Plants

Naturally, the unique climatic conditions, elevations, the degree of slopes, aspects, and soil properties affect the distribution and diversity of medicinal plant species across the Himalayan range (Kala 2005c). Moreover, in the high-altitude Himalayan areas, the major precipitation in the form of the snow and subsequent cool climatic conditions creates a unique environment for the growth of some exceptional medicinal plant species. The Indian Himalaya is sharply differentiated into the northwest Himalaya and the eastern Himalaya based on the amount of rainfall as the northwest Himalaya receives less rainfall than the eastern Himalaya. In addition, the Indian trans-Himalaya receives the lowest rainfall and hence is called the cold desert areas (Aswal and Mehrotra 1994; Kala 2003). Apart from temperature, elevation, aspects, and soil characteristics, the amount of moisture determines the distribution of medicinal plant species across the Himalayan region.

The localized availability of moisture contents in soil within the cold desert gives way to patchy distribution of medicinal plants with high species diversity in comparison to adjacent low moisture localities. Besides, various levels of anthropogenic pressures influence the diversity and distribution of plants within their natural habitats. The heavy influx of livestock grazing, over-harvesting, unregulated tourism, and habitat alteration are some of the man-made pressures altering natural populations of medicinal plants, even in the high-altitude Himalayan areas (Kala and Mathur 2002; Kala 2005c). The distribution of some prominent medicinal plants across the northwest, northeast, and trans-Himalaya is shown in Table 4.2.

4.4 Northwest Himalaya

The northwest Himalaya is endowed with high medicinal plants' diversity. The high-altitude areas hold high moisture contents; hence during the summer season, when temperature is optimum for the growth of species, the alpine meadows in northwest Himalaya remain thickly packed with medicinal plant species. Many of these medicinal plants are ephemeral and complete their lifecycle within 1 or 2 weeks. Depending on the seasonality, one set of medicinal plants is replaced by other set of medicinal plants within a growing season. The prominent medicinal plants, which grow in the high-altitude areas here, include *Potentilla atrosanguinea, Rheum australe, Rhododendron anthopogon, Selinum tenuifolium, Heracleum*

Table 4.2 Distribution of some prominent medicinal plants across the northwest, northeast, and trans-Himalaya

Area	Species
Northwest Himalaya	*Nardostachys grandiflora, Pricrorhiza kurrooa, Origanum vulgare, Aconitum heterophyllum, Fritillaria roylei, Dactylorhiza hatagirea, Rhododendron anthopogon, Angelica glauca, Saussurea costus, Saussura obvallata, Thymus linearis, Rheum australe*
Eastern Himalaya	*Swertia hookeri, Silene himalayensis, Rheum nobile, Gentiana stylophora, Ajania tibetica, Nardostachys jatamansi, Saussurea obvallata, Saussurea tridactyla, Eriophyton wallichii*
Trans-Himalaya	*Ephedra gerardiana, Hyoscymus niger, Lancea tibetica, Lloydia serotina, Physochlaina praealta, Arnebia euchroma, Hippophae rhamnoides, Ferula jaeschkeana, Waldheimia tomentosa, Saussurea gnaphaloides, Saussurea bracteata*

wallichii, Saussurea obvallata, Saussurea simsoniana, Saussurea gossypiphora, Corydalis meifolia, C. govaniana, Iris kumaonensis, Fritillaria roylei, Polygonum affine, Rheum moorcroftianum, Rheum webbianum, Dactylorhiza hatagirea, Picrorhiza kurrooa, Aconitum heterophyllum, Aconium rotundifolium, Aconitum violaceum, Delphinium cashmerianum, and *Nardostachys grandiflora* (Samant et al. 2007; Kala 1998, 2010).

The unique Ayurvedic medicine—the Astavarga, which is made up by mixing tubers and roots of eight medicinal plants—is the product of exclusive Himalayan environment. All these eight medicinal plants such as *Habenaria intermedia, Habenaria edgeworthii, Malaxis acuminata, Malaxis muscifera, Polygonatum cirrhifolium, Polygonatum verticillatum, Fritillaria roylei,* and *Lilium polyphyllum* are found in the high altitude of the Himalaya. *Fritillaria roylei* has a good population density in the Valley of Flowers of Uttarakhand state (Kala 2005c). Similarly, *Trillidium govanianum,* one of the high value medicinal plants, is reported to have high population density in Dhauladhar mountain range of Himachal Pradesh (Sharma and Kala 2018).

4.5 Trans-Himalaya

Studies conducted in the Indian trans-Himalaya report sparse distribution of plant species and so that the low diversity of medicinal plant species. However, a total of 335 medicinal plant species are known to grow in the Indian trans-Himalaya those are used by amchis for curing various ailments. Of these medicinal plants, 45 are placed in various threat categories, such as critically endangered (8 species), and 7 each in endangered and vulnerable (Kala 2003).

Plant species are rarely distributed uniformly across the Indian trans-Himalaya and, thus, most medicinal plants are localized and restricted to a few localities and habitat types. A study reports that the highest numbers of medicinal plant species (92 species) within the trans-Himalaya are restricted to marshy and moist areas, followed by dry scrub (46 species), rocks (39 species), boulders (32 species), and undulating lands (30 species). Similarly, most of the rare medicinal plant species (25 species) grow in marshy and moist areas. Two species (*Inula racemosa and Saussurea lappa*) are found in cultivated farmlands in the Lahual valley of Himachal Pradesh. The highest percentage of the medicinal plant is found in Zanskar (33%), followed by the Pin valley (31%) and Leh valley (26%). The highest numbers of rare medicinal plants (23 species) are found in the Pin valley, followed by Zanskar valley (21 species), high passes/pastures in the Leh valley (18 species), Nubra (12 species), Malari (5 species), and 4 species each in Kargil, Lahaul, and Changthang areas (Kala 2003, 2006).

A study reports that due to habitat diversity, relatively higher soil moisture and humidity, a few localities within borderline of trans-Himalaya and Himalaya sustain rich medicinal plants' diversity. These localities include Tarbak and Pilshur in the Pin valley of Spiti region and Padum, Rangdum, Sapi Nala, Ralakunb, and Sinku-la in the Zanskar region of Ladakh. These areas also sustain good populations of rare and endangered medicinal plants such as *Dactylorhiza hatagirea, Bergenia stracheyi, Delphinium cashmerianum, Saussurea bracteata, Podophyllum hexandrum,* and *Picrorhiza kurrooa* (Kala 2003). Many medicinal plants are common in both trans-Himalaya and northwestern Himalaya, which include *Bergenia stracheyi, Dactylorhiza hatagirea, Juniperus communis, Malaxis muscifera, Thymus linearis, Waldheimia tomentosa,* and *Ferula jaeschkeana* (Kala 2005c).

4.6 Eastern Himalaya

Medicinal plants of high-altitude regions, especially above 3500 mamsl in the eastern Himalaya, are less studied due to inhospitable climatic conditions, tough terrain, and lack of proper local informants. Monpa community is one of the high landers communities of the eastern Himalaya, mainly inhabiting Tawang and West Kameng districts of Arunachal Pradesh, and is known to use many high-altitude medicinal plant species. Besides herbs, the high-altitude areas in these districts harbor low-lying shrubs such as Rhododendron and Juniper. The major medicinal plants known to occur here in the high-altitude areas are *Swertia hookeri, Silene himalayensis, Podophyllum hexandrum, Rheum nobile, Rheum acuminatum, Gentiana kurroo, Gentiana stylophora, Ajania tibetica, Aconitum hookeri, Aconitum heterophyllum,* and *Aconitum ferox* (Tsering and Tag 2015; Bhuyan and Pangu 2018). Within the high-altitude areas, different species do have their preferences for specialized habitats. For instance, *Aconitum ferox* and *Aconitum heterophyllum* are inhabited the alpine meadows and grasslands, while *Aconitum hookeri* grows near lakes and moist soil of alpine region (Tsering and Tag 2015).

The eastern Himalayan region has a unique assemblage of endemic plant species, and many of them are known for their medicinal values, including *Rheum nobile*. *Rheum nobile* is a 1–2 m high herbaceous plant distributed above 4000–6000 m altitude, mostly on open screes. It has large translucent cream-colored bracts concealing the inflorescence that produce a greenhouse effect against extreme cold in high mountains (Omori and Ohba 1999). It blooms during early June to early

July, and the inflorescence lasts for 10 days. It is a threatened monocarpic perennial, which has an extended juvenile phase, saving resources for a single and fatal act of reproduction during which all available resources are consumed (Schaffer 1974). It is endemic to the Himalayas and distributed across northeast Afghanistan, India, Nepal, Bhutan, Tibet, Myanmar to the Hengduan Mountains in southwest China (Song et al. 2020). Yak grazing is common in its habitat, and it is suffered from overgrazing in its natural habitat range.

The amplitude of cattle grazing is one of the most destructive activities in the high-altitude areas, leading to a decline in natural population of medicinal plants such as *Nardostachys, Picrorrhiza,* and *Aconitum.* A study reports that some medicinal plant species in the high altitude of the Himalaya are positively influenced by livestock grazing, to a certain level. However, livestock grazing in combination with a high level of medicinal plants harvesting exerts a negative impact on species diversity (Ghimire et al. 2006). Besides livestock grazing, the commercial forces are the major factors as extraction of medicinal plants depends on the demand and so that the population and distribution of species are impacted. For instance, extraction of two high Himalayan herbs such as *Trillidium govanianum* for its rhizomes and *Fritillaria roylei* for its bulbs has increased in Himachal Pradesh, Jammu & Kashmir, and Uttarakhand (Goraya and Ved 2017; Sharma and Kala 2018). Therefore, the population of such species has gone down in the natural habitats.

The traditional knowledge on high-altitude medicinal plants and the natural distribution of such an important healthcare commodity need to be examined in the present era of global climate change, which may not only affect the diversity and distribution of medicinal plants but may influence their very inherent properties, including phytochemicals and phenophases. The change in distribution of medicinal plant species due to a number of anthropogenic pressures, including climate change and overexploitation of useful species, may be counterproductive to the livelihood of local people and also to the sustainability of associated traditional knowledge. Therefore, concerted efforts are required to be implemented in the restoration of high-altitude areas in view of maintaining the sustained flow of various ecosystem services, including medicinal plants.

References

Aswal BS, Mehrotra BN (1994) Flora of Lahaul-Spiti (A cold desert in north west Himalaya). Bishen Singh Mahendra Pal Singh, Dehradun

Aswani S, Lemahieu A, Sauer WHH (2018) Global trends of local ecological knowledge and future implications. PLoS One 13:01954400–0e195519. https://doi.org/10.1371/journal.pone.0195440

Bhuyan LR, Pangu Y (2018) High altitude medicinal plants of Tawang district, Arunachal Pradesh. Bull Arunachal For Res 33(1):35–43

Devkota S, Chaudhary RP, Werth S, Scheidegger C (2017) Indigenous knowledge and use of lichens by the lichenophilic communities of the Nepal Himalaya. J Ethnobiol Ethnomed 13:1–10

Fei Y, Wang J, Peng B, Peng J, Hu JH, Zeng ZP, Chen J, Li QE, Gao ZP, Yan XL (2017) Phenolic constituents from Rheum nobile and their antioxidant activity. Nat Prod Res 31(24):2842–2849

Ghimire SK, Mckey D, Aumeeruddy-Thomas Y (2006) Himalayan medicinal plant diversity in an ecologically complex high altitude anthropogenic landscape, Dolpo, Nepal. Environ Conserv 33(2):128–140

Goraya GS, Ved DK (2017) Medicinal plants in India: an assessment of their demand and supply. National Medicinal Plants Board, Ministry of AYUSH, Government of India, New Delhi

Kala CP (1998) Ecology and conservation of alpine meadows in the Valley of Flowers National Park, Garhwal Himalaya. PhD thesis, Forest Research Institute (Deemed University), Dehradun, India

Kala CP (2005a) Health traditions of Buddhist community and role of amchis in trans-Himalayan region of India. Curr Sci 89(8):1331–1338

Kala CP (2005b) Current status of medicinal plants used by traditional Vaidyas in Uttaranchal state of India. Ethnobot Res Appl 3:267–278

Kala CP (2003) Medicinal plants of Indian Trans-Himalaya: focus on Tibetan use of medicinal resources. Bishen Singh Mahendra Pal Singh, Dehradun

Kala CP (2005c) Indigenous uses, population density, and conservation of threatened medicinal plants in protected areas of the Indian Himalayas. Conserv Biol 19(2):368–378

Kala CP (2006) Medicinal plants of the high altitude cold desert in India: diversity, distribution and traditional uses. Int J Biodivers Sci Ecosyst Serv Manage 2(1):43–56

Kala CP (2009) Ethnobotanical and ecological approaches for conservation of medicinal and aromatic plants. Acta Hortic 860:19–26. https://doi.org/10.17660/ActaHortic.2010.860.1

Kala CP (2010) Medicinal plants of Uttarakhand: diversity, livelihood and conservation. Biotech Books, New Delhi

Kala CP (2023) Medicinal plants used for the management of eye diseases in Garhwal region of Uttarakhand state in India. Stud Ethno-Med 17:14–18

Kala CP, Mathur VB (2002) Patterns of plant species distribution in the Trans-Himalayan region of Ladakh, India. J Veg Sci 13(6):751–754

Kaul MK (2010) High altitude botanicals in integrative medicine: case studies from Northwest Himalaya. Indian J Tradit Knowl 9(1):18–25

Khakurel D, Uprety Y, Ahn G, Cha JY, Kim WY, Lee SH, Rajbhandary S (2022) Diversity, distribution, and sustainability of traditional medicinal plants in Kaski district, Western Nepal. Front Pharmacol 13:1076351

Nayaka S, Upreti DK, Khare R (2010) Medicinal lichens of India. In: Trivedi PC (ed) Drugs from plants. Avishkar Publishers, Distributors, Jaipur

Omori Y, Ohba H (1999) Thermal condition of the inflorescence of a glasshouse plant, Rheum nobile Hook. f. and Thoms., and microclimatic features of its habitat in Jaljale Himal, East Nepal. Newslet Himalayan Bot 25:5–11

Samant, Pant S, Singh M, Lal M, Singh A, Sharma A, Bhandari S (2007) Medicinal plants in Himachal Pradesh, north western Himalaya, India. Int J Biodivers Sci Manag 3(4):234–251

Schaffer WM (1974) Selection for optimal life histories: the effects of age structure. Ecology 55:291–303

Sharma N, Kala CP (2018) Patterns in distribution, population density and uses of medicinal plants along the altitudinal gradient in Dhauladhar mountain range of Indian Himalayas. Curr Sci 114(11):2323–2328

Song B, Stoll P, Peng D, Sun H, Stöcklin J (2020) Demography of the giant monocarpic herb Rheum nobile in the Himalayas and the effect of disturbances by grazing. Ann Bot 125(3):447–458

Tali BA, Khuroo AA, Ganie AH, Nawchoo IA (2019) Diversity, distribution and traditional uses of medicinal plants in Jammu and Kashmir (J&K) state of Indian Himalayas. J Herb Med 17:100280

Tsarong TJ (1994) Tibetan medicinal plants. Tibetan Medical Publications, New Delhi

Tsering J, Tag H (2015) High altitude ethnomedicinal plants of Western Arunachal Himalayan landscape. Pleione 9(1):18–25

Website

https://www.flowersofindia.net/catalog/slides/Snow%20Lotus.html

Available Knowledge, Trends in Research, and Future Prospects of High-Altitude Medicinal Plants

5

Indra Dutt Bhatt, Laxman Singh, and Deep Chandra Tiwari

Abstract

The alpine region of the Himalaya is a rich repository of medicinally important plants used in both traditional and modern therapeutics. These species are the predominant floral wealth of the Himalaya bestowed by unique geographical location, climatic conditions, and edaphic and altitudinal variations. The continuous and often indiscriminate collections (i.e., well beyond their natural regenerative capacity), coupled with habitat destruction, have led to many species being listed in the IUCN Red Data Book and/or in various threat categories. However, as an issue of global concern, conservation and management of these valuable genetic resources are essential. Although extensive research has been conducted, assessing these scientific publications needs a thorough investigation. The study, therefore, aims to identify and analyze the central themes and research trends through a bibliometric analysis and highlights available knowledge on high-altitude Himalayan medicinal plants. A total of 1236 scientific documents (obtained from the Scopus database) from 1990 to 2024 were analyzed. The parameters assessed for the results include (i) publication trends, (ii) analysis of contributing countries, (iii) thematic evolution, (iv) co-occurrence of authors' keywords, (v) prominent species studied, and (vi) organizations. The study identified an evident upward trend in research on high-altitude Himalayan medicinal plants. The country-wise analysis indicates that the highest number of publications is attributed to India (226), followed by China (54), while Pakistan and Nepal contribute 27 and 17, respectively. Similarly, the author's keyword analysis identified the keyword "medicinal plant" with the highest occurrences (56 hits) and a total link strength of 122, followed by the keyword "plant extract" (49 hits; a link strength of 116), "traditional medicine," and "ethnobotany" with 22

I. D. Bhatt (✉) · L. Singh · D. C. Tiwari
G. B. Pant National Institute of Himalayan Environment, Almora, Uttarakhand, India

S. Chandra, H. Lata (eds.), *High Altitude Medicinal Plants*,
https://doi.org/10.1007/978-3-031-99406-7_5

hits and 53 as link strength. Among the species, *Picrorhiza kurroa* was the most extensively studied species, with 33 publications, followed by *Saussurea costus* and *Swertia chirayita*, with 27 and 24 documents. Thematic-wise, most of the publications were from the biological sciences. In contrast, documents from chemistry, immunology, microbiology, and social sciences were comparatively lesser in number, thus suggesting emerging or interdisciplinary research in these areas. While this study provided a systematic overview of productivity and visibility of research work on high-altitude medicinal plants, it has a limitation in that only documents from the Scopus database were included in the analysis. Although Scopus encompasses a substantial number of peer-reviewed journals, it may exclude some relevant research on the topic. Furthermore, while the main research themes were identified and discussed briefly, more comprehensive information on each central theme is still required.

5.1 Introduction

It is evident that medicinal and aromatic plants (MAPs) have played a significant role in human well-being's economic, social, cultural, and ecological aspects for centuries and continue to do so today. The basis of most modern drug discovery is often based on ethnobotanical case studies passed down between generations. To an estimate, between 50,000 and 80,000 flowering plant species worldwide are used for medicinal purposes (Chen et al. 2016). The ingenuity of these herbal formulations is such that they are perceived and branded as natural, intrinsically safe, and cost-effective. Globally, the herbal market is increasing and is expected to grow with a compound annual growth rate (CAGR) of 10.6% from $169.1 billion (2023) to $279.8 billion by the end of 2028 (BCC Research 2023). In India—a trove of medicinal plants repository and one of the leading exporters—during the last 3 years the exports of botanical raw materials stand at 139380.46 MT (2020–2021), 133264.35 MT (2021–2022), 129856.76 MT (2022–2023), and 22279.61 MT for the year 2023–2024 (till May 2023; https://www.dgciskol.gov.in). Meanwhile, the export value of ayurvedic and herbal products is reported to have increased from (*values in million US dollars*) $354.68 (2015) to $628.48 for the year 2023 and is projected to stand at $651.17 for the fiscal year 2024 (https://www.statista.com). This growth is fueled by the increasing demand for herbal remedies globally and a greater social acceptance. Thus, contributing significantly to several Sustainable Development Goals (SDGs) such as poverty reduction (SDG-1), good health and well-being (SDG-3), responsible consumption and production (SDG-12), and protect, restore, and promote sustainable use of biodiversity elements (SDG-15).

Though the exact number of the documented (i.e., 48,655 species including lower taxonomic groups) plant species in medicinal practices remains a tantalizing question, nearly 9500 species are reported to have ethnobotanical significance, while around 7500 species are such that are exploited in both traditional and modern pharmacopeia (Gowthami et al. 2021). While exceptions prevail, most of these resources are extensively exploited from the wild (to meet the demand and supply) to the extent that many are now listed in the prominent threat categories of the IUCN. Data retrieved and synthesized from the IUCN database (IUCN 2024) highlights a total of 3336 species of the country are evaluated, among which 02 species are now extinct in the wild, 148 are critically endangered, 274 are endangered, and 222 species are in the vulnerable category (IUCN 2024). Nevertheless, conservation biologists argue that the number reflected remains small and assert that it would increase if the country's red list were prepared. This is primarily because the majority (70–90%) of the species are exclusively wild-harvested, despite continuous efforts to mainstream these species (Singh et al. 2021, 2022).

This holds true for the MAPs of the Indian Himalayan region with 1748 medicinal species, where compounding threats (biotic and abiotic, including disasters and climate warming) work in tandem (Singh et al. 2017). The root cause of most of the regional emanating threats is being catalyzed by anthropogenic disturbances (mainly habitat loss and fragmentation), overexploitation, and illegal trade. Other than these, the atypical biology of the Himalayan plants, that is, habitat specificity, low seed set and germination, prolonged dormancy, narrow niche, low population size, etc., has aggravated the problem. Even though many of these plants are prohibited from collection/and trade by the different state governments, the Wildlife Protection Act 1972, and the Convention on International Trade in Endangered Species (CITES), illicit trade and unsustainable harvesting practices do not seem to halt (Singh et al. 2023). Therefore, conservation and sustainable utilization of the Himalayan MAPs warrant priority attention. Developing a more comprehensive taxonomic catalog, conservation plans, and protocols is essential to prevent species extinctions. Unfortunately, there remains a paucity of information even after three decades of research on medicinal plants of the Himalayan region. Conversely, although a research gap persists, the current knowledge base is sufficient to implement practical conservation measures and mainstreaming. Herein, we have reviewed and analyzed the current state of knowledge regarding high-altitude Himalayan medicinal plants, specifically emphasizing their ecological, pharmacological, and conservation aspects, and have investigated the general trends in medicinal plant utilization patterns over time.

5.2 Data Extraction and Processing

A comprehensive systematic search was conducted using the Scopus database for data collection. The Scopus database was selected over the Web of Science due to its broader coverage of the research topics pertinent to the present study. Search terms were formulated to retrieve a wide spectrum of research on high-altitude Himalayan medicinal plants. Data were aggregated through the Boolean information retrieval method, employing species names along "AND"/"OR" and "NOT" operators (Singh et al. 2020a, 2020b). The strings used include "high-altitude Himalayan medicinal plant," "medicinal plant," "medicinal herbs," and "medicinal flora," combined with regional identifiers such as "Himalaya," "Himalayan region," "Indian Himalayan region," "Nepal Himalaya," "Bhutan Himalaya" "Tibet,"

"Xinjiang," "China," "Pakistan Himalaya." Likewise, subject keywords include "ecology," "pharmacology," "physiology," "ethnobotany," "plant tissue culture," "propagation," "genetic diversity" "threats," "biodiversity conservation," etc. The bibliometric analysis was conducted using the Preferred Reporting Items for Systematic Reviews and Meta-Analyses (PRISMA), which involved four key steps: identification, screening, eligibility, and inclusion. In the identification process, a total of 2011 papers were retrieved. At first, a total of 27 documents pertained to other languages (other than English) were excluded. Thereafter, documents not relevant were excluded, that is, (i) documents from outside the Himalayan region (878), and (ii) not relevant documents (35). After the removal of the irrelevant documents, 1071 documents were processed during the screening process. Here, 672 documents were such that were found outside the subject area and were thus excluded. In the eligibility phase, 399 documents were assessed, out of which 86 were not relevant to the topic and excluded. Finally, following a comprehensive inspection, 313 papers were considered appropriate and included in the study. The methodology followed a strict and lucid selection procedure, offering a strong basis for the ensuing study (Fig. 5.1).

5.3 Bibliometric Analysis of Publications on Himalayan Medicinal Plants

5.3.1 Year-Wise Trend

From 1990 to 2024, a total of 313 studies related to high-altitude Himalayan medicinal plants have been published. As shown in Fig. 5.2, a significant increase in the publication is seen—year-wise (1990–2024), with a steady rise, particularly after the year 2005. In the subsequent years, this upward trend continued (except on some occasions), and the maximum number of studies was found in the years 2021 and 2022 (37 no), while the minimum (2 no) in the years between 1997 and 2002. This trend suggests increased scientific interest and knowledge of the value of Himalayan medicinal plants, reflecting the region's high biodiversity and the potential advantages these plants might provide in terms of traditional medicine, pharmaceuticals, and conservation efforts.

5.3.2 Country-Wise Contributions

Analysis of country-wise distributions of the publications demonstrated significant research interest from countries in the Himalayan region. Figure 5.3a illustrates the countries that have published on the subject and the intensity of their publications. The data analysis shows that India, with over 200 publications, is notably the most prominent country that has contributed significantly to the research on high-altitude medicinal plants. In detail, the country-wise analysis indicates that the highest number of publications is attributed to India (226 no), followed by China (54 no);

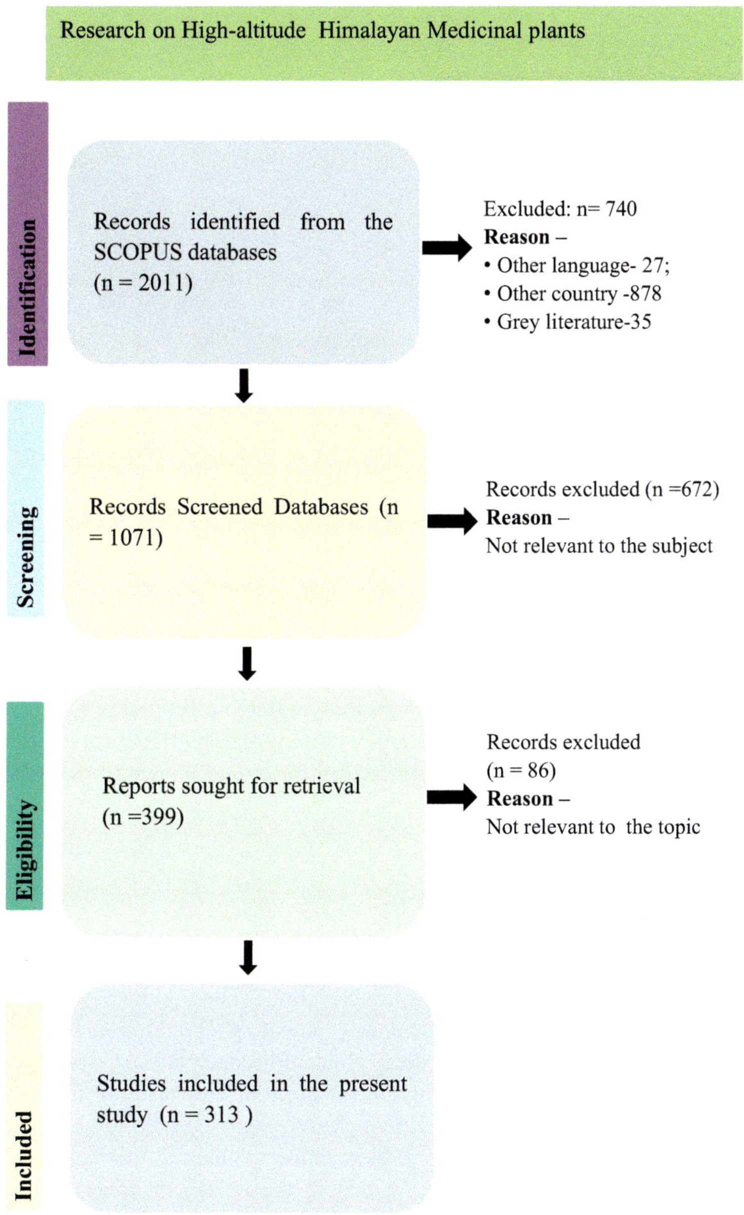

Fig. 5.1 Flow chart describing selection and screening of literature using Preferred Reporting Items for Systematic Reviews and Meta-Analyses (PRISMA)

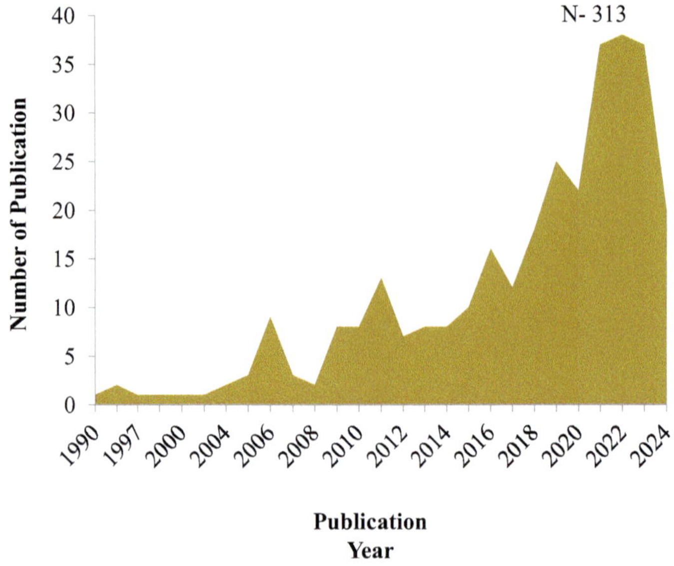

Fig. 5.2 Year-wise scientific document published during the period 1990–2024

likewise, Pakistan has 27 publications, and Nepal has 17 publications, while Bhutan and Bangladesh's contributions were minimal, indicating limited research activity in these countries. This distribution suggests that research efforts are concentrated in India, likely due to the extensive Himalayan range and increased research funding in this area.

5.3.3 Subject-Wise

The subject area analysis reveals that research on high-altitude Himalayan medicinal plants encompasses diverse scientific disciplines. The majority of publications are in the agriculture and biological sciences, indicating a predominant focus on the botanical and ecological aspects of these plants. Pharmacological, ecological, and ethnobotanical topics constitute a significant portion, emphasizing the medicinal and pharmacological significance of these plants (Fig. 5.3b). Following biological sciences, notable publications were in biochemistry, genetics, molecular biology, and plant physiology. The author's interest in these subject areas reflects their emphasis on biochemical studies. Chemistry, immunology, microbiology, and social sciences have comparatively fewer publications, suggesting emerging or interdisciplinary research in these areas. This comprehensive overview underscores the multifaceted nature of research on high-altitude Himalayan medicinal plants conducted by regional and international scientific communities across various disciplines.

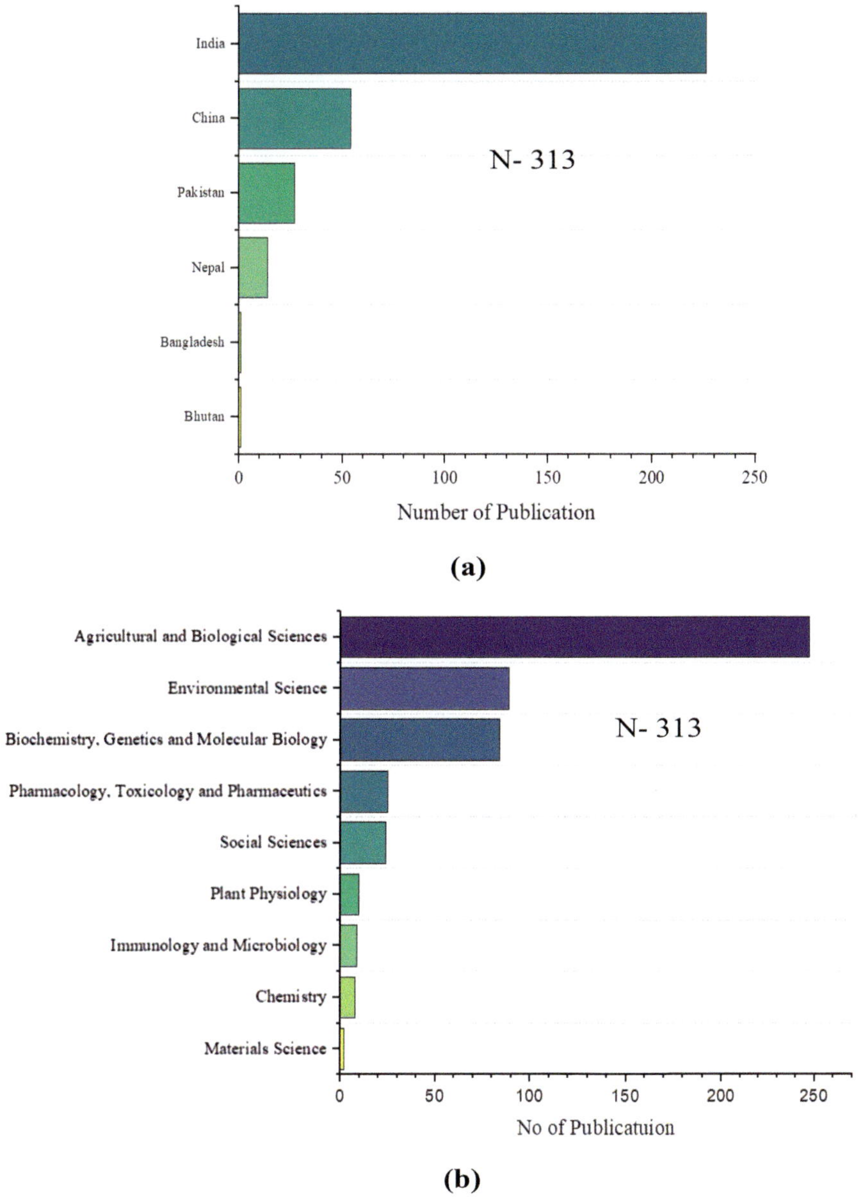

Fig. 5.3 Categorization of publications into (**a**) country-wise contribution and (**b**) thematic evolution of publications into different research domains

5.3.4 Organization Contribution

In the bibliometric analysis of the organization, we examine the publication output of the top 15 institutions with the highest number of publications in the field (Fig. 5.4). The G. B. National Institute of Himalayan Environment, Almora, emerged as the leading institution, with a total of 35 publications, reflecting its over three-decade-long extensive research on Himalayan ecology and conservation studies. This was closely followed by the University of Kashmir and the Chinese Academy of Sciences, with 33 and 21 publications, respectively. Institutes like CSIR—Institute of Himalayan Bioresource Technology and CSIR—Indian Institute of Integrative Medicine focus on biotechnological and pharmacological research. Concurrently, Kumaun University, H.N.B. Garhwal University, and Tribhuvan University significantly contributed to ecology, ethnobotanical, and pharmacological-related studies. This distribution of research among top organizations on high-altitude Himalayan medicinal plants underscores the prominent role of these institutions in advancing research within the field. This information is essential for comprehending the distribution of research efforts and identifying leading organizations.

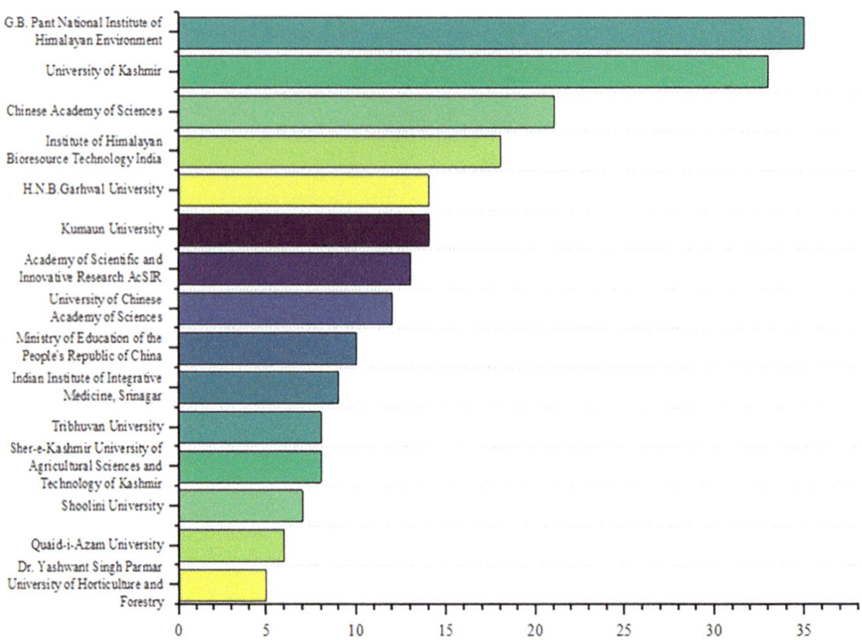

Fig. 5.4 Categorization of publications into prominent research organizations that published extensively on alpine medicinal plants

5.3.5 Keyword Analysis

This bibliometric analysis delves into author keywords related to high-altitude medicinal plants in the Himalayan region from 2005 to 2024 (Fig. 5.6). To establish a relationship among the different keywords used and types of keywords used over the period, a terminological keyword map was prepared using VOS viewer (https://www.vosviewer.com/) to specify distinct thematic clusters. The map identified 7 clusters and 125 keywords, providing a visual and quantitative overview of key terms. "Medicinal plant" emerged as the most frequently used keyword (56 occurrences, link strength of 122), followed by "medicinal plants" (49 occurrences, link strength of 116), and "plant extract," "traditional medicine," and "ethnobotany" (22 occurrences, link strength of 53). Keywords that appear at least five times in the sample were chosen to construct a keyword map. From Fig. 5.5, the diameter of the circle designates how frequently the keyword is used. The distance between the circles is a proxy for the strength of the relationship between the terms, and it gets stronger the closer they get.

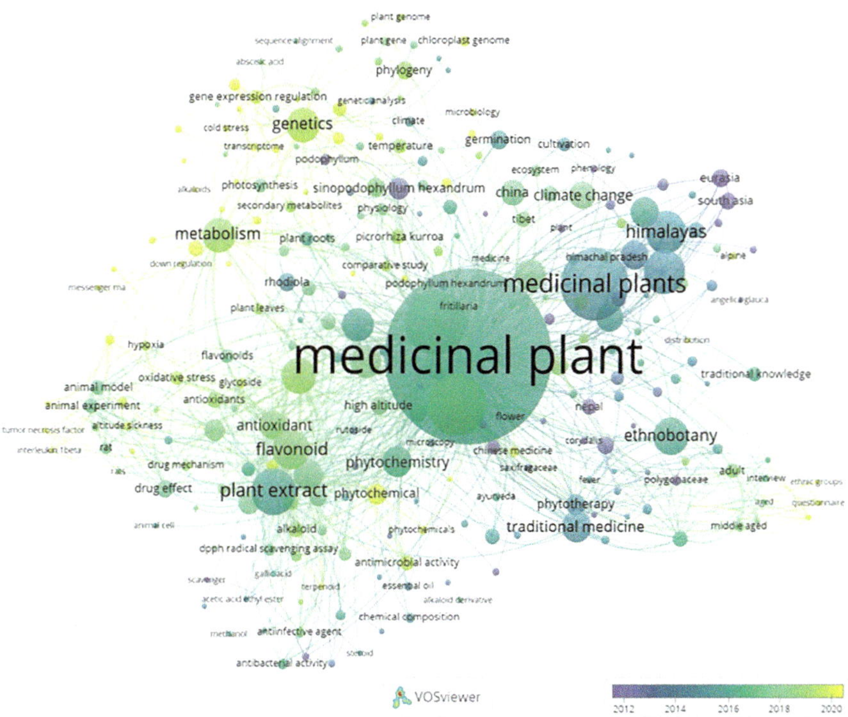

Fig. 5.5 Overlay map of the co-occurrence network based on the author's keyword during the period 1990–2024

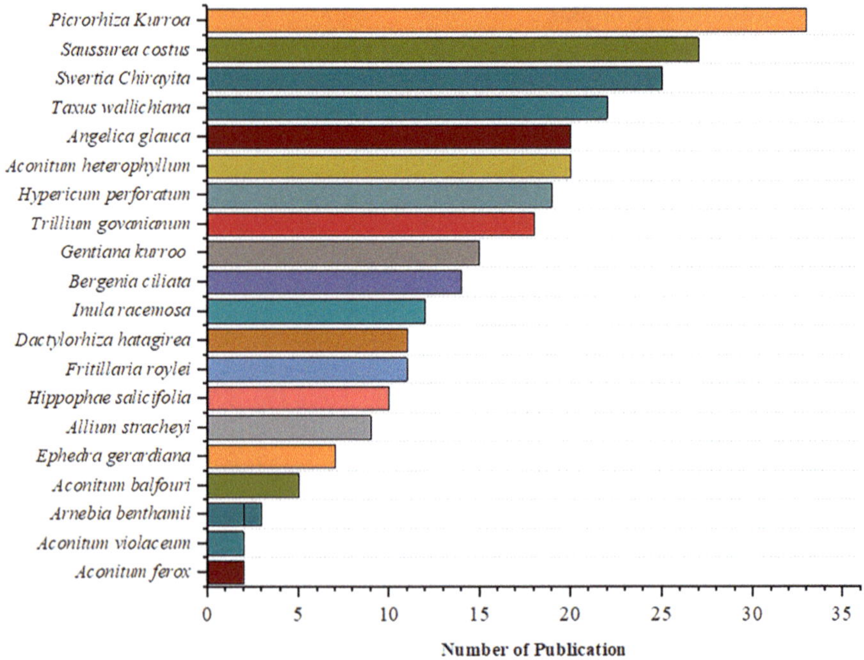

Fig. 5.6 Cataloging species with species-wise publication numbers

5.3.6 Species-Wise Analysis

The species-wise trend reveals a distinct pattern in publication frequency for various species. Based on the available literature on alpine Himalayan medicinal plants, *Picrorhiza kurroa* emerges as the most extensively studied species, with 33 publications, followed by *Saussurea costus* and *Swertia chirayita*, with 27 and 24 publications, respectively (Fig. 5.6). Similarly, *Taxus wallichiana* and *Angelica glauca* are among the species that were extensively studied with more than 15 publications each. At the same time, *Aconitum heterophyllum* and *Hypericum perforatum* are closely followed, indicating substantial research engagement. Furthermore, other prominent species such as *Trillium govanianum*, *Gentiana kurroo*, *Bergenia ciliata*, *Inula racemosa, Dactylorhiza hatagirea*, *Fritillaria roylei*, *Hippophae salicifolia*, and *Allium stracheyi* showed moderate research interest with more than 10 publications each. This pattern highlights well-researched and under-researched species within this field of study, underscoring the differences in the amount of attention paid to various alpine Himalayan medicinal plants.

5.4 Available Knowledge of High-Altitude Himalayan Medicinal Plants

5.4.1 Diversity and Distribution

Species richness and community structure in alpine ecosystems are a function of several factors, including soil moisture and composition, habitat heterogeneity, precipitation patterns during summer and winter, topographical variability, elevation, aspect, and human-induced pressures. For example, the alpine zones of the Greater Himalayas demonstrate rich species diversity and vegetation coverage, particularly in humid, south-facing areas, while the adjacent trans-Himalayan region exhibits a spectrum of unique ecological variations within the terrain (Rawat 2007). In Himachal Pradesh, Badola (2001) reported species such as *Centella asiatica*, *Dioscorea deltoidea*, *Jurinea dolomiaea*, *Podophyllum hexandrum*, *Taxus baccata*, and *Valeriana jatamansi* were more pervasive at an altitudinal range of 1000–3000 m. Furthermore, elucidated the altitudinal range between 1500–2000 m and 2500–3000 m as the richest reservoir of medicinal plants, with 305 species and 297 species, respectively (Badola 2001). At the family level, in the western Himalayas, the prominent families are Asteraceae, Poaceae, Brassicaceae, Scrophulariaceae, Ranunculaceae, Fabaceae, Rosaceae, and Ranunculaceae. Among the dominant genera with more than 20 genera include *Astragalus*, *Potentilla*, *Polygonum*, *Pedicularis*, *Corydalis*, *Primula*, *Saussurea*, *Gentiana*, and *Carex* (Rau 1975). Likewise, Singh and Hajra (1996) reviewed 8000 species of angiosperms in the Himalaya and reported Berberidaceae, Saxifragaceae, Ranunculaceae, Rosaceae, and Umbelliferae are some of the leading Himalayan families with a high degree of endemism, that is, 98.7%, 92.3%, 72.7%, 70.4%, and 68.8%, respectively. Despite the abundance of eco-floristic research conducted in the Western Himalayan alpine region, numerous ecological aspects remain insufficiently explored. Notably, a comprehensive understanding of plant species diversity patterns, community structure, composition, and successional trajectories in the region is yet to be fully elucidated. Population assessment studies in *Dactylorhiza hatagirea* by Singh et al. (2021) highlight that the mean value range for density in the present study (0.70–2.43 ind/m^2) and was in broad consonance with the reported range (0.15–4.70 ind/m^2) in the earlier reports from the region (Uniyal et al. 2002; Nautiyal et al. 2004; Bhatt et al. 2005a, 2005b; Pandey et al. 2019). Similar reports were presented by Kala (2000) and Kala (2005) in the Spiti Valley (Himachal Pradesh) and Indian trans-Himalaya region, wherein the density of the species varied between 5.0–30.0 ind/m^2 and 0.37–18.0 ind/m^2, respectively. In another study, Rinchen et al. (2012) report differences in species diversity from Jammu and Kashmir, which varied between 0.44 and 8.0 ind/m^2. In Nepal, the diversity of the species ranges between 0.67 and 4.06 ind/m^2 (Ghimire et al. 1999; Bhattarai et al. 2014; Khadka et al. 2016); however, going too Far East in Bhutan, the diversity of the species reported was 0.40 ind/m^2, which is least from the Himalayan region (Lakey and Dorji 2016). In the case of *Lillium polyphyllum,* Dhyani et al. (2018) revealed the density of the species to vary between 0.2 and 1.2 ind/m^2 at an altitude range of

2200–3200 m asl and under the canopy of Himalayan cedar (*Cedrus deodara*). Likewise, in the western Himalayas, the density of *Rheum emodi* is reported to vary between 1.93 and 4.24 (ind/m^2) (Rawat et al. 2016; Uniyal et al. 2006), while in Jammu and Kashmir, Wani et al. (2021) reported the density of *Rheum webbianum* to vary between 0.1 and 1.23 (ind/m^2). In the case of *Polygonatum verticillatum*, Suyal et al. (2020) reported the mean value range for density was 0.43–4.80 ind/m^2 from the western Himalayas and was comparable to the values (0.11–4.40 ind/m^2) reported in earlier studies from low- and high-altitude habitats of west Himalaya (Bhatt et al. 2014). For *Polygonatum cirrhifolium*, the mean value range for density was 1.20–6.33 ind/m^2 and was in broad consonance with the reported range (0.12–7.20 ind/m^2) in the earlier reports from the region (Bhatt et al. 2014; Pandey et al. 2019). Further, in some sites (Sunderdhunga, Kafni, Phurkia, Khati, Vanasur, and Khetikhan), the entire population was suspected to have been lost either due to overharvesting or due to poor regeneration (Suyal et al. 2020). Ecological observations of *Trillium govanianum* were reported by Chauhan et al. (2018) and highlighted the density of the species to vary between 1.20 and 5.25 ind/m^2 (for one leaf individual) and from 0.70 to 4.15 ind/m^2 for three leaf plants in the northwestern Himalaya (Himachal Pradesh and Uttarakhand). At the landscape level, Pandey et al. (2019) studied 22 threatened and high-value medicinal plants in cold desert areas of Johar Valley, western Himalaya. The study report, in the region, shrubs such as *Rhododendron anthopogon*, *Juniperus indica*, *Ephedra intermedia*, *Juniperus communis*, and *Ephedra gerardiana* were most abundant, while the concentration of medicinal herbs varied with density <20 ind ha^{-1} (*Juniperus semiglobosa*) to 38972.83 ind ha^{-1} for *Picrorhiza kurrooa*.

5.4.2 Ethnomedicinal Knowledge

The Himalayan region demonstrates a profound connection between traditional knowledge and biodiversity, stemming from the reciprocal relationship between plant and cultural diversity interplayed by cultural reminiscences, ecological awareness, and established social norms (Haq et al. 2023). Ethnomedicinal knowledge, therefore, is fascinating because of the wide range of medicinal plants used in different traditional medical systems (Table 5.1). Over the years, various researchers have documented ethno-medico-botanical information and traditional use of the high-altitude Himalayan medicinal plants. Ballabh et al. (2008) documented that the cold desert medicinal plants of Ladakh were used against kidney and urinary disorders. A total of 68 medicinal plants belonging to 29 families and 58 genera were documented to be used either alone or in combination with some exotic species. Among others, *Gentiana squarrosa* roots combined with *Dactylorhiza hatagirea* (tubers) or *Rubia cordifolia* (roots) in the form of tablets for 12–15 days helped regulate urine discharge and checked burning sensation. Likewise, *Malaxis muscifera* roots combined with minerals and medicinal stones extraction were prescribed for 8–10 days to revitalize kidney function. Furthermore, *Picrorhiza kurrooa* (roots), *Polygonatum verticillatum* (rhizomes), and *Podophyllum hexandrum* (fruits)

Table 5.1 Ethnobotanical categorization of selected high-altitude medicinal plants in the Indian Himalayan region

Species name	Common name	Family	Elevation (m asl)	State-wise distribution	Traditional uses	Parts used
Aconitum ferox	Indian aconite	Ranunculaceae	2500–4500	UK, J&K, HP, SK	Fever.	Root, tuber
Aconitum balfourii	Balchar	Ranunculaceae	3200–3800	UK, J&K, HP, SK, MG	Fever, pain, and inflammation	Root
Aconitum heterophyllum	Atis	Ranunculaceae	2500–4500	HP, UK	Anti-inflammatory, fever	Root, tuber
Aconitum lacinniatum	Mohra vish	Ranunculaceae	3800–4600	UK, J&K, HP, SK, AP	Fever and respiratory	Root
Allium stracheyi Baker	Ganderan	Pinaceae	2500–4000	UK, HP	Anti-inflammatory, digestive	Leaves
Aconitum vioaceum	Tilla	Ranunculaceae	3500–4000	UK, J&K, HP, SK	Anti-inflammatory, digestive	Root
Angelica glauca	Choru	Apiaceae	3000–4000	UK, HP	Digestive aid, tonic	Root
Beberis jasechkiana	Kilmor	Berberidaceae	3000–5200	UK, J&K, HP, SK	Diabetes, digestive	Root
Bergenia stracheyi	SIlphod	Saxifragaceae	2500–4000	UK, HP	Stones, respiratory issues	Rhizome, root
Bunium persicum	Black Cumin	Apiaceae	2500–4000	UK, J&K	Digestive aid, aromatic	Seed
Dactylorhiza hatagirea	Salampanja	Orchidaceae	3000–4200	UK, J&K, HP	Aphrodisiac, urinary	Tuber
Fritillaria cirrhosa	Kakoli	Liliaceae	3000–4500	UK, HP, SK, AP	Respiratory issues	Bulb
Gentiana kurroo	Indian Gentian	Gentianaceae	1500–3500	UK, HP	Stomach ailments, fever	Root
Hippophae rhamnoides	Seabuckthorn	Elaeagnaceae	2000–3800	HP, UK, SK, AP	Immune booster, digestive	Fruit
Inula racemosa	Pushkarmool	Asteraceae	2500–3500	HP, J&K	Digestive aid	Root
Jurinea dolomiaea	Dhoop	Asteraceae	2800–4200	UK, HP	Aromatic and digestive issues	Root
Morchella esculenta	Guchhi	Morchellaceae	2500–3500	HP, UK	Immune booster, digestive	Fruit
Nardostachys jatamansi	Jatamansi	Caprifoliaceae	3000–5000	UK, J&K, HP	Nervine tonic, stress relief	Rhizome, root
Paris polyphylla	Satuwa	Melanthiaceae	1500–3000	UK, SK	Pain relief, wound healing	Rhizome

(continued)

Table 5.1 (continued)

Species name	Common name	Family	Elevation (m asl)	State-wise distribution	Traditional uses	Parts used
Picrorhiza kurroa	Kutki	Plantaginaceae	2700–4500	UK, HP	Liver disorders, fever	Root, rhizome
Potentilla atrosanguinea	Bajradanti	Rosaceae	3000–4600	UK, HP, SK, AP	Diarrhea and skin diseases	Leaves, root
Podophyllum hexandrum	Van kakari	Berberidaceae	2000–4000	UK, J&K, HP, SK, MG, AP	Liver issues, digestive tonic	Rhizome
Polygonatum cirrhifolium	Mahameda	Asparagaceae	3000–4500	UK, HP	Anti-inflammatory	Rhizome
Rheum emodi	Dolu	Polygonaceae	3000–4500	HP, UK	Laxative, digestive ailments	Root, stem
Rheum webbianum	Indian Rhubarb	Polygonaceae	2500–4500	UK, HP	Digestive health	Root, stem
Rhododendron campanulatum	Buransh	Ericaceae	2500–4000	UK, HP	Anti-inflammatory, cardiac	Flower, leaves
Saussurea chrysanthemoides	Chrysanthemum	Asteraceae	3000–4500	UK, HP	Wound healing, digestive	Root
Saussurea costus	Kuth	Asteraceae	2500–4000	HP, J&K	Skin conditions, digestive	Root
Saussurea obvallata	Brahmkamal	Asteraceae	3800–4600	UK, HP	Anti-inflammatory, cardiac	Flower, leaves
Saussurea gossypiphora	Kasturi kamal	Asteraceae	4000–5500	UK, HP	Cold relief, respiratory	Whole plant
Selinum candollei	Bhootkesh	Apiaceae	3000–3800	UK, HP	Digestive health	Root
Senecio kraschennikowii	Kraschennikov's	Asteraceae	3000–4500	UK, HP	Anti-inflammatory, used in	Whole plant
Swertia chirayita	Chirata	Gentianaceae	1200–3000	UK, J&K, HP, SK	Diabetes, blood purifier	Whole plant
Tanacetum dolichophyllum	Long-leaved	Asteraceae	2500–4000	UK, HP, SK	Anti-inflammatory	Leaves
Tanacetum gracile	Slender Tansy	Asteraceae	3000–4200	UK, HP, SK	Used in respiratory issues	Leaves, stem
Tanacetum tenuifolium	Fine-leaved Tansy	Asteraceae	3000–4500	UK, HP, SK	Anti-inflammatory, digestive	Leaves, flower
Tanacetum tomentosum	Woolly Tansy	Asteraceae	2500–4000	UK, HP, SK	Skin ailments, wound	Leaves, root

Taraxacum officinale	Dandelion	Asteraceae	1500–3500	UK, HP, SK	Digestive health, liver tonic	Leaves, root
Taxus wallichiana	Himalayan Yew	Taxaceae	1800–3300	UK, HP, SK	Respiratory diseases	Bark, leaves
Trillium govanianum	Nagchhatri	Melanthiaceae	2500–4000	UK, HP, SK	Reproductive, wound-healing	Rhizome, root
Valeriana jatamansi	Tagar	Caprifoliaceae	1500–3500	UK, J&K	Anxiety, insomnia	Rhizome
Viola odorata	Nirbishi	Violaceae	3000–4000	UK, HP, J&K	Sore throats	Flower, leaves
Rheum australe	Dolu	Polygonaceae	3000–4200	UK, HP, J&K	Laxative, digestive ailments	Root
Mecnopsis aculeata	Popy	Papaveraceae	3200–3800	UK, HP, J&K	Anti-inflammatory, cardiac	Leaves, root
Polygonatum multiflorum	Mahmeda	Asparagaceae	3000–4000	UK, HP, J&K	Anti-inflammatory	Leaves
Fritillaria roylei	Kshirakakoli	Liliaceae	2800–3500	UK, HP, J&K	Fever, cough, and cold	Root
Pleurospermum densiflorus		Asparagaceae	3000–4000	UK, HP, SK	Anti-inflammatory	Root
Rheum webbianum		Polygonaceae	2400–4200	SK	Laxative, digestive ailments	Root

Note: *UK* Uttarakhand, *HP* Himachal Pradesh, *J&K* Jammu and Kashmir, *SK* Sikkim, *MG* Meghalaya, *AP* Arunachal Pradesh

combined with *Emblica officinalis* (fruits) or with *Mangifera indica* (seeds)/*Bergenia ligulata* (roots) blended in tablet form for 10–15 days helped relieve all kinds of kidney and urinary disorders, control inflammation (Ballabh et al. 2008). Singh et al. (2021) reported that the tuber powder extract of *Dactylorhiza hatagirea* is very useful in cuts and wound healing, stomach-ache, immunity enhancer and energy tonic, seminal debility, diabetes, diarrhea, and chronic pain. Also, when consumed, the tuber of the species is reported to overcome impotency, low sperm count, spermatorrhea, and increased vigor (Wangchuk and Samten 2009; Thakur and Dixit 2007). The rhizomes of *Paris polyphylla* are reported to be used as anthelmintic, against stomach-aches, as antispasmodic, digestive, vermifuge, expectorant, and ethno-pediatrics for diarrhea (Nautiyal and Nautiyal 2004). Also, the decoction of *P. polyphylla* rhizomes is reported to help treat snake and insect bites and boils (Rawat et al. 2023). Similarly, species such as *Polygonatum cirrhifolium*, *P. verticillatum,* and *Malaxis muscifera* are important ingredients of Astavarga (a group of eight medicinal plants in Ayurveda medicine) considered very good rasayana with rejuvenating and health-promoting properties and are known to strengthen the immune system with immense cell regeneration capacity (Giri et al. 2017). Further, these species are useful for healing fractures, seminal weakness, fever, abnormal thirst, and diabetic conditions and as a cure for vata, pitta, and rakta doshas (Acharya Balkrishna et al. 2012). The use of species such as *Nardostachys jatamansi* dates back to the Vedic period (500–1000 BC) and is well documented in authoritative Ayurvedic texts, such as "Sushruta Samhita," "Nighantus Chikitsa Granthas," and "Charaka Samhita." Since then, the dried rhizomes and roots of the species have been an integral component of "havan," a ritualistic practice in Hindu worship (Dhiman and Bhattacharya 2020). A widespread belief persists that the combustion of these herbs may offer therapeutic benefits for various ailments. Other than this, traditionally, the species have been used to treat various disorders and ailments of the nervous, digestive, circulatory, respiratory, urinary, and reproductive systems (Dhiman and Bhattacharya 2020). The inhabitants of western Nepal use powdered infusions of the rhizomes of *N. jatamansi* to cure cough, fever, food poisoning, cholera, stomach disorders, intestinal worms, headaches, high altitude sickness, leprosy, etc. Elsewhere, decoctions of the roots are taken early in the morning to serve as a tonic for memory enhancement (Rokaya et al. 2010; Joshi et al. 2016). In Tibet, the rhizomes of the species are used for curing wounds, chronic fever, inflammation, intestinal worms, high blood pressure, food poisoning, and gastritis. In Amchis (agro-pastoralist Tibetan community), leaves of *N. jatamansi* are used for curing headaches, high-altitude sickness, fever, and wounds (Ghimire and Thomas 2009). As the region harbors a rich diversity of ethno-medical plants, their uses against different ailments and across different phytogeographical regions are mentioned in Table 5.1.

5.4.3 Conservation Planning, Propagation, and Multiplication Strategies

The guiding principles of systematic conservation planning involve knowledge of the species vulnerability status and irreplaceability extent (i.e., the degree to which the geographical possibilities for conservation objectives are limited if the site is destroyed) (Rodrigues et al. 2006). Yearly evidence on the Himalayan medicinal and aromatic plants (MAPs) indicates that most of these medicinal plants exhibit a continuous decreasing trend over the years (Singh et al. 2023). Some of the high-stakes species promptly figured in various threat categories yet highly traded are *Angelica glauca*, *Aconitum violaceum*, *Aconitum heterophyllum*, *Aquilegia nivalis*, *Coptis teeta*, *Dactylorhiza hatagirea*, *Gymnocladus assamicus*, *Magnolia pleiocarpa*, *Lilium polyphyllum*, *Arnebia euchroma*, *Corydalis cashmeriana*, *Ephedra gerardiana*, *Hypericum gaitii*, *Paris polyphylla*, *Malaxis muscifera*, *Meconopsis aculeta*, *Nardostachys jatamansi*, *Rheum austral*, *Rheum webbianum*, *Roscoea purpurea*, *Polygonatum cirrhifolium*, *Polygonatum verticillatum*, *Trillium govanium*, etc. (Dhyani et al. 2018; Barik et al. 2018; Mehta et al. 2020; Singh et al. 2021, 2023). These species contain unique bioactive compounds with high therapeutic potential, making them a special candidature of global interest. Being narrowly distributed and destructive harvesting practice emphasizes the conservation implication needs of such species (Dhar et al. 2000; Rai et al. 2000; Mehta et al. 2020).

Given the gravity of the issues, it is necessary to establish a conservation framework surrounding habitat restoration and promote cultivation, thereby reducing pressure on wild populations. Various propagation protocols involving macro-propagation and micro-propagation techniques have been used for high-value endemic and threatened MAPs in the region. These techniques have significantly improved plant propagation strategy, mainly multiplication and large-scale availability of planting materials. In this context, several vegetative propagation protocols (Table 5.2) have been developed. These include *Aconitum heterophyllum* (Pandey et al. 2005), *Heracleum candicans* (Butola and Badola 2007), *Arnebia benthamii* (Manjkhola et al. 2003), *Cornus capitata* (Airi et al. 2005), *Swertia agustifolia* (Bhatt et al. 2005a, 2005b), *Angelica glauca* (Bisht et al. 2006), *Selinum wallichianum* (Joshi et al. 2006), *Swertia chirayita* (Pradhan and Badola 2011), *Podophyllum hexandrum* (Kharkwal et al. 2008), *Picrorhiza kurrooa* (Sharma et al. 2012), *Polygonatum verticillatum* (Lohani et al. 2012), *Lilium polyphyllum* (Dhyani et al. 2022), *Hippophae rhamnoides* (Dolkar et al. 2016), and *Trillium govanianum* (Chauhan et al. 2020). This is very important because seed germination in Himalayan medicinal plants comprises long dormancy periods, poor seed germination, high seedling mortality, etc., which inhibit the success in obtaining desired results. In the last three decades, tissue culture-mediated biotechnological interventions have gained widespread recognition. It has achieved great success in alpine MAPs in addressing the increasing demand for plant propagules in a short period (Table 5.3). Some of the propagation milestones achieved in this direction include *Podophyllum hexandrum* (Nadeem et al. 2000; Rajesh et al. 2014), *Saussurea obvallata* (Joshi and Dhar 2003), *Aconitum balfourii* (Pandey et al. 2004), *Arnebia benthamii*

Table 5.2 Vegetative propagation studies on selected high-altitude medicinal plants of the Indian Himalayan region

Name of plant species	Explant	Effective treatment			Sprouting%	Rooting %	References
		IBA	IAA	GA_3			
Paris polyphylla	Rz bd	100 mg/L		100 mg/L	83.33	73.33	Danu et al. (2015)
	Rz bd				93.33	53.33	Puwien and Thomas (2019)
Picrorhiza kurroa	Br A Ct	50.0μM				87	Chandra et al. (2006)
Dactylorhiza hatagirea	Tb	50.0μM			61.1	38.88	Giri and Tamta (2012)
Gentiana kurroo	Rz	200 ppm					Tomar et al. (2011)
Angelica glauca			0.25 mM		93.33	90	Butola and Badola (2007)
Hippophae rhamnoides	Sm	1000 ppm				98.8	Dolkar et al. (2016)
Hippophae salicifolia	Sm	400 ppm					Rao et al. (2010)
Podophyllum hexandrum	Rz	100μM				70	Nadeem et al. (2000)
	Rz		400 ppm			32 roots	Kharkwal et al. (2008)
Trillium govanianum	Rz	100 ppm					Chauhan et al. (2020)

Note: *Ap Rz* apical part of rhizome, *Rz ct* rhizome cutting, *Rz bd* rhizome bud, *Rz* rhizome, *Br A Ct* Br A Ct, *Ct* cutting, *Tb* tuber, *Sm* stem cutting, *Sh* shoot, *Br* branch

(Manjkhola et al. 2005), *Angelica glauca* (Bisht et al. 2015; Rawat et al. 2018), *Hedychium spicatum* (Dhyani et al. 2024), *Habenaria edgeworthii* (Giri et al. 2012), *Lilium polyphyllum* (Dhyani et al. 2014), *Dactylorhiza hatagirea* (Warghat et al. 2014), *Valeriana jatamansi* (Nazir et al. 2022), *Paris polyphylla* (Raomai et al. 2015), *Nardostachys jatamansi* (Bose et al. 2016; Dhiman et al. 2021), *Berberis asiatica* (Bisht et al. 2022), *Gentiana kurroo* (Sharma et al. 2014), *Swertia chirayita* (Kumar et al. 2018), *Mahonia jaunsarensis* (Bisht et al. 2024), *Aconitum ferox* (Singh et al. 2020a, b), *Rhodiola imbricata* (Bhardwaj et al. 2018), *Aconitum chasmanthum* (Rafiq et al. 2021), *Rheum emodi* (Tabin et al. 2018), and *Rheum spiciforme* (Khan et al. 2022).

Table 5.3 In vitro propagation studies undertaken in prominent high-altitude medicinal plants of the Himalayan region

Species name	Explants used	PGRs concentration used for callus induction (*mg/L and # μM)	PGRs concentration used for shoot multiplication (* mg/L and # μM)	PGRs concentration used for rooting (* mg/L and # μM)	References
Aconitum heterophyllum	L	1.0# NAA and 5.0# BAP	1.0# NAA and 5.0# BAP	11.4# IBA	Nandi et al. (2016)
Aconitum balfourii	L	4.5# 6 BAP and 26.9# NAA	22.2# BAP + 5.4# NAA	12.3# IBA	Nandi et al. (2016)
Aconitum violaceum	Cd		1.0* TDZ		Rawat et al. (2013)
	L, R	2.5# IBA 2.5# BA	1.0* IBA and 2.5* BAP	2.0* IBA	Hadi et al. (2023)
Angelica glauca	St			2.5* NAA and 1.25* BAP	Butola et al. (2010)
	Rz		8.0# BAP and 0.1# IAA	0.5# IAA 0.1# NAA	Rawat et al. (2018)
Arnebia euchroma	Cd		4.5* TDZ	4.9* IBA	Jiang et al. (2005)
	L	1.5# IBA and 1# BAP	2.5# BAP		Majkhola et al. (2004)
	Bb	1.0* Kn + 0.5* NAA	3# and 2# BAP		Kumar et al. (2011)
Aconitum ferox	Rt	2.26# 2,4-D	6# BAP	1# IBA	Singh et al. (2020)
Bunium persicum	S	2* NAA and 2* Kn		2* NAA + 2* Kn	Vlizadeh et al. (2006)
	Rt	0.5* 2 4 D			Otroshy et al. (2013)
	Hc	2* BAP	2* BAP 1* NAA	1.5* IBA	Bashir et al. (2014)
	S	1* 2,4-D	1* 2,4-D		Valizadeh et al. (2009)
	Cd	1* BAP and 3* IAA			Bashir et al. (2014)

(continued)

Table 5.3 (continued)

Species name	Explants used	PGRs concentration used for callus induction (*mg/L and # μM)	PGRs concentration used for shoot multiplication (* mg/L and # μM)	PGRs concentration used for rooting (* mg/L and # μM)	References
Dactylorhiza hatagirea	T		10# TDZ	50.0# IBA	Giri and Tamta (2012)
Gentiana kurrooa	St		0.5* IAA + 1.0* BAP		Kaushal et al. (2014)
	Em c		2* BAP + 0.5 mg/L IAA		Alphonse et al. (2022)
	Pt	1* BA and 3*NAA	0.10* NAA + 1.0* TDZ	1/2 MS 1* IBA	Sharma et al. (2014)
	Sm		8.9# BAP		Sharma et al. (1993)
Hippophae rhamnoides	L		4.5* TDZ and 2.6* BAP		Dolkar et al. (2016)
Lilium polyphyllum	Bb	25# IBA			Dhyani et al. (2014)
Nardostychys jatamansi	Rz		0.5* 1 Kn + 0.75* TDZ	1* IAA	Sharma et al. (2016)
	L	3* NAA and 0.25* L Kn		0.25* to 1* IAA	Mathur (1992)
	N		2.0* BAP + 1.5* Kn	1.5 NAA 1.5* BAP	Pant et al. (2021)
	L	21.1# NAA and 3.16 # Kn	23.1# NAA + 4.0# BAP		Rawat et al. (2020)
Polygonatum verticillatum	Rz		1* BAP + 0.5* NAA	0.5* IBA	Bisht et al. (2012)
	S	8* Kn and 2* IAA	4* BAP + 1* NAA	2* IBA	Qadir et al. (2020)
	Ac		0.25* Kn + 0.25* BAP	0.25* IBA	Bisht et al. (2017)
	Ns		0.5# TDZ	MS	Patial et al. (2012)
	Rn			2.5# IBA + 1.0# IBA	Chandra et al. (2006)
	L	0.75* TDZ and 0.75* IBA	0.50* BAP + 0.75* Kn		Pratap et al. (2022)

(continued)

Table 5.3 (continued)

Species name	Explants used	PGRs concentration used for callus induction (*mg/L and # μM)	PGRs concentration used for shoot multiplication (* mg/L and # μM)	PGRs concentration used for rooting (* mg/L and # μM)	References
Paris polyphylla	Rz		5* BAP NAA 2*		Puwein and Thomas (2022)
Podophyllum hexandrum	Em		1# IAA and 1# IBA	4.6# IBA	Nadeem et al. (2000)
	Rt	0.5–1.5*Kn and BAP 0.2–1*	0.5* BAP 5* IAA	0.5–2* NAA	Sultan et al. (2006)
	Rt	0.1* BAP + 1.0* NAA	2.0* NAA 0.2* BAP	BAP 0.5* + 2,4-D 0.5*	Sharma et al. (2022)
Saussurea costus	St		1.14# TDZ 2.68# NAA	2.46* IBA	Sharma et al. (2019)
	L	0.5* 2,4-D and 1.0* Kn	2.0* BAP 1.0* NAA	0.5* BAP 0.5* IBA	Khan et al. (2021)
	Ab		5.0* BAP + 1.0* IAA		Mishra and Shukla (2010)
	As		1.0* BAP	1.0* IBA	Nisa et al. (2019)
Sassurea ovaballata	L	2.5# BAP and 10.0# NAA	5# BAP + 1# NAA	1/2MS + 2.5# IBA	Dhar and Joshi (2005)
	Ec		1# Kn and 0.25# α-NAA	Half MS 2.5* IBA	Joshi and Dhar (2003)
Swertia chirayita	L	1.0* 2, 4 D + 0.5 BAP	1.5* + IAA 2.0* Kn + 0.2* GA3		Kumar et al. (2014)
	St		2* Kn + 3* GA3 + 2* IBA		Kumar et al. (2014)
	Bd		4.44# BAP + 2.85# IAA	4.90# IBA	Pant et al. (2010)
	L	15.0# 2,4-D	5.0# BAP	5.0# IBA	Kumar et al. (2018)
	L	1* NAA and 3.0* BAP	0.1* NAA and 3.0* BAP	0.1*NAA	Shailja et al. (2017)

Another aspect of tissue culture studies that has become vital to address the increasing demand for pharmaceutical industries is optimizing the plant growth regulators (PGRs) concentrations, optimal photoperiods, temperature range, and identifying the optimum stage for collection of propagules, etc., aiming to support commercial cultivation. Studies have demonstrated that specific PGR combinations are crucial for the effective in vitro propagation of the above-mentioned species. For instance, *Aconitum heterophyllum* and *A. balfourii* are reported to exhibit positive responses to naphthaleneacetic acid (NAA) and 6-benzylaminopurine (BAP) combinations for callus induction and shoot formation, while indolebutyric acid (IBA) facilitates root development (Nandi et al. 2016). Likewise, thidiazuron (TDZ) has been shown to enhance shoot initiation in *Angelica glauca* and *Arnebia euchroma*, whereas *Bunium persicum* displays efficient callus formation and rooting when treated with NAA and kinetin (Kn) combinations (Bashir et al. 2014). The cytokinin–auxin balance plays a significant role in the development of *Dactylorhiza hatagirea* and *Gentiana kurroo*, with TDZ and IBA promoting shoot and root growth, respectively (Giri and Tamta 2012; Kaushal et al. 2014). In *Picrorhiza kurroa* and *Polygonatum verticillatum*, IBA and TDZ support root formation, while Kn and BAP induce substantial shoot development (Bisht et al. 2012; Patial et al. 2012). Further, NAA and BAP have been found to enhance overall plantlet development in *Podophyllum hexandrum* and *Paris polyphylla* (Nadeem et al. 2000; Puwein and Thomas 2022). For *Saussurea costus* and *Swertia chirayita*, balanced PGR combinations promote consistent growth across various explants, achieving stable callus, shoot, and root initiation, which is essential for large-scale propagation (Kumar et al. 2014). These findings underscore the importance of tailored PGR combinations in the successful in vitro propagation of Himalayan medicinal plants.

5.4.4 Good Agricultural and Collection Practices

The National Medicinal Plants Board (NMPB), under the Department of AYUSH, Government of India, has instituted and implemented various initiatives focusing on Good Agricultural and Collection Practices (GACPs). A notable example is the "Central Scheme for Development of Agro-techniques and Cultivation of Medicinal Plants," which has established several protocols and agro-techniques for cultivating and harvesting MAPs from wild sources, including Himalayan species. Table 5.4 outlines some of the promising agro-techniques developed for high-altitude Himalayan MAPs. In general, medicinal plant cultivation depends on several parameters, such as soil types, propagation materials, nursery techniques, and harvesting methods, each influencing productivity and market value. For example, *Aconitum balfourii* and *Aconitum heterophyllum* flourish in sandy loam with mildly acidic soils (pH 5.1–6.0). Their propagation involves seeds and tuber segments, with approximately 2 kg of seeds necessary to generate 50,000 seedlings for a hectare of land, planted at a depth of 0.5–0.7 cm in mist chambers or polyhouses. Harvesting occurs in August–September in lower elevations and October–November in higher regions, yielding 450 kg per hectare with a market rate of ₹172 per kg for *A. balfourii*, and 518 kg per hectare with a higher market value of ₹2800–3600 per kg for

Table 5.4 Agro-techniques developed for the cultivation of different Himalayan medicinal plants

Name of medicinal plants	Soil	Propagation material	Nursery technique	Harvesting	Production (Kg/ha)	Market rate
Aconitum balfourii	Sandy loam and slightly acidic soils (pH 5.1–5.5)	Seeds and tuber segments	Seeds are sown in a nursery at a soil depth of 0.5–0.7 cm in mist chambers or shade houses/nearly 2 kg seeds are required for raising seedlings for 1 hectare of land, that is, for planting 50,000 seedlings at a spacing of 45 cm × 45 cm.	October–November	450 kg	172 per kg
Aconitum heterophyllum	Sandy loam and slightly acidic soil, with a pH of about 6	Seeds and tuber segments	They are sown in the sand and FYM (farmyard manure)/compost mixture (1:2) at a depth of 0.5 cm in styrofoam trays in a mist house/approximately 1.5 kg seeds are required for obtaining seedlings for 1 hectare.	May–June	518 kg	Rs 2800–3600 per kg
Inula racemosa	Well-drained, clay-loam soils	Seeds and rootstocks	In nursery in November or early March by broadcasting the seeds in well-prepared nursery beds/ about 1 kg of seeds are required to raise about 40,000 seedlings required for 1 hectare of land.	October–November	80 quintals	
Nardostachys jatamansi	Loamy porous soil rich in organic matte	Seeds	Seeds may be sown manually in small plots of 1 m × 1 m size or in 1 Agro techniques study carried out by High Altitude Plant Physiology Research Centre, HNB Garhwal University, Srinagar, Garhwal, Uttarakhand styrofoam trays in the polyhouse/ approximately 600 g of seeds are required for raising seedlings for transplanting in 1 hectare of land.	October–November	835 kg	Rs 170 per kg
Picrorhiza kurroa	Sandy clay textured soi	Rhizomes/ stolons	The nursery beds are planted using rhizomes/stolons/ offsets in October–November. The planting stock may be raised in polybags, styrofoam trays, or motherbeds/ about 165,000 seedlings per hectare are needed, which means 1–1.5 kg seeds are required for raising saplings for 1 hectare of land.	August–September	11 quintals	Rs. 230–265 per kg dry roots

(continued)

Table 5.4 (continued)

Name of medicinal plants	Soil	Propagation material	Nursery technique	Harvesting	Production (Kg/ha)	Market rate
Rheum emodi	Well-drained, porous, humus-rich soil	Seeds/ rhizome	Seeds are sown in lines 15 cm apart. In temperate conditions, seeds are sown in the spring season (March–early April)/ about 600 g of seeds are required to raise a nursery for planting at a spacing of 50 cm × 50 cm in 1 hectare of land.	September and October	37 quintals	
Swertia chirata	Loamy to sand-loam, friable, and well-drained soils	Seeds	Seeds are sown in rows 10–15 cm apart and covered with a 0.5-cm-thick layer of sand or fine soil/about 200 g of seeds are required for raising planting stock for 1 hectare of land.	October–November	3.75 tons	

Source: NMPB (2014)

A. heterophyllum. In the case of *Nardostachys jatamansi*, the species is reportedly grown in loamy, porous, organic-rich soil and requires about 600 grams of seeds to produce seedlings for a hectare. Harvested primarily in October–November, it yields 835 kg of dry roots per hectare, valued at ₹170 per kg. While *Picrorhiza kurroa* prefers sandy clay-textured soils and is propagated using rhizomes or stolons, with 1–1.5 kg of seeds needed for seedlings covering a hectare. Planted in polybags or mother beds in October–November and harvested in August–September, it produces 11 quintals of dry roots per hectare, with market prices ranging from ₹230 to ₹265 per kg. These cultivation methodologies highlight the significance of developing agro-techniques and conservation practices to enhance the sustainability of high-altitude medicinal plant cultivation in the Himalayas (NMPB 2014).

5.4.5 Listing of Medicinal Plant Species in Trade Through Mandis and Price Analysis

Approximately 7500 plant species are utilized in various traditional and herbal medicinal practices (Dubey et al. 2004), with 242 species consumed at volumes exceeding 100 metric tons annually (Goraya and Ved 2017; Table 5.5). Furthermore, 650–700 species are commercially exploited for phytopharmaceutical formulations (Maitra and Satya 2015). This extensive utilization has established India as a crucial hub for medicinal plants in Asia, generating an estimated trade value of US$140 million (Mishra et al. 2016) and supporting roughly 35 million workdays of employment (Traffic—the Wildlife Trade Monitoring Network 2018). The domestic demand for these plants was estimated at 195,000 tons, while export demand reached 134,500 tons. Notably, medicinal plant exports experienced substantial growth, increasing from Rs. 345.80 crore in 2005–2006 to Rs. 3211 crores in 2014–2015, representing a ninefold expansion within a decade (www.nmpb.nic.in). The market dynamics of high-value medicinal plants in India demonstrate considerable variation in market value, sourcing, and estimated annual trade volumes (Table 5.5; NMPB). For example, *Swertia chirayita* stands out with its exceptional market value range (320–80,000 per kg) and substantial estimated trade volume (500–1000 MT), indicative of its strong demand in both cultivated and wild-sourced markets. Certain species, such as *Picrorhiza kurroa* (Rs. 1370–2100 per kg) and *Delphinium denudatum* (Rs. 1000 per kg), are primarily wild-sourced and have low annual trade volumes (<10 MT), signaling potential overexploitation and conservation concerns. Conversely, *Picrorhiza kurroa* and *Berberis aristata* (Rs. 500–600 kg) exhibit high annual trade volumes of 1000–2000 MT, underscoring their economic significance. Rare and endangered species like *Aconitum ferox* (Rs. 900–1250 per kg) and *Delphinium denudatum* (1000 per kg), predominantly sourced from the wild, have limited trade volumes (<10 MT), further emphasizing their vulnerability. While species such as *Polygonatum verticillatum* (Rs. 16–50 per kg) demonstrate restricted trade volumes (<10 MT), highlighting the necessity for sustainable harvesting methods. Cultivated species such as *Gloriosa superba* (Rs. 200–2500 per kg) and *Taxus wallichiana* (Rs. 40–50 per kg) are traded at moderate levels (100–200

Table 5.5 Market value and estimated trade of high-altitude medicinal plants as per Goraya and Ved (2017)

Botanical name	Current market value in India (Rs/kg)	Source	Estimated annual trade (MT)
Aconitum ferox	900–1250	W	50–100
Aconitum heterophyllum	450–6300	W	100–200
Abrus precatorius	250–1000	C/W	200–500
Angelica glauca	5750–9500	W	Oct-50
Swertia chirata	320–80,000	C/W	500–1000
Oroxylum indicum	60–100	W	500–1000
Podophyllum hexandrum	60–100	W	Oct-50
Taxus wallichiana	40–50	C/W	100–200
Rheum australe	150–575	W	100–200
Polygonatum verticillatum	16–50	W	<10
Rauwolfia serpentina	700–1000	C/W	200–500
Delphinium denudatum	1000	W	<10
Nardostachys jatamansi	550–1750	W	500–1000
Berberis aristata	500–600	W	1000–2000
Ephedra gerardiana	80–100	W	100–200
Picrorhiza kurroa	1370–2100	W	1000–2000
Plumbago zeylanica	120–1100	W	500–1000
Betula utilis	215–400	W	100–200
Valeriana hardwickii	380–2360	W	<10
Gloriosa superba	200–2500	C/W	100–200

Note: *C* cultivated, *W* wild source

MT annually), reflecting efforts to reduce reliance on wild sources. Species such as *Rheum australe* (Rs. 150–575 per kg) and *Ephedra gerardiana* (Rs. 80–10 per kg) show moderate trade volumes of 100–200 MT. Some of the prominent Himalayan species with high trade volume and that draw high monetary value include *Angelica glauca* (Rs. 5750–9500 per kg) and *Nardostachys jatamansi* (Rs. 550–1750 per kg) remain crucial in traditional and commercial medicinal systems. However, this heightened demand has led to conservation concerns as 70–90% of these species are sourced from wild populations. These findings underscore the critical need for targeted conservation strategies, sustainable harvesting protocols, and improved cultivation practices to address overexploitation and ensure the long-term availability of these essential medicinal resources. Therefore, cultivating these high-value medicinal plants is useful not only for reducing the pressures on the wild population but also for improving the overall rural economy.

5.4.6 Organization Working Toward Conservation and Sustainable Utilization of Plants

At the international level, organizations such as the International Union for Conservation of Nature (IUCN), the Convention on International Trade in Endangered Species of Wild Fauna and Flora (CITES), and Trade Record Analysis of Flora and Fauna in Commerce (Traffic) are engaged in the conservation of threatened plant species. In India, the Forest Conservation Act of 1980, the Forest (Conservation) Amendment Act of 2023, and the National Wildlife Act of 1972 regulate trade and stipulate that any activities detrimental to such plants must be subject to legal sanctions. Various ministries of the central governments, viz., Ministry of AYUSH, Ministry of Environment, Forest and Climate Change (MoEF&CC), Ministry of Science and Technology, Ministry of Agriculture and Farmer's Welfare, Ministry of Rural Development, etc., and their statutory autonomous departments like the National Medicinal Plant Board of India (NMPB), Botanical Survey of India (BSI), G. B. Pant National Institute of Himalayan Environment (NIHE), National Bureau of Plant Genetic Resources (NBPGR), Indian Council for Agricultural Research (ICAR), Council for Scientific and Industrial Research (CSIR), etc. Other than these, every state has devised guidelines for the cultivation, trade, and conservation of medicinal plants, and these guidelines are often formulated in collaboration with the National Medicinal Plant Board of India and other relevant central government agencies.

5.5 Conclusion and Future Prospects

Over the years, considerable efforts have been made to investigate alpine medicinal plants from ethno-botany to advance "omics" involving proteomics, genomics, and metabolomics. Apart from their application in the folk/traditional systems of medicine, these plants are central pillars of most modern medicine—one-fourth of all prescribed medicines have herbal constituents as active therapeutic agents. As a result, the utilization of herbal remedies by means of a dietary supplement is seeing an upsurge, given the general belief that herbs are natural, intrinsically safe, and cost-effective. Very recently, alpine herbs have shown significant potential—in silico studies, in the possible treatment of novel SARS coronavirus (SARS-CoV2), thus holding the trust that herbal medicine has a role in treating and preventing an epidemic disease. Against this backdrop and considering several other epidemiological diseases, the global market is now occupied with many herbal products; however, time and forth, serious concerns have been raised regarding the originality of the product. Also, most herbal products fail to demonstrate their safety and efficacy as they are not properly evaluated. Conducting in-depth research and developing a holistic understanding of the pharmacological and pharmacokinetics of Himalayan herbs are vital for advancing the efficacy of herbal blends, providing strong scientific support regarding their safety and toxicity, thus overcoming trust-deficiency issues, and market expectations—meeting national/international standards, and ensuring quality. In the long run, innovations in herbal medications and collaborations between academics, industries, and traditional practitioners are essential for the herbal market to thrive. Unfortunately, many of these organizations are hesitant to share knowledge or reluctant to share the benefits equitably—access and benefit sharing. Contrarily, many of the Himalayan plant species are now threatened due to destructive harvesting practices and climate change. Although several preventive initiatives (by both the state and the central governments) are being implemented to tackle this issue, the outcomes still need improvement. Some of the key recommendations for strengthening research on alpine medicinal plants could include the following:

- Digitizing the available information and compiling it into a comprehensive database for informed decision-making and halting biodiversity loss.
- Standardized plant identification system to overcome taxonomic ambiguity.
- Detailed know-how of species in trade, including population, habitat and ecology, threats, and kind of conservation actions needed or strategies for mainstreaming.
- Overcoming variability in post-harvest processing methods, adulteration, and contamination (chemical or biological) checks.
- Mainstreaming and market value-chain development.
- Value addition and product diversification.
- Regulatory approvals for herbal medicines ensuring safety and quality assurance.

Acknowledgments The authors are thankful to the director (NIHE) for providing the necessary facilities and encouragement.

Conflicts of Interest The authors declare no conflict of interest.

Funding Financial support from NMHS (NMHS 2023–2024/SC-XI/LG13/03) funded project is gratefully acknowledged.

References

Acharya Balkrishna AB, Anupam Srivastava AS, Mishra RK, Patel SP, Vashistha RK, Ajay Singh AS et al (2012) Astavarga plants-threatened medicinal herbs of the North-West Himalaya. Int J Med Arom Plants 2(4):661–676

Airi S, Rawal RS, Dhar U (2005) Presowing treatment effects on germination of Cornus capitata seeds. Seed Sci Technol 33(1):77–86

Alphonse M, Chandrasekaran R, Ramamoorthy S, Fulzele DP, Raina R, Thiagarajan K (2022) Optimizing shoot formation in *Gentiana kurroo* Royle for gentiopicroside production. J Plant Growth Regul 41(3):1–10

Badola HK (2001) Medicinal plant diversity of Himachal Pradesh. In: Himalayan medicinal plants: potential and prospects. Gyanodaya Prakashan, Nainital, pp 87–116

Ballabh B, Chaurasia OP, Ahmed Z, Singh SB (2008) Traditional medicinal plants of cold desert Ladakh—used against kidney and urinary disorders. J Ethnopharmacol 118(2):331–339

Barik SK, Tiwari ON, Adhikari D, Singh PP, Tiwary R, Barua S (2018) Geographic distribution pattern of threatened plants of India and steps taken for their conservation. Curr Sci 114(3):470–503

Bashir I, Kaloo ZA, Singh S, Rashid S (2014) Comparative efficiency of various explants for callus production in *Bunium persicum* (Boiss.) B. Fedtsch. J Sci Innov Res 3(6):572–577

BCC Research Report (2023) Global Herbal Medicine Market. https://www.bccresearch.com/marketresearch/pharmaceuticals/herbal-medicine-market.html?

Bhardwaj AK, Singh B, Kaur K, Roshan P, Sharma A, Dolker D et al (2018) In vitro propagation, clonal fidelity and phytochemical analysis of *Rhodiola imbricata* Edgew: a rare trans-Himalayan medicinal plant. Plant Cell Tissue Organ Cult 135:499–513

Bhatt A, Joshi SK, Gairola S (2005a) *Dactylorhiza hatagirea* (D. Don) Soo—a west Himalayan orchid in peril. Curr Sci 89(4):610–612

Bhatt A, Rawal RS, Dhar U (2005b) Germination improvement in *Swertia angustifolia*: a high value medicinal plant of Himalaya. Curr Sci 89(6):1008–1012

Bhatt D, Kumar R, Tewari LM, Joshi GC (2014) *Polygonatum cirrhifolium* Royle and *Polygonatum verticillatum* (L.) Allioni: status assessment and medicinal uses in Uttarakhand, India. J Med Plant Res 8(5):253–259

Bhattarai P, Pandey B, Gautam RK, Chhetri R (2014) Ecology and conservation status of threatened orchid *Dactylorhiza hatagirea* (D. Don) Soo in Manaslu conservation area, Central Nepal. Am J Plant Sci 5(23):3483

Bisht AK, Bhatt A, Rawal, Dhar U (2006) Assessment, extraction pattern and restoration through propagation of *Angelica glauca*, a critically endangered Himalayan medicinal herb. Int J Biodivers Sci Manag 2(2):105–113

Bisht S, Bisht NS, Bhandari S (2012) In vitro micropropagation in *Polygonatum verticillatum* (L.) All. an important threatened medicinal herb of Northern India. Physiol Mol Biol Plants 18:89–93

Bisht AK, Bhatt A, Dhar U, Bhatt A, Dhar U (2015) Note on somatic embryogenesis and synthetic seed production in *Angelica glauca*: a valuable medicinal plant of Himalaya. J Med Plant Res 9(12):419–425

Bisht SS, Bisht AS, Chauhan RS (2017) In-vitro mutagenesis induction to improve abiotic stress in tissue cultured plantlet of *Picrohiza kurroa* Royle ex. *Benth*: an endangered plant of Western Himalayas, India. Med Aromat Plants (Los Angel) 6(287):2167–0412

Bisht A, Giri L, Belwal T, Pandey A, Bahukhandi A, Bhatt ID, Rawal RS (2022) In vitro propagation and antioxidant potential of *Berberis asiatica* from Western Himalaya. Plant Biosyst 156(2):490–496

Bisht A, Singh L, Singh B, Bisht M, Pandey A, Bhatt ID, Pande V (2024) In vitro propagation of endemic species *Mahonia Jaunsarensis* Ahrendt through callus culture. Appl Biochem Biotechnol 196(1):113–128

Bose B, Kumaria S, Choudhury H, Tandon P (2016) Assessment of genetic homogeneity and analysis of phytomedicinal potential in micropropagated plants of *Nardostachys jatamansi*, a critically endangered, medicinal plant of alpine Himalayas. Plant Cell Tissue Organ Cult 124:331–349

Butola JS, Badola HK (2007) Vegetative propagation of *Angelica glauca* and *Heracleum candicans*. J Trop Med Plants 8(1):85–91

Butola JS, Vashistha RK, Samant SS, Malik AR (2010) Technology for propagation and cultivation of *Angelica glauca* Edgew.: a threatened high value Himalayan medicinal cum edible herb. Med Plant 2(1):67–72

Chandra B, Palni LMS, Nandi SK (2006) Propagation and conservation of *Picrorhiza kurrooa* Royle ex Benth.: an endangered Himalayan medicinal herb of high commercial value. Biodivers Conserv 15:2325–2338

Chauhan HK, Bisht AK, Bhatt ID, Bhatt A, Gallacher D, Santo A (2018) Population change of *Trillium govanianum* (Melanthiaceae) amid altered indigenous harvesting practices in the Indian Himalayas. J Ethnopharmacol 213:302–310

Chauhan HK, Bisht AK, Bhatt ID, Bhatt A (2020) Protocol for vegetative propagation of *Trillium govanianum* Wall ex D. Don. J Appl Res Med Aromat Plants 16:100233
Chen SL, Yu H, Luo HM, Wu Q, Li CF, Steinmetz A (2016) Conservation and sustainable use of medicinal plants: problems, progress, and prospects. Chin Med 11:1–10
Danu K, Adhikari RS, Pande V, Singh MK, Rawal P (2015) Vegetative propagation of an endangered medicinal plant of Himalayan region, *Paris polyphylla* smith. Int J Curr Microbiol App Sci 4(6):660–665
Dhar U, Joshi M (2005) Efficient plant regeneration protocol through callus for *Saussurea obvallata* (DC.) Edgew. (Asteraceae): effect of explant type, age and plant growth regulators. Plant Cell Rep 24:195–200
Dhar U, Rawal RS, Upreti J (2000) Setting priorities for conservation of medicinal plants—a case study in the Indian Himalaya. Biol Conserv 95(1):57–65
Dhiman N, Bhattacharya A (2020) *Nardostachys jatamansi* (D. Don) DC.—challenges and opportunities of harnessing the untapped medicinal plant from the Himalayas. J Ethnopharmacol 246:112211
Dhiman N, Devi K, Bhattacharya A (2021) Development of low cost micropropagation protocol for *Nardostachys jatamansi*: a critically endangered medicinal herb of Himalayas. S Afr J Bot 140:468–477
Dhyani A, Sharma G, Nautiyal BP, Nautiyal MC (2014) Propagation and conservation of *Lilium polyphyllum* D. Don ex Royle. J Appl Res Med Aromat Plants 1(4):144–147
Dhyani A, Nautiyal BP, Nautiyal MC (2018) Distribution, status, and conservation of *Lilium polyphyllum* (Liliaceae), a critically endangered medicinal plant from India. Plant Biosyst 152(4):608–611
Dhyani A, Nautiyal BP, Nautiyal MC (2022) Seedling performance, phenotypic traits and growing media for domestication of *Lilium polyphyllum*. Plants People Planet 4(6):685–695
Dhyani V, Bhatt ID, Pande V (2024) Efficient plant regeneration through callus culture in *Hedychium spicatum* Buch. Ham. Ex. D. Don using response surface methodology. Plant Cell Tissue Organ Cult 158:37. https://doi.org/10.1007/s11240-024-02831-4
Dolkar P, Dolkar D, Angmo S, Srivastava RB, Stobdan T (2016) An improved method for propagation of seabuckthorn (*Hippophae rhamnoides* L.) by cuttings. Nat Acad Sci Lett 39(5):323–326
Dubey NK, Kumar R, Tripathi P (2004) Global promotion of herbal medicine: India's opportunity. Curr Sci 86(1):37–41. https://www.jstor.org/stable/24109515
Ghimire SK, Aumeeruddy-Thomas Y (2009) Ethnobotanical classification and plant nomenclature system of high altitude agro-pastoralists in Dolpo, Nepal. Plant Sci J 6:56–68
Ghimire SK, Sah JP, Shrestha KK, Bajracharya D (1999) Ecological study of some high altitude medicinal and aromatic plants in the Gyasumdo valley, Manang, Nepal. Ecoprint 6(1):17–25
Giri D, Tamta S (2012) Propagation and conservation of *Dactylorhiza hatagirea* (D. Don) Soo, an endangered alpine orchid. Afr J Biotechnol 11(62):12586–12594
Giri L, Jugran A, Rawat S, Dhyani P, Andola H, Bhatt ID et al (2012) In vitro propagation, genetic and phytochemical assessment of *Habenaria edgeworthii*: an important Astavarga plant. Acta Physiol Plant 34:869–875
Giri L, Belwal T, Bahukhandi A, Suyal R, Bhatt ID, Rawal RS, Nandi SK (2017) Oxidative DNA damage protective activity and antioxidant potential of Ashtvarga species growing in the Indian Himalayan Region. Ind Crop Prod 102:173–179
Goraya GS, Ved DK (2017) Medicinal plants in India: an assessment of their demand and supply. National Medicinal Plants Board, Ministry of AYUSH, Government of India, New Delhi and Indian Council of Forestry Research & Education, Dehradun, pp 1–430
Gowthami R, Sharma N, Pandey R, Agrawal A (2021) Status and consolidated list of threatened medicinal plants of India. Genet Resour Crop Evol 68(6):2235–2263
Haq SM, Waheed M, Khoja AA, Amjad MS, Bussmann RW, Ali K (2023) A cross-cultural study of high-altitude botanical resources among diverse ethnic groups in Kashmir Himalaya, India. J Ethnobiol Ethnomed 19(1):12
https://www.bccresearch.com/market-research/pharmaceuticals/herbal-medicine-market.html. Assessed on 05 Sept 2024

https://www.dgciskol.gov.in. Assessed on 25 Sept 2024

https://www.statista.com/statistics/652418/export-value-of-ayurvedic-and-herbal-products-india/. Assessed on 25 Sept 2024

IUCN Red List of Threatened Species (2024). https://www.iucnredlist.org/. Accessed on September 2024

Jiang BO, Yang YG, Guo YM, Guo ZC, Chen YZ (2005) Thidiazuron-induced in vitro shoot organogenesis of the medicinal plant *Arnebia euchroma* (Royle) Johnst. In Vitro Cell Dev Biol Plant 41:677–681

Joshi M, Dhar U (2003) In vitro propagation of *Saussurea obvallata* (DC.) Edgew.—an endangered ethnoreligious medicinal herb of Himalaya. Plant Cell Rep 21:933–939

Joshi M, Rawal RS, Dhar U (2006) Breaking seed dormancy of *Selinum wallichianum*, a medicinal plant of the Himalaya. Seed Technol 28(1):87–92

Joshi RK, Satyal P, Setzer WN (2016) Himalayan aromatic medicinal plants: a review of their ethnopharmacology, volatile phytochemistry, and biological activities. Medicine 3(1):6

Kala CP (2000) Status and conservation of rare and endangered medicinal plants in the Indian trans-Himalaya. Biol Conserv 93(3):371–379

Kala CP (2005) Indigenous uses, population density, and conservation of threatened medicinal plants in protected areas of the Indian Himalayas. Conserv Biol 19(2):368–378

Kaushal S, Sidana A, Dev K (2014) In vitro plant production through apical meristem culture of *Gentiana kurroo* Royle. J Med Plant Stud 3(1):4–9

Khadka CB, Hammet AL, Singh A, Balla MK, Timilsina YP (2016) Ecological status and diversity indices of Panchaule (*Dactylorhiza hatagirea*) and its associates in Lete village of Mustang district, Nepal. Banko Janakari 26(1):45–52

Khan A, Shah AH, Ali N (2021) In-vitro propagation and phytochemical profiling of a highly medicinal and endemic plant species of the Himalayan region (*Saussurea costus*). Sci Rep 11(1):23575

Khan MI, Shahzad A, Ganie IB, Pandith SA (2022) Organogenesis, direct somatic embryogenesis, and shoot proliferation of *Rheum spiciforme* Royle: an endemic and vulnerable medicinal herb from Indian Trans Himalayas. In Vitro Cell Dev Biol Plant 58(1):35–50

Kharkwal AC, Kushwaha R, Prakash O, Ogra RK, Bhattacharya A, Nagar PK, Ahuja PS (2008) An efficient method of propagation of *Podophyllum hexandrum*: an endangered medicinal plant of the Western Himalayas under ex situ conditions. J Nat Med 62:211–216

Kumar R, Sharma N, Malik S, Bhushan S, Sharma UK, Kumari D et al (2011) Cell suspension culture of *Arnebia euchroma* (Royle) Johnston—a potential source of naphthoquinone pigments. J Med Plant Res 5:6048–6054

Kumar V, Singh S, Bandopadhyay R, Sharma M, Chandra S (2014) In vitro organogenesis secondary metabolite production and heavy metal analysis in *Swertia chirayita.* Open Life Sci 9(7):686–698

Kumar RR, Purohit VK, Prasad P, Nautiyal AR (2018) Efficient in vitro propagation protocol of *Swertia chirayita* (Roxb. ex Fleming) Karsten: a critically endangered medicinal plant. Nat Acad Sci Lett 41:123–127

Lakey, Dorji K (2016) Ecological status of high altitude medicinal plants and their sustainability: Lingshi, Bhutan. BMC Ecol 16:1–14

Lohani N, Kumar R, Tewari LM, Joshi GC (2012) Ex-situ conservation of *Polygonatum verticillatum* (L.) Allioni under different types of organic treatments. Int J Biodivers Conserv 4:22–31

Maitra S, Satya P (2015) Value added products from medicinal plants. In: Sharangi A, Datta S (eds) Value addition of horticultural crops: recent trends and future directions. Springer, New Delhi. https://doi.org/10.1007/978-81-322-2262-0_7

Manjkhola S, Dhar U, Rawal RS (2003) Treatments to improve seed germination of *Arnebia benthamii*: an endangered medicinal herb of high altitude Himalaya. Seed Sci Technol 31(3):571–577

Manjkhola S, Dhar U, Joshi M (2005) Organogenesis, embryogenesis, and synthetic seed production in *Arnebia euchroma*—a critically endangered medicinal plant of the Himalaya. In Vitro Cell Dev Biol Plant 41:244–248

Mathur J (1992) In vitro morphogenesis in *Nardostachys jatamansi* DC.: shoot regeneration from callus derived roots. Ann Bot 70(5):419–422

Mehta P, Sekar KC, Bhatt D, Tewari A, Bisht K, Upadhyay S et al (2020) Conservation and prioritization of threatened plants in Indian Himalayan Region. Biodivers Conserv 29:1723–1745

Mishra P, Kumar A, Nagireddy A, Mani DN, Shukla AK, Tiwari R, Sundaresan V (2016) DNA barcoding: an efficient tool to overcome authentication challenges in the herbal market. Plant Biotechnol J 14(1):8–21

Misra P, Shukla PK (2010) In vitro approaches for mass production of medicinal plants. In: Chemistry biochemistry and ayurveda of Indian medicinal plants. International E-Publication

Nadeem M, Palni LMS, Purohit AN, Pandey H, Nandi SK (2000) Propagation and conservation of *Podophyllum hexandrum* Royle: an important medicinal herb. Biol Conserv 92(1):121–129

Nandi SK, Palni LMS, Pandey H, Chandra B, Nadeem M (2016) Selection of elites and in vitro propagation of selected high-value Himalayan medicinal herbs for sustainable utilization and conservation. In: Plant tissue culture: propagation, conservation and crop improvement. Springer, Singapore, pp 15–44

National Medicinal Plants Board (2014) Agro-techniques of selected medicinal plants, vol 2. Department of AYUSH, Ministry of Health and Family Welfare, Government of India, New Delhi. ISBN 978-81-909121-3-6

Nautiyal MC, Nautiyal BP (2004) Agrotechniques for high altitude medicinal & aromatic plants. High Altitude Plants Physiology Research Centre

Nautiyal MC, Nautiyal BP, Prakash V (2004) Effect of grazing and climatic changes on alpine vegetation of Tungnath, Garhwal Himalaya, India. Environmentalist 24:125–134

Nazir U, Gul Z, Shah GM, Khan NI (2022) Interaction effect of auxin and cytokinin on in vitro shoot regeneration and rooting of endangered medicinal plant *Valeriana jatamansi* Jones through tissue culture. Am J Plant Sci 13(2):223–240

Otroshy M, Pakravan V, Enteshary S, Mokhtari A, Roozbeh S (2013) Micropropagation of *Bunium Persicum* (Boiss.) B. Fedtsch. via indirect somatic embryogenesis. J Agric Eng Biotechnol 1(3):48

Pandey H, Nandi SK, Kumar A, Palni UT, Chandra B, Palni LMS (2004) In vitro propagation of *Aconitum balfourii* Stapf.: an important aconite of the Himalayan alpines. J Hortic Sci Biotechnol 79(1):34–41

Pandey S, Kushwaha R, Prakash O, Bhattacharya A, Ahuja PS (2005) Ex situ conservation of *Aconitum heterophyllum* Wall.—an endangered medicinal plant of the Himalaya through mass propagation and its effect on growth and alkaloid content. Plant Genet Res 3(2):127–135

Pandey A, Chandra Sekar K, Joshi B, Rawal RS (2019) Threat assessment of high-value medicinal plants of cold desert areas in Johar valley, Kailash Sacred Landscape, India. Plant Biosyst 153(1):39–47

Pant M, Bisht P, Gusain MP (2010) In vitro propagation through axillary bud culture of *Swertia chirata* Buch.-Ham ex Wall: an endangered medicinal herb. Int J Integr Biol 10(1):48–53

Pant HC, Pant HV, Kumar A, Tomar H, Dev M, Gaurav N (2021) In vitro clonal propagation of *Nardostachys jatamansi*: a traditional Himalayan medicinal plant. J Mt Res 16(3):87–98

Partap M, Chhimwal J, Kumar P, Kumar D, Padwad Y, Warghat AR (2022) Growth dynamics and differential accumulation of picrosides and its precursor metabolites in callus cell lines of *Picrorhiza kurroa* with distinct anti-steatotic potential. Process Biochem 120:85–100

Patial V, Devi K, Sharma M, Bhattacharya A, Ahuja PS (2012) Propagation of *Picrorhiza kurroa* Royle ex Benth: an important medicinal plant of Western Himalaya. J Med Plant Res 6(34):4848–4860

Pradhan BK, Badola HK (2011) Assessment of seedling emergence and vigour for quality planting material in thirteen populations of *Swertia chirayita*—a high value endangered medicinal herb, using substrate combinations. Z Arznei Gewurzpfla 16(1):34–41

Puwein A, Thomas SC (2019) A suitable vegetative propagation technique for effective fruiting and conservation of a vulnerable medicinal plant, Paris polyphylla smit. Med Plant 11(4):475–480

Puwein A, Thomas SC (2022) Effect of plant growth regulators on in vitro propagation of *Paris Polyphylla*. Int J Geogr Inf Syst 28(2):206–216

Qadir J, Singh S, Kour S, Kaloo ZA, Ganai BA (2020) In vitro propagation of *Polygonatum verticillatum* all. A threatened medicinal herb through seed explant. J Sci Res 64:111–117
Rafiq S, Wagay NA, Bhat IA, Kaloo ZA, Rashid S, Lin F et al (2021) In vitro propagation of *Aconitum chasmanthum* Stapf Ex Holmes: an endemic and critically endangered plant species of the western Himalaya. Horticulturae 7(12):586
Rai LK, Prasad P, Sharma E (2000) Conservation threats to some important medicinal plants of the Sikkim Himalaya. Biol Conserv 93(1):27–33
Rajesh M, Sivanandhan G, Subramanyam K, Kapildev G, Jaganath B, Kasthurirengan S et al (2014) Establishment of somatic embryogenesis and podophyllotoxin production in liquid shake cultures of *Podophyllum hexandrum* Royle. Ind Crop Prod 60:66–74
Rao VK, Yadav VK, Sharma SK, Sah VK (2010) Effect of season and plant bio-regulators on hardwood cutting of seabuckthorn (*Hipophae salicifolia*) under Garhwal Himalayas. Indian J Agric Sci 80(10):910–912
Raomai S, Kumaria S, Kehie M, Tandon P (2015) Plantlet regeneration of *Paris polyphylla* Sm. via thin cell layer culture and enhancement of steroidal saponins in mini-rhizome cultures using elicitors. Plant Growth Regul 75:341–353
Rau MA (1975) High altitude flowering plants of west Himalaya. BSI, Howrah, 214pp
Rawat GS (2007) Alpine vegetation of the western Himalaya: species diversity, community structure, dynamics and aspects of conservation. Doctoral dissertation, DSc thesis, Kumaun University, Nainital, 239pp
Rawat GS, Adhikari BS, Tiwari UK et al (2016) Medicinal plants of Garhwal region Uttarakhand: a baseline on the status and distribution. Wildlife Institute of India and Uttarakhand Forest Development Corporation, Dehradun
Rawat JM, Bhandari A, Mishra S, Rawat B, Dhakad AK, Thakur A, Chandra A (2018) Genetic stability and phytochemical profiling of the in vitro regenerated plants of *Angelica glauca* Edgew.: an endangered medicinal plant of Himalaya. Plant Cell Tissue Organ Cult 135:111–118
Rawat V, Ghildiyal A, Singh L, Jugran AK, Bhatt ID, Nandi SK, Pande V (2020) *Methyl jasmonate* induced polyphenols and antioxidant production in callus suspension culture of *Nardostachys jatamansi*. Plant Biosyst 154(6):851–859
Rawat JM, Pandey S, Rawat B, Rai N, Preeti P, Thakur A et al (2023) Traditional uses, active ingredients, and biological activities of *Paris polyphylla* Smith: a comprehensive review of an important Himalayan medicinal plant. J Chem 2023(1):7947224
Rinchen T, Pant S, Anwar M (2012) Population census of critically endangered *Dactylorhiza hatagirea* (D. Don) Soo in Suru valley, (cold desert region, Jammu and Kashmir, India). Int J Biodivers Conserv 4(9):332–335
Rodrigues AS, Pilgrim JD, Lamoreux JF, Hoffmann M, Brooks TM (2006) The value of the IUCN Red List for conservation. Trends Ecol Evol 21(2):71–76
Rokaya MB, Münzbergová Z, Timsina B (2010) Ethnobotanical study of medicinal plants from the Humla district of western Nepal. J Ethnopharmacol 130(3):485–504
Shailja, Kanwar K, Soni M, Singh A (2017) In vitro propagation and conservation of an endangered high value medicinal herb *Swertia chirayita* of temperate Himalayas. Indian J Plant Physiol 22:247–257
Sharma N, Chandel KPS, Paul A (1993) In vitro propagation of *Gentiana kurroo*—an indigenous threatened plant of medicinal importance. Plant Cell Tissue Organ Cult 34:307–309
Sharma S, Negi PS, Singh J (2012) Vegetative propagation of *Picrorhiza kurrooa* Royle ex Benth and *Valeriana jatamansi* Jones through macroproliferation technique for large scale nursery production. Int J Farm Sci 2(1):91–97
Sharma A, Kaur R, Sharma N (2014) In vitro morphogenic response of different explants of *Gentiana kurroo* Royle from Western Himalayas—an endangered medicinal plant. Physiol Mol Biol Plants 20:249–256
Sharma N, Sharma AR, Patel BD, Shrestha K (2016) Investigation on phytochemical, antimicrobial activity and essential oil constituents of *Nardostachys jatamansi* DC. in different regions of Nepal. J Coast Life Med 4(1):56–60

Sharma S, Sharma R, Sharma P, Thakur K, Dutt B (2019) Direct shoot organogenesis from seedling derived shoot tip explants of endangered medicinal plant *Saussurea costus* (Falc.) Lipsch. Proc Natl Acad Sci India B Biol Sci 89:755–764

Sharma N, Thakur M, Sharma P, Sharma YP, Dutt B (2022) In vitro propagation from rhizomes, molecular evaluation and podophyllotoxin production in Himalayan May Apple (*Sinopodophyllum hexandrum* Royle TS Ying): an endangered medicinal plant. Plant Cell Tissue Organ Cult 149(1):159–173

Singh DK, Hajara PK (1996) Changing perspectives of biodiversity status in the Himalaya. British Council Division, New Delhi, pp 23–38

Singh M, Chettri A, Pandey A, Sinha S, Singh KK, & Badola HK (2020) In vitro propagation and phytochemical assessment of Aconitum ferox wall: a threatened medicinal plant of Sikkim Himalaya. Proceedings of the National Academy of Sciences, India Section B: Biological Sciences 90(2):313–321

Singh L, Tariq M, Sekar KC, Bhatt ID, Nandi SK (2017) Ecological niche modelling: an important tool for predicting suitable habitat and conservation of the Himalayan medicinal herbs. ENVIS Bull Himal Ecol 25:154–155

Singh L, Joshi T, Tewari D, Echeverría J, Mocan A, Sah AN et al (2020a) Ethnopharmacological applications targeting alcohol abuse: overview and outlook. Front Pharmacol 10:1593

Singh M, Chettri A, Pandey A, Sinha S, Singh KK, Badola HK (2020b) In vitro propagation and phytochemical assessment of *Aconitum ferox* wall: a threatened medicinal plant of Sikkim Himalaya. Proc Natl Acad Sci India B Biol Sci 90:313–321

Singh L, Bhatt ID, Negi VS, Nandi SK, Rawal RS, Bisht AK (2021) Population status, threats, and conservation options of the orchid *Dactylorhiza hatagirea* in Indian Western Himalaya. Reg Environ Chang 21:1–16

Singh L, Kanwar N, Bhatt ID, Nandi SK, Bisht AK (2022) Predicting the potential distribution of *Dactylorhiza hatagirea* (D. Don) Soo-an important medicinal orchid in the West Himalaya, under multiple climate change scenarios. PLoS One 17(6):e0269673

Singh L, Nandi SK, Bhatt ID, Bisht AK (2023) Inventorization of ecology, ethnobotany, and conservation status of *Dactylorhiza hatagirea*: problems, progress, and prospects. In: Jha S, Halder M (eds) Medicinal plants: biodiversity, biotechnology and conservation, Sustainable development and biodiversity, vol 33. Springer, Singapore. https://doi.org/10.1007/978-981-19-9936-9_12

Sultan P, Shawl AS, Ramteke PW, Jan A, Chisti N, Jabeen N, Shabir S (2006) In vitro propagation for mass multiplication of *Podophyllum hexandrum*: a high value medicinal herb. Asian J Plant Sci 5(2):179–184

Suyal R, Bhatt D, Rawal RS, Tewari LM (2020) Status of two threatened Astavarga herbs, *Polygonatum cirrhifolium* and *Malaxis muscifera*, in west Himalaya: conservation implications. Proc Natl Acad Sci India B Biol Sci 90(3):695–704

Tabin S, Kamili AN, Gupta RC, Parray JA, Bansal A (2018) In vitro culture of *Rheum emodi* Wall: an endangered medicinal plant of Northwestern Himalaya. Proc Natl Acad Sci India B Biol Sci 88:995–1006

Thakur M, Dixit VK (2007) Aphrodisiac activity of *Dactylorhiza hatagirea* (D. Don) Soo in male albino rats. Evid Based Complement Alternat Med 4:29–31

Tomar A, Manhas RK, Srivastava RK (2011) Macroproliferation of *Gentiana kurroo* royle. Int J Green Pharm 5(2):138

TRAFFIC (2018) Wild at home. Retrieved from https://www.traffic.org/site/assets/files/9241/wild-athome.pdf on 24 July 2019

Uniyal SK, Awasthi A, Rawat GS (2002) Current status and distribution of commercially exploited medicinal and aromatic plants in upper Gori valley, Kumaon Himalaya, Uttaranchal. Curr Sci 82(10):1246–1252

Uniyal SK, Singh KN, Jamwal P et al (2006) Traditional use of medicinal plants among the tribal communities of Chhota Bhangal, Western Himalaya. J Ethnobiol Ethnomed 2:14

Vlizadeh M, Safarnejad A, Nematzadeh GA, Kazemitabar SK (2006) Regeneration of plantlets from embryo explants of *Bunium persicum* (Boiss.) B. Fedtsch. Indian J Crop Sci 1(1 and 2):93–96

Valizadeh M, & Kazemi TS (2009) Investigation of Plant Growth Regulators Effects onCallus Induction and Shoot Regeneration of Bunium persicum (Boiss.) B. Fedtsch

Wangchuk P, Samten U (2009) High altitude medicinal plants of Bhutan. An illustrated guide for practical use. Pharmaceutical and Research Unit, Institute of Traditional Medicine Services, Ministry of Health, Thimphu, pp 112–119

Wani IA, Verma S, Kumari P, Charles B, Hashim MJ, El-Serehy HA (2021) Ecological assessment and environmental niche modelling of Himalayan rhubarb (*Rheum webbianum* Royle) in north-west Himalaya. PLoS One 16(11):e0259345

Warghat AR, Bajpai PK, Srivastava RB, Chaurasia OP, Chauhan RS, Sood H (2014) In vitro protocorm development and mass multiplication of an endangered orchid, *Dactylorhiza hatagirea*. Turk J Bot 38(4):737–746

Zaib-Un-Nisa ZUN, Shah SH, Ghulam Farooq GF, Sajad MA, Khan MAS (2019) Identification, quantification and comparative assessment of costunolide in wild, cultivated and in vitro callus culture of *Saussurea lappa* by HPLC technique. Pure Appl Biol 8(1):910–919

Ethnobotany of Medicinal Plants from High Mountains: The Case Study of Lombardy (Italy)

6

Claudia Giuliani, Fabrizia Milani, Martina Bottoni, Lorenzo Colombo, Paola Sira Colombo, Piero Bruschi, and Gelsomina Fico

Abstract

Ethnobotanical research in Alpine regions, especially in Lombardy (Northern Italy), has undergone a noticeable process of evolution from different points of view. This chapter guides the reader in this evolutionary path through the analysis of the ethnobotanical investigations published in this region in the last 15 years. Special attention is paid to two aspects of ethnobotany: the analysis and presentation of the primary data as a useful tool for selecting target species for the subsequent phytochemical and pharmacological research and the description of the last and often neglected phase of a survey, namely the return to the territory and its population with the thorough dissemination of the scientific results. Through the critical analysis of the current scientific literature in the ethnobotanical field and the proposal of specific examples based on the investigations conducted in Lombardy, this journey unfolds along these two parallel tracks to finally merge in the cyclic course of ethnobotany, in which scientific research goes forward and looks to the future, starting from the traditions of a territory and returning to the same territory, ever moving, ever evolving, never forgetting its roots.

C. Giuliani · F. Milani (✉) · M. Bottoni · L. Colombo · P. S. Colombo · G. Fico
Department of Pharmaceutical Sciences, University of Milan, Milan, Italy

Botanical Garden G.E. Ghirardi, Department of Pharmaceutical Sciences, University of Milan, Toscolano Maderno, Brescia, Italy
e-mail: fabrizia.milani@unimi.it

P. Bruschi
Department of Agricultural, Environmental, Food and Forestry Science and Technology, University of Florence, Florence, Italy

S. Chandra, H. Lata (eds.), *High Altitude Medicinal Plants*,
https://doi.org/10.1007/978-3-031-99406-7_6

6.1 Ethnobotany as a Tool to Protect the Biocultural Diversity of Alpine Areas

The unique intertwining of several factors, such as climatic heterogeneity, geographical and topographical variety, and isolation, is what has since always characterized the sub-Alpine and Alpine biomes around the world, from the Europeans Alps to the Rocky Mountains of North America and the Andes in South America, and facilitated the development of its peculiar floristic and faunistic biodiversity (Inouye 2020; Verrall and Pickering 2020). Unfortunately, this richness is as extraordinary as it is extremely fragile. The climate crisis we are currently experiencing, with the rising temperatures, declining snowfall, melting glaciers, and consequent shifts of seasonal cycles, poses a direct threat on indigenous Alpine plant species and their pollinators (Inouye 2020; Verrall and Pickering 2020), and is counted among the many factors that lead to the ever-increasing presence of alien invasive species in Alpine areas (Carboni et al. 2018), to the extent of altering the entire ecosystem stability.

Furthermore, the inevitable bond that humans create with the natural environment in which they are immersed cannot be dismissed. For example, human beings have learned long ago how to benefit from the plants, how to use them in different aspects of their everyday lives. They have the possibility to forge a traditional knowledge that is naturally and inevitably influenced by the plant species that they can find around them and their seasonality (Kupika et al. 2019). At the same time, human activities themselves, as well as the abrupt lack thereof, can mold the surrounding environment. For instance, on one side the indiscriminate use of plant species might be the cause for their endangerment, but on the other the connection between humans and territory is so deep rooted that the abandonment of lands, especially in mountainous, sub-Alpine, and Alpine areas, might influence the ecosystem stability of a territory (i.e., by facilitating the advance of alien species) (Carboni et al. 2018). This relationship is like an ongoing loop: if it is true that the diversity of plant species of a territory influences the development of a richer ecological knowledge, it is equally true that to lose plant species would also mean to lose the traditions associated with them. As an outcome, to lose this knowledge, which is already in danger due to its declining oral transmission, would mean for a territory and its inhabitants to lose a significant portion of their identity.

Furthermore, species that have been traditionally used for medicinal purposes for centuries are *reservoirs* of potential active compounds, ready to be studied. If we consider the high variety of medicinal plant species that can be found in mountainous areas and the traditional therapeutic uses associated with them, the loss of local knowledge might translate into loss of these potential active compounds as well (Dutfield 2010; Balick and Cox 2021).

Ethnobotany, as "*the study of the interactions between plants and people, including the influence of plants on human culture*" (Balick and Cox 2021), has been proven to be a pivotal tool to study, enhance, and preserve the traditional uses of spontaneous and cultivated species of a territory for different purposes: that is, medicinal, veterinary, culinary, domestic, religious, artisanal, etc. Ethnobotanical surveys can thus help staunch the erosion of traditional knowledge and the loss of cultural diversity, as well as protect the plant biodiversity of a territory (i.e., by raising awareness concerning the use of protected species, educating younger generations, preserving and maintaining species, etc.) and the potential scientific research development that comes with it.

For these reasons, particular effort must be made in ethnobotanical research in Alpine territories. Taking into consideration the European Alpine and trans-Alpine regions, some field works could be found, especially in Switzerland (Abbet et al. 2014; Dal Cero et al. 2014; Disler et al. 2014; Cero et al. 2015; Mayer et al. 2017; Stucki et al. 2019; Mertenat et al. 2020; Poncet et al. 2021) and in Austria (Schunko et al. 2012; Grasser et al. 2016), but also at the border among Croatia, Slovenia, and Italy (Pieroni and Giusti 2009). Concerning the Italian Alps, ethnobotanical surveys were conducted in Liguria (Cornara et al. 2014), Piedmont (Bellia and Pieroni 2015; Fontefrancesco and Pieroni 2020; Sulaiman et al. 2024), Valle D'Aosta (Danna et al. 2022), and Trentino Alto Adige (Cavalloro et al. 2022; Mattalia et al. 2023). Finally, considering the Alpine valleys of the Lombardy region, focus of this work, one ethnobotanical investigation was performed in the province of Brescia, specifically in Valvestino (Vitalini et al. 2009), while nine throughout the province of Sondrio, in Val San Giacomo (Vitalini et al. 2013), Valfurva (Dei Cas et al. 2015), Parco dello Stelvio (Vitalini et al. 2015), Valtellina (Bottoni et al. 2022b), and Valmalenco (Bottoni et al. 2020, 2022a, 2024; Milani et al. 2023, 2024) (Fig. 6.1).

The aim of the work presented herein is to guide the readers through an evolutionary journey in the ethnobotany of Alpine Lombardy, from the first survey published in 2009 to the most recent works. The focus will be on two key elements that can illustrate this evolution: (1) the presentation and analysis of the primary data for

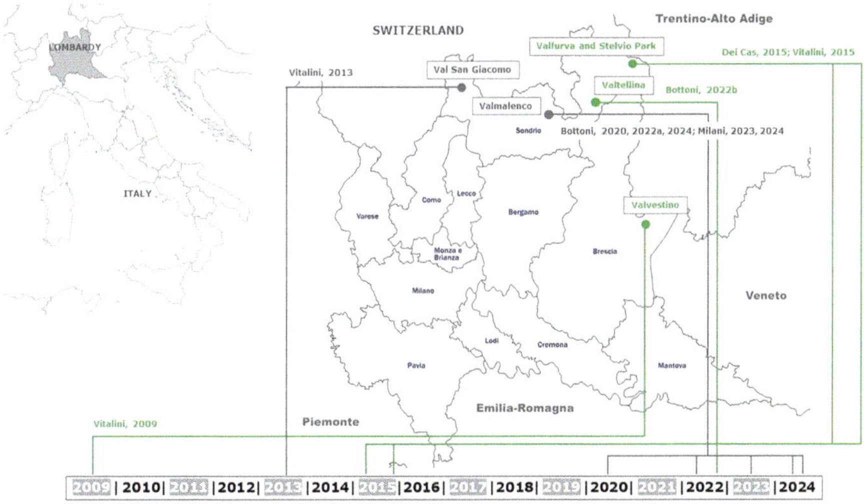

Fig. 6.1 Study area and timeline of the different investigations conducted on the Alpine territories of the Lombardy region (North of Italy)

the selection of target medicinal species for further examinations and (2) the effort put in place to return the scientific results to the investigated territory, with important practical and economic implications.

6.2 Ethnobotany of Medicinal Species in Lombardy: An Evolving Path

As mentioned in Sect. 6.1, ethnobotanical research in Lombardy, and specifically along the Alpine territory, has been widely developed in the last 15 years, with 10 works published between 2009 and 2024. This development is highlighted not only by the increasing number of papers, but also by the undergoing process of evolution that can be observed while analyzing the concerning literature. In fact, taking into account the works cited before, some common features can be found, along with some variations and integrations that characterized a journey meant to elevate ethnobotanical investigations to pivotal tools for the preservation and strong valorization of the traditional knowledge of a territory, as well as for the undertaking of in-depth phytochemical and pharmacological analyses on medicinal plant species intrinsically linked to and inextricably guided by said traditional knowledge.

Specifically, Vitalini et al. (2009, 2013, 2015), Dei Cas et al. (2015), Bottoni et al. (2020, 2022b), and Milani et al. (2024) described only the field phase of an ethnobotanical investigation, with the subsequent analyses of the data, focusing on different sectors of use of spontaneous and cultivated plant species of the territory, with particular attention to the medicinal, veterinary, and culinary uses. The remaining works presented the field phase as a starting point for the further in-depth analysis of a single or few selected species from a micromorphological, phytochemical, nutritional, and pharmacological point of view (Table 6.1).

Vitalini et al. (2009)—Valvestino (BS)

The survey presented in Vitalini et al. (2009) was conducted in Valvestino, a fairly isolated Alpine valley in the province of Brescia, as part of the "*LEADER 2000–2006 Lombardy Region*" project, aimed at promoting and sustaining fringe rural areas. The ethnobotanical investigation meant to abide by these goals through the valorization of traditional knowledge concerning plant species and the study of intergenerational transfer. Additionally, an attempt at preserving some of the plant species surfaced during the survey was made ex situ at the "G.E. Ghirardi" Botanical Garden in Toscolano Maderno, Brescia (Department of Pharmaceutical Sciences, University of Milan).

The work was developed into three main phases: (a) preliminary study of the flora, with the collection of photographs and *herbarium* specimens of the plants; (b) open interviews conducted on 54 informants (45–94 years old); and (c) basic data analysis and browsing of the scientific literature in order to evaluate the traditional uses collected. From the survey, the uses of 58 species (belonging to 30 botanical families) were reported, mainly concerning the medicinal field.

Table 6.1 Breakdown of the 10 ethnobotanical investigations published in Alpine Lombardy from 2009 to 2024

Title, authors, year, journal	Area of study and (province)	Number of informants	Type of sampling and interviews	Fields of use	Number of species (botanical families)	Projects	Strategies for the return to the population phase
Traditional uses of medicinal plants in Valvestino (Italy). Vitalini S., Tomè F., Fico G. 2009 Journal of Ethnopharmacology	Valvestino (BS)	54 (45–94 years old)	Snowballing Open interviews Audio recordings	Medicinal Some information also on Food and Veterinary	58 (30)	Leader Plus Program 2000–2006 To promote in a sustainable way the local endogenous potential of peripheral rural areas	Ex situ preservation of some of the species at the “G.E. Ghirardi” BG
Traditional knowledge on medicinal and food plants used in Val San Giacomo (Sondrio, Italy)—An alpine ethnobotanical study. Vitalini S., Iriti M., Puricelli C., Ciuchi D., Segale A., Fico G. 2013 Journal of Ethnopharmacology	Valle San Giacomo (SO)	104 (35–89)	Snowballing Open interviews Audio recordings (only for group interviews)	All the fields	66 (35)	Projects linked to MU.VI.S (Museo della Via Spluga e della Val San Giacomo) To promote spontaneous species	Not specified

(continued)

Table 6.1 (continued)

Title, authors, year, journal	Area of study and (province)	Number of informants	Type of sampling and interviews	Fields of use	Number of species (botanical families)	Projects	Strategies for the return to the population phase
Tradition of use on medicinal species in Valfurva (Sondrio, Italy). Dei Cas L., Pugni F., Fico G. 2015 Journal of Ethnopharmacology	Valfurva (SO)	92 (23–97)	Snowballing Semi-structured interviews Audio recordings	All the fields	126 (48)	No	Not specified
Plants, people and traditions: ethnobotanical survey in the Lombard Stelvio National Park and neighbouring areas (Central Alps, Italy). Vitalini S., Puricelli C., Mikerezi I., Iriti M. 2015 Journal of Ethnopharmacology	Stelvio National Park (SO)	328 (28–88)	Snowballing	All the fields	192 (65)	No	Not specified
Using Medicinal Plants in Valmalenco (Italian Alps): From Tradition to Scientific Approaches. Bottoni M., Milani F., Colombo L., Nallio K., Colombo P.S., Giuliani C., Bruschi P., Fico G. 2020 Molecules	Valmalenco—Municipality of Caspoggio (SO)	137	Snowballing Open interviews	Medicinal and veterinary	59 (30)	Interreg Italy/Switzerland B-ICE & Heritage	Yes (but not specified in the paper)

Alpine ethnobotanical knowledge in Sondalo (SO, Lombardy, Italy) Bottoni M., Colombo L., Gianoli C., Milani F., Colombo P.S., Bruschi P., Giuliani C., Fico G. 2022 Ethnobotany Research & Applications	Sondalo (Valtellina, SO).	101 (25–98 anni)	Snowballing Open interviews	All the fields	112 (52)	No	No
Achillea moschata Wulfen: From Ethnobotany to Phytochemistry, Morphology, and Biological Activity. Bottoni M., Baron G., Gado F., Milani F., Santagostini L., Colombo L., Colombo P.S., Caporali E., Spada A., Biagi M., Giuliani C., Bruschi P., Aldini G., Fico G. 2022 Molecules	Valmalenco—Municipality of Chiesa in Valmalenco (SO)	125 (18–96)	Snowballing Open interviews	Medicinal	148	Interreg Italy/ Switzerland B-ICE & Heritage	Yes (but not specified in the paper)

(continued)

Table 6.1 (continued)

Title, authors, year, journal	Area of study and (province)	Number of informants	Type of sampling and interviews	Fields of use	Number of species (botanical families)	Projects	Strategies for the return to the population phase
Alpine Diet in Valmalenco (Lombardy, Italy): Nutritional Features of Spontaneous Plants and Traditional Dishes. Milani F., Bottoni M., Giuliani C., Colombo L., Casiraghi M.C., Colombo P.S., Bruschi P., Erba D., Fico G. 2023 Nutrients	Valmalenco—all 5 municipalities (SO)	401 (11–95)	Snowballing Open interviews	Food	154 (54)	Interreg Italy/ Switzerland B-ICE & Heritage and GEMME projects	Yes (but not specified in the paper)
From Primary Data to Ethnopharmacological Investigations on *Achillea erba-rotta* subsp. *moschata* (Wulfen) I. Richardson as a Remedy against Gastric Ailments in Valmalenco (Italy). Bottoni M., Martinelli G., Maranta N., Sabato E., Milani F., Colombo L., Colombo P.S., Piazza S., Sangiovanni E., Giuliani C., Bruschi P., Vistoli G., Dell'Agli M., Fico G. 2024 Plants	Valmalenco—all 5 municipalities (SO)	401 (11–97)	Snowballing Open interviews	Medicinal, Food, Veterinary	Only *Achillea erba-rotta* subsp. *moschata*	Interreg Italy/ Switzerland B-ICE & Heritage and GEMME projects	Yes (but not specified in the paper)

Beyond the Border: comparative ethnobotany in Valmalenco (SO, Italy) and Valposchiavo (Canton of Grisons, Switzerland). Milani F., Bottoni M., Colombo L., Colombo P.S., Bruschi P., Giuliani C., Fico G. 2024 Journal of Ethnobiology and Ethnomedicine	Valmalenco—all 5 municipalities (SO) + Valposchiavo—all 3 municipalities (Canton of Grisons, SW)	401 Valm/70 Valp	Snowballing Open interviews	All the fields	Valm 227 (77); Valp 226 (65)	Interreg Italy/Switzerland B-ICE & Heritage and GEMME projects	Yes (but not specified in the paper)

Vitalini et al. (2013)—Val San Giacomo (SO)

This work, conducted in Val San Giacomo (province of Sondrio), was loosely linked to the MU.VI.S, *Museo della* Via *Spluga e della Val San Giacomo* (Museum of Via Spluga and Val San Giacomo), in the effort of valorizing and preserving the spontaneous flora of the territory through the restoration of traditional knowledge. In 2013, the ethnobotanical landscape in Lombardy, specifically on the Alps, could have been described as fragmentary compared to what was published in Central and Southern Italy, so that this survey was considered one of the first attempts at enriching and expanding it. Just as for Vitalini et al. (2009), the pre-field and field phases of the survey were conducted in a similar manner, with a focus on all the fields of use of plants, whilst the data analysis steered toward a more quantitative approach. In fact, the authors used an extensive number of ethnobotanical indices and then accompanied the results with what information could be found in scientific literature regarding the medicinal uses. In total, 104 informants (35–89 years old) were involved in the survey, with a total of 66 species mentioned (35 botanical families), mainly by elderly people.

Dei Cas et al. (2015)—Valfurva (SO)

This work, conducted in Valfurva in the province of Sondrio, was not part of a wider project but it should be highlighted that one of the authors was native to the area of study and that the survey itself originated by their strong wish to deepen their own understanding of the folk traditions of the territory and to value them. The area belongs in its entirety to the Stelvio National Park, where a strict regulation on the collection of spontaneous plants is maintained. The first phase of the survey was carried out as in the previous works cited, while semi-structured interviews were conducted during the field phase involving 92 informants (23–97 years old). The approach to the pool of respondents was definitely eased by the constant presence on the territory of a native researcher. The analysis of the dataset was performed through the calculation of a more limited number of ethnobotanical indices. Though all the fields of use were considered, 91 species were used for medicinal purposes and the bibliographic research highlighted scientific studies conducted on 41 out of these. Among the medicinal species cited, *Achillea erba-rotta* subsp. *moschata* (Wulfen) I. Richardson, the use of which had already been mentioned in Val San Giacomo, turned out to be known and used by the whole pool of informants in the medicinal, veterinary, and food fields.

Vitalini et al. (2015)—Stelvio National Park (SO)

The study area of this survey was the Stelvio National Park, in the province of Sondrio, an Alpine territory that includes the municipalities of Bormio, Livigno, Sondalo, Valdidentro, Valdisotto, and Valfurva. Given the broader territory, a total number of 328 informants (28–88 years old), both native to the area and resident for at least 15 years, was involved in semi-structured interviews. As in Vitalini et al. (2013), a large number of different ethnobotanical indices was used for the quantitative analysis of data concerning 192 species (65 families and 149 genera). The authors also stressed the importance of ethnobotanical surveys of this kind in the discovery process of potentially active compounds obtained from spontaneous medicinal species, especially in a territory as rich in biodiversity as the Alpine range, and provided extensive bibliographic research meant to preliminarily validate the traditional uses.

Bottoni et al. (2020)—Caspoggio, Valmalenco (SO)
The work presented in Bottoni et al. (2020) was the first of a series of papers published on the ethnobotanical survey conducted in Valmalenco (province of Sondrio) as part of the European Interreg Italy–Switzerland B-ICE & Heritage and the subsequent Interreg GEMME projects. These two projects covered a period of time of 5 years (2018–2023) and were aimed at the sustainable development of the environmental and cultural resources of a territory threatened by the ongoing climate crisis in order to create new opportunities of tourism attractiveness. In Bottoni et al. (2020), the authors presented the partial results of the investigation, concerning the medicinal and veterinary uses in one of the municipalities of the valley, Caspoggio. In addition to the phases already described in the previous works, in this paper special attention was given to the comparison of species and traditional uses recorded in Caspoggio and the ones reported in the neighboring areas. Furthermore, a clear example of preliminary validation of traditional use was given not only through the presentation of the bibliographic research collected on the ethnobotany, phytochemistry, and pharmacology of each species, but also with considerations on the types of extracts described in literature and comparable to the traditional preparations of the valley. Finally, a box was specifically dedicated to the case of *Pinus mugo* Turra, which was one of the most cited plants. The box unfolded through the evaluation of phytochemical and pharmacological investigations previously published on this species, in order to highlight its potential activities in the treatment of upper airways ailments.

Bottoni et al. (2022a)—Chiesa in Valmalenco, Valmalenco (SO)
This work proposed the ethnobotanical results obtained from the interviews conducted on 125 informants (18–95 years old) in Chiesa in Valmalenco, as part of the Interreg project described in the previous paragraph on Bottoni et al. (2020). Specifically, the medicinal field was the main focus of the paper, and the survey could be considered the starting point from which underlining the importance of the traditional use of *A. erba-rotta* subsp. *moschata* in the treatment of digestive problems. The multidisciplinary laboratory investigation that stemmed from the field phase described, developed through micromorphological observations of the species collected during its balsamic time, according to the tradition of the valley, phytochemical LC-MS/MS analyses on aqueous extracts from dried flowerheads, and in vitro evaluations regarding the antioxidant and anti-inflammatory activity of the same extracts.

Bottoni et al. (2022b)—Sondalo (SO)
As it happened in Dei Cas et al. (2015), the survey conducted in Sondalo (province of Sondrio) was not a part of a wider project, but it stemmed from the thesis work

of one of the authors, native to the study area. Sondalo is located in the heart of the Stelvio National Park, in the territory of upper Valtellina. The structure of the work followed some of the previous surveys, such as Dei Cas et al. (2015) and Bottoni et al. (2020), focusing on all the use fields, from the medicinal to the artisanal, through the domestic, food, veterinary, ludic, and religious. A total of 112 plant species (52 families) were mentioned by the 101 informants interviewed (25–98 years old). The phase of data analysis developed through the calculation of only two ethnobotanical indices, which were mainly aimed at evaluating the consensus among the 101 respondents concerning the use of medicinal species.

Milani et al. (2023)—Valmalenco (SO)

The ethnobotanical results on the food field of use were the main focus of the survey phase presented in Milani et al. (2023). In fact, based on the data obtained from the entire pool of informants (n = 401) six spontaneous edible species were selected among the most mentioned and subsequently analyzed regarding their nutritional features, such as their proximate composition and minerals concentration, the content in total polyphenols and the total carotenoids. Specifically, the selected species were *Aruncus dioicus* (Walter) Fernald (young sprouts, raw and cooked), *Chenopodium bonus-henricus* L. (leaves, raw and cooked), *Rumex acetosa* L. (leaves and young stems, raw), *Silene vulgaris* (Moench) Garcke (leaves, raw and cooked), *Taraxacum officinale* Weber (leaves, raw and cooked; leaves, flowerheads, and roots, raw), and *Urtica dioica* L. (leaves, raw and cooked). Furthermore, nutritional analysis on traditional recipes, green gnocchi with *U. dioica* or *C. bonus-henricus*, was carried out in order to evaluate the in vitro effect of the presence of the vegetable ingredients on the starch digestibility. This represents one of the few works that captured the nutritional value of species collected in the Alps and of plant-based typical dishes of the area.

Bottoni et al. (2024)—Valmalenco (SO)

Like in Bottoni et al. (2022b), in the framework of the European Interreg Italy–Switzerland B-ICE & Heritage project, the laboratory investigation described in this 2024 paper stemmed from the description of the field phase conducted on the entire territory of Valmalenco, specifically focusing on the species *A. erba-rotta* subsp. *moschata* and its traditional use for the treatment of digestive problems, such as dyspepsia. The chosen approach involved the presentation of the raw primary data instead of focusing on the use of indices, for example, by reporting sentences directly recorded from the informants. Special attention was given to the evaluation of the anti-inflammatory potential at gastrointestinal level of the traditional decoction of flowerhead. Moreover, the concentration-dependent inhibition of *H. pylori* growth and the anti-adhesive effect of the same extracts were assessed. Finally, the bitter taste of the aqueous preparation was investigated through molecular docking techniques.

Milani et al. (2024)—Valmalenco (SO)
As final step of the European Interreg Italy–Switzerland Gemme project, the ethnobotanical results regarding every field of use, recorded throughout the entire territory and on the entire pool of informants (*n* = 401 in 2019–2022), were presented in this paper alongside their comparison to the ones obtained in 2023 the neighboring Swiss valley of Valposchiavo (number of informants = 70). The qualitative and quantitative comparison between the two valleys was then enriched by the one with the published ethnobotanical literature concerning both the Italian and Swiss neighboring valleys.

6.2.1 Overview of the Main Results

The studies were conducted on individuals ranging from 60 to 401 in number, most of which were elderly. Asteraceae turned out to be the most cited botanical family in all of the studies considered, while other families commonly mentioned were Lamiaceae and Rosaceae. As for the plant species, 146 were cited in at least two areas, while it is interesting to highlight that 16 species were commonly cited along all the territories considered, specifically *Abies alba* L., *A. erba-rotta* subsp. *moschata*, *Achillea millefolium* L., *Arnica montana* L., *Artemisia absinthium* L., *Hypericum perforatum* L., *Juniperus communis* L., *Larix decidua* Mill., *Matricaria chamomilla* L., *Melissa officinalis* L., *Pinus mugo* Turra, *Plantago major* L., *Rosa canina* L., *Sambucus nigra* L., *Taraxacum officinale* Weber, and *U. dioica*. In terms of traditional medicinal uses, *A. montana* and *A. erba-rotta* subsp. *moschata* were among the most relevant species throughout the entire area, especially in the medicinal sector of use. *A. montana* was deemed a useful remedy for muscle-skeletal problems, whereas *A. erba-rotta* subsp. *moschata* was mentioned in Valfurva and Valmalenco for digestive purposes, but also as a hemostatic agent and in wound healing, or as an analgesic and expectorant, as well as for veterinary purposes (Dei Cas et al. 2015; Bottoni et al. 2020, 2022b; Milani et al. 2024). In Val San Giacomo and in Valtellina, for example, it was mainly known for its digestive properties (Vitalini et al. 2013; Bottoni et al. 2022b).

The "exclusive" species are also worth mentioning. In fact, while in the majority of these ethnobotanical surveys the number ranges from 4 to 26, a total of 64 plant species were cited exclusively during the investigation conducted on the territory of Stelvio National Park (Vitalini et al. 2015). This higher number might be attributable to the wider territory examined in the work. However, despite the high number of exclusive species in this paper, if we consider the medicinal sector, the area of Stelvio National Park has one of the highest levels of similarity, according to the Jaccard's Indices in Table 6.2, specifically with Valfurva (JI = 0.40). A JI of 0.40 was also found between Valfurva and Val San Giacomo and, interestingly enough given the different provinces, between Val San Giacomo and Valvestino. On the other hand, it is quite striking that the lowest similarity (JI = 0.21) was found between the Stelvio Park and Sondalo, considering that Sondalo is actually part of the park itself. Overall, the JI (range 0.21–0.40/1.00) are rather low considering that

Table 6.2 Jaccard's Similarity Index calculated, according Pitman et al. (2005), for the medicinal uses in the different areas considered

	Valvestino	Val San Giacomo	Valfurva	Stelvio Park	Sondalo	Valmalenco
	Vitalini et al. (2009)	Vitalini et al. (2013)	Dei Cas et al. (2015)	Vitalini et al. (2015)	Bottoni et al. (2022a)	Bottoni et al. (2020) and Milani et al. (2024)
Valvestino	1.00	0.40	0.29	0.26	0.26	0.26
Valle San Giacomo	0.40	1.00	0.40	0.28	0.30	0.26
Valfurva	0.29	0.40	1.00	0.40	0.33	0.27
Stelvio Park	0.26	0.28	0.40	1.00	0.21	0.28
Sondalo	0.26	0.30	0.33	0.21	1.00	0.35
Valmalenco	0.26	0.26	0.27	0.28	0.35	1.00

JI values may range from 0 to 1, where 1 represents the highest level of similarity between two areas

some of the areas described in the papers not only border, sometimes they even overlap.

These discrepancies may be justified by the challenge of comparing datasets belonging to different works published over the years (a span of 15 years, in this specific case). Nonetheless, this might also represent further evidence of the high heterogeneity and depth of the traditional knowledge of the Alpine area.

6.2.2 Ethnobotanical Indices: Just a Whim or Actually Useful?

One aspect of the evolution of this ethnobotanical journey in Alpine Lombardy is the analysis and presentation of the primary data. The majority of the surveys presented herein developed the field phase through the quantitative analysis of the data collected. One of the most common ways of this examination involves the use of ethnobotanical indices, according to the reference scientific literature (Table 6.3).

As can be observed in Table 6.3, in Vitalini et al. (2009) the authors chose not to calculate ethnobotanical indices in the presentation of the primary data. In fact, they opted for presenting only the number of citations/use reports. The medicinal uses of the species were then collected in a dedicated table, while the food and veterinary sectors were briefly described in the text. Overall, this kind of presentation of field data was clear and concise. Nonetheless, though the authors dedicated the introduction of the work to the importance of wild medicinal species in discovering new potential active compounds, the raw primary data did not seem designed to guide and support the readers and the authors themselves in selecting a species of interest for further laboratory investigations.

As diametrically opposed to this approach as it could get, we find Vitalini et al. (2013, 2015), in which a massive use of indices could be found: ethnophytonomic index (EPI), ethnobotanicity index (EI), relative frequency of citation (RCF), relative importance (RI), use value (UV), and informant consensus factor (ICF) were

Table 6.3 Use of indices as a means to present primary data in ethnobotanical surveys along the Lombardy Alps

Index	Significance of use	Vitalini et al. (2009)	Vitalini et al. (2013)	Vitalini et al. (2015)	Dei Cas et al. (2015)	Bottoni et al. (2020)	Bottoni et al. (2022a)	Bottoni et al. (2022b)	Milani et al. (2023)	Bottoni et al. (2024)	Milani et al. (2024)
EPI Bonet et al. (1999)	Defines the correlation between the number of species known by their vernacular names and the number of species constituting the flora of the studied territory	–	+	+	–	–	–	–	–	–	–
EI Portères (1970)	Defines the correlation between the plant species with at least one known traditional use and the whole flora of the territory	–	+	+	–	–	–	–	–	–	–
RCF Tardio and Pardo-De-Santayana (2008)	Defines whether and how much a species is used in a territory	–	+	–	+	–	–	–	–	–	–
RI Bennet and Prance (2000)	Gives information on the versatility of uses for a species	–	+	–	+	–	–	–	–	–	–
UV Phillips and Gentry (1993) and Tardio and Pardo-de-Santayana (2008)	Defines the mean of the uses per informant for every species	–	+	–	–	–	–	–	–	–	–

(continued)

Table 6.3 (continued)

Index	Significance of use	Vitalini et al. (2009)	Vitalini et al. (2013)	Vitalini et al. (2015)	Dei Cas et al. (2015)	Bottoni et al. (2020)	Bottoni et al. (2022a)	Bottoni et al. (2022b)	Milani et al. (2023)	Bottoni et al. (2024)	Milani et al. (2024)
ICF Trotter and Logan (1986)	Studies the agreement among the informants on the use of species for the treatment of an ailment	–	+	+	+	+	–	+	–	–	+
FL Friedman (1986)	Defines the consensus among the informants on medicinal uses	–	–	+	–	+	–	+	–	–	–
JI Pitaman et al. (2005)	Compares data of neighboring territories based on the presence/absence of the reported uses	–	–	–	–	+	–	–	–	–	+

EPI ethnophytonomic index, *EI* ethnobotanicity index, *RFC* relative frequency of citation, *RI* relative importance, *UV* use value, *ICF* informant consensus factor, *FL* fidelity level, *JI* Jaccard's Index among the selected indices. Additionally, at the end of Vitalini et al. (2013), the statistical correlation among the different indices of relative importance was calculated, along with the correlation between the number of informants and citations and the quantitative indices.

Regarding EPI and EI, although these indices may be useful and are still calculated in most recent works (Petelka et al. 2020; Gras et al. 2021), they may also fall short when it comes to territories in which the whole pool of local flora has not been wholly mapped or has not been mapped altogether (Dei Cas et al. 2015). Moreover, in the case of EPI specifically, the number of species that are known by their vernacular names in a given territory might not automatically reflect their use and relevance in said territory (Signorini et al. 2013).

The authors proceeded to rank the first 20 species obtained by number of informants, number of citations, RFC, UV, and RI. While most of the species in the rankings were comparable, it is interesting to note that the position of many of them changed. Not only that, but for the RI ranking there were at least seven exclusive species. The perfect correspondence between the lists obtained by number of informants and number of citations with the ones by RFC and UV was justified by the fact that RFC and UV are actually directly proportional to number of informants and citations, respectively. The authors themselves found a strong statistical correlation between these numbers and the indices, thus further solidifying the issue. Therefore, there seems to be no benefit in calculating two indices when the same information could be obtained by just observing the raw primary data (in this case, the simple number of informants and citations).

Moreover, basing an index, such as RI, on the variety of uses, seems furtherly confounding: species like *A. erba-rotta* subsp. *moschata* or *A. montana*, which are used extensively in these territories but in fewer manners, would be in lower positions in a ranking based on RI or, like in the case of *A. montana* in Vitalini et al. (2013), would not even be included at all.

Even ICF, which was utilized not only in Vitalini et al. (2013, 2015), but also in the majority of the other surveys analyzed herein (Dei Cas et al. 2015; Bottoni et al. 2020, 2022b; Milani et al. 2024), and FL (Vitalini et al. 2015; Bottoni et al. 2020, 2022b) should be used carefully. In fact, these indices should disclose the consensus among the informants on medicinal uses and the agreement among the informants on the use of species for the treatment of an ailment, respectively. However, when comparing ICFs or FLs of different investigations, the problem of the comparability of different datasets must be considered, especially in the definition of citation/use report and in the classification of the different categories of pathology (apparatuses). As already mentioned, the same issue could prove to be a problem when using Jaccard's similarity Index (JI) to compare data of neighboring territories based on the presence/absence of the reported species or uses.

Although indices such as these represented an attempt at bringing rigor to the quantitative analysis of ethnobotanical datasets, which was first defined in the 1980s (Prance et al. 1987), in recent times they have not been exempted from criticism even by some of the authors that once supported them, stating that such indices are not built upon solid statistical principles, or that many of these indices may be effectively replaced by the solid quantitative and statistical analysis of the use reports (Verpoorte 2012; Weckerle et al. 2018; Leonti 2022).

Even from the papers discussed so far it is quite clear that the selection of medicinal species of interest to be further investigated during the laboratory phase cannot be based on the mere use of ethnobotanical indices and that, sometimes, they might even prove to be misleading. Some of the indices, such as ICF, might be useful when conveniently supported by the primary data. Not only by the number of informants or the number of citations, but also by other qualitative information that could never be translated into numbers or indices: the use of a vernacular name for a specific plant, the personal tales of everyday life thoroughly intertwined to a species, etc.

Surely, all of this must be then supported by extensive bibliographic research on the selected species or on the respective genus and botanical family and, also during the laboratory investigation, every single step must consistently guide each subsequent phase.

When the selection of a species of interest is carried out with these premises, even the limited number of informants involved in ethnobotanical investigations becomes less of a problem. With the quantitative analysis being supported by the other aforementioned elements, the analysis of new or little-studied species and of their extracts, as well as the research of new active compounds may move forward. The progress of the scientific research on new plant-based active compounds is actually possible precisely thanks to the field phase of ethnobotany.

This evolution in data analysis methodologies is reflected in the most recent Valmalenco articles: Bottoni et al. (2022a, 2024) and Milani et al. (2023). Specifically, in Milani et al. (2023), the selection of six target species in the food sector was not performed by use of specific indices, but by considering the plants with at least 50 use reports and the numerous local culinary recipes and samples that were collected during the field investigation with the ever-present help and the guidance from the inhabitants of the valley. This allowed not only for the analysis of these six species from a nutritional point of view, but also for the assessment of nutritional features and their potential impact on human health of traditional dishes in which some of these species are the main ingredients.

Also, in both Bottoni (2022a, 2024), the selection of *A. erba-rotta* subsp. *moschata* as a symbol species of the territory, thus slated for further investigations in a laboratory setting, was firmly based on the valorization of the raw primary data gathered during the field-work phase. This plant was specifically used against ailments typical of the gastrointestinal tract, such as stomachache, dyspepsia, gastritis, and even ulcers, alone or as support to the traditional pharmacological treatment.

Therefore, the types of analysis performed on the species were aligned with the actual traditional uses of the species and were heavily influenced by the raw data itself.

Other than by the investigation conducted in Valmalenco and described in both the papers also through the account of anecdotal evidence recorded during the field work, the traditional use of this species for the treatment of gastrointestinal ailments was confirmed by the other works conducted on the Alpine range. Furthermore, the choice was also guided by the scarce scientific literature on the species (Vitalini et al. 2016, 2022; Argentieri et al. 2020, on the phytochemistry and biological activity; Apel et al. (2021), with the phytochemical and antimicrobial activity investigation on different species of the genus; Fusani et al. (2022), with the study on the variability in essential oil composition), which was not always based on the analysis of traditional preparations and on the validation of the traditional use.

The laboratory phases described in Bottoni et al. (2022a, 2024) developed through different steps, starting with the micro-morphological and histochemical investigations. The observation of the number and the distribution pattern of the glandular trichomes responsible for the production and secretion of active compound effectively confirmed the juxtaposition of the time of the year when the species is traditionally collected and its actual balsamic time.

This was followed by an attentive phytochemical investigation, aimed at characterizing the decoction of *A. erba-rotta* subsp. *moschata* as it is prepared by the informants themselves. The analysis was furtherly enriched by the assessment of the bitter taste of the decoction by means of molecular docking techniques. Finally, the in vitro tests that were carried out on the aqueous preparations allowed us to evaluate the different mechanisms behind its antioxidant and anti-inflammatory activities and to underline an interesting anti-bacterial and anti-adhesion activity against *Helicobacter pylori*.

This preliminary validation of the traditional use of the species may be just the starting point from which a 360-degree investigation could be conducted. In fact, the interest toward this species seems to have spiked in the last few years as demonstrated, other than by the papers already mentioned, by some of the most recent works on the species (Aiello and Fusani (2023), with the morphological and quantitative characteristics of five accessions of cultivated musk yarrow and Salamone et al. (2023), with the specific analysis of the nonvolatile terpenoids and lipophilic flavonoids of the species). The different lines of research are really numerous and extremely interesting, especially when the centenarian tradition revolving around this species is taken as a reference point.

6.2.3 "Render Unto Caesar...", Even Better If More Is Returned!

Seven out of the ten papers considered herein were part of a wider framework (Table 6.1). However, while Vitalini et al. (2009, 2013) very briefly referred to projects aimed at enhancing rural areas through the valorization of the local flora, the investigation conducted in Valmalenco was entirely the result of the pluriannual Interreg Italy–Switzerland B-ICE & Heritage and Interreg Gemme projects (2018–2023). These were designed to enhance the Italian and Swiss Alpine valleys located on both sides of the Pizzo Bernina (4048 m a.s.l.), specifically Valmalenco in Italy and Valposchiavo in Switzerland. Through the retrieval and preservation of the Local Ecological Knowledge revolving around traditional uses of wild plant species, the efforts were focused on finding viable alternatives to winter tourism, which is now in danger due to the ever more severe consequences of the ongoing climate crisis on snow precipitations and glaciers stability. As the important economic inflow of winter tourism is put at risk, cultural tourism might be the better alternative, since it is not bound by climate conditions but instead rooted in the local traditions (Consorzio turistico del mandamento di Sondrio https://www.sondrio-evalmalenco.it/it/progetto-b-ice-heritage).

Being part of a wider project is also reflected by the possibility to dedicate part of the endeavors of an investigation to give something back to the population, be it better knowledge, more awareness, or economic return. These concepts are all part of the Open Science methodology. After all, an ethnobotanical investigation is always deeply rooted in a territory and, as it takes its first steps among its people, it is only fair for its last steps to be taken in the same territory. Valuable local traditions shall be further enriched of scientific knowledge and return to the population, as in the cycle described in Fig. 6.2.

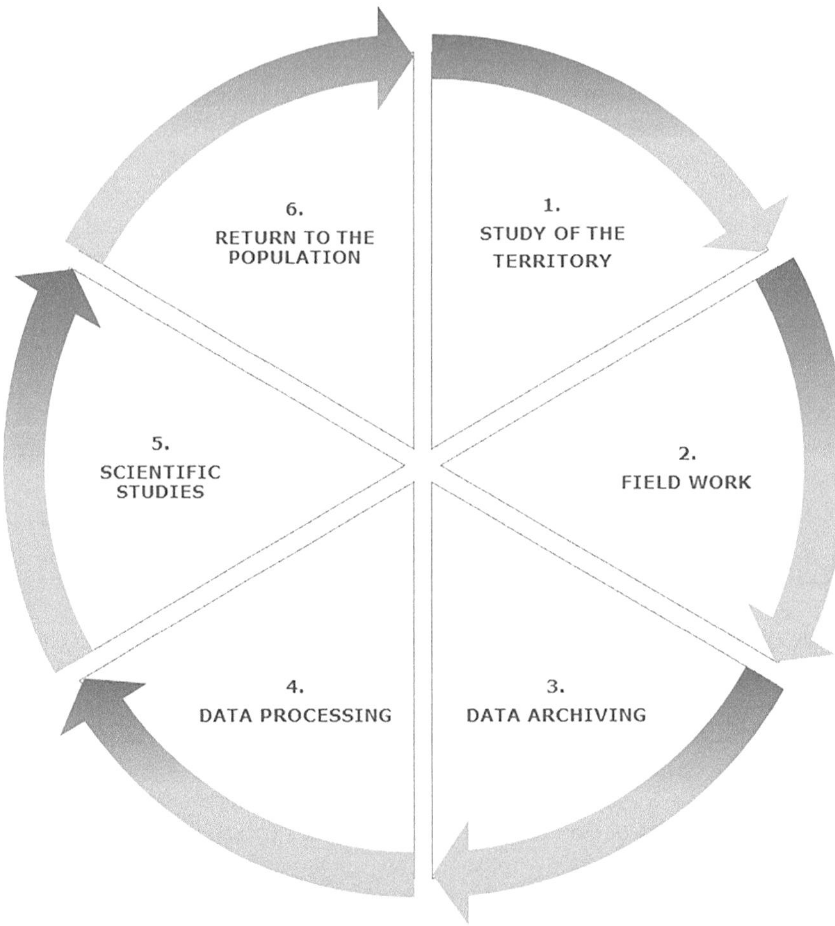

Fig. 6.2 Cyclic representation of an ethnobotanical survey, from the territory to the territory and its inhabitants

This "return to the population" phase can be articulated in a variety of ways. After the investigation in Valvestino, for example, the strategy was the ex situ cultivation of some of the most relevant species selected during the field work in a dedicated flowerbed at the "G.E. Ghirardi" Botanical Garden in Toscolano Maderno (Brescia). The species were then accompanied by an ad hoc didactic and informative apparatus born from the investigation itself.

Things were different in Valfurva (Dei Cas et al. 2015) and Sondalo (Bottoni et al. 2022b). These two surveys were not part of any project but were strongly desired and supported by two of the authors, native to the two territories, who in the end dedicated their dissertations to the valorization of their own culture and traditions.

It wasn't until a more complex and articulated project, such as the aforementioned Interreg in Valmalenco, that implementing several different return strategies became feasible, in an unprecedented effort to abide by the Open Science guidelines.

The greatest achievement in this phase was without a doubt the conceptualization, design, and realization of the "E. Dioli" Botanical Garden in Sant'Antonio di Caspoggio (1300 m a.s.l.), which was inaugurated in July 2023 (Orto Botanico "E. Dioli" 2023, https://www.ortobotanicocaspoggio.it/). As mentioned before, this territory has been severely affected by the ongoing climate change. The area of Sant'Antonio di Caspoggio, specifically, used to be a sought-after ski destination up until 10 years ago, when the ski resorts were closed due to insufficient snow precipitation (Calufetti 2013).

The creation of the "E. Dioli" BG represented a cultural Renaissance for the area because it is the first high-altitude botanical garden completely dedicated to the wild plant species of a territory and their traditional uses, which were first identified during the field work phase of an ethnobotanical investigation.

The garden was named after Erminio Dioli, a local nineteenth-century architect and artist. Additionally, the name of the garden was chosen to foster the connection with the "Dioli" Museum in Caspoggio, which in turn allowed for the creation of a cultural circuit dedicated to the artist, his accomplishments, and his native territory.

Almost 100 selected plant species were arranged in eight flowerbeds, each dedicated to a traditional sector of use. Experts were consulted in order to consider several aspects of the design of the garden, such as total extension of the area, height reached by the plants, potential species incompatibility, special needs in terms of soil, watering, and pedo-climatic characteristics, level of endangerment, and aesthetics. New illustrative panels were created (Fig. 6.3), as well as labels that identified each flowerbed through ad hoc drawings specifically designed for the garden (Fig. 6.4). Also, new labels were dedicated to the plant species providing the following information: botanical family, scientific name, common and vernacular name, botanical characters, habitat and altitudinal zone, protection, balsamic time, traditional uses identified during the field investigation, and part of the plant used (Fig. 6.5).

Accessibility of the spaces and sustainability of the materials used were also major concerns in the realization of the Dioli Garden. Inclusiveness is a core value of the BG, and plans have already been made to include interpretative apparatuses, which provide information in Braille, and to translate the content at least in English.

Fig. 6.3 Illustrative panel situated at the entrance of the "E. Dioli" Botanical Garden. (Designing of the logo: Fabrizio Chierichetti. Graphic design: Arch. Patrizia Berera. Original drawings: Carlo Grimoldi)

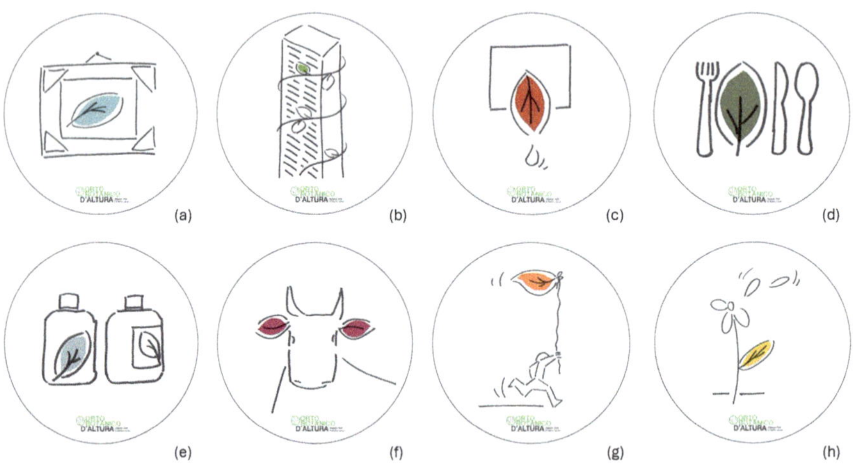

Fig. 6.4 Drawings dedicated to each flowerbeds/sector of uses: (**a**) flowerbed of beauty; (**b**) construction; (**c**) medicinal; (**d**) culinary; (**e**) domestic and cosmetic; (**f**) veterinary; (**g**) ancient games and toys; (**h**) ritualistic/religious. (Original drawings: Carlo Grimoldi)

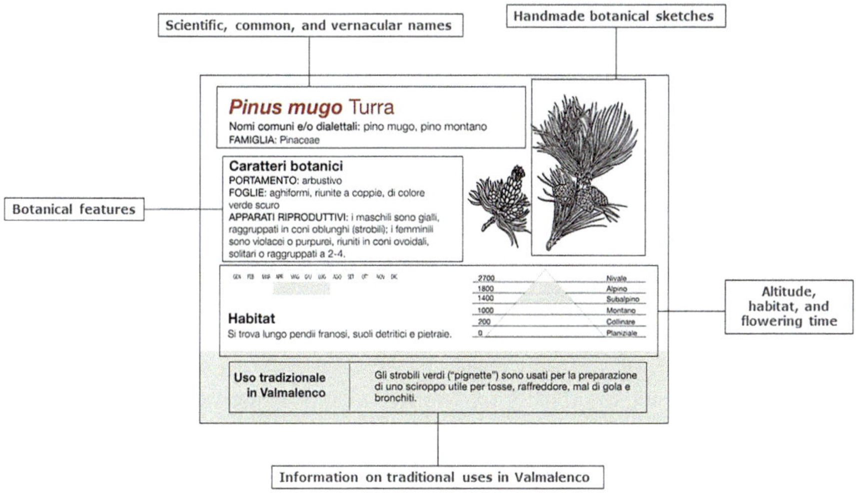

Fig. 6.5 The new labels dedicated to the plant species of the "E. Dioli" Botanical Garden. (Graphic design: Arch. Patrizia Berera. Original drawings: Lorenzo Colombo)

As for local resources development, and as part of a Citizen Science framework, local guides and volunteers were trained through a three-part course to manage the yearly summer opening. Along with these selected guides and other experts, weekly activities, laboratories, guides, and tours were implemented for both children and adults, with special attention to people with disabilities.

Finally, booklets for children, leaflets, bookmarks, and gadgets were created with the aid of professional illustrators in order to help the self-support of the garden. The creation of this Garden, which is the first of its kind in Italy, would have not been possible without the active participation from the inhabitants of the valley. Needless to say, every information concerning the traditional uses of plant species came from the inhabitants themselves, but this is not where their involvement stopped: the local administrations also played a pivotal role in the realization of the project, and even the schools of Valmalenco and the neighboring valleys have begun including the laboratories and activities offered by the Garden in their educational proposals.

The creation of a botanical garden is not the only effort made in this last phase of the survey. In fact, for the past 3 years, an event called "Festival of medicinal plants" has been held in the valley, in which the scientific results of the investigations were divulged back to the local population. Additionally, this dissemination was carried out in creative ways, such as the publication of educational booklets for both adults and children, with the extraordinary participation of illustrators and graphic designers (Fico et al. 2023a, b), the organization of various public seminars, etc.

It is thus evident that, in the framework of a cyclic ethnobotanical investigation, both the "E. Dioli" Botanical Garden and the other strategies put in place in Valmalenco could represent an effective example of Open Science concepts put into practice.

6.3 Conclusions

In the last few years, ethnobotanical research in Lombardy has evolved considerably. This progress is evident since the very first investigations performed in this region, particularly on the Alpine range. As a matter of fact, the primary objective of these first works was to fill gaps in scientific literature concerning the complex mosaic of traditional knowledge in this area, which had never been investigated before from an ethnobotanical standpoint.

However, as the years went by, this approach became wholly insufficient. Even though these works actually depicted the general picture of traditional knowledge heterogeneity, taking next steps had become crucial: starting with the traditions of the land thus enhancing scientific research in several different ways.

This work focused specifically on two main aspects: the way the primary data gathered in the field is presented and highlighted and how a target species is subsequently investigated in a laboratory setting with tradition as main guidance; the way the return to the population phase, the "closing" and often disregarded phase of the cycle, is conducted.

The project in Valmalenco stood out among the other investigations in Lombardy in both these aspects, and it is actually emblematic of the evolutionary process described. First, by giving the utmost importance to the raw primary data directly coming from the informants and second by putting into place strategies of dissemination that lead to the opening of a botanical garden dedicated to the traditions of the valley.

The dualistic nature of the new approaches adopted finally merged in the completion of the ethnobotany cycle, where every phase is as important as the previous one: from the field work, through the laboratory, arriving at concrete solutions for the preservation of local traditional knowledge and valorization of the territory.

References

Abbet C, Mayor R, Roguet D et al (2014) Ethnobotanical survey on wild alpine food plants in Lower and Central Valais (Switzerland). J Ethnopharmacol 151:624–634. https://doi.org/10.1016/j.jep.2013.11.022

Aiello N, Fusani P (2023) Morpho-quantitative characteristics of cultivated musk yarrow (*Achillea erba-rotta* subsp. moschata). Italus Hortus 30:53–61. https://doi.org/10.26353/j.itahort/2023.2.5361

Apel L, Lorenz P, Urban S et al (2021) Phytochemical characterization of different yarrow species (Achillea sp.) and investigations into their antimicrobial activity. Z Naturforsch C J Biosci 76:55–65. https://doi.org/10.1515/znc-2020-0149

Argentieri MP, Madeo M, Avato P et al (2020) Polyphenol content and bioactivity of *Achillea moschata* from the Italian and Swiss Alps. Z Naturforsch C J Biosci 75:57–64. https://doi.org/10.1515/znc-2019-0214

Balick MJ, Cox PA (2021) Plants, people, and culture. The science of ethnobotany, 2nd edn. CRC Press/Taylor & Francis Group—A Garland Science Book

Bellia G, Pieroni A (2015) Isolated, but transnational: the glocal nature of Waldensian ethnobotany, Western Alps, NW Italy. J Ethnobiol Ethnomed 11:37. https://doi.org/10.1186/s13002-015-0027-1

Bennett BC, Prance GT, Url S (2011) Introduced Plants in the Indigenous Pharmacopoeia of Northern South America. Econ Bot 54:90–102

Bonet MÀ, Parada M, Selga A, Vallès J (1999) Studies on pharmaceutical ethnobotany in the regions of L'Alt Emporda and Les Guilleries (Catalonia, Iberian Peninsula). J Ethnopharmacol 68:145–168. https://doi.org/10.1016/S0378-8741(99)00083-5

Bottoni M, Milani F, Colombo L et al (2020) Using medicinal plants in Valmalenco (Italian Alps): from tradition to scientific approaches. Molecules 25:1–27. https://doi.org/10.3390/molecules25184144

Bottoni M, Baron G, Gado F et al (2022a) *Achillea moschata* Wulfen: from ethnobotany to phytochemistry, morphology, and biological activity. Molecules 27:8318. https://doi.org/10.3390/molecules27238318

Bottoni M, Colombo L, Gianoli C et al (2022b) Alpine ethnobotanical knowledge in Sondalo (SO, Lombardy, Italy). Ethnobot Res Appl 24:1–63. https://doi.org/10.32859/era.35.9.1-63

Bottoni M, Martinelli G, Maranta N et al (2024) From primary data to ethnopharmacological investigations on remedy against gastric ailments in Valmalenco (Italy). Plants (Basel) 13(4):539

Calufetti P (2013) Valmalenco, chiude la stazione sciistica di Caspoggio. https://www.montagna.tv/44933/valmalenco-chiude-la-stazione-sciistica-di-caspoggio/

Carboni M, Guéguen M, Barros C et al (2018) Simulating plant invasion dynamics in mountain ecosystems under global change scenarios. Glob Chang Biol 24:e289–e302. https://doi.org/10.1111/gcb.13879

Cavalloro V, Robustelli della Cuna FS, Quai E et al (2022) Walking around the Autonomous Province of Trento (Italy): an ethnobotanical investigation. Plan Theory 11:1–18. https://doi.org/10.3390/plants11172246

Cero MD, Saller R, Weckerle CS (2015) Herbalists of today's Switzerland and their plant knowledge. A preliminary analysis from an ethnobotanical perspective. Complement Med Res 22:238–245. https://doi.org/10.1159/000438809

Consorzio turistico del mandamento di Sondrio B-ICE & Heritage. https://www.sondrioevalmalenco.it/it/progetto-b-ice-heritage

Cornara L, La Rocca A, Terrizzano L et al (2014) Ethnobotanical and phytomedical knowledge in the North-Western Ligurian Alps. J Ethnopharmacol 155:463–484. https://doi.org/10.1016/j.jep.2014.05.046

Dal Cero M, Saller R, Weckerle CS (2014) The use of the local flora in Switzerland: a comparison of past and recent medicinal plant knowledge. J Ethnopharmacol 151:253–264. https://doi.org/10.1016/j.jep.2013.10.035

Danna C, Poggio L, Smeriglio A et al (2022) Ethnomedicinal and ethnobotanical survey in the Aosta Valley side of the Gran Paradiso National Park (Western Alps, Italy). Plan Theory 11:170. https://doi.org/10.3390/plants11020170

Dei Cas L, Pugni F, Fico G (2015) Tradition of use on medicinal species in Valfurva (Sondrio, Italy). J Ethnopharmacol 163:113–134. https://doi.org/10.1016/j.jep.2014.12.054

Disler M, Ivemeyer S, Hamburger M et al (2014) Ethnoveterinary herbal remedies used by farmers in four north-eastern Swiss cantons (St. Gallen, Thurgau, Appenzell Innerrhoden and Appenzell Ausserrhoden). J Ethnobiol Ethnomed 10:1–22. https://doi.org/10.1186/1746-4269-10-32

Dutfield G (2010) Why traditional knowledge is important in drug discovery. Future Med Chem 2:1405–1409. https://doi.org/10.4155/fmc.10.210

Fico G, Giuliani C, Bottoni M et al (2023a) Scienza & Gusto, Quaderno B-ICE Gemme. Toscolano Maderno (Brescia, Italy)

Fico G, Giuliani C, Bottoni M et al (2023b) Quaderno Bimbi, vol 1. Caspoggio (Sondrio, Italy)

Fontefrancesco MF, Pieroni A (2020) Renegotiating situativity: transformations of local herbal knowledge in a Western Alpine valley during the past 40 years. J Ethnobiol Ethnomed 16:1–20. https://doi.org/10.1186/s13002-020-00402-3

Friedman J, Yaniv Z, Dafni A, Palewitch D (1986) A preliminary classification of the healing potential of medicinal plants, based on a rational analysis of an ethnopharmacological field survey among Bedouins in the Negev Desert, Israel. J Ethnopharmacol 16:275–287. https://doi.org/10.1016/0378-8741(86)90094-2

Fusani P, Aiello N, Biazzi E, Tava A (2022) Variability in the essential oil composition of *Achillea erba-rotta* subsp. *moschata* (Wulfen) I. Richardson growing in the Eastern Italian Alps. Chem Biodivers 19:e202200628. https://doi.org/10.1002/cbdv.202200628

Gras A, Parada M, Vallès J, Garnatje T (2021) The role of traditional plant knowledge in the fight against infectious diseases: a meta-analytic study in the Catalan linguistic area. Front Pharmacol 12:1–8. https://doi.org/10.3389/fphar.2021.744616

Grasser S, Schunko C, Vogl CR (2016) Children as ethnobotanists: methods and local impact of a participatory research project with children on wild plant gathering in the Grosses Walsertal Biosphere Reserve, Austria. J Ethnobiol Ethnomed 12:46. https://doi.org/10.1186/s13002-016-0119-6

Inouye DW (2020) Effects of climate change on alpine plants and their pollinators. Ann N Y Acad Sci 1469:26–37. https://doi.org/10.1111/nyas.14104

Kupika OL, Gandiwa E, Nhamo G, Kativu S (2019) Local ecological knowledge on climate change and ecosystem-based adaptation strategies promote resilience in the Middle Zambezi Biosphere Reserve, Zimbabwe. Scientifica (Cairo) 2019:3069254. https://doi.org/10.1155/2019/3069254

Leonti M (2022) The relevance of quantitative ethnobotanical indices for ethnopharmacology and ethnobotany. J Ethnopharmacol 288:115008. https://doi.org/10.1016/j.jep.2022.115008

Mattalia G, Graetz F, Harms M et al (2023) Temporal changes in the use of wild medicinal plants in Tyrol, South Trentino. Plants 12(12):2372

Mayer M, Zbinden M, Vogl CR et al (2017) Swiss ethnoveterinary knowledge on medicinal plants—a within-country comparison of Italian speaking regions with north-western German speaking regions. J Ethnobiol Ethnomed 13:1. https://doi.org/10.1186/s13002-016-0106-y

Mertenat D, Cero MD, Vogl CR et al (2020) Ethnoveterinary knowledge of farmers in bilingual regions of Switzerland—is there potential to extend veterinary options to reduce antimicrobial use? J Ethnopharmacol 246:112184. https://doi.org/10.1016/j.jep.2019.112184

Milani F, Bottoni M, Giuliani C et al (2023) Alpine diet in Valmalenco (Lombardy, Italy): nutritional features of spontaneous plants and traditional dishes. Nutrients 15(8):1988

Milani F, Bottoni M, Colombo L et al (2024) Beyond the border: comparative ethnobotany in Valmalenco (SO, Italy) and Valposchiavo (Canton of Grisons, Switzerland). J Ethnobiol Ethnomed 20:21. https://doi.org/10.1186/s13002-024-00718-4

Orto Botanico "E. Dioli" (2023). https://www.ortobotanicocaspoggio.it/

Petelka J, Plagg B, Säumel I, Zerbe S (2020) Traditional medicinal plants in South Tyrol (Northern Italy, Southern Alps): biodiversity and use. J Ethnobiol Ethnomed 16:74. https://doi.org/10.1186/s13002-020-00419-8

Phillips O, Gentry AH (1993) The useful plants of Tambopata, Peru: I. Statistical hypotheses tests with a new quantitative technique. Econ Bot 47:15–32. https://doi.org/10.1007/BF02862203

Pieroni A, Giusti ME (2009) Alpine ethnobotany in Italy: traditional knowledge of gastronomic and medicinal plants among the Occitans of the Upper Varaita Valley, Piedmont. J Ethnobiol Ethnomed 5:32. https://doi.org/10.1186/1746-4269-5-32

Pitman NCA, Cerón CE, Reyes CI, et al (2005) Catastrophic natural origin of a species-poor tree community in the world's richest forest. J Trop Ecol 21:559–568. https://doi.org/10.1017/S0266467405002713

Portères R (1970) Ethno-Botanique Générale. Paris: Muséum National d'Histoire Naturelle/ Faculté des Lettres.

Poncet A, Schunko C, Vogl CR, Weckerle CS (2021) Local plant knowledge and its variation among farmer's families in the Napf region, Switzerland. J Ethnobiol Ethnomed 17:1–19. https://doi.org/10.1186/s13002-021-00478-5

Prance GT, Balée W, Boom BM, Carneiro RL (1987) Quantitative ethnobotany and the case for conservation in ammonia. Conserv Biol 1:296–310. https://doi.org/10.1111/j.1523-1739.1987.tb00050.x

Salamone S, Aiello N, Fusani P et al (2023) Non-volatile terpenoids and lipophilic flavonoids from *Achillea erba-rotta* Subsp. *moschata* (Wulfen) I. Richardson. Plan Theory 12:1–9. https://doi.org/10.3390/plants12020402

Schunko C, Grasser S, Vogl CR (2012) Intracultural variation of knowledge about wild plant uses in the Biosphere Reserve Grosses Walsertal (Austria). J Ethnobiol Ethnomed 8:1. https://doi.org/10.1186/1746-4269-8-23

Signorini M, Bruschi P, Camangi F et al (2013) Metodi della ricerca etnobotanica. In: Etnobotanica. Conservazione di un patrimonio culturale come risorsa per uno sviluppo sostenibile. EdiPuglia, Italy, Santo Spirito, Bari, pp 43–68

Stucki K, Cero MD, Vogl CR et al (2019) Ethnoveterinary contemporary knowledge of farmers in pre-alpine and alpine regions of the Swiss cantons of Bern and Lucerne compared to ancient and recent literature—is there a tradition? J Ethnopharmacol 234:225–244. https://doi.org/10.1016/j.jep.2018.12.022

Sulaiman N, Zocchi DM, Borrello MT et al (2024) Traditional knowledge evolution over half of a century: local herbal resources and their changes in the Upper Susa Valley of Northwest Italy. Plan Theory 13:13010043. https://doi.org/10.3390/plants13010043

Tardío J, Pardo-De-Santayana M (2008) Cultural importance indices: A comparative analysis based on the useful wild plants of southern Cantabria (northern Spain). Econ Bot 62:24–39. https://doi.org/10.1007/s12231-007-9004-5

Trotter RT, Logan MH (1986) Informant consensus: a new approach for identifying potentially effective medicinal plants. In: Etkin NL (ed) Plants in Indigenous Medicine & Diet: Biobehavioral Approaches. Redgrave Publishing Company, Bedford Hills, New York, pp 91–112

Verpoorte R (2012) Primary data are the basis of all science! J Ethnopharmacol 139:683–684. https://doi.org/10.1016/j.jep.2011.09.054

Verrall B, Pickering CM (2020) Alpine vegetation in the context of climate change: a global review of past research and future directions. Sci Total Environ 748:141344. https://doi.org/10.1016/j.scitotenv.2020.141344

Vitalini S, Tomè F, Fico G (2009) Traditional uses of medicinal plants in Valvestino (Italy). J Ethnopharmacol 121:106–116. https://doi.org/10.1016/j.jep.2008.10.005

Vitalini S, Iriti M, Puricelli C et al (2013) Traditional knowledge on medicinal and food plants used in Val San Giacomo (Sondrio, Italy)—an alpine ethnobotanical study. J Ethnopharmacol 145:517–529. https://doi.org/10.1016/j.jep.2012.11.024

Vitalini S, Puricelli C, Mikerezi I, Iriti M (2015) Plants, people and traditions: ethnobotanical survey in the Lombard Stelvio National Park and neighbouring areas (Central Alps, Italy). J Ethnopharmacol 173:435–458. https://doi.org/10.1016/j.jep.2015.05.036

Vitalini S, Madeo M, Tava A et al (2016) Chemical profile, antioxidant and antibacterial activities of *Achillea moschata* Wulfen, an endemic species from the alps. Molecules 21:1–15. https://doi.org/10.3390/molecules21070830

Vitalini S, Garzoli S, Sisto F et al (2022) Digestive and gastroprotective effects of *Achillea erba-rotta* subsp. *moschata* (Wulfen) I. Richardson (syn. *A. moschata* Wulfen) (Asteraceae): from traditional uses to preclinical studies. J Ethnopharmacol 298:115670. https://doi.org/10.1016/j.jep.2022.115670

Weckerle CS, de Boer HJ, Puri RK et al (2018) Recommended standards for conducting and reporting ethnopharmacological field studies. J Ethnopharmacol 210:125–132. https://doi.org/10.1016/j.jep.2017.08.018

7 Pharmaceutical Potentials of Alpine Plants: A Special Reference to Their Antioxidant Properties

Vijay Laxmi Trivedi, Pallavi Sati, Pradeep Dobhal, Akanksha Bisht, and M. C. Nautiyal

Abstract

Alpine plants, subjected to harsh environmental conditions like intense temperature extremes, high light intensity, and high winds, have evolved robust antioxidant defense systems to mitigate oxidative stress caused by reactive oxygen species (ROS). These systems comprise both enzymatic and nonenzymatic antioxidants. Nonenzymatic antioxidants include carotenoids, α-tocopherol, ascorbic acid, zeaxanthin, and glutathione, while enzymatic antioxidants include superoxide dismutase (SOD), catalase (CAT), and glutathione peroxidase (GPx). The production of secondary metabolites, such as flavonoids, tannins, terpenoids, and alkaloids, further contributes to alpine plants' antioxidant capacity. These plants offer a rich source of phytochemicals with potential therapeutic applications in various disorders linked to oxidative stress. This chapter explores the pharmaceutical potential of alpine plants, focusing on their remarkable antioxidant properties. The chapter details the role of alpine plant antioxidants in addressing conditions such as cancer, cardiovascular diseases, neurodegenerative diseases, diabetes, obesity, and fatigue. Specific examples of alpine plants and their bioactive compounds are presented, including *Rhodiola rosea*, *Saussurea lappa*, *Berberis aristata*, and *Gentiana lutea*. The antioxidant mechanisms of action of these compounds are also discussed in this chapter.

V. L. Trivedi (✉) · P. Sati · P. Dobhal · A. Bisht · M. C. Nautiyal
High Altitude Plant Physiology Research Centre (HAPPRC), H.N.B. Garhwal University, Srinagar Garhwal, Uttarakhand, India

S. Chandra, H. Lata (eds.), *High Altitude Medicinal Plants*,
https://doi.org/10.1007/978-3-031-99406-7_7

7.1 Introduction

Reactive oxygen species are some of the extreme challenges the alpine plants face, primarily due to the harsh environmental factors. High light intensities, diurnal and seasonal temperature extremes, high-speed wind exposure, and higher UV radiations are some factors that are exposed to plants. Such extreme events linked with plant phenological events can further change the ROS dynamics in alpine plants. Even alpine plants completed their life cycle somehow and withstood millions of years of evolutionary events. The credit goes to the strong antioxidative system of the alpine plants (Trivedi and Nautiyal 2021). The control of reactive oxygen species (ROS) production and their detoxification is essential for alpine plant viability because oxidative stress is brought on by low temperature circumstances and high photosynthetic photon flux density (PFD). Since oxidative stress is induced at high PFD and low temperature, control of ROS formation and ROS detoxification might be of key importance for the survival of alpine plants. The plants achieve this protection by synchronizing enzymatic and nonenzymatic antioxidant systems (Ort and Baker 2002).

The stress-driven factors in the alpines subjected the plants to various biological modifications that led to the accumulation of secondary metabolites and other compounds, most of which are part of the nonenzymatic system. Nonenzymatic antioxidants include carotenoids, α-tocopherol, and ascorbic acid, while zeaxanthin and glutathione assist scavenge singlet oxygen and protect against reactive oxygen species (ROS) (Streb et al. 2003; Wang et al. 2009). Phenolic compounds, such as flavonoids and tannins, have been extensively investigated because of their powerful antioxidant properties (Trivedi and Nautiyal 2021). Free radicals are extremely reactive compounds with one unpaired electron in their outer shell. The most prevalent in biological systems are reactive oxygen species (ROS) and reactive nitrogen species (RNS), which can facilitate cell signaling but, in excess, can also result in oxidative stress. ROS or RNS are generated during electron transfer reactions through the loss or gain of electrons (Halliwell and Gutteridge 2015). Cellular processes, including division, differentiation, and proliferation, depend on ROS. ROS and RNS promote physiological activities, control important transcription factors, and preserve redox balance at low to moderate levels. However, high concentrations can harm all vital biomolecules, such as proteins, membranes, and DNA, through oxidative stress and cellular malfunction (Liguori et al. 2018).

Plants and plant-based products have been used for centuries to treat the complications of free radicals and maintain overall health. Since alpine plants are rich sources of antioxidant compounds, studies showed they have wide application in redox homeostasis (Trivedi and Nautiyal 2021). Antioxidants are compounds that, even in little amounts, help prevent or decrease the oxidation of compounds that are extremely oxidative. Antioxidants can trap metal ions, block specific oxidizing enzymes, as well as react with oxidizing compounds before they damage other molecules, and even fix systems like proteins that transport iron (Brewer 2011). According to Bhatt et al. (2013), all plant-based antioxidants, phenolics, and flavonoids are suggested to be most potent in quenching all types of free radicals.

Similarly, the antioxidant activity of polyphenols has been extensively studied in relation to cardiovascular diseases, as well as their hepatoprotective, antiviral, anti-inflammatory, anticarcinogenic, and antimicrobial effects are well investigated. Anthocyanins act as antioxidants by inhibiting chemically induced cancer and suppressing gene associated in angiogenesis, inflammation, and proliferation (Karlsen et al. 2007). In addition to aiding in division of cell and protein modification, carotenoids and ascorbic acid are essential for reducing the oxidative stress brought by photosynthesis. According to McGregor and Biesalski (2006), ascorbic acid has both pro-oxidant and antioxidant properties, its conversion from ascorbate to dehydroascorbate in vitro has been shown to reduce transition metals, including ferric ions (Fe^{3+}) to ferrous (Fe^{2+}) and cupric ions (Cu^{2+}) to cuprous (Cu^{1+}) (Satoh and Sakagami 1997). Additionally, vitamin C is an antioxidant that is necessary for both scurvy prevention and treatment. Furthermore, many phytochemicals found in plants have for a unique way to prevent and cure diseases.

7.2 Antioxidant Defense Systems

Antioxidants are critical for avoiding or allowing down the breakdown of target molecules caused by ROS, which can lead to oxidative stress. These compounds function as antioxidant by donating electrons to free radicals, thereby neutralizing their damaging effects on protein, DNA lipids, and other biomolecules. Acting as scavengers within biological systems, they play a crucial role in defending against oxidative stress (Jomova et al. 2024). Antioxidants can be classified based on their source of origin as either exogenous, which are primarily derived from food, and, on the other hand, endogenous antioxidants are those that are naturally produced within the body (Panova and Tatikolov 2023). Endogenous antioxidants are further divided into enzymatic and nonenzymatic categories. Enzymatic antioxidants proteins in the human body that possess antioxidant properties. These enzymes can modify their valence state, allowing them to donate electrons to free radicals, thereby promoting their degradation and neutralization (Mironczuk-Chodakowska et al. 2018). Glutathione reductase (GR), superoxide dismutase (SOD), catalase (CAT), and glutathione peroxidase (GPx) are examples of enzymatic antioxidants. Glutathione reductase plays a key role in generating reduced glutathione and lessen oxidative damage brought by ROS. Likewise, SOD plays a vital role in detoxifying free radicals by converting superoxide radicals into hydrogen peroxide (Kwon et al. 2019). The body spontaneously produces nonenzymatic endogenous antioxidants through a variety of physiological and metabolic mechanism; these antioxidants are essential for protecting cells from oxidative stress and neutralizing ROS. Glutathione (GSH), alpha-lipoic acid, bilirubin, uric acid, and melatonin are a few examples (Rudenko et al. 2024). Exogenous antioxidants, on the other hand, originate from external sources and are primarily obtained through diet or supplements. These include essential nutrients such as vitamin E, vitamin C, omega-6 fatty acids, and omega-3. Exogenous antioxidants encompass a wide range of plant-derived phytochemicals such as flavonoids and polyphenols, as well as essential trace elements like manganese and zinc. Additionally, synthetic antioxidants such as butyl hydroxyanisole (BHA) may also be classified as exogenous antioxidants as they prevent lipid oxidation (Ponnampalam et al. 2022). Low-molecular-weight, nonenzymatic bioactive compounds produced by plants, known as phytochemicals, are well known for their therapeutic properties and medicinal use (Ashraf et al. 2024). By efficiently interacting with reactive radicals and counteracting their harmful effects through a variety of metabolic pathways, certain phytochemicals demonstrate antioxidant potential. By efficiently interacting with reactive radicals and counteracting their harmful effects through a variety of metabolic pathways, certain phytochemicals demonstrate antioxidant potential.

Various groups of phytochemicals, including ascorbic acid, carotenoids, and flavonoids, display a range of antioxidant activities targeting different ROS. Medicinal plants adapted to high-altitude environment possess intrinsic defense mechanism against the damaging results of ROS. Kumar et al. (2021) reported that high-altitude medicinal plant possess a strong antioxidants defense system by utilizing enzymatic (SOD and CAT), as well as nonenzymatic antioxidants (tannins, flavonoids, and ascorbic acid) to mitigate oxidative stress. Despite their potential, enzymatic antioxidants derived from plants are rarely utilized for therapeutic application due to the risk of denaturation. Certain plant species have genetic capabilities that let them synthesize phytochemicals with strong antioxidant capabilities that effectively counteract toxic ROS. The synthesis of these beneficial molecules is controlled by key metabolic pathways, ensuring cellular defense against oxidative stress. These phytochemicals promote plant resilience by reducing the oxidation of essential macromolecules like proteins, lipids, and DNA (Ponnampalam et al. 2022; Ashraf et al. 2024).

7.2.1 Alpine Plants Have a Specialized Defense Mechanism

Stressful environmental conditions such as extreme diurnal and seasonal temperature shifts, high UV radiation and light intensities, and other associated factors subjected the alpine plants to complex oxidative stress. Alpine plants produce various antioxidants to cope with ROS stress, which protects them enzymatically and nonenzymatically. The antioxidant can be defined as a molecule that can neutralize ROS oxidation before it reacts with other cellular biomolecules. This ROS protection in plants worked on two levels, viz., primary and secondary. The primary mechanism involves the endogenous enzymes, which scavenge the ROS directly before damaging intracellular biomolecules (Ighodaro and Akinloye 2018). Plant cells control ROS production and scavenging via organelles-specific systems found in mitochondria, peroxisomes, and chloroplasts.

These pathways operate cooperatively to keep the redox equilibrium. The nonenzymatic antioxidative defense system comprises primary cellular redox buffers, along with low-molecular-weight antioxidants such as glutathione, tocopherol, ascorbic acid, and carotenoids, and higher-molecular-weight phenolic compounds. These compounds interact with other cellular structures, serving crucial roles in defense and as enzyme cofactors, and providing second line of antioxidant protection (Sharma et al. 2012). Furthermore, secondary metabolites contribute to stress resistance via two mechanisms: passive resistance, in which metabolites are always available despite the presence of stressors, and second being active resistance, in which metabolites production is driven by specific stressors (Korkina 2007). These metabolites, including phenolics, terpenoids, alkaloids, and flavonoids, are crucial in mitigating oxidative damage and protecting cellular integrity. Alpine plants, such as *Rhododendron*, *Primula*, and *Artemisia* species, are rich in phenolic compounds that act as potent antioxidants, scavenging ROS, and counterbalancing free radicals. Flavonoids, commonly found in alpine plants, provide UV protection and reduce oxidative stress through their free radical-scavenging ability (Ashraf et al. 2024). Terpenoids, such as carotenoids and essential oils, are also abundant in alpine flora. For instance, the carotenoids in *Rhododendron anthopogon* and *Artemisia annua* provide photoprotection and combat lipid peroxidation (Trivedi and Nautiyal 2021). Alkaloids, such as berberine and other nitrogen-containing compounds that are found in plant species such as *Berberis* and *Gentiana*, have a high antioxidant effect because they reduce ROS levels and enhance the action of antioxidant enzymes (Ashraf et al. 2024). Furthermore, glucosinolates in alpine cruciferous plants induce detoxification enzymes, thereby providing additional defense mechanisms against oxidative damage (Hayes et al. 2008). This potential of the plants can be utilized for maintaining overall health and for the treatment of various illnesses because most of the health complications involve ROS (Fig. 7.1).

7.3 Disorders Related to ROS and the Role of Antioxidants from Alpine Plants

Free radicals generate oxidative stress, which can result in cellular damage. It also impairs DNA repair pathways and increases protein aggregation, which contributes to the onset of different illnesses. The cellular source of the free radicals involves plasma membrane-associated NADPH oxidases, lipoxygenases, quinone oxidases, endoplasmic reticulum-associated flavoproteins, and cytochrome P450; peroxisomes-associated oxidases and flavoproteins, and mitochondria-associated electron transport chain (Jomova et al. 2023). External factors like exposure to radiation, cancer treatment therapy, alcohol consummation, and specific medication such as bleomycin, gentamycin, and others are known to increase oxidative stress (Bhattacharyya et al. 2014). They function as signaling molecules in various physiological processes at low to moderate levels. However, excessive ROS production leads to oxidative stress, damaging cellular components like DNA, lipids, and proteins, which contributes to the pathogenesis of numerous human disorders (Valko et al. 2007). Reactive oxygen species (ROS) are associated with neurodegenerative illnesses, genetic mutations, chronic inflammation, tumor formation, diabetes, aging, and age-related ailments such as cardiovascular decline, osteoporosis, and sarcopenia. While free radicals serve an important function in maintaining cellular homeostasis and acting as signaling molecules, their excessive accumulation causes

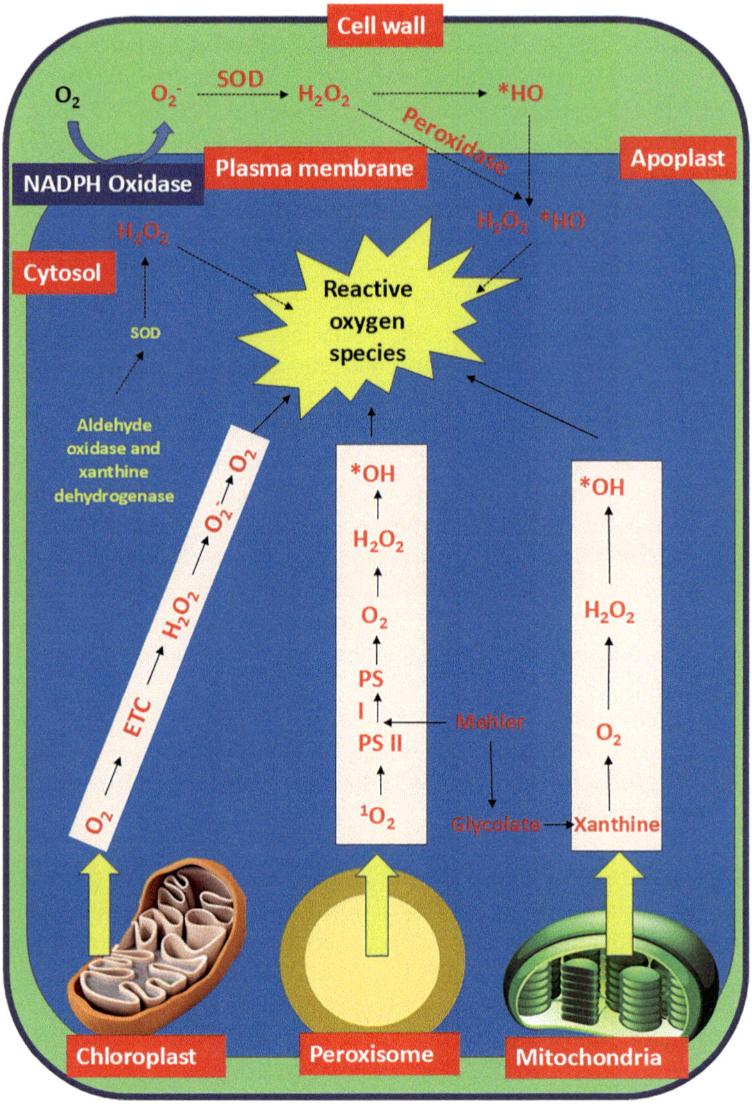

Fig. 7.1 Reactive oxygen species and antioxidant defense system in plants

oxidative stress (Sies 2020), which contributes to a variety of degenerative illnesses. In this context, antioxidants play an important role in safeguarding the body by limiting ROS production, neutralizing free radicals, and repairing cellular enzymes. In several experimental studies, both natural and synthetic antioxidants were found to help prevent diseases, while epidemiological evidence suggests that plant-derived antioxidants may play an important role in lowering the incidence of oxidative stress–associated degenerative diseases (Miguel et al. 2021).

Phytoconstituents have showed promise as therapeutic agents for free radical scavenging and treating oxidative stress-related diseases. Compared to traditional medications, they are frequently more efficient and less harmful. Several plant-based antioxidants have shown promising results in the treatment of oxidative stress-related illnesses. Plant compounds from high-altitude medicinal herbs have received attention due to their powerful antioxidant effects. These substances serve to neutralize free radicals, alleviate inflammation, control cell signaling processes, and improve immunological function. The harsh climatic circumstances of high-altitude locales, such as lack of oxygen, severe ultraviolet (UV) light, and temperature variations, cause these plants to develop unique bioactive chemicals, making them valuable sources of medicinal medicines. Exploring high-altitude medicinal herbs not only helps conserve cultural traditions, but also reveals their immense therapeutic potential for contemporary medicine, particularly in the treatment of oxidative stress-related disorders and the discovery of novel pharmaceutical chemicals. These locations, which are abundant with plant species and genetic variety, are relatively unexplored, providing enormous opportunities for finding new bioactive chemicals and investigating evolutionary adaptations to severe settings (Kasote et al. 2015) (Fig. 7.2).

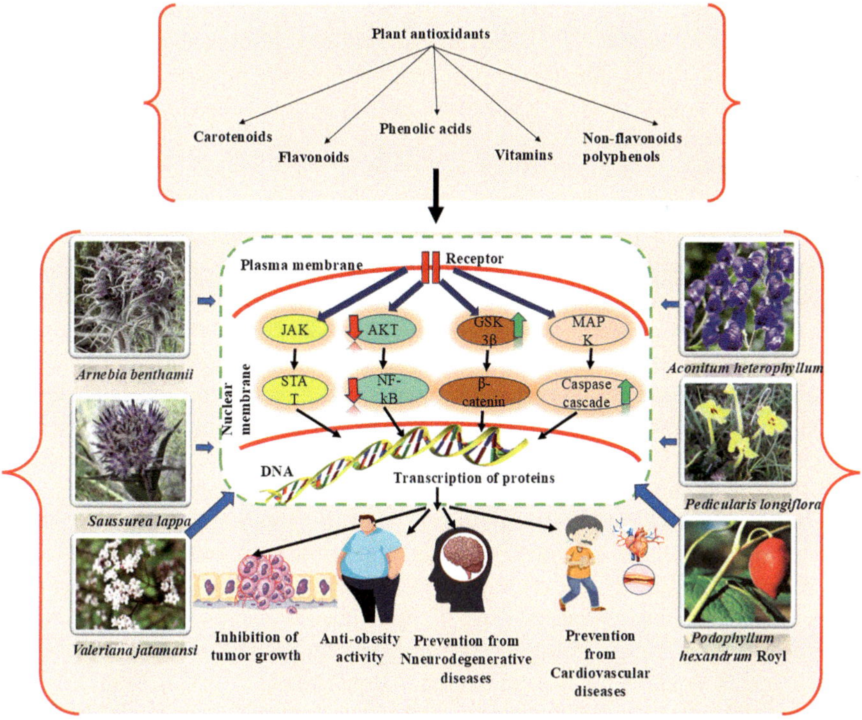

Fig. 7.2 Antioxidant-based pharmaceutical properties of some alpine medicinal plants

While the advantages and limitations of plant-based antioxidants are still being debated, boosting the body's antioxidant defenses—whether through dietary supplementation or encouraging naturally occurring antioxidant production—remains a promising strategy to combat diseases associated with free radical damage. Plant-derived antioxidants like flavonoids and ascorbic acid are among the most powerful exogenous antioxidants. Indeed, these chemicals reduce ROS generation by scavenging free radicals and contribute to the body's innate antioxidant defenses.

7.3.1 Cancer ROS and Antioxidants from Alpine Plants

Cancer has become quite a common disease nowadays, with approximately millions of lives lost per year worldwide. Oxidative stress-induced DNA damage remains one of the primary reasons for cancer as it can cause significant damage to DNA, leading to base modifications, strand breaks, or even DNA–protein crosslinks. Unrepaired damage has the potential to cause mutations in tumor suppressor and oncogene genes, which will ultimately aid in the development and spread of cancer. Furthermore, ROS are essential for the activation of several cellular signaling pathways, including nuclear factor-kappa B (NF-κB) and mitogen-activated protein kinases (MAPKs), which further encourage tumor development, invasion, and metastasis. ROS stimulate VEGF and HIF-1α synthesis, leading to tumor angiogenesis (Arfin et al. 2021). These elements promote the development of new vessels, guaranteeing a consistent flow of nutrients and oxygen to sustain tumor growth and metastasis. Furthermore, ROS promote neovascularization by suppressing anti-angiogenic processes and activating pro-angiogenic pathways, which both aid in the advancement of cancer (Khan and Kerbel 2018). Additionally, they promote the formation of cancer by inducing the production of inflammatory mediators that facilitate the growth and invasion of tumors. The development of cancer is significantly influenced by oxidative stress, and inflammatory cells are a key source of ROS. These ROS and reactive nitrogen species (RNS) accelerate the development of disease by oxidatively damaging essential cellular components. Furthermore, angiogenesis is triggered by a number of pro-inflammatory mediators, including cytokines, prostaglandins, and chemokines, which activate pathways largely controlled by vascular endothelial growth factors (VEGFs). Immune suppression is also associated with inflammation in cancer, which enables tumor cells to elude the body's defenses and proliferate unchecked. Various antioxidant mechanisms have evolved to maintain genomic stability in the organisms and counter oxidative damage from ROS. These processes include DNA repair enzymes, other cellular systems of genomic surveillance, phase II detoxification, and other antioxidation enzymes that function in cellular defense, such as GPx, SOD, GST, and catalase and other constitutive and inducible antioxidants (Rajendran et al. 2014). Many remarkable phytochemicals from alpine meadows can cure cancer by quenching ROS; for example, costunolide, a phytochemical obtained from the roots of the *Saussurea lappa* (Decne.), an alpine plant, known for its exceptional antioxidant properties, is believed to help combat cancer. This impact is mostly owing to the presence of

N-acetylcysteine, a molecule that plays an important role in neutralizing damaging reactive oxygen species (ROS). It accomplishes this by promoting the synthesis of important enzymes such as catalase and superoxide dismutase, which aid in the protection of cells from oxidative stress (Ansari et al. 2021a, b). With its potent antioxidant qualities, costus is a viable natural treatment that can help prevent and treat a wide range of illnesses. Shikonin, a potent substance found in *Arnebia benthamii* (Wall. ex G. Don), another well-known alpine herb, aids in shielding cells from oxidative damage. Shikonin protects DNA from oxidative stress, stops lipid peroxidation, and scavenges dangerous free radicals (Ganie et al. 2014). Its protective properties are further enhanced by its ability to neutralize ferrous ions as a quinone derivative. Shikonin's role in boosting general cellular health is further supported by its occurrence in Himalayan Arnebia, which emphasizes its medicinal significance and possible therapeutic advantages. Strong phytochemicals like podophyllotoxin, which are present in the rhizome, leaves, and other components of *Podophyllum hexandrum* Royle, also referred to as the Himalayan May apple, provide it a high antioxidant content (Motyka et al. 2023). This plant's extracts efficiently counteract dangerous superoxide and hydrogen peroxide radicals, preventing lipid peroxidation and increasing the activity of antioxidant enzymes. Similarly, a special class of terpenoids known as valepotriates and lignans is responsible for *Valeriana jatamansi*'s therapeutic qualities (Huong et al. 2022). Its rhizome is rich in flavonoids and phenols, which neutralize reactive oxygen species (ROS) and donate hydrogen atoms, making them potent antioxidants. Furthermore, by binding with specific metal ions, it can stop dangerous oxidative chemicals from forming. A promising natural treatment for several illnesses, *Pedicularis longiflora* is another priceless medicinal herb that is known for its potent anti-inflammatory and antioxidant qualities (Yatoo et al. 2017). Beneficial phytochemicals, including flavonoids and phenols, which are abundant in *Pedicularis longiflora*, efficiently scavenge superoxide radicals and hence lower lipid peroxidation. This herb's capacity to fight oxidative stress is further strengthened by the fact that it increases the activity of two important antioxidant enzymes, catalase (CAT) and superoxide dismutase (SOD). Similar to this, *Aconitum heterophyllum* is a potent medicinal plant whose roots, stems, and leaves contain flavonoids and alkaloids that are essential for the body's detoxification of reactive oxygen species (ROS). These substances enhance general health by preserving cellular equilibrium and shielding tissues from oxidative damage (Saravanan et al. 2023).

7.3.2 Fatigue-Related Disorder, ROS, and Antioxidants from Alpine Plants

Free radicals are organic by-products of metabolism, including superoxide and hydroxyl anion radicals. Their removal and production are balanced under typical circumstances. But vigorous exercise can upset this balance, causing an overabundance of reactive oxygen species (ROS) to build up. It is thought that this oxidative stress plays a role in fatigue and muscle injury (Kumar et al. 2023). The body's natural defensive systems must be supplemented with exogenous antioxidant sources to combat weariness caused by oxidative stress (González-Peña et al. 2023). High-altitude plants that are rich in antioxidants are essential for repairing oxidative damage, balancing the body by scavenging reactive oxygen species, and reducing fatigue. Furthermore, by regulating the central nervous system, restoring essential energy, and improving muscle function, traditional Chinese medicines (TCM) with high-altitude plants have shown remarkable anti-fatigue capabilities, providing promising natural remedies (Ashraf et al. 2024).

Alpine plants have demonstrated exceptional promise in controlling physical exhaustion, especially those utilized in Tibetan medicine. For example, *Rhodiola rosea* L. roots and rhizomes have long been prized for their capacity to momentarily alleviate signs of stress, such as weakness and exhaustion (Lekomtseva et al. 2017). Other medicinal plants, such as *Crocus sativus* L. (Khazdair et al. 2015), *Lepidium meyenii* (Walp.) (Zhu et al. 2021), and *Hippophae rhamnoides* L. (Zheng et al. 2012), have been widely used to prevent and treat fatigue-related symptoms, including weakness, slowed movement (bradykinesia), and low mood. *Saussurea involucrata* is another noteworthy example, which has potent anti-fatigue qualities because of its rich flavonoid, coumarin, lactone, steroid, phenylpropanoid, and lignan composition. These bioactive ingredients help make it more effective at preventing fatigue and encouraging overall physical resilience (Li et al. 2011).

7.3.3 Diabetes ROS and Antioxidant

Oxidative stress occurs when free radical synthesis exceeds the capacity of the body to eliminate them, leading to either an excess of free radicals or a decline in antioxidant mechanisms of defense, or both. Hyperglycemia has been associated with increased endothelial cell death in terms of vitro and in vivo (Johansen et al. 2005). A form of diabetes, whether type 1 or type 2, has been linked to increased amounts of free radicals and a reduced ability for antioxidant defense, resulting in oxidative damage to cells and tissues (Matough et al. 2012). In diabetes, reactive oxygen species (ROS) are produced by both mitochondria and non-mitochondrial sources, and they help accelerate four essential molecular pathways implicated in hyperglycemia-induced tissue damage. Protein kinase C (PKC) activation, increased flux through the hexosamine route, accumulation of advanced glycation end-products (AGEs), and increased polyol pathway activity are some of these processes (Coughlan and Sharma 2016). Oxidative stress can affect insulin signaling by enhancing the serine phosphorylation of insulin receptor substrate 1 (IRS-1), preventing it from binding to the insulin receptor. This prevents phosphatidylinositol 3-kinase (PI3K) and Akt from being activated downstream, which lowers glucose absorption, decreases glycogen synthesis, and increases gluconeogenesis and lipolysis. As a result, this contributes to increased insulin resistance and hyperglycemia, which are hallmarks of type 2 diabetes (Bhatti et al. 2022).

Several high-altitude plants with aromatic and medicinal properties have shown potential in diabetes treatment. *Rhodiola rosea*, specifically its active component

Salidroside, has been shown to safeguard β-cells from oxidative injury caused by hyperglycemia (Ju et al. 2017). Similarly, *Vaccinium vitisidaea* (lingonberry), which contains flavonoids such as quercetins, phenolic acids, anthocyanins, catechins, lignans, proanthocyanidins, catechins, and stilbenes such as resveratrol, is important in diabetes management due to its antidiabetic properties (Pärnänen et al. 2024). *Dipsacus fullonum, Dipsacus sylvestris* (Oszmiański et al. 2020), and *Juniperus recurva* proved to be very effective antidiabetics due to the antioxidative activities of their *Polyphenols* and *Flavonoids* (Raina et al. 2019). The carotenoids (Carotenes, Xanthophylls) are present in other plants such as *Gentiana algida* Pall. (Olennikov et al. 2015), *Rhododendron ferrugineum* L. (Chosson et al. 1998) ascorbic acid, Tocopherols, in *Vaccinium macrocarpon* Aiton. (Ghosh et al. 2024) (Mountain cranberry), *Sorbus aucuparia* Poir., *Sorbus scopulina* Greene (Sarv et al. 2020), *Carum carvi* (Ahmad et al. 2020), *Swertia* sp. (Roy et al. 2015), *Neopicrorhiza scrophulariiflora* and *Rheum* sp. with their Chrysophanol, physcion, rhein, emodin, and aloe emodin (Kumari et al. 2023), *Picrorhiza kurroa* along with *Berberis aristata* and others (Table 7.1).

7.3.4 Cardiovascular Diseases ROS and Antioxidants

Atherosclerosis is a chronic inflammatory disorder that contributes significantly to coronary heart disease and stroke in people. Natural and synthetic antioxidants play important roles in both avoiding and treating this disease via several methods. These methods include inhibiting LDL oxidation, reducing reactive oxygen species (ROS), and suppressing cytokine production. Antioxidants also serve to prevent the formation of plaques at the site of atherosclerosis, reduce platelet aggregation, and decrease mononuclear cell infiltration. They can also improve endothelial function and vasodilation, increase nitric oxide (NO) bioavailability, and modulate the level of expression of adhesion molecules such as vascular adhesion molecule-1 (VCAM-1) and intercellular attachment molecule-1 (ICAM-1) on endothelial cells, thus hindering the development of foam cells (Malekmohammad et al. 2019). Medicinal plants are especially useful for preventing and controlling atherosclerosis since they often have less adverse effects than synthetic medicines (Kajal et al. 2016). Many of these plants have significant antioxidant characteristics, which are thought to be the primary mechanism for their effects (Trivedi et al. 2022).

However, the medicinal activity of some plants may be due to the combined action of additional bioactive components. The processes by which plant-based bioactive chemicals help prevent or treat atherosclerosis are numerous and complex. These chemicals have endothelial-protective properties, reduce blood lipid levels, and modulate inflammation. They also play an important function in preventing foam cell formation, monocyte migration/activation, and vascular smooth muscle cell (VSMC) migration/proliferation. These chemicals also decrease platelet aggregation and coagulation, as well as having antiplatelet and anti-lipid properties. Plant-derived bioactive chemicals, as natural molecules, represent a viable technique for decreasing the progression of atherosclerosis. They accomplish this by

Table 7.1 Phytochemical composition, antioxidant mechanism, and pharmaceutical properties of some alpine medicinal plants

Sl. no.	Plant name	Family	Phytochemical constituents	Antioxidant mechanism	Pharmaceutical properties	References
1	*Saussurea lappa*	Asteraceae	Costunolide, N-acetylcysteine, phenolics, and flavonoids	Neutralizing ROS by easing key enzymes production like SOD and CAT	Anticancerous and antiobesity	Patel et al. (2020)
2	*Arnebia benthamii*	Boragineacae	Shikonin	Prevention of oxidative DNA damage	Anticancerous	Ashraf et al. (2024)
3	*Podophyllum hexandrum*	Berberidaceae	Podophyllotoxin	Neutralizing superoxide and hydrogen peroxide radicals preventing lipid peroxidation, and stimulation of antioxidant enzymes	Anticancerous	Ashraf et al. (2024)
4	*Valeriana jatamansi*	Valerianaceae	Valepotriates, phenols, and flavonoids	Donation of hydrogen atoms or quenching singlet oxygen species, and chelating metal ions	Anticancerous	Ashraf et al. (2024)
5	*Pedicularis longiflora*	Orobanchaceae	Flavonoids and phenols	Reduce lipid peroxidation enhances CAT and SOD activity	Anticancerous and anti-inflammatory	Ashraf et al. (2024)
6	*Aconitum heterophyllum*	Ranunculaceae	Alkaloids and flavonoids	Detoxifying ROS	Prevention of gastrointestinal problems such as liver inflammation, anticancerous, antiobesity, and antidiabetic	Ashraf et al. (2024)
7	*Nardostachys jatamansi*	Caprifoliaceae	Polyphenols and flavonoids, terpenoids, actinidine, and glaziovine	Nitric oxide scavenging activity and iron chelation activity	Anticancerous and neurodegenerative diseases protectant	Razack et al. (2017) and Anupama et al. (2022)

8	*Aconogonon alpinum*	Polygonaceae	Polyphenol	Activate the NRF2/ARE pathway, which upregulates antioxidant enzymes	Hepatoprotective	Mhalla et al. (2018), Khan et al. (2020) and Dufour et al. (2022)
9	*Selinum carvifolia*	Apiaceae	Polyphenol	Neutralize ROS, reducing oxidative stress, upregulate antioxidant enzymes (SOD, CAT, GPx), enhancing cellular defense, and inhibit lipid peroxidation, protecting cell membranes	Cytotoxic	Srivastava et al. (2023)
12	*Gentiana lutea*	Gentianeae	Secoiridoid bitter compounds	Antioxidant	Microbial infections, effective against inflammation, obesity, and atherosclerosis	Park et al. (2020)
16	*Rhodiola* spp. *(Rhodiola rosea L)*	Crassulaceae	Polyphenols flavonoids stilbenes, phenolic acids, and phytosterols	Direct scavenging of ROS, modulation of antioxidant enzyme activities (such as SOD and POD), and potential effects on energy pathways linked to the pentose phosphate pathway	Antiobesity, effective in neurodegenerative diseases, type 2 diabetes, and cardiovascular diseases, levated cholesterol level, inflammation, and immunosuppression	Bernatoniene et al. (2023)
17	*Vaccinium vitis-idaea*	Ericaceae	Polyphenols flavonoids stilbenes, and phenolic acids	Plant extract and its fractions have been shown to inhibit apoptosis in H9c2 cells during simulated ischemia-reperfusion injury. This protective effect is associated with the inhibition of several markers of apoptosis, including nuclei condensation, caspase-3 activation, and MAP kinase signaling	Antiobesity, effective in neurodegenerative diseases, type 2 diabetes, and cardiovascular diseases	Ashraf et al. (2024)

(continued)

Table 7.1 (continued)

Sl. no.	Plant name	Family	Phytochemical constituents	Antioxidant mechanism	Pharmaceutical properties	References
18	*Dipsacus fullonum*	Caprifoliaceae	Polyphenols flavonoids stilbenes, and phenolic acids	The active constituents of plant prevent oxidative cellular damage and DNA damage	Antiobesity, effective in neurodegenerative diseases, type 2 diabetes, and cardiovascular diseases	Ashraf et al. (2024)
19	*Dipsacus sylvestris*	Caprifoliaceae	Polyphenols, flavonoids stilbenes, and phenolic acids	Transferring a hydrogen atom or an electron to reactive species, thereby deactivating them	Antiobesity, effective in neurodegenerative diseases, type 2 diabetes, and cardiovascular diseases	Ashraf et al. (2024)
20	*Juniperus recurva*	Cupressaceae	Polyphenols, flavonoids, stilbenes, ascorbic acid, tocopherols, tocotrienols, and phytosterols	Hepatoprotective effects against CCl4-induced oxidative damage in rats by maintaining antioxidant status and preventing alterations in hepatic glutathione levels and enzyme activities	Antiobesity, effective in neurodegenerative diseases, type 2 diabetes, and cardiovascular diseases, age-related muscular degeneration, cataract, immunosuppression, cancer, obesity, elevated cholesterol level, and inflammation	Ashraf et al. (2024)
21	*Picrorhiza scrophulariiflora*	Plantaginaceae	Luteolin and picroside II	Phosphorylation of Akt, p38, ERK, and JNK, indicating its inhibitory effect on PI3K/Akt and MAPK signaling pathways	Neurodegenerative	Rokaya et al. (2020)
22	*Rosa pendulina*	Rosaceae	Ascorbic acid	Enhances insulin sensitivity, facilitating the uptake of glucose into cells; reduces risk of hyperglycemia; and modulates lipid metabolism by reducing lipid peroxidation	Antidiabetic	Oprica et al. (2015)

23	*Rhododendron ferrugineum*	Ericaceae	Dehydrins proteins, polyphenols, triterpenoids, and diterpenes	Reactive oxygen species (ROS) scavenging and maintaining cellular hydration	Cardiovascular diseases, type 2 diabetes, cancer, skin and eye diseases, ageing, inflammation, and antiaging	Hubert et al. (2022) and Ashraf et al. (2024)
24	*Ranunculus glacialis*	Ranunculaceae	Flavonoids, polyphenolics, carotenoids, and anthocyanins	Enzymatic detoxification, ROS scavenging, UV protection, and redox signaling	Cardiovascular diseases, type 2 diabetes, cancer, skin and eye diseases, ageing, and inflammation	Goo (2022)
25	*Saxifraga oppositifolia*	Saxifragaceae	Flavonoids, polyphenolics, carotenoids, anthocyanins, triterpenoids, and saponins	Enzymatic detoxification, ROS scavenging, UV protection, and stress-adaptive mechanisms	Cardiovascular diseases, type 2 diabetes, cancer, skin and eye diseases, aging, and inflammation	Ashraf et al. (2024)
26	*Primula hirsuta*	Primulaceae	Flavonoids, polyphenolics, carotenoids, anthocyanins, triterpenoids, and saponins	Enzymatic detoxification, ROS scavenging, UV protection, and stress-adaptive mechanisms	Cardiovascular diseases, type 2 diabetes, cancer, skin and eye diseases, aging, and inflammation	Ashraf et al. (2024)
27	*Vaccinium macrocarpon*	Ericaceae	Ascorbic acid, tocopherols, and tocotrienols	ROS scavenging, enzymatic detoxification, and lipid protection	Age-related muscular degeneration, cataract, cardiovascular diseases, immunosuppression, cardiovascular diseases, cancer, obesity, and diabetes	Ashraf et al. (2024)
28	*Sorbus aucuparia, Sorbus scopulina*	Rosaceae	Ascorbic acid, tocopherols, and tocotrienols	ROS scavenging, enzymatic detoxification, and lipid protection	Age-related muscular degeneration, cataract, cardiovascular diseases, immunosuppression, cardiovascular diseases, cancer, obesity, and diabetes	Ashraf et al. (2024)

(continued)

Table 7.1 (continued)

Sl. no.	Plant name	Family	Phytochemical constituents	Antioxidant mechanism	Pharmaceutical properties	References
29	*Betula utilis*	Betulaceae	Pentacyclic triterpenoids, betulin, betulinic acid, lupeol, ursolic acid (UA), oleanolic acid, and β-amyrin	Activate the Nrf2-Keap1 pathway, upregulating the cell's endogenous antioxidant capacity, and/or by preventing the activation of the pro-oxidant and pro-inflammatory NF-κB pathway	Antiobesity	Kumar et al. (2020)
30	*Bunium periscum*	Apiaceae	Polyphenolics and flavonoids	NF-E2-related factor 2 (Nrf2) pathway, as observed in studies with other spice-derived phytochemicals	Antiobesity	Kumar et al. (2020)
31	*Angelica glauca*	Apiaceae	Polyphenolics, flavonoids, and terpenoids	Free radical scavenging, metal chelation, and boosting the body's antioxidant defense systems	Antiobesity	Kumar et al. (2020)
32	*Podophyllum hexandrum*	Berberidaceae	Polyphenolics, flavonoids, and terpenoids	Free radical scavenging, metal chelation, and boosting the body's antioxidant defense systems	Antiobesity	Kumar et al. (2020)
33	*Bergenia ciliate*	Bergenia ciliate	Polyphenolics and flavonoids	Free radical scavenging, metal chelation, and boosting the body's antioxidant defense systems	Antiobesity	Kumar et al. (2020)
34	*Rumex acetosa*	Polygonaceae	Polyphenolics and flavonoids	Free radical scavenging, metal chelation, and boosting the body's antioxidant defense systems activity	Antiobesity	Kumar et al. (2020)

35	*Carum carvi*	Apiaceae	Essential oil (mainly contains carvone, limonene, anethole and carveol)	Polyphenols act by donating hydrogen atoms or electrons to free radicals, chelating metal ions, and inhibiting lipid peroxidation	Anthropometric and metabolic indices in overweight and type 2 diabetic subjects	Haidari et al. (2011)
36	*Swertia barunensis, Swertia multicaulis*	Gentianaceae	Xanthones and amarogentin	Enzyme inhibition, insulin sensitization, free radical scavenging, and anti-inflammatory pathways	Antidiabetic and antioxidant activities	Roy et al. (2015)
37	*Neopicrorhiza scrophulariiflora*	Plantaginaceae	Glycosides such as caffeoyl, cucurbitacin, hydroquinone, iridoid glucoside, cyclopentanoid monoterpenes, nonglycosidic iridoid, phenol, and secoiridoid glycoside	Protection against oxidative stress and inflammation, and pancreatic β-cell protection	Antidiabetic activity	Rokaya et al. (2020)
38	*Rheum australe*	Polygonaceae	Polyphenols, flavonoids, quinones, saponins and tannins, chrysophanol, physcion, rhein, emodin, and aloe emodin	ROS scavenging and cytokine modulation	Antidiabetic activity	Kumai et al. (2023)
39	*Picrorhiza kurroa*	Plantaginaceae	Picroside-I, picroside-II and kutkin	Neutralizes ROS, upregulating SOD, CAT, GPx and reducing glucose absorption	Antidiabetic activity	Mohammed et al. (2014)
40	*Berberis aristata*	Berberidaceae	Berberine and polyphenols	Prevent oxidative stress-induced β-cell damage	Antidiabetic activity	Singh and Kakkar (2009)

(continued)

Table 7.1 (continued)

Sl. no.	Plant name	Family	Phytochemical constituents	Antioxidant mechanism	Pharmaceutical properties	References
41	*Malaxis acuminata*	Orchidaceae	Dietary fatty acids, α-hydroxy acids, phenolic acids, sterols, amino acids, sugars, and glycosides	Scavenge free radicals (ROS) and protecting pancreatic β-cells	Anti-inflammatory potential and in vitro photoprotective and skin aging	Bose et al. (2017)
42	*Delphinium uncinatum*	Ranunculaceae	Flavonoid compounds, rutoside, nicotiflorin, narcissoside, and violanthin	Scavenge free radicals (ROS)	Antioxidant, antimicrobial, cytotoxic, anti-inflammatory, and antiaging	Rehman et al. (2023)
43	*Leontopodium alpinum*	Asteraceae	Flavonoid compounds	Neutralize free radicals (ROS) and reduce oxidative stress, cellular longevity, and DNA	Antiaging	Cho et al. (2020)
44	*Dendrobium sulcatum*	Orchidaceae	Alkaloids (dendrobine) flavonoids, phenolic compounds glycosides, and polysaccharides	Neutralize ROS, upregulate SOD, CAT, GPx, and prevent oxidative stress, prevent neuronal apoptosis, and reducing chronic inflammation	Antioxidant, anti-lipid peroxidation, antiglycation, anti-inflammatory, and anti-tyrosinase properties	Rungsang et al. (2023)

lowering LDL cholesterol, inflammation, and oxidative stress, all while causing few negative effects (Scolaro et al. 2018). Many of these plant-based chemicals are not only beneficial, but have long been utilized in traditional medicine, with numerous studies documenting their therapeutic potential (Liwa et al. 2017). Due to research showing that a balanced diet, regular exercise, and lifestyle modifications can help lower the risk of coronary heart disease (CHD) by addressing factors like elevated LDL, inflammation, cholesterol, triglycerides, and cellular oxidative stress, maintaining a healthy diet and lifestyle is essential (Köhler et al. 2017). Certain plants, such as *Gentiana lutea*, have been demonstrated to have substantial anti-atherosclerosis properties. Gentisin, bellidifolin-8-O-glucoside, demethylbellidifolin-8-O-glucoside, isovitexin, and amarogentin are phytochemicals found in the plant that effectively reduce endothelial inflammation and inhibit smooth muscle cell migration, both of which are important processes in the development of atherosclerosis (Kesavan et al. 2016).

Elevated ROS levels contribute significantly to the evolution of hypertension by inducing endothelial damage, increasing cardiovascular remodeling, decreasing renal function, stimulating the sympathetic nervous system, and activating immunological responses that result in systemic inflammation. Furthermore, oxidative stress reduces NO bioavailability, a critical element in regulating vascular tone, aggravating endothelial dysfunction and hypertension. Antioxidants are vital in counteracting these detrimental effects, neutralizing ROS and restoring vascular homeostatic. Enzymatic and nonenzymatic antioxidants, including α-lipoic acid, glutathione peroxidase, catalase, superoxide dismutase, vitamins, and coenzyme Q10, are essential for neutralizing oxidative damage (Młynarska et al. 2024). Oxidative stress is a major factor in the onset and progression of atherosclerosis, leading to plaque production and instability, increasing the possibility of plaque rupture and thrombosis. Beyond vascular problems, high ROS can directly injure cardiomyocytes, impairing cardiac contractility. Additionally, oxidative stress activates inflammatory pathways, which promote cardiac inflammation and fibrosis. These processes work together to cause cardiac repair and dysfunction, which exacerbates the progression of cardiovascular disease (Moris et al. 2017).

Several alpine medicinal plants effectively treat various cardiovascular diseases. However, most of them are effective for myocardial infarction. *Rhodiola rosea*, a well-known adaptogenic plant, has been shown to have strong cardioprotective qualities due to its high phytochemical content, which includes tyrosol, rosavin, salidroside, rosaline, and gallic acid (Chen et al. 2022). According to studies, salidroside modulates vascular endothelial growth factor (VEGF) expression, which improves vasodilation and stimulates angiogenesis in ischemic myocardium. This process promotes the healing of ischemic cardiac tissue and protects against myocardial ischemia-reperfusion damage. Furthermore, Gupta et al. (2021) found that *Rhodiola rosea* extracts effectively reduce oxidative stress by decreasing (MDA) and lactate dehydrogenase (LDH) levels in blood, liver, and muscle tissues while increasing the activity of key antioxidants such as reduced glutathione and SOD. These results show *Rhodiola rosea*'s potential as a natural medicinal agent for cardiovascular health and oxidative stress management. Several studies showed

Vaccinium vitis-idea berries have and the role of its flavonoids as cardio-protectant is well known (Ghosh et al. 2024) and associated with high antioxidant activity. *Dipsacus* sp., mainly reported to be effective in cardiovascular diseases (atherosclerosis), leaves of *D. sativus* (L.) Honck. are used as an infusion for treating cardiovascular diseases (Skała and Szopa 2023). *Juniperus* sp. leaves are rich in alpha-pinene, sabinene, and δ-3-Carene, and flavonoids like amentoflavone and rutin, effective in cardiovascular treatments (Tavares and Seca 2018). Various other alpine plants, such as the role of the genus *Rhododendron*, with its terrific polyphenolic profile, are also well known for their atherosclerosis and other cardiovascular diseases (Hussain et al. 2024). Other species include *Primula hirsuta*, *Ranunculus glacialis* L., *Saxifraga oppositifolia* L., *Sorbus* sp., and *Vaccinium macrocarpon* (Table 7.1).

7.4 Neurodegenerative Diseases, ROS, and Antioxidants

Neurodegenerative diseases are a varied collection of conditions characterized by the slow degeneration and loss of certain neuron populations, which result in distinct clinical symptoms. These disorders are frequently associated with aberrant protein aggregation, which impairs a variety of biological processes. Mitochondrial malfunction, excessive ROS production, excitotoxicity, poor synaptic signaling, faulty degradation of proteins, the endoplasmic reticulum stress, damage to DNA, prolonged inflammation, and aberrant cell cycle re-entry are all important pathogenic factors (Chi et al. 2018). Unintentional consequences of the mitochondria electron transport cycle include superoxide radicals (O_2^-).

This oxidative imbalance causes severe harm to neurons and plays an important role in the course of neurodegenerative diseases. However, some reactive species play important physiological roles. Nitric oxide is important for neurotransmission, and activated microglia produce superoxide (O_2^-) during immunological responses. Because of its high ATP requirement, the central nervous system utilizes oxygen at a quick pace, rendering it especially sensitive to mitochondrial dysfunction. Such deficiencies can increase superoxide generation and exacerbate oxidative stress. To combat oxidative damage, the brain possesses several antioxidant defense systems, including mitochondrial manganese-containing superoxide dismutase (Mn-SOD) and reduced glutathione (GSH) playing particularly important roles. Nonetheless, studies have repeatedly found higher oxidative degradation of lipids, DNA, and proteins in postmortem brain tissues of people with neurodegenerative disorders like as Parkinson's, Alzheimer's, and amyotrophic lateral sclerosis (Jomova et al. 2024).

As a result, antioxidants have gained popularity as potential preventive and therapeutic agents for Alzheimer's disease. Natural antioxidants have shown therapeutic potential in preclinical investigations, and combination antioxidant therapy is being studied as a novel method for preventing and treating Alzheimer's disease (Collins et al. 2022). Endogenous and exogenous antioxidants protect cells from oxidative damage induced by free radicals. Endogenous antioxidants, such as enzymes, directly neutralize free radicals, while exogenous antioxidants, derived from the diet, donate electrons to stabilize them. *Nardostachys jatamansi* has long been as key ingredient Medhya Rasayana a formulation of traditional Indian Ayurveda medicine to treat neurological problems and improve memory. Jatamansinol from the *N. jatamansi* against Alzheimer's disease (AD) showed that it could be a multitargeted drug against AD. *N. jatamansi*, with its actinidine and glaziovine bioactive compounds, are reportedly effective for neurodegenerative diseases. Some other compounds, like Nardostachin, also protect the nervous system with their neuroinflammatory effects (Krishnan et al. 2023).

Another plant, *Rhodiola rosea,* has also been identified as a potential treatment for neurodegenerative illnesses such as Alzheimer's disease (AD), owing to its anti-inflammatory and neuroprotective characteristics. The active component, salidroside, is thought to produce neuroprotective benefits by lowering oxidative stress and increasing antioxidant enzyme activity (Wang et al. 2015; Bernatoniene et al. 2023). Similarly, *Vaccinium vitis-idaea,* sometimes known as lingonberry, is considered a "superfruit" due to its extraordinarily high antioxidant content and several health advantages. Numerous studies have demonstrated its anti-inflammatory, antioxidant, and even anticancer effects (Kowalska 2021). *Picrorhiza scrophulariiflora*, another promising plant, has been found to reduce neuropathic pain through its active components, including luteolin and picroside II. According to Nong and Lan (2018), these bioactive components function by suppressing spinal reactive astrocyte-mediated neuroinflammation through NF-κB signaling pathway.

7.5 Antiobesity Activity

Oxidative stress activates pro-inflammatory signaling pathways, including NF-κB and JNK, which contribute to adipocyte malfunction and chronic inflammation. ROS trigger the production of pro-inflammatory cytokines, including TNF-α and IL-6, from adipocytes and macrophages in adipose tissue. Chronic inflammation leads to resistance to insulin, dyslipidemia, and metabolic abnormalities (Čolak and Pap 2021). Studies show that particular high-altitude medicinal herbs, such as *Gentiana lutea* (also known as yellow gentian), have strong antiobesity potential. The plant's significant antioxidant capacity has been found to reduce oxidative stress associated with obesity, inhibit adipocyte development, and may aid in weight management. *G. Lutea* that also have important effects such as iridoidloganic acid and xanthone glycosides compassing its antioxidant capability, can be responsible for *Gentiana lutea's* antiobesity activities (Park et al. 2020). Several other medicinal herbs, including *Rhodiola rosea*, have been found as safer natural treatments for overweight and type 2 diabetes, or T2D (Loodu and Rupasinghe 2019). These plants not only help control the level of blood glucose and prevent hypertension, but they also boost the human body's antioxidant defenses and aid in insulin management. *R. rosea* contains bioactive chemicals, including salidroside and rosavines, that have considerable metabolic advantages. Extracts containing both salidroside and rosavines are more effective at regulating the molecular and cellular processes of fat cell growth, whereas salidroside-rich extracts primarily accelerate fat breakdown and induce apoptosis in maturing adipocytes (Pomari et al. 2015). Similarly, *Vaccinium vitis*-idaea (lingonberry) has shown the capacity to diminish moderate inflammation while avoiding diet-induced obesity in diabetic animals. Preliminary research suggests that certain plants' favorable metabolic benefits may be related to high levels of total polyphenols, antioxidants, and bioactive substances such as proanthocyanidins, anthocyanins, resveratrol, and kaempferol. Ryyti et al. (2020, 2024) found that air-dried lingonberry powder significantly enhanced the metabolic results in mice with high-fat (HF) diet-induced obesity, highlighting its potential significance in weight management. Furthermore, some alpine medicinal herbs, such as *Betula utilis*, *Bunium persicum*, *Angelica glauca*, *Saussurea lappa*, *Podophyllum hexandrum*, *Bergenia ciliata*, *Rumex acetosa*, *Aconitum heterophyllum*, and *Juniperus communis*, have demonstrated promising antiobesity capabilities. Kumar et al. (2020) discovered their capacity to regulate the breakdown of lipids and oxidative stress, which makes them potential therapies for obesity management.

7.6 Anti-aging Activity

Aging undermines efforts to prevent or halt the progression of age-related diseases such as diabetes, cardiovascular disease, neurodegenerative disorders, and cancer, reducing human independence, general well-being, and morbidity (Davalli et al. 2016). Telomere shortening, oxidative stress, and cellular senescence are all important aspects in the aging process, and they are caused by abnormalities in cellular metabolism. Notably, mitochondrial malfunction and a breakdown of communication involving mitochondria and the nucleus are associated with age-related alterations. Inadequate regulation of reactive oxygen species (ROS) within mitochondrial super complexes disrupts ROS signaling, prompting cellular stress responses that hasten aging-related damage. As cells age, their ability to neutralize ROS diminishes, causing them to become pro-oxidants (Guo et al. 2022). Given this, antioxidant supplementation has been advocated as a viable treatment method, in line with the widely accepted Free Radical Theory of Aging (FRTA) (Harman 2006).

Several natural substances, either alone or in combination, have been investigated for their possible use in supplemental therapy. Notable among these are resveratrol, hydroxytyrosol, vitamin C, vitamin E, curcumin, and coenzyme Q10, all of which have shown promise in preventing age-related deterioration (Conti et al. 2016). Plant-derived polyphenols, such as stilbenes, lignans, flavonoids, phenolic acids, and coumarins, have also been shown to improve aging symptoms (Hano and Tungmunnithum 2020). Alpine plants, noted for their high polyphenol content and other bioactive chemicals, are especially useful in anti-aging treatments. Several species, including *Rhodiola rosea*, tend to slow age-related decline in physical activity and immunological function, as well as improve stress tolerance (Li et al. 2017). Similarly, polysaccharides, the principal pharmacological components of *Polygonatum kingianum*, have been discovered to exhibit excellent antioxidant, anti-aging, and antibacterial capabilities, indicating their potential in age-related treatments (Di et al. 2023). Natural antioxidants, such as the saponins found in *Polygonatum kingianum*, show promise for therapeutic use in anti-aging therapies. *Anthriscus sylvestris*, often known as wild chervil, is another plant with noteworthy anti-aging qualities. It has long been utilized in traditional medicine, particularly in Korea and other places. The plant's primary chemical composition comprises organic acids, fatty acids, flavonoids, phenylpropanoids, and steroids, which contribute to its wide range of pharmacological actions, including anticancer, antioxidant, and anti-aging properties (Zhang et al. 2024). Furthermore, *A. sylvestris* polysaccharides have shown considerable anti-aging capabilities due to their combination antioxidant and immune-regulating actions, laying a solid theoretical foundation for producing anti-aging products from this plant (Xiaoyun et al. 2024). *Koenigia alpina* extracts and fractions show antioxidant, anti-inflammatory activities, whitening, and anti-aging suggesting that they could be used as anti-aging cosmeceuticals (Lee et al. 2024). Lee et al. (2024) found that the 70% ethanol extract, ethyl acetate fraction, and aglycone fraction of *K. alpina* can reduce ROS, limit the activity of enzymes involved in extracellular matrix degradation, and reduce inflammation. Similarly, the anti-aging and skin-rejuvenating properties of the alpine rose, *Rhododendron ferrugineum*, are well recognized. This plant has demonstrated interesting in vivo activity, such as the reduction of carbonylated proteins, a significant indicator of skin aging (Gafner et al. 2012). The extract is primarily composed of flavonoids, with hyperoside, naringenin, farrerol, and taxifolin being the most likely components responsible for the senolytic and skin rejuvenating effecting of plant extract (Hubert et al. 2022).

7.7 Pharmaceuticals Potential of Some Alpine Plants with Special Reference to Their Antioxidant Activities

1. *Aconitum species:* Different *Aconitum* species exhibited antioxidant activity such as *A. carmichaelii* Debx. (Yu et al. 2023), *A. kusnezoffii* Reichb (Su et al. 2021), and *A. heterophyllum* Wall. ex-Royle. Research has identified aconitine as a potential active content in these species. *Aconitine* exposure in *Danio rerio* embryos resulted in increased reactive oxygen species levels and triggered cell apoptosis in the heart and brain regions. It also reduced antioxidant enzyme activities like superoxide dismutase and caused the accumulation of lipid peroxidation products. In oxidative stress experiments, expression of oxidative stress-related genes, such as the protein Nrf2 and the HO-1, was downregulated, while pro-apoptotic proteins were increased (Xia et al. 2021). In cardiac cells, aconitine-induced intracellular calcium buildup altered mitochondrial membrane potential and enhanced the expression of autophagy and apoptosis-related genes, leading to oxidative stress and cell death (Ma et al. 2018). Neuronal cell studies revealed that aconitine toxicity was connected to glutamate and aspartate buildup, resulting in intracellular calcium excess, oxidative stress, and death (Zhang et al. 2023). These findings imply that *A. chasman-*

thum and *A. heterophyllum* may be useful in the prevention and treatment of oxidative stress-related illnesses.

2. *Rheum emodi: R. emodi* wall exhibited possible antioxidant compounds such as myricitrin, myricetin 3-galloyl rhamnoside, and myricetin. Recent research indicates that myricetin could potentially be exploited in the biological processes related to cardiovascular safeguarding, diabetes, and intrinsic resistance to carcinogens due to its significant iron-chelating, antioxidant, and free-radical scavenging characteristics (Yao et al. 2014). These compounds were initially found in *R. emodi*, and it was indicated that the antioxidant compounds would help use *R. emodi* as a medical herb (Domitrović et al. 2015). Another study found that stilbenes and Rheum-derived extracts improved blood plasma's non-enzymatic antioxidant capacity and successfully decreased the oxidative and nitrative alterations of lipids and proteins (Liudvytska et al. 2023). Additionally, studies using the isolated fibrinogen demonstrated the protective properties of the compounds under investigation against the oxidative stress-induced changes of this protein, such as the production of tyrosine nitration, protein aggregates, and tryptophan residue modification (Nowak et al. 2007; Mira et al. 2002). The *R. emodi* extracts' ability to defend against changes in fibrinogen molecules brought on by oxidative stress may be essential for preserving hemostatic equilibrium. Although fibrinogen is an essential protein for hemostasis, it is extremely vulnerable to oxidative and nitrative changes, especially damage caused by peroxynitrite (Martinez et al. 2013; Ponczek et al. 2014; Liudvytska et al. 2023).
3. *Saussurea lappa: S. lappa* ethanolic extract significantly reduced malondialdehyde concentrations and increased superoxide dismutase as well as glutathione peroxidase activities in rat lung and spleen tissues (Abd El-Rahman et al. 2020). This protective effect was most pronounced when *S. lappa* was administered prophylactically. *S. costus* is distinguished by the presence of costunolide, a phytochemical found largely in its roots (Mlejnek et al. 2021). Costunolide possesses remarkable antioxidant capability, which has been linked to its capacity to prevent cancer development. The chemical contains N-acetylcysteine, which helps neutralize ROS by boosting the manufacture of critical enzymes including CAT and SOD (Elshaer et al. 2022). These help the plant's antioxidant defense system by detoxifying harmful free radicals. Because of its antioxidant properties, *S. costus* has the potential to be employed as a medicinal agent to prevent and treat numerous diseases (Ashraf et al. 2024).
4. *Allium humile*: Allicin, the principal ingredient of *Allium humile*, has been shown to have significant antioxidant properties that help reduce oxidative stress in plants. Studies have shown that allicin can reduce intracellular reactive oxygen species (ROS) generation and prevent cells from H_2O_2-induced damage (Chan et al. 2014; Tu et al. 2016). It was shown that allicin increased the glutathione/glutathione disulfide ratio in retinal pigmented epithelium cells while decreasing intracellular ROS as well as malondialdehyde levels (Tu et al. 2016). Instead of scavenging free radicals or extracellular H_2O_2, allicin's protective action appears to be more concentrated on preventing the generation of ROS

inside cells (Chan et al. 2014). Allicin's mode of action sets it apart from other antioxidants. Furthermore, allicin has been shown to increase the activity of nuclear factor erythroid 2-related factor 2 in H_2O_2-stimulated cells and alter the expression levels of ROS-associated enzymes like superoxide dismutase, NAD(P)H dehydrogenase quinone 1, and NADPH oxidase (Tu et al. 2016).

5. *Berberis aristata* DC: Berberine, a compound found in *Berberis aristata*, has demonstrated significant potential in reducing human oxidative stress. Research has found that berberine can scavenge free radicals and improve oxidative stress indicators such as reduced glutathione, superoxide dismutase, nitric oxide, and malondialdehyde (Naz et al. 2022). According to Hojs et al. (2020), berberine often reduces MDA content. It raised SOD, GSH, along with GSH-Px contents, which might aid in scavenging excess free radicals and overcoming oxidative stress. Another study (Lao-ong et al. 2012) found that berberine medication significantly increased the mRNA level of SOD in diabetic mice while decreasing the levels of glutathione as well as glutathione peroxidase. Lao-ong et al. reported that although SOD was decreased in diabetic mice, glutathione levels were elevated. This was likely caused by the acute phase in experimental diabetes mellitus (Majithiya and Balaraman 2005), which typically happened 3–4 weeks after streptozotocin injection (Naz et al. 2022). Berbarine may therefore control the antioxidant mechanism of glutathione peroxidase differently depending on the stage of diabetes mellitus.
6. *Gastrodia elata:* Gastrodin, the main bioactive component of *G. elata* Blume, has demonstrated significant potential in reducing oxidative stress in human cells and tissues. Multiple studies have shown its antioxidant properties and protective effects against oxidative stress-induced conditions. According to studies, gastrodin increased the activities of SOD and total antioxidant capacity in the brain of mice subjected to lead. It also increased the expression of HO-1 along with NQO1 and brought back the nuclear translocation of Nrf2. These findings suggest that gastrodin may exhibit antioxidant effects by controlling the Wnt/Nrf2 signaling pathway (Liu et al. 2020). Treatment with gastrondin can control the Nrf2/HO-1 pathway, prevent the rise in oxidative stress markers like MDA, 3-nitrotyrosine, and 8-hydroxy-2-desoxyguanosine brought on by subarachnoid hemorrhage, and raise SOD levels, all of which have an antioxidant effect (Wang et al. 2019). According to Jiang et al. (2020), gastrodin has been demonstrated to protect C6 cells from H_2O_2-induced ferroptosis and reduce oxidative damage by increasing glutathione as well as glutathione peroxidase activity, promoting the expression of GPX4, Nrf2, and HO-1, and lowering ROS levels. Gastrodin treatment may increase cerebral ATP generation, preserve and restore mitochondrial function, and protect cells from oxidative damage (Jiang et al. 2020; Dai et al. 2024). Another study found that gastrodin can prevent glutamate-induced cell death in rat pheochromocytoma (PC12) cells by reducing ROS formation and decreasing oxidative stress via the CaMKII/ASK-1/p38 MAPK/p53 signaling pathway (Wang et al. 2019). Thus, the Nrf2 signaling pathway may be responsible for gastrodin's antioxidant

activity by modulating the expression of antioxidant proteins such as SOD and protecting mitochondrial function (Dai et al. 2024).

7. *Sophora flavescens* Aiton: Matrine (MAT), a naturally occurring alkaloid component derived from the herb *Sophorae flavescentis*, has a molecular weight of 258.43 ($C15H24N2O$) and has been widely used in the treatment of human hepatitis B and leukocytopenia with little adverse effects (Feng 2022). Matrine lowers oxidative stress and encourages mitochondrial production, as per earlier research. Nrf2, a major transcription factor of the mitochondrial endogenous antioxidant system, influences the expression of various antioxidant pathways, including glutathione and cytoprotective genes (Liang et al. 2017). Song et al. (2022), that when Matrine was administered to experimental autoimmune encephalomyelitis, the expression of Nrf2 increased dramatically. This shows that Matrine improves mitochondrial production and reduces oxidative stress, which could be linked to Nrf2 activation. Additionally, EX527, a Sirtuin 1 inhibitor, blocked Nrf2 expression, indicating that Matrine activated the Sirtuin 1 pathway to upregulate Nrf2 expression in retinal ganglion cells (Huang et al. 2015).
8. *Millettia nitida* Benth: Orientin, a flavonoid component with antioxidant capacity, has shown significant potential in reducing oxidative stress in human tissues and cells. Studies have shown that orientin can alleviate liver damage by improving antioxidant ability and decreasing levels of oxidative stress markers (Li et al. 2022). It promotes the Nrf2/ARE pathway through PI3K/Akt and P38/MAPK signaling, both of which are required for cellular oxidative stress defense mechanisms (Li et al. 2022; An et al. 2016). In erythrocytes, orientin has been found to protect against oxidative damage by reducing oxidative products, increasing antioxidative enzyme activities, and preserving the integrity of erythrocyte structure (An et al. 2016).
9. *Angelica glauca: A. glauca* Edgew, a valuable medicinal and aromatic herb, has been discovered to have high antioxidant capabilities that can help lower oxidative stress in humans. The root extracts of *A. glauca* contain high levels of active compounds, particularly polyphenols, and flavonoids, which contribute to its antioxidant potential (Devi et al. 2018). According to the previous study, angelicin, which is one of the important active compounds of the plant, has shown promising results in inhibiting the proliferation of glioblastoma cells by causing cell cycle arrest and inhibiting cell migration. Angelicin inhibited the expression of YAP and β-catenin, which are crucial proteins in cancer growth. These findings indicate that angelicin may be a promising chemical for treating GBM, a malignant tumor with a dismal survival rate (Wang et al. 2023).
10. *Nardostachys jatamansi*: *Nardostachys jatamansi* has demonstrated significant promise for reducing oxidative stress and its associated effects. The plant extract has strong antioxidant and anti-inflammatory properties, making it useful in treating oxidative stress and related inflammatory diseases (Rajyalakshmi et al. 2018). In a *Drosophila* model of neurological illnesses such as Alzheimer's, jatamansinol from *N. jatamansi* demonstrated neuroprotective properties by increasing antioxidant enzyme activity and decreasing Tau-induced oxidative

stress (Anupama et al. 2022). The extract also had chemopreventive properties, reducing benzoyl peroxide-induced cutaneous oxidative stress and damage in mice (Ali et al. 2005). Furthermore, *N. jatamansi* extract protected C6 glioma cells from oxidative stress caused by hydrogen peroxide, raising antioxidant enzyme levels and lowering lipid peroxidation (Khan et al. 2012; Dhuna et al. 2013). In chronic stress scenarios, the extract protected rats against stress-induced deficits in spatial learning and memory.

11. *Potentilla species*: *Potentilla* species contain various bioactive compounds that significantly combat oxidative stress in humans. These substances are largely phenolic, consisting of flavonoids, tannins, and phenolic acids. These chemicals have been shown to effectively scavenge free radicals and reactive oxygen species (ROS), shielding cells from oxidative damage (Hao et al. 2015; Muscolo et al. 2024). Potentilla extracts' antioxidant activities have been connected to their potential to prevent and treat oxidative stress-related disorders such as tumors, autoimmune diseases, and diabetes (Hao et al. 2015). *Potentilla* species contain various terpenes and terpenoids, contributing to their medicinal properties and biological activities. Terpenes are commonly found in *Potentilla* species; some studies have focused on other compound classes. *Potentilla erecta*, for example, has yielded coumarins and flavonoids (Goncharov et al. 1989). *Potentilla fulgens* produces chemicals such as rosamultic acid, hyptadienic acid, and tormentic acid (Kumar et al. 2013). Among these, hyptadienic acid, an A-ring contracted triterpenoid, displayed the highest inhibition of α-glucosidase, suggesting its potential relevance in diabetes control. Furthermore, linalool, an active component found in various *Potentilla* species, has been demonstrated to lower reactive oxygen species (ROS) levels and oxidative stress. Linalool intake was found to lower Aβ-induced ROS levels and oxidative stress in a fruit fly Alzheimer's disease model (Yuan et al. 2021). Similarly, in rat hippocampus, linalool attenuated Aβ1-42-induced oxidative stress (Yuan et al. 2021). In neuronal HT-22 cells, linalool reduced mitochondrial ROS and preserved mitochondrial membrane potential under glutamate-induced oxidative stress conditions (Sabogal-Guáqueta et al. 2019).
12. *Juniperus communis* L.: *Juniperus communis* (J. communis) essential oil has shown significant antioxidant activity in a variety of in vitro experiments. According to research, the oil successfully scavenges radicals such as DPPH, ABTS, hydroxyl, and superoxide radicals, as well as scavenging hydrogen peroxide and inhibiting xanthine oxidase. Juniper berries essential oil's principal antioxidant action is attributed to electron transfer processes, making it a powerful antioxidant; however, its activity via hydrogen atom transfer is less potent (Elmastaş et al. 2006; Höferl et al. 2014). Both water and ethanol extracts of juniper fruit have strong antioxidant capabilities, such as reducing power, free radical scavenging, and metal chelation activities (Elmastaş et al. 2006). In vivo studies have validated the antioxidant properties of *J. communis* essential oil, revealing its potential to inhibit oxidation processes in yeast cells. This is accomplished through enhanced activity of antioxidant enzymes such as glutathione peroxidase, SOD, and POD (Höferl et al. 2014). These findings indicate

that *J. communis* essential oil has promise as a natural antioxidant in both in vitro and in vivo systems, providing protection against oxidative stress and associated cellular damage.

13. *Hippopahe rhamnoides*: *Hippophae rhamnoides* L., or sea buckthorn, is known for its health-promoting characteristics. Flavones and catechins in the species are thought to be powerful flavonoids that defend against reactive oxygen species (ROS) (Trivedi et al. 2022). Quercetin, the most abundant dietary flavonol, has been shown to have antioxidant, anti-inflammatory, antiallergic, antiviral, and anticancer properties, as well as protection against liver and cardiovascular diseases. Sea buckthorn leaves, fruits, and oils contain high levels of vitamins A, C, and E, as well as lipids, carotenoids, unsaturated fatty acids, amino acids, microelements, and tannins. *H. rhamnoides* contains carotenoids, phytosterols, tocopherols, and flavonoids that can protect against the harmful effects of cisplatin and radiation treatment, while the plant's pectins can stimulate the immune system. According to pharmacological studies, crude extracts, or compounds from *H. rhamnoides* L., have a variety of health benefits, including antibacterial, anticancer, antioxidant, cardiovascular, hypoglycemic, hypolipidemic, neuroprotective, hepatoprotective, and anti-inflammatory properties.
14. *Rhodolia rosea: Rhodiola rosea* has recently received a lot of interest from scientists because of its potential as an adaptogen. This herb is also known as an ergogenic aid, which is a chemical used to improve both physical and mental performance. *R. rosea*'s adaptogenic and ergogenic benefits include increased performance, decreased weariness, and relief of depressive symptoms, the majority of which are due to its antioxidant capability. *R. rosea,* often known as Golden Root, is a valuable functional and medicinal food plant that is widely utilized in China and other Asian countries to maintain health. It is available in the United Kingdom as well as across Europe as a food supplement used to treat stress-induced weariness and fatigue (Booker et al. 2016). Salidroside, a bioactive molecule produced from *R. rosea*, has strong hepatoprotective, antioxidant, antiviral, and properties neuroprotective (Yu et al. 2007). Salidroside has been found to enhance defective hippocampus neurogenesis in a rat Alzheimer's disease model by preserving neural stem cells (NSCs) and scavenging intracellular reactive oxygen species (ROS) (Qu et al. 2012). *R. rosea*'s anti-inflammatory and neuroprotective characteristics may assist to treat neurodegenerative illnesses such as Alzheimer's disease (AD). The flavonoids in *R. rosea*, which are recognized for their antioxidant and anti-inflammatory properties, may contribute to the plant's general health benefits by supporting cellular health, lowering oxidative stress, and controlling inflammation. *R. rosea* is distinguished by its high concentration of cinnamic alcohol glucosides and phenylpropanoids such as rosavin, rosarin, and rosin (Ma et al. 2006).

7.8 Conclusion and Future Remarks

Alpine plants, facing constant oxidative stress from harsh environmental conditions, have evolved strong antioxidant defense systems. These systems, encompassing enzymatic and nonenzymatic antioxidants and a diverse array of secondary metabolites, provide significant protection against reactive oxygen species (ROS). The remarkable antioxidant capacity of these plants holds immense promise for therapeutic applications in various ROS-related diseases, including cancer, cardiovascular diseases, neurodegenerative disorders, diabetes, obesity, and fatigue. Many specific examples of alpine plants and their bioactive compounds with significant antioxidant potential have been discussed, highlighting their therapeutic potential. Future research should focus on several key areas. Firstly, more comprehensive studies are needed to fully elucidate the mechanisms of action of these phytochemicals at a molecular level. Secondly, optimized extraction and purification techniques need to be developed to ensure the consistent delivery of bioactive compounds with high bioavailability. Rigorous preclinical and clinical trials are essential to assess the safety and efficacy of alpine plant extracts and isolated compounds in treating human diseases. Additionally, further exploration of the vast biodiversity of alpine flora is crucial to identify novel compounds with even greater therapeutic potential.

References

Abd El-Rahman GI, Behairy A, Elseddawy NM, Batiha GES, Hozzein WN, Khodeer DM, Abd-Elhakim YM (2020) *Saussurea lappa* ethanolic extract attenuates triamcinolone acetonide-induced pulmonary and splenic tissue damage in rats via modulation of oxidative stress, inflammation, and apoptosis. Antioxidants 9(5):396

Ahmed MS, Khan AU, Kury LTA, Shah FA (2020) Computational and pharmacological evaluation of carveol for antidiabetic potential. Front Pharmacol 11:919. https://doi.org/10.3389/fphar.2020.00919

Ali A, Dua Y, Siddiqui AW, Sultana S, Rafiullah MRM (2005) Inhibition of benzoyl peroxide–induced cutaneous oxidative stress, toxicity, and ear edema in mice by *Nardostachys jatamansi*. Pharm Biol 43(6):533–539

Ansari S, Hasan K, Bhat S (2021a) Anticancer, antioxidant, and hepatoprotective activity of *Saussurea lappa*, C.B. Clarke (qust) on human hepatoma cell line. J Cancer Res Ther 17(2):499–503. https://doi.org/10.4103/jcrt.JCRT_571_19. PMID: 34121698

Ansari S, Maaz M, Ahmad I, Hasan SK, Bhat SA, Naqui SK, Husain M (2021b) Quality control, HPTLC analysis, antioxidant and antimicrobial activity of hydroalcoholic extract of roots of qust (*Saussurea lappa*, CB Clarke). Drug Metab Pers Ther 36(2):145–153

Anupama KP, Antony A, Shilpa O, Raghu SV, Gurushankara HP (2022) Jatamansinol from *Nardostachys jatamansi* ameliorates tau-induced neurotoxicity in drosophila Alzheimer's disease model. Mol Neurobiol 59(10):6091–6106

Arfin S, Jha NK, Jha SK, Kesari KK, Ruokolainen J, Roychoudhury S, Rathi B, Kumar D (2021) Oxidative stress in cancer cell metabolism. Antioxidants 10(5):642. https://doi.org/10.3390/antiox10050642

Ashraf MV, Khan S, Misri S, Gaira KS, Rawat S, Rawat B, Khan MH, Shah AA, Asgher M, Rawat B, Ahmad S (2024) High-altitude medicinal plants as promising source of phytochemical antioxidants to combat lifestyle-associated oxidative stress-induced disorders. Pharmaceuticals 17(8):975. https://doi.org/10.3390/ph17080975

Bernatoniene J, Jakstas V, Kopustinskiene DM (2023) Phenolic compounds of *Rhodiola rosea* L. as the potential alternative therapy in the treatment of chronic diseases. Int J Mol Sci 24(15):12293. https://doi.org/10.3390/ijms241512293

Bhatt ID, Rawat S, Rawal RS (2013) Antioxidants in medicinal plants. In: Chandra S, Lata H, Verma A (eds) Biotechnology for medicinal plants. Springer, Berlin, pp 296–326

Bhattacharyya A, Chattopadhyay R, Mitra S, Crowe SE (2014) Oxidative stress: an essential factor in the pathogenesis of gastrointestinal mucosal diseases. Physiol Rev 94(2):329–354. https://doi.org/10.1152/physrev.00040.2012

Bhatti JS, Sehrawat A, Mishra J, Sidhu IS, Navik U, Khullar N, Kumar S, Bhatti GK, Reddy PH (2022) Oxidative stress in the pathophysiology of type 2 diabetes and related complications: current therapeutics strategies and future perspectives. Free Radic Biol Med 184:114–134. https://doi.org/10.1016/j.freeradbiomed.2022.03.019

Brewer MS (2011) Natural antioxidants: sources, compounds, mechanisms of action and potential applications. Compr Rev Food Sci Food Saf 10(4):221–247. https://doi.org/10.1111/j.1541-4337.2011.00156

Chosson E, Chaboud A, Chulia AJ, Raynaud J (1998) Dihydroflavonol glycosides from *Rhododendron ferrugineum*. Phytochemistry 49(5):1431–1433. https://doi.org/10.1016/S0031-9422(98)00080-6

Čolak E, Pap D (2021) The role of oxidative stress in the development of obesity and obesity-related metabolic disorders. J Med Biochem 40(1):1–9. https://doi.org/10.5937/jomb0-24652. PMID: 33584134; PMCID: PMC7857849

Collins AE, Saleh TM, Kalisch BE (2022) Naturally occurring antioxidant therapy in Alzheimer's disease. Antioxidants 11(2):213

Conti V, Izzo V, Corbi G, Russomanno G, Manzo V, De Lise F, Di Donato A, Filippelli A (2016) Antioxidant supplementation in the treatment of aging-associated diseases. Front Pharmacol 7:24. https://doi.org/10.3389/fphar.2016.00024

Coughlan MT, Sharma K (2016) Challenging the dogma of mitochondrial reactive oxygen species overproduction in diabetic kidney disease. Kidney Int 90(2):272–279. https://doi.org/10.1016/j.kint.2016.02.043

Davalli P, Mitic T, Caporali A, Lauriola A, D'Arca D (2016) ROS, cell senescence, and novel molecular mechanisms in aging and age-related diseases. Oxidative Med Cell Longev 2016(1):3565127. https://doi.org/10.1155/2016/3565127

Devi K, Samant SS, Puri S, Kundra R, Kumari P (2018) Investigation of antioxidant and radical scavenging potential of *Angelica glauca* Edgew. and *Aralia cachemirica* Decne: a high value medicinal plants from Kanawar Wildlife Sanctuary in Himachal Pradesh of north western Himalayas. Med Plants Int J Phytomed Relat Ind 10(4):312–319

Dhuna K, Dhuna V, Bhatia G, Singh J, Kamboj S (2013) Cytoprotective effect of methanolic extract of *Nardostachys jatamansi* against hydrogen peroxide induced oxidative damage in C6 glioma cells. Acta Biochim Pol 60(1):21–31

Domitrović R, Rashed K, Cvijanović O, Vladimir-Knežević S, Škoda M, Višnić A (2015) Myricitrin exhibits antioxidant, anti-inflammatory and antifibrotic activity in carbon tetrachloride-intoxicated mice. Chem Biol Interact 230:21–29

Elmastaş M, Gülçin I, Beydemir Ş, İrfan Küfrevioğlu Ö, Aboul-Enein HY (2006) A study on the in vitro antioxidant activity of juniper (*Juniperus communis* L.) fruit extracts. Anal Lett 39(1):47–65

Elshaer SE, Hamad GM, Hafez EE, Baghdadi HH, El-Demerdash FM (2022) Root extracts of *Saussurea costus* as prospective detoxifying food additive against sodium nitrite toxicity in male rats. Food Chem Toxicol 166:113225

Feng Q (2022) Gastrodin attenuates lipopolysaccharide-induced inflammation and oxidative stress, and promotes the osteogenic differentiation of human periodontal ligament stem cells through enhancing sirtuin3 expression. Exp Ther Med 23(4):1–11

Gafner F, Schmid D, Zülli F (2012) Ecocert certified alpine rose active protects skin proteins. Acta Hortic 955:351–355. https://doi.org/10.17660/ActaHortic.2012.955.53

Ganie SA, Dar TA, Hamid R, Zargar O, Abeer SU, Masood A, Amin S, Zargar MA (2014) In vitro antioxidant and cytotoxic activities of *Arnebia benthamii* (Wall ex. G. Don): a critically endangered medicinal plant of Kashmir valley. Oxidative Med Cell Longev 2014(1):792574. https://doi.org/10.1155/2014/792574

Ghosh A, Debnath SC, Igamberdiev AU (2024) Effects of *Vaccinium*-derived antioxidants on human health: the past, present and future. Front Mol Biosci 11:1520661. https://doi.org/10.3389/fmolb.2024.1520661

Goncharov NF, Stupakova ĖP, Komissarenko NF (1989) Polyphenol composition of the epigeal part of *Potentilla erecta*. Chem Nat Compd 25:375–376

González-Peña MA, Ortega-Regules AE, Anaya de Parrodi C, Lozada-Ramírez JD (2023) Chemistry, occurrence, properties, applications, and encapsulation of carotenoids—a review. Plan Theory 12:313

Guo J, Huang X, Dou L, Yan M, Shen T, Tang W, Li J (2022) Aging and aging-related diseases: from molecular mechanisms to interventions and treatments. Signal Transduct Target Ther 7:391. https://doi.org/10.1038/s41392-022-01251-0

Halliwell B, Gutteridge JMC (2015) Free radials in biology and medicine. Oxford University Press, p 896

Hano C, Tungmunnithum D (2020) Plant polyphenols, more than just simple natural antioxidants: oxidative stress, aging and age-related diseases. Medicine 7(5):26. https://doi.org/10.3390/medicines7050026

Hao DA, Gu XJ, Xiao PG (2015) Potentilla and rubus medicinal plants: potential non-camellia tea resources. In: Medicinal plants, chemistry, biology and omics. Woodhead Publishing, Cambridge, MA, pp 373–430

Hayes JD, Kelleher MO, Eggleston IM (2008) The cancer chemopreventive actions of phytochemicals derived from glucosinolates. Eur J Nutr 2:73–88. https://doi.org/10.1007/s00394-008-2009-8

Höferl M, Stoilova I, Schmidt E, Wanner J, Jirovetz L, Trifonova D, Krastanov A (2014) Chemical composition and antioxidant properties of Juniper berry (*Juniperus communis* L.) essential oil. Action of the essential oil on the antioxidant protection of *Saccharomyces cerevisiae* model organism. Antioxidants 3(1):81–98

Hojs NV, Bevc S, Ekart R, Hojs R (2020) Oxidative stress markers in chronic kidney disease with emphasis on diabetic nephropathy. Antioxidants 9:925

Huang Q, Shi J, Gao B, Zhang HY, Fan J, Li XJ, Liu J (2015) Gastrodin: an ancient Chinese herbal medicine as a source for anti-osteoporosis agents via reducing reactive oxygen species. Bone 73:132–144

Hubert J, Kotland A, Henes B, Poigny S, Wandrey F (2022) Deciphering the phytochemical profile of an alpine rose (*Rhododendron ferrugineum* L.) leaf extract for a better understanding of its senolytic and skin-rejuvenation effects. Cosmetics 9(2):37. https://doi.org/10.3390/cosmetics9020037

Huong TT, Tram LH, Van Thong N, Minh NH, Minh TT, Quang TT, Ha TT, Hung ND, Van Don D, Nguyen TM, Park JT (2022) *Furofuran lignans* from *Valeriana jatamansi* with their antioxidant and anticancer properties. Vietnam J Chem 60(2):157–163. https://doi.org/10.1002/vjch.202100084

Hussain A, Azam S, Maqsood R, Anwar R, Akash MS, Hussain H, Wang D, Imran M, Kotwica-Mojzych K, Khan S, Hussain S (2024) Chemistry, biosynthesis, and theranostics of antioxidant flavonoids and polyphenolics of genus Rhododendron: an overview. Naunyn Schmiedeberg's Arch Pharmacol 14:1–44

Ighodaro OM, Akinloye OA (2018) First line defence antioxidants-superoxide dismutase (SOD), catalase (CAT) and glutathione peroxidase (GPX): their fundamental role in the entire antioxidant defence grid. Alexandria J Med 54:287–293. https://doi.org/10.1016/j.ajme.2017.09.001

Johansen JS, Harris AK, Rychly DJ, Ergul A (2005) Oxidative stress and the use of antioxidants in diabetes: linking basic science to clinical practice. Cardiovasc Diabetol 4:1–11. https://doi.org/10.1186/1475-2840-4-5

Jomova K, Raptova R, Alomar SY, Alwasel SH, Nepovimova E, Kuca K, Valko M (2023) Reactive oxygen species, toxicity, oxidative stress, and antioxidants: chronic diseases and aging. Arch Toxicol 97(10):2499–2574. https://doi.org/10.1007/s00204-023-03562-9

Jomova K, Alomar SY, Alwasel SH, Nepovimova E, Kuca K, Valko M (2024) Several lines of antioxidant defense against oxidative stress: antioxidant enzymes, nanomaterials with multiple enzyme-mimicking activities, and low-molecular-weight antioxidants. Arch Toxicol 98:1323–1367. https://doi.org/10.1007/s00204-024-03696-4

Ju L, Wen X, Wang C, Wei Y, Peng Y, Ding Y, Feng L, Shu L (2017) Salidroside, a natural antioxidant, improves β-cell survival and function via activating AMPK pathway. Front Pharmacol 8:749. https://doi.org/10.3389/fphar.2017.00749

Kajal A, Kishore L, Kaur N, Gollen R, Singh R (2016) Therapeutic agents for the management of atherosclerosis from herbal sources. Beni Suef Univ J Basic Appl Sci 5(2):156–169. https://doi.org/10.1016/j.bjbas.2016.02.004

Karlsen A, Retterstøl L, Laake P, Paur I, Kjølsrud-Bøhn S, Sandvik L, Blomhoff R (2007) Anthocyanins inhibit nuclear factor-kB activation in monocytes and reduce plasma concentrations of pro-inflammatory mediators in healthy adults. J Nutr 137:1951–1954. https://doi.org/10.1093/jn/137.8.1951

Kasote DM, Katyare SS, Hegde MV, Bae H (2015) Significance of antioxidant potential of plants and its relevance to therapeutic applications. Int J Biol Sci 11(8):982–991. https://doi.org/10.7150/ijbs.12096

Kesavan R, Chandel S, Upadhyay S, Bendre R, Ganugula R, Potunuru UR, Giri H, Sahu G, Kumar PU, Reddy GB, Joksic G, Bera AK, Dixit M (2016) *Gentiana lutea* exerts anti-atherosclerotic effects by preventing endothelial inflammation and smooth muscle cell migration. Nutr Metab Cardiovasc Dis 4:293–301

Khan KA, Kerbel RS (2018) Improving immunotherapy outcomes with anti-angiogenic treatments and vice versa. Nat Rev Clin Oncol 15:310–324. https://doi.org/10.1038/nrclinonc.2018.9

Khan MB, Hoda MN, Ishrat T, Ahmad S, Moshahid Khan M, Ahmad A, Islam F (2012) Neuroprotective efficacy of *Nardostachys jatamansi* and crocetin in conjunction with selenium in cognitive impairment. Neurol Sci 33:1011–1020

Khazdair MR, Boskabady MH, Hosseini M, Rezaee R, Tsatsakis AM (2015) The effects of Crocus sativus (saffron) and its constituents on nervous system: a review. Avicenna J Phytomed 5(5):376–391. PMID: 26468457; PMCID: PMC4599112

Korkina LG (2007) Phenylpropanoids as naturally occurring antioxidants: from plant defense to human health. Cell Mol Biol 53:15–25

Kowalska K (2021) Lingonberry (*Vaccinium vitis-idaea* L.) fruit as a source of bioactive compounds with health-promoting effects—a review. Int J Mol Sci 22(10):5126. https://doi.org/10.3390/ijms22105126

Krishnan N, Mariappanadar V, Dhanabalan AK, Devadasan V, Gopinath SC, Raman P (2023) Purification, identification and in silico models of alkaloids from *Nardostachys jatamansi*—bioactive compounds for neurodegenerative diseases. Biomass Convers Biorefinery 13:14889–14900. https://doi.org/10.1007/s13399-022-03237-y

Kumai GP, Baskota M, Aryal P, Subba B (2023) Study of phytochemical constituent and biological activities of methanolic extract of rhizomes of *Neopicrorhiza scrophulariiflora* and roots of *Rheum australe* collected from the Alpine Region of Nepal. J Nepal Chem Soc 43(2):130–140. https://doi.org/10.3126/jncs.v43i2.53359

Kumar D, Ghosh R, Pal BC (2013) α-Glucosidase inhibitory terpenoids from *Potentilla fulgens* and their quantitative estimation by validated HPLC method. J Funct Foods 5(3):1135–1141

Kumar M, Guleria S, Chawla P, Khan A, Modi VK, Kumar N, Kaushik R (2020) Anti-obesity efficacy of the selected high altitude Himalayan herbs: in vitro studies. J Food Sci Technol 57(8):3081–3090. https://doi.org/10.1007/s13197-020-04341-5

Kumar A, Guleria S, Ghosh D, Dogra V, Kumar S (2021) Managing reactive oxygen species—some learnings from high altitude extremophytes. Environ Exp Bot 189:104525

Kumar P, Rana V, Singh AN (2022) *Angelica glauca* Edgew.—a comprehensive review. J Appl Res Med Aromat Plants 31:100397

Kumar APN, Kumar M, Jose A, Tomer V, Oz E, Proestos C, Zeng M, Elobeid TKS, Oz F (2023) Major phytochemicals: recent advances in health benefits and extraction method. Molecules 28(2):887

Kwon DH, Cha HJ, Lee H, Hong SH, Park C, Park SH, Kim GY, Kim S, Kim HS, Hwang HJ, Choi YH (2019) Protective effect of glutathione against oxidative stress-induced cytotoxicity in RAW 264.7 macrophages through activating the nuclear factor erythroid 2-related factor-2/heme oxygenase-1 pathway. Antioxidants 8(4):82

Lao-ong T, Chatuphonprasert W, Nemoto N, Jarukamjorn K (2012) Alteration of hepatic glutathione peroxidase and superoxide dismutase expression in streptozotocin-induced diabetic mice by berberine. Pharm Biol 50(8):1007–1012. https://doi.org/10.3109/13880209.2012.655377

Lee KW, Mun SH, Kim YA, Kim HR, Jin Q, Lee MK, Park SN (2024) Functional properties and components of *Koenigia alpina* extract. Skin Res Technol 30:e70102. https://doi.org/10.1111/srt.70102

Lekomtseva Y, Zhukova I, Wacker A (2017) *Rhodiola rosea* in subjects with prolonged or chronic fatigue symptoms: results of an open-label clinical trial. Complement Med Res 24(1):46–52. https://doi.org/10.1159/000457918

Li Y, Pham V, Bui M, Song L, Wu C, Walia A, Uchio E, Smith-Liu F, Zi X (2017) *Rhodiola rosea L.*: an herb with anti-stress, anti-aging, and immunostimulating properties for cancer chemoprevention. Curr Pharmacol Rep 3(6):384–395. https://doi.org/10.1007/s40495-017-0106-1

Liang WZ, Jan CR, Hsu SS (2017) Cytotoxic effects of gastrodin extracted from the rhizome of *Gastrodia elata* Blume in glioblastoma cells, but not in normal astrocytes, via the induction of oxidative stress-associated apoptosis that involved cell cycle arrest and p53 activation. Food Chem Toxicol 107:280–292

Liguori I, Russo G, Curcio F, Bulli G, Aran L, Della-Morte D, Gargiulo G, Testa G, Cacciatore F, Bonaduce D, Abete P (2018) Oxidative stress, aging, and diseases. Clin Interv Aging 13:757–772. https://doi.org/10.2147/CIA.S158513

Liu CM, Tian ZK, Zhang YJ, Ming QL, Ma JQ, Ji LP (2020) Effects of gastrodin against lead-induced brain injury in mice associated with the Wnt/Nrf2 pathway. Nutrients 12:1805

Liudvytska O, Bandyszewska M, Skirecki T, Krzyżanowska-Kowalczyk J, Kowalczyk M, Kolodziejczyk-Czepas J (2023) Anti-inflammatory and antioxidant actions of extracts from *Rheum rhaponticum* and *Rheum rhabarbarum* in human blood plasma and cells in vitro. Biomed Pharmacother 165:115111

Ma G, Li W, Dou D, Chang X, Bai H, Satou T, Li J, Sun D, Kang T, Nikaido T, Koike K (2006) Rhodiolosides A-E, monoterpene glycosides from *Rhodiola rosea*. Chem Pharm Bull 54(8):1229–1233

Ma LQ, Yu Y, Chen H, Li M, Ihsan A, Tong HY, Huang XJ, Gao Y (2018) Sweroside alleviated aconitine-induced cardiac toxicity in H9c2 cardiomyoblast cell line. Front Pharmacol 9:1138

Majithiya JB, Balaraman R (2005) Time-dependent changes in antioxidant enzymes and vascular reactivity of aorta in streptozotocin-induced diabetic rats treated with curcumin. J Cardiovasc Pharmacol 46(5):697–705. https://doi.org/10.1097/01.fjc.0000183720.85014.24

Martinez M, Weisel JW, Ischiropoulos H (2013) Functional impact of oxidative posttranslational modifications on fibrinogen and fibrin clots. Free Radic Biol Med 65:411–418

Matough FA, Budin SB, Hamid ZA, Alwahaibi N, Mohamed J (2012) The role of oxidative stress and antioxidants in diabetic complications. Sultan Qaboos Univ Med J 12(1):5–18. https://doi.org/10.12816/0003082

McGregor GP, Biesalski HK (2006) Rationale and impact of vitamin C in clinical nutrition. Curr Opin Clin Nutr Metabol Care 9(6):697–703

Miguel CA, Noya-Riobó MV, Mazzone GL, Villar MJ, Coronel MF (2021) Antioxidant, anti-inflammatory and neuroprotective actions of resveratrol after experimental nervous system insults. Special focus on the molecular mechanisms involved. Neurochem Int 150:105188. https://doi.org/10.1016/j.neuint.2021.105188

Mira L, Fernandez MT, Santos M, Rocha R, Florencio MH, Jenninqs KR (2002) Interactions of flavonoids with iron and copper ions: a mechanism for their antioxidant activity. Free Radic Res 36:1199–1208

Mironczuk-Chodakowska I, Witkowska AM, Zujko ME (2018) Endogenous non-enzymatic antioxidants in the human body. Adv Med Sci 63:68–78

Mlejnek P, Dolezel P, Kriegova E, Pastvova N (2021) N-acetylcysteine can induce massive oxidative stress, resulting in cell death with apoptotic features in human leukemia cells. Int J Mol Sci 22:12635

Młynarska E, Biskup L, Możdżan M, Grygorcewicz O, Możdżan Z, Semeradt J, Uramowski M, Rysz J, Franczyk B (2024) The role of oxidative stress in hypertension: the insight into antihypertensive properties of vitamins A, C and E. Antioxidants 13(7):848. https://doi.org/10.3390/antiox13070848

Moris D, Spartalis M, Spartalis E, Karachaliou GS, Karaolanis GI, Tsourouflis G, Tsilimigras DI, Tzatzaki E, Theocharis S (2017) The role of reactive oxygen species in the pathophysiology of cardiovascular diseases and the clinical significance of myocardial redox. Ann Transl Med 5(16):326. https://doi.org/10.21037/atm.2017.06.27

Motyka S, Jafernik K, Ekiert H, Sharifi-Rad J, Calina D, Al-Omari B, Szopa A, Cho WC (2023) Podophyllotoxin and its derivatives: potential anticancer agents of natural origin in cancer chemotherapy. Biomed Pharmacother 158:114145

Naz I, Masoud MS, Chauhdary Z, Shah MA, Panichayupakaranant P (2022) Anti-inflammatory potential of berberine-rich extract via modulation of inflammation biomarkers. J Food Biochem 46(12):e14389

Nong X, Lan Y (2018) Picroside II attenuates CCI-induced neuropathic pain in rats by inhibiting spinal reactive astrocyte-mediated neuroinflammation through the NF-κB pathway. Neurochem Res 43:1058–1066. https://doi.org/10.1007/s11064-018-2518-7

Nowak P, Zbikowska HM, Ponczek M, Kolodziejczyk J, Wachowicz B (2007) Different vulnerability of fibrinogen subunits to oxidative/nitrative modifications induced by peroxynitrite: functional consequences. Thromb Res 121(2):163–174

Olennikov DN, Kashchenko NI, Chirikova NK, Koryakina LP, Vladimirov LN (2015) Bitter gentian teas: nutritional and phytochemical profiles, polysaccharide characterisation and bioactivity. Molecules 20(11):20014–20030. https://doi.org/10.3390/molecules201119674. PMID: 26556333; PMCID: PMC6331966

Ort DR, Baker NR (2002) A photoprotective role for O_2 as an alternative electron sink in photosynthesis? Curr Opin Plant Biol 5:193–198

Oszmiański J, Wojdyło A, Juszczyk P, Nowicka P (2020) Roots and leaf extracts of *Dipsacus fullonum* L. and their biological activities. Plan Theory 9:78. https://doi.org/10.3390/plants9010078

Panova IG, Tatikolov AS (2023) Endogenous and exogenous antioxidants as agents preventing the negative effects of contrast media (contrast-induced nephropathy). Pharmaceuticals 16(8):1077. https://doi.org/10.3390/ph16081077

Park E, Lee CG, Kim J, Yeo S, Kim JA, Choi CW, Jeong SY (2020) Antiobesity effects of *Gentiana lutea* extract on 3T3-L1 preadipocytes and a high-fat diet-induced mouse model. Molecules 25(10):2453. https://doi.org/10.3390/molecules25102453

Pärnänen P, Niikko S, Lähteenmäki H, Räisänen IT, Tervahartiala T, Sorsa T, Ranki A (2024) Lingonberry (*Vaccinium vitis-idaea* L.) fruit phenolic bioactivities—a review of in vitro and in vivo human studies. Microorganisms 12:1850. https://doi.org/10.3390/microorganisms12091850

Pomari E, Stefanon B, Colitti M (2015) Effects of two different *Rhodiola rosea* extracts on primary human visceral adipocytes. Molecules 20(5):8409–8428. https://doi.org/10.3390/molecules20058409

Ponczek M, Bijak M, Saluk J, Kolodziejczyk-Czepas J, Nowak P (2014) The comparison of peroxynitrite action on bovine, porcine and human fibrinogens. Open Life Sci 9(3):233–241

Ponnampalam EN, Kiani A, Santhiravel S, Holman BWB, Lauridsen C, Dunshea FR (2022) The importance of dietary antioxidants on oxidative stress, meat and milk production, and their preservative aspects in farm animals: antioxidant action, animal health, and product quality—invited review. Animals (Basel) 12(23):3279. https://doi.org/10.3390/ani12233279

Raina R, Verma PK, Peshin R, Kour H (2019) Potential of *Juniperus communis* L as a nutraceutical in human and veterinary medicine. Heliyon 5(8):e02376. https://doi.org/10.1016/j.heliyon.2019.e02376

Rajakrishnan R, Alfarhan A, Al-Ansari A, Lekshmi R, Sreelakshmi R, Benil P, Kim Y, Tack J, Na S, Kim H (2020) Therapeutic efficacy of the root tubers of *Aconitum heterophyllum* and its substitute *Cyperus rotundus* in the amelioration of pylorus ligation induced ulcerogenic and oxidative damage in rats. Saudi J Biol Sci 27(4):1124–1129. https://doi.org/10.1016/j.sjbs.2019.12.025

Rajendran P, Nandakumar N, Rengarajan T, Palaniswami R, Gnanadhas EN, Lakshminarasaiah U, Gopas J, Nishigaki I (2014) Antioxidants and human diseases. Clin Chim Acta 436:332–347. https://doi.org/10.1016/j.cca.2014.06.004. Epub 2014 Jun 13. PMID: 24933428

Rajyalakshmi I, Rao VA, Potti RB, Murty KVGS (2018) Effects of *Nardostachys jatamansi* extract on inflammatory and oxidative stress markers in hypertensive, hyperglycaemic patients: an open-label, prospective study. Int J Res Pharm Sci 9(4):1096–1098

Roy P, Abdulsalam FI, Pandey DK, Bhattacharjee A, Eruvaram NR, Malik T (2015) Evaluation of antioxidant, antibacterial, and antidiabetic potential of two traditional medicinal plants of India: *Swertia cordata* and *Swertia chirayita*. Pharm Res 7(Suppl 1):S57–S62. https://doi.org/10.4103/0974-8490.157997

Rudenko AY, Mariasina SS, Bolikhova AK, Nikulin MV, Ozhiganov RM, Vasil'ev VG, Ikhalaynen YA, Khandazhinskaya AL, Khomutov MA, Sergiev PV, Khomutov AR, Polshakov VI (2024) Organophosphorus *S*-adenosyl-*L*-methionine mimetics: synthesis, stability, and substrate properties. Front Chem 12:1448747. https://doi.org/10.3389/fchem.2024.1448747

Ryyti R, Hämäläinen M, Peltola R, Moilanen E (2020) Beneficial effects of lingonberry (*Vaccinium vitis-idaea* L.) supplementation on metabolic and inflammatory adverse effects induced by high-fat diet in a mouse model of obesity. PLoS One 15:e0232605

Sakagami H, Satoh K (1997) Prooxidant action of two antioxidants: ascorbic acid and gallic acid. Anticancer research. 1;17(1A):221–4

Saravanan S, Hari R, Sekar K (2023) Anti-proliferative potentials of *Aconitum heterophyllum* root extract in human breast cancer (MDA-MB-231) cell lines-genetic and antioxidant enzyme approach. Avicenna J Med Biotechnol 15(3):188–195. https://doi.org/10.18502/ajmb.v15i3.12929

Sarv V, Venskutonis PR, Bhat R (2020) The *Sorbus* spp.—underutilised plants for foods and nutraceuticals: review on polyphenolic phytochemicals and antioxidant potential. Antioxidants (Basel) 9(9):813. https://doi.org/10.3390/antiox9090813. PMID: 32882984; PMCID: PMC7555345

Sekhon-Loodu S, Rupasinghe HPV (2019) Evaluation of antioxidant, antidiabetic and antiobesity potential of selected traditional medicinal plants. Front Nutr 6:53. https://doi.org/10.3389/fnut.2019.00053

Sharma P, Jha AB, Dubey RS, Pessarakli M (2012) Reactive oxygen species, oxidative damage, and antioxidative defense mechanism in plants under stressful conditions. J Bot 2012(1):217037. https://doi.org/10.1155/2012/217037

Sies H (2020) Oxidative stress: concept and some practical aspects. Antioxidants 9:852. https://doi.org/10.3390/antiox9090852

Skała E, Szopa A (2023) *Dipsacus* and *Scabiosa* species—the source of specialized metabolites with high biological relevance: a review. Molecules 28(9):3754. https://doi.org/10.3390/molecules28093754

Song Y, Wang M, Zhao S, Tian Y, Zhang C (2022) Matrine promotes mitochondrial biosynthesis and reduces oxidative stress in experimental optic neuritis. Front Pharmacol 13:936632

Streb P, Aubert S, Gout E, Bligny R (2003) Reversibility of cold- and light-stress tolerance and accompanying changes of metabolite and antioxidant levels in the two high mountain plant species *Soldanella alpina* and *Ranunculus glacialis*. J Exp Bot 54:405–418. https://doi.org/10.1093/jxb/erg048

Su X, Gao X, Shang L, Li Z, Zhao W, Chi B, Sun J (2021) Cytotoxicity of mongolian medicine *Aconitum kusnezoffii*. Chin Med 12:10–18

Tavares WR, Seca AML (2018) The current status of the pharmaceutical potential of *Juniperus* L. metabolites. Medicine 5(3):81. https://doi.org/10.3390/medicines5030081

Trivedi VL, Nautiyal MC (2021) Antioxidant properties of alpine plant. IntechOpen. https://doi.org/10.5772/intechopen.96735. https://www.intechopen.com/online-first/antioxidant-properties-of-alpine-plant

Trivedi VL, Attri DC, Bahukhandi A, Nautiyal MC, Sati J (2022) Seabuckthorn (*Hippophae salicifolia*) leaves, a good source of natural antioxidant compounds. Natl Acad Sci Lett 45(2):195–198

Valko M, Leibfritz D, Moncol J, Cronin MT, Mazur M, Telser J (2007) Free radicals and antioxidants in normal physiological functions and human disease. Int J Biochem Cell Biol 39:44–84. https://doi.org/10.1016/j.biocel.2006.07.001

Wang Y, He W, Huang H, An L, Wang D, Zhang F (2009) Antioxidative responses to different altitudes in leaves of alpine plant *Polygonum viviparum* in summer. Acta Physiol Plant 31:839–848. https://doi.org/10.1007/s11738-009-0300-9

Wang S, He H, Chen L, Zhang W, Zhang X, Chen J (2015) Protective effects of salidroside in the MPTP/MPP(+)-induced model of Parkinson's disease through ROS-NO-related mitochondrion pathway. Mol Neurobiol 51:718–728. https://doi.org/10.1007/s12035-014-8755-0

Wang X, Li S, Ma J, Wang C, Chen A, Xin Z, Zhang J (2019) Effect of gastrodin on early brain injury and neurological outcome after subarachnoid hemorrhage in rats. Neurosci Bull 35:461–470

Wang M, Xing S, Jia J, Zeng W, Lei J, Qian Y, Huang Z (2023) Angelicin impedes the progression of glioblastoma via inactivation of YAP signaling pathway. Biomed Pharmacother 161:114462

Xia Q, Gao S, Rapael Gnanamuthu SR, Zhuang K, Song Z, Zhang Y, Wang X, Tu P, Li J, Liu K (2021) Involvement of Nrf2-HO-1/JNK-Erk signaling pathways in aconitine-induced developmental toxicity, oxidative stress, and ROS-mitochondrial apoptosis in zebrafish embryos. Front Pharmacol 12:642480

Yao Y, Lin G, Xie Y, Ma P, Li G, Meng Q, Wu T (2014) Performulation studies of myricetin: a natural antioxidant flavonoid. Pharmazie 69:19–26

Yatoo MI, Dimri U, Gopalakrishnan A, Karthik K, Gopi M, Khandia R, Saminathan M, Saxena A, Alagawany M, Farag MR, Munjal A (2017) Beneficial health applications and medicinal values of Pedicularis plants: a review. Biomed Pharmacother 95:1301–1313

Yu J, Xia J, Xu J, Chen S, Zhang Y, Yin F, Fang J, Cai L, Zhang B, Zhan Y, Zhang X (2023) Chemical profile, anti-hepatoma activity, anti-acetylcholinesterase and antioxidant activity of aerial part of *Aconitum carmichaeli* Debx. Nat Prod Res 37(22):3884–3888

Yuan C, Shin M, Park Y, Choi B, Jang S, Lim C, Yun HS, Lee IS, Won SY, Cho KS (2021) Linalool alleviates Aβ42-induced neurodegeneration via suppressing ROS production and inflammation in fly and rat models of Alzheimer's disease. Oxidative Med Cell Longev 2021(1):8887716

Zhang Y, Chen S, Fan F, Xu N, Meng XL, Zhang Y, Lin JM (2023) Neurotoxicity mechanism of aconitine in HT22 cells studied by microfluidic chip-mass spectrometry. J Pharm Anal 13(1):88–98

Zhang M, Ji X, Li Y, Chen X, Wu X, Tan R, Jiang H (2024) *Anthriscus sylvestris*: an overview on bioactive compounds and anticancer mechanisms from a traditional medicinal plant to modern investigation. Mini-Rev Med Chem 24(12):1162–1176. https://doi.org/10.2174/0113895575271848231116095447

Zheng X, Long W, Liu G, Zhang X, Yang X (2012) Effect of seabuckthorn (*Hippophae rhamnoides* ssp. sinensis) leaf extract on the swimming endurance and exhaustive exercise-induced oxidative stress of rats. J Sci Food Agric 92(4):736–742

Zhu H, Xu W, Wang N, Jiang W, Cheng Y, Guo Y, Yao W, Hu B, Du P, Qian H (2021) Anti-fatigue effect of *Lepidium meyenii* Walp. (Maca) on preventing mitochondria-mediated muscle damage and oxidative stress in vivo and vitro. Food Funct 12(7):3132–3141. https://doi.org/10.1039/d1fo00383f. Epub 2021 Mar 17. PMID: 33729250

8 Traditional Use, Phytochemistry and Pharmacological Potential of *Crocus sativus* and *Emblica officinalis*: High Value Medicinal Plants from North Western Himalayan Mountains

Jaspreet Kour

Abstract

One of the 36 worldwide biodiversity hotspots, the Himalaya is well-known for its benefits on both an ecological and financial basis. Indian Himalayan region covers about 16.2% of India's land area, 100% of the alpine cover, and roughly 32.1% of the nation's forests. India is one of the world's largest exporters of raw herbal medications, and the Himalayas are renowned for their vast variety of medicinal plants. Jammu & Kashmir, located in the North-Western Himalayas, is gifted with a remarkable array of medicinal plants. Among the many plants found in this beautiful state, *Emblica officinalis* and *Crocus sativus* have been used for thousands of years in traditional medical systems. The fruit of *Emblica officinalis* is rich in tannins, flavonoids, pectin, gallic acid, quercetin, and other polyphenolic compounds. Numerous phytochemicals, including tannins, alkaloids, flavonoids, terpenoids, and carbohydrates, have been shown to produce beneficial biological effects. On the other hand, the stigmas of *Crocus sativus* have been widely used due to their diverse chemical components, including crocetin, crocin, and other flavonoids. These components offer a range of medicinal benefits, such as anticancer, antioxidant, anti-inflammatory, anticonvulsant, anti-Alzheimer's, and antitussive properties, which help treat various ailments. This chapter focuses on the traditional uses, phytochemistry, and pharmacological properties of these medicinal plants.

J. Kour (✉)
Dendrimers and Heterochemistry Team, Laboratoire de Chimie de Coordination, UPR CNRS 8241, Toulouse, France

S. Chandra, H. Lata (eds.), *High Altitude Medicinal Plants*,
https://doi.org/10.1007/978-3-031-99406-7_8

Abbreviations

IHR	Indian Himalayan Region
WHO	World Health Organization
TMS	Traditional System of Medicine
EO	Emblica officinalis
GSH	Glutathione
SOD	Superoxide Dismutase
CAT	Catalase
GPX	Glutathione Peroxidase
MAO	Monoamine Oxidase
HDL	High-Density Lipoprotein
LDL	Low-Density Lipoprotein
LPO	Lipid Peroxide
EOT	Tannoids of *E. officinalis*
ESRD	End-Stage Renal Disease
OTA	Ochratoxin A
NK	Natural Killers
DNA	Deoxyribonucleic Acid
RNA	Ribonucleic Acid
EEO	Ethanol Emblica Officinalis
AEO	Aqueous Emblica Officinalis
HAEEO	Hydroalcoholic extract of Emblica Officinalis
ROS	Reactive Oxygen Species
AD	Alzheimer's Disease
AChE	Acetyl Cholinesterase
STZ	Streptozotocin
CCl_4	Carbon Tetrachloride

8.1 Introduction

Traditional herbal remedies can be utilized to combat a wide range of ailments with minimal to no adverse effects. They have made a significant impact on health systems worldwide. Herbal plants are employed as natural remedies for various ailments, including pharyngitis, asthma, diabetes mellitus, heart disease, wound healing, and tuberculosis. Bioactive phytochemicals found in these plants, such as alkaloids, flavonoids, tannins, and polyphenols, contribute to their therapeutic properties. Over the past few decades, many emerging countries have increased research into and usage of herbal medicines. According to the World Health Organization (WHO), traditional medicines, including herbal drugs, are recognized as forms of treatment that have been used for centuries and remain in practice today. The development of herbal medicines is grounded in traditional healing practices that involve therapeutic plant mixtures. India, long known for its rich diversity of medicinal plants, is considered a hub of "living traditions" in herbal medicine. The traditional system of medicine (TMS) aims to prevent or alleviate illnesses, often through various herbal formulations made from medicinal plants.

Siddha, homeopathy, and traditional Chinese medicine (TCM) practitioners have developed highly effective herbal formulations with few side effects, using plants from the Himalayan region. The Himalayas are a rich source of information about plant-based medicine. The diverse topography, soil types, rainfall, and weather conditions of the Himalayan Forest systems contribute to the growth of unique and valuable flora. The vegetation in the Himalayas changes with altitude, ranging from tropical deciduous forests at lower elevations to temperate forests at mid-altitudes. At higher elevations, one can find coniferous, subalpine, and alpine forests, transitioning to alpine meadows, grasslands, scrublands, and permanent snow lines. The flora exhibits increasing diversity with higher altitudes, and the region is home to a wide variety of aromatic, medicinal, and wild food plants, along with orchids, ferns, bamboo, and ornamental species. The Himalayan forests are a primary source of medicinal plants that are vital to the livelihoods of local communities. Plants from higher altitudes tend to have a rich variety of phytochemicals due to the challenging environmental conditions, including higher radiation, poor soil, moisture stress, and extreme temperatures. These plants play a crucial role in supporting the wellbeing of people worldwide.

This chapter discusses two well-known medicinal plants from the Indian North Western Himalayan region, *Emblica officinalis* and *Crocus sativus,* along with their extracts and associated pharmacological properties, such as antioxidant, anti-diabetic, antimicrobial, and anticancer effects. The goal of this chapter is to provide a thorough scientific review of the key phytochemicals in these plants and their pharmacological actions, potentially contributing to the development of novel ethnomedicines in the future.

8.2 Role of Medicinal Plants to Economic Development

Researchers have documented 1123 plant species with medicinal applications in Jammu and Kashmir, which are also a significant component of the rich biodiversity of the Himalayan region. In addition to being utilized in the conventional medicinal system to treat a variety of illnesses, these medicinal plants also give an opportunity to move up on the socioeconomic scale for households. The socioeconomic profile of the local population suggests that these individuals lead challenged lives. Medicinal plants of great commercial value, such as *Emblica officinalis* and *Crocus sativus*, among others, can be used as a source of income. These plant species are rich in phytochemicals and can help create different herbal remedy compositions. *Emblica officinalis* has produced a significant socioeconomic influence in a significant portion of the country. Through commercial orcharding, nursery production, and processing of herbs with consideration for fruit size, maturity period, and yield per plant, it has made a substantial contribution to improve the economic position of farming communities.

Additionally, saffron possesses an orange stigma that, when combined with style-tops, produces saffron of commerce. It is regarded as the most valuable therapeutic food that can be eaten or taken as medication. One of the priciest and oldest spices in the world is saffron. It is estimated that between 225,000 and 500,000 dried stigmas make up one pound of saffron, meaning that 75,000 flowers must be manually picked. Many career opportunities arise from the cultivation and processing of saffron, especially for women in the Jammu and Kashmir region who perform 80% of the labor-intensive jobs such as harvesting, refining, and packing. Due to its numerous unique qualities and applications, as well as the substantial economic impact it exerts on farmers in the Jammu and Kashmir region, saffron requires specific attention in order to resolve production, export, and retailing difficulties. In addition to providing jobs, it may discourage people from moving from rural to urban regions. In contrast to other agricultural products, it has the potential to generate enormous profits for both farmers and the country. Saffron is important for the food industry, the economy, healthcare, and employment creation. In 226 villages in J&K, over 30,000 families are being convinced to cultivate saffron. The only sole state in the nation where saffron is farmed is Jammu & Kashmir, which ranks as the second-largest saffron-producing region in the world.

8.3 *Emblica officinalis*

Emblica officinalis usually referred as Indian gooseberry, Amla or Phyllanthus Emblica is a medicinal plant in the Indian traditional system of medicine, the Ayurveda which is a member of large genus comprising about 550–750 species and family Euphorbiaceae. According to mythology originating in ancient India, it is thought to be the universe's first tree. Originally India is the species' habitat of origin. and also found growing in South East Asia, China, Pakistan, Uzbekistan, Malaysia, Sri Lanka, and other tropical a well as subtropical regions. Two types of Amla are present: the wild (vanya) and the cultivated (gramya). *Emblica officinalis* is a small to medium-sized deciduous tree with a maximum height of 8–18 m. The leaves of EO are simple, fluffy, light green, oblong, subsessile, and resemble pinnate leaves. They are three to five inches long. Their branches extend out and their trunk is crooked. The branchlets are 10–20 cm long, usually deciduous, and glabrous or slightly pubescent. The axillary fascicles contain the unisexual flowers. There are many yellow male flowers with three stamens, an oblong shape with an entire edge, an obtuse apex, and short, slender pedicels. The anthers are oblong and upright. Less flowers, sub-sessile, rectangular to spathulate, and with three stigmas make up the female species. The ovary has three loci or trilocular (Shanmugasundaram et al. 1983; Unander et al. 1990). Different growing stages of this plant or phenology can be studied from October to March. The fruit has six vertical stripes or furrows and is almost spherical, soft, and hard to the touch. Its color is pale greenish yellow. When fruit is young, it is green; when it reaches maturity, it turns yellow. The fruit is typically 20–25 g in weight. Its wood is reddish and its bark is gray. Its wood has a challenging texture. The fruit is made up of mesocarp and epicarp in addition to pericarp. The single layer of oblong epidermal cells with a cuticle covering them makes up the epicarp. The mesocarp, which is composed of polyhedral parenchymal cells, is the largest portion of the fruit. Vascular bundles that are dispersed throughout the mesocarp are made up of xylem and phloem elements. Except for completely sandy soil, amla can be grown well in both light and heavy soils; calcareous soil with a rocky substrate can also be beneficial. For a better and larger yield, fertile loamy soil that drains effectively is the ideal type. The plant may grow successfully in soils that are somewhat alkaline and can adapt to arid environments. Additionally, it grows exceptionally well in tropical climates where the plant requires an annual rainfall of between 650 and 800 mm. Amla has demonstrated a significant output under these settings. The plants require sufficient protection from the hotter May and June winds as well as from frost because they are not extremely frost hardy after they reach an average height of three feet or more. An Amla tree that is well-grown and mature can withstand freezing temperatures as well. During summer periods, the amla plants need to be watered regularly for around 20 days until they get well established. After that, no irrigation is recommended until the very beginning of the monsoon season. Mature fruit-bearing plants should be watered every 2 weeks during the height of summer to increase fruit production and decrease fruit drop. February is the ideal month to harvest because it is when the fruits are fully mature and contain the most ascorbic acid. Fruit is available all year round in South India, and when it reaches maturity, it is sufficiently hard to fall from the tree without causing the tree to tremble.

8.3.1 Chemical Constituents of *Emblica officinalis*

E. officinalis is one of the most explored medicinal plants and outcomes shown that it includes phenolic compounds, alkaloids, and tannins. According to findings, the fruits of *E. officinalis* have larger amounts of most minerals, protein, and amino acids, including glutamic acid, proline, aspartic acid, alanine, cysteine, and lysine, as well as a higher quantity of vitamin C. There is more ascorbic acid or vitamin C in them than in oranges, tangerines, or lemons (Scartezzini et al. 2006; Jain and Khurdiya 2004; Poltanov et al. 2009). Higher concentrations of hydrolyzable tannins, such as emblicanin A and B, punigluconin, and pedunculagin, are found in the fresh pericarp of *E. officinalis*. The phytochemicals found in *E. officinalis* were discovered by an activity-directed fractionation and purification procedure. Using chromatography and spectroscopy, they have determined the presence of gallic acid, methyl gallate, corilagin, furosin, and geraniin in *E. officinalis*. According to phytochemical studies, *E. officinalis* has higher concentrations of flavonoids including quercetin. Fruits' alkaloidal content was also examined. IR spectrum investigations and chromatography were used to demonstrate the presence of alkaloids such as phyllantine and phyllantidine. Gallic acid, ellagic acid, chebulinic acid, chebulagic acid, emblicanin A, emblicanin B, punigluconin, pedunculagin, citric acid, ellagotannin, trigallayl glucose, pectin, 1,6-di-O-galloyl-β-D-glucose, corilagin, 1,6-di-O-galloyl-β-D-glucose, 3 ethylgallic acid (3 ethoxy 4,5 dihydroxy benzoic acid), and

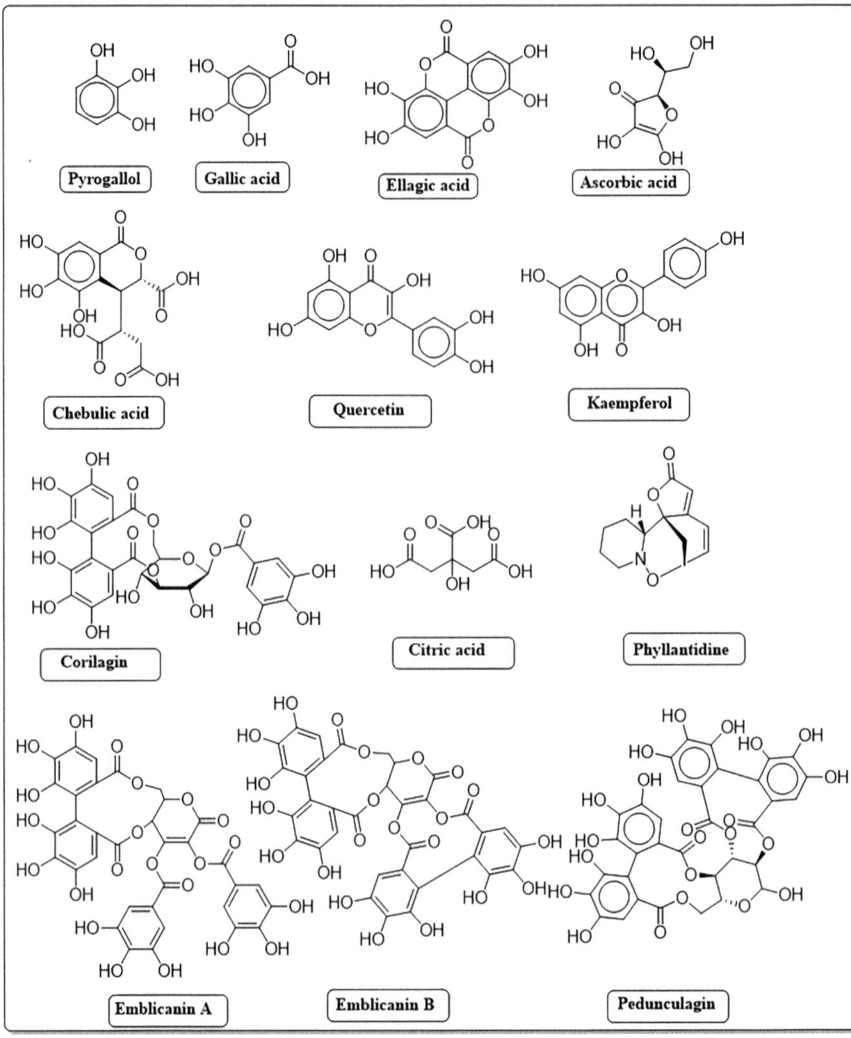

Fig. 8.1 Structures of chemical constituents present in fruit of *Emblica officinalis*

isostrictiniin are also present in the fruit. (Zhang et al. 2001a, 2004; Nisha et al. 2004; Kumaran and Karunakaran 2006; Bajpai et al. 2005; Bhattacharya et al. 2002; Anila and Vijayalakshmi 2002). It also contains flavonoids such as quercetin, kaempferol 3 O-α-L (6″ methyl) rhamnopyranoside and kaempferol 3 O-α-L (6″ ethyl) rhamnopyranoside shown in (Fig. 8.1). *Embilica Officinalis* roots have tannins and glycosides in them (Zhang et al. 2000, 2001b). The fruit pulp of Embilica Officinalis has the following average percentage composition: Nicotinic acid (0.2 mg/100 gm), Vitamin C (600 mg/100 gm) (81.2%), protein (0.5%), fat (0.1%), mineral matter (0.7%), fiber (3.4%), carbohydrate (14.1%), calcium (0.05%), phosphorous (0.02%) and Iron (1.2 mg/100 gm). When compared to apples, the edible fruit tissue has three times the protein content and 160 times the ascorbic acid content. It also has a significantly larger concentration of most minerals and amino acids. Of the total amino acids, glutamic acid, proline, aspartic acid, alanine, and lysine make up 29.6%, 14.6%, 8.1%, 5.4%, and 5.3%, respectively (Krishnaveni and Mirunalini 2010).

8.3.2 Molecular Structure of *Emblica officinalis*

8.3.3 Traditional Uses of *Emblica officinalis*

E. officinalis is used in the Siddha, Unani, Tibetan, Sri Lankan, and Chinese medical systems. *E. officinalis* is thought to be a potent rasayana (rejuvenator), helpful in postponing both the senescence and degenerative processes. It is well-known for its use in Chyawanprash, an Indian herbal concoction or polyherbal jam, and Triphala, a blend of dried powdered fruits of *E. officinalis*, *Terminalia bellirica*, and *Terminalia chebula*. It is thought that Maharshi Chyavan created the chyawanprash recipe after his marriage to a young woman. It helps in constipation relief, digestion improvement, and longevity. According to the ayurvedic medical system, it also reduces fever, purifies the blood, lessens coughing, soothes asthma, strengthens the heart, is good for the eyes, promotes hair development, energizes the body, and enhances intelligence. The astringent fruits are used in many traditional medicines to cure a variety of conditions, including gastritis, hyperacidity, constipation, colitis, dyspepsia, hemorrhoids, hematuria, menorrhagia, asthma, anemia, diabetes, cough, osteoporosis, premature graying of the hair, weakness, and exhaustion. Moreover, hepatoprotective, cardioprotective, laxative, refrigerant, stomachic, diuretic, restorative, alterative, antipyretic, and anti-inflammatory qualities are claimed for *E. officinalis*. In addition to its use as a hair tonic, *E. officinalis* is a digestive aid that helps avoid peptic ulcer dyspepsia (Baliga and Dsouza 2011; Calixto et al. 1998; Vasudevan and Parle 2007).

8.3.4 Pharmacological Importance of *Emblica officinalis*

8.3.4.1 Antioxidant Activity

Natural antioxidants and their function in preserving human health and preventing disease have garnered increased attention nowadays. In general, antioxidants keep a balance between the creation of free radicals and their scavenging in healthy persons. Depletion of antioxidants causes the generation of free radicals to rise and oxidative stress, which in turn causes several disruptions throughout the organism. In fact, oxidative stress causes oxidation at the molecular and cellular levels. It also damages cells and tissues, which is linked to a host of illnesses, including cancer, metabolic, inflammatory, neurological, and cardiovascular diseases (Aprioku 2013). Together with a variety of dietary components, nature has endowed us with shielding antioxidant mechanisms such as catalase, glutathione (GSH), reductase, superoxide dismutase, GSH peroxidases, vitamin C, and vitamin E (tocopherols and tocotrienols). It is well recognized that natural food ingredients, such as Indian spices and medicinal plants, have antioxidant properties. The antioxidant activity of the tannoid active principles of *E. officinalis*, consisting of emblicanin A (37%), emblicanin B (33%), punigluconin (12%), and pedunculagin (14%), was investigated based on their effects on the concentrations of the oxidative free radical scavenging enzymes, superoxide dismutase (SOD), catalase (CAT), and glutathione peroxidase (GPX), as well as lipid peroxidation in terms of thiobarbituric acid-reactive products, in the rat brain frontal cortical and striatal. The outcomes were contrasted with those caused by deprenyl, a monoamine oxidase (MAO) B inhibitor that is selective and has a strong track record of antioxidant efficacy. When given once daily for 7 days, the active tannoids of *E. officinalis* (EOT) at doses of 5 and 10 mg/kg, i.p., and deprenyl (2 mg/kg, i.p.) caused a decrease in lipid peroxidation and an increase in SOD, CAT, and GPX activity in the frontal cortical and striatal brain areas. The effects of acute single dosing of deprenyl and EOT were negligible. The findings also suggest that rather than vitamin C itself, the antioxidant property of *E. officinalis* may be found in the tannoids, which are the plant's fruits and have qualities similar to those of vitamin C. The results of the experiment showed that the low molecular weight hydrolysable tannins (gallo-ellagi) (65–70%) were responsible for the fruits' strong vitamin C-like activity. From the fresh pericarp, four such compounds—emblicanin A, emblicanin B, punigluconin, and pedunculagin—were extracted, and spectroscopic investigations and chemical transformations were used to determine their chemical structures. Additionally, it has been demonstrated that emblicanins A and B have a strong antioxidant action in vitro. It was shown that the *Emblica officinalis* extract, which is high in emblicanin A and B, considerably raises the concentrations of the antioxidant enzymes SOD, CAT, and GPX in the cortical and striatal regions of the rabbit brain while also lowering lipid peroxidation (Bhattacharya et al. 1999; Adil et al. 2010).

8.3.4.2 Anti-inflammatory Activity

Limited anti-inflammatory effects are shown in animal models of acute and chronic inflammation, with reduced granulomatous tissue and edema at higher dosages (Dang et al. 2011; Muthuraman et al. 2011). There are no clinical research results available on the potential of Emblica as an anti-inflammatory drug. Nicolis et al. 2008 conducted in vitro experiments using bronchial epithelial cells obtained from a cystic fibrosis patient, which revealed a suppression of pro-inflammatory cytokine expression. Further, in vitro studies utilizing donor cartilage from osteoarthritic patients showed that dried emblica fruit extract inhibited the enzymes hyaluronidase and collagenase (Sumantran et al. 2008). Additionally, crude extracts have caused mature osteoclasts to undergo apoptosis without influencing osteoclastogenesis; this may have happened through the modification of transcription factor (Piva et al. 2009; Penolazzi et al. 2008). These days, a number of research findings suggest that hydroalcoholic extract of *Emblica officinalis* (HAEEO) has strong anti-inflammatory properties and may 1 day be used therapeutically to treat both acute and chronic inflammatory diseases.

8.3.4.3 Antimicrobial Activity

Both alcoholic and aqueous extracts of Emblica demonstrated efficaciousness against common human infections such as viruses, bacteria, and fungi. The efficacy against fungi is modest, and the action against gram-positive bacteria seems to be higher (Rani and Khullar 2004; Dutta et al. 1998; Rahman et al. 2009; Liu et al. 2009b). The content of the phenol has been associated with activity against herpes simplex viruses 1 and 2 (Xiang et al. 2011). While phyllaemblicin B separated from the plant's roots showed efficacy against the coxsackie virus (Liu et al. 2009a; Wang et al. 2009). Mice with induced Klebsiella pneumoniae responded, during a 30-day period, to dietary supplementation of powdered fruit. In the short term, however, colonization was not stopped (15 days) (Saini et al. 2008).

8.3.4.4 Antidiabetic Activity

Numerous research investigations and experiments have demonstrated amla's potential as a promising drug for decreasing blood glucose levels in both healthy individuals and diabetic patients. An age-old herb called amla has been utilized in traditional medicine to treat a range of illnesses, including diabetes. Regular foods such amla, garlic, onions, turmeric, and others have a beneficial effect on lowering blood sugar levels by blocking the production of glucose and glycogenolysis. A daily dose of two to three grams of amla powder can effectively lower LDL cholesterol and raise HDL cholesterol. Similarly, amla fruit has been used to treat diabetes patients who are developing neuropathy. Numerous studies have shown that amla's high vitamin C concentration is beneficial in regulating blood glucose levels, which in turn helps people manage their diabetes. Indian medicine makes extensive use of the methanolic extract (75%) of Terminalia chebula, Terminalia belerica, EO, and their combination known as "Triphala" (equal share of above three plant extracts). They were found to scavenge hydroxyl and superoxide radicals in vitro and to limit the development of lipid peroxide (LPO). Plant extracts with concentrations of 85.5, 27, 74, and 69 μg/ml were shown to prevent 50% of lipid peroxidation precipitated with Fe $(^{2+})$/ascorbate, respectively. The desired concentrations to limit hydroxyl radical scavenging were found to be 165, 71, 155.5, and 151 μg/ml, as well as 20.5, 40.6, 6.5, and 12.5 μg/ml for superoxide scavenging activity, respectively. Within 4 h, oral treatment of the extracts (100 mg/kg body weight) significantly reduced the blood sugar level in both common and alloxan (120 mg/kg) diabetic rats. Daily, continuous medication administration resulted in a long-lasting effect (Sabu and Kuttan 2002).

8.3.4.5 Anticancer Activity

The possibility of using Emblica to cure cancer is a topic of great interest, yet no clinical trials or epidemiological data have been published. Rats administered Emblica extracts demonstrated a decrease in chromosomal abnormalities, a rise in damaged cells, a decrease in the frequency of micronuclei in bone marrow cells, a decrease in free radical production, and an increase in cell survival in response to heavy metal carcinogens (arsenic, chromium, and nickel) (Biswas et al. 1999; Sai Ram et al. 2002, 2003; Dhir et al. 1991). In response to tumor cells (lymphoma and mammary carcinoma), rats given Emblica extracts demonstrated increased natural killer cell activity, antibody-dependent cellular cytotoxicity, and survival (Jeena et al. 1999; Suresh and Vasudevan 1994; Veena et al. 2006). One of these tests demonstrated a decrease in tumor volume but no effect on tumor development (Biswas et al. 1999). It has been proven that hydrolyzable tannins and organic acid gallates are cytotoxic to tumor cells (Zhang et al. 2004). According to one published study, emblica had no effect on the characteristics associated with lung cancer in mice (Menon et al. 1997). Mice exposed to radiation were shielded from radiation illness by emblica extracts, which also improved their 30-day survival rate and reduced overall mortality (Singh et al. 2005; Jagetia et al. 2002; Jindal et al. 2009). There is no clinical study information available addressing the use of Emblica in malignant diseases. Nonetheless, research has been done in vitro with human cancer cell lines, such as those from the lung, liver, cervical, ovarian, and breast (Sharma et al. 2000; Arora et al. 2003; Yang et al. 2009; Ngamkitidechakul et al. 2010). Different emblica preparations reduced the risk of hepatocarcinogenesis based on liver damage indicators, enzyme levels, and tumor incidence (Jose et al. 2001; Pinmai et al. 2008; Sultana et al. 2008; Jeena et al. 1999).

8.3.4.6 Hepatoprotective Activity

In rat tests, alcoholic and aqueous extracts of *Emblica officinalis* fruits revealed hepatoprotective effects. An extract of the fruit of *Emblica officinalis* that is 50% hydroalcoholic lessen the degree of hepatic fibrosis caused by thioacetamide and carbon tetrachloride (*Emblica officinalis*-50). *Emblica officinalis*-50 successfully reversed the fibrogenic processes because of its antioxidative action. Ascorbic acid, tannins, and flavanols—phenols found in amla—protect against liver cell disruption and harmful effects. Amla water extract reduces lipid oxidation, increases serum glutamic oxaloacetic transaminase and pyruvic transaminase levels, and positively affects the liver and nephron in oxidative stress caused by ochratoxin A (OTA) in all important organs. Chronic CCl_4 and thioacetamide treatment is the source of the abnormality in the histology showing pre-fibrogenic events. Amla possesses the ability to counteract these changes with significant regenerative advancements, demonstrating its protective function in the pre-fibrogenesis of the liver. Hepatotoxic substances come in a wide range of forms, including heavy metals, ochratoxins, ethanol, paracetamol, carbon tetrachloride, hexachlorocyclohexane, and anti-tubercular medications. Along with reducing hyperlipidemia, metabolic syndrome, hepatocellular cancer, and hepatotoxicity brought on by iron overload, the phytochemical components found in amla, such as gallic acid, quercetin, ellagic acid, and corilagin have also been shown to be hepatoprotective against such agents. The primary phytochemical constituent's quercetin, gallic acid, corilagin, and ellagic acid were also found to have hepatoprotective properties opposing the toxicity of lipopolysaccharide, paracetamol, micro cystins, and galactosamine. These studies showed protective and restorative effects that were enzymatic and/or histological (Tasduq et al. 2005a, b; Haque et al. 2001; Jose and Kuttan 2000; Sultana et al. 2004; Pramyothin et al. 2006; Verma and Chakraborty 2008; Chakraborty and Verma 2010; Sharma et al. 2009; Reddy et al. 2009). There are no clinical studies on the use of Emblica in renal disease; nevertheless, Emblica decreased oxidative stress indicators in individuals with uremia as a result of end-stage renal disease (ESRD). Hepatic, renal, and diabetes indices showed no change (Chen et al. 2009).

8.3.4.7 Hypolipidemic Activity

A number of animal investigations that reveal improved lipid profiles (Anila and Vijayalakshmi 2000; Kim et al. 2010; Yokozawa et al. 2007). Flavonoid extracts from Emblica fruits increased hepatic HMG-CoA reductase, which in turn prevented the synthesis of cholesterol and promoted its breakdown (Anila and Vijayalakshmi 2002). When rats with generated metabolic syndrome were given fresh Emblica fruit juice, their induced aortic plaques regressed to almost normal, serum and tissue lipid levels dropped, and the ethanol extract of Emblica enhanced the lipid profile and decreased hypertension (Kim et al. 2010). In a human clinical research, serum cholesterol levels decreased in hypercholesterolemic, healthy men (35–55 years old) who took Emblica supplementation for 28 days. When the supplement was stopped, the condition was reversible (Jacob et al. 1988). Two to three grams of powdered, dried Emblica fruit per day, both in individuals with and without type 2 diabetes, improved the lipid profile at 21 days (higher-density lipoprotein; lowered total cholesterol, low-density lipoprotein, and triglycerides). At the 3 g daily dose, total lipids decreased only in the participants who had diabetes (Akhtar et al. 2011).

8.3.4.8 Gastroprotective Activity

Although phytochemical studies have primarily focused on preventing cancer, a number of plant-based products are being investigated or utilized to treat a range of gastrointestinal issues (Romano 2014). Beyond its culinary value, Phyllanthus emblica has the potential to be employed as a gastroprotective agent in cases of gastropathy caused by non-steroidal anti-inflammatory drug use. It has been proposed that Phyllanthus emblica's antioxidative quality is the secret to its medicinal benefits (Chatterjee et al. 2011). Furthermore, rats with indomethacin-induced ulceration showed increased production of stomach mucus and hexosamine after pretreatment with a butanol extract of the aqueous fraction of Phyllanthus emblica fruits (Sairam et al. 2002). The methanolic extract of *E. officinalis* was also evaluated for its ability to prevent ulcers in various acute gastric ulcer models in rats. In rats, acetic acid was used to cause chronic stomach ulcers, whereas aspirin, ethanol, cold restraint stress, and pyloric ligation were used to cause ulcers. In all of the aforementioned acute ulcer models, *E. officinalis* demonstrated dose-dependent protective effects against ulcers in addition to a notable healing effect after 5–10 days of treatment. The findings demonstrated that *E. officinalis* had a strong anti-ulcer and healing effect, which may have been brought about by its effects on both defensive and offensive mucosal components. A study by (Bandyopadhyay et al. 2000) aimed to provide the likely pathways for Phyllanthus emblica's medicinal application in diarrhea. The findings suggest that the fruit extract of Phyllanthus emblica has antidiarrheal and spasmolytic properties, which may be mediated by dual blocking of muscarinic receptors and Ca($^{2+}$) channels. This explains the fruit extract's medical application in the treatment of diarrhea (Mehmood et al. 2011).

8.3.4.9 Immunostimulant

Numerous plants exhibit immunostimulatory properties (Kumar et al. 2011). It has been discovered that *E. officinalis*, a great source of vitamin C (ascorbate), enhances the activity of natural killer (NK) cells and antibody-dependent cellular cytotoxicity. Splenic NK cell activity was induced to proliferate double by *E. officinalis*. In tumor-bearing mice treated with *E. officinalis*, a 35% increase in life duration was seen (Suresh and Vasudevan 1994). Furthermore, employing chromium (VI) as an immunosuppressive agent (Sai Ram et al. 2002), examined the anti-oxidant and immunomodulatory qualities of *E. officinalis*. Chromium is responsible for cytotoxicity, the generation of free radicals, lipid peroxidation, reduced GPx activity, and decreased GSH levels. The proliferation of lymphocytes induced by lipopolysaccharide and concanavalin-A was likewise markedly reduced. Inhibiting concanavalin-A induced interleukin-2 and gamma-interferon synthesis was another effect of chromium. Cr increased DNA fragmentation and apoptosis. *E. officinalis* effectively decreased the generation of free radicals generated by Cr and returned the antioxidant capacity to the control level. Moreover, *E. officinalis* prevented DNA fragmentation and apoptosis brought on by Cr. It is interesting to note that *E. officinalis* significantly restored the production of gamma-IFN and IL-2 while also mitigating the immunosuppressive effects of Cr on lymphocyte proliferation (Sai Ram et al. 2002). Despite its many therapeutic benefits, amla also exhibits immunostimulatory effects on lymphocyte activity, especially in immunosuppressive conditions. The vitamin C found in amla fruit pulp stimulates the immune system. Many items made from amla, such as syrups, dried extracts, tonics, etc., have the benefit of boosting immunity, giving you energy, and warding off illness. Because it enhances cryoprotection and reduces apoptosis and DNA disintegration, these compounds are also known as immunomodulators. Amalaki churna and Amalaki swaras fruit extracts from *E. officinalis* have been shown to have strong pharmacological effect as well as immunostimulant, immunomodulatory, and mild cytoprotective properties.

8.3.4.10 Analgesic/Antipyretic Effects of *Emblica officinalis*

The antipyretic analgesic and analgesic properties of alcohol and aqueous extracts of Emblica fruit were examined in mice. Except for emblica's lack of activity in the heat pain model, the results were comparable to those of aspirin (Perianayagam et al. 2004). The current study aimed to investigate the antipyretic and analgesic effects of ethanol (EEO) and aqueous (AEO) extracts of *Emblica officinalis* fruits utilizing a range of experimental models. Rats' brewer's yeast-induced hyperthermia was greatly decreased by a single oral dosage of EEO and AEO (500 mg/kg, i.p.). In the analgesic test, mice's acetic acid-induced writhing reaction was likewise markedly inhibited by EEO and AEO. The tail-immersion test revealed no appreciable analgesic efficacy from either AEO or EEO. These results imply that *Emblica officinalis* fruit extracts had strong analgesic and antipyretic properties. Alkaloids, tannins, phenolic compounds, amino acids, and carbohydrates were detected in the extracts during a preliminary phytochemical screening process. These substances may be in charge of the extracts' antipyretic and analgesic properties.

8.3.5 Other Uses of *Emblica officinalis*

8.3.5.1 Indigestion

It is recognized that green, fresh leaves combined with curds have a comparable outcome to the soft amla shoots used in buttermilk therapy for indigestion. Fruit has stomachic and carminative properties. It is known that mixing fresh green leaves with curds produces a comparable outcome to the mild shot used in buttermilk therapy for dyspepsia. Tannins, which are the main component of EO fruit and are highly astringent by nature, are extremely effective in treating intestinal conditions including dysentery and diarrhea. In tests using charcoal meals on rats, the methanol extract of EO demonstrated enormous inhibitory effects on magnesium sulfate- and castor oil-induced diarrhea as well as a decrease in gastrointestinal motility. Additionally, it reduced the rat's production of prostaglandin E2. In piles, amla is frequently used as a laxative to treat constipation (Perianayagam et al. 2005).

8.3.5.2 Skin Protection

Studies employing human skin fibroblasts in vitro have shown enhanced collagen formation and cell proliferation at wound and ultraviolet B light (UBV) photo-aged sites (Sumitra et al. 2009; Fujii et al. 2008; Adil et al. 2010). These investigations have been carried out in conjunction with animal research. As an alternative to hydroquinone, Emblica extracts have been the subject of a small number of clinical trials when combined with additional ingredients in skin-lightening creams (Costa et al. 2010). Amla appears to have the potential to energize and strengthen the body. Amla may also be beneficial to the health of the skin and hair. Amla may therefore contribute to youth, health, and vitality in some way. Although more investigation is need to support these assertions. Indian gooseberry includes an ingredient that is particularly beneficial in avoiding aging and retaining strength in old age, which is why it has revitalizing effects. It strengthens the body's defenses against infection and increases resistance. It fortifies the body's various glands, heart, and hair.

8.3.5.3 Ophthacare/Eye Disorder

One ingredient in a blended herbal eye drop composition called Emblica (Ophthacare) shown efficacy in treating minor infections and inflammatory eye diseases in a clinical research, although the study's methodology was of low quality (Biswas et al. 2001). Patients with a variety of ocular disorders, such as conjunctival xerosis (dry eye), conjunctivitis, acute dacryocystitis, degenerative conditions (pterygium or pinguecula), and postoperative cataract patients, participated in an open prospective multicenter clinical trial using a herbal eye drop preparation called Ophthacare®. This preparation contains the fundamental ingredients of several herbs that have been traditionally used in the Ayurvedic medical system for ages. These consist of *Rosa damascena, Ocimum sanctum, Curcuma longa, Carum copticum, Terminalia belirica, Emblica officinalis*, and *Melampodium paludosum*. It is said that these herbs have anti-inflammatory and anti-infective qualities. The goal of this study is to clarify how this herbal substance affects different types of eye conditions. During the study, no side effects were documented. After using the herbal eye drop for therapy, there was a noticeable improvement in most cases.

8.3.5.4 Oils and Spots

The pericarp of the fruit of amla is usually applied externally on boils with cow ghee to induce suppuration, and used in decoctions with other sections of the fruit.

8.3.5.5 Natural Cure for Anemia

EO (amla) is exceptionally abundant in ascorbic acid, often known as vitamin C, which is necessary for the absorption of iron. Amla supplements can be extremely helpful for people with iron deficiencies Anemia.

8.3.5.6 Insecticidal Activity

Essential components of EO (amla), saponins, have properties that make them either cytotoxic or insecticidal to specific types of insects. Despite the fact that saponins with insecticidal properties were previously gathered from conventional sources rather than EO (amla). However, since EO (amla) also contains saponins, which are bioactive substances, it is evident that EO (amla) may have insecticidal activity. Further testing might be done to provide a more thorough study (Chaieb 2010).

8.3.5.7 Larvicidal and Mosquitocidal

Analysis tests revealed that the methanol extract of EO (amla) had larvicidal and pupicidal effects against the malarial vector, Anopheles stephensi, with a 98% fatality rate at 100 ppm. The EO (amla) methanol and EEO extracts also demonstrated 100% mortality (no hatchability) at 400 ppm, and the EO (amla) ethyl acetate leaf extracts were shown to be larvicidal. The research found that, when compared to Aedes aegypti larvae, the ethyl acetate extract of EO (amla) demonstrated the highest larvicidal activity (99.6%) with a lethal concentration of 50% (lethal awareness results in 50% mortality) of 78.89 ppm (Jeyasankar et al. 2012).

8.4 *Crocus sativus*

Crocus sativus L. (saffron) also known as Red gold, Golden spice or "Kesar" is a spice that is highly valued worldwide and belongs to the Iridaceae family (Giorgi et al. 2017). The Arabic term Zafran, which signifies yellow, is the origin of the English word saffron (Winterhalter and Straubinger 2000). The main nations that grow saffron include India, Iran, Greece, Morocco, China, Italy, Turkey, France,

Switzerland, Pakistan, Japan, Spain, and Australia. Iran is the world's top producer of saffron, meeting around 80% of the global demand (Menia et al. 2018). India is the world's third-highest producer of saffron, behind Iran and Spain, yet its percentage of the worldwide market is a tiny 5%. Ninety percent of the country's saffron is grown in Jammu and Kashmir (J&K), mostly in the districts of Budgam and Pulwama (Tantry and Dar 2017). The rapid growth of *Crocus sativus* widespread cultivation was driven by the plant's ability to produce the stigmas, which are used to generate the costlier and well-known spice saffron. Saffron has around 150 volatile compounds that produced aroma, mostly terpenes, terpene alcohol, and their esters. Saffron is a widely used spice, coloring and flavoring component in culinary and cosmetic applications due to its distinctive color, scent and flavor. It also offers a lot of health advantages and has been used to both classical and modern medical techniques (Srivastava et al. 2010). *Crocus sativus*, commonly called as saffron, is a perennial herb that grows 10–25 cm tall from bulbs. The bulb varies in size and shape and has a sub-ovoid morphology. Its structure is massive and covered in several concentric spathes. From the apical buds of each mother bulb, one to three big daughter bulbs are produced, and from the lateral buds, several small bulbs (Zubor et al. 2004). Saffron has two different types of roots: contractile roots that grow at the base of the lateral buds and fibrous (absorbing) and thin roots at the base of the mother bulb (Khalesi et al. 2004). A great source of vitamins B1, B2, B6, and C are saffron roots. It is abundant in minerals, including iron, calcium, and magnesium. Further the roots of saffron are extensively used in medicine. There are five to eleven leaves per bud. They have a dark green shades and measure between 1.5 and 2.5 mm, making them extremely narrow. Their length ranges from 20 to 60 cm, and they have a rib on the outside and a white band inside. The purple flowers of *Crocus sativus*, which have six tepals total—three internal and three externals—start to bloom in the early fall around the end of September. The tepals meet at a long tube that protrudes from the upper part of the ovary. The flowers are protected when they first emerge by white membrane bracts. The pistil is made up of an inferior ovary that gives rise to a slender style that is between 9 and 10 cm long. The style terminates in a single stigma made up of three deeply red filaments that are longer than the tepals, the portion of the plant that is most fascinating to humans from an aesthetic point of view (Molina et al. 2004). The meristematization of parenchyma cells at the base of terminal, lateral buds, and the corm forms the contractile roots of *Crocus sativus*. The plant may thrive in low-nutrient soil, but it favors light (sandy) and medium (loamy) soils that drain well. The saffron plant and flower flourish in a variety of climates with different humidity levels, temperature ranges, and altitudes. Light precipitation in mid-September is helpful to start plant growth and encourage flowering; excessive rains during bloom have a negative impact on the quality of saffron. Corm formation and subsequent productivity depend on adequate soil moisture during early vegetative growth (November–December) and maximal photosynthetic activity (March–April) (Tantry and Dar 2017). The hillside plateaus and mountain valleys with an altitudinal range of 600–1700 m above sea level are the most favorable to the effective growth of this crop. The plant benefits from being able to survive in arid and semi-arid areas with severe summertime water shortages.

With an average elevation of 1850 m above sea level, the Kashmir valley nearly meets all the necessary geographical requirements for the successful production of saffron, earning it the distinction of being one of the few places in the world where high-quality saffron is grown. Saffron seeds are referred to as bulbs or corms. Saffron corms are sowed once a year for several years instead of annually. The slow-growing perennial saffron plant regenerates by the vegetative division of its subterranean corms. Depending on the size of the corm, the mother corm's surface has a variety of scars, which resemble buds covered with scaly leaves. The daughter corms at the base are produced by the aerial shots produced by the buds.

8.4.1 Chemical Constituents of *Crocus sativus*

Numerous studies on the phytochemical components of saffron have showed the existence of over 150 volatile and non-volatile chemicals. More than forty components make up the volatiles; the majority are monoterpenes, their alcohol and ester derivatives, with safranal (2,6,6-trimethyl1,3-cyclohexadiene-1-carboxaldehyde, accounting for 60–70% of the volatiles) being the principal constituent. The non-volatile constituents consist of flavonoids, crocetin, crocin, and picrocrocin (Mykhailenko et al. 2019). However, fewer than 50 constituents have been found thus far (Winterhalter and Straubinger 2000), Of these, three major constituents—safranal, picrocrocin, and crocetin glycosides (crocin)—are responsible for the characteristic spicy aroma, flavor, bitter taste, and yellow-orange color respectively (Fig. 8.2). A relatively rare carotenoid found in nature, crocin ($C_{44}H_{64}O_{24}$) is readily soluble in water (cis and trans glucosyl esters of crocetin). Because of its high solubility, crocin offers a broader range of applications as a food and medicine as

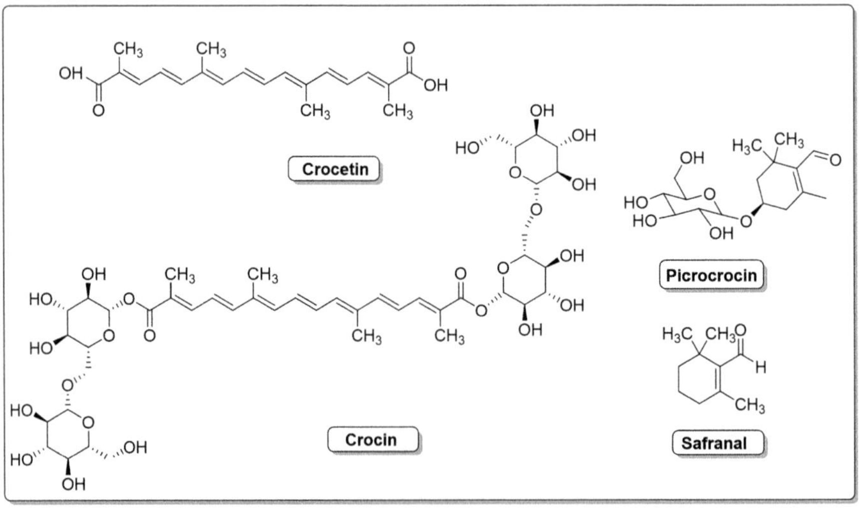

Fig. 8.2 Structures of biologically active chemical constituents present in *Crocus sativus*

compared to other carotenoids. The crocins' red coloration is due to the presence of long-conjugated system of the crocetin and can be detected by using UV-VIS spectra. Here UV-VIS spectra, which show a characteristic peak VIS absorbance in the 400–500 nm range, are useful for both qualitative and quantitative studies. Furthermore, crocetin and crocins only have stereochemistry derived from a single, strongly orientated double bond conformation, either trans or cis, with the trans form often being more stable than the cis form. The primary component influencing saffron's bitter flavor is the monoterpene aldehydes picrocrocin ($C_{16}H_{26}O_7$), which can crystallize by hydrolysis (Moghaddasi 2010). Fresh stigmas may or may not contain the deglycosylated derivative safranal (dehydro-β-cyclocitral, $C_{10}H_{14}O$), which gives saffron its fragrance and makes up 70% of the volatile component. The concentration of this derivative is dependent on how the saffron is dried and preserved. Apart from crocin and picrocrocin, the primary constituents of saffron are anthocyanins and flavonoids (such as kaempferol). It also contains significant amounts of starch, vitamins, amino acids, proteins, carbohydrates, gums, and minerals (Winterhalter and Straubinger 2000). Additionally, it contains a large number of non-volatile substances, many of which are carotenoids, such as lycopene, zeaxanthin, and other α- and β-carotenes. More than 34 components namely Terpenes, terpene alcohols, and their esters make up the vast majority of the volatiles' and give off a pungent odor (Pitsikas 2021). In addition to this, they have 5–8% fat and wax, 12–13% protein, and a small amount of essential oils, which give saffron its potent scent (Hosseinzadeh and Younesi 2002). The amount of each of these three primary metabolites—which give stigmas their distinct flavor and color—determines the quality of saffron. The environment and tradition influence the contents of saffron (Gresta et al. 2009). The chemical content of saffron samples from various nations shows that the mentioned values heavily depend on the extraction, drying, and stigma analysis techniques (Kanakis et al. 2004; Zareena et al. 2001). It has been suggested that the highest quality saffron should contain roughly 20% crocin, 5% picrocrocin, and 0.5% safranal. Several techniques for testing saffron components have been reported, and the profile of crocin, picrocrocin, and safranal helps to manage saffron quality (Tarantilis et al. 1995).

8.4.2 Molecular Structure of *Crocus sativus*

8.4.3 Traditional Uses of *Crocus sativus*

The earliest known cultures to employ saffron were those of Mesopotamia; they mostly used it as a condiment, in ceremonial and religious contexts, and as a fabric dye (Yasmin and Nehvi 2013). It is estimated that 3000–4000 years ago was when saffron was first utilized as a spice and food coloring. Generally saffron has long been utilized in herbal remedies, ayurvedic medications, and at-home cures for specific illnesses. Aphrodisiac, stimulant, anti-poison, livotonic, lactogogue, nervine tonic, cardiac tonic, diaphoretic, diuretic, sedative, emmenagogue, relaxing, febrifuge, anti-stress, and anti-anxiety are some of the other important uses of saffron. As a result, it can be used to treat a wide range of illnesses and ailments, including insect bites and stings, edema, general debility, diabetes, inflammation, alcoholism, and disorders in children whose root causes are not clear. Additionally, studies on saffron have showed promise in treating skin conditions, measles, smallpox, scarlet fever, respiratory problems, and gastrointestinal disorders (Mousavi and Bathaie 2011). Saffron and its metabolites show a significant anticancer effect against the breast, lung, pancreatic, and leukemic cancer cells in different in vivo and in vitro models (Milajerdi et al. 2016). The carotenoids in saffron are not extracted from the raw material before being added to food; nonetheless, some dishes—such as Indian pulav rice—are made with whole or powdered saffron stigmas, and the water-soluble crocins in the stigmas give the dish its characteristic crimson red color. The use of pigment from the dried stigma in animal-themed cave paintings is one of the earliest examples that is known to exist. A prominent component in many Indian dishes, including kheer, biryani, Kashmiri pulao, and other sweet foods, is saffron. Without saffron, "kehwa," a traditional beverage consumed in Kashmir, is incomplete. Saffron is used in many rituals and head markings by religions like Hinduism. Buddhism also views saffron as auspicious. It has antibacterial, antidepressant, antispasmodic, anticancer, and carminative properties in modern as well as traditional medicine. It is also used as an herbal remedy to treat respiratory infections such coughs, colds, scarlet fever, and asthma. Numerous other industries, including textile dyes, perfumes, and cosmetics, use it (Menia et al. 2018).

8.4.4 Pharmacological Importance of *Crocus sativus*

8.4.4.1 Antioxidant Activity

Several investigations have looked into using saffron and its constituents, which have potent biological and antioxidant properties, to treat neurodegenerative illnesses. Carotenoids comprising crocin and crocetin, play an extremely important role in health by functioning as natural antioxidants. They shield tissues and cells from the damaging effects of reactive oxygen species (ROS) and free radicals. When it comes to saffron's antioxidant qualities, crocin has been researched more than any other active component. Nevertheless, it functions in combination with other elements including safranal, dimethylcrocetin, and flavonoids rather than acting alone (Shamsa et al. 2009). Additional research concentrated on the adverse effects of oxidative stress on the brain, as it is the organ most exposed to oxidation. This is because of the high phospholipid content of neuronal membranes and the connection between oxidative stress and the development of neurodegenerative pathologies like Alzheimer's disease. Saffron treatment has been shown to inhibit the aggregation and deposition of amyloid β peptide in the human brain, making it useful in the treatment of Alzheimer's disease (Bolhassani et al. 2014). It has been observed that the methanolic extract of *crocus sativus* and its constituents, such as safranal and crocin, exhibit radical scavenging properties. This suggests that the extract may find application as a food supplement, cosmetic to treat age-related illnesses etc. (Hosseinzadeh and Sadeghnia 2007). Crocin was discovered to have a higher antioxidant capacity than alpha-tocopherol in neuronally differentiated pheochromocytoma cells lacking of glucose; the absence of glucose resulted in peroxidation of the cell membrane lipids and decreased intracellular superoxide dismutase activity. Crocin showed potential as a special and strong antioxidant that fights oxidative stress in neurons by reversing these effects. It was also reported to maintain the functional levels of other antioxidants and raise the levels of several enzymes, including glutathione reductase and glutathione-S-transferase, indicating its potential as an antioxidant.

8.4.4.2 Anticancer Activity

Cancer is the second most serious disease involving abnormal cell growth with the potential to invade or spread to other parts of the body and leading to the death of approximately 6 million people per year from worldwide (Abdullaev et al. 2000). Chemoprevention, which is defined as the use of either natural or synthetic chemicals (alone or in combination) to prevent the development of cancer in humans, is currently a very promising method of cancer prevention. Several research organizations have been interested in the chemical constituents of saffron in recent decades because of its three primary active ingredients (crocins, picrocrocin, and safranal). Strong evidence collected from large number of studies suggests that saffron and its constituents have anticarcinogenic and antitumor effects both in vitro *and* in vivo (Escribano et al. 2000; Geromichalos et al. 2014; Nair et al. 1991). Excellent outcomes have been shown against skin neoplasia (retardation or even inhibition) when saffron extract is applied topically, taken intravenously or orally. The presence of crocin, crocetin, picrocrocin, and safranal in saffron in quantities ranging from 2 μM to 3 mM is responsible for its anticarcinogenic qualities (Abdullaev 2002). Water insoluble carotenoids represent the majority of carotenoids from different sources utilized in chemotherapy and chemoprevention. Saffron, on the other hand, has water-soluble carotenoids that are easy to take orally. Most notably, the skin texture of every mouse treated with saffron showed a noticeable improvement. When comparing the skin histology of the saffron-treated mice to that of normal control mice, the appearance was almost normal (Das et al. 2010). Furthermore, it was discovered that using crocetin for lung cancer reduced lipid peroxidation, glutathione-metabolizing enzymes, and the histopathological alterations linked to the growth of tumors, indicating its antitumor (Inoue et al. 2005; Samarghandian et al. 2014). Using in-vivo experimental models, saffron extract has demonstrated the ability to suppress and/or retard a range of neoplasias. For instance, crocetin and saffron aqueous extract both prevent the growth of stomach cancer in rats in a dose-dependent way (Bathaie et al. 2013). It is believed that *Crocus sativus* triggers the process of cell death in a variety of cancer types, including bladder, colorectal, and pancreatic cancer (Bathaie et al. 2013; Bakshi et al. 2010; Sun et al. 2013). Saffron's antitumor or anticarcinogenic properties might originate from a variety of processes. It has been suggested that saffron and its constituents hinder the synthesis of cellular DNA and RNA, but not protein (Nair et al. 1991, 1995). Since the majority of carotenoids are lipid-soluble and function as membrane-associated, highly effective free radical scavengers (antioxidant properties), their inhibition of the free radical chain reaction indicates another potential antitumor strategy (Molnár et al. 2000). The metabolic conversion of retinoids from naturally occurring carotenoids is a third putative anticarcinogenic mechanism. Nevertheless, it was recently reported that anticancer action does not require the conversion of carotenoids to vitamin A (Smith 1998). Lastly, it has also been proposed that topoisomerase II, an enzyme that controls the synthesis of proteins and DNA in cells, and carotenoids interact to provide the cytotoxic effect of saffron (Molnár et al. 2000).

8.4.4.3 Anti-inflammatory and Analgesic Activity

The pharmaceutical industry is paying a lot of attention to traditional medicinal herbs, which were once used to treat a wide range of illnesses as first aid measures. Natural substances that have been used for generations to treat pain and inflammation, such as nutritional supplements and herbal medicines, are gaining a lot of attention (Tamaddonfard et al. 2013). Saffron extracts and tinctures have been utilized to treat fever, wounds, abscesses, gingivitis, and pain associated with an infant's first tooth eruption (Premkumar et al. 2003). Their antinociceptive and anti-inflammatory properties are demonstrated by the effects of aqueous and alcoholic extracts of stigmas and saffron petals, as well as by xylene-induced ear edema in mice and formalin-induced edema in rat paws for both temporary and long-term pain (Poma et al. 2012).

8.4.4.4 Antitussive Activity

The antitussive activity of *C. sativus* stigma and petal extracts and its constituents, safranal and crocin, was determined using the nebolized solution of citric acid 20% in guinea pigs. The number of coughs in guinea pigs was decreased by safranal (0.25–0.75 ml/kg) and the ethanolic extract of *C. sativus* (100–800 mg/kg). There was no antitussive effect demonstrated by the aqueous ethanolic and ethanolic extracts of petal and crocin (Hosseinzadeh and Ghenaati 2006).

8.4.4.5 Anticonvulsant Activity

According to recent research, safranal may have anticonvulsant effects. After pentylene tetrazole was administered to rats peripherally, it was demonstrated that the coloring component of saffron, safranal, reduced the incidence of both generalized tonic-clonic seizures and minor clonic seizures in a dose-dependent manner. Safranal's action was believed to be at least partially mediated by the GABA (A)-benzodiazepine receptor complex (Hosseinzadeh and Ghenaati 2006). Crocin, another saffron ingredient, had no effect on mice's convulsions induced by the drug pentylenetetrazole (Sheng et al. 2006).

8.4.4.6 Anti-Alzheimer Activity

The stigmas of the *Crocus sativus* plant, which are often known as saffron, are made up of particularly polar carotenoids. Amyloid β-peptide (Aβ) fibril deposition is the pathological description of Alzheimer's disease (AD). In a concentration- and time-dependent manner, the aqueous-ethanolic (50:50, v/v) extract of *C. sativus* stigmas exhibited good antioxidant effects, surpassing those of carrot and tomato. This was accompanied by suppression of Aβ fibrillogenesis. The primary carotenoid component that prevented Aβ fibrillogenesis was trans-crocin-4, which is the digentibiosyl ester of crocetin (Papandreou et al. 2006). Streptozotocin (STZ) injections intracerebroventricularly (ICV) into mice have been widely utilized as an animal model for sporadic AD (Lannert and Hoyer 1998; Veerendra Kumar and Gupta 2003). Previous research has shown that treating rats with *C. sativus* extract (30 mg/kg) for 3 weeks can considerably repair cognitive deficits induced on by ICV injection of STZ (Khalili et al. 2010). Additionally, it has been demonstrated that crocin (30 mg/kg) has antagonizes the cognitive deficits in rats generated by STZ (Khalili and Hamzeh 2010). It has been demonstrated that the saffron extract prevented acetylcholine breakdown, the primary treatment strategy for Alzheimer's disease AD, and exhibited a moderate (up to 30%) inhibitory efficacy on acetyl-cholinesterase (AChE) (Geromichalos et al. 2012).

8.4.4.7 Effects on Ocular Blood Flow and Retinal Function

The two most common visual conditions that result in blindness are age-related macular degeneration and ischemic retinopathy. The decrease in blood flow to the retina and/or choroid is one of the causes of these disorders. Analogs of crocin, which are extracted from *Crocus sativus*, have been shown to improve blood flow to the retina and choroid through vasodilation. They also aid in the restoration of retinal function, which wards off age-related macular degeneration and ischemic retinopathy, which can lead to blindness. An interesting structure-activity connection was observed between the potency of crocin disaccharide analogs, such as crocin-1 and crocin-2, and its monosaccharide analogs, such as crocin-3 and crocin-4 (Xuan et al. 1999). Different studies showed that crocin reduced glucose levels in diabetic mice and inhibited the pro-inflammatory response in retinal cells. Additionally, saffron shielded photoreceptors in retinal cells from light-induced damage, and crocetin stopped retinal degeneration.

8.4.4.8 Effects on Blood Pressure

The ethanolic and aqueous extracts of *Crocus sativus* petals revealed a dose-dependent decline in blood pressure in anesthetized rats, isolated rat vas deferens, guinea pig ileum etc. When electrical stimulation was used to trigger chemical processes. Postsynaptic mediation has been hypothesized as the mechanism for this drop in blood pressure (Abe and Saito 2000; Soeda et al. 2001).

8.4.4.9 Effects on Learning Behavior and Long-Term Potentiation

Folk medicine utilizes *Crocus sativus* L., or saffron, for many different kinds of purposes. Saffron extracts have been shown in current pharmacological research to possess antitumor, radical scavenger, and hypolipemic qualities. The primary component of saffron extract that is accountable for these pharmacological effects is crocetin. Herbal remedies have been utilized for decades for treating Alzheimer's disease-related symptoms and to enhance cognitive abilities. One of the powerful medications that grows as a herb is saffron. In Chinese traditional medicine, *Crocus sativus* L. is used to treat several different central nervous system problems. *Crocus sativus* L. contains a component called crocin, which is ethanol-extractable and has been demonstrated to protect mice's learning and memory against ethanol-induced damage. Furthermore, it has been shown by current behavioral and electrophysiological research that saffron extract influences memory and learning in test animals. Saffron extract reduced the deficiencies of learning behaviors in mice caused by ethanol and blocked the reduction of hippocampus long-term potentiation, an activity-dependent synaptic plasticity that may sustain memory and learning. It is believed that crocin (crocetin di-gentiobiose ester), not crocetin, is responsible for this effect of saffron extract. The active ingredients of saffron, crocetin and crocin, may be able to help cure neurodegenerative conditions that accompany memory loss. Oral administration of saffron in formulation form is an option for treating neurogenerative illnesses and their associated memory impairment (Abe and Saito 2000; Soeda et al. 2001).

8.4.4.10 Treating Sexual Dysfunction

During ancient Egyptian, Greek, Roman and other communities, saffron and other spices were regarded as aphrodisiacs. Saffron is traditionally used as a sexual stimulant by Chinese, Muslims, and Phoenicians (Dwyer et al. 2011). The aphrodisiac properties of safranal and crocin, two of the constituents of the aqueous extract of *C. sativus* stigmas, were assessed. It seems that saffron can safely treat several fluoxetine-induced sexual issues in women, like lubrication, excitation, or pain (Moshiri et al. 2006). Even after just 10 days of treatment, saffron has been demonstrated to improve sexual function in erectile dysfunction patients by increasing the frequency and length of erections.

8.4.5 Other Uses of *Crocus sativus*

8.4.5.1 Culinary Use

Saffron is common and has been utilized in cookery from ancient times all throughout the world. Chefs and saffron specialists say it smells like honey with metallic notes (Basker and Negbi 1983). Certain nations, like Spain, Iran, and India, employ saffron as a rice condiment. Fish-based zarzuela and rice-based Paella Valenciana are examples of Spanish cuisine. Iran's national dish, chelow kabab, is made with saffron. Saffron is primarily used in Indian cuisine to make traditional rice-based dishes, desserts, and especially biryanis. Additionally, when creating desserts like kulfi and gulab jamun (Tsatsaroni et al. 1998). Saffron is a key component of the chermoula herb blend used in Moroccan cooking, which gives the food its unique scent (Modaghegh et al. 2008). In Morocco, saffron is substituted for mint in tea and used as a spice in a number of traditional meals, such as koftas, which are prepared from any meatballs as well as tomatoes and mrouzia, which is a sweet and salty dish cooked with mutton or dill.

8.4.5.2 Coloring Power

Some nations have banned synthetic food dyes due to their negative effects, forcing people to switch back to natural coloring. The great solubility of crocine in water makes saffron an ideal alternative color for the agro-food industry (Ramadan et al. 2012). Thus, butter, pasta, cheeses, and oleomargarines have long been colored using saffron's strong coloring ability, which is also useful for beauty purposes. Saffron's golden yellow color is utilized in textiles and painting. The saffron solutions maintain well in both acidic and alkaline media. The acid dissociation constant (pKa) of crocin, dicarboxylic acids, esters, and nitrogen compounds are responsible for this characteristic. Solutions containing saffron buffers reduce the oxidation of cellulose. Silk, wool, Oriental carpets, and Buddhist monks' garments are still dyed with saffron. Compared to some synthetic colors, natural dyes are less poisonous, more environmentally friendly, biodegradable, and less allergic.

8.4.5.3 Color Complexation

Traditionally, it is believed that pregnant women who consume saffron daily with milk can give their unborn child a fair complexion.

8.5 Need for Conservation

The North-Western Himalayan region of Jammu and Kashmir is home to an astonishing variety of therapeutic plants. Kashmir is well-known worldwide and is frequently referred to as a "Terrestrial Paradise." Many medicinal plants that have been used for countless years in ancient medical systems can be found in the stunning state of Jammu & Kashmir. Health care knowledge is passed down from generation to generation. However, in recent decades, it has been deteriorating as a result of swift socio-cultural shifts. Furthermore, the overuse of therapeutic plants is occurring at a startling rate. Comprehensive documentation on the utilization of traditional medicinal herbs to cure ailments, and scientific recommendations for the preservation and management of Jammu & Kashmir's medicinal plant resources are urgently needed. Numerous governmental and non-governmental organizations (NGOs) are engaged in the conservation of these plant species in light of the depletion of medicinal plants. In Jammu and Kashmir, various organizations such as Shri Mata Vaishno Devi University, Katra, University of Jammu, Indian Institute of Integrative Medicine, Jammu (IIIM-Jammu) (formerly known as RRL, Jammu), Centre for Biodiversity Studies, BGSB University, Rajouri, Sher-e-Kashmir Agriculture University of Science and Technology, Jammu and Srinagar, University of Kashmir, Srinagar, State Forest Research Institute, J&K and Defense Institute of High Altitude Research, Leh are keenly involved in developing conservation technologies for medicinal plants of the State. The demands of the conventional medicine markets, which are present in both local and international markets, have been significantly reduced by medicinal and aromatic plants (MAPs). With the expansion of human needs, population and commerce, there is a growing requirement for a broad range of wild species of aromatic and medicinal plants. Overexploitation of several wild species of medicinal plants has been reported. Many organizations are advocating for the commercial cultivation of wild species of medicinal plants. An urgent evaluation is required for the state's medicinal and aromatic plant resources, as there are many high-demand and highly valuable commercial varieties in Jammu and Kashmir. It has been noted that the state's raw medicinal plant industry does not operate under strong authoritarian control. The highly desired varieties of aromatic and medicinal plants are unique in that they are readily available locally, and the popularity of herbs is closely correlated with their availability. The identification of these herbs requires experience. In order to encourage the widespread growth of natural herbs in the state, significant efforts are required. The World Health Organization (WHO) estimates that the current market for ethno-medical plants is worth US$14 billion annually. The market for raw materials derived from medicinal plants is growing at a pace of 15–25% per year, and there will probably be further growth in demand for ethno-medicinal plants in the future. In order to improve the socioeconomic standing of the farmers in Jammu and Kashmir, the Indian government has introduced the Jammu and Kashmir Aroma Arogya Gram (JAAG) project, which aims to motivate them to grow aromatic and medicinal plants that have been proven to be both commercially advantageous and scientifically tested. Under this program, the state's farmers receive professional advice from the government's premier research and development organization, the Council for Scientific and Industrial Research (CSIR), as well as saplings of up to nine different types of fragrant and medicinal plants. It is envisioned that these strategies may aid in the preservation of Jammu and Kashmir's aromatic and medicinal plants.

8.6 Conclusion and Future Perspective

The Himalayan state of Jammu and Kashmir's hilly terrain is vital to both economic growth and human welfare since it offers a variety of basic essentials, such as food, energy, fresh water, and medicinal plants. The Ayurvedic physician's inventory of medicinal plants from the high-altitude area accounts for 35.7% of all plant species found in the alpine and subalpine regions of the Himalaya. Various ethnic groups in the Himalayan region have distinct indigenous healthcare systems. Different applications of therapeutic plants exist for these tribes, dependent on the ecology and terrain of their respective locales. Recently, there has been a revived interest on medicinal plant research. Because plant-based medication is natural, it doesn't cause any significant issues. In this chapter, two medicinal plants have been the subject of a review and an attempt to compile all the relevant information available on these two medicinal plants. Even though *Emblica officinalis* and *Crocus sativus* have a variety of medical uses, it is urgent to investigate their medicinal potential at the molecular level using a range of biotechnological instruments and methodologies. To clarify the molecular mechanisms behind the interactions between diverse plant-based medications and the human body in various disorders, further research needs to be done. Based on our study's analysis, we can conclude that in addition to contributing significantly to the state economy, it is also rife with a variety of issues.

Several disadvantages have been brought to light, including the preservation of therapeutic herbs. As a result, widespread knowledge is required to ensure the survival of medicinal plant production. Additionally, participation from all societal segments is necessary. The government's guidance, additional instruction, training and encouragement for medicinal plant cultivators are urgently needed. It is important to employ a variety of conservation techniques in order to prevent the extinction of medicinal plants. In addition, in order to prevent the extinction of this commercial crop, it is imperative that the financial, technological, and infrastructure issues facing this sector be addressed more quickly.

Acknowledgements The author would like to acknowledge Prof. Vania Bernardes-Genisson (Laboratoire de chimie de Coordination (LCC) du CNRS, France) for her endless motivation as well as support and CEFIPRA for financial support to work in LCC-CNRS.

References

Abdullaev FI (2002) Cancer chemopreventive and tumoricidal properties of saffron (*Crocus sativus* L.). Exp Biol Med (Maywood) 227(1):20–25. https://doi.org/10.1177/153537020222700104

Abdullaev FI, Rivera-Luna R, Roitenburd-Belacortu V, Espinosa-Aguirre J (2000) Pattern of childhood cancer mortality in Mexico. Arch Med Res 31(5):526–531. https://doi.org/10.1016/s0188-4409(00)00094-1

Abe K, Saito H (2000) Effects of saffron extract and its constituent crocin on learning behaviour and long-term potentiation. Phytother Res 14(3):149–152. https://doi.org/10.1002/(sici)1099-1573(200005)14:3<149::aid-ptr665>3.0.co;2-5

Adil MD, Kaiser P, Satti NK, Zargar AM, Vishwakarma RA, Tasduq SA (2010) Effect of *Emblica officinalis* (fruit) against UVB-induced photo-aging in human skin fibroblasts. J Ethnopharmacol 132(1):109–114. https://doi.org/10.1016/j.jep.2010.07.047

Akhtar MS, Ramzan A, Ali A, Ahmad M (2011) Effect of Amla fruit (*Emblica officinalis* Gaertn.) on blood glucose and lipid profile of normal subjects and type 2 diabetic patients. Int J Food Sci Nutr 62(6):609–616. https://doi.org/10.3109/09637486.2011.560565

Anila L, Vijayalakshmi NR (2000) Beneficial effects of flavonoids from *Sesamum indicum*, *Emblica officinalis* and *Momordica charantia*. Phytother Res 14(8):592–595. https://doi.org/10.1002/1099-1573(200012)14:8<592::aid-ptr772>3.0.co;2-#

Anila L, Vijayalakshmi NR (2002) Flavonoids from *Emblica officinalis* and *Mangifera indica*-effectiveness for dyslipidemia. J Ethnopharmacol 79(1):81–87. https://doi.org/10.1016/s0378-8741(01)00361-0

Aprioku JS (2013) Pharmacology of free radicals and the impact of reactive oxygen species on the testis. J Reprod Infertil 14(4):158–172

Arora S, Kaur K, Kaur S (2003) Indian medicinal plants as a reservoir of protective phytochemicals. Teratog Carcinog Mutagen Suppl 1:295–300. https://doi.org/10.1002/tcm.10055

Bajpai M, Pande A, Tewari SK, Prakash D (2005) Phenolic contents and antioxidant activity of some food and medicinal plants. Int J Food Sci Nutr 56(4):287–291. https://doi.org/10.1080/09637480500146606

Bakshi H, Sam S, Rozati R, Sultan P, Islam T, Rathore B, Lone Z, Sharma M, Triphati J, Saxena RC (2010) DNA fragmentation and cell cycle arrest: a hallmark of apoptosis induced by crocin from Kashmiri saffron in a human pancreatic cancer cell line. Asian Pac J Cancer Prev 11(3):675–679

Baliga MS, Dsouza JJ (2011) Amla (*Emblica officinalis* Gaertn), a wonder berry in the treatment and prevention of cancer. Eur J Cancer Prev 20(3):225–239. https://doi.org/10.1097/CEJ.0b013e32834473f4

Bandyopadhyay SK, Pakrashi SC, Pakrashi A (2000) The role of antioxidant activity of *Phyllanthus emblica* fruits on prevention from indomethacin induced gastric ulcer. J Ethnopharmacol 70(2):171–176. https://doi.org/10.1016/s0378-8741(99)00146-4

Basker D, Negbi M (1983) Uses of saffron. Econ Bot 37(2):228–236. https://doi.org/10.1007/BF02858789

Bathaie SZ, Miri H, Mohagheghi M-A, Mokhtari-Dizaji M, Shahbazfar A-A, Hasanzadeh H (2013) Saffron aqueous extract inhibits the chemically-induced gastric cancer progression in the Wistar albino rat. Iran J Basic Med Sci 16(1):27–38

Bhattacharya A, Chatterjee A, Ghosal S, Bhattacharya SK (1999) Antioxidant activity of active tannoid principles of *Emblica officinalis* (amla). Int J Exp Biol 37(07):676–680

Bhattacharya SK, Bhattacharya A, Sairam K, Ghosal S (2002) Effect of bioactive tannoid principles of *Emblica officinalis* on ischemia-reperfusion-induced oxidative stress in rat heart. Phytomedicine 9(2):171–174. https://doi.org/10.1078/0944-7113-00090

Biswas S, Talukder G, Sharma A (1999) Protection against cytotoxic effects of arsenic by dietary supplementation with crude extract of *Emblica officinalis* fruit. Phytother Res 13(6):513–516. https://doi.org/10.1002/(sici)1099-1573(199909)13:6<513::aid-ptr525>3.0.co;2-6

Biswas NR, Gupta SK, Das GK, Kumar N, Mongre PK, Haldar D, Beri S (2001) Evaluation of Ophthacare® eye drops—a herbal formulation in the management of various ophthalmic disorders. Phytother Res 15(7):618–620. https://doi.org/10.1002/ptr.896

Bolhassani A, Khavari A, Bathaie SZ (2014) Saffron and natural carotenoids: biochemical activities and anti-tumor effects. Biochim Biophys Acta 1845(1):20–30. https://doi.org/10.1016/j.bbcan.2013.11.001

Calixto JB, Santos AR, Cechinel Filho V, Yunes RA (1998) A review of the plants of the genus Phyllanthus: their chemistry, pharmacology, and therapeutic potential. Med Res Rev 18(4):225–258. https://doi.org/10.1002/(sici)1098-1128(199807)18:4<225::aid-med2>3.0.co;2-x

Chaieb I (2010) Saponins as insecticides: a review. Tunis J Plant Prot 5(1):39–50

Chakraborty D, Verma R (2010) Ameliorative effect of *Emblica officinalis* aqueous extract on ochratoxin-induced lipid peroxidation in the kidney and liver of mice. Int J Occup Med Environ Health 23(1):63–73. https://doi.org/10.2478/v10001-010-0009-4

Chatterjee A, Chattopadhyay S, Bandyopadhyay SK (2011) Biphasic effect of *Phyllanthus emblica* L. extract on NSAID-induced ulcer: an antioxidative trail weaved with immunomodulatory effect. Evid Based Complement Alternat Med 2011:146808. https://doi.org/10.1155/2011/146808

Chen T-S, Liou S-Y, Chang Y-L (2009) Supplementation of *Emblica officinalis* (Amla) extract reduces oxidative stress in uremic patients. Am J Chin Med 37(1):19–25. https://doi.org/10.1142/S0192415X09006680

Costa A, Moisés TA, Cordero T, Alves CRT, Marmirori J (2010) Association of emblica, licorice and belides as an alternative to hydroquinone in the clinical treatment of melasma. An Bras Dermatol 85(5):613–620. https://doi.org/10.1590/s0365-05962010000500003

Dang GK, Parekar RR, Kamat SK, Scindia AM, Rege NN (2011) Antiinflammatory activity of *Phyllanthus emblica*, *Plumbago zeylanica* and *Cyperus rotundus* in acute models of inflammation. Phytother Res 25(6):904–908. https://doi.org/10.1002/ptr.3345

Das I, Das S, Saha T (2010) Saffron suppresses oxidative stress in DMBA-induced skin carcinoma: a histopathological study. Acta Histochem 112(4):317–327. https://doi.org/10.1016/j.acthis.2009.02.003

Dhir H, Agarwal K, Sharma A, Talukder G (1991) Modifying role of *Phyllanthus emblica* and ascorbic acid against nickel clastogenicity in mice. Cancer Lett 59(1):9–18. https://doi.org/10.1016/0304-3835(91)90129-6

Dutta BK, Rahman I, Das TK (1998) Antifungal activity of Indian plant extracts. Mycoses 41(11–12):535–536. https://doi.org/10.1111/j.1439-0507.1998.tb00718.x

Dwyer AV, Whitten DL, Hawrelak JA (2011) Herbal medicines, other than St. John's Wort, in the treatment of depression: a systematic review. Altern Med Rev 16(1):40–49

Escribano J, Díaz-Guerra MJ, Riese HH, Alvarez A, Proenza R, Fernández JA (2000) The cytolytic effect of a glycoconjugate extracted from corms of saffron plant (*Crocus sativus*) on human cell lines in culture. Planta Med 66(2):157–162. https://doi.org/10.1055/s-2000-11127

Fujii T, Wakaizumi M, Ikami T, Saito M (2008) Amla (*Emblica officinalis* Gaertn.) extract promotes procollagen production and inhibits matrix metalloproteinase-1 in human skin fibroblasts. J Ethnopharmacol 119(1):53–57. https://doi.org/10.1016/j.jep.2008.05.039

Geromichalos GD, Lamari FN, Papandreou MA, Trafalis DT, Margarity M, Papageorgiou A, Sinakos Z (2012) Saffron as a source of novel acetylcholinesterase inhibitors: molecular docking and in vitro enzymatic studies. J Agric Food Chem 60(24):6131–6138. https://doi.org/10.1021/jf300589c

Geromichalos GD, Papadopoulos T, Sahpazidou D, Sinakos Z (2014) Safranal, a *Crocus sativus* L constituent suppresses the growth of K-562 cells of chronic myelogenous leukemia. In silico and in vitro study. Food Chem Toxicol 74:45–50. https://doi.org/10.1016/j.fct.2014.09.001

Giorgi A, Pentimalli D, Giupponi L, Panseri S (2017) Quality traits of saffron (*Crocus sativus* L.) produced in the Italian Alps. Open Agric 2(1):52–57. https://doi.org/10.1515/opag-2017-0005

Gresta F, Avola G, Lombardo GM, Siracusa L, Ruberto G (2009) Analysis of flowering, stigmas yield and qualitative traits of saffron (*Crocus sativus* L.) as affected by environmental conditions. J Sci Hortic 119:320–324. https://doi.org/10.1016/j.scienta.2008.08.008

Haque R, Bin-Hafeez B, Ahmad I, Parvez S, Pandey S, Raisuddin S (2001) Protective effects of *Emblica officinalis* Gaertn. in cyclophosphamide-treated mice. Hum Exp Toxicol 20(12):643–650. https://doi.org/10.1191/096032701718890568

Hosseinzadeh H, Ghenaati J (2006) Evaluation of the antitussive effect of stigma and petals of saffron (*Crocus sativus*) and its components, safranal and crocin in guinea pigs. Fitoterapia 77(6):446–448. https://doi.org/10.1016/j.fitote.2006.04.012

Hosseinzadeh H, Sadeghnia HR (2007) Protective effect of safranal on pentylenetetrazol-induced seizures in the rat: involvement of GABAergic and opioids systems. Phytomedicine 14(4):256–262. https://doi.org/10.1016/j.phymed.2006.03.007

Hosseinzadeh H, Younesi HM (2002) Antinociceptive and anti-inflammatory effects of *Crocus sativus* L. stigma and petal extracts in mice. BMC Pharmacol 2:7. https://doi.org/10.1186/1471-2210-2-7

Inoue E, Shimizu Y, Shoji M, Tsuchida H, Sano Y, Ito C (2005) Pharmacological properties of N-095, a drug containing red ginseng, polygala root, saffron, antelope horn and aloe wood. Am J Chin Med 33(1):49–60. https://doi.org/10.1142/S0192415X05002655

Jacob A, Pandey M, Kapoor S, Saroja R (1988) Effect of the Indian gooseberry (amla) on serum cholesterol levels in men aged 35-55 years. Eur J Clin Nutr 42(11):939–944

Jagetia GC, Baliga MS, Malagi KJ, Sethukumar Kamath M (2002) The evaluation of the radioprotective effect of Triphala (an ayurvedic rejuvenating drug) in the mice exposed to gamma-radiation. Phytomedicine 9(2):99–108. https://doi.org/10.1078/0944-7113-00095

Jain SK, Khurdiya DS (2004) Vitamin C enrichment of fruit juice based ready-to-serve beverages through blending of Indian gooseberry (*Emblica officinalis* Gaertn.) juice. Plant Foods Hum Nutr 59(2):63–66. https://doi.org/10.1007/s11130-004-0019-0

Jeena KJ, Joy KL, Kuttan R (1999) Effect of *Emblica officinalis*, *Phyllanthus amarus* and *Picrorrhiza kurroa* on N-nitrosodiethylamine induced hepatocarcinogenesis. Cancer Lett 136(1):11–16. https://doi.org/10.1016/s0304-3835(98)00294-8

Jeyasankar A, Premalatha, Elumalai K (2012) Larvicidal activity of *Phyllanthus emblica* Linn. (Euphorbiaceae) leaf extracts against important human vector mosquitoes (Diptera: Culicidae). Asian Pac J Trop Dis 2(1):399–403. https://doi.org/10.1016/S2222-1808(12)60190-X

Jindal A, Soyal D, Sharma A, Goyal PK (2009) Protective effect of an extract of *Emblica officinalis* against radiation-induced damage in mice. Integr Cancer Ther 8(1):98–105. https://doi.org/10.1177/1534735409331455

Jose JK, Kuttan R (2000) Hepatoprotective activity of *Emblica officinalis* and Chyavanaprash. J Ethnopharmacol 72(1–2):135–140. https://doi.org/10.1016/s0378-8741(00)00219-1

Jose JK, Kuttan G, Kuttan R (2001) Antitumour activity of *Emblica officinalis*. J Ethnopharmacol 75(2–3):65–69. https://doi.org/10.1016/s0378-8741(00)00378-0

Kanakis CD, Daferera DJ, Tarantilis PA, Polissiou MG (2004) Qualitative determination of volatile compounds and quantitative evaluation of safranal and 4-hydroxy-2,6,6-trimethyl-1-cyclohexene-1-carboxaldehyde (HTCC) in Greek saffron. J Agric Food Chem 52(14):4515–4521. https://doi.org/10.1021/jf049808j

Khalesi M, Behboodi BS, Ebrahimzadeh H (2004) Development and contraction of contractile roots in *Crocus sativus*. Acta Hortic 650:55–58. https://doi.org/10.17660/ActaHortic.2004.650.4

Khalili M, Hamzeh F (2010) Effects of active constituents of *Crocus sativus* L., crocin on streptozocin-induced model of sporadic Alzheimer's disease in male rats. Iran Biomed J 14(1–2):59–65

Khalili M, Kiasalari Z, Rahmati B, Narenjkar J (2010) Behavioral and histological analysis of *Crocus sativus* effect in intracerebroventricular streptozotocin model of Alzheimer disease in rats. Iran J Pathol 5(1):27–33

Kim HY, Okubo T, Juneja LR, Yokozawa T (2010) The protective role of amla (*Emblica officinalis* Gaertn.) against fructose-induced metabolic syndrome in a rat model. Br J Nutr 103(4):502–512. https://doi.org/10.1017/S0007114509991978

Krishnaveni M, Mirunalini S (2010) Therapeutic potential of *Phyllanthus emblica* (amla): the ayurvedic wonder. J Basic Clin Physiol Pharmacol 21(1):93–105. https://doi.org/10.1515/jbcpp.2010.21.1.93

Kumar S, Gupta P, Sharma S, Kumar D (2011) A review on immunostimulatory plants. Zhong Xi Yi Jie He Xue Bao 9(2):117–128. https://doi.org/10.3736/jcim20110201

Kumaran A, Karunakaran RJ (2006) Nitric oxide radical scavenging active components from *Phyllanthus emblica* L. Plant Foods Hum Nutr 61(1):1–5. https://doi.org/10.1007/s11130-006-0001-0

Lannert H, Hoyer S (1998) Intracerebroventricular administration of streptozotocin causes long-term diminutions in learning and memory abilities and in cerebral energy metabolism in adult rats. Behav Neurosci 112(5):1199–1208. https://doi.org/10.1037/0735-7044.112.5.1199

Liu Q, Wang Y-F, Chen R-J, Zhang M-Y, Wang Y-F, Yang C-R, Zhang Y-J (2009a) Anti-coxsackie virus B3 norsesquiterpenoids from the roots of *Phyllanthus emblica*. J Nat Prod 72(5):969–972. https://doi.org/10.1021/np800792d

Liu X, Zhao M, Luo W, Yang B, Jiang Y (2009b) Identification of volatile components in *Phyllanthus emblica* L. and their antimicrobial activity. J Med Food 12(2):423–428. https://doi.org/10.1089/jmf.2007.0679

Mehmood MH, Siddiqi HS, Gilani AH (2011) The antidiarrheal and spasmolytic activities of *Phyllanthus emblica* are mediated through dual blockade of muscarinic receptors and Ca2+ channels. J Ethnopharmacol 133(2):856–865. https://doi.org/10.1016/j.jep.2010.11.023

Menia M, Iqbal S, Zahida R, Tahir S, Kanth RH, Saad AA, Hussian A (2018) Production technology of saffron for enhancing productivity. J Pharmacogn Phytochem 7(1):1033–1039

Menon LG, Kuttan R, Kuttan G (1997) Effect of rasayanas in the inhibition of lung metastasis induced by B16F-10 melanoma cells. J Exp Clin Cancer Res 16(4):365–368

Milajerdi A, Djafarian K, Hosseini B (2016) The toxicity of saffron (*Crocus sativus* L.) and its constituents against normal and cancer cells. J Nutr Intermed Metab 3:23–32. https://doi.org/10.1016/j.jnim.2015.12.332

Modaghegh M-H, Shahabian M, Esmaeili H-A, Rajbai O, Hosseinzadeh H (2008) Safety evaluation of saffron (*Crocus sativus*) tablets in healthy volunteers. Phytomedicine 15(12):1032–1037. https://doi.org/10.1016/j.phymed.2008.06.003

Moghaddasi MS (2010) Saffron chemicals and medicine usage. J Med Plant Res 4:427–430

Molina RV, García-Luis A, Valero M, Navarro Y, Guardiola JL (2004) Extending the harvest period of saffron. Acta Hortic 650:219–225. https://doi.org/10.17660/ActaHortic.2004.650.25

Molnár J, Szabó D, Pusztai R, Mucsi I, Berek L, Ocsovszki I, Kawata E, Shoyama Y (2000) Membrane associated antitumor effects of crocine-, ginsenoside- and cannabinoid derivates. Anticancer Res 20(2A):861–867

Moshiri E, Basti AA, Noorbala A-A, Jamshidi A-H, Hesameddin Abbasi S, Akhondzadeh S (2006) *Crocus sativus* L. (petal) in the treatment of mild-to-moderate depression: a double-blind, randomized and placebo-controlled trial. Phytomedicine 13(9–10):607–611. https://doi.org/10.1016/j.phymed.2006.08.006

Mousavi SZ, Bathaie SZ (2011) Historical uses of saffron: identifying potential new avenues for modern research. Avicenna J Phytomed 1(2):57–66. https://doi.org/10.22038/ajp.2011.122

Muthuraman A, Sood S, Singla SK (2011) The antiinflammatory potential of phenolic compounds from *Emblica officinalis* L. in rat. Inflammopharmacology 19(6):327–334. https://doi.org/10.1007/s10787-010-0041-9

Mykhailenko O, Kovalyov V, Goryacha O, Ivanauskas L, Georgiyants V (2019) Biologically active compounds and pharmacological activities of species of the genus Crocus: a review. Phytochem Rev 162:56–89. https://doi.org/10.1016/j.phytochem.2019.02.004

Nair SC, Pannikar B, Panikkar KR (1991) Antitumour activity of saffron (*Crocus sativus*). Cancer Lett 57(2):109–114. https://doi.org/10.1016/0304-3835(91)90203-t

Nair SC, Kurumboor SK, Hasegawa JH (1995) Saffron chemoprevention in biology and medicine: a review. Cancer Biother 10(4):257–264. https://doi.org/10.1089/cbr.1995.10.257

Ngamkitidechakul C, Jaijoy K, Hansakul P, Soonthornchareonnon N, Sireeratawong S (2010) Antitumour effects of *Phyllanthus emblica* L.: induction of cancer cell apoptosis and inhibition of in vivo tumour promotion and in vitro invasion of human cancer cells. Phytother Res 24(9):1405–1413. https://doi.org/10.1002/ptr.3127

Nicolis E, Lampronti I, Dechecchi MC, Borgatti M, Tamanini A, Bianchi N, Bezzerri V, Mancini I, Giri MG, Rizzotti P, Gambari R, Cabrini G (2008) Pyrogallol, an active compound from the medicinal plant *Emblica officinalis*, regulates expression of pro-inflammatory genes in bronchial epithelial cells. Int Immunopharmacol 8(12):1672–1680. https://doi.org/10.1016/j.intimp.2008.08.001

Nisha P, Singhal RS, Pandit AB (2004) A study on degradation kinetics of ascorbic acid in amla (*Phyllanthus emblica* L.) during cooking. Int J Food Sci Nutr 55(5):415–422. https://doi.org/10.1080/09637480412331321823

Papandreou MA, Kanakis CD, Polissiou MG, Efthimiopoulos S, Cordopatis P, Margarity M, Lamari FN (2006) Inhibitory activity on amyloid-beta aggregation and antioxidant properties of *Crocus sativus* stigmas extract and its crocin constituents. J Agric Food Chem 54(23):8762–8768. https://doi.org/10.1021/jf061932a

Penolazzi L, Lampronti I, Borgatti M, Khan MTH, Zennaro M, Piva R, Gambari R (2008) Induction of apoptosis of human primary osteoclasts treated with extracts from the medicinal plant *Emblica officinalis*. BMC Complement Altern Med 8:59. https://doi.org/10.1186/1472-6882-8-59

Perianayagam JB, Sharma SK, Joseph A, Christina AJM (2004) Evaluation of anti-pyretic and analgesic activity of *Emblica officinalis* Gaertn. J Ethnopharmacol 95(1):83–85. https://doi.org/10.1016/j.jep.2004.06.020

Perianayagam JB, Narayanan S, Gnanasekar G, Pandurangan A, Raja S, Rajagopal K, Rajesh R, Vijayarajkumar P, Vijayakumar SG (2005) Evaluation of antidiarrheal potential of *Emblica officinalis*. Pharm Biol 43(4):373–377. https://doi.org/10.1080/13880200590951856

Pinmai K, Chunlaratthanabhorn S, Ngamkitidechakul C, Soonthornchareon N, Hahnvajanawong C (2008) Synergistic growth inhibitory effects of *Phyllanthus emblica* and *Terminalia bellerica* extracts with conventional cytotoxic agents: doxorubicin and cisplatin against human hepatocellular carcinoma and lung cancer cells. World J Gastroenterol 14(10):1491–1497. https://doi.org/10.3748/wjg.14.1491

Pitsikas N (2021) *Crocus sativus* L. extracts and its constituents crocins and safranal; potential candidates for schizophrenia treatment? Molecules 26(5):1237. https://doi.org/10.3390/molecules26051237

Piva R, Penolazzi L, Borgatti M, Lampronti I, Lambertini E, Torreggiani E, Gambari R (2009) Apoptosis of human primary osteoclasts treated with molecules targeting nuclear factor-kappaB. Ann N Y Acad Sci 1171:448–456. https://doi.org/10.1111/j.1749-6632.2009.04906.x

Poltanov EA, Shikov AN, Dorman HJD, Pozharitskaya ON, Makarov VG, Tikhonov VP, Hiltunen R (2009) Chemical and antioxidant evaluation of Indian gooseberry (*Emblica officinalis* Gaertn., syn. *Phyllanthus emblica* L.) supplements. Phytother Res 23(9):1309–1315. https://doi.org/10.1002/ptr.2775

Poma A, Fontecchio G, Carlucci G, Chichiricco G (2012) Anti-inflammatory properties of drugs from saffron crocus. Antiinflamm Antiallergy Agents Med Chem 11(1):37

Pramyothin P, Samosorn P, Poungshompoo S, Chaichantipyuth C (2006) The protective effects of *Phyllanthus emblica* Linn. extract on ethanol induced rat hepatic injury. J Ethnopharmacol 107(3):361–364. https://doi.org/10.1016/j.jep.2006.03.035

Premkumar K, Abraham SK, Santhiya ST, Ramesh A (2003) Protective effects of saffron (*Crocus sativus* Linn.) on genotoxins-induced oxidative stress in Swiss albino mice. Phytother Res 17(6):614–617. https://doi.org/10.1002/ptr.1209

Rahman S, Akbor MM, Howlader A, Jabbar A (2009) Antimicrobial and cytotoxic activity of the alkaloids of Amlaki (*Emblica officinalis*). Pak J Biol Sci 12(16):1152–1155. https://doi.org/10.3923/pjbs.2009.1152.1155

Ramadan A, Soliman G, Mahmoud SS, Nofal SM, Abdel-Rahman RF (2012) Evaluation of the safety and antioxidant activities of *Crocus sativus* and *Propolis* ethanolic extracts. J Saudi Chem Soc 16(1):13–21. https://doi.org/10.1016/j.jscs.2010.10.012

Rani P, Khullar N (2004) Antimicrobial evaluation of some medicinal plants for their anti-enteric potential against multi-drug resistant *Salmonella typhi*. Phytother Res 18(8):670–673. https://doi.org/10.1002/ptr.1522

Reddy VD, Padmavathi P, Varadacharyulu NC (2009) *Emblica officinalis* protects against alcohol-induced liver mitochondrial dysfunction in rats. J Med Food 12(2):327–333. https://doi.org/10.1089/jmf.2007.0694

Romano M (2014) ADDENDUM: nutraceuticals for protection and healing of gastrointestinal mucosa. Romano M, Vitaglione P, Sellitto S, D'Argenio G. Curr Med Chem 2012, 19: 109-117. Curr Med Chem 21(28):3310. https://doi.org/10.2174/092986732128140825173622

Sabu MC, Kuttan R (2002) Anti-diabetic activity of medicinal plants and its relationship with their antioxidant property. J Ethnopharmacol 81(2):155–160. https://doi.org/10.1016/s0378-8741(02)00034-x

Sai Ram M, Neetu D, Yogesh B, Anju B, Dipti P, Pauline T, Sharma SK, Sarada SKS, Ilavazhagan G, Kumar D, Selvamurthy W (2002) Cyto-protective and immunomodulating properties of Amla (*Emblica officinalis*) on lymphocytes: an in-vitro study. J Ethnopharmacol 81(1):5–10. https://doi.org/10.1016/s0378-8741(01)00421-4

Sai Ram M, Neetu D, Deepti P, Vandana M, Ilavazhagan G, Kumar D, Selvamurthy W (2003) Cytoprotective activity of Amla (*Emblica officinalis*) against chromium (VI) induced oxidative injury in murine macrophages. Phytother Res 17(4):430–433. https://doi.org/10.1002/ptr.1157

Saini A, Sharma S, Chhibber S (2008) Protective efficacy of *Emblica officinalis* against Klebsiella pneumoniae induced pneumonia in mice. Indian J Med Res 128(2):188–193

Sairam K, Rao CV, Babu MD, Kumar KV, Agrawal VK, Goel R (2002) Antiulcerogenic effect of methanolic extract of *Emblica officinalis*: an experimental study. J Ethnopharmacol 82(1):1–9. https://doi.org/10.1016/S0378-8741(02)00041-7

Samarghandian S, Shoshtari ME, Sargolzaei J, Hossinimoghadam H, Farahzad JA (2014) Anti-tumor activity of safranal against neuroblastoma cells. Pharmacogn Mag 10(Suppl 2):S419–S424. https://doi.org/10.4103/0973-1296.133296

Scartezzini P, Antognoni F, Raggi MA, Poli F, Sabbioni C (2006) Vitamin C content and anti-oxidant activity of the fruit and of the Ayurvedic preparation of *Emblica officinalis* Gaertn. J Ethnopharmacol 104(1–2):113–118. https://doi.org/10.1016/j.jep.2005.08.065

Shamsa A, Hosseinzadeh H, Molaei M, Shakeri M (2009) Evaluation of *Crocus sativus* L. (saffron) on male erectile dysfunction: a pilot study. Phytomedicine 16:690–693. https://doi.org/10.1016/j.phymed.2009.03.008

Shanmugasundaram KR, Seethapathy PG, Shanmugasundaram ERB (1983) Anna Pavala Sindhooram—an antiatherosclerotic indian drug. J Ethnopharmacol 7(3):247–265. https://doi.org/10.1016/0378-8741(83)90001-6

Sharma N, Trikha P, Athar M, Raisuddin S (2000) In vitro inhibition of carcinogen-induced mutagenicity by *Cassia occidentalis* and *Emblica officinalis*. Drug Chem Toxicol 23(3):477–484. https://doi.org/10.1081/dct-100100129

Sharma A, Sharma MK, Kumar M (2009) Modulatory role of *Emblica officinalis* fruit extract against arsenic induced oxidative stress in Swiss albino mice. Chem Biol Interact 180(1):20–30. https://doi.org/10.1016/j.cbi.2009.01.012

Sheng L, Qian Z, Zheng S, Xi L (2006) Mechanism of hypolipidemic effect of crocin in rats: crocin inhibits pancreatic lipase. Eur J Pharmacol 543(1–3):116–122. https://doi.org/10.1016/j.ejphar.2006.05.038

Singh I, Sharma A, Nunia V, Goyal PK (2005) Radioprotection of Swiss albino mice by *Emblica officinalis*. Phytother Res 19(5):444–446. https://doi.org/10.1002/ptr.1600

Smith TA (1998) Carotenoids and cancer: prevention and potential therapy. Br J Biomed Sci 55(4):268–275

Soeda S, Ochiai T, Paopong L, Tanaka H, Shoyama Y, Shimeno H (2001) Crocin suppresses tumor necrosis factor-alpha-induced cell death of neuronally differentiated PC-12 cells. Life Sci 69(24):2887–2898. https://doi.org/10.1016/s0024-3205(01)01357-1

Srivastava R, Ahmed H, Dixit RK, Dharamveer, Saraf SA (2010) *Crocus sativus* L.: a comprehensive review. Pharmacogn Rev 4(8):200–208. https://doi.org/10.4103/0973-7847.70919

Sultana S, Ahmed S, Sharma S, Jahangir T (2004) *Emblica officinalis* reverses thioacetamide-induced oxidative stress and early promotional events of primary hepatocarcinogenesis. J Pharm Pharmacol 56(12):1573–1579. https://doi.org/10.1211/0022357044931

Sultana S, Ahmed S, Jahangir T (2008) *Emblica officinalis* and hepatocarcinogenesis: a chemopreventive study in Wistar rats. J Ethnopharmacol 118(1):1–6. https://doi.org/10.1016/j.jep.2007.04.021

Sumantran VN, Kulkarni A, Chandwaskar R, Harsulkar A, Patwardhan B, Chopra A, Wagh UV (2008) Chondroprotective potential of fruit extracts of *Phyllanthus emblica* in osteoarthritis. Evid Based Complement Alternat Med 5(3):329–335. https://doi.org/10.1093/ecam/nem030

Sumitra M, Manikandan P, Gayathri VS, Mahendran P, Suguna L (2009) *Emblica officinalis* exerts wound healing action through up-regulation of collagen and extracellular signal-regulated kinases (ERK1/2). Wound Repair Regen 17(1):99–107. https://doi.org/10.1111/j.1524-475X.2008.00446.x

Sun Y, Xu H-J, Zhao Y-X, Wang L-Z, Sun L-R, Wang Z, Sun X-F (2013) Crocin exhibits antitumor effects on human leukemia HL-60 cells in vitro and in vivo. Evid Based Complement Alternat Med 2013:690164. https://doi.org/10.1155/2013/690164

Suresh K, Vasudevan DM (1994) Augmentation of murine natural killer cell and antibody dependent cellular cytotoxicity activities by *Phyllanthus emblica*, a new immunomodulator. J Ethnopharmacol 44(1):55–60. https://doi.org/10.1016/0378-8741(94)90099-x

Tamaddonfard E, Farshid A-A, Eghdami K, Samadi F, Erfanparast A (2013) Comparison of the effects of crocin, safranal and diclofenac on local inflammation and inflammatory pain responses induced by carrageenan in rats. Pharmacol Rep 65(5):1272–1280. https://doi.org/10.1016/s1734-1140(13)71485-3

Tantry DMA, Dar PBA (2017) An economic analysis of production and marketing of saffron in Jammu and Kashmir. Int J Soc Relevant Concern 5(10):12–19

Tarantilis PA, Tsoupras G, Polissiou M (1995) Determination of saffron (*Crocus sativus* L.) components in crude plant extract using high-performance liquid chromatography-UV-visible photodiode-array detection-mass spectrometry. J Chromatogr A 699(1–2):107–118. https://doi.org/10.1016/0021-9673(95)00044-n

Tasduq SA, Kaisar P, Gupta DK, Kapahi BK, Maheshwari HS, Jyotsna S, Johri RK (2005a) Protective effect of a 50% hydroalcoholic fruit extract of *Emblica officinalis* against anti-tuberculosis drugs induced liver toxicity. Phytother Res 19(3):193–197. https://doi.org/10.1002/ptr.1631

Tasduq SA, Mondhe DM, Gupta DK, Baleshwar M, Johri RK (2005b) Reversal of fibrogenic events in liver by *Emblica officinalis* (fruit), an Indian natural drug. Biol Pharm Bull 28(7):1304–1306. https://doi.org/10.1248/bpb.28.1304

Tsatsaroni E, Liakopoulou-Kyriakides M, Eleftheriadis I (1998) Comparative study of dyeing properties of two yellow natural pigments—effect of enzymes and proteins. Dyes Pigments 37(4):307–315. https://doi.org/10.1016/S0143-7208(97)00069-7

Unander DW, Webster GL, Blumberg BS (1990) Records of usage or assays in Phyllanthus (Euphorbiaceae) I. Subgenera Isocladus, Kirganelia, Cicca and Emblica. J Ethnopharmacol 30(3):233–264. https://doi.org/10.1016/0378-8741(90)90105-3

Vasudevan M, Parle M (2007) Effect of *Anwala churna* (*Emblica officinalis* GAERTN.): an ayurvedic preparation on memory deficit rats. J Pharm Soc Jpn 127(10):1701–1707. https://doi.org/10.1248/yakushi.127.1701

Veena K, Shanthi P, Sachdanandam P (2006) Anticancer effect of Kalpaamruthaa on mammary carcinoma in rats with reference to glycoprotein components, lysosomal and marker enzymes. Biol Pharm Bull 29(3):565–569. https://doi.org/10.1248/bpb.29.565

Veerendra Kumar MH, Gupta YK (2003) Effect *of Centella asiatica* on cognition and oxidative stress in an intracerebroventricular streptozotocin model of Alzheimer's disease in rats. Clin Exp Pharmacol Physiol 30(5–6):336–342. https://doi.org/10.1046/j.1440-1681.2003.03842.x

Verma R, Chakraborty D (2008) Alterations in DNA, RNA and protein contents in liver and kidney of mice treated with ochratoxin and their amelioration by *Emblica officinalis* aqueous extract. Acta Pol Pharm 65(1):3–9

Wang Y-F, Wang X-Y, Ren Z, Qian C-W, Li Y-C, Kaio K, Wang Q-D, Zhang Y, Zheng L-Y, Jiang J-H, Yang C-R, Liu Q, Zhang Y-J, Wang Y-F (2009) Phyllaemblicin B inhibits coxsackie virus B3 induced apoptosis and myocarditis. Antivir Res 84(2):150–158. https://doi.org/10.1016/j.antiviral.2009.08.004

Winterhalter P, Straubinger M (2000) Saffron—renewed interest in an ancient spice. Food Rev Int 16(1):39–59. https://doi.org/10.1081/FRI-100100281

Xiang Y, Pei Y, Qu C, Lai Z, Ren Z, Yang K, Xiong S, Zhang Y, Yang C, Wang D, Liu Q, Kitazato K, Wang Y (2011) In vitro anti-herpes simplex virus activity of 1,2,4,6-tetra-O-galloyl-β-D-glucose *from Phyllanthus emblica* L. (Euphorbiaceae). Phytother Res 25(7):975–982. https://doi.org/10.1002/ptr.3368

Xuan B, Zhou YH, Li N, Min ZD, Chiou GC (1999) Effects of crocin analogs on ocular blood flow and retinal function. J Ocul Pharmacol Ther 15(2):143–152. https://doi.org/10.1089/jop.1999.15.143

Yang C-J, Wang C-S, Hung J-Y, Huang H-W, Chia Y-C, Wang P-H, Weng C-F, Huang M-S (2009) Pyrogallol induces G2-M arrest in human lung cancer cells and inhibits tumor growth in an animal model. Lung Cancer 66(2):162–168. https://doi.org/10.1016/j.lungcan.2009.01.016

Yasmin S, Nehvi FA (2013) Saffron as a valuable spice: a comprehensive review. Afr J Agric Res 8(3):234–242

Yokozawa T, Kim HY, Kim HJ, Okubo T, Chu D-C, Juneja LR (2007) Amla (*Emblica officinalis* Gaertn.) prevents dyslipidaemia and oxidative stress in the ageing process. Br J Nutr 97(6):1187–1195. https://doi.org/10.1017/S0007114507691971

Zareena AV, Variyar PS, Gholap AS, Bongirwar DR (2001) Chemical investigation of gamma-irradiated saffron (*Crocus sativus* L.). J Agric Food Chem 49(2):687–691. https://doi.org/10.1021/jf000922l

Zhang YJ, Tanaka T, Iwamoto Y, Yang CR, Kouno I (2000) Novel norsesquiterpenoids from the roots of *Phyllanthus emblica*. J Nat Prod 63(11):1507–1510. https://doi.org/10.1021/np000135i

Zhang YJ, Abe T, Tanaka T, Yang CR, Kouno I (2001a) Phyllanemblinins A-F, new ellagitannins from *Phyllanthus emblica*. J Nat Prod 64(12):1527–1532. https://doi.org/10.1021/np010370g

Zhang YJ, Tanaka T, Iwamoto Y, Yang CR, Kouno I (2001b) Novel sesquiterpenoids from the roots of *Phyllanthus emblica*. J Nat Prod 64(7):870–873. https://doi.org/10.1021/np010059z

Zhang Y-J, Nagao T, Tanaka T, Yang C-R, Okabe H, Kouno I (2004) Antiproliferative activity of the main constituents from *Phyllanthus emblica*. Biol Pharm Bull 27(2):251–255. https://doi.org/10.1248/bpb.27.251

Zubor AA, Surányi G, Győri Z, Borbély G, Prokisch J (2004) Molecular biological approach of the systematics of *Crocus sativus* L. and its allies. Acta Hortic 650:85–93. https://doi.org/10.17660/ActaHortic.2004.650.8

Medicinal Potential of Himalayan High-Altitude Plants: A Significant Resource for Healthcare and Their Socioeconomic Significance

9

Muhammad Manzoor, Mushtaq Ahmad, Syed Waseem Gillani, Shazia Sultana, Muhammad Kafeel Ansari, Mumna Munir, Sanakulov Akmal Lapasovich, Fayziev Vakhid Bahramovich, Nizomova Maksoda Usmankulovna, and Shabir Ahmad

Abstract

The Himalayan Mountain ecosystem, rich in biodiversity, is the homeland of an extensive spectrum of plant species that have overcome obstacles to thrive in a harsh environment. The study explores the medicinal value of high-altitude plants from the Himalayan region and their therapeutic uses, and modern medicinal roles, which also discuss challenges to conservation and sustainable harvest-

M. Manzoor · S. W. Gillani · S. Sultana · M. Munir · S. Ahmad
Department of Plant Sciences, Quaid-i-Azam University, Islamabad, Pakistan

M. Ahmad (✉)
Department of Plant Sciences, Quaid-i-Azam University, Islamabad, Pakistan

Department of Medicinal Plants, Tashkent State Agrarian University, Tashkent, Uzbekistan
e-mail: mushtaq@qau.edu.pk

M. K. Ansari
Department of Biology, Faculty of Natural Sciences, University of Guyana, GeogeTown, Guyana

S. A. Lapasovich
Department of Plant Physiology and Microbiology, Samarkand State University Named After Sharof Rashidov,
Samarkand, Uzbekistan

F. V. Bahramovich
Department of Biology, Chirchik State Pedagogical University, Chirchik, Uzbekistan

N. M. Usmankulovna
Department of Medicinal Plants, Tashkent State Agrarian University, Tashkent, Uzbekistan

S. Chandra, H. Lata (eds.), *High Altitude Medicinal Plants*,
https://doi.org/10.1007/978-3-031-99406-7_9

ing practices. The region's resilient environment yields specialized plant species that exhibit potential for healing and pharmaceutical applications. The Himalayas, particularly the western Himalayas, hold significant medicinal knowledge, with 80% of the developing population relying on plant resources for healthcare. The rapid shift toward modernization has depleted traditional knowledge, which may influence pharmacological research. Traditional herbal medicine in tribal areas utilizes medicinal plants and essential oils, but information on their biological activities and ethnobotanical uses is limited due to non-volatile components. Western Himalayan plants offer alternative medicine due to their mineral content, alkaloid, saponin, and flavonoid content. In a tribal-dominated area of the high-altitude Himalayas, the investigation seeks to emphasize the medicinal plants used in traditional healthcare practices as well as the influence of developmental activities and changing socioeconomic conditions on traditional knowledge by utilizing both modern research and medicinal or healing knowledge. Documentation of native understanding through ethnobotanical and ethnomedicine research constitutes crucial evidence for the conservation and implementation of the natural assets in the world's richest region.

9.1 Introduction

The Himalaya, the youngest mountain ecosystem globally recognized as a biodiversity hotspot, is home to a rich repository of unique biodiversity (Mallen-Cooper and Pickering 2008). Mountains are rich in biodiversity, providing ecosystem services like climate regulation, water availability, soil conservation, and recreation (Kokkoris et al. 2018). The vast mountain range, reaching up to 8000 masl, exhibits wide bioclimatic regimes, with monsoonal and non-monsoonal climates (Vetaas and Grytnes 2002). The region's extreme conditions, including temperature fluctuations, rainfall, wind velocity, and snowbound conditions, result in a variety of ecological traits and specialized habits (Körner 2007). Tree species in high elevations struggle to adapt to environmental stress, leading to shorter, lower-density treelines. This zone includes alpine scrubs and alpine meadows. The Himalayan ecosystem is fragile and threatened due to anthropogenic pressure and overexploitation (Bisht et al. 2022). Changes in biotic and abiotic components can cause imbalance. Due to remoteness, inaccessibility, and harsh climates, these areas have largely escaped attention.

The Himalayas, particularly the western Himalayas, hold significant medicinal knowledge, with 80% of the developing population relying on plant resources for healthcare. However, information on plant medicinal uses in interior areas is lacking due to remoteness and lack of modern health facilities (Farnsworth et al. 1985; Uniyal et al. 2006). The region's rapid transformation as an ideal paragliding site has depleted traditional knowledge, potentially losing important leads to drug discovery. Recent investigations have focused on plant diversity and composition in these high-altitude regions (Acharya et al. 2011).The Himalayan mountain range, spanning Afghanistan, Pakistan, India, China, Nepal, Tibet, and Bhutan, is home to a diverse flora and medicinal plants, including *Artemisia*, *Rhododendron*, *Cinnamomum*, *Juniperus*, *Cymbopogon*, *Aegle*, *Swertia*, *Pinus*, *Origanum*, and *Saussurea* (Painuli et al. 2021).

The Himalayan Region is a rich reservoir of biological diversity, used for medicine, food, fuel, fodder, timber, agricultural tools, fiber, and religious purposes (Adnan and Hölscher 2011). The herbaceous layer decreases with altitude, suggesting higher elevation communities harvest more medicinal plants due to lack of healthcare facilities (Kumar et al. 2011). Traditional therapies are increasingly being used due to their cultural acceptability, compatibility with the human body, and lesser side effects. Herbal medicine is the primary health care for 75–80% of the world's population, mainly in developing countries (Kamboj 2000). Forests are crucial for local healers who gather and sell medicinal plants for various ailments (Vidyarthi et al. 2013). *Artemisia dubia*, a plant native to Nepal, is used as an antiseptic and pesticide in Dolpa and Newar communities for wound healing and cytotoxic activity against MCF-7 breast tumor cells. Its essential oil contains chrysanthenone, coumarins, and camphor (Satyal et al. 2012). *Artemisia* gmelinii, a plant native to Himalayan India, is used externally for headache, boils, and pimples (Rokaya et al. 2010). However, these compounds are not particularly effective against bacteria or fungal infections (Joshi et al. 2016).

Traditional herbal medicine in tribal areas utilizes medicinal plants and essential oils, but information on their biological activities and ethnobotanical uses is limited due to non-volatile components (Badola and Aitken 2003). Many phytochemical researchers have overlooked bioactivity screening for ethnopharmacological uses of medicinal plants, which have commercial potential in flavoring agents, fragrances, cosmetics, and pesticides (Joshi et al. 2016). Western Himalayan plants like *Acer pictum*, *Acer caecium*, *Betula utilis*, *Oxalis corniculata*, *Euphorbia pilosa*, *Heracleum lanatum*, *Urtica dioica*, *Berberis lycium*, *Berberis asiaticaand*, and *Quercus ilex* offer alternative medicine due to their mineral content in addition to alkaloid, saponin, and flavonoid (Radha et al. 2021).

The study explores the medicinal plants used in traditional healthcare practices in a tribal-dominated region of high-altitude Himalaya. The study also explores the impact of developmental activities and evolving socio-economic conditions on traditional knowledge. Himalayan region faces significant challenges in conservation and sustaining ecological balance due to anthropogenic pressure on mountainous forests. Forests provide livelihoods for 500 million people and are crucial for tribal communities (Prakash 2015). Overexploitation, commercialization of agriculture, and the commercialization of medicinal sectors have led to the decline of important medicinal plant species. Traditional uses of medicinal plants may also decline due to increased tourism. Records of indigenous knowledge through ethnobotanical and ethno-medicine studies are essential for conservation and utilization of biological resources. Conservation of medicinal plants is crucial for long-term availability, requiring detailed data on individual species. Local forest departments and NGOs must prioritize conservation efforts.

9.2 Materials and Methods

9.2.1 Study Area

The State of Azad Jammu and Kashmir (AJK) is a regarded as biodiversity hotspot, with a hilly, mountainous topography and diverse agroclimatic zones. An ethnobotanical survey was conducted in six districts at higher elevations, focusing on subalpine and alpine pastures. The climate is extreme cold, with heavy snowfall and freezing temperatures from November to April. The area receives about 1000 mm of precipitation annually, mostly as snow in winter.

9.2.2 Data Collection

The study used a multistage random sampling method to select high-altitude sites in the Himalayan region. In the studied literature, informants, preferably older people, were surveyed to document indigenous ethno-medicinal knowledge and their experience using medicinal plants for primary healthcare management. Field visits were conducted from June to September to document ethno-medicinal plants in the western Himalayan region of Kashmir, as this season is the growing season in the alpine zone (Manzoor et al. 2023). Participants were briefed on the research's objectives and purposes, and their contributions were documented using open-ended, semi-structured, and pre-tested questionnaire methods.

The study surveyed informants from different ethnic groups in the upper temperate to alpine zones, focusing on the harvesting and utilization of high-altitude plants in their local habitats used for the treatment of different types of ailments. The interviews were conducted in local languages, and participants were asked about local names, collection seasons, parts used, mode of utilization, and recipes. Most informants in the Himalayan high-altitude areas were illiterate but had extensive expertise in the wide range and utilization of wild plants in the study area. The collected specimens in the studied scenarios were dried, pressed, and mounted on herbarium sheets. Identification of the specimens using taxonomic literature and online databases was employed.

9.2.3 Ethnographic Information

The region of Division Muzaffarabad, Azad Kashmir is diverse and ethnically diverse, with Gujjar tribes being the most influential. Most residents are Muslims, and their languages include Hindko, Gojri, Pahari, and Kashmiri (Abdullah et al. 2021). The region is remote, with poor infrastructure and limited access to modern healthcare facilities. Agriculture relies on rain-fed techniques, with maize being the primary crop. Health services are limited, and indigenous ethno-medicinal knowledge is primarily held by elders and health practitioners (Gillani et al. 2024a).

The study analyzed the consumption of Himalayan medicinal plants among distinct ethnic groups, revealing the variability and difference in plant cultural markers. The Gujjar ethnic group was observed to utilize the largest percentage of indigenous plants, likely due to increasing population density, social contacts, and occupations related to wild food plants (Awan et al. 2021). The sociocultural differences among the groups may contribute to variations in species usage across different geographical areas. Tourism has significantly improved the region's socioeconomic conditions by creating employment opportunities (Gillani et al. 2024b).

9.3 Results and Discussion

9.3.1 Healthcare

9.3.1.1 Traditional Healing Practices of Himalayan High-Altitude Plants

Medicinal plants and traditional healing systems are national heritage, requiring recognition and visibility in policies for long-term sustainability (Lakey and Dorji 2016). The Himalayan Mountain region has 70% wild medicinal plants and animals, and 70–80% of the local population relies on traditional remedies for healthcare (Pie and Manandhar 1987). Ethno-ecology documents ecological knowledge, aiding in long-term biodiversity management and conservation (Balick 1996). Traditional medicines are frequently employed in impoverished nations for basic healthcare management, as well as in industrialized countries because people believe that "natural is healthier" (Lewis 2003). Wounds, injuries, and boils are frequent skin conditions seen in traditional dermatology techniques in India and Pakistan (Kumar et al. 2013; Wagh and Jain 2020). *Aconitum* species are used in traditional medicine for minor ailments like fever, stomachic, tonsillitis, throat pain, and cheilitis. It is also used to keep yak products safe in herds. *Aconitum* species are found in sub-alpine grassland, lakes, and moist soil. Wild Allium species are considered more effective than cultivated crops and are used for medicinal purposes only in North East India and Arunachal Himalayan Region (Tsering and Tag 2015). Rhizomes of various plants such as *Aconitum, Saussurea, and Corydalis govianiana* were used for various purposes, including killing lice, tooth cleaning, respiratory diseases, and for various ailments. Leaves of *Arnebia benthamii* were used for internal injuries, brain tonics, muscle sprain remedies, dye production, drug addiction relief, and stomach issues (Haq et al. 2023).

9.3.1.2 Pharmacological Properties

Ethno-pharmacological sciences are critical for detecting, analyzing, monitoring, and treating adverse effects of pharmaceutical drugs, as well as aiding in genetic resource conservation and land use management strategies (Grumbine and Xu 2021). *Arnebia* species are found in the Himalayas and are used to treat toothaches, earaches, eye problems, and fire burns (Hu et al. 2006). The roots contain naphthoquinone pigments, terpenes, and sterols, and are highly prized in Asia and Northern

Africa (Ishida and Sakaguchi 2007). Seabuckthorn (SBT) is a deciduous shrub or tree found in subtropical mountainous zones of Asia and Europe. Its phyto-chemistry is rich in proteins, fatty acids, and flavonoids, with higher Vitamin E content than wheat embryo, safflower, maize, and soybean (Dhyani et al. 2007). SBT has historical medicinal references, including use in ancient Greece and China, and has been used in Ayurveda for treating stomach malfunctioning, circulatory disorders, burns, and wound healing (Tiitinen et al. 2006). Studies show that seabuckthorn extracts promote immune function and have antimutagenic properties in mice (Nersesyan and Muradyan 2004). *Rheum emodi*, a Polygonaceae plant, belonging to Himalayan is known for its medicinal properties, including secondary metabolites like aloe-emodin, chrysophanol, and rhein. Anthraquinones, anthracene ring-like compounds, have shown anticancer activities (Bringmann et al. 2006). *Rheum emodi* is the main source of Himalayan rhubarb and has been found to increase DNA repair, inhibit Hep G2 cell growth, induce apoptosis, and block cell cycle progression (Bhatia et al. 2011; Cao et al. 2005). Major pharmacological activities of medicinal plants from Himalayas are shown in Fig. 9.1.

9.3.1.3 Clinical Applications

The diverse uses of plant parts for medicinal preparations, with leaves being the most frequently used part. Leaves are preferred due to their sustainability and longer medicinal demand over an annual cycle, making them a more sustainable use of natural resources (Alamgeer et al. 2018). The mode of preparation and formulation is crucial, with the majority of drugs being made into pastes, followed by juice, decoction, and powder. Multiple uses of drugs from a species and modes of preparation are recorded, indicating potential use of the entire spectrum of metabolites (Sharma et al. 2014). Most species with clear records of administration routes are used externally, while internal uses account for 16%, including both external and internal uses. For example, the *Aconitum* species and *Allium species ,various plant parts like* stems, leaves, flowers, and seeds were used in a decoction for treating different diseases (Heinrich et al. 2021). *Achyranthes species*, used for pimples, boils, and bites, have potential applications in skin diseases due to their steroidal content and tissue regeneration potential (Chen et al. 2004; Rao 1978).

9.3.1.4 Integrative Approaches in Modern Healthcare

Traditional medicine is a holistic approach combining indigenous knowledge, skills, and cultural practices from various cultures to treat and improve physical and mental well-being (Shewamene et al. 2020). The Himalayas, spanning eight countries, are home to a vast array of valuable medicinal plant species, including 1748 used medicinally. However, the demand for herbal products and plant-based drugs has led to heavy exploitation, resulting in habitat degradation, unscientific harvesting, and overexploitation (Malik et al. 2014). Over 90% of plant species used in herbal industries are extracted from the wild, primarily from the sub-alpine and alpine regions of the Himalaya. These plant-derived drugs were discovered through traditional cures and folk knowledge, and some cannot be substituted despite advancements in synthetic chemistry (Malik et al. 2015). Developed countries have access

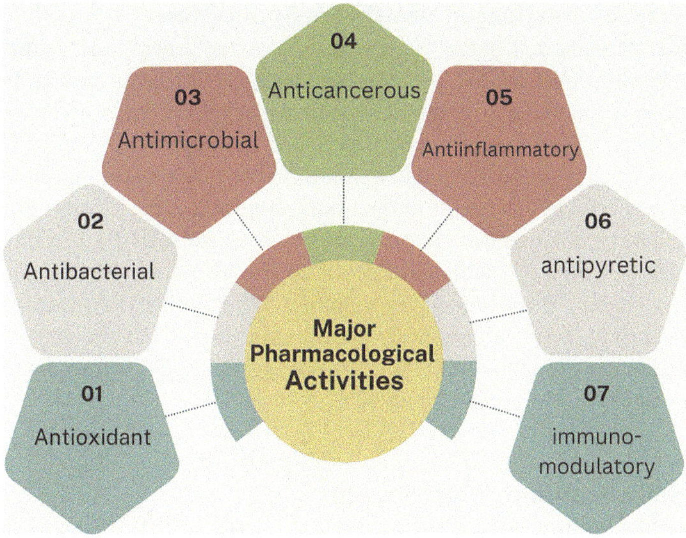

Fig. 9.1 Major pharmacological activities of medicinal plants from Himalayan altitudes

to modern medicines, but still rely heavily on traditional herbal medicine for minor illnesses (Chen et al. 2016). Traditional knowledge, such as ethnobotanical knowledge (TK), is being endangered due to the loss of knowledge. Ethnopharmacological studies can help preserve this knowledge by transferring local popular knowledge to the rest of the population (Sen et al. 2011). Disseminating ethnobotanical knowledge promotes knowledge exchange and cultural networks, despite the potential loss of TK in many societies (Singh et al. 2020).

9.3.1.5 Preservation of Ethno-medicinal Knowledge

The Himalayan Mountain region is a biocultural hub with diverse medicinal and food plants, particularly crucial for indigenous communities. They rely on non-timber forest products for livelihood and traditional healthcare systems (Haq et al. 2021). However, rapid globalization and modernization have led to a decline in traditional ecological knowledge (Fernández-Giménez and Fillat Estaque 2012). Ethnobotanical documentation aids in conserving plant diversity, culture, and interactions, aiding in the identification of new ingredients for medications (Malhotra et al. 2021), and symbiotically relating to cultural diversity and religious traditions in the Himalayas, providing direct economic benefits and alleviating poverty (Olofsson et al. 2021). The Himalayan region is at the edge of an herbal resurgence, providing medicinal plant wealth to meet global market requirements. However, this knowledge is increasingly threatened by factors like biodiversity destruction, loss of indigenous knowledge, bioprospecting, climate change, and improper government policies. Therefore, sustainable conservation and management are crucial for this valuable resource (Sharma et al. 2024).

9.3.1.6 Future Directions in Healthcare Applications

Knowledge of medicinal plants is passed down over generations, particularly in rural communities (Verma 2014). Herbal practice in Himalayas efficiently handles health problems, particularly in remote locations. Insight of medicinal plant conservation connects environmental conservation and indigenous knowledge (Cameron 2008). However, this knowledge and cultural practices are declining. Investigating these habits is critical for future research and improving the cultural practices of nomads and other communities who rely on plants (Jardine 2019). It is important to highlight the potential of traditional knowledge for affordable healthcare and conservation practices. There should be scientific validation of medicinal plants for efficacy against diseases, highlighting the importance of local abundance and seasonal diseases. Legal regulations and ecosystem restoration measures are needed to protect biodiversity and protect plant taxa.

9.3.2 Socioeconomic Significance

9.3.2.1 Economic Value of High-Altitude Plants

Plants have long been used for physical and spiritual well-being, and medicinal plants continue to gain prominence (Awoyemi et al. 2012). Around 40,000–50,000 plant species are used for traditional healthcare and drugs worldwide. Sixty-five to eighty percent of the world's population, particularly in developing countries, depend on plants for healing due to poverty and lack of access to modern medicine. Traditional herbal medicine has a long history and is considered safe and non-toxic. The market for herbal drugs is estimated at 40 billion, with an expected 16% increase in the next 3–4 years (Semwal et al. 2019). Thus, recognizing the significance of medicinal plant documentation for future conservation is critical before it disappears from nature. Aside from safeguarding these rich plant resources of great economic significance, valuable plant information from the Himalayas must be systematically collected and documented for future generations.

9.3.2.2 Community-Based Initiatives

The Himalayan region's mountain ecosystems are crucial for economic development and human well-being, providing necessities like fresh water, energy, food, and medicinal plants (Liu et al. 2021). Ethnobotanical knowledge is derived from interactions between humans and their environment, influenced by factors (Maffi 2005) like local classification systems, language, cultural history, beliefs, religion, social networks, and information access (Bandiera and Rasul 2006). The diverse ethnic communities in the region, rely on forests for their primary healthcare and livelihoods. Ethnic groups in the region have unique indigenous healthcare systems using medicinal plants. Rural communities rely on herbal resources for disease treatment, resulting in folk medicine and traditional systems. Scientific research has yielded significant medications in Western medicine (Thakur et al. 2014).

9.3.2.3 Impact on Livelihoods

The diverse ethnic communities in the region, rely on forests for their primary healthcare and livelihoods (Uniyal 2003). Natural Forest Products (NTFPs) play a crucial role in the Himalayan region, where rural populations rely on forests for livelihood needs. These products provide food, nutrition, fodder, fiber, medicine, and other uses, generating substantial revenue (Sundriyal and Sundriyal 2004). Livelihood security depends on the condition of natural resources, with many village communities generating 10–50% of their income from forest products (Abbasi et al. 2013). The Food and Agriculture Organization estimates that 80% of developing countries rely on NTFPs for nutritional and health needs (Saha and Sundriyal 2012). A Venn diagram illustrating the synergy between sustainable livelihoods and medicinal plants is shown in Fig. 9.2.

9.3.2.4 Policy and Governance

Allopathic medicine, on the other hand, remains out of reach for the vast majority of villagers who live in remote areas with little access to medical facilities. Documenting traditional healthcare practices can assist planners and politicians in better managing and sustaining these local resources. Operational plans focus on forest management for timber, fuelwood, and fodder, but lack specific guidelines for NTFP development. Existing forest rules lack clarity on scope and opportunities. Government bans on threatened species have not enhanced conservation, leading to illegal trade and disincentives for communities to conserve valuable species (Subedi 2004). In the case of Non-Timber forest products, poor communities heavily rely on NTFPs for domestic and market needs, causing pressure (Masoodi and Sundriyal 2020). State agencies lack a robust conservation plan. Long-term management requires species-specific strategies, proper harvesting, cultivation, quality material supply, product diversification, value chain development, and market assurance.

9.4 Challenges and Opportunities

The agricultural sector in high-elevation Himalayan locations struggles to give enough pay to people due to the lack of access to market centers, motorable roads, and variables such as medicinal and fragrant plants. Local people thus depend on natural resources for subsistence and income, but population pressure, ecosystem disturbances, and overexploitation by industries limit natural regeneration of endangered flora and fauna (Nautiyal et al. 2005). The commercial trade of natural products, particularly plant medicines and crafts, has led to overexploitation of many species, particularly commercially viable ones (Joshi et al. 2018). Scientific documentation on diversity, distribution, and economic importance can help conserve and sustainably use these resources, while quantitative understanding of biological and cultural diversity is crucial for local population conservation (Kufer et al. 2005). The trade of medicinal plants has been studied extensively, to improve rural livelihoods. However, patterns of trade, particularly in threatened high-altitude species, are rarely available. Empirical data does not depict unfair harvester exploitation,

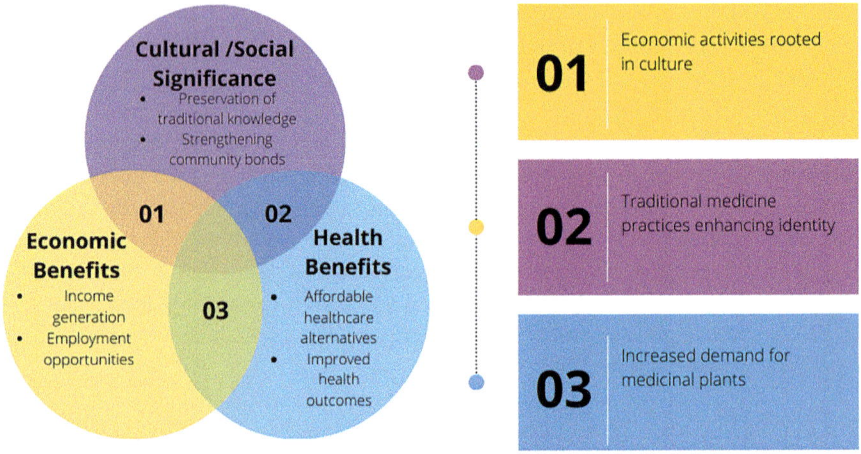

Fig. 9.2 Venn diagram depicting synergy of benefits: Sustainable live hood and medicinal plants

and more stringent price data collection is needed (Chauhan et al. 2013). There should be clear plans regarding household income, trade patterns, and policy management for the conservation of medicinal plant species in hill people. It is a great opportunity that ethno-medicine utilizes plant therapeutic properties for modern medications, with 70,000 plant species used globally, especially in developing nations due to affordability and safety compared to conventional therapies (Aziz et al. 2018).

References

Abbasi AM, Khan MA, Shah MH, Shah MM, Pervez A, Ahmad M (2013) Ethnobotanical appraisal and cultural values of medicinally important wild edible vegetables of Lesser Himalayas-Pakistan. J Ethnobiol Ethnomed 9:1–13

Abdullah A, Khan SM, Pieroni A, Haq A, Haq ZU, Ahmad Z, Sakhi S, Hashem A, Al-Arjani A-BF, Alqarawi AA (2021) A comprehensive appraisal of the wild food plants and food system of tribal cultures in the Hindu Kush Mountain Range: a way forward for balancing human nutrition and food security. Sustain For 13:5258

Acharya BK, Chettri B, Vijayan L (2011) Distribution pattern of trees along an elevation gradient of Eastern Himalaya, India. Acta Oecol 37:329–336

Adnan M, Hölscher D (2011) Medicinal plants in old-growth, degraded and re-growth forests of NW Pakistan. For Ecol Manag 261:2105–2114

Alamgeer, Sharif A, Asif H, Younis W, Riaz H, Bukhari IA, Assiri AM (2018) Indigenous medicinal plants of Pakistan used to treat skin diseases: a review. Chin Med 13:1–26

Awan AA, Akhtar T, Ahmed MJ, Murtaza G (2021) Quantitative ethnobotany of medicinal plants uses in the Jhelum valley, Azad Kashmir, Pakistan. Acta Ecol Sin 41:88–96

Awoyemi O, Abdulkarim I, Ewa E, Aduloju A (2012) Ethnobotanical assessment of herbal plants in South-Western Nigeria. Acad Res Int 2:50

Aziz MA, Adnan M, Khan AH, Shahat AA, Al-Said MS, Ullah R (2018) Traditional uses of medicinal plants practiced by the indigenous communities at Mohmand Agency, FATA, Pakistan. J Ethnobiol Ethnomed 14:1–16

Badola HK, Aitken S (2003) The Himalayas of India: a treasury of medicinal plants under siege. Biodiversity 4:3–13

Balick MJ (1996) Transforming ethnobotany for the new millennium. Ann Mo Bot Gard 83:58–66

Bandiera O, Rasul I (2006) Social networks and technology adoption in northern Mozambique. Econ J 116:869–902

Bhatia A, Arora S, Singh B, Kaur G, Nagpal A (2011) Anticancer potential of Himalayan plants. Phytochem Rev 10:309–323

Bisht M, Chandra Sekar K, Mukherjee S, Thapliyal N, Bahukhandi A, Singh D, Bhojak P, Mehta P, Upadhyay S, Dey D (2022) Influence of anthropogenic pressure on the plant species richness and diversity along the elevation gradients of Indian Himalayan high-altitude protected areas. Front Ecol Evol 10:751989

Bringmann G, Noll TF, Gulder TA, Grüne M, Dreyer M, Wilde C, Pankewitz F, Hilker M, Payne GD, Jones AL (2006) Different polyketide folding modes converge to an identical molecular architecture. Nat Chem Biol 2:429–433

Cameron M (2008) Modern desires, knowledge control, and physician resistance: regulating Ayurvedic medicine in Nepal. Asian Med 4:86–112

Cao Y, Xia Q-H, Meng H, Zhong A-P (2005) Antitumor and synergistic effect of Chinese medicine "bushen huayu jiedu recipe" and chemotherapy on transplanted animal hepatocarcinoma. World J Gastroenterol: WJG 11:5218

Chauhan R, Nautiyal B, Nautiyal M (2013) Trade of threatened Himalayan medicinal and aromatic plants-socioeconomy, management and conservation issues in Garhwal Himalaya, India. Global J Pathol Microbiol 13:9–18

Chen JK, Chen TT, Crampton L (2004) Chinese medical herbology and pharmacology. Art of Medicine Press City of Industry, Williams

Chen S-L, Yu H, Luo H-M, Wu Q, Li C-F, Steinmetz A (2016) Conservation and sustainable use of medicinal plants: problems, progress, and prospects. Chin Med 11:1–10

Dhyani D, Maikhuri R, Rao K, Kumar L, Purohit V, Sundriyal M, Saxena K (2007) Basic nutritional attributes of *Hippophae rhamnoides* (Seabuckthorn) populations from Uttarakhand Himalaya, India. Curr Sci 92:1148–1152

Farnsworth N, Akerele O, Bingel A (1985) Medicinal plants in therapy. Bull World Health Organ 63(6):965–981

Fernández-Giménez ME, Fillat Estaque F (2012) Pyrenean pastoralists' ecological knowledge: documentation and application to natural resource management and adaptation. Hum Ecol 40:287–300

Gillani SW, Ahmad M, Zafar M, Haq SM, Waheed M, Manzoor M, Shaheen H, Sultana S, Rehman FU, Makhkamov T (2024a) An insight into indigenous ethnobotanical knowledge of medicinal and aromatic plants from Kashmir Himalayan region. Ethnobot Res Appl 28:1–21

Gillani SW, Ahmad M, Zafar M, Manzoor M, Shah GM, Shaheen H, Zaman W, Sultana S, Sadia B, Khishlatovna KK (2024b) Ethnobotanical exploration of traditional medicinal plants among the rural inhabitants of district Muzaffarabad, Kashmir Himalayan Region. Plant Sci Today 11. https://horizonepublishing.com/journals/index.php/PST/article/view/3265

Grumbine RE, Xu J (2021) Mountain futures: pursuing innovative adaptations in coupled social–ecological systems. Front Ecol Environ 19:342–348

Haq SM, Yaqoob U, Calixto ES, Rahman IU, Hashem A, Abd-Allah EF, Alakeel MA, Alqarawi AA, Abdalla M, Hassan M (2021) Plant resources utilization among different ethnic groups of Ladakh in Trans-Himalayan Region. Biology 10:827

Haq SM, Waheed M, Khoja AA, Amjad MS, Bussmann RW, Ali K (2023) A cross-cultural study of high-altitude botanical resources among diverse ethnic groups in Kashmir Himalaya, India. J Ethnobiol Ethnomed 19:12

Heinrich M, Jiang H, Scotti F, Booker A, Walt H, Weckerle C, Maake C (2021) Medicinal plants from the Himalayan region for potential novel antimicrobial and anti-inflammatory skin treatments. J Pharm Pharmacol 73:956–967

Hu Y, Jiang Z, Leung KS-Y, Zhao Z (2006) Simultaneous determination of naphthoquinone derivatives in Boraginaceous herbs by high-performance liquid chromatography. Anal Chim Acta 577:26–31

Ishida T, Sakaguchi I (2007) Protection of human keratinocytes from UVB-induced inflammation using root extract of Lithospermum erythrorhizon. Biol Pharm Bull 30:928–934

Jardine TD (2019) Indigenous knowledge as a remedy for shifting baseline syndrome. Front Ecol Environ 17:13–14

Joshi RK, Satyal P, Setzer WN (2016) Himalayan aromatic medicinal plants: a review of their ethnopharmacology, volatile phytochemistry, and biological activities. Medicine 3:6

Joshi S, Ballabh B, Negi P, Dwivedi S (2018) Diversity, distribution, use pattern and evaluation of wild edible plants of Uttarakhand, India. Def Life Sci J 3:126–135

Kamboj VP (2000) Herbal medicine. Curr Sci 78:35–39

Kokkoris IP, Drakou EG, Maes J, Dimopoulos P (2018) Ecosystem services supply in protected mountains of Greece: setting the baseline for conservation management. Int J Biodivers Sci Ecosyst Serv Manage 14:45–59

Körner C (2007) The use of 'altitude' in ecological research. Trends Ecol Evol 22:569–574

Kufer J, Heinrich M, Förther H, Pöll E (2005) Historical and modern medicinal plant uses—the example of the Ch'orti 'Maya and Ladinos' in eastern Guatemala. J Pharm Pharmacol 57:1127–1152

Kumar M, Sheikh MA, Bussmann RW (2011) Ethnomedicinal and ecological status of plants in Garhwal Himalaya, India. J Ethnobiol Ethnomed 7:1–13

Kumar A, Pandey VC, Singh AG, Tewari DD (2013) Traditional uses of medicinal plants for dermatological healthcare management practices by the Tharu tribal community of Uttar Pradesh, India. Genet Resour Crop Evol 60:203–224

Lakey, Dorji K (2016) Ecological status of high altitude medicinal plants and their sustainability: Lingshi, Bhutan. BMC Ecol 16:1–14

Lewis WH (2003) Pharmaceutical discoveries based on ethnomedicinal plants: 1985 to 2000 and beyond. Econ Bot 57:126–134

Liu F, Gao C, Chen M, Tang G, Sun Y, Li K (2021) The impacts of flowering phenology on the reproductive success of the narrow endemic Nouelia insignis Franch (Asteraceae). Ecol Evol 11:9396–9409

Maffi L (2005) Linguistic, cultural, and biological diversity. Annu Rev Anthropol 34:599–617

Malhotra A, Nandigama S, Bhattacharya KS (2021) Food, fields and forage: a socio-ecological account of cultural transitions among the Gaddis of Himachal Pradesh in India. Heliyon 7(7):e07569

Malik ZA, Bhat JA, Bhatt A (2014) Forest resource use pattern in Kedarnath wildlife sanctuary and its fringe areas (a case study from Western Himalaya, India). Energy Policy 67:138–145

Malik ZA, Bhat JA, Ballabha R, Bussmann RW, Bhatt A (2015) Ethnomedicinal plants traditionally used in health care practices by inhabitants of Western Himalaya. J Ethnopharmacol 172:133–144

Mallen-Cooper J, Pickering CM (2008) Linear declines in exotic and native plant species richness along an increasing altitudinal gradient in the Snowy Mountains, Australia. Austral Ecol 33:684–690

Manzoor M, Ahmad M, Zafar M, Gillani SW, Shaheen H, Pieroni A, Al-Ghamdi AA, Elshikh MS, Saqib S, Makhkamov T (2023) The local medicinal plant knowledge in Kashmir Western Himalaya: a way to foster ecological transition via community-centred health seeking strategies. J Ethnobiol Ethnomed 19:56

Masoodi HUR, Sundriyal R (2020) Richness of non-timber forest products in Himalayan communities—diversity, distribution, use pattern and conservation status. J Ethnobiol Ethnomed 16:1–15

Nautiyal S, Rajan K, Shibasaki R (2005) Interaction of biodiversity and economic welfare—a case study from the Himalayas of India. J Environ Inf 6:111–119

Nersesyan A, Muradyan R (2004) Sea-buckthorn juice protects mice against genotoxic action of cisplatin. Exp Oncol 26:153–155

Olofsson M, Ros-Tonen M, Gupta J, de Steenhuijsen Piters B, Van Leynseele Y (2021) Rethinking the divide: exploring the interdependence between global and nested local markets. J Rural Stud 83:60–70

Painuli S, Semwal P, Cruz-Martins N, Bachheti RK (2021) Medicinal plants of Himalayan forests. In: Non-timber forest products: food, healthcare and industrial applications. Springer, pp 175–212

Pie S, Manandhar N (1987) Sources of some local medicines in the Himalayan Regions. Himalayan Ecosyst 97:112

Prakash R (2015) Medicinal plants used by tribal communities: a study of Uttarakhand Himalayan region. Int J Humanit Soc Sci Invent 4:55–61

Radha, Kumar M, Puri S, Pundir A, Bangar SP, Changan S, Choudhary P, Parameswari E, Alhariri A, Samota MK (2021) Evaluation of nutritional, phytochemical, and mineral composition of selected medicinal plants for therapeutic uses from cold desert of Western Himalaya. Plan Theory 10:1429

Rao KR (1978) Effects of ecdysterone, inokosterone and eyestalk ablation on limb regeneration in the fiddler crab, Uca pugilator. J Exp Zool 203:257–269

Rokaya MB, Münzbergová Z, Timsina B (2010) Ethnobotanical study of medicinal plants from the Humla district of western Nepal. J Ethnopharmacol 130:485–504

Saha D, Sundriyal R (2012) Utilization of non-timber forest products in humid tropics: implications for management and livelihood. Forest Policy Econ 14:28–40

Satyal P, Paudel P, Kafle A, Pokharel SK, Lamichhane B, Dosoky NS, Moriarity DM, Setzer WN (2012) Bioactivities of volatile components from Nepalese Artemisia species. Nat Prod Commun 7:1934578X1200701228

Semwal DK, Chauhan A, Kumar A, Aswal S, Semwal RB, Kumar A (2019) Status of Indian medicinal plants in the International Union for Conservation of Nature and the Future of Ayurvedic Drugs: shouldn't think about Ayurvedic fundamentals? J Integr Med 17:238–243

Sen S, Chakraborty R, De B (2011) Challenges and opportunities in the advancement of herbal medicine: India's position and role in a global context. J Herb Med 1:67–75

Sharma J, Gairola S, Sharma YP, Gaur R (2014) Ethnomedicinal plants used to treat skin diseases by Tharu community of district Udham Singh Nagar, Uttarakhand, India. J Ethnopharmacol 158:140–206

Sharma A, Patel SK, Singh GS (2024) Revitalizing traditional knowledge of the Himalayan indigenous tribes: an ethnoecological and ethnobotanical approach, learning 'from' and 'with' the locals: traditional knowledge systems for environmental sustainability in the Himalayas. Springer, pp 111–133

Shewamene Z, Dune T, Smith CA (2020) Use of traditional and complementary medicine for maternal health and wellbeing by African migrant women in Australia: a mixed method study. BMC Complement Med Ther 20:1–12

Singh B, Singh B, Kishor A, Singh S, Bhat MN, Surmal O, Musarella CM (2020) Exploring plant-based ethnomedicine and quantitative ethnopharmacology: medicinal plants utilized by the population of Jasrota Hill in Western Himalaya. Sustain For 12:7526

Subedi BP (2004) Linking plant-based enterprises and local communities to biodiversity conservation in Nepal Himalaya. Unpublished doctoral Dissertation, Kumaun University, India

Sundriyal M, Sundriyal R (2004) Wild edible plants of the Sikkim Himalaya: marketing, value addition and implications for management. Econ Bot 58:300–315

Thakur KS, Kumar M, Bawa R, Bussmann RW (2014) Ethnobotanical study of herbaceous flora along an altitudinal gradient in Bharmour Forest Division, District Chamba of Himachal Pradesh, India. Evid Based Complement Alternat Med 2014:946870

Tiitinen KM, Yang B, Haraldsson GG, Jonsdottir S, Kallio HP (2006) Fast analysis of sugars, fruit acids, and vitamin C in sea buckthorn (*Hippophae rhamnoides* L.) varieties. J Agric Food Chem 54:2508–2513

Tsering J, Tag H (2015) High altitude ethnomedicinal plants of Western Arunachal Himalayan landscape. Pleione 9:18–25

Uniyal B (2003) Utilization of medicinal plants by the rural women of Kullu, Himachal Pradesh. Indian J Tradit Knowl 2(4):366–370

Uniyal SK, Singh K, Jamwal P, Lal B (2006) Traditional use of medicinal plants among the tribal communities of Chhota Bhangal, Western Himalaya. J Ethnobiol Ethnomed 2:1–8

Verma RK (2014) An ethnobotanical study of plants used for the treatment of livestock diseases in Tikamgarh District of Bundelkhand, Central India. Asian Pac J Trop Biomed 4:S460–S467

Vetaas OR, Grytnes JA (2002) Distribution of vascular plant species richness and endemic richness along the Himalayan elevation gradient in Nepal. Glob Ecol Biogeogr 11:291–301

Vidyarthi S, Samant SS, Sharma P (2013) Traditional and indigenous uses of medicinal plants by local residents in Himachal Pradesh, North Western Himalaya, India. Int J Biodivers Sci Ecosyst Serv Manage 9:185–200

Wagh VV, Jain AK (2020) Ethnopharmacological survey of plants used by the Bhil and Bhilala ethnic community in dermatological disorders in Western Madhya Pradesh, India. J Herb Med 19:100234

High Altitude Medicinal Plants and Integrative Medicine in Bhutan: Identification, Phytochemical and Pharmacological Properties

10

Thinley Gyeltshen and Phurpa Wangchuk

Abstract

Traditional medicine of Bhutan was once the primary provider of health care services in Bhutan and continues to deliver services to the Bhutanese population as part of the integrated medical system. This chapter reviewed the phytochemicals present in the high-altitude medicinal plants of Bhutan isolated through natural products isolation studies and the pharmacological properties of the extracts and phytochemicals. Phytochemicals were isolated from 11 medicinal plants, and these represent less than 2% of medicinal plants known in Bhutan. More than 50 metabolites were isolated belonging to alkaloids, flavonoids, terpenes, phenylpropanoids, and furanocoumarins. Primarily, diterpenoid alkaloids were isolated from two *Acotinium* species, whereas isoquinoline alkaloids were obtained from *Meconopsis simplicifolia* and three *Corydalis* species. Anthocyanins and flavonol glycosides were isolated as constituents of blue and yellow flowers of four *Meconopsis* species that feature in BTM. The crude extracts and selected isolated metabolites from medicinal plants *A. laciniatum, A. nubigena, C. bhutanica, C. crispa, C. dubia, M. simplicifolia* and *H. amabile* reportedly exhibited anti-inflammatory and immunomodulatory activities, as well as antiplasmodial properties. These reported pharmacological properties were consistent with the traditional uses and validate their uses in BTM.

T. Gyeltshen (✉)
University of Tasmania, Sandy Bay, TAS, Australia
e-mail: thinley.gyeltshen@utas.edu.au

P. Wangchuk (✉)
James Cook University, Smithfield, QLD, Australia
e-mail: phurpa.wangchuk@jcu.edu.au

S. Chandra, H. Lata (eds.), *High Altitude Medicinal Plants*,
https://doi.org/10.1007/978-3-031-99406-7_10

10.1 Introduction

Humans have long recognized plants for their beneficial and medicinal properties and for thousands of years have been used to treat an array of diseases. Such knowledge concerning the use of plants have evolved into many distinct practices in different regions around the world and has been passed down for many generations that continues to thrive today as traditional medicinal practices (Yuan et al. 2016). Many of these knowledge and practices of using plants as medicines have been recorded, such as the Traditional Chinese Medicine (TCM), Indian Ayurvedic medicine, Japanese Campo medicine, Traditional Korean Medicine, Tibetan Sowa Rigpa medicine and Unani (Moreira et al. 2014; Yuan et al. 2016). These traditional practices continues to serve a large portion of global population, mainly in the developing countries, as an alternative or additional service to modern medicine (Ekor 2014). Today, this traditional knowledge serves as an important lead towards discovery of secondary metabolites that are used as drugs in modern medicine (Ekor 2014). A classic example is artemisinin, one of the many drugs used in treatment of malaria caused by *Plasmodium falciparum*, and was isolated from the plant *Artemisia annua* based on the information obtained from the Chinese Traditional Medicine (Pan et al. 2013; Yuan et al. 2016).

Tibetan *Sowa Rigpa* medicine, which originated in central Tibet between seventh and twelfth century (Kloos et al. 2020) has emerged as one of the prominent scholarly medical knowledge of the century with strong influences from Aryuvedic and Chinese medicine practices. It has spread to surrounding regions together with the Tibetan Buddhism (Tobgay et al. 2011; Kloos et al. 2020). Presently, besides its wide practice in the autonomous region of Tibet (China), *Sowa Rigpa* is also widely practiced in the Bhutan, Nepal, India, Mongolia, the Buryat and Tuva regions and Kalmykia of the Russian Federation (Gurmet 2004; Ulanov and Badmaev 2018; Yeshi et al. 2019; Ghimire et al. 2021). It is one of the vibrant scholarly traditional medical systems of the world that not only provides primary health care services but also has potential application in modern pharmaceuticals.

In Bhutan, the Tibetan *Sowa Rigpa* medicine was formally introduced by Lama Zhabdrung Ngawang Namgyal in the seventeeth century along with Kagyupa sect of Buddhism (Wangchuk et al. 2013e). *Sowa rigpa* in Bhutan was initially taught in the monastic institutions (Wangchuk et al. 2016a). It was Tenzing Drugyal, the personal physician of Zhabdrung, who is attributed for spreading the teaching of *Sowa rigpa* in Bhutan (Wangchuk et al. 2007b; Steiner 2017; McKay and Wangchuk 2018). The medical philosophy, disease concepts, the diagnosis of diseases, and the treatment regimens remained same as Tibetan medicine (McKay and Wangchuk 2018). However, over hundreds of years that followed, the Bhutanese traditional medicine has been gradually tailored by the traditional physicians according to the local needs, surrounding biodiversity, and historical prevalence of diseases in the country (Yeshi et al. 2019). The differences have been highlighted by McKay and Wangchuk (2018). For this reason, it is locally known as Bhutanese traditional medicine (BTM). The BTM continues to play key role in providing health care services to Bhutanese population. An estimate of 30,000 patients continue to avail the services annually (Lhamo and Nebel 2011).

The BTM uses three main ingredients in the current formulations, which are derived from vegetables, animals, and minerals. The medicinal plants that belong to the category of vegetable origin are the bulk ingredient and are divided into two sub-categories: high altitude medicinal plants (2000–6000 m above sea level) and low altitude medicinal plants (95–2000 m above sea level). This chapter describes how these high-altitude medicinal plants of Bhutan are identified and used in the integrated health care system of Bhutan. It also describes the reported phytochemicals and the biological activities of selected plants, which grows in the higher altitude Himalayan mountains of Bhutan.

10.2 Integrated Medical System of Bhutan

In Bhutan, for hundreds of years, the Bhutanese traditional medicine was the primary provider of health care services to the Bhutanese population, until the western allopathic medicine arrived in the 1960s (Tobgay et al. 2011; Wangchuk et al. 2013e; Yangchen et al. 2016). Today, both modern allopathic and traditional medical systems co-exist and comprise the national health care systems in Bhutan. This is as a result of the integration of the traditional medical services into the national health system in 1967, to maximize public healthcare services and promote conservation of traditional practices (Lhamo and Nebel 2011; Wangchuk et al. 2016a; McKay and Wangchuk 2018).

The traditional Bhutanese medicine was formally integrated into the national medical system of Bhutan through a Royal Decree issued in 1967 (Wangchuk et al. 2007b; Lhamo and Nebel 2011; Yangchen et al. 2016; McKay and Wangchuk 2018). It began from a single dispensary unit that was first established in Thimphu in 1968 and underwent multiple upgradations until becoming the Institute of Traditional Medicine Services (ITMS) in 1998 (Wangchuk et al. 2007b; Lhamo and Nebel 2011). Today there are three separate organizational entities: (i) Department of Traditional Medicine Services (DTMS), (ii) Faculty of Traditional Medicine (FoTM), and (iii) the Menjong Sorig Pharmaceuticals Corporation Limited (MSPCL) under the Ministry of Health. While the DTMS is responsible for providing the medical care and therapies through different district hospitals and basic health units in the country, the FoTM serves as a premier training institute for producing highly trained traditional physicians and health assistants. Under umbrella of the Ministry of Health, the traditional hospitals and units were established in all 20 districts of Bhutan by the end of 2001, and attached to all the hospitals in the 20 districts and 29 Basic Health Units (Lhamo and Nebel 2011; Wangchuk et al. 2013e). Presently, there are more than 80 traditional medicine units established across all districts in Bhutan, providing the traditional medical services to their communities. These health service centers provide the primary healthcare services to 20–30% of the outpatients in the country.

The MSPCL is tasked with the collection of raw materials, produce traditional medicines, perform research to assess efficacy and quality assurances of the medicines. Presently, MSPCL produces 108 essential traditional medicines in different dosage forms such as tablets, capsules, syrup, ointment, powder, and herbal mixtures for the Ministry of Health. These essential medicines list is reviewed every after 2 years. Depending upon the disease trend and the requirement of the medicines, the new essential medicines are added, or the old ones are removed from the list. These essential medicines are all prescription medicines and cannot be accessed as over the counter drugs through pharmacy or chemist shops in Bhutan. In addition to essential medicines, the MSPCL produces 19 health supplements and herbal products. Bhutan's rich biodiversity supports this important health system.

10.3 Medicinal Plants Diversity of Bhutan

Geographically, Bhutan is a small mountainous country, nestled between China and India. Situated in the Eastern region of the Himalayas, the mountainous terrain arises from the Indo-Gangetic plains with elevation points that are as low as 100 m above sea level (masl) in the south to mountainous peaks as high as 7570 masl adjoining the Tibetan plateau in the north (Gyeltshen et al. 2020). These varying landscapes provide a diverse habitat for a wide range of flora and fauna. This is further aided by the strong policies Bhutan has in place which has promoted environmental conservation and allowed more than 70.46% of the land area to remain covered with forest (Gyeltshen et al. 2020). As of 2017, it was reported that 5369 plant species are found in Bhutan (Gyeltshen et al. 2020). Of these higher plant species, more than 600 medicinal plants have been identified for using in BTM. However, only 116 high altitude medicinal plants and 108 low altitude medicinal plants were botanically described to date (Wangchuk 2009a, b). There are two separate collection and drying centers in Bhutan—Lingzhi for high altitude plants and Langthel for low altitude plants. The MSPCL organize the annual collection of plants with the respective farmers and the farmers get paid for whatever collection they make. These farmers are trained on the sustainable collection medicinal plants.

10.4 High Altitude Medicinal Plants and Identification

High-altitude medicinal plants are collected between the months of May and September, and are collected from their pristine mountain wild habitat (Fig. 10.1a). Lingzhi under Thimphu district has been a permanent collection sites for high altitude medicinal plants (Fig. 10.1b). Identification of these high altitude medicinal plants are based on the traditional medicinal texts and are performed by a team of the traditional physicians and plant taxonomists (Wangchuk et al. 2008). The process of identification also involves physical verification in the field by experienced team members through frequent field visits and studies. Some of the commonly collected high altitude medicinal plants are given in Fig. 10.1c–f.

Fig. 10.1 Himalayan high altitude medicinal plants of Bhutan. (**a**) Vegetation type. (**b**) Lingzhi drying center. (**c**) *Meconopsis horridulla*. (**d**) *Meconopsis simplicifolia*. (**e**) *Aconitum laciniatum*. (**f**) *Aconitum orochryseum*

Due to continuous collection program, the pressure on medicinal plants has been increasing in this collection region. In addition, yak grazing and climate change are constantly changing the landscape and the medicinal plant density in the region, undermining the long-term sustainability of not only the income sources of the nomadic yak herders of the region but also the production of medicines that depend on these high-altitude medicinal plants. Recently, to relive pressure on the population density of medicinal plants that are under constant collection program, the Department of Traditional Medicinal Services under the Ministry of Health in 2020 has identified more than 530 high-altitude medicinal plant species (belonging to 87 Families) from other parts of the country (Wangchuk 2020).

10.5 Phytochemicals of Selected High-Altitudinal Medicinal Plants of Bhutan

A handful of studies performed are available on the medicinal plant species from Bhutan. As many as 25 high-altitude medicinal plants were screened for major class of phytochemicals such as alkaloids, flavonoids, tannins, terpenoids and saponins, that resulted in selection of seven medicinal plants which were subjected to further isolation and pharmacological studies (Wangchuk et al. 2011). Altogether, a total of 17 high altitude medicinal plants of Bhutan have been studied for their phytochemical and biological activities (see Table 10.1). These studies had reported on the isolation of more than 50 different phytochemicals belonging to alkaloids, flavonol glycosides, terpenes, phenylpropanoids and coumarins. For example, as many as 40 phytochemicals were isolated from five Bhutanese *Sowa Rigpa* medicinal plants: *Aconitum laciniatum*, *Ajania nubegina*, *Corydalis crispa*, *Corydalis dubia,* and *Pleurospermum amabile* (Wangchuk et al. 2018). Overall, phytochemicals were isolated from a total of 11 high-altitude medicinal plants. This represent less than 2% of the total medicinal plant species which are found in Bhutan that were subjected to natural products isolation studies.

10.5.1 Alkaloids

Alkaloids are a class of secondary metabolites and are of profound interest as they possess diverse biological properties, many of which has beneficial effects and are used as drugs for treatment of many diseases. Indeed, most drugs (for instance many chemotherapy drugs) from natural sources approved for use are alkaloids. Traditional practices have served as important source of information for drug discovery searches. As a result, countless alkaloids were isolated from medicinal plants used in the many traditional practices. Alkaloids have been isolated from few high-altitude medicinal plants of Bhutan.

Alkaloids have been isolated from five high-altitude medicinal plants of Bhutan, two species of genus *Aconitum*, three *Corydalis* species and from *M. simplicifolia*. The two species of *Aconitum* have yielded diterpenoid alkaloids, again the types of which were further distinctive between the two species. C_{19}-diterpenoid alkaloids, such as pseudaconitine, 14-veratroylpseudaconine, 14-*O*-acetylneoline, neoline, and senbusine A, were isolated for the first time from the roots of *A. laciniatum* (Wangchuk et al. 2015). Alkaloids isolated from *A. orochryseum* were of hetisine-type diterpenoid alkaloids, orochrine, 2-*O*-acetylorochrine, and 2-*O*-acetyl-7α-hydroxyorochrine, atisinium chloride and virescenine (Wangchuk et al. 2007a).

Table 10.1 A list of 16 high altitude medicinal plants of Bhutan studied for their phytochemical and biological activities

Botanical name	Family	Traditional uses	Isolated phytochemicals	Reported biological activities
Aconitum laciniatum (Brühl) Stapf	Ranunculaceae	As anthelmintic; towards gout and arthritis; in treatment of mumps and leprosy, and chronic infections	Alkaloids: pseudaconitine, 14-veratroylpseuda-conine, 14-*O*-acetylneoline, neoline, senbusine A (Wangchuk et al. 2015)	Anti-inflammatory (Wangchuk et al. 2013c, 2015); immunomodulatory properties (Wangchuk et al. 2018)
Aconitum orochryseum Stapf	Ranunculaceae	As an anti-dote for snake bites; bile inflammation and jaundice; in dysentery and food poisonings	Alkaloids: orochrine, 2-*O*-acetylorochrine, and 2-*O*-acetyl-7*R*-hydroxyorochrine, atisinium chloride, virescenine (Wangchuk et al. 2007a, 2010b)	Antiplasmodial (Wangchuk et al. 2010b)
Ajania nubigena (Wall. ex DC.) Muldashev (synonym *Tanacetum nubigenum* Wall. ex DC.)	Asteraceae	For alleviating abscess; kidney related disorders and swollen limbs; in haemorrhages and benign tumours	Terpenes: (3*R*,6*R*)-linalool oxide acetate, chamazulene, (*E*)-2-(2,4-hexadiynylidene)-1,6-dioxaspiro[4,4]non-3-ene, and (*Z*)-2-(2,4-hexadinylidene)-1,6-dioxas-piro[4,4]non-3-ene, 1-(4-hydroxyphenyl)propan-1-one (Wangchuk et al. 2016b, 2018; Shepherd et al. 2018); flavonoids: oxyanin B, luteolin, luteolin-7-*O*-β-D-glucoside (Wangchuk et al. 2013a, 2016b, 2018)	Anthelminthic (Wangchuk et al. 2013c, 2016b); immunomodulatory properties (Wangchuk et al. 2018); antimicrobial, antimalarial and cytotoxicity (Wangchuk et al. 2013a)
Aletris pauciflora (Klotzsch) Hand.-Mazz.	Nartheciaceae	Handling of hepatic disorders; respiratory issues including pneumonia; concerning blood infections		Antioxidant (Yeshi et al. 2017)

(continued)

Table 10.1 (continued)

Botanical name	Family	Traditional uses	Isolated phytochemicals	Reported biological activities
Codonopsis bhutanica Ludlow	Campanulaceae	Concerning kidney disorder; relieving of tingling and numbness; in gout and leprosy		Anti-inflammatory activity (Wangchuk et al. 2013c)
Corydalis calliantha D. G. Long	Papaveraceae	Treating of malaria, cough and cold, and biliary inflammation	Alkaloids: protopine, stylopine, cheilanthifoline, scoulerine (Wangchuk et al. 2010a)	Antiplasmodial (Wangchuk et al. 2010a)
Corydalis crispa Prain	Papaveraceae	As a febrifuge; for treating blood, liver and bile infections	Alkaloids: protopine, 13-oxoprotopine, 13-oxocry-ptopine, stylopine, coreximine, rheagenine, ochrobirine, sibiricine and bicuculline (Wangchuk et al. 2012a)	Anti-inflammatory (Wangchuk et al. 2013c); immunomodulatory properties (Wangchuk et al. 2018); acetylcholinesterase inhibitory and antiplasmodial (Wangchuk et al. 2012a)
Corydalis dubia Prain	Papaveraceae	As a febrifuge; for treating blood, liver and bile infections (malarial and antimicrobial)	Alkaloids: dubiamine, scoulerine, cheilanthifoline, protopine, capnoidine, bicuculline, corydecumbine, hydrastine (Wangchuk et al. 2012b; Shepherd et al. 2018)	Anti-inflammatory (Wangchuk et al. 2013c; Shepherd et al. 2018); immunomodulatory properties (Wangchuk et al. 2018); antimalarial (Wangchuk et al. 2012b)

Erysimum pachycarpum Hook. f. & Thomson (synonym *Erysimum bhutanicum* W. W. Sm.)	Brassicaceae	As an antispasmodic and antidepressant agent; concerning lung, heart and blood disorders; for relieving meat related food poisoning		Antioxidant (Yeshi et al. 2017)
Hymenidium amabile (Craib & W. W. Sm.) Pimenov & Kljuykov (synonym *Pleurospermum amabile* Craib ex W. W. Sm.)	Apiaceae	Treating dyspepsia, poisoning and fever	Phenylpropanoids: (*E*)-isomyristicin, (*E*)-isoapiol, methyl eugenol, (*E*)-isoelemicin (Wangchuk et al. 2013d, 2014; Shepherd et al. 2018); furanocoumarins: psoralen, bergapten, isoimperatorin, isopimpinellin, oxypeucedanin hydrate, oxypeucedanin methanolate (Wangchuk et al. 2014)	Anti-inflammatory activity (Wangchuk et al. 2013c); immunomodulatory properties (Wangchuk et al. 2018); antiplasmodial and antibacterial (Wangchuk et al. 2013d, 2014)
Koenigia tortuosa (D. Don) T. M. Schust. & Reveal (synomym *Aconogonon tortuosum* (D. Don) H. Hara)	Polygonaceae	Abating fever; in diarrhea and dysentery; in relieving haemorrhoids		Antioxidant (Yeshi et al. 2017)

(continued)

Table 10.1 (continued)

Botanical name	Family	Traditional uses	Isolated phytochemicals	Reported biological activities
Meconopsis horridula Hook. f. & Thomson	Papaveraceae	In cases of bone fractures including skull fractures; revitalizing of synovial fluid; treat wounds	Anthocyanins: cyanidin 3-*O*-sambubioside-7-*O*-glucoside, cyanidin 3-*O*-[xylosyl-(1 → 2)-(6″-malonylglucoside)]-7-*O*-glucoside (Iwashina et al. 2019); flavonol glycosides: kaempferol 3-*O*-glucoside, kaempferol 3-*O*-gentiobioside (Iwashina et al. 2019)	
Meconopsis paniculata (D. Don) Prain	Papaveraceae	Managing fever related to respiratory and hepatic disorders; promoting digestion	Flavonol glycosides: 6-hydroxykaempferol 3-*O*-glucoside, 6-hydroxykaempferol 3-*O*-sophoroside, kaempferol 3-*O*-glucoside, kaempferol 3-*O*-sopho-roside (Yokoyama et al. 2018)	
Meconopsis primulina Prain	Papaveraceae	For managing bleeding and blood regulation; removing of placenta	Anthocyanins: cyanidin 3-*O*-[xylosyl-(1 → 2)-(6″-malonylglucoside)], cyanidin 3-*O*-sambubioside, quercetin 3-*O*-glucoside, isorhamnetin 3-*O*-glycoside (Iwashina et al. 2019); flavonol glycosides: quercetin 3-*O*-glucoside, isorhamnetin 3-*O*-glycoside (Iwashina et al. 2019)	
Meconopsis simplicifolia (D. Don) Walp.	Papaveraceae	As an antipyretic agent against cough and cold, and fever associated with respiratory and liver inflammation	Anthocyanins: cyanidin 3-*O*-sambubioside-7-*O*-glucoside, cyanidin 3-*O*-[xylosyl-(1 → 2)-(6-malonylglucoside)]-7-*O*-glucoside, cyanidin 3-*O*-[xylosyl-(1 → 2)-(6-malonylglucoside)], cyanidin 3-*O*-sambubioside (Iwashina et al. 2019); flavonol glycosides: kaempferol 3-*O*-sophoroside, kaempferol 3-*O*-gentiobioside, kaempferol 3-*O*-xylosylgentiobioside (Iwashina et al. 2019); alkaloids: protopine, norsanguinarine, dihydro-sanguinarine, 6-methoxydihydrosanguinarine, oxysanguinarine (Wangchuk et al. 2013b)	Antiplasmodial (Wangchuk et al. 2013b); anti-inflammatory activity (Wangchuk et al. 2013c)

Thalictrum chelidonii DC.	Ranunculaceae	As an antidote, antimalarial, antipyretic, antiseptic, febrifuge		Antioxidant (Yeshi et al. 2017)
Thlaspi arvense L.	Brassicaceae	Alleviating respiratory and renal inflammation, pneumonia; reduce joints fluid accumulation		Antioxidant (Yeshi et al. 2017)

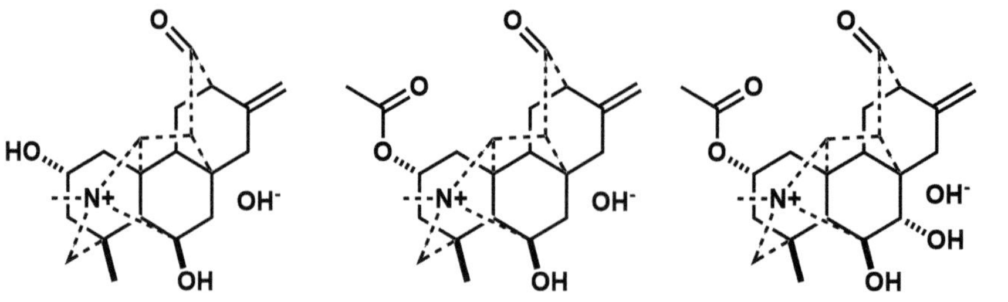

orochrine 2-*O*-acetylorochrine 2-*O*-acetyl-7α-hydroxyorochrine

Fig. 10.2 Three novel hetisine-type diterpenoid alkaloids isolated from *A. orochryseum*

Three of these hetisine-type diterpenoid alkaloids were previously not reported (Fig. 10.2).

Isoquinoline alkaloids were isolated from the three *Corydalis* species and *M. simplicifolia*. As many as 15 isoquinoline alkaloids have been isolated from three *Corydalis* species (Wangchuk et al. 2020). *C. crispa* provided nine known isoquinoline alkaloids, protopine, 13-oxoprotopine, 13-oxocryptopine, stylopine, coreximine, rheagenine, ochrobirine, sibiricine, and bicuculline (Wangchuk et al. 2012a). Propopine in *C. crispa* was the major alkaloid with 0.05% yield (w/w dry weight) (Wangchuk et al. 2012a). Likewise, *C. dubia* had yielded seven known isoquinoline alkaloids, scoulerine, cheilanthifoline, protopine, capnoidine, bicuculline, corydecumbine and hydrastine, together with a novel alkaloid, dubiamine (Wangchuk et al. 2012b). Alkaloids, capnoidine and scoulerine, were reisolated from the same plant in a subsequent study (Shepherd et al. 2018). Alkaloids obtained from *C. calliantha* were protopine, scoulerine, cheilanthifoline, and stylopine (Wangchuk et al. 2010a).

Among the four species of *Meconopsis* used in BTM, *M. simplicifolia* is the sole species from which alkaloids were isolated. A new protoberberine type isoquinoline alkaloid, simplicifolianine, together with five known alkaloids: protopine, norsanguinarine, dihydrosanguinarine, 6-methoxydihydrosanguinarine, and oxysanguinarine were obtained from the ariel parts of *M. simplicifolia* (Wangchuk et al. 2013b) (Fig. 10.3).

10.5.2 Terpenes

Terpenes are a class of natural products derived from isoprene units and widely distributed in plants. Terpenes in plants are the major constituent in essential oils (EO) (Masyita et al. 2022). Consistent to it, purification of *A. nubigena* EO had yielded (*3R,6R*)-linalool oxide acetate, chamazulene, (*E*)-2-(2,4-hexadiynylidene)-1,6-dioxaspiro[4,4]non-3-ene, and (*Z*)-2-(2,4-hexadinylidene)-1,6-dioxas-piro[4,4]non-3-ene, and 1-(4-hydroxyphenyl)propan-1-one (Wangchuk et al. 2013a). With 75.8% composition in the EO, (*3R,6R*)-linalool oxide acetate (1.2 g; 0.12% w/w dry weight) was identified as the major constituent in *A. nubigena* (Wangchuk et al. 2013a, 2018).

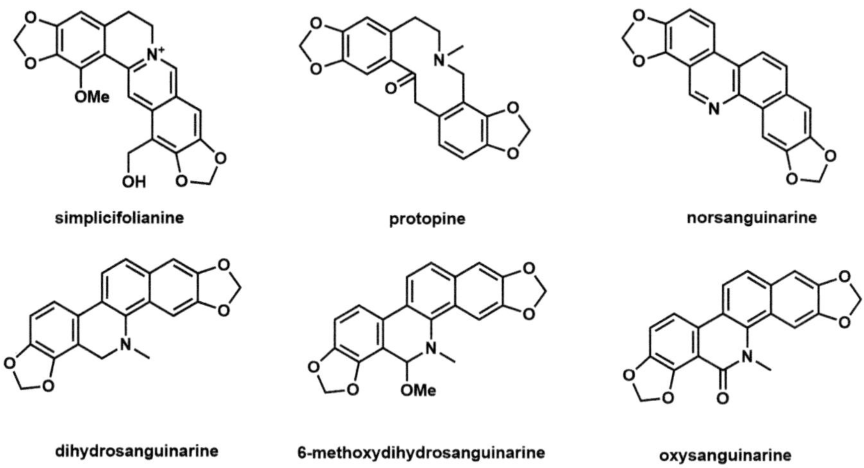

Fig. 10.3 Structures of isoquinoline alkaloids isolated from *M. simplicifolia*

This major molecule was isolated again from the ariel parts of the same plant species (Shepherd et al. 2018).

10.5.3 Flavonoids

Flavonoids are polyphenolic secondary metabolites which are widely distributed in plants. They are often consumed through plant dietary sources such as fruits and vegetables and many flavonoid molecules are known for their beneficial properties. It is unsurprising that these molecules have been isolated from high-altitude medicinal plants of Bhutan.

Alongside anthocyanins, a total of 12 flavonol glycosides were isolated from the blue flowers of six Bhutanese *Meconopsis* speces (Iwashina et al. 2019). Among the isolated flavonols, kaempferol 3-*O*-glucoside and kaempferol 3-*O*-gentiobioside were isolated from *M. horridula*, quercetin 3-*O*-glucoside and isorhamnetin 3-*O*-glycoside, from *M. primilina*, and kaempferol 3-*O*-sophoroside, kaempferol 3-*O*-gentiobioside and kaempferol 3-*O*-xylosylgentiobioside, from *M. simplicifolia* (Iwashina et al. 2019). Another similar study was conducted on the yellow flowers of *M. paniculata* and *M. integrifolia* (Yokoyama et al. 2018). Again, of the two *Meconopsis* species which produce yellow flowers, *M. paniculata* feature among the high-altitude medicinal plants. The yellow flowers of *M. paniculata* yielded two major flavonol glycosides, 6-hydroxykaempferol 3-*O*-glucoside and 6-hydroxykaempferol 3-*O*-sophoroside, together with two minor kaempferol 3-*O*-glucoside and kaempferol 3-*O*-sophoroside (Yokoyama et al. 2018). The two major flavonol constituents of *M. paniculata* yellow flowers were reported for the first time. Thus, together with anthocyanins, flavonol glycosides are the constituents in the blue- and yellow-colored flowers of *Meconopsis* species.

Besides the flavonol glycosides, flavones such as oxyanin B, luteolin, and luteolin-7-*O*-*β*-D-glucoside were obtained from the methanolic extract of *A. nubigena* (Wangchuk et al. 2013a). Luteolin in *A. nubigena* is a major constituent, followed by its glycoside, luteolin-7-*O*-*β*-D-glucoside, obtained in reasonable quantity (Wangchuk et al. 2018).

10.5.4 Furanocoumarins

Furanocoumarins are a class of secondary metabolites produced by a variety of plants, particularly in the families Apiaceae and Rutaceae. The high-altitude medicinal plant, *H. amabile* belong to Apiaceae family and six furanocoumarins were obtained from this medicinal plant. The isolated molecules were linear furanocoumarins, psoralen, bergapten, isoimperatorin, isopimpinellin, oxypeucedanin hydrate, and oxypeucedanin methanolate (Wangchuk et al. 2014). These furanocoumarins are the major phytochemicals in *H. amabile*, with bergapten (2.5 g, 0.1% w/w yield) obtained as the primary constituent from 2 kg of dried plant material and followed by others with reasonably good quantities.

10.5.5 Anthocyanins

Anthocyanins are colorful pigments found in red, blue, and purple colored fruits and flowers. Iwashina and group had surveyed the anthocyanin composition in six Bhutanese *Meconopsis* species that give blue flowers, among which three species, *M. horridula*, *M. primulina* and *M. simplicifolia*, feature in the BTM (Iwashina et al. 2019). In the study, a total of seven anthocyanins were isolated and the identities of four were determined. These anthocyanins were cyanidin 3-*O*-sambubioside-7-*O*-glucoside, cyanidin 3-*O*-[xylosyl-(1 → 2)-(6″-malonylglucoside)]-7-*O*-glucoside, cyanidin 3-*O*-[xylosyl-(1 → 2)-(6″-malonylglucoside)] and cyanidin 3-*O*-sambubioside (Iwashina et al. 2019). By respective species, all four anthocyanins were isolated from *M. simplicifolia*; the first two anthocyanins, cyanidin 3-*O*-sambubioside-7-*O*-glucoside and cyanidin 3-*O*-[xylosyl-(1 → 2)-(6″-malonylglucoside)]-7-*O*-glucoside, were isolated from *M. horridula*; and the latter two anthocyanins, cyanidin 3-*O*-[xylosyl-(1 → 2)-(6″-malonylglucoside)] and cyanidin 3-*O*-sambubioside, were isolated from *M. primulina* (Iwashina et al. 2019). However, the identity of additional three anthocyanins isolated in the study was not established.

10.5.6 Phenylpropanoids

Phenylpropanoids in plants are simple secondary metabolites derived from phenylalanine in the Shikimic pathway. Phenylpropenes are phenylpropanoids with allylbenzene (or 3-phenylpropene) unit derived from monolignols. These are aromatic metabolites and primary constituents in the essential oils of many aromatic plants.

H. amabile is an aromatic high-altitude plant used in BTM for treating dyspepsia, poisoning and fever (Wangchuk et al. 2014). From this plant, four phenylpropanes, (*E*)-isomyristicin, (*E*)-isoapiol, methyl eugenol and (*E*)-isoelemicin, were isolated together with furanocoumarins (Wangchuk et al. 2014). In comparison to furanocoumarins, these phenylpropenes are minor constituents in *H. amabile*, with only (*E*)-isomyristicin obtained in reasonable amount.

10.6 Pharmacological Properties

Several of the high-altitude medicinal plants of Bhutan that were subjected to phytochemical investigation were further evaluated for pharmacological properties of both crude extracts and metabolites obtained from them. In this way, few of the high-altitude plants were investigated for their anti-inflammatory, antiparasitic, and immunomodulatory activities. For examples, *A. laciniatum, A. nubigena, C. bhutanica, C. crispa, C. dubia, M. simplicifolia,* and *H. amabile*, which are used in BTM for management of various ailments including disorders, have been studied by the Bhutanese researchers to find scientific basis for their claimed effects of traditional uses in BTM.

10.6.1 Anti-inflammatory

The crude extracts from these medicinal plants were evaluated for in vitro anti-inflammatory activity by evaluating the inhibition of TNF-α production in LPS-activated THP-1 monocytic cells (Wangchuk et al. 2013c). Except for *A. laciniatum*, the crude extracts of the six medicinal plants displayed TNF-α inhibitory activity (Wangchuk et al. 2013c). Interestingly, despite the crude extract not exhibiting anti-inflammatory activity, diterpenoid alkaloids were isolated from *A. laciniatum* (Wangchuk et al. 2015). Two of the isolated alkaloids were tested, and 14-*O*-acetylneoline had displayed significant anti-inflammatory properties in the TNBS-induced colitis model in mice (Wangchuk et al. 2015). Similarly, capnoidine is an alkaloid isolated from *C. dubia*, and conferred protection against TNBS-induced colitis in mice (Shepherd et al. 2018). The findings of these studies in both in vitro and in vivo conditions validate the traditional uses of these high-altitude medicinal plants for management of inflammatory conditions.

10.6.2 Immunomodulatory

As many as 40 phytochemicals were isolated from five medicinal plants, *A. laciniatum, A. nubegina, C. crispa, C. dubia* and *P. amabile*. Selectively, 14 of the isolated compounds were tested for their immunomodulatory properties using a murine dendritic cell (DC) line (Wangchuk et al. 2018). Among the tested compounds, scoulerine exhibited adjuvant-like properties and significantly upregulated MHC-I gene and protein expression, and bergapten had demonstrated immunosuppressive properties and significantly downregulated MHC-I gene and protein expression and other co-stimulatory molecules (Wangchuk et al. 2018). Both molecules had exhibited low cytotoxicity against normal healthy cells, validating the traditional uses of these medicinal plants towards boosting of immune response.

10.6.3 Antimicrobial

Essential oils of *A. nubigena* had exhibited strong antibacterial activity against *Bacillus subtilis* and moderate antifungal activity against *Candida albicans* with MIZ values of 13 mm and 11 mm, respectively (Wangchuk et al. 2013a). However, (*Z*)-spiroether isolated from EO displayed weak and selective antimicrobial activity against *B. subtilis, S. epidermis,* and *C. albicans* (Wangchuk et al. 2013a). Interestingly, luteolin, a flavonoid purified from the methanol extract of *A. nubigena*, had exhibited antimicrobial properties against all gram-positive bacteria (*B. subtilis*, *Staphylococcus aureus*, methicillin resistant *S. aureus* (MRSA), *S. epidermidis*) and fungi (*C. albicans*) (Wangchuk et al. 2013a). Another study on the methanol extract obtained from *H. amabile* had afforded oxypeucedanin methanolate and had exhibited weak antimicrobial activity against *B. subtilis* (Wangchuk et al. 2013d, 2014).

10.6.4 Antimalarial

A few high-altitude medicinal plants had displayed antiparasitic properties. In the evaluation of Bhutanese medicinal plants for classes of phytochemicals and biological activities, five medicinal plants (*C. crispa, C. dubia, A. nubigena, M. simplicifolia, and H. amabile*) were identified to possess antiplasmodial properties (Wangchuk et al. 2011). Further isolation studies had yielded metabolites that exhibited in vitro antiplasmodial activity against the *P. falciparum* strains, TM4/8.2 (chloroquine–antifolate sensitive strain) and K1CB1 (multidrug resistant strain). These metabolites include coreximine from *C. crispa* (Wangchuk et al. 2012a); scoulerine, cheilanthifoline and protopine from *C. dubia* (Wangchuk et al. 2012b); luteolin and luteolin-7-*O*-*β*-D-glucoside from *A. nubigena* (Wangchuk et al. 2013a); simplicifolianine from *M. simplicifolia* (Wangchuk et al. 2013b); and isoimperatorin, isopimpinellin, oxypeucedanin hydrate and oxypeucedanin methanolate from *H. amabile* (Wangchuk et al. 2014). Similarly, alkaloids atisinium chloride isolated from *A. orochryseum* (Wangchuk et al. 2010b), and protopine and cheilanthifoline obtained from *C. calliantha* (Wangchuk et al. 2010a), had exhibited in vitro antiplasmodial activity against the *P. falciparum* strains. Atisinium chloride has since become an antimalarial structural lead.

10.6.5 Anthelmintic

Four major constituents that were isolated from *A. nubigena* were further investigated for anthelminthic activity against nematode whipworm *Trichuris muris* and platyhelminth blood fluke *Schistosoma mansoni* using the xWORM assay technique (Wangchuk et al. 2016b). Two compounds, luteolin and (3*R*,6*R*)-linalool oxide acetate, exhibited anthelmintic activity against *S. mansoni* and *T. muris* with IC_{50} range values of 5.8–36.9 μg/mL and 9.7–20.4 μg/mL, respectively (Wangchuk et al. 2016b). Additional test against schistosomula of *S. mansoni* revealed luteolin had lethal effect at lowest dilution of 3.91 μg/mL and achieved 98–100% killing at dilution of 31.3 μg/mL (Wangchuk et al. 2016b). In vivo assessment of luteolin against *T. muris* infection in mice resulted in 27.6% reduction in worm burden, but was markedly weaker than mebendazole (93.1%) (Wangchuk et al. 2016b). Additionally, moderate anti-*Trypanosoma brucei rhodesiense* activity were reported for extracts obtained from *C. crispa* and *H. amabile* (Wangchuk et al. 2011).

10.6.6 Anti-oxidant

Antioxidant activities were investigated of methanol and aqueous extracts obtained from five high-altitude medicinal plants, *Aconogonon tortuosum, Thlaspi arvense, Erysimum bhutanicum, Thalictrum chelidonii,* and *Aletris pauciflora* (Yeshi et al. 2017). Aqueous extract of *T. chelidonii* and methanol extract of *E. bhutanicum* exhibited free radical scavenging activity, and strong ferric reducing antioxidant potential by the aqueous and methanol extracts of *A. tortuosum* (Yeshi et al. 2017).

10.7 Conclusion

The Bhutanese Traditional Medicine originates from the Tibetan *Sowa Rigpa* and over centuries has evolved through incorporation of the local biodiversity and the prevalence of diseases in the country. BTM was once the primary health care provider in the country and continues to provide medical services through the integrated medical system of Bhutan. Medicinal plants play a fundamental role in providing the medical services. Phytochemical analysis of 11 high-altitude medicinal plants has yielded more than 50 metabolites belonging to alkaloids, flavonoids, terpenes, phenylpropanoids, and furanocoumarins. Primarily, hetisine-type and C_{19}-diterpenoids alkaloids were obtained from two of the *Aconitium* species, whereas isoquinoline alkaloids were isolated from *M. simplicifolia* and *Corydalis* species. The *Meconopsis* species had afforded anthocyanins and flavonol glycosides as the pigments of the blue and yellow color flowers. Several alkaloids and flavonoids (luteolin and luteolin glycoside) isolated from the high-altitude plant species had showed, immunomodulatory, anti-inflammatory, anti-helminthic, and antiplasmodial activities. These properties were consistent with the traditional uses of these medicinal plants.

However, phytochemical studies of Bhutanese medicinal plants represent less than 2% of medicinal plants known in Bhutan. These high-altitude medicinal plants have the huge potential for discovery of novel compounds and for applications as molecular scaffolds and towards drug discovery. A good example is atisinium chloride isolated from *A. orochryseum* which has become an antimalarial structural lead. There is even more urgency as the climate change has shown impact on these high-altitude resident medicinal plants. Especially, of great concern is of high-altitude endemic species. For this reason, a preliminary review of phytochemicals found in these high-altitude medicinal plants from studies of species that shares geographical distribution with neighboring countries would provide directions for future studies and should be of focus in the future.

References

Ekor M (2014) The growing use of herbal medicines: issues relating to adverse reactions and challenges in monitoring safety. Front Neurol 4(JAN):1–10. https://doi.org/10.3389/fphar.2013.00177

Ghimire SK, Bista AG, Lama NS, Craig SR (2021) Sowa Rigpa, ethnobotany, and conservation of threatened species in Nepal. WWF Nepal, p 428

Gurmet P (2004) "Sowa—Rigpa ": Himalayan art of healing. Indian J Tradit Knowl 3:212–218

Gyeltshen C, Prasad K, Dema S (2020) Number of species in Bhutan. Conserv Sci Pract 2:2–4. https://doi.org/10.1111/csp2.146

Iwashina T, Yokoyama K, Yangzom R, Mizuno T, Devkota HP, Murai Y, Dorji K, Wangmo C, Gyeltshen C (2019) Anthocyanins and flavonols from the blue flowers of six Meconopsis species in Bhutan. Biochem Syst Ecol 86:103925. https://doi.org/10.1016/j.bse.2019.103925

Kloos S, Madhavan H, Tidwell T, Blaikie C, Cuomu M (2020) The transnational Sowa Rigpa industry in Asia: new perspectives on an emerging economy. Soc Sci Med 245:112617. https://doi.org/10.1016/j.socscimed.2019.112617

Lhamo N, Nebel S (2011) Perceptions and attitudes of Bhutanese people on Sowa Rigpa, traditional Bhutanese medicine: a preliminary study from Thimphu. J Ethnobiol Ethnomed 7:3. https://doi.org/10.1186/1746-4269-7-3

Masyita A, Mustika Sari R, Dwi Astuti A, Yasir B, Rahma Rumata N, Bin ET, Nainu F, Simal-Gandara J (2022) Terpenes and terpenoids as main bioactive compounds of essential oils, their roles in human health and potential application as natural food preservatives. Food Chem X 13:100217. https://doi.org/10.1016/j.fochx.2022.100217

McKay A, Wangchuk D (2018) Traditional medicine in Bhutan. Asian Med 1:204–218. https://doi.org/10.1163/157342105777996737

Moreira D de L, Teixeira SS, Monteiro MHD, De-Oliveira ACAX, Paumgartten FJR (2014) Traditional use and safety of herbal medicines. Rev Bras 24:248–257. https://doi.org/10.1016/j.bjp.2014.03.006

Pan SY, Zhou SF, Gao SH, Yu ZL, Zhang SF, Tang MK, Sun JN, Ma DL, Han YF, Fong WF, Ko KM (2013) New perspectives on how to discover drugs from herbal medicines: CAM'S outstanding contribution to modern therapeutics. Evid Based Complement Alternat Med 2013:627375. https://doi.org/10.1155/2013/627375

Shepherd C, Giacomin P, Navarro S, Miller C, Loukas A, Wangchuk P (2018) A medicinal plant compound, capnoidine, prevents the onset of inflammation in a mouse model of colitis. J Ethnopharmacol 211:17–28. https://doi.org/10.1016/j.jep.2017.09.024

Steiner RP (2017) Tibetan medicine part I: introduction to Tibetan medicine and the rGyud-bzi (Fourth Tantra). Am J Chin Med 15:80–88

Tobgay T, Dorji T, Pelzom D, Gibbons RV (2011) Progress and delivery of health care in Bhutan, the land of the thunder dragon and gross national happiness. Trop Med Int Health 16:731–736. https://doi.org/10.1111/j.1365-3156.2011.02760.x

Ulanov MS, Badmaev VN (2018) Kalmyk traditional medicine: a historical and cultural analysis. New Res Tuva 7:138–160. https://doi.org/10.25178/nit.2018.4.8

Wangchuk T (2020) A pictorial guide to high altitude medicinal flora of Bhutan. Department of Traditional Medicine Services Ministry of Health, Royal Government of Bhutan

Wangchuk P, Bremner JB, Samosorn S (2007a) Hetisine-type diterpenoid alkaloids from the Bhutanese medicinal plant *Aconitum orochryseum*. J Nat Prod 70:1808–1811. https://doi.org/10.1021/np070266k

Wangchuk P, Wangchuk D, Aagaard-Hansen J (2007b) Sowarigpa medicine as the integrated health care system of Bhutan. Southeat Asian J Trop Med Public Health 38:161–167

Wangchuk P, Samten U, Thinley J, Afaq SH (2008) High altitude plants used in Bhutanese traditional medicine (gSo-ba-rig-pa). Ethnobotany 20:54–64

Wangchuk P, Samten, and Ugyen (2009a) High altitude medicinal plants of Bhutan: an illustrated guide for practical use. Pharmaceutical and Research Unit, Institute of Traditional Medicine Services, Ministry of Health, Thimphu, Bhutan

Wangchuk, P, and Samten (2009b) Low altitude medicinal plants of Bhutan: an illustrated guide for practical use. Pharmaceutical and Research Unit, Institute of Traditional Medicine Services, Ministry of Health, Thimphu, Bhutan

Wangchuk P, Bremner JB, Samten, Rattanajak R, Kamchonwongpaisan S (2010a) Antiplasmodial agents from the Bhutanese medicinal plant *Corydalis calliantha*. Phyther Res 24:481–485. https://doi.org/10.1002/ptr.2893

Wangchuk P, Bremner JB, Samten, Skelton BW, White AH, Rattanajak R, Kamchonwongpaisan S (2010b) Antiplasmodial activity of atisinium chloride from the Bhutanese medicinal plant, *Aconitum orochryseum*. J Ethnopharmacol 130:559–562. https://doi.org/10.1016/j.jep.2010.05.057

Wangchuk P, Keller PA, Pyne SG, Taweechotipatr M, Tonsomboon A, Rattanajak R, Kamchonwongpaisan S (2011) Evaluation of an ethnopharmacologically selected Bhutanese medicinal plants for their major classes of phytochemicals and biological activities. J Ethnopharmacol 137:730–742. https://doi.org/10.1016/j.jep.2011.06.032

Wangchuk P, Keller PA, Pyne SG, Sastraruji T, Taweechotipatr M, Rattanajak R, Tonsomboon A, Kamchonwongpaisan S (2012a) Phytochemical and biological activity studies of the Bhutanese medicinal plant *Corydalis crispa*. Nat Prod Commun 7:575–580. https://doi.org/10.1177/1934578x1200700507

Wangchuk P, Keller PA, Pyne SG, Willis AC, Kamchonwongpaisan S (2012b) Antimalarial alkaloids from a Bhutanese traditional medicinal plant *Corydalis dubia*. J Ethnopharmacol 143:310–313. https://doi.org/10.1016/j.jep.2012.06.037

Wangchuk P, Keller PA, Pyne SG, Korth J, Samten, Taweechotipatr M, Rattanajak R, Kamchonwongpaisan S (2013a) Antimicrobial, antimalarial and cytoxicity activities of constituents of a Bhutanese variety of *Ajania nubigena*. Nat Prod Commun 8:733–736. https://doi.org/10.1177/1934578x1300800613

Wangchuk P, Keller PA, Pyne SG, Lie W, Willis AC, Rattanajak R, Kamchonwongpaisan S (2013b) A new protoberberine alkaloid from *Meconopsis simplicifolia* (D. Don) Walpers with potent antimalarial activity against a multidrug resistant *Plasmodium falciparum* strain. J Ethnopharmacol 150:953–959. https://doi.org/10.1016/j.jep.2013.09.052

Wangchuk P, Keller PA, Pyne SG, Taweechotipatr M (2013c) Inhibition of TNF-α production in LPS-activated THP-1 monocytic cells by the crude extracts of seven Bhutanese medicinal plants. J Ethnopharmacol 148:1013–1017. https://doi.org/10.1016/j.jep.2013.05.055

Wangchuk P, Keller PA, Pyne SG, Taweechotipatr M, Kamchonwongpaisan S (2013d) GC/GC-MS analysis, isolation and identification of bioactive essential oil components from the Bhutanese medicinal plant, *Pleurospermum amabile*. Nat Prod Commun 8:1305–1308. https://doi.org/10.1177/1934578x1300800930

Wangchuk P, Pyne SG, Keller PA (2013e) An assessment of the Bhutanese traditional medicine for its ethnopharmacology, ethnobotany and ethnoquality: textual understanding and the current practices. J Ethnopharmacol 148:305–310. https://doi.org/10.1016/j.jep.2013.04.030

Wangchuk P, Pyne SG, Keller PA, Taweechotipatr M, Kamchonwongpaisan S (2014) Phenylpropanoids and furanocoumarins as antibacterial and antimalarial constituents of the Bhutanese medicinal plant *Pleurospermum amabile*. Nat Prod Commun 9:957–960. https://doi.org/10.1177/1934578x1400900719

Wangchuk P, Navarro S, Shepherd C, Keller PA, Pyne SG, Loukas A (2015) Diterpenoid alkaloids of *Aconitum laciniatum* and mitigation of inflammation by 14-O-acetylneoline in a murine model of ulcerative colitis. Sci Rep 5:1–10. https://doi.org/10.1038/srep12845

Wangchuk P, Gyeltshen T, Tobgay T (2016a) Bhutanese traditional medical education. In: Schuelka MJ, Maxwell TW (eds) Education in Bhutan: culture, schooling, and gross national happiness. Springer, pp 91–108

Wangchuk P, Pearson MS, Giacomin PR, Becker L, Sotillo J, Pickering D, Smout MJ, Loukas A (2016b) Compounds derived from the Bhutanese daisy, *Ajania nubigena*, demonstrate dual anthelmintic activity against Schistosoma mansoni and *Trichuris muris*. PLoS Negl Trop Dis 10:1–18. https://doi.org/10.1371/journal.pntd.0004908

Wangchuk P, Apte SH, Smout MJ, Groves PL, Loukas A, Doolan DL (2018) Defined small molecules produced by Himalayan medicinal plants display immunomodulatory properties. Int J Mol Sci 19:3490. https://doi.org/10.3390/ijms19113490

Wangchuk P, Yeshi K, Vennos C, Mandal SC, Kloos S, Nugraha AS, Tashi, Samten (2020) Three medicinal *Corydalis* species of the Himalayas: their ethnobotany, pharmacognosy, phytochemistry and pharmacology. J Herb Med 23:100384. https://doi.org/10.1016/j.hermed.2020.100384

Yangchen S, Tashi T, Bjorn M (2016) Bhutanese health and the health care system: past, present, and future. Druk J 2(1):80–93

Yeshi K, Yangdon P, Kashyap S, Wangchuk P (2017) Antioxidant activity and the polyphenolic and flavonoid contents of five high altitude medicinal plants used in Bhutanese Sowa rigpa dedicine. J Biol Act Prod Nat 7:18–26. https://doi.org/10.1080/22311866.2017.1287593

Yeshi K, Gyal Y, Sabernig K, Phuntsho J, Tidwell T, Jamtsho T, Dhondup R, Tokar E, Wangchuk P (2019) An integrated medicine of Bhutan: Sowa Rigpa concepts, botanical identification, and the recorded phytochemical and pharmacological properties of the eastern Himalayan medicinal plants. Eur J Integr Med 29:100927. https://doi.org/10.1016/j.eujim.2019.100927

Yokoyama K, Yangzom R, Mizuno T, Murai Y, Dorji K, Wangmo C, Gyeltshen C, Iwashina T (2018) Flavonol glycosides in the flowers of the Himalayan *Meconopsis paniculata* and *Meconopsis integrifolia* as yellow pigments. Biochem Syst Ecol 81:102–104. https://doi.org/10.1016/j.bse.2018.10.006

Yuan H, Ma Q, Ye L, Piao G (2016) The traditional medicine and modern medicine from natural products. Molecules 21:559. https://doi.org/10.3390/molecules21050559

11 Drugs Derived from High Altitude Plants

K. Swathi, K. Suganya, R. Aswathy, Chalos Angel, Balraj Sudha, S. Sumathi, and Sanjeeb Kumar Mandal

Abstract

High-altitude environments harbour a diverse array of plants equipped with remarkable adaptations to extreme conditions, including low oxygen levels and intense ultraviolet radiation. In India, only 20% of high-altitude medicinal plants contribute to the drug trade, revealing an untapped reservoir of diversity. Thriving in harsh climates, these plants deploy unique defence mechanisms against abiotic stresses crucial for survival. Environmental factors significantly impact secondary metabolite production, crucial for maintaining physiological functions and acting as antioxidants. Spanning countries, the high-altitude region integrates Ayurveda, Traditional Chinese Medicine, and Unani medicine into robust traditional medicine systems. High-altitude flora exhibits rich biochemical diversity and addresses major ailments contributing to drug development and ethno-

K. Swathi (✉)
Department of Biochemistry, Dr. N. G.P. Arts and Science College, Coimbatore, Tamilnadu, India
e-mail: swathi.k@drngpasc.ac.in

K. Suganya · R. Aswathy · C. Angel · S. Sumathi
Department of Biochemistry, Biotechnology & Bioinformatics, Avinashilingam Institute for Home Science and Higher Education for Women, Coimbatore, Tamilnadu, India

B. Sudha
Department of Biochemistry, Biotechnology & Bioinformatics, Avinashilingam Institute for Home Science and Higher Education for Women, Coimbatore, Tamilnadu, India

Department of Biotechnology, Dr. G.R.Damodaran College of Science, Coimbatore, Tamilnadu, India

S. K. Mandal (✉)
Department of Biotechnology, Chaitanya Bharathi Institute of Technology (CBIT), Hyderabad, Telangana, India
e-mail: sanjeebkumar_biotech@cbit.ac.in

S. Chandra, H. Lata (eds.), *High Altitude Medicinal Plants*,
https://doi.org/10.1007/978-3-031-99406-7_11

botanical uses. Studies reveal compounds with therapeutic applications like *Rhodiola rosea's* stress-relieving adaptogens and *Artemisia annua*, the source of the antimalarial drug artemisinin. These kinds of high-altitude-derived drugs show promise for infectious diseases, cardiovascular conditions, respiratory disorders, neurodegenerative diseases, inflammatory conditions, and immunomodulatory effects, providing inspiration for ongoing drug discovery. Despite their ethnobotanical importance, many high-altitude plant species face extinctions due to prolonged exploitation. Government agencies, non-governmental organizations (NGOs), and local farmers forefront conservation initiatives, employing a combination of ex situ and in situ methods. The exploration of high-altitude plants not only opens doors to innovative solutions but also expands our understanding of the intricate relationships between nature and human well-being, particularly in the realm of drug discovery.

11.1 Introduction

Plants are used by humans for many different purposes, one of which being their medical usefulness, which is well known worldwide. One of the main sources of revenue for people living in the Himalayan highlands is medicinal herbs (Qian et al. 2016). The destruction of unique natural habitats, especially in regions of vulnerability, is one of the key concerns surrounding the effects of climate change since these changes affect the dynamics of plant development, distribution, and population structure (Li et al. 2019). Droughts, nutrient shortages, and temperature increases are predicted to become more frequent as a result of climate change, and their combined impacts will be more detrimental to species survival than any one of them alone. Plants modify their photosynthetic capacity or their physiological and morphological traits in response to environmental cues in order to maintain a homeostatic level of photosynthetic efficiency (Jakada et al. 2019; Yousefzadeh Najafabadi and Ehsanzadeh 2017).

Traditional medical systems frequently employ traditional knowledge to treat a wide range of illnesses, which is crucial for the preservation of floral diversity. The medicinal plant uses of many ethnic groups in the Himalayan areas vary according to location and environment, and each group has its own indigenous healthcare system. In order to prioritize conservation efforts and fully appreciate how climate change affects plant phenological features, it is crucial to understand plant ecological behavior (Yang et al. 2018a). The hard and complex climatic conditions imposed by the high altitude of the mountains limit plant growth, diversity, and spread. These extreme environmental circumstances include low CO_2 and O_2 levels, severe winds, high illumination, extremely thin soils with minimal water and nutrient contents, and low temperatures (El-Keblawy and Khedr 2017). Many plants have persevered and advanced despite the possibility that these stressors may impede plant growth. In response, they have developed a range of adaptive defence systems (Wonsick and Pinker 2014; Elkelish et al. 2020).

Plant secondary metabolites are a valuable range of natural compounds that are also associated with the activation and enhancement of plant defence mechanisms against biotic and abiotic environmental challenges (Gong et al. 2018). Similarly, the quality and quantity of soil microorganisms produced by plant populations are significantly influenced by environmental factors such as soil microorganisms, water availability, soil pH, nutrients, high and low temperatures, drought, high altitudes, and light conditions. In actuality, these secondary metabolites help plants retain their morphological and physiological characteristics. As an example, they boost the antioxidant capacity of phenols, tannins, and flavonoids (Nawrot-Chorabik et al. 2022; Wang et al. 2019; Soliman et al. 2018).

High altitude environmental conditions such as low temperature, dryness, and high light intensity increase the production of reactive oxygen species (ROS) like singlet oxygen, hydroxyl radicals, and superoxide, which can cause significant oxidative damage to plants (Moussa et al. 2019). This damage includes impairing photosynthesis, closing stomata, and reducing other physiological functions like antioxidant activity. While ROS generation is generally low and balanced by antioxidant molecules, achieving ROS homeostasis, excessive ROS can overwhelm this balance (Khan et al. 2019). Plants use two antioxidant systems to scavenge ROS: the enzymatic system (including catalase, peroxidase, superoxide dismutase, and ascorbate peroxidase) and the non-enzymatic system (comprising compounds like glutathione, flavonoids, phenolic compounds, carotenoids, and water-soluble α-tocopherols). These systems help mitigate oxidative damage and maintain cellular function (Rahimi et al. 2018).

Gaining knowledge about the characteristics and advantages of high-altitude medicinal plants not only helps with new drug discovery but also emphasizes how crucial it is to protect these special ecosystems (Fig. 11.1). Given the challenges posed by climate change, habitat degradation, and overharvesting, conservation activities are essential. Conserving high altitude medicinal plants is crucial for future generations, as these plants offer invaluable benefits to both biodiversity and medicine.

11.2 Habitat and Adaptation

High altitude medicinal plants, or HAMPs, are distinguished by a particular habitat and survival adaptations against harsh environmental circumstances. These plants are found in high altitude areas, usually above 1700 m, where they have developed special adaptations to withstand harsh winters, dry climates, little precipitation, strong winds, snowfall, blizzards, and intense UV radiation. These areas' vegetation demonstrates specific responses to harsh environments, like the compression of thermal living zones and the division of the terrain into distinct ecological niches (Kour et al. 2020). These adaptations contribute to the great variety and therapeutic potential of high-altitude medicinal plants by enabling them to flourish in harsh conditions. High altitude regions are considered a repository of diverse medicinal plant species, with a significant proportion being herbs, gymnosperms, and

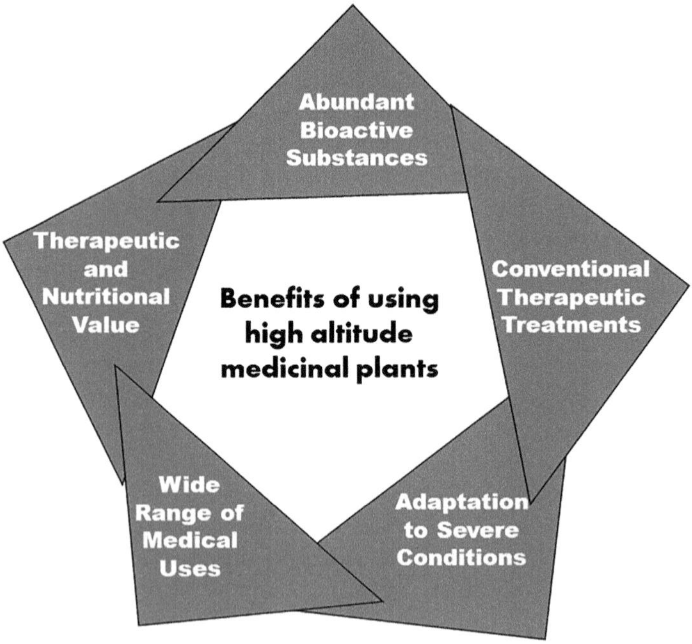

Fig. 11.1 Figure showing the various benefits of HAMP

pteridophytes (Bargali et al. 2022). It is estimated that over 50% of the identified medicinal plants in the Indian Himalayan Region (IHR) occur as herbs in temperate or alpine regions, with more than 20% being native and over-exploited (Dar and Khuroo 2020).

The ecological state of high-altitude medicinal plants includes *Ephedra gerardiana*, this alpine plant is known for its stimulant and decongestant qualities, among other therapeutic uses (Choudhary et al. 2021; Rinchen et al. 2021). *Podophyllum hexandrum*, known for its antineoplastic qualities, this herb grows at high altitudes and is used medicinally. It has proven therapeutic claims and is used to treat testicular cancer (Kalam et al. 2021). *Panax* sp., known for its adaptogenic and therapeutic qualities, Panax species, including *Panax ginseng*, are farmed in high altitude areas (Walia et al. 2023). *Ginkgo biloba* is a high-altitude plant known for its antioxidant and cognitive-improving qualities with therapeutic significance (Wahid et al. 2020). *Echinacea* are grown in high altitude areas; echinacea species are among the most popular herbal remedies in developed nations. Their ability to strengthen the immune system makes them valuable (Kaur et al. 2016).

HAMP and biodiversity have a complicated interaction that is influenced by many different circumstances. According to the research findings, there appears to be a strong positive link between species richness and human population density (HPD), suggesting that human settlements and species-rich areas frequently coexist. The sustainability of ecosystems and conservation efforts may be impacted by this geographic congruence of people and biodiversity. Furthermore, the biodiversity of these areas is enhanced by HAMP's adaptive responses to high altitude conditions, which include increased production of antioxidant chemicals and secondary metabolites. All things considered, the existence of HAMP at high altitude regions enhances the biodiversity of these areas and highlights the complex interactions that exist between the health of the ecosystem, human activity, and medicinal plants (Hazarika et al. 2023; Rosso et al. 2023).

Due to their ability to promote wound healing, HAMPs like *Ficus religiosa* are useful in the treatment of skin conditions and wounds (Murugesu et al. 2021). It has anti-inflammatory qualities that can aid in lowering inflammation and easing associated symptoms. Antioxidant-rich bioactive chemicals found in HAMP can help prevent oxidative stress and shield cells from harm. Because of their therapeutic potential and importance in ethnobotany, it has been traditionally utilized to treat a variety of health issues. By providing a natural approach to health and wellness, the bioactive components in HAMP have the ability to prevent diseases and disorders caused by oxidative stress (Kukreti et al. 2023).

11.3 Acclimatization of Medicinal Flora to High Altitude Stress

Medicinal plants have developed a variety of adaptation responses to high altitude, including coping mechanisms for the harsh climate found in these areas. The regulation of antioxidants and secondary metabolites, proteomic studies, morphological, physiological, and biochemical adaptations, and protein-level adaptations are a few important adaptive responses (Pan et al. 2023). Antioxidant enzymes like catalase and ascorbate peroxidase are more active in medicinal plants at high altitudes to counteract oxidative stress brought on by strong light. In order to live in harsh environments, high altitude medicinal plants use strategies like deep super-cooling or spontaneous ice formation in extracellular spaces in response to low temperatures and freezing stress (Kumar et al. 2021). These plants have changed how they grow, develop, and distribute resources in order to adapt tolerance mechanisms that allow them to flourish in high altitude settings. The adaptive responses of medicinal plants to high altitude conditions have been linked to stress-responsive proteins such as heat shock proteins (HSPs), antifreeze proteins (AFPs), and late embryogenesis abundant (LEA) proteins, according to proteomic studies (Shantappa et al. 2022). For the purpose of adjusting to the demanding high altitude environment, the plants may go through post-translational alterations of proteins, such as phosphorylation (Sharma et al. 2023). There are a number of significant ramifications for researching how medicinal plants adapt to high altitude environments:

Identification of Bioactive Compounds: As an adaptive reaction, medicinal plants produce and accumulate secondary metabolites in response to the high altitude environment. Gaining insight into how these secondary metabolites are regulated may help identify new bioactive substances with potential medical uses (Semwal et al. 2023).

Knowing Stress Tolerance Mechanisms: Studies on how medicinal plants adapt to high altitude stresses, such as oxidative stress, low temperature, and a lack of nutrients or water, can shed light on the underlying physiological, molecular, and biochemical processes that enable these plants to survive in harsh environments. The development of ways to increase the stress tolerance of economically significant crops and medicinal plants can benefit from this information (Hashim et al. 2020).

Preservation and Ecological Application: Researching the adaptable reactions of medicinal plants at high elevations can help guide conservation initiatives aimed at shielding these priceless genetic assets from the effects of climate change and other environmental disturbances. Sustainable harvesting and production methods can be guided by this knowledge to guarantee the long-term supply of these medicinal plants (Chen et al. 2016b).

The main therapeutic qualities of high-altitude medicinal plants include chemopreventive and radioprotective actions. Because they are exposed to intense UV radiation, high altitude medicinal plants have been shown to have chemo preventive and radioprotective qualities that can be helpful in the treatment of cancer. These plants' bioactive substances may be able to lessen the consequences of DNA damage and oxidative stress (Painuli and Kumar 2016). Antioxidants and cytoprotectants is another property, as an adaptive response to oxidative stress, high altitude medicinal plants show increased production of antioxidant chemicals such as flavonoids, phenolic acids, and terpenoids. These plants are useful for treating oxidative stress-related illnesses because of their antioxidant components, which aid in preventing oxidative damage to cells and tissues. The research has also thrown light on the effects that are analgesic and anti-inflammatory, in which high altitude medicinal plants such as *Ficus religiosa* have been shown to have analgesic that is pain-relieving and anti-inflammatory effects in their extracts. The treatment of inflammatory and pain-related conditions may benefit from the existence of bioactive substances having these therapeutic properties (Ahmad et al. 2023).

Another study has demonstrated antimicrobial properties against common pathogens, indicating their potential utility in the treatment of skin infections and the encouragement of wound healing. These plants' antibacterial and wound-healing qualities are attributed to certain compounds, including tannins and volatile oils. Additionally, studies have documented the immunomodulatory and adaptogenic effects, the immune system is supported and the body is able to adapt to stress because of the adaptogenic qualities of several high-altitude medicinal plants, such as the Panax species. These plants might help treat ailments like immune system malfunction, exhaustion, and stress (Suryakumar et al. 2019).

Tanacetum sinaicum, Echinops spinosissimus, Crataegus x sinaica, and *Cotoneaster orbicularis* are the four medicinal plants that were researched; of these, *Cotoneaster orbicularis* showed the greatest adaptive response for handling high-altitude pressures. They revealed buildup of osmoprotectants such as proline and soluble carbohydrates, along with notable increases in antioxidant enzymes such as catalase and ascorbate peroxidase. Elevated levels of antioxidant enzymes and secondary metabolites, such as flavonoids, tannins, and phenols, were shown to be adaptive responses in *Crataegus sinaica* to height. Proline and soluble sugars are examples of compatible solutes that *Echinops spinosissimus* produced in greater quantities to counteract oxidative stressors at high altitudes. Like the other species, *Tanacetum sinaicum* exhibited similar adaptive responses at higher altitudes, with increased secondary metabolite buildup and antioxidant enzyme activity (Ibrahim et al. 2022).

11.4 Pharmacognosy and Phytochemistry of High-Altitude Medicinal Flora

High-altitude medicinal plants and fruits have garnered increasing attention due to their diverse pharmacological and biological activities. These high-altitude plants are rich sources of bioactive compounds such as flavonoids, polysaccharides, phenols, triterpenes, alkaloids, and glycosides. These compounds contribute to a wide range of pharmacological effects, including scavenging free radicals, antioxidant properties, anti-inflammatory effects, as well as neuromodulation and immune stimulation. The pharmacological diversity of high-altitude plants underscores their potential for various therapeutic applications.

Coleus forskohlii, found at higher altitudes, exhibits significant therapeutic potential due to its phenol, flavonoid, and terpenoid components. Research indicates that variations in growing altitudes influence the quantity of secondary metabolites in *Coleus forskohlii*, impacting its antioxidant potential (Rana et al. 2020). Furthermore, cold stress poses a significant challenge to plants by inducing excessive production of Reactive Oxygen Species (ROS). *Picrorhiza kurroa*, a high-altitude medicinal plant, demonstrates higher antioxidant potential, particularly in its leaf part. Research conducted in regions of varying altitudes in Uttarakhand, India, suggests that non-enzymatic antioxidants play a crucial role in scavenging ROS and enabling adaptation to cold climatic conditions (Agarwal et al. 2023).

Selinum carvifolia, a perennial herb native to the high altitudes of the Indian Western Himalayas, has demonstrated valuable therapeutic effects primarily attributed to its phenolic, flavonoid, and terpenoid constituents. In traditional Indian medicine, known as Bhutkeshi, it serves various purposes including as an insecticide, nervine sedative, anti-spasmodic, and digestive aid. Eight phenolic compounds have been identified in *S. carvifolia*, contributing to its antimicrobial, cytotoxic, and antioxidant properties (Srivastava et al. 2023).

Aconitum heterophyllum Wall, also known as "Atis," is a biennial herb native to the Himalayan regions of the Indian subcontinent. Studies have revealed its diverse pharmacological properties, including analgesic, anti-inflammatory, antiarrhythmic, antiparasitic, anticancer, and central nervous system effects (Wani et al. 2022). These are just a few examples of the myriad medicinal plants found in high-altitude regions, each offering a unique array of therapeutic benefits. In addition to this, various other medicinal plants from high-altitude regions are listed in Table 11.1.

Table 11.1 Pharmacological activities of high-altitude medicinal plants

S. no	Plant/species name	Phytochemistry	Pharmacological activity	References
1	*Allardia tridactylites*	Monoterpenes, sesquiterpenes, α-pinene, α-thujone, valeranone and α-bisabolol	Phytotoxic, antibacterial, and cytotoxic activities	Zomba et al. (2024)
2	*Aconitum orochryseum* Stapf	Aconitine, mesaconitine, and hypaconitine	Antiseptic, analgesic, anthelminthic properties, and are also useful against leprosy and bone disorders	Jamtsho et al. (2022)
3	*Chrysosplenium* species	Hydroxylated and methoxylated flavonoids and triterpenoids	Anti-tumor, antibacterial, anti-viral, hepatoprotective, and insecticidal properties	Zhao et al. (2022)
4	*Fritillariae species*	Alkaloids, terpenoids, nucleosides, organic acids, saponins, carbohydrates, amines, and sterols	Anti-tussive effect, expectorant effect, anti-asthmatic effect	Wang et al. (2021)
5	*Meconopsis horridula*	Alkaloids (oleracein E) and flavonoids	Anti-inflammatory, antitumor and anti-Shigella activities	Poudeyal et al. (2020)
6	*Neopicrorhiza scrophulariiflora*	Caffeoyl glycoside, cucurbitacin glycoside, cyclopentanoid monoterpenes, hydroquinone glycoside, iridoid glucoside, non-glycosidic iridoid, phenol, phenyl glycoside, phenylpropanoid, phenylethyl glycoside and secoiridoid glycoside	Anti-atherosclerotic, antidiabetic and anti-inflammatory, antimicrobial, antimalarial, antioxidative, hepatoprotective, immunomodulatory and nerve growth factor potentiating activity	Rokaya et al. (2020)
7	*Saussurea costus*	*Sesquiterpene terpenes* (ST), anthraquinones, alkaloids and flavonoids	CNS depressant, anti-microbial, antiulcer, anti-inflammatory, hepato-protective and anti-nematodial, anti-protozoal and cytotoxic activity	Mujammami (2020)

11.5 Traditional Therapeutic Uses of High Altitude Medicinal Plants

Traditional knowledge of high-altitude plants encompasses the accumulated wisdom, practices, and beliefs of indigenous communities living in mountainous regions and other elevated terrains. In recent years, the scientific community has increasingly recognized the value of traditional knowledge in pharmacology and drug discovery (Mahizan et al. 2018). Integrating traditional knowledge with modern pharmacological research holds great promise for developing new drugs derived from high-altitude plants. This approach respects the cultural heritage and practices of indigenous communities and contributes to the discovery of novel therapeutic agents that can address various health challenges. As research continues to explore the rich biodiversity of high-altitude regions, these plants will likely play an increasingly important role in future drug development and biotechnological applications (Zhang et al. 2017). Table 11.2 shows some examples of traditional uses of high-altitude plants.

Indigenous communities have utilized these plants for centuries, drawing upon the rich ethnobotanical knowledge passed down through generations. Furthermore, the traditional uses of these plants often extend beyond individual health remedies to encompass holistic approaches to well-being (Pan et al. 2019).

11.6 Natural Medicine for High Altitude Ailments

High altitude illness encompasses a spectrum of conditions stemming from insufficient oxygen levels at altitudes surpassing 2500 m. Although the body can gradually adjust to low oxygen levels, adapting to extreme environments takes time and often leads to the development of high-altitude illness. These conditions, collectively known as High Altitude Illness (HAI), present diverse manifestations such as acute mountain sickness (AMS), high-altitude cerebral edema (HACE), high-altitude pulmonary edema (HAPE), chronic mountain sickness (CMS), and notably, high-altitude pulmonary hypertension (HAPH). Of particular concern, HAPH can progress to right ventricle hypertrophy and heart failure, posing significant risks to individuals at high altitudes. The pathogenesis of these conditions involves complex molecular mechanisms, with some linked to oxidative stress. As the body copes with reduced oxygen levels at high altitudes, oxidative stress may occur, leading to cellular damage and dysfunction (Li et al. 2023; Pena et al. 2022).

Throughout history, plants have been humanity's primary source of medicinal remedies, offering solutions to a myriad of ailments. Their resilience is particularly evident in extreme environments, such as high altitudes. In these regions, over a hundred varieties of medicinal plants thrive, thanks to the diverse biological niches and altitude ranges spanning from 100 to 7500 m above sea level. High altitude plants have evolved specialized adaptations to withstand the harsh environmental conditions prevalent at higher elevations, including intense ultraviolet (UV) radiation. They produce a plethora of protective compounds, such as UV filters, potent antioxidants, free radical scavengers, and antifreeze carbohydrates. Through these mechanisms, high altitude plants synthesize a wide array of metabolites, enabling them to thrive in their challenging habitats. The biochemical machinery of high-altitude plants has undergone significant evolution to cope with their harsh surroundings, resulting in the development of novel biosynthetic pathways and molecular structures absent in plants at lower altitudes. As a result, high-altitude

Table 11.2 List of high-altitude plants and their traditional applications

Plant name	Traditional uses	References
Rhodiola rosea	Adaptogen, fatigue reduction, altitude sickness	Panossian and Wikman (2010)
Hippophae rhamnoides	Skin diseases, digestive disorders, wound healing	Beveridge et al. (1999)
Artemisia absinthium	Antiparasitic, digestive aid, anti-inflammatory	Mancuso and Santangelo (2017)
Saussurea lappa	Anti-inflammatory, analgesic, respiratory conditions	Kumar and Mishra (2011)
Swertia chirayita	Bitter tonic, fever, liver disorders	Pandey and Pant (2014)
Berberis aristata	Antimicrobial, digestive aid, eye infections	Singh et al. (2010)
Picrorhiza kurroa	Hepatoprotective, anti-inflammatory, liver disorders	Chander et al. (2013)
Aconitum heterophyllum	Antipyretic, digestive issues, respiratory conditions	Singhuber et al. (2009)
Nardostachys jatamansi	Sedative, anxiety, mental disorders	Bimakr et al. (2011)
Dactylorhiza hatagirea	Aphrodisiac, general tonic, respiratory issues	Ballabh et al. (2008)
Allium wallichii	Cardioprotective, digestive aid, antimicrobial	Purohit and Samant (1995)
Gentiana kurroo	Antipyretic, anti-inflammatory, digestive aid	Kala (2005)
Podophyllum hexandrum	Anticancer, antiviral, purgative	Bragg and Branham (1999)
Meconopsis aculeate	Pain relief, anti-inflammatory, respiratory conditions	Rawat et al. (2010)
Inula racemosa	Cardioprotective, respiratory aid, digestive aid	Singh and Singh (2005)
Thymus linearis	Antiseptic, digestive aid, respiratory aid	Gaur et al. (1992)
Polygonum viviparum	Antidiarrheal, anti-inflammatory, wound healing	Manandhar and Plants (2002)
Valeriana jatamansi	Sedative, digestive aid, anxiety	Singh and Singh (2006)
Acorus calamus	Digestive aid, respiratory aid, antimicrobial	Dhar et al. (1968)
Cordyceps sinensis	Immunomodulatory, aphrodisiac, anti-fatigue	Holliday and Cleaver (2008)
Gaultheria fragrantissima	Anti-inflammatory, pain relief, respiratory aid	Chhetri et al. (2006)
Paris polyphylla	Anticancer, antimicrobial, anti-inflammatory	Liu et al. (2010)
Juniperus communis	Antiseptic, respiratory aid, digestive aid	Mathew et al. (2004)
Betula utilis	Antiseptic, anti-inflammatory, wound healing	Negi et al. (2011)
Bergenia ciliate	Antipyretic, analgesic, anti-inflammatory	Bisht et al. (2011)

(continued)

Table 11.2 (continued)

Plant name	Traditional uses	References
Bupleurum falcatum	Anti-inflammatory, liver tonic, respiratory aid	Lee et al. (2006)
Astragalus membranaceus	Anti-inflammatory	Cho and Yoon (2011)
Rheum austral	Digestive aid, laxative, anti-inflammatory	Purohit and Vyas (2004)
Ephedra sinica	Bronchodilator, stimulant, respiratory conditions	Kletter and Kriechbaum (2001)
Morina longifolia	Digestive aid, tonic, anti-inflammatory	Kunwar and Bussmann (2006)
Urtica dioica	Anti-inflammatory, diuretic, arthritis relief	Upton (2013)
Geranium wallichianum	Anti-inflammatory, astringent, wound healing	Watanabe et al. (2005)
Anaphalis triplinervis	Respiratory conditions, anti-inflammatory, wound healing	Kala (2005)
Elsholtzia stachyodes	Antimicrobial, digestive aid, respiratory conditions	Singh et al. (2013)
Paeonia emodi	Anti-inflammatory, analgesic, liver tonic	Nisar and Khan (2017)
Rhododendron anthopogon	Respiratory conditions, antiseptic, anti-inflammatory	Pandey et al. (2004)
Gentiana tibetica	Digestive aid, tonic, anti-inflammatory	Joshi and Joshi (2000)
Potentilla anserine	Anti-diarrheal, anti-inflammatory, wound healing	Martin and Chambers (2001)
Plantago major	Anti-inflammatory, wound healing, respiratory aid	Samuelsen (2000)
Leontopodium alpinum	Anti-inflammatory, antioxidant, skin care	Szabo et al. (2009)

plants hold both economic and medicinal value, presenting unique opportunities for research and therapeutic applications (Kour et al. 2020).

Acute Mountain Sickness (AMS) is characterized by a range of symptoms including headache, nausea or vomiting, dyspnea, fatigue, poor appetite, dizziness, and difficulty sleeping. Research suggests that *Rhodiola rosea*, a traditional Tibetan medicinal plant in China, may alleviate AMS symptoms by inhibiting the HIF-1 degradation pathway. During hypoxia, the degradation of HIF-1α is suppressed, leading to its accumulation. Subsequently, HIF-1α forms a heterodimer with HIF-1β, known as HIF-1, which possesses transcriptional activity. This complex binds to hypoxia response elements (HREs) and activates the transcription of approximately 60 target genes involved in adaptation to hypoxic conditions, including vascular endothelial growth factor (VEGF), inducible nitric oxide synthase (iNOS), and erythropoietin (EPO) (Liang et al. 2018). *Ginkgo biloba* extract emerges as a promising remedy for Acute Mountain Sickness (AMS), with studies demonstrating that its flavonoids competitively inhibit carbonic anhydrase II (CAII), potentially mitigating AMS symptoms. This inhibition may alleviate the increase in systolic pressure and respiratory alkalosis associated with AMS. Consequently, *Ginkgo biloba* extract holds potential as a substitute for acetazolamide, a synthetic drug currently used commercially, aiming to reduce potential side effects and enhance bioavailability (Wiboonkiat et al. 2023). *Salvia miltiorrhiza Bunge*, commonly known as red sage or Danshen, is frequently utilized in Chinese herbal medicine for its diverse biological effects. Regulatory bodies such as the Food and Drug Administration in China and other Asian countries endorse *Salvia miltiorrhiza* therapy for cardiovascular protection, particularly in individuals with peripheral arterial disease experiencing intermittent claudication, as well as for the treatment of Acute Mountain Sickness (AMS) (Ou et al. 2020).

High Altitude Cerebral Edema (HACE) typically occurs at altitudes ranging between 4500 and 5500 m, resulting in brain tissue damage and potentially fatal outcomes if left untreated. It is characterized by worsening headaches, vomiting, lethargy, coma, behavioral abnormalities, ataxia, papilledema, confusion, and brain edema. Failure to promptly recognize and treat HACE can lead to severe consequences (Yang et al. 2023). Meanwhile, High Altitude Pulmonary Edema (HAPE) refers to a form of non-cardiogenic pulmonary edema that develops in high-altitude environments exceeding 2500 meters. It carries a mortality rate ranging from 36% to 90%. HAPE patients may rapidly develop dyspnea, reduced arterial oxygen saturation, exacerbated tachycardia, lethargy, sputum production, cyanosis, and profound hypoxemia (Yang et al. 2023). *Gynostemma pentaphyllum*, widely used in traditional Asian medicine as a functional food and tea, shows promise in treating HACE. Specifically, Gypenoside-14 (GP-14), a bioactive fraction of *G. pentaphyllum* ethanol extract, exhibits significant protective effects against neuro-inflammation and blood-brain barrier disruption, suggesting its potential as a future treatment for HACE (Geng et al. 2022). *Phlomis younghusbandii Mukerjee*, known as "Lu Mu Er" in Tibetan medicine, contains phenylethanoid glycosides (PhGCs) that have been shown to suppress inflammatory responses and oxidative stress in acute hypobaric hypoxia-stimulated high altitude cerebral edema rats. PhGCs present a potential therapeutic option for HACE treatment (Luan et al. 2019). Medicinal plants used for treating HAPE include *Brassica rapa* Linn. (Brassicaceae), *Conioselinum anthriscoides* (H.Boissieu) *Pimenov* and *Kljuykov* (Apiaceae), and *Polygonum cuspidatum Sieb.* et Zucc. (Polygonaceae). These plants can modulate hemodynamic changes by enhancing the release of nitric oxide (NO) or inhibiting the release of endothelin-1 (ET-1). Other plant such as *Hippophae rhamnoides* Linn. (Elaeagnaceae) is also utilized for their therapeutic properties in HAPE (Yang et al. 2023).

Chronic mountain sickness (CMS) is a clinical syndrome commonly observed in populations residing in high-altitude areas, typically above 2500 m. It is characterized by excessive proliferation of erythrocytes, severe hypoxemia, and pulmonary hypertension. If left untreated, CMS can progress to pulmonary heart disease, eventually leading to congestive heart failure. Glycyrrhiza, derived from the dried roots and stems of leguminous plants, contains key components such as glycyrrhizic acid and glycyrrhetinic acid. Studies have shown that glycyrrhetinic acid can ameliorate heart injury in rats with chronic mountain sickness (Tao et al. 2021).

Oxidative stress and endothelial cell dysfunction, along with the proliferation of pulmonary artery smooth muscle cells (PASMCs), are key pathological changes that drive the progression of High-Altitude Pulmonary Hypertension (HAPH). *Terminalia bellirica tannin* has demonstrated the ability to alleviate or slow pulmonary vascular remodeling, inhibit reactive oxygen species (ROS) production, and mitigate oxidative stress. These effects are attributed to its regulation of the Nrf2/HO-1 signaling pathway. *Terminalia bellirica tannin* shows a promising potential for use in the treatment of HAPH (Aimaier et al. 2023). *Salvia przewalskii Maxim* (SPM), native to northwestern China, exhibits antioxidant properties by inhibiting the redox system and modulating fructose metabolism through downregulation of Khk and AldoB protein activity, aiding in hypoxia adaptation. SPM extracts enhance the activities of superoxide dismutase and lactate dehydrogenase in acute hypoxic conditions. Furthermore, they downregulate hypoxia-inducible factor-1α, proliferating cell nuclear antigen, Bcl-2, cyclin-dependent kinase 4, CyclinD1, and P27Kip1, inhibit monocyte chemoattractant protein-1 and nuclear factor-kappaB, and regulate the RhoA-Rho-associated protein kinase signaling pathway. These actions promote the repair of chronic hypoxia-induced lung injury, rendering SPM beneficial for the treatment of HAPH (Li et al. 2023). Tsantan Sumtang (TS), derived from the Four Tantras, is an empirical Tibetan medicine prescription with the ability to upregulate the AKT/eNOS signaling pathway, thereby protecting pulmonary vascular endothelial function. This property underscores its potential in safeguarding against the development of HAPH (Li et al. 2024).

Recent advancements in research have provided valuable insights into the pathophysiological mechanisms driving high altitude illness, revealing the roles of oxidative stress, endothelial dysfunction, and smooth muscle cell proliferation. In response to these challenges, traditional medicinal plants and plant-derived compounds have emerged as promising avenues for prevention and treatment. The exploration of natural compounds and traditional medicines offers a beacon of hope for more effective strategies in managing high altitude illness. Continued research into the pharmacological mechanisms of these natural remedies holds the promise of developing novel interventions that can enhance adaptation to high-altitude environments and improve the well-being of individuals residing in or traveling to such regions.

11.7 Protective Effects of High-Altitude Plant Compounds on Oxidative Stress

Oxidative stress is a biochemical process that affects proteins, lipids, DNA, and other components of cells and tissues. It is caused by an abnormal build-up and production of reactive oxygen species (ROS), also known as free radicals, which the human body's antioxidants are unable to neutralize (Hajam et al. 2022; Checa and Aran 2020). Inadequate management can hasten the aging process and contribute to numerous chronic and degenerative diseases and disorders, as well as acute pathologies, including stroke, trauma, and other conditions (Kunst et al. 2023).

Bioactive chemicals are extra-nutritional substances typically present in various foods, including grains, fruits, and vegetables, and they have health advantages in addition to their conventional nutritional value (Banwo et al. 2021). Many bioactive components have been extracted from plants, and these plants have been used for their medicinal properties since ancient times.

High-altitude medicinal plants are noted for their ethnobotanical significance and have been used by numerous populations worldwide to cure many diseases and problems due to their healing powers. These plants produce unique bioactive metabolites due to their diverse geographic and climatic conditions, which include biotic and abiotic variables (Ashraf et al. 2023; Khan et al. 2023). Many possible metabolites with antibacterial, antiviral, cardioprotective, anti-inflammatory, anticancer, neuroprotective, nephroprotective, and chemoprotective properties have been found in natural sources in recent years. These include flavonoids, polysaccharides, phenols, triterpenes, alkaloids, glycosides, and other main active ingredients, which are summarized in Table 11.3. Numerous pieces of evidence point to these phytochemicals' potential epidemiological and clinical health benefits (Hazarika et al. 2022; Semwal et al. 2023; Melfi et al. 2023; Chan et al. 2023).

11.8 Drugs Discovered from High Altitude Medicinal Plants

Drug discovery from medicinal plants involves a variety of disciplines and research approaches. The plants of interest are frequently gathered and identified by botanists, ethnobotanists, ethnopharmacologists, and plant ecologists (Balunas and Kinghorn 2005). Plants can now be used as "factories" to produce natural medicinal chemicals for use in the creation of biotech medicines, therapies, and treatments due to recent technical advancements (Paul and Ma 2011). One of the most notable accomplishments in the history of medicine was the discovery of the early pharmaceuticals, which included quinine, digitoxin, cocaine, pilocarpine, codeine, and quilicoxin (Kong et al. 2003),

Plant-based pharmaceutical approaches provide a safe, efficient, and economical substitute for traditional methods involving animal cell cultures or microbial fermentation in the drug-development process. Thus, pharmaceuticals made from plant-based natural chemicals can provide patients with more and faster access to treatment (Subramoniam 2014; Veeresham 2012). Natural products' mostly untapped structural variety is what makes them the most amazing in terms of their ongoing significance in medication discovery (Chaachouay and Zidane 2024). Modern pharmaceutical research is increasingly focusing on the study of chemicals produced by plants because of their therapeutic properties. Considerable advancement in this field has been made possible by the use of synthetic biology and biotechnology to produce these chemicals under controlled circumstances (Ausländer et al. 2017). This process allows for the significant production of uncontaminated compounds, which enhances the safety and efficacy of employing these substances. Phytochemicals are widely used in modern medicine because many medicines and therapies are sourced from natural sources. Synthetic biology and biotechnology have opened up new avenues for medication discovery and development (Zajtchuk 1999). This makes it possible to produce novel compounds with enhanced

Table 11.3 Bioactive compounds from high altitude medicinal plants

S. no	Bioactive molecule	Source plant	Biological property of bioactive molecule	Treated disease	References
1.	Phenolic compounds	*Rhodiola rosea* L	Antioxidant, anti-inflammatory properties, and anticancer activity	Chronic diseases (metabolic, cardiovascular diseases, ischemic stroke and CNS diseases)	Bernatoniene et al. (2023)
2.	Limonin	*Rutaceae* plants	Anti-cancer, anti-inflammatory, analgesic, antibacterial and antiviral, anti-oxidant, liver-protecting properties	Non-alcoholic fatty liver disease	Li et al. (2022)
3.	Phenol	*Dioscorea japonica var*; *Japonica* [Dioscoreaceae]; *Paeonia suffruticosa Andrews* [Paeoniaceae]; *Arisaema erubescens* (Wall.) Schott [Araceae], and *Paeonia lactiflora Pall* [Paeoniaceae].	Antitumor, anti-inflammatory, neuroprotective, cardioprotective, nephroprotective activities	Diabetic retinopathy	Adki and Kulkarni (2022)
4.	Salidroside	*Rhodiola* species	Adaptogenic, antifatigue, antidepressant, antioxidant, anti-inflammatory, antinoception, and anticancer activities	Ischemia-related diseases	Han et al. (2022)

(continued)

Table 11.3 (continued)

S. no	Bioactive molecule	Source plant	Biological property of bioactive molecule	Treated disease	References
5.	Saffron (aqueous extract)	*Crocus sativus L*	Free radical scavenging and antioxidant properties	Reducing oxidative stress in ischemic strokes	Gudarzi et al. (2022)
6.	Aqueous extract	*Brassica rapa*	Antidiabetic, anti-inflammatory, and antioxidant properties	Type 2 diabetes mellitus	Berdja et al. (2016)
7.	Phenolic fraction	*Hippophae rhamnoides*	Antioxidant and antiplatelet activity	Oxidative stress-associated diseases	Olas et al. (2016)
8.	Polysaccharides	*Polygonatum kingianum*	Anti-oxidative stress, anti-inflammation, anti-fibrosis, lipid regulation, and anti-atherosclerosis	Cardiovascular diseases	Lin et al. (2024)
9.	Flavonoid				
10.	Saponin				

medicinal properties. This method has the potential to revolutionize the pharmaceutical industry by generating cutting-edge and potent treatments for a wide range of diseases and conditions.

High-throughput screening, computer modelling, and bioinformatics are all having a significant impact on the discovery process. High-throughput screening is an automated procedure that checks a large number of isolated chemicals or plant extracts against predetermined biological targets (Ismail et al. 2024). This makes it possible to find active chemicals quickly. The interactions between chemicals originating from plants and biological targets are predicted by computational modelling techniques like virtual screening and molecular modelling. Molecular modelling is a computational technique that predicts molecular behavior and interactions with other molecules using computer simulations (Dhoundiyal and Alam 2023). These contemporary methods combine powerful elements to solve environmental concerns and speed up the search for and creation of novel drugs.

Natural chemicals originating from plants are expected to continue to play a significant role in the development of novel drugs and therapeutic approaches. There is a large variety of compounds with potential use in medicine due to the wide range of chemical changes present in plants (Lautie et al. 2020). Drug development and discovery offer substantial prospects for the use of compounds originating from plants. However, in order to maximize advantages and minimize negative consequences on the environment, society, and healthcare, it is imperative that these problems be approached appropriately.

Native Americans used herbs, and this led to the development of several pharmacopoeias in scientific medicine during the twentieth century. Many pharmaceuticals used today, such as taxol, ergot, codeine, morphine, atropine, hyoscyamine, ephedrine, quinine, colchicine, digoxin, strychnine, reserpine, and artemisinin, are derived from plants. In the USA, community pharmacies dispense prescription medications with at least one active ingredient derived from plant material in about 25% of cases. Modern medicine research is primarily based on indigenous cures in folklore in one culture or another (Kaul 2010). The utilization of traditional folklore of some medicinal plants grown in high altitude and temperate regions to produce medications containing a single chemical substance has been reported (Table 11.4).

Table 11.4 Modern drugs from high altitude medicinal plants

S. no	Plant name	Drugs	References
1.	*Artemisia annua Linn.*	Artemisinin	Wang et al. (2024)
2.	*Atropa belladonna Linn.*	Atropine	Owais et al. (2014)
3.	*Atropa acuminata Royle*	Atropine	Maqbool et al. (2016)
4.	*Colchicum luteum Baker*	Colchicine	Khan et al. (2011)
5.	*Digitalis lanata Ehrh*	Digoxin	van Wietmarschen et al. (2016)
6.	*Ephedra gerardiana wall*	Ephedrine	Choudhary et al. (2021)
7.	*Hyoscyamus niger Linn*	Hyoscyamine	Al-Snafi (2018)
8.	*Papaver somniferum L.,*	Codeine	Bajpai et al. (2001)
9.	*Papaver somniferum L.*	Morphine	Bajpai et al. (2001)
10.	*Polygonatum verticillatum; Cinchona officinalis*	Quinine	Virk et al. (2016)
11.	*Taxus baccata*	Taxol (Paclitaxel)	Kaul (2010)
12.	*Rauvolfia serpentina*	Reserpine	Kaul (2010)
13.	*Podophyllum hexandrum*	Podophyllotoxin	Pandey et al. (2007)
14.	*Silybum marianum*	Silymarin	Khan et al. (2022)
15.	*Galanthus nivalis*	Galantamine	Juneja et al. (2017)
16.	*Berberis aristate*	Berberine	Malhotra et al. (2021)
17.	*Nothapodytes nimmoniana*	Camptothecin	Suhas et al. (2007)
18.	*Coleus forskohlii*	Forskolin	Kaushal et al. (2013)
19.	*Pilocarpus microphyllus*	Pilocarpine	Alves-Lameira et al. (2021)

11.9 Challenges and Opportunities in Altitude Plant-Derived Drug Development

11.9.1 Challenges

- High-altitude regions are characterized by low temperatures, which can limit plant growth and affect metabolic processes. Cold stress can trigger the production of secondary metabolites in plants as a defence mechanism against environmental challenges. Increased UV radiation at high altitudes can lead to oxidative stress in plants, prompting the synthesis of protective compounds such as flavonoids and phenolics. Reduced oxygen availability at high altitudes affects plant respiration and metabolism, leading to altered biochemical pathways and the accumulation of specific secondary metabolites (Paul et al. 2016).
- Many high-altitude regions are remote and inaccessible, requiring researchers to overcome logistical challenges such as difficult terrain, lack of infrastructure, and limited transportation options. Researchers conducting fieldwork in high-altitude areas often encounter harsh weather conditions such as strong winds, snowstorms, and unpredictable weather patterns, disrupting research activities and posing safety risks (Bhattarai et al. 2004).
- High-altitude regions harbour diverse plant species adapted to extreme environmental conditions. These regions are considered biodiversity hotspots due to many endemic and rare species. Identifying and studying the diverse flora of high-altitude regions requires taxonomic expertise and extensive fieldwork involving botanical surveys, specimen collection, and documentation of plant distributions (Grytnes and Vetaas 2002).
- Fluctuations in temperature, precipitation, and other environmental factors in high-altitude regions result in seasonal variations in plant physiology and biochemistry. Seasonal changes can influence the synthesis and accumulation of secondary metabolites in plants, leading to chemical composition and bioactivity variability. Standardizing extraction methods and ensuring consistency in the quality and efficacy of plant-derived drugs have become challenging due to this seasonal variation (Chen et al. 2012).
- Some high-altitude plants may be endangered or protected species, necessitating ethical considerations in their collection and use for pharmacological research. Researchers must adhere to ethical guidelines and regulations governing plant materials' sustainable harvest and conservation (Shrestha and Bawa 2014).

11.9.2 Opportunities

- High-altitude plants often involve the production of bioactive compounds. These compounds, including alkaloids, flavonoids, and terpenoids, possess various pharmacological activities such as antioxidant, anti-inflammatory, and anticancer effects (Wang et al. 2017). Alkaloids, for instance, have been found in plants growing at high altitudes and are known for their diverse medicinal properties, including pain relief and antimicrobial activity. Flavonoids, another group of bioactive compounds, contribute to the health benefits of high-altitude plants, such as improving cardiovascular health and reducing inflammation. Terpenoids exhibit a wide range of biological activities, including antiviral, antifungal, and insecticidal properties, making them valuable targets for drug discovery (Yang et al. 2018b).

- Exploring the chemical diversity of high-altitude plants presents an opportunity to discover novel therapeutic agents with unique mechanisms of action. These plants often produce secondary metabolites in their adaptive response to environmental stressors, offering a rich source of bioactive compounds for drug development (Ticktin et al. 2012).
- Researchers can identify lead compounds with promising pharmacological activities by systematically screening plant extracts using bioassays and high-throughput screening techniques (Atanasov et al. 2021).
- Indigenous communities living in high-altitude regions often possess valuable traditional knowledge about the medicinal properties of local plants. Collaborating with these communities can provide insights into plant selection, preparation methods, and therapeutic uses, facilitating the development of culturally appropriate medicines. Indigenous knowledge systems offer a holistic understanding of plant-based medicines, including information about plant harvesting, processing, and usage practices passed down through generations (Rumpf et al. 2018).
- Studying high-altitude plants can provide insights into the effects of climate change on plant physiology and biochemistry. Researchers can investigate how environmental stressors such as temperature changes and water scarcity influence the production of bioactive compounds in plants. Understanding these mechanisms is crucial for predicting how climate change will impact the availability and efficacy of medicinal plants in high-altitude regions and identifying new targets for drug development (Ncube et al. 2014).
- Advances in biotechnology offer new opportunities for the sustainable production of bioactive compounds from high-altitude plants. Techniques such as tissue culture, genetic engineering, and metabolic engineering can enhance the yield and quality of target compounds, reduce production costs, and minimize environmental impact. Biotechnological approaches also enable the production of rare or endangered plant compounds in controlled environments, reducing the need for wild harvesting and preserving biodiversity (Sharifi-Rad et al. 2018).

By addressing these challenges and leveraging the opportunities of high altitude plants, researchers can unlock the full potential of these natural resources for drug discovery and development, benefiting both human health and biodiversity conservation.

11.10 Conservation and Sustainability of High Altitude Medicinal Plants

High altitude medicinal plants are crucial resources with immense cultural, economic, and ecological value. Thriving in mountainous regions above 1500 m, these plants possess unique biochemical compositions vital for traditional medicine systems and pharmaceutical industries globally. However, they face significant threats to their sustainability, including habitat loss, overharvesting, climate change, and unsustainable harvesting practices. These threats jeopardize not only the availability of medicinal resources but also the delicate ecological balance in mountain ecosystems, where these plants play key roles in stabilizing soil, regulating water flow, and supporting biodiversity (Mishra et al. 2023).

In the high altitude regions of the Himalayas, particularly in countries like Nepal and Bhutan, medicinal plants constitute a significant source of income for local inhabitants. However, indiscriminate harvesting practices pose a serious threat to their sustainability. The increasing popularity of Medicinal and Aromatic Plants (MAPs) for pharmaceuticals and traditional healthcare systems worldwide exacerbates this issue. Despite the promotion of green consumerism, the exploitation of these resources persists due to a lack of emphasis on conservation and sustainable development (Phondani et al. 2016). To address this, there's a pressing need to promote the cultivation and long-term utilization of high-altitude medicinal and aromatic plants for socioeconomic development, aiming to generate rural employment, strengthen state economies, foster industrial development, and counter population decline, particularly in steep districts (Gupta et al. 2023).

Globally, over 422,000 plant species are recognized for their medicinal value, with approximately 52,885 species traded internationally. The majority of these species (80–90%) are sourced from the wild. The relentless expansion of the global botanical market further strains plant resources. This escalating demand underscores the urgent need for sustainable practices to alleviate pressure on wild populations and ensure the long-term viability of medicinal plant resources.

The sustainability of high altitude medicinal plants in regions like the Himalayas is under threat from various factors such as excessive harvesting, overgrazing, climate change, and habitat destruction. This is exemplified by the precarious situation of *Angelica glauca*, an endangered perennial medicinal plant, which faces extinction due to unscientific harvesting practices, overgrazing, and the impacts of climate change (Gautam et al. 2023). Despite these challenges, the Himalayan region is recognized globally as a center for medicinal plants, with Bhutan alone boasting around 600 species out of its reported 5603 higher plant species with medicinal properties. The diverse ecological conditions in the region, spanning altitudes from 100 to 7500 m above sea level, provide favorable growing conditions for a wide variety of medicinal plants. These plants not only contribute directly to healthcare systems but also serve as a primary source of income for local communities, thereby contributing to the concept of Gross National Happiness (Wangchuk and Tobgay 2015).

Certain species within the Himalayan flora have highly specific ecological requirements, reflecting their restricted distribution. For example, *Delphinium glaciale*, occurring at only three sites and exhibiting limited range of slope angles, highlights the vulnerability of such species. Conversely, species like *Corydalis crispa, Nardostachys grandiflora, Dactylorhiza hatagirea*, and *Meconopsis simplicifolia* show broader ranges of altitude and slope conditions, indicating a relatively wider distribution across the region. Species like *Nardostachys grandiflora, Neopicrorhiza scrophulariiflora, Onosma hookeri,* and *Fritillaria delavayi*, valued in traditional medicine, show rapid regeneration through vegetative propagation, emphasizing the critical need for conservation for potential drug discovery (Airi et al. 2000).

The continuation of traditional healing systems ensures the ongoing wild harvest of medicinal plants, underscoring the profound implications for the long-term sustainability of these resources. A study highlights the importance of conservation of key species, such as *Corydalis crispa, Nardostachys grandiflora, Rhodiola crenulata, Meconopsis simplicifolia, Neopicrorhiza scrophulariiflora*, and *Onosma hookeri,* being identified as most threatened due to high exploitation levels. Additionally, species like *Dactylorhiza hatagirea* and *Gentiana urnula* are noted to have become relatively rare and inaccessible, emphasizing the urgent need for conservation action to ensure their survival (Lakey and Dorji 2016).

Indiscriminate and non-systematic collection practices have further exacerbated the pressure on medicinal plant populations, leading to rarity, threats, or endangerment of species like *Aconitum heterophyllum, Podophyllum hexandrum, Nardostachys jatamansi, Picrorhiza kurrooa, Swertia chirata,* and *Bergenia ciliata.* Conservation efforts recommend in-depth study and monitoring of low-density species, establishment of wild nurseries, and adoption of scientific techniques for cultivation to ensure sustainability and provide sustainable livelihood options for local communities in hilly areas (Nainwal and Singh 2020).

To safeguard high altitude medicinal plants, a comprehensive approach addressing ecological and socioeconomic factors is essential. Strategies such as establishing protected areas, promoting habitat restoration, and encouraging sustainable harvesting practices are critical. Additionally, initiatives focusing on cultivation, research, and community involvement are vital for long-term conservation success. Successful examples like the Himalayan Plant Conservation Program in Nepal, the Andean Medicinal Plant Conservation Project in South America, and the Mount Kinabalu Medicinal Plant Sanctuary in Malaysia highlight the effectiveness of community-based conservation efforts and the importance of integrating indigenous knowledge into conservation strategies (Larsen and Olsen 2007; Kor et al. 2021; Bidder 2023). Through collaborative action, it's possible to ensure the continued availability and sustainable use of high altitude medicinal plants for generations to come.

Effective management plans for sustainable harvesting, developed collaboratively with local communities and incorporating capacity building on the reproductive biology of priority species, are crucial for the conservation of high altitude medicinal plants. Stakeholders, including legislative and governmental bodies,

NGOs, traditional medicine practitioners, and local communities, need a comprehensive understanding of the ecological status and significance of these plants to achieve sustainable harvest practices (Lakey and Dorji 2016). However, challenges persist in domesticating and cultivating high altitude species due to difficulties in replicating their natural growing environment and economic feasibility concerns, alongside cultural beliefs.

In mountain systems, agrodiversity faces challenges such as marginality, inaccessibility, and fragility, hindering the sustainability of resilience. Sikkim Himalayan farmers uphold four crucial components-biophysical diversity, management diversity, agrobiodiversity, and organizational diversity-leveraging their agro-ecological expertise to promote sustainable production. Enhancing traditional farming in the Himalayas supports biodiversity conservation and sustainable development while also offering opportunities for drug discovery from high altitude medicinal plants. By leveraging the region's unique ecological characteristics and agricultural practices, we can bridge ecological services with healthcare needs, benefiting both current and future generations (Sharma and Rai 2012).

Sustainability, in the context of high altitude medicinal plants, entails the responsible use and management of resources to meet present needs while ensuring the ability of future generations to meet their own needs. This encompasses ecological, social, and economic considerations, emphasizing the importance of balancing environmental conservation, community welfare, and economic prosperity for sustainable outcomes.

In the realm of high altitude medicinal plants, ecological sustainability encompasses preserving ecosystem health and resilience by conserving habitats, protecting biodiversity, and implementing sustainable harvesting practices. This ensures the continued availability of valuable plant species for drug discovery while minimizing negative impacts on ecosystems. Social sustainability of high altitude medicinal plants prioritizes the well-being of local communities and indigenous peoples who rely on these plants. This involves engaging communities in decision-making, providing alternative livelihood opportunities, and preserving traditional knowledge essential for drug discovery endeavors. Economic sustainability in the realm of high altitude medicinal plants focuses on ensuring equitable benefits from their harvesting and trade. This includes promoting fair trade practices, enhancing the value of medicinal plant products through drug discovery efforts, and facilitating market access to sustainably sourced products, thereby ensuring the long-term viability of these resources (Hamilton 2004; Suneetha 2010; Lakey and Dorji 2016).

Several successful initiatives, such as the Sustainable Harvesting Program for Tibetan Medicinal Plants in China, the Sustainable Wild Collection Program for Himalayan Medicinal Plants in India, and the FairWild Standard, demonstrate these principles by involving local communities, promoting sustainable harvesting practices, and improving market access for wild-collected medicinal plants (Brinckmann et al. 2018). Preserving high altitude medicinal plants for drug development requires a holistic approach, considering ecological, social, and economic factors. Collaboration among stakeholders is vital to sustainably harness the potential of high altitude plants for drug discovery.

11.11 Future Avenues: Potential New Drugs from High Altitude Medicinal Plants

The future prospects for potential new drugs from high altitude plants are promising, offering opportunities for the discovery and development of novel therapeutic agents. High altitude plants produce bioactive compounds with diverse chemical structures and pharmacological activities. These compounds have the potential to target a wide range of biological pathways and molecular targets implicated in various diseases, including cancer, cardiovascular disorders, neurological conditions, and infectious diseases (Kumar et al. 2020). Here are some potential areas of focus and the prospects they hold:

11.11.1 Anti-cancer Agents

Several bioactive compounds isolated from high altitude plants have demonstrated anticancer properties in preclinical studies. For example, alkaloids, flavonoids, and terpenoids from plants like *Taxus wallichiana*, *Rhodiola rosea*, and Aconitum species have shown cytotoxic effects against cancer cells and have been investigated for their potential as anticancer agents (Chen et al. 2016a; Pan et al. 2019). Further research is needed to elucidate the mechanisms of action of these compounds and evaluate their safety and efficacy in clinical trials.

11.11.2 Cardiovascular Health

Flavonoids and other polyphenolic compounds found in high altitude plants have been associated with cardiovascular health benefits, including anti-inflammatory, antioxidant, and vasodilatory effects. These compounds have the potential to prevent or alleviate conditions such as hypertension, atherosclerosis, and thrombosis (Yang et al. 2018b). Future research could focus on identifying specific compounds with cardioprotective properties and developing them into therapeutics for cardiovascular diseases.

11.11.3 Neurological Disorders

Some high altitude plants contain bioactive compounds that may have neuroprotective and neuroregenerative properties. For example, alkaloids such as huperzine A from *Huperzia serrata* have shown potential for treating Alzheimer's disease by inhibiting acetylcholinesterase activity and enhancing cholinergic neurotransmission (Zhang et al. 2017). Continued investigation into the neuropharmacological effects of compounds derived from high altitude plants could lead to the development of novel treatments for neurodegenerative disorders and cognitive impairment.

11.11.4 Infectious Diseases

High altitude plants produce secondary metabolites with antimicrobial, antiviral, and antiparasitic activities, which could be explored as potential treatments for infectious diseases. For instance, alkaloids from plants such as Berberis species have shown antibacterial effects against drug-resistant pathogens (Mahizan et al. 2018). Research efforts may focus on identifying and characterizing novel antimicrobial compounds from high altitude plants to combat emerging infectious diseases and address the growing problem of antimicrobial resistance.

11.11.5 Immunomodulatory Agents

High altitude plants produce compounds with immunomodulatory properties that could be explored for the treatment of immune-related disorders. For example, polysaccharides from medicinal mushrooms found in high altitude regions have shown immunostimulatory effects and potential applications in cancer immunotherapy and autoimmune diseases (Wasser 2011). Research efforts may focus on identifying novel immunomodulatory compounds from high altitude plants and elucidating their mechanisms of action to develop therapeutic interventions for immune disorders.

11.11.6 Metabolic Syndrome and Diabetes Management

Bioactive compounds from high altitude plants, such as polyphenols and alkaloids, have shown potential for managing metabolic syndrome and diabetes by modulating glucose metabolism, insulin sensitivity, and lipid profiles (Wang et al. 2018). Further investigation into the molecular mechanisms underlying the metabolic effects of these compounds could lead to the development of new therapies for metabolic disorders and diabetes-related complications.

11.11.7 Wound Healing and Tissue Regeneration

High altitude plants produce compounds with wound healing and tissue regenerative properties, which could be harnessed for developing novel therapies for acute and chronic wounds, burns, and tissue injuries (Kumar et al. 2020). For example, flavonoids and terpenoids from plants like *Rhodiola rosea* and *Hippophae rhamnoides* have demonstrated wound healing and tissue repair-promoting effects through various mechanisms, including antioxidant, anti-inflammatory, and angiogenic activities. These properties can be used in future to produce effective ointments and bandages.

11.11.8 Neuropharmacology and Mental Health

Compounds from high altitude plants have the potential to modulate neurotransmitter systems and neurotrophic factors, offering therapeutic opportunities for neurological and psychiatric disorders such as depression, anxiety, and cognitive impairment (Sarris and Wardle 2013). Future research may focus on identifying natural compounds with neuroprotective and neuroregenerative properties from high altitude plants and evaluating their efficacy in preclinical and clinical studies.

11.11.9 Environmental and Nutraceutical Applications

High altitude plants produce compounds with antioxidant, antimicrobial, and anti-inflammatory properties that could find applications in environmental remediation, food preservation, and nutraceutical formulations (Sharifi-Rad et al. 2020). Natural antioxidants derived from high altitude plants can be used as food additives to extend shelf life and enhance nutritional quality, while antimicrobial compounds can be employed in eco-friendly pesticides and antimicrobial coatings.

11.11.10 Personalized Medicine and Pharmacogenomics

High altitude plants may contain genetic variants and chemotypes that influence their chemical composition and pharmacological activities. Integrating pharmacogenomic approaches with traditional knowledge and ethnopharmacology could enable the development of personalized medicine tailored to individual genetic profiles and therapeutic needs (Narahari et al. 2016). Pharmacogenomic studies could help identify genetic markers associated with drug response and adverse reactions, optimizing drug efficacy and safety in diverse populations.

Thus, high altitude plants represent a valuable source of bioactive compounds with diverse pharmacological activities. Continued research into these plants holds great promise for the discovery of novel drugs to treat a wide range of diseases and address global health challenges.

11.12 Conclusion

In conclusion, high-altitude environments are a vital source of unique medicinal plants with significant pharmacological potential due to their extraordinary adaptations to extreme conditions. Medicinal plants from high altitudes have proven to have a rich composition of pharmacologically active compounds. While these plants contribute only a small proportion to drug trafficking in India, they offer significant potential for treating a range of diseases, including anti-inflammatory, anticancer, antimicrobial, and neuroprotective properties. A sustainable resource for novel drugs lies in regions such as the Himalayas and Andes. However, translating these compounds into clinically viable drugs presents challenges such as sustainable harvesting, ecological impact, and standardization. For these issues to be addressed and these plants to be tapped to their full potential, interdisciplinary collaboration between ethnobotanists, pharmacologists, and biotechnologists is critical. However, overexploitation threatens their existence, underscoring the need for conservation efforts. Exploring and preserving these plants helps us discover innovative drug solutions and better understand nature's relationship with human health, thus contributing to the advancement of global medical knowledge.

References

Adki KM, Kulkarni YA (2022) Paeonol attenuates retinopathy in streptozotocin-induced diabetes in rats by regulating the oxidative stress and polyol pathway. Front Pharmacol 13:891485

Agrawal S, Bhandari N, Saklani P (2023) Effect of altitude on the non-enzymatic antioxidant potential of the leaves of a high-altitude medicinal plant, *Picrorhiza kurroa*. Biochem Cell Arch 23(2):1085–1090

Ahmad S, Ali MD, Alotibi AM, Aljedaie SM, Alotaibi KS, Alam N et al (2023) Phytochemical profiling and anti-inflammatory, analgesic activity of *Ficus religiosa* L. and molecular docking study against iNOS, TNF-α enzyme. Pak J Pharm Sci 36(5):1467–1481

Aimaier S, Tao Y, Lei F, Yupeng Z, Wenhui S, Aikemu A, Maimaitiyiming D (2023) Protective effects of the *Terminalia bellirica* tannin-induced Nrf2/HO-1 signaling pathway in rats with high-altitude pulmonary hypertension. BMC Complement Med Ther 23(1):150. https://doi.org/10.1186/s12906-023-03981-2

Airi S, Rawal RS, Dhar U, Purohit AN (2000) Assessment of availability and habitat preference of Jatamansi—a critically endangered medicinal plant of west Himalaya. Curr Sci 79(10):1467–1471

Al-Snafi AE (2018) Therapeutic importance of Hyoscyamus species grown in Iraq (*Hyoscyamus albus*, *Hyoscyamus niger* and *Hyoscyamus reticulates*)—a review. IOSR J Pharm 8(6):18–32

Alves-Lameira O, Santos-Ribeiro FN, Teixeira-Rocha T, Mesquita-Germano C, Alves de Assis RM (2021) Branch regeneration of jaborandi submitted to diferentes types of pruning and cropping systems. Acta Agron 70(3):285–291

Ashraf MV, Pant S, Khan MH, Shah AA, Siddiqui S, Jeridi M et al (2023) Phytochemicals as antimicrobials: prospecting Himalayan medicinal plants as source of alternate medicine to combat antimicrobial resistance. Pharmaceuticals 16(6):881

Atanasov AG, Waltenberger B, Pferschy-Wenzig EM, Linder T, Wawrosch C, Uhrin P, Rollinger JM (2021) Discovery and resupply of pharmacologically active plant-derived natural products: a review. Biotechnol Adv 45:107657

Ausländer S, Ausländer D, Fussenegger M (2017) Synthetic biology—the synthesis of biology. Angew Chem Int Ed 56(23):6396–6419

Bajpai S, Gupta AP, Gupta MM, Kumar S (2001) Inter-relationships between morphine and codeine in the Indian genetic resources of opium poppy. Int J Geogr Inf Syst 8(4):75–81

Ballabh B, Chaurasia OP, Ahmed Z, Singh SB (2008) Traditional medicinal plants of cold desert Ladakh-used in treatment of cold and kidney disorders. J Ethnopharmacol 118(2):331–339

Balunas MJ, Kinghorn AD (2005) Drug discovery from medicinal plants. Life Sci 78(5):431–441

Banwo K, Olojede AO, Adesulu-Dahunsi AT, Verma DK, Thakur M, Tripathy S et al (2021) Functional importance of bioactive compounds of foods with potential health benefits: a review on recent trends. Food Biosci 43:101320

Bargali H, Kumar A, Singh P (2022) Plant studies in Uttarakhand, Western Himalaya—a comprehensive review. Tree For People 8:100203

Berdja S, Smail L, Saka B, Neggazi S, Haffaf EM, Benazzoug Y et al (2016) Glucotoxicity induced oxidative stress and inflammation in vivo and in vitro in *Psammomys obesus*: involvement of aqueous extract of *Brassica rapa rapifera*. Evid Based Complement Alternat Med 2016:3689208

Bernatoniene J, Jakstas V, Kopustinskiene DM (2023) Phenolic compounds of *Rhodiola rosea* L. as the potential alternative therapy in the treatment of chronic diseases. Int J Mol Sci 24(15):12293

Beveridge T, Li TS, Oomah BD, Smith A (1999) Sea buckthorn products: manufacture and composition. J Agric Food Chem 47(9):3480–3488

Bhattarai KR, Shrestha NS, Kushwaha BP (2004) High altitude medicinal plants of Nepal Himalaya: chemistry, ethnobotany and pharmacology. Kathmandu Univ J Sci Eng Technol 1(1):1–13

Bidder C (2023) Sacred natural sites in the face of global changes: the case of Mount Kinabalu in Malaysian Borneo. Malays J Sustain Environ 10(2):21–44

Bimakr M, Rahman RA, Taip FS, Ganjloo A (2011) Supercritical carbon dioxide extraction of bioactive flavonoid compounds from spearmint (*Mentha spicata* L.) leaves. Food Bioprocess Technol 4(8):1287–1293

Bisht VK, Negi JS, Bhatt VP (2011) Traditional utilization of *Bergenia ciliata* in Sikkim Himalaya. Indian J Tradit Knowl 10(2):312–315

Bragg R, Branham SE (1999) Podophyllotoxin production by plant cell cultures. Biotechnol Adv 17(3):207–223

Brinckmann JA, Luo W, Xu Q, He X, Wu J, Cunningham AB (2018) Sustainable harvest, people and pandas: assessing a decade of managed wild harvest and trade in *Schisandra sphenanthera*. J Ethnopharmacol 224:522–534

Chaachouay N, Zidane L (2024) Plant-derived natural products: a source for drug discovery and development. Drugs Drug Candidate 3(1):184–207

Chan WJJ, Adiwidjaja J, McLachlan AJ, Boddy AV, Harnett JE (2023) Interactions between natural products and cancer treatments: underlying mechanisms and clinical importance. Cancer Chemother Pharmacol 91(2):103–119

Chander R, Kapoor NK, Dhawan BN (2013) Hepatoprotective activity of *Picrorhiza kurroa* against carbon tetrachloride induced hepatotoxicity. Pharmacogn Mag 9(36):289–295

Checa J, Aran JM (2020) Reactive oxygen species: drivers of physiological and pathological processes. J Inflamm Res 13:1057–1073

Chen X, Qiu F, Yu BY (2012) Seasonal variation in phenolic compounds and antioxidant activity in natural populations of common buckwheat (*Fagopyrum esculentum*) from high and low altitude. Food Chem 133(2):161–169

Chen X, Wang W, Liu S, Wang M, Dong X (2016a) Flavonoids isolated from Tibetan medicine *Halenia elliptica* inhibit inflammatory responses in RAW264.7 macrophages and experimental colitis mice. Int Immunopharmacol 39:21–28

Chen SL, Yu H, Luo HM, Wu Q, Li CF, Steinmetz A (2016b) Conservation and sustainable use of medicinal plants: problems, progress, and prospects. Chin Med 11:37. https://doi.org/10.1186/s13020-016-0108-7

Chhetri DR, Sharma P, Gajurel PR (2006) Medicinal plants used by traditional healers in Sikkim Himalaya. Ethnobot Leafl 2006(5):1

Cho WC, Yoon JH (2011) *Astragalus membranaceus* and its immune modulatory functions. Am J Chin Med 39(4):785–807

Choudhary S, Kaurav H, Chaudhary G (2021) Medicinal importance of in ayurveda and modern sciences: a *Ephedra gerardiana*. Asian J Pharm Pharmacol 7(3):110–117

Dar GH, Khuroo AA (2020) An introduction to biodiversity of the Himalaya: Jammu and Kashmir state. In: Biodiversity of the Himalaya: Jammu and Kashmir state. Springer, pp 3–26

Dhar U, Manjkhola S, Joshi M, Bisht D (1968) Aspects of ethnobotany of Kailash sacred landscape. Indian J Tradit Knowl 7(3):13–22

Dhoundiyal S, Alam MA (2023) Advances in pharmacokinetic modelling and computational approaches for nanoparticles in drug delivery systems. Recent Adv Drug Deliv Formul 17(3):210–227

El-Keblawy AA, Khedr AHA (2017) Population structure and ecological role of *Moringa peregrina* (Forssk.) Fiori. at its northwestern range edge in the Hajar Mountains. Plant Biosyst 151(1):29–38

Elkelish AA, Alhaithloul HAS, Qari SH, Soliman MH, Hasanuzzaman M (2020) Pretreatment with *Trichoderma harzianum* alleviates waterlogging-induced growth alterations in tomato seedlings by modulating physiological, biochemical, and molecular mechanisms. Environ Exp Bot 171:103946

Gaur RD, Bisht NS, Sheth PK (1992) Ethnomedicinal plants of Uttar Pradesh hills: a study. Ethnobotany 4:55–60

Gautam K, Raina R, Dikshit N (2023) Current knowledge on sustainability and conservation of endangered Himalayan medicinal herb *A. glauca* Edgew.—a review. J Herb Med 2023:100764

Geng Y, Yang J, Cheng X, Han Y, Yan F, Wang C et al (2022) A bioactive gypenoside (GP-14) alleviates neuroinflammation and blood brain barrier (BBB) disruption by inhibiting the NF-κB signaling pathway in a mouse high-altitude cerebral edema (HACE) model. Int Immunopharmacol 107:108675

Gong J, Zhang Z, Zhang C, Zhang J, Ran A (2018) Ecophysiological responses of three tree species to a high-altitude environment in the southeastern Tibetan plateau. Forests 9(2):48

Grytnes JA, Vetaas OR (2002) Species richness and altitude: a comparison between null models and interpolated plant species richness along the Himalayan altitudinal gradient, Nepal. Am Nat 159(3):294–304

Gudarzi S, Jafari M, Pirzad Jahromi G, Eshrati R, Asadollahi M, Nikdokht P (2022) Evaluation of modulatory effects of saffron (*Crocus sativus* L.) aqueous extract on oxidative stress in ischemic stroke patients: a randomized clinical trial. Nutr Neurosci 25(6):1137–1146

Gupta S, Jasrotia S, Purohit MC (2023) Conservation, cultivation, and sustainable use of high altitude medicinal and aromatic plants for socio-economic development: a review. J Mt Res 18(1):201–209

Hajam YA, Rani R, Ganie SY, Sheikh TA, Javaid D, Qadri SS et al (2022) Oxidative stress in human pathology and aging: molecular mechanisms and perspectives. Cells 11(3):552

Hamilton AC (2004) Medicinal plants, conservation and livelihoods. Biodivers Conserv 13:1477–1517

Han J, Luo L, Wang Y, Wu S, Kasim V (2022) Therapeutic potential and molecular mechanisms of salidroside in ischemic diseases. Front Pharmacol 13:974775

Hashim AM, Alharbi BM, Abdulmajeed AM, Elkelish A, Hozzein WN, Hassan HM (2020) Oxidative stress responses of some endemic plants to high altitudes by intensifying antioxidants and secondary metabolites content. Plants (Basel, Switzerland) 9(7):869. https://doi.org/10.3390/plants9070869

Hazarika TK, Tayeng B, Ngurthankhumi R, Lalruatsangi E, Upadhyaya K, Lyngdoh N (2022) Unlocking wild edible fruits of Indo-Burma biodiversity hot spot, Arunachal Pradesh, India, to support food security and sustainable rural livelihood. Sustain For 14(23):16088

Hazarika TK, Varte L, Mathipi V, Khawlhring L, Lalruatsangi E, Debbarma P, Senthil Kumar N (2023) Phytochemicals constituents, antioxidant activities and cytotoxicity assays of few wild edible fruits of North-East India. Int J Food Prop 26(1):1020–1035

Holliday JC, Cleaver MP (2008) Medicinal value of the caterpillar fungi species of the genus *Cordyceps* (Fr.) Link (Ascomycetes): a review. Int J Med Mushrooms 10(3):219–234

Ibrahim IA, Jabbour AA, Abdulmajeed AM, Elhady ME, Almaroai YA, Hashim AM (2022) Adaptive responses of four medicinal plants to high altitude oxidative stresses through the regulation of antioxidants and secondary metabolites. Agronomy 12(12):3032

Ismail FM, Nahar L, Sarker SD (2024) High throughput screening of phytochemicals: application of computational methods. In: Computational phytochemistry. Elsevier, pp 219–253

Jakada BH, Aslam M, Fakher B, Greaves JG, Li Z, Li W et al (2019) Identification of SWI2/SNF2-related 1 chromatin remodeling complex (SWR1-C) subunits in pineapple and the role of pineapple SWR1 COMPLEX 6 (Ac SWC6) in biotic and abiotic stress response. Biomol Ther 9(8):364

Jamtsho T, Yeshi K, Wangchuk P (2022) Comparative analysis of two Himalayan Aconitum species for their phytopharmaceutical properties. J Herb Med 32:100497

Joshi K, Joshi R (2000) The gentians of the Himalayas: a review. Himal J Sci 2(4):56–62

Juneja K, Saini SS, Gaid M, Beerhues L, Sircar D (2017) Production of life-saving drugs from Himalayan herbs. In: Sustainable utilization of natural resources. CRC Press, pp 333–351

Kala CP (2005) Indigenous uses, population density, and conservation of threatened medicinal plants in protected areas of the Indian Himalayas. Conserv Biol 19(2):368–378

Kalam MA, Malik AH, Ganie AH, Butt TA (2021) Medicinal importance of Papra (*Podophyllum hexandrum* Royle) in Unani system of medicine. J Complement Integr Med 18(3):485–490

Kaul MK (2010) High altitude botanicals in integrative medicine—case studies from Northwest Himalaya. Indian J Tradit Med 9(1):18–25

Kaur T, Bhat R, Vyas D (2016) Effect of contrasting climates on antioxidant and bioactive constituents in five medicinal herbs in Western Himalayas. J Mt Sci 13:484–492

Kaushal K, Shinde A, Gahunge P, Nagar L (2013) A comprehensive review: *Coleus forskohlii*. Int J Ayurvedic Herb Med 3(1):1106–1113

Khan H, Tariq SA, Khan MA (2011) Biological and phytochemical studies on corms of *Colchicum luteum* Baker. J Med Plant Res 5(32):7031–7035

Khan MN, Zhang J, Luo T, Liu J, Ni F, Rizwan M et al (2019) Morpho-physiological and biochemical responses of tolerant and sensitive rapeseed cultivars to drought stress during early seedling growth stage. Acta Physiol Plant 41:1–13

Khan N, Ullah R, Ali K, Jones DA, Khan MEH (2022) Invasive milk thistle (*Silybum Marianum* (L.) Gaertn.) causes habitat homogenization and affects the spatial distribution of vegetation in the semi-arid regions of Northern Pakistan. Agriculture 12(5):687

Khan S, Masoodi TH, Pala NA, Islam MA, Raja A, Rizvi SZ (2023) Cultural significance of Western Himalayan wild food plants. Acta Ecol Sin. https://doi.org/10.1016/j.chnaes.2023.07.011

Kletter C, Kriechbaum M (eds) (2001) Tibetan medicinal plants. CRC Press

Kong JM, Goh NK, Chia LS, Chia TF (2003) Recent advances in traditional plant drugs and orchids. Acta Pharmacol Sin 24(1):7–21

Kor L, Homewood K, Dawson TP, Diazgranados M (2021) Sustainability of wild plant use in the Andean Community of South America. Ambio 50(9):1681–1697

Kour J, Balgotra S, Rajput P, Kour H, Verma PK, Sawant SD (2020) Medicinal value of high-altitude plants of Indian Himalaya. In: Singh B (ed) Botanical leads for drug discovery. Springer, Singapore, pp 295–324. https://doi.org/10.1007/978-981-15-5917-4_14

Kukreti N, Chitme HR, Varshney VK, Abdel-Wahab BA, Khateeb MM, Habeeb MS (2023) Antioxidant properties mediate nephroprotective and hepatoprotective activity of essential oil and hydro-alcoholic extract of the high-altitude plant *Skimmia anquetilia*. Antioxidants 12(6):1167

Kumar S, Mishra A (2011) Costunolide, a sesquiterpene lactone, inhibits inflammatory response by suppressing the activation of NF-κB. Biol Pharm Bull 34(7):1058–1063

Kumar S, Sharma A, Gupta A (2020) Rhodiola rosea: a versatile adaptogen. Complement Ther Med 49:102353

Kumar A, Guleria S, Ghosh D, Dogra V, Kumar S (2021) Managing reactive oxygen species—some learnings from high altitude extremophytes. Environ Exp Bot 189:104525

Kunst C, Schmid S, Michalski M, Tümen D, Buttenschön J, Müller M, Gülow K (2023) The influence of gut microbiota on oxidative stress and the immune system. Biomedicine 11(5):1388

Kunwar RM, Bussmann RW (2006) Indigenous use of medicinal plants in central Nepal. J Ethnobiol Ethnomed 2(1):1–6

Lakey, Dorji K (2016) Ecological status of high altitude medicinal plants and their sustainability: Lingshi, Bhutan. BMC Ecol 16:1–14

Larsen HO, Olsen CS (2007) Unsustainable collection and unfair trade? Uncovering and assessing assumptions regarding Central Himalayan medicinal plant conservation. Biodivers Conserv 16:1679–1697

Lautie E, Russo O, Ducrot P, Boutin JA (2020) Unraveling plant natural chemical diversity for drug discovery purposes. Front Pharmacol 11:397

Lee JH, Cho SH, Lee IS (2006) Anti-inflammatory effects of *Bupleurum falcatum* in mice. J Ethnopharmacol 106(1):24–28

Li X, Guo C, Ahmad S, Wang Q, Yu J, Liu C, Guo Y (2019) Systematic analysis of MYB family genes in potato and their multiple roles in development and stress responses. Biomol Ther 9(8):317

Li Y, Yang M, Lin H, Yan W, Deng G, Ye H et al (2022) Limonin alleviates non-alcoholic fatty liver disease by reducing lipid accumulation, suppressing inflammation and oxidative stress. Front Pharmacol 12:801730

Li L, Lin L, Wen B, Zhao PC, Liu DS, Pang GM, Wang ZR, Tan Y, Lu C (2023) Promising natural medicines for the treatment of high-altitude illness. High Alt Med Biol 24(3):175–185. https://doi.org/10.1089/ham.2022.0139

Li N, Su S, Xie X, Yang Z, Li Z, Lu D (2024) Tsantan Sumtang, a traditional Tibetan medicine, protects pulmonary vascular endothelial function of hypoxia-induced pulmonary hypertension rats through AKT/eNOS signaling pathway. J Ethnopharmacol 320:117436

Liang ZL, Zhang XY, Wang F, Zhang K, Liu HF, Liu HL (2018) Understanding molecular mechanisms of *Rhodiola rosea* for the treatment of acute mountain sickness through computational approaches (a STROBE-compliant article). Medicine 97(39):e11886

Lin H, Wang W, Peng M, Kong Y, Zhang X, Wei X, Shang H (2024) Pharmacological properties of Polygonatum and its active ingredients for the prevention and treatment of cardiovascular diseases. Chin Med 19(1):1

Liu J, Yang F, Ye LB, Yang XW, He K (2010) *Paris polyphylla*: chemical and biological assessment. Nat Prod Bioprospect 1(2):72–80

Luan F, Li M, Han K, Ma Q, Wang J, Qiu Y, Yu L, He X, Liu D, Lv H (2019) Phenylethanoid glycosides of *Phlomis younghusbandii* Mukerjee ameliorate acute hypobaric hypoxia-induced brain impairment in rats. Mol Immunol 108:81–88. https://doi.org/10.1016/j.molimm.2019.02.002

Mahizan NA, Teh SS, Mahyudin NA (2018) Berberine from *Coptidis rhizoma*, *Rhizoma coptidis*, *Lonicera japonica* and *Phellodendron amurense*: a review of its traditional uses, phytochemistry, pharmacology and toxicology. Molecules 23(9):2326

Malhotra B, Kulkarni GT, Dhiman N, Joshi DD, Chander S, Kharkwal A et al (2021) Recent advances on *Berberis aristata* emphasizing berberine alkaloid including phytochemistry, pharmacology and drug delivery system. J Herb Med 27:100433

Manandhar NP, Plants SM (2002) People of Nepal. Timber Press Inc

Mancuso C, Santangelo R (2017) Panax ginseng and *Panax quinquefolius*: from pharmacology to toxicology. Food Chem Toxicol 107:362–372

Maqbool F, Singh S, Kaloo ZA, Jan M (2016) A rapid micropropagation protocol of *Atropa acuminata* Royle ex Lindl.—a threatened medicinal plant species of Kashmir Himalaya. Indian J Biotechnol 15:576–580

Martin JT, Chambers TC (2001) Potentilla: a genus of traditional remedies. J Ethnopharmacol 76(2):145–165

Mathew S, Abraham TE, Zakaria ZA (2004) Antimicrobial activity of juniper berry essential oil: comparison with antibiotics. J Appl Microbiol 97(2):1156–1162

Melfi F, Carradori S, Mencarelli N, Campestre C, Gallorini M, Di Giacomo S, Di Sotto A (2023) Natural products as a source of new anticancer chemotypes. Expert Opin Ther Pat 33(11):721–744

Mishra AP, Kumar A, Yadav SN (2023) Ecology and conservation of threatened medicinal plants in the Trans-Himalayan region of Nanda Devi Biosphere Reserve, Western Himalaya. Tree For People 14:100451

Moussa Z, Judeh ZM, Ahmed SA (2019) Nonenzymatic exogenous and endogenous antioxidants. In: Free radical medicine and biology, vol 1. IntechOpen, pp 11–22

Mujammami M (2020) Clinical significance of *Saussurea Costus* in thyroid treatment. Saudi Med J 41(10):1047–1053. https://doi.org/10.15537/smj.2020.10.25416

Murugesu S, Selamat J, Perumal V (2021) Phytochemistry, pharmacological properties, and recent applications of *Ficus benghalensis* and *Ficus religiosa*. Plan Theory 10(12):2749

Nainwal P, Singh N (2020) Sustainable use of high altitude medicinal and aromatic plants for socio-economic development in Uttarakhand: a review. Int J Pharm Sci Res 11(9):4238–4243

Narahari A, Maher AR, Wetherington JD (2016) Pharmacogenomics: the future of personalized medicine? Perm J 20(1):94–100

Nawrot-Chorabik K, Sułkowska M, Gumulak N (2022) Secondary metabolites produced by trees and fungi: achievements so far and challenges remaining. Forests 13(8):1338

Ncube B, Finnie JF, Van Staden J (2014) Biotechnological production of plant secondary metabolites: recent advances and future challenges. Plant Cell Tissue Organ Cult 118(1):1–17

Negi JS, Bisht VK, Bhandari AK (2011) Medicinal uses of "Bhojpatra" (*Betula utilis*): an ethnopharmacological review. Asian J Tradit Med 6(2):41–47

Nisar M, Khan I (2017) Biological activities of *Paeonia emodi* Royle. J Med Plant Res 5(4):575–581

Olas B, Kontek B, Malinowska P, Żuchowski J, Stochmal A (2016) *Hippophae rhamnoides* L. fruits reduce the oxidative stress in human blood platelets and plasma. Oxidative Med Cell Longev 2016:4692486

Ou C, Geng T, Wang J, Gao X, Chen X, Luo X et al (2020) Systematically investigating the pharmacological mechanism of Dazhu Hongjingtian in the prevention and treatment of acute mountain sickness by integrating UPLC/Q-TOF-MS/MS analysis and network pharmacology. J Pharm Biomed Anal 179:113028

Owais F, Anwar S, Muhammad S, Ishtiaque S, Mohiuddin O (2014) Analgesic, anti-inflammatory and neuropharmacological effects of Atropa belladonna. Pak J Pharm Sci 27(6):2183–2187

Painuli S, Kumar N (2016) Prospects in the development of natural radioprotective therapeutics with anti-cancer properties from the plants of Uttarakhand region of India. J Ayurveda Integr Med 7(1):62–68

Pan L, Kardono LB, Riswan S, Chai H (2019) Alkaloids from natural herbs as anticancer agents. Adv Exp Med Biol 1140:297–338

Pan L, Yang N, Sui Y, Li Y, Zhao W, Zhang L, Mu L, Tang Z (2023) Altitudinal variation on metabolites, elements, and antioxidant activities of medicinal plant *Asarum*. Metabolites 13(12):1193. https://doi.org/10.3390/metabo13121193

Pandey A, Pant DR (2014) *Swertia chirayita*: a traditional herb and its medicinal uses. J Ethnopharmacol 158:331–342

Pandey MM, Rastogi S, Rawat AKS, Negi KS (2004) Traditional uses and medicinal potential of *Rhododendron anthopogon*. J Ethnopharmacol 92(1):123–131

Pandey H, Nandi SK, Kumar A, Palni UT, Palni LMS (2007) Podophyllotoxin content in *Podophyllum hexandrum* Royle plants of known age of seed origin and grown at a lower altitude. Acta Physiol Plant 29:121–126

Panossian A, Wikman G (2010) Evidence-based efficacy of adaptogens in fatigue, and molecular mechanisms related to their stress-protective activity. Curr Clin Pharmacol 5(3):198–219

Paul M, Ma JKC (2011) Plant-made pharmaceuticals: leading products and production platforms. Biotechnol Appl Biochem 58(1):58–67

Paul A, Malla G, Ghimire SK, Bhattarai KR (2016) Impact of climate change on the distribution of plant species and communities in the Himalaya. In: Climate change impacts on high-altitude ecosystems. Springer, pp 15–32

Pena E, El Alam S, Siques P, Brito J (2022) Oxidative stress and diseases associated with high-altitude exposure. Antioxidants 11(2):267

Phondani PC, Bhatt ID, Negi VS, Kothyari BP, Bhatt A, Maikhuri RK (2016) Promoting medicinal plants cultivation as a tool for biodiversity conservation and livelihood enhancement in Indian Himalaya. J Asia Pac Biodivers 9(1):39–46

Poudeyal MR, Kunwar RM, Bussmann RW, Paniagua-Zambrana NY (2020) *Meconopsis aculeata* Royle *Meconopsis horridula* Hook. f. and *Thomson Meconopsis latifolia* (Prain) *Prain Meconopsis quintuplinervia* Regel *Meconopsis simplicifolia* (D. Don) Walpers Papaveraceae. In: Ethnobotany of the Himalayas. Springer, Cham, pp 1–9

Purohit VK, Samant SS (1995) Medicinal plants used by the tribals of Mandi district, Himachal Pradesh. Ethnobotany 7:107–110

Purohit SS, Vyas SP (2004) Aegle marmelos Correa ex Roxb. (Bael), medicinal plant cultivationa scientific approach. Agrobios, Jodhpur, pp 498–504

Qian C, Yin H, Shi Y, Zhao J, Yin C, Luo W et al (2016) Population dynamics of *Agriophyllum squarrosum*, a pioneer annual plant endemic to mobile sand dunes, in response to global climate change. Sci Rep 6(1):26613

Rahimi Y, Taleei A, Ranjbar M (2018) Long-term water deficit modulates antioxidant capacity of peppermint (*Mentha piperita* L.). Sci Hortic 237:36–43

Rana PS, Saklani P, Chandel C (2020) Research article influence of altitude on secondary metabolites and antioxidant activity of *Coleus forskohlii* root extracts. Res J Med Plant 14(43–52):2020

Rawat S, Pant G, Rawal RS, Pangtey YPS (2010) Meconopsis aculeata Royle: a high value medicinal plant of Kumaon Himalaya. Ethnobot Leafl 2010(5):10

Rinchen T, Gurmet P, Dolker P, Stobgais T (2021) Cultivation and population status of Ephedra (*Ephedra gerardiana* Wall. Ex Stapf) critically endangered medicinal plant for the conservation in cold desert of Ladakh, India. Research Square

Rokaya MB, Parajuli B, Bhatta KP, Timsina B (2020) *Neopicrorhiza scrophulariiflora* (Pennell) Hong: a comprehensive review of its traditional uses, phytochemistry, pharmacology and safety. J Ethnopharmacol 247:112250

Rosso L, Cantamessa S, Bergante S, Biselli C, Fricano A, Chiarabaglio PM et al (2023) Responses to drought stress in poplar: what do we know and what can we learn? Life 13(2):533

Rumpf SB, Hülber K, Wessely J, Willner W, Moser D, Gattringer A, Dullinger S (2018) Range dynamics of mountain plants decrease with elevation. Proc Natl Acad Sci 115(8):1848–1853

Samuelsen AB (2000) The traditional uses, chemical constituents and biological activities of Plantago major L. J Ethnopharmacol 71(1–2):1–21

Sarris J, Wardle J (2013) Plant-based medicines for anxiety disorders, part 2: a review of clinical studies with supporting preclinical evidence. CNS Drugs 27(4):301–319

Semwal P, Rauf A, Simal-Gandara J (2023) High altitude medicinal plants and their bioactive compounds for the prevention of oxidative stress-induced diseases and disorders. Front Pharmacol 14:1257018

Shantappa T, Navi V, Shet RM, Hongal S, Athani SI (2022) Low temperature stress induced changes in plants and their management. In: Abiotic & biotic stress management in plants. CRC Press, pp 95–120

Sharifi-Rad M, Mnayer D, Morais-Braga MFB, Carneiro JNP, Bezerra CF, Coutinho HDM, Salehi B, Martorell M, Del Mar Contreras M, Soltani-Nejad A et al (2018) Echinacea plants as antioxidant and antibacterial agents: from traditional medicine to biotechnological applications. Phytother Res 32:1653–1663. https://doi.org/10.1002/ptr.6101

Sharifi-Rad M, Anil Kumar NV, Zucca P, Varoni EM, Dini L, et al (2020) Lifestyle, oxidative stress, and antioxidants: back and forth in the pathophysiology of chronic diseases. Front physiol 2(11):552535

Sharma G, Rai LK (2012) Climate change and sustainability of agrodiversity in traditional farming of the Sikkim Himalaya. In: Climate change in Sikkim: patterns, impacts, initiatives. Information and Public Relations Department, Government of Sikkim, Gangtok, pp 193–218

Sharma M, Sidhu AK, Samota MK, Gupta M, Koli P, Choudhary M (2023) Post-translational modifications in histones and their role in abiotic stress tolerance in plants. Proteome 11(4):38

Shrestha UB, Bawa KS (2014) Impact of climate change on potential distribution of Chinese caterpillar fungus (*Ophiocordyceps sinensis*) in Nepal Himalaya. PLoS One 9(9):e106405

Singh B, Singh VP (2005) Anti-inflammatory and antimicrobial activities of *Inula racemosa* root extract. J Ethnopharmacol 96(3):305–309

Singh H, Singh PP (2006) Pharmacological studies of *Valeriana jatamansi*. J Ethnopharmacol 107(3):469–475

Singh A, Duggal S, Kaur N, Singh J (2010) Antimicrobial activity of *Berberis aristata*. J Ethnopharmacol 132(3):504–508

Singh R, Singh UK, Singh KK (2013) Ethnomedicinal plants used for primary health care among the indigenous people of Manang district, central Nepal. J Ethnopharmacol 150(1):221–227
Singhuber J, Zhu M, Prinz S (2009) Aconitum in traditional Chinese medicine: a valuable drug or an unpredictable risk? J Ethnopharmacol 126(1):18–30
Soliman MH, Alayafi AA, El Kelish AA, Abu-Elsaoud AM (2018) Acetylsalicylic acid enhance tolerance of *Phaseolus vulgaris* L. to chilling stress, improving photosynthesis, antioxidants and expression of cold stress responsive genes. Bot Stud 59:1–17
Srivastava RP, Kumar S, Singh L, Madhukar M, Singh N, Saxena G, Pandey S, Singh A, Devkota HP, Verma PC, Shiva S, Malik S, Rustagi S (2023) Major phenolic compounds, antioxidant, antimicrobial, and cytotoxic activities of *Selinum carvifolia* (L.) collected from different altitudes in India. Front Nutr 10:1180225. https://doi.org/10.3389/fnut.2023.1180225
Subramoniam A (2014) Present scenario, challenges and future perspectives in plant based medicine development. Ann Phytomed 3(1):31–36
Suhas S, Ramesha BT, Ravikanth G, Gunaga RP, Vasudeva R, Ganeshaiah KN, Shaanker RU (2007) Chemical profiling of *Nothapodytes nimmoniana* populations in the Western Ghats, India for anti-cancer compound, camptothecin. Curr Sci 92(8):1142–1147
Suneetha MS (2010) Sustainability issues for biodiversity business. Sustain Sci 5:79–87
Suryakumar G, Rathor R, Agrawal A, Singh SN, Kumar B (2019) Optimizing performance under high-altitude stressful conditions using herbal extracts and nutraceuticals. In: Nutraceuticals and natural product pharmaceuticals. Academic, pp 141–166
Szabo E, Schuehly W, Hohmann J (2009) *Leontopodium alpinum* (Edelweiss) as a cosmetic ingredient. Int J Cosmet Sci 31(4):333–347
Tao Y, Siyiti A, Minmin W, Maimaiyiming D, Aikemu A (2021) Protective effects of glycyrrhetnic acid on a rat model of chronic mountain sickness. Int J Clin Exp Med 14(4):1581–1593
Ticktin T, Ganesan R, Paramesha M, Setty S (2012) Disentangling the effects of multiple anthropogenic drivers on the decline of two tropical dry forest trees. J Appl Ecol 49(4):774–784
Upton R (2013) Stinging nettles leaf (*Urtica dioica* L.): extraordinary vegetable medicine. J Herb Med 3(1):9–38
van Wietmarschen HA, Hagels H, Peters R, Heistek J, van der Greef J, Wang M (2016) Optimizing growth conditions for digoxin production in *Digitalis lanata* Ehrh. World J Tradit Chin Med 2(2):24–35
Veeresham C (2012) Natural products derived from plants as a source of drugs. J Adv Pharm Technol Res 3(4):200–201
Virk JK, Kumar S, Singh R, Tripathi AC, Saraf SK, Gupta V, Bansal P (2016) Isolation and characterization of quinine from *Polygonatum verticillatum*: a new marker approach to identify substitution and adulteration. J Adv Pharm Technol Res 7(4):153–158
Wahid M, Ali A, Saqib F, Aleem A, Bibi S, Afzal K et al (2020) Pharmacological exploration of traditional plants for the treatment of neurodegenerative disorders. Phytother Res 34(12):3089–3112
Walia S, Kumar P, Kumar D, Kumar R (2023) A preliminary study on suitability of growing ginseng (*Panax ginseng* Meyer) in the Western Himalayan region. Plant Soil Environ 69(2):71–80
Wang S, Li J, Tian Y (2017) Alkaloids from plants in the genus *Amaryllis*. Molecules 22(9):1534
Wang S, Li J, Ma J, Fu J (2018) Natural products with anti-diabetic potential from high altitude medicinal plants: a review. Phytochem Rev 17(4):795–812
Wang Y, Gao S, He X, Li Y, Li P, Zhang Y, Chen W (2019) Growth, secondary metabolites and enzyme activity responses of two edible fern species to drought stress and rehydration in Northeast China. Agronomy 9(3):137
Wang Y, Hou H, Ren Q, Hu H, Yang T, Li X (2021) Natural drug sources for respiratory diseases from Fritillaria: chemical and biological analyses. Chin Med 16(1):40
Wang Y, Yuan X, Ren M, Wang Z (2024) Ferroptosis: a new research direction of artemisinin and its derivatives in anti-cancer treatment. Am J Chin Med 2024:1–21
Wangchuk P, Tobgay T (2015) Contributions of medicinal plants to the Gross National Happiness and Biodiscovery in Bhutan. J Ethnobiol Ethnomed 11:1–13

Wani TA, Kaloo ZA, Dangroo NA (2022) *Aconitum heterophyllum* Wall. ex Royle: a critically endangered medicinal herb with rich potential for use in medicine. J Integr Med 20(2):104–113

Wasser SP (2011) Current findings, future trends, and unsolved problems in studies of medicinal mushrooms. Appl Microbiol Biotechnol 89(5):1323–1332

Watanabe K, Kinoshita M, Hasegawa T, Umemoto K (2005) Medicinal plants of Nepal-Himalaya. Jpn J Pharmacogn 59(2):99–109

Wiboonkiat K, Pornthanamongkol C, Intarapradit S, Vatanashevanopakorn C, Dangprapai Y (2023) Ginkgo biloba flavonoid extracts as a carbonic anhydrase II inhibitors for acute mountain sickness prevention: a molecular docking study. J Asian Med Stud Assoc. https://doi.org/10.52629/jamsa.vi.499

Wonsick MM, Pinker RT (2014) The radiative environment of the Tibetan Plateau. Int J Climatol 34(7):2153–2162

Yang L, Wen KS, Ruan X, Zhao YX, Wei F, Wang Q (2018a) Response of plant secondary metabolites to environmental factors. Molecules 23(4):762

Yang X, Zhao Y, Zhou Y, Zhao T (2018b) Flavonoids from medicinal plants: a potential therapeutic treasure for cardiovascular disease. Pharmacol Res 119:13–23

Yang L, Li Y, Yue T, Zhang Z, Hu Y, Hou J (2023) Comprehensive prospects of high-altitude cognitive dysfunction and the positive role of traditional Chinese medicine. Pharmacol Res Mod Chin Med 9:100317

Yousefzadeh Najafabadi M, Ehsanzadeh P (2017) Photosynthetic and antioxidative upregulation in drought-stressed sesame (*Sesamum indicum* L.) subjected to foliar-applied salicylic acid. Photosynthetica 55:611–622

Zajtchuk R (1999) New technologies in medicine: biotechnology and nanotechnology. Dis Mon 45:453–495

Zhang HY, Tang XC, Bai DL (2017) Natural alkaloids and their derivatives as potent anti-Alzheimer's agents. Chem Biol Drug Des 90(5):724–735

Zhao J, Qiu X, Zhao Y, Wu R, Wei P, Tao C, Wan L (2022) A review of the genus Chrysosplenium as a traditional Tibetan medicine and its preparations. J Ethnopharmacol 290:115042

Zomba D, Sharma M, Jandrotia R, Singh HP, Batish DR (2024) Chemical profiling, phytotoxicity, cytotoxicity, and antibacterial activity of the essential oil of a high-altitude plant, *Allardia tridactylites*, from the Trans-Himalayan Region, Ladakh. Chem Pap 78(3):1887–1896

Cultivation and Conservation of High Value Medicinal and Aromatic Plants from the Himalayas

12

Yashwant Singh Tariyal, R. S. Chauhan, and B. P. Nautiyal

Abstract

Traditional Indian systems of medicine like Ayurveda, Siddha, and Unani have long relied on Medicinal and Aromatic Plants (MAPs), forming a rich foundation of knowledge and practices.

India's diverse climate and geography have fostered a vast array of MAPs, making it a global hotspot for plant-based traditional knowledge. Uttarakhand, nestled in the Himalayas, boasts a particularly rich flora, including many MAPs adapted to temperate ecosystems. With society's shift towards preventive healthcare and personalized medicine, there's a growing demand for nature-based products with fewer side effects. India stands out as a major contributor to the global herbal market with its 16 agro-climatic zones and immense biodiversity. Despite its advantages, India faces challenges such as habitat loss, overexploitation, and unsustainable harvesting practices. Conservation efforts are crucial, including establishing protected areas, habitat restoration, promoting sustainable harvesting, engaging local communities, conducting research, and enacting supportive policies. In Uttarakhand, where traditional agriculture faces challenges, cultivation of Maps presents an opportunity to meet market demands, improve farmers' livelihoods, and conserve endangered species. Cultivating MAPs organically not only meets market demands but also conserves biodiversity and fetches premium prices. This approach contributes to both economic prosperity and ecological sustainability. By cultivating threatened and endangered species, farmers can play a vital role in conserving these plants while meeting market demands.

Y. S. Tariyal · R. S. Chauhan · B. P. Nautiyal (✉)
Department of Plantation, Spices, Medicinal & Aromatic Plants, College of Horticulture, VCSG Uttarakhand University of Horticulture and Forestry, Bharsar, Uttarakhand, India

S. Chandra, H. Lata (eds.), *High Altitude Medicinal Plants*,
https://doi.org/10.1007/978-3-031-99406-7_12

12.1 Introduction

Medicinal and aromatic plants (MAPs) have been integral part of traditional Indian systems of medicine such as Ayurveda, homeopathy, Siddha, and Unani for millennia. Various indigenous knowledge and traditional medicinal system (TSMs) form a strong base and vast history for the utilization of medicinal plants. Apart from the medicinal uses, these plants are also a part of various traditional ceremonies, religious offerings, ornamental and aesthetic purpose (Verma et al. 2018). India's diverse climatic and geographical conditions have contributed to the proliferation of a wide array of MAPs, making it one of the world's richest repositories of plant-based traditional knowledge. Uttarakhand, situated in the Himalayan region, is particularly renowned for its rich flora, including numerous medicinal and aromatic species adapted to its ecosystems (Rao and Smitha 2018).

Currently society is moving from illness to wellness, treatment of aliments to prevention and early diagnosis. The pharmaceutical industry is now moving toward personalized medicines from generalized medicines. The market is now a day moving toward nature-based products causing lesser side effects at comparatively low cost. It has been reported that around 80–85% of the total world population relies on herbal based medicines and health products (WHO). The rapid growth in the herbal market has been observed in recent years. WHO projected that the global market of herbal medicines would grow to $5 trillion from current market value of over $62 billion (Gautam et al. 2020; Yadav 2019).

India with 16 agro-climatic zones is the center of origin for many plant species. India is one of the seventeen mega-biodiversity countries and ecological hotspots contributing about 7% of world's total biodiversity. Wide variety of climate and topography, India has been considered treasure house or botanical garden of plant genetic resources. There are about 45,000 different plant species across the country, out of which 18,000 are of medicinal use. The traditional health care practitioners of India are using about 4500–5000 species of plants for medicinal purpose. The demand of herbal raw material was estimated to be around 195,000 MT during 2014–2015 and total domestic consumption was estimated to be around 12,000 MT. It was also reported that only 22% of total herbal raw material is procured through cultivation (NMPB 2019). Instead of having all the advantages India is second largest exporter of herbal raw products and herbal medicines after China. India and China account for about 70% of global demand of herbal medicines. In 2017–2018 India exported raw herbal drugs of worth approximately $330.18 million and of value-added herbal products was around $456.12 million with growth rate of 14.22% over the year 2016–2017 and export of value-added extracts of medicinal plant were valued around $456.12 million with the growth rate of 12.23% over the year 2016–2017. Out of 4500–5000 medicinal plants species used in traditional medicinal system across the country, only 1178 plant species are currently under trade. Out of which 42% are herbs, 31% shrub and climbers and 27% trees. Only 242 species share major share in trade annually (>100MT) (Goraya and Ved 2017).

The temperate regions of Uttarakhand, characterized by high altitude, rugged terrain, and extreme climatic conditions, harbor unique plant communities adapted to survive in such harsh environments. These plants have evolved diverse mechanisms to cope with cold temperatures, intense solar radiation, and nutrient-poor soils. Additionally, they play crucial ecological roles, such as stabilizing soil, regulating water flow, and providing habitat and food for diverse flora and fauna. Indigenous communities inhabiting the Himalayan region have long relied on MAPs for their healthcare needs (Joshi and Kumar 2013). The traditional knowledge associated with these plants has been passed down through generations, encompassing a wide range of medicinal preparations and therapeutic practices. Common uses include the treatment of digestive disorders, respiratory ailments, skin conditions, and rheumatic pains. Moreover, many aromatic plants are used in religious rituals, perfumery, and culinary applications (Sharma and Samant 2014).

Scientific studies have validated the pharmacological efficacy of numerous medicinal plants found in Uttarakhand's alpine regions. These plants contain bioactive compounds such as alkaloids, flavonoids, terpenoids, and polyphenols, which exhibit diverse biological activities including anti-inflammatory, antimicrobial, antioxidant, anticancer, and immunomodulatory effects. Examples of well-studied species include *Picrorhiza kurrooa*, *Aconitum heterophyllum*, *Valeriana jatamansi*, *Saussurea costus*, etc.

12.2 Conservation and Sustainable Utilization Strategies

Despite ecological and socio-economic implications, many temperate MAPs face threats due to narrow distribution, habitat loss, endemism, overexploitation, unsustainable harvesting practices, habitat fragmentation, and climate change. Additionally, the commercialization of herbal products has led to increased pressure on wild populations, raising concerns about genetic erosion and loss of traditional knowledge. We need to accept conservation means of MAPs with commercial systems as one of the main sub-sectors. Therefore, conservation efforts must focus on threats assessment, habitat protection, sustainable harvesting practices, community-based resource management, ex-situ conservation and cultivation, and documentation of traditional knowledge as highlighted below:

1. *Assessment of Threats:* Identify key threats to temperate MAPs (e.g., climate change, habitat destruction, overexploitation). Conduct scientific studies to understand the extent and impact of each threat (Pant et al. 2020).
2. *Prioritization of Species:* Prioritize medicinal plant species based on their ecological significance, economic value, and conservation status. Focus conservation efforts on species at higher risk of extinction or those with limited distribution with ecosystem approach.
3. *Protected Areas:* Establishing protected areas and conservation reserves in high altitude regions can help safeguard the habitats of MAPs from encroachment, habitat degradation, and unsustainable harvesting. These protected areas should be carefully managed to balance conservation objectives with the needs of local communities to ensure active participation and cooperation (Negi et al. 2002).

4. *Habitat Restoration:* Implementing habitat restoration initiatives, e.g., reforestation, seed broadcasting of inhabitant species, soil conservation, and erosion control measures, can enhance the resilience of high-altitude ecosystems and promote the regeneration of MAPs populations (Tewari et al. 2018).
5. *Sustainable Harvesting Practices:* Adopting sustainable harvesting practices, guided by scientific research and traditional knowledge, is essential for ensuring the long-term viability of MAPs populations. Harvesting methods should be selective, ensuring the regeneration of plants and minimizing ecological impact. In Uttarakhand, sustainable harvesting from wilds can be effective by following steps:

   - The entire natural habitats may be divided into different divisions for rotational harvesting.
   - Harvesting on a 5-year basis as per conservation, development, and harvesting (CDH) plan (Fig. 12.1).
   - Categorization of the species into five groups:

- Banned species: CITES included, Negative List of Export and Threatened species.
- Species identified for rotational system (rare, endemic).
- Species for limited harvesting (Conservation dependent).

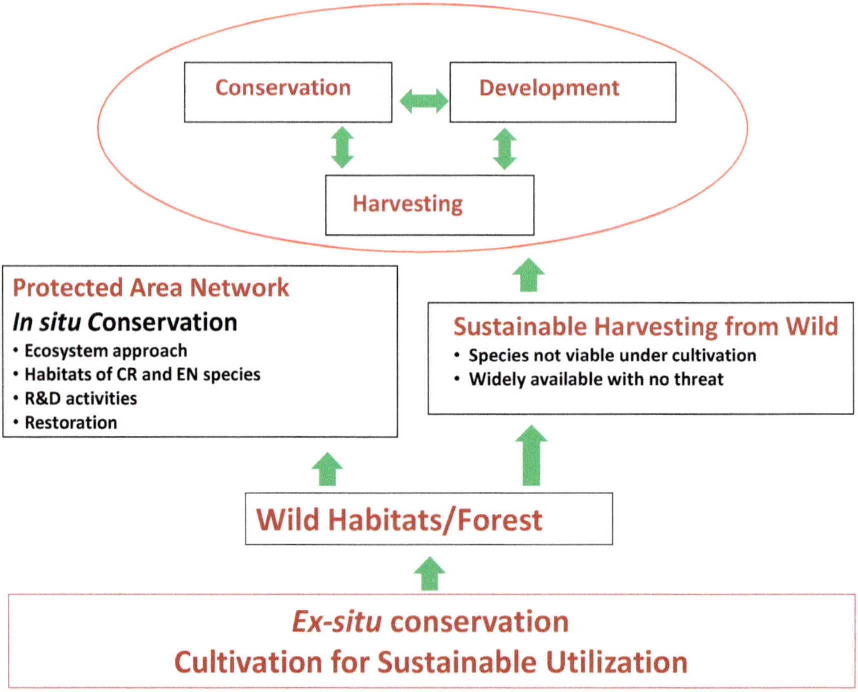

Fig. 12.1 Conservation and sustainable utilization of MAPs

- Species permitted for large scale harvesting (Common, Roadside).
- Exotic species (No Ban, except for CITES).

6. *Community Participation:* Engaging local communities in conservation efforts through participatory approaches, community-based management schemes, and capacity building initiatives fosters a sense of ownership and stewardship towards high altitude ecosystems and their biodiversity. Provide training and support for local communities to diversify their livelihoods away from reliance on medicinal plant harvesting. Promote eco-tourism, sustainable agriculture, and other income-generating activities that reduce pressure on natural resources (Pandey and Tripathi 2017).
7. *Establish seed banks, botanical/herbal gardens, and germplasm repositories:* to conserve genetic diversity of MAPs.
8. *Research and Monitoring:* Conducting scientific research to assess the population dynamics, distribution patterns, growth pattern and ecological requirements of MAPs for conservation decision. Long-term monitoring programs can track changes in plant populations and ecosystem health, facilitating adaptive management strategies (Upreti et al. 2018).
9. *Policy and Legislation:* Enacting and enforcing policies and legislation that regulate the harvesting, trade, and utilization of MAPs is essential for combating illegal harvesting, overexploitation, and habitat destruction. Incorporating traditional knowledge systems into legal frameworks promotes culturally sensitive conservation practices (Dhyani and Dhyani 2021).
10. *Collaboration and Partnerships:* Foster collaboration among government agencies, NGOs, research institutions, and local communities to coordinate conservation efforts. Establish partnerships with international organizations and funding agencies to mobilize resources and expertise for conservation projects (Singh and Sureja 2008).
11. *Trade chain:* Government is just one player in the diverse landscape of stakeholders. Producers, private enterprises, knowledge centers, and community representatives form a human chain with government that allows for the flow of information, capital, and produce (ICP). This chain needs to be strengthened for steady flow of ICPs (Aggarwal and Panwar 2020).

The Uttarakhand Himalayan region is rich in MAPs. A total of more than 2000 species of medicinal plants are found in region which includes majority of herbs out of which currently about 960 plants are active in trade (Ved and Gorya 2007). The total cultivation of medicinal plants in the state accounts about 200 MT/year by volume. The demand for plant-based medicine is continuously increasing and the productivity is low in comparison to the market demand. In hilly states like Uttarakhand, there are many constrains in conventional crop production which ultimately ends with decrease in interest of farmers in agriculture leads to unemployment, lower living standard and migration. Many of the medicinal plants are cultivated substantially and are not exploited to commercial level. Their cultivation will not only be helpful to meet the market demand but also improves the living standard of the farmers (Nainwal and Singh 2020). Studies conducted on economics and employment generation in medicinal and aromatic plant sector shows that the number of registered farmers engaged in the cultivation of MAPs increasing year by year. The number which was 301 in year 2003–2004 increases to 2714 in year 2006–2007 and increases to approximately 8936 people during 2008–2013 (Pangriya 2015).

12.3 History of MAP Cultivation

In Uttarakhand, the cultivation of Allium stracheyi, a spice, is believed to have begun between 1915 and 1920 as an experimental effort, according to the residents of Tolma village in Niti Valley, Chamoli district. Initially, the crop was grown on a small scale, primarily as part of local kitchen gardens. Local elders recall that before the Indo-Tibet border trade was closed in the early 1960s, A. stracheyi was traded with India by Tibetan Lamas and other traders in exchange for daily necessities. It is likely that commercial cultivation of A. stracheyi began in the region after the 1960s (Nautiyal and Nautiyal 2004a). Over time, the cultivation spread to other villages in Niti Valley, as well as to Mana Valley in Chamoli and Johaar Valley in Pithoragarh. Currently, 20 families in Tolma village are cultivating A. stracheyi across a total area of 1.10 hectares (Kuniyal and Negi 2018). However, villages such as Malari, Jhelum, Gamshali, and Niti in Chamoli district, and Pachhu, Martoli, and Milam in Pithoragarh district, have only limited cultivation (Bhatt 1999). The trade of medicinal and aromatic plants (MAPs) largely depends on wild harvesting, which is often illegal and unsustainable.

To promote the cultivation of medicinal and aromatic plants (MAPs), a seminar titled "*Harvesting Herbs-2000 for Medicinal and Aromatic Plants—An Action Plan for Uttarakhand*" was organized in 1997 at the High-Altitude Plant Physiology Research Centre (HAPPRC), HNB Garhwal University, Srinagar Garhwal. The seminar proceedings addressed various issues related to the cultivation and trade of high-altitude MAPs (Purohit 1997). In a pioneering study, Nautiyal et al. (2001, 2003a, b) examined the yield and related parameters of *Picrorhiza kurrooa* and *Rheum emodi* for the first time, and found that cultivating these species is commercially viable at lower altitudes, up to 2000 m. They also outlined the essential agro-technological factors for cultivating MAPs (Nautiyal et al. 2002). The institute (HAPPRC) later published a book on agro-techniques for high-altitude medicinal and aromatic plants, providing detailed information on the cultivation of 20 such species (Nautiyal and Nautiyal 2004a).

Following this, a systematic and planned initiative was undertaken by HAPPRC, Srinagar Garhwal, under the leadership of Prof. A.N. Purohit and Prof. M.C. Nautiyal, to kickstart the commercial cultivation of *P. kurrooa* in the remote village of Ghesh, Chamoli district (Nautiyal and Nautiyal 2004b). A dedicated team of villagers (primarily ex-army personnel) and scientists from HAPPRC, including B.P. Nautiyal and Vinay Prakash, demonstrated the cultivation of *P. kurrooa* on village farms. This effort was followed by training sessions and a tripartite agreement between the villagers, HAPPRC, and the industry (Dhawan International), which included a buy-back provision. The first harvest took place in 2005. Later, *P. kurrooa* cultivation expanded to other villages in Chamoli, Rudraprayag, and Pithoragarh districts, although the cultivated area has yet to meet expectations. In addition, significant contributions were made by HAPPRC in promoting the cultivation of other high-altitude medicinal plants such as *Nardostachys jatamansi* (Chauhan and Nautiyal 2005, 2007), *Angelica glauca* and *A. archangelica* (Vishashita et al. 2007, 2008), *Aconitum* species (Nautiyal et al. 2006; Attri et al. 2021; Dhyani et al. 2022).

The Ghesh and neighboring villages in Chamoli district, as well as the upper Johaar valley, have frequently been highlighted in recent literature as areas where significant cultivation of medicinal and aromatic plants (MAPs) is taking place. However, in reality, only 1–2% of the total cultivated land in the entire region is dedicated to MAP cultivation. In contrast, substantial profits have been generated from the sale of wild MAPs, primarily *Yar Tsa Gumba* or *Keeda Jadi* (Cordyceps sinensis), along with other species such as *P. kurrooa*, *A. heterophyllum*, and *N. jatamansi*. It is clear that while some MAPs are grown in home gardens or on small plots of land, this is often used as a cover for the unsustainable wild harvesting that is actually taking place, leading to the depletion of important species in the area. Additionally, the cultivation of *A. stracheyi* is seen as a potential strategy for conserving its genetic resources (Pandey et al. 2008). Similar conservation-focused approaches are urgently needed for other threatened species of commercial value.

12.4 Cultivation Status

Besides sustainable utilization, the aim of cultivation of medicinal plants is to increase the uniformity and concentration of bioactive ingredients of plants. In recent years, the demand for herbal based medicines and herbal raw products increases rapidly due to their utilization in the value addition of food products and acceleration in pharmaceutical interests. The National Medicinal Plants Board (NMPB), GOI, New Delhi and the Food and Agriculture Organization recommended that all the MAPs to be cultivated organically. Effective nutrient management techniques will not only help in fulfilling nutritional requirements of plants but also in maintaining the biological, physical, and chemical ecosystem. Organically grown medicinal and aromatic products are not only readily acceptable in the global market but also fetch premium prices than those grown with conventional farming. It has been aggravated that quality of MAPs deteriorates with chemical fertilizers.

As most of the high-altitude plants are threatened, the cultivation of these plants will not only help in increasing production and meet the market demand, but it will also help in conservation of threatened species in farmer's field as well as in wild habitats. Therefore, important MAPs from high altitude regions are described in following Tables 12.1, 12.2 and 12.3 with brief description, status of cultivation, and information related to various agronomic practices:

Table 12.1 Notably, very few species are under cultivation with subsidiary yield / no record compared to market demand. Perusal of information depicted in Tables 12.1, 12.2 and 12.3 clearly indicates that besides low area under cultivation and not

Table 12.1 Species under cultivation

Common name	Kutki	Jamboo Pharan, Dhungar	Black Caraway, Caraway, Kalajira	Kuth
Botanical name	*Picrorhizakurroa* Royle ex Benth.	*Allium stracheyi* baker	*Carum carvi* L.	*Saussurealappa* C.B. Clarke
Family	Scrophulariaceae	Alliaceae	Apiaceae	Asteraceae
Center of origin	India	Himalayan region	Europe	
Distribution	It occurs in temperate to alpine Himalayas, from Kashmir to Sikkim, between 3000 and 4500 m altitudes.	This species occurs in Jammu Kashmir, Himachal Pradesh and Uttarakhand (India), Nepal and Pakistan at the elevation of 2500–3625 m	It is cultivated in countries such as Jamaica, India, Canada, United States and Australia. In India, this spice is known as Kashmiri jeera	Naturally grows at altitudes from 2500 to 3500 masl. In Uttar Pradesh Himanchal Pradesh, Uttarakhand, Arunachal Pradesh, Sikkim, Jammu and Kashmir
Chromosome number	$2n = 34$	$X = 7, 8$ or 9	$2n = 2x = 20$	$2n = 30$
Inflorescence	Flowers are small, white or pale blue purple, in dense terminal spikes. Bracts are oblong or lanceolate, equalling the calyx. Sepals are lanceolate. Corolla units are five in number and have a cleft in the middle.	Inflorescence is of umbel type. It has rosy flowers. Sepals are oblong, blunt, stamen filaments are threadlike, much protruding out.	It has a high vernalisation requirement to initiate the production of flowering stems in the second year. The flowers are produced on umbels, are white and 2–3 mm across, the outer ones larger than the inner ones. They open from late April onwards	It features a capitulumin florescence, consists of numerous small florets arranged on a common receptacle. Capitula are typically grouped together in dense clusters, forming a corymbose or paniculate arrangement. Flowering & fruits occurs during August–September
Fruit type	Fruit is a two-celled spherical capsule. Seeds are many, white, with an oblong curved raphae. Flowering and fruiting occur from June to August.	Capsule	Fruit which Schizocarp is 3–6 mm long, and light brown, ripening from early June and July	The fruit is a small, dry, one-seeded achene. Each achene contains a single seed. The achene is typically crowned with a pappus (aid in dispersal by wind), which consists of fine bristles or scales. Seeds collected during Sep–Oct
Economic part	Roots and rhizomes	Dried flowers and herb	Seeds	Tuberous root

Major chemical constituent	Rhizome contains a brown resinous glucoside, picrorhizin, and a glycone, picrorhizetin. Root contains bitter principle Kutkin.	Phenolic acids, flavonoids, thiosulfinates, alkaloids, fixed oils, phytosterols, sulfur containing compounds 1,2-*bis*(methylthio) ethene, 2,4-dimethylthiophene, dimethyl disulfide and dimethyl trisulfide, as the major volatile components	Seeds contain between 0.6% and 4.4% essential oil, rich inmonoterpenes carvacrol, carvone, α-pinene, limonene, γ-terpinene, linalool, carvenone, and p-cymene	Inulin, Costunolide, Saussureal, Dehydrocostuslactone, Caffeic Acid, Cynaropicrin, Quercetin, Kaempferol, Chlorogenic Acid
Uses	It is anti-pyretic, anthelmintic, carminative, stomachic, anti-malarial, and laxative in small doses butcathartic in large doses. It improves appetite and stimulates gastric secretions. It is useful for asthma, cold, and cough, and also administered in liver complaints, anaemia, and jaundice.	It has anti-inflammatory, hypoglycemic, antioxidant, antimicrobial, hemolytic properties	It has anti-microbial, anti-diabetic, hepatoprotective, antioxidant activity	Anti-inflammatory, anticancer, gastroprotective, antimicrobial, immunomodulatory
Soil and climate	The plant grows well in cool and moist climate. Sandy clay textured porous soil is the best which facilitate horizontal spreading of the rhizomes underneath, that produce aerial sprouts from the nodes. Prefers sufficiently moist and shaded locations. It suffers heavy mortality during prolonged and intense rains	It thrives best in temperate climates, with temperatures ranging between 10 and 20 °C. Adequate sunlight and moderate moisture are essential for its optimal growth. It benefits from the monsoon rains, although excessive moisture can lead to bulb rot. Typically found at altitudes ranging from 2500 to 3625 m. It requires well-drained, loamy soils with a pH between 6.0 and 7.5	Caraway thrives in well-drained, loamy soil with a pH between 6.0 and 7.5. It prefers a temperate climate and can tolerate cold, making it suitable for cultivation in regions with mild summers and cold winters. It grows best in cool temperate climates with average temperatures ranging from 15 to 25 °C with annual rainfall between 400 and 600 mm. It can tolerate frost and temperatures as low as −4 °C to −6 °C during the vegetative stage	It prefers well-drained sandy loam and clay loam, loamy soils that are rich in organic matter. The ideal soil pH for cultivation is between 6.0 and 7.5. It thrives in cool temperate climates. The ideal temperature range is between 15 and 25 °C. It can tolerate lower temperatures and even light frost, which is typical of its native high-altitude habitats. The plant requires moderate rainfall, ideally between 600 and 1200 mm annually. Consistent moisture is essential, but excessive rainfall can lead to waterlogging and root diseases

(continued)

Table 12.1 (continued)

Common name	Kutki	Jamboo Pharan, Dhungar	Black Caraway, Caraway, Kalajira	Kuth
Propagation	The planting stock can initially be raised through seedlings, but can also be propagated through rhizomes/stolons/offsets. Seeds may be collected in August–September for raising nursery. Viability of seeds is more than 60% for a period of about 6 months. The nursery beds are planted using rhizomes/stolons/offsets in Oct–Nov. The planting stock may be raised in polybags, or mother beds. About 165,000 seedlings per hectare are needed, which means 1–1.5 kg seeds are required for raising saplings for 1 hectare of land	Seeds are collected from mature plants in late summer or early autumn. Seeds are dried and stored in a cool, dry place to maintain viability. Seeds are sown in nursery beds or directly in the field. Nursery beds should be prepared with a fine tilth and adequately manured. Seeds are sown at a depth of 1–2 cm and spaced 5–10 cm apart. Generally, propagated by using bulbils. Bulbils are planted in prepared beds at a depth of 2–3 cm, spaced 10–15 cm apart. Mature plants can be divided into smaller clumps and replanted. This method ensures genetic uniformity and quicker establishment	Seeds can be sown in either late spring (March–May) or late summer to early autumn (August–September). In temperate regions, sowing in autumn allows the plants to establish before winter. Seed rate is 8–10 kg per hectare	Propagated by both seed and vegetative methods. The recommended seed rate is approximately 5–6 kg/hectare. Seeds benefit from pre-sowing treatments by soaking in water for 12–24 h to enhance germination. Nursery is sown during March–April at spacing of 10–15 cm. Seeds germinate within 10–15 days under optimal conditions. When seedlings reach 6–8 cm in height and have 4–6 true leaves, they are ready for transplanting. For vegetative method collect root cuttings (10–15 cm in length) from mature, healthy plants during the dormant season (late autumn to early spring). Ensure cuttings have at least one bud or shoot. Treat cuttings with rooting hormone to enhance root development

Method of planting	Transplanting is done at a spacing of 20 cm× 30 cm or 30 cm× 30 cm. If the agronomic practices are carried out properly, plant density may increase to 300,000 after the first growth season	Transplant seedlings or bulbils in early spring or autumn. Spacing should be maintained at 15–20 cm between plants and 30–40 cm between rows	Sowing is done at spacing of 20–30 cm between plants and 30–40 between rows.	Plant root cuttings directly in the field or in nursery beds, maintaining proper spacing. Incorporate well-decomposed organic matter (compost or farmyard manure) into the soil. Transplant seedlings to the main field during the onset of the monsoon (June–July) to ensure adequate moisture for establishment. Maintain a spacing of 30–40 cm between plants and 60–75 cm between rows to provide sufficient space for growth and airflow. Direct sowing is also preferred, sow seeds directly into prepared beds or fields. Maintain row spacing of 30–40 cm to facilitate weeding and other cultural practices.
Nutrient management	Compost is mixed with the soil at the rate of 6–8 tonnes/ hectare/year	Incorporate well-rotted farmyard manure or compost during soil preparation. Apply balanced NPK fertilizers based on soil test recommendations	Incorporate well-decomposed organic matter or compost into the soil during land preparation. Add 10–20 tons per hectare of well-rotted farmyard manure or compost. 100–120 kg N per hectare, applied in split doses. 40–60 kg P_2O_5 and K_2O each per hectare, applied at the time of sowing	Farmyard manure or compost, 10–15 tons per hectare during land preparation. Additionally, 60–80 kg N 40–60 kg P_2O_5 and K_2O per hectare is applied, with one-third of nitrogen at planting and the rest in split doses during the growing season

(continued)

Table 12.1 (continued)

Common name	Kutki	Jamboo Pharan, Dhungar	Black Caraway, Caraway, Kalajira	Kuth
Inter-cultural operation	Field should be irrigated on alternate days during summer and as and when required during winter. The field should be kept sufficiently moist at all times. Manual weeding must take place frequently at an interval of 5–7 days during the first growing season. At later stages, weeding along with hoeing at an interval of 1 month is recommended	Light irrigation is required after planting to help establishment of seedlings. Moderate irrigation should be provided, avoiding waterlogging which can cause bulb rot. Regular weeding is necessary to reduce competition for nutrients. Mulching with organic materials helps retain soil moisture and control weeds	Thinning is done 30–40 days after sowing to maintain plant population. Maintain adequate soil moisture to facilitate nutrient uptake but avoid waterlogging, which can cause nutrient leaching and root diseases. Regular irrigation is essential, especially during dry spells, to support nutrient absorption and plant growth	Ensure regular irrigation, especially during the initial establishment phase. Avoid waterlogging as it can lead to root diseases. In areas with insufficient rainfall, provide supplementary irrigation during dry spells. Perform regular weeding to reduce competition for nutrients, water, and sunlight. Mulching can help suppress weed growth and retain soil moisture
Harvesting and yield	This crop has a life cycle of 3 years. After the completion of flowering, the fruiting starts in August and continues up to September. The plant needs 3 years for complete maturity of the seed. The roots and rhizomes are manually harvested in September when the shoots or the aerial parts begin to wither and dry. To get higher active contents, plants must be collected before flowering occurs. The plant gives about 15–18 quintals/hectare of dry roots and rhizomes after the third-year growth in Bharsar (2000 masl)	Herbage got ready for harvest 6–8 months after planting when the 50% flowering complete and lower leaves turn yellow and start drying	Under optimal conditions, it produce an average seed yield of 800–1200 kg per hectare. In some cases, yields can reach up to 1500 kg per hectare with improved agronomic practices. Seeds typically contain 0.6–4.4% essential oil, with carvone being the principal component. The oil yield depends on the variety and growing conditions	The plants are harvested when it reaches maturity, which is second year of growth. The roots are the main part harvested, as they contain the highest concentration of active compounds. Harvesting is often done during the autumn months (Sept-Nov) when the plant's energy is concentrated in the roots as per ayurveda. However, roots may be harvested after 50% flowering. Carefully dig out the roots using a spade or fork to avoid damaging the roots. Remove soil and debris from the roots, and trim off aerial parts. The yield of roots can vary based on cultivation practices, soil fertility, and climatic conditions. On average, under good management practices, the yield can range from 16 to 20 Q of dried roots per hectare

Post harvest processing	The roots and rhizomes should be dried in shade, and after proper drying, the material should be packed in polythene-lined jute bags to ensure protection from moisture	Arieal part of plant is harvested after 50% of flowering and shade dried to maintain color. After drying stored in polythene lined bags till marketing	Seeds should be dried to a moisture content of around 10% to prevent mould growth during storage. After threshing, seeds are cleaned to remove chaff and other impurities. Store seeds in a cool, dry place to maintain their quality and viability	Roots are usually dried immediately after harvest to preserve their medicinal properties. Spread them in a well-ventilated area away from direct sunlight. Clean, cut, and sort the roots based on size and quality before drying. Roots are dried until they reach a moisture content of about 10–12%. This can take 5–7 days depending on drying conditions. Store dried roots in a cool, dry place in airtight containers to maintain quality and prevent moisture absorption

Table 12.2 Domesticated species but cultivation technology needs to be standardized

Common name	Chorak, Chora, Choru, Gandrayan, Angelica	Meetha Vish, IndianAconite	Ativisha, Atees, Atees, Atis	Chirata, Chirayata, Chirayita	Vanyakarkati, Papra, Bankakri, Indian Podophyllum
Botanical name	*Angelica glauca* Edgew. *Angelica archangelica* L.	*Aconitum balfourii* syn. *Aconitum atrox*	*Aconitum heterophyllum* L.	*Swertia chirayita* (Roxb. ex Flem.) Karsten	*Podophyllum hexandrum* Royle
Family	Apiaceae	Ranunculaceae	Ranunculaceae	Gentianaceae	Podophyllaceae
Center of origin	Eastern Afghanistan Western Himalayas to Western Tibet	Alpine Himalayan Region, Southern China	Alpine Himalayan Region, Southern China	Himalayan region	Himalayan region
Distribution	The plant is endangered, which is distributed in Western Himalaya from Kashmir to Uttarakhand, in alpine scrub and forest shades between 2700 and 3700 m	The plant is found in the temperate alpine in the Himalayas with the *Rhododendron* at altitudes of 2800 m and 4200 m	The species is found in grassy slopes of alpine Himalayan region, between 3000 and 4200 masl altitude	Distributed throughout the temperate Himalayas, between 1200 and 3000 m altitudes, extending from Kashmir to Bhutan	It occurs in Indian Himalayas
Chromosome number	$2n = 22$	$2n = 16$	$2n = 16$	$2n = 26$	$2n = 12$
Inflorescence	Inflorescence is compound umbel with numerous rays. Flowers are white, yellow or purple in color, bracteate; florets white or purple	Compound raceme, flowers are bluish-violet with and attractive five carpels	Slender racemes or lax leafy panicles with large, hooded, white or violet-colored flowers. Corolla hairy corolla, carpels five in number. Flowering and fruiting occur from August to October after the third year of growth	Inflorescence is a large leafy panicle. Flowers are numerous, greenish yellow, and tinged with purple. Sepals and petals are four in number.	Flowers are white to pinkish in color, 4 cm across, appear in the fork of the stem as terminal cymose. Sepals are 3 in number and petaloid. Stamens are usually 6

Fruit type	Fruit is schizocarp 8–12 mm long, elliptic to oblong, lateral ridges broadly winged. Seeds are small in size and winged	The fruit is an achene containing 16–25 seeds, and is obpyramidal in shape. Flowering and fruiting occur during Sept–Nov.	The fruit is an achene containing 16–25 seeds, and is obpyramidal in shape. Fruiting occurs during September to Oct.	Capsules are minute, ovate, about 6 mm in diameter, and sharp pointed. Seeds are smooth and many angled.	Fruits are fleshy berry ovoid, pulpy 5 cm long and scarlet when ripe.
Economic part	Rhizomes	Tuber	Tuber	Whole plant	Roots and rhizomes
Major chemical constituent	Roots contain 1–1.5% volatile oil, valeric acid, angelic acid, lactones, sesquiterpenes, cadinene, umbelliprenin, terpene alcohol and angelisineresin	Pseudo-aconitine (0.4–0.5%), Aconine, Aconitine, Picroaconine, Hemonapelline and Benzyl Aconine	Atisine (0.19–0.27%). 0.79% of total alkaloids of the plant are found in tubers. Heteratisine (0.3%), Histisine, Atidine, and Hitidine, Aconitic acid	Glycoside chiratin, which on hydrolysis yields two bitter principles: ophelic acid, anamorphous bitter hygroscopicprinciple, and chiratogenin, a yellow bitter glycoside, insoluble in water. Amatroswerin, Amarozentin and mangiferin, etc	The primary constituents are lignin glycosides, podophyllotoxin, podophyllic acid and picropodophyllin, α-peltalin and β-peltalin
Uses	Used in dyspepsia, constipations, ulcer of palate, infantile atrophy, dysentery, menorrhagia and rinderpest. Rhizomes are used for wounds and gastric pains. The powdered root is administered in warm water for children's stomach ailments; it also checks vomiting. It is also used for flavoring confectionery and liquors	Diaphoretic, diuretic is analgesic, febrifuge, anti-inflammatory, anti-rheumatic, anti-pyretic, and vermifuge. It is used in all types of pains and inflammations. In large doses, it acts as powerful sedative, narcotic and poison	Cooling in potency used as expectorant, febrifuge, anthelmintic, anti-diarrhoeal, anti-emetic, and anti-inflammatory. Also used against poisoning due to scorpion or snake bite and to cure fever and contagious diseases, root induces hypertension through action on the sympathetic nervous system and in higher doses, it becomes lethal	It is a bitter tonic, carminative, laxative, anti-pyretic, febrifuge, anti-periodic, anti-inflammatory, stomachic, and antihelmintic. It is used in treating piles, skin diseases, ulcers, fever and diabetes.	It is used for typhoid fever, jaundice, dysentery, chronic hepatitis, rheumatism, tumerous growth, kidney and bladder problems, is also used for gonorrhoea, and syphilis. The podophyllumis used as a purgative and also for treatment of cancer and vaginal warts

(continued)

Table 12.2 (continued)

Common name	Chorak, Chora, Choru, Gandrayan, Angelica	Meetha Vish, IndianAconite	Ativisha, Atees, Atees, Atis	Chirata, Chirayata, Chirayita	Vanyakarkati, Papra, Bankakri, Indian Podophyllum
Soil and climate	It requires cool and temperate climate. It can be cultivated between 2000 and 3000 m above msl. It requires deep rich porous and moist soil with shady situations. For its ideal cultivation, the plenty of organic manure is required	Sandy loam and slightly acidic soils (pH 5.1–5.5). Partially shaded areas, thick soil, and moist conditions provide a healthy environment for the plants	Sandy (10–15 cm deep) soils with rich in organic matter. Sandy loam and slightly acidic soil, with pH of 6. The plant prefers open, sunny sites, and abundant air and soil moisture during summer months	The plant inhabits temperate regions in the Himalayas. Loamy to sandy loam, friable, and well-drained soils are suitable for its cultivation. The soil should be enriched with FYM, and if soils are clayey, addition of sand is recommended. The crop can be grown in areas having mild rainfall (100 cm) in rainy season and in areas with long cold winter, receiving snowfall frequently.	The species thrives best as undergrowth as well as in forests in well drained humus rich soil in temperate and subalpine zones

Propagation	Seeds and roots splits. 6.2 kg seeds are required for its cultivation in one hectare land area. Seeds are sown immediately after harvesting during Nov-Dec. Seed viability is very low and moist seeds have a better germination potential than dry seeds. Germination takes 25–40 days. Approximately 50,000 root splits are required to plant a hectare area	Propagation is done through seeds and tuber segments. The capsules are collected in late Oct-Nov for better germination of seeds. Stem cuttings are successful for multiplication. Seed sowing is done during Oct–Nov or March–April in temperature-controlled chambers at middle altitudes (1800–2200 m). At lower elevations, sowing is done in February–March in glass houses and during May–June at alpine sites in open beds or hot house	Seeds, tuber segments or young leafy stems can be used as propagules. The capsules that turn light brown are collected during late Oct to mid-November for better germination of seeds. The seeds should preferably be collected during morning hours. Seeds are sown during Oct–Nov or March–April in polyhouses at middle altitudes (1800–2200 m), during February–March in glass houses at lower altitudes (600–1000 m), and during May–June at alpine sites in open beds or in hot houses	The plant is successfully propagated by seeds. Completely mature seeds may be collected in autumn. About 10–20 g of seeds are required for raising planting stock for 1 hectare area. Sowing is done in Octo–Nov, seeds sown in March–April show poor germination and low survival rate	The plant is propagated by seeds and rhizome cuttings. Seeds are to be sown in spring season. The germination is up to 76%. Seedlings exhibits low growth rate and even after two seasons of growth. They could attain a height of 6.5–7.2 cm. The survival rate of this plants ranges from 85.0% to 92.6%. 3–4 kg seeds are required for an area of one hectare land. Pretreatment of seed is done by GA or by chilling

(continued)

Table 12.2 (continued)

Common name	Chorak, Chora, Choru, Gandrayan, Angelica	Meetha Vish, IndianAconite	Ativisha, Atees, Atees, Atis	Chirata, Chirayata, Chirayita	Vanyakarkati, Papra, Bankakri, Indian Podophyllum
Method of planting	After 4–6 months of growth, seedlings are transplanted at the beginning of rainy season. Raised beds are better for growth. If the site is moist or has good irrigation facilities, transplanting can be done during April and May. Seedlings are transplanted 45 cm × 45 cm apart. Apical portions of roots are also transplanted during the rainy season	800–1000 g seeds are required for raising seedlings for 1 hectare of land, that is, for planting 50,000 seedlings. Seedlings are transplanted after 3 months of the first true leaf initiation during March–April at lower altitudes and during May–August at alpine sites at a spacing of 45 cm × 45 cm	They are sown in sand and FYM/compost mixture (1:2) at a depth of 0.5 cm. Nursery beds of 2 m × 1.25 m or even smaller size is better for raising seedlings. 800–1000 g seeds are required for 1 hectare. 111,000 tuber segments are needed for plantation on 1 hectare. Propagules/ seedlings are transplanted after 3 months of the first true leaf initiation during March–April at middle altitudes and during May–August at alpine sites. A spacing of 30 cm× 30 cm in field is considered optimum for better vegetative growth and resultant high tuber yield	Seeds germinate well under nursery conditions in the media having FYM, sand, and soil in 2:2:1 ratio. Seeds are sown in rows 10–15 cm apart and covered with 0.5-cm thick layer of sand or fine soil. It takes about 25–28 days for complete germination. Chilling treatment of seeds at 3 °C or below for 15 days is required for good germination in the nursery. About 50,000 plants would be planted in 1 hectare. Transplanting of seedlings is done in March–April, at a spacing of 45 cm× 45 cm	Seedlings of 10–12 cm height are transplanted at a spacing of 30 cm× 30 cm or 40 cm× 40 cm. It takes about 15 days for establishment

Nutrient management	Approximately 15–20 tonnes of manure are required for one hectare of land initially. And also applied during winter season. Forest litter can also be used	FYM/compost at the rate of 25–30 quintals/hectare/year as basal dose at the time of land preparation. During second year onward application of forest litter at the rate of 100–150 quintals/hectare/year for the entire crop period (5–7 years) during Feb–March of each year	Forest leaf litter/FYM from different sources/leaf compost can be used to provide nutrition to the growing plants at the rate of 20–30 Q/ha/year. Leaves may turn slightly yellowish and the plants may wither due to waterlogging in beds. To overcome this problem, the crop should be planted in well-drained beds	FYM 10 tonnes/hectare, Vermicompost 3.75 tonnes/hectare and forest leaf litter at the rate of 2 tonnes/hectare at the time of field preparation is recommended	Well-decomposed Farm Yard Manure at the rate of 60 t/ha is to be mixed thoroughly with the soil in three split dosages
Inter-cultural operation	The intercultural operation like weeding/hoeing is carried out periodically or when required. Irrigation twice a week during the dry season is required. Weeding once a month and earthing up every month during the rainy season and every 2–3 months during the dry season is essential	In the winter season and after the establishment of seedlings or sprouting of tuber cuttings in early growth period, mulching with leaf litter is recommended as also the manuring before the initiation of new plantlets. Irrigation depends on the location of sites and texture of soil. Soil humidity should be maintained constantly	During the rainy season, weeding is required every 15 days. In other seasons, weeding is done at an interval of 30–40 days during the entire winter season. Planted beds need irrigation during early summer. The retention of soil moisture is necessary to decrease seedling mortality. Irrigation requirement depends on the texture and porosity of the soil. In dry season, irrigation at least once in a week is necessary to retain soil moisture	Weeding along with hoeing is done manually once in a month. A proper drainage system should be ensured by digging channels around the fields, especially during rains, to protect the plant against stagnating water. Alternatively, raised beds should be prepared. The field should be irrigated as and when required, maybe every alternate day during summers and fortnightly in winters	Weeding-cum-hoeing may be carried out regularly after 4 weeks interval. Crop normally prefers cool, shady environments and light irrigation during hot season in April–June. Regular weeding, hoeing operations are required during the months from March to May when the crop is grown at lower altitude.

(continued)

Table 12.2 (continued)

Common name	Chorak, Chora, Choru, Gandrayan, Angelica	Meetha Vish, IndianAconite	Ativisha, Atees, Atees, Atis	Chirata, Chirayata, Chirayita	Vanyakarkati, Papra, Bankakri, Indian Podophyllum
Harvesting and yield	Harvesting can be done after 3 years. Roots are harvested during September and October when seeds become partially mature. Harvesting can be done after every 2 years, once the cultivation is well established and gives the maximum yield. At high altitudes nearly 593–600 kg/ha of rhizomes can be obtained	In nature, plants require 5–7 years to complete the reproductive phase. In comparison, at lower altitudes, plants mature within 3–5 years when cultivated through seedlings. Generally, the plants from alpine areas complete their reproductive phase by Oct or first week of Nov, while the plants at lower altitudes complete reproductive phase during the Oct. Highest quantity of active ingredients is found during Nov–Dec. However, percentage of pseudoaconitine and aconitine is found to be maximum during May–June when the plant becomes 6 years old. 300–400 Kg/ha yield at the end of 3rd year can be obtained	Harvesting of tubers is recommended after the completion of reproductive phase and ripening of seeds during Oct–Nov. However, active content (atisine) and alkaloid contents have been found to be maximum when tubers/plants are harvested in July–August at the budding stage. Actual production of 518–579 kg/hectare, has been recorded after third year of cultivation	Few leaves of the plants can be harvested by the autumn of first year. Whole plants are harvested by the autumn of second year when seed become mature in Dec. The whole plant is collected and dried. About 3.75 tonnes/hectare of dried herbage yield is estimated in cultivation in 2 years	It is a perennial crop. The plants remain in vegetative stage till 2nd year. Flowering start from 3rd year onwards. Rootstocks are harvested in spring season. Rootstocks are found to be rich in resin contents during this season. 3.0–4.0 t of dried rootstock and 10 kg of seeds from one hectare after 5 years can be collected

Post-harvest processing	After harvesting the rhizomes, an apical portion is transplanted in a field for future crops. The remaining portion is washed with water to remove soil, and roots are cut into small pieces and put in partial shade for drying. After complete drying, roots are stored and packed in cloth bags	The tubers without rhizome are cut into 4–6 cm long slices. They should be dried in shade at room temperature. After complete drying, when moisture content is not more than 10% in slices, tubers can be stored in damp-proof containers in dark, dry, and preferably cool places for not more than 6 months, after which the quality starts deteriorating	Tubers are harvested by digging the field. Whole tuber is harvested. Topmost segments of tubers can be used for further multiplication, as they have better survival and growth rate. Whole tubers or tubers without the top segment should be dried in partial shade or at room temperature. After complete drying, tubers can be stored in wooden boxes or airtight polythene bags	The collection/harvesting of seeds is difficult due to their very small size. Therefore, a piece of cloth should be placed below the plant while harvesting. After harvesting, the plants should be dried in shade and packed accordingly. Plants harvested post-fruiting are considered to be of inferior quality with reduced active principles	Roots and rhizomes are harvested after drying of above ground parts. Roots and rhizomes are then cut into pieces of 15–20 cm long, washed and dried in shade and stored in moisture proof bags

Table 12.3 Species in which cultivation technology is under standardization

Common name	Jatamansi, Jatamansi, Jatamansi, Spikenard	Salampunja, Himalayan Marsh Orchid
Botanical name	*Nardostachy grandiflora* DC Syn. *Nardostachy jatamansi* DC	*Dactylorhiza hatagirea* (D. Don) Soo
Family	Valerianaceae	Orchidaceae
Center of origin	Himalayan region	North western Himalayan region
Distribution	The species occurs in montane and sub-alpine zones of the Himalayas, from Kumaon to Sikkim, Nepal, and Bhutan, from 3000 to 5000 m altitude in steep open areas.	This attractive terrestrial orchid grows in damp pastures of Himalayas from Kashmir to Nepal at the altitude of 2500–5000 masl.
Chromosome number	$2n = 32$	–
Inflorescence	Flowering shoot is 30–40 cm high and bears 40–50 flowers in three to seven condensed cymes. Flowers are campanulate, pinkish red to bluish white. Corolla tube is 6 mm long and hairy.	Inflorescence is raceme, 5.0–15.0 cm long, crowded with many flowers. Flowers are purple and the bracts green, narrowly lance shaped, lower bract longer than the flowers, upper slightly shorter. Flowers are about 1.8 cm long
Fruit type	Fruit is covered with 4-mm-long hairs and crowned by the acute, often dentate, calyx teeth. Flowering occurs from June to August, according to elevation, while fruiting occurs from September to October. Seeds are collected in October when they turn pale brown in color	Dry dehiscent capsule
Economical part	Rhizomes and roots	Tubers
Major chemical constituent	Dried rhizomes are steam-distilled to yield 1.5–1.9% of a pale-yellow essential oil, commercially known as spikenard oil, emitting a pleasant odour. It contains patchouli alcohol, β-patchoulene, α-patchoulene, seychellene, α-selinene, and δ-guaiene	Tubers contain 45% mucilage, 2.7%starch, 1% sugar, 2% phosphate and chloride. Tubers also containing lucoside-loroglossin
Uses	It is a nervine tonic, hypotensive, antiseptic, stomachic, carminative, tranquilizer, sedative, anti-bacterial, antispasmodic, and stimulant. It is useful in the treatment of insomnia, hysteria, convulsions, vertigo, and is also used as a cardiac tonic, etc	It is used as a farinaceous food, nervine tonic, used as an aphrodisiac, and in diarrhea, dysentery, chronic fever and in leucorrhea. It is also considered a very good tonic, especially for women after delivery, with suitable drugs in liquid form. The tubers are used as food, nervine tonic etc

(continued)

Table 12.3 (continued)

Common name	Jatamansi, Jatamansi, Jatamansi, Spikenard	Salampunja, Himalayan Marsh Orchid
Soil and climate	The plant generally grows in steep hills with 40°–70° inclined slopes and is more frequent on open, stony, and grassy slopes. It is more abundant in the western cooler slopes. It occurs on litter and soil deposits as pioneer species. For the cultivation loamy porous soil rich in organic matter like humus is considered the best. Thick humus layer promotes rapid growth and profuse branching.	Plants grow in the alpine climate of central and Western Himalayas where average summer temperature ranges between 15 and 18 °C and winters are very harsh generally below the freezing point. Plants grow best in moist meadow soils. These soils are usually dark grey, granular, sandy loam, micaceous sandy soils at greater depth
Propagation	Seeds are the best propagation material; vegetative propagation may be undertaken if sufficient mother plants are available. Initially the planting stock has to be raised through seeds only. Seeds have more than >80% germination rate when sown in a mixture of soil, sand, and compost in equal quantities. The crop can be grown by raising a nursery in March–May from seeds or vegetative rhizomes separated from the mother plants. Approximately 800–1000 g of seeds are required for raising seedlings for transplanting in 1 hectare of land. Both seeds and rhizomes are treated with GA_3 100 ppm for 48 h for rapid germination/sprouting	Propagation is done by seeds and tuber cuttings. Seeds did not show germination. It requires thawing under the snow for about 4–5 months and mycorrhizal association for germination. It is reported to be an inherently slow growing and poorly regenerating species because of pollinator specificity
Method of planting	Raised beds are suitable for good biomass production at high altitudes. 40 q/ha compost as basal dose is required for the better growth of plants and good biomass production. The seedlings may be transplanted to the main field about 90 days after germination at a spacing of 20 × 20 cm or 20 × 30 cm	Vegetative propagation through tuber cutting is quite successful in this plant. Small slices of tuber of 4.0 mm size, with meristematic tissues develop plantlets when transplanted at 5.0–7.0 cm depth at a spacing of 15 cm × 15 cm. Plant raised from tuber cutting develops tubers and roots after 1 year and it becomes about 4.0–8.0 cm tall
Nutrient management	Manure or forest litter (80 quintals) is recommended for 1 hectare of land in three split dosages. Fifty percent manure is used during the first year and rest is applied in two divided doses during the second and third years	Approximately 5 t manure is required for one hectare of land. Manure requirement increases at lower altitudes depending upon soil fertility

(continued)

Table 12.3 (continued)

Common name	Jatamansi, Jatamansi, Jatamansi, Spikenard	Salampunja, Himalayan Marsh Orchid
Inter-cultural operation	Initially, watering should be done on alternate day still proper rooting is developed. Later, watering is done at weekly intervals during dry season. Constant humidity should be maintained in the soil avoiding waterlogging. Manual weeding is carried out (fortnightly) during early growth season, and later at monthly intervals or as and when required to keep the crop weed-free	Frequent weeding at every 7–10 days especially during the rainy season favors optimal growth. At the early stage, 80–90% humidity is required for plant development of rooting and leaf initiation. Irrigation at every 12 h is needed, especially at lower altitude during this stage
Harvesting and yield	Maturity period of plant depends on propagules used. Plants raised through seeds take 3–4 years to mature, while plants raised vegetatively mature in 2–3 years and are ready for harvesting. To obtain good active constituent, plants must be collected after senescence in Oct. At an elevation of 3600 m, the recorded yield was 835 kg/hectare dry roots in the plantation raised through seedlings. At lower altitudes (2200 m), the recorded yield was 670 kg/hectare dry weight after third year of cultivation in the plantation raised through seedlings	Tubers are usually harvested after 5 years to fetch a good yield. Tubers should be collected after seed maturity at late September. Its approximate yield is 1.8 t/ha
Post-harvest processing	After harvesting in October, the roots should be washed and well dried in shade to reduce their moisture content to 8–10%. Dried material should be filled in jute bags or wooden boxes, which can be stored in dry places. During storage, hairs on root separate due to rubbing, and are often used as dhoop	Old tubers are separated from young and then steeped in hot water for 1 h. Outer membranes of tubers are removed through this process, and tubers turn light yellow in color. These are sun dried and can be preserved for a long time. In some regions tubers are boiled with buffalo milk before drying in the sun, which is considered to keep its contents active for long time

many species, packages for cultivations are still deficit in terms of propagation protocol, no information on nutrients and water requirements of the plant, harvesting, post-harvest handling and quality issues. Habitat suitability and spatial planning is not available. Choru (*Angelica glauca*), ban kakri (*Podophyllum hexandrum*) as it is not possible to cultivate these species in native habitats, i.e., subalpine region (The entire alpine /subalpine region in Uttarakhand under Protected Area Network) beside extreme climate. Nevertheless, notice must be taken of the people's experiences in successfully regenerating medicinal herbs such as Kutki (*Picrorrhiza kurrooa*), Kuth (*Saussurea lappa*), Kala jeera (*Carum carvi*) and domestication of

Archa (*Rheum emodi*), Atis (*Aconitum heterophyllum*), Choru (*Angelica glauca*), Ban kakri (*Podophyllum hexandrum*) etc. However, the area under cultivation is still undersized. The principal factors behind low area under cultivation of MAPs at present are (1) lack of undemanding and standard agro-techniques and availability of quality planting material; (2) high risk involved in the cultivation of medicinal plants; (3) a long growing period, often 3 years or more; (4) frequent low yields; (5) a lack of financial and technological support in terms of harvesting, storage, transportation, and quality improvement; (6) lack of financial support to construct fences to protect the crop from wild animals, irrigation, etc.; (7) lack of institutional support for regular monitoring; (8) a wide gulf between the prices of cultivated MAPs and those of MAPs collected from the wild, with the traders preferring the wild harvested MAPs, and (9) lack of market chain and trade related issues viz., *S. lappa* is easy to cultivate and yield is higher per plant / per unit area, the source of crude drug is cultivation only at present but still included in CITES Appendix I and trade is not permitted.

12.5 Challenges Ahead

One of the foremost challenges impacting alpine medicinal plants is climate change. Rising temperatures, altered precipitation patterns, and shifting snowfall regimes are disrupting the delicate ecological balance of entire ecosystems. These changes directly affect the distribution, phenology, and abundance of medicinal plant species. Additionally, invasive species favored by warmer temperatures pose a threat to native flora, further exacerbating the problem. Anthropogenic as well as natural factors lead to a change in the climate over a period of time. Anthropogenic activities such as the rapid pace of infrastructure development, tourism, and land-use change in Uttarakhand have led to extensive habitat destruction in natural habitat. Deforestation, road construction, and urbanization encroach upon the habitats of medicinal plants, diminishing their population size and genetic diversity. Fragmentation of habitats also disrupts ecological processes such as pollination and seed dispersal, hampering the natural regeneration of plant populations. Over a period of time, with the loss of the natural population, demand for these high value MAPs is also rising, which increases the overexploitation from natural habitats. Unsustainable harvesting practices driven by the increase demand for MAPs pose a significant threat to their survival. Overexploitation, often exacerbated by illegal harvesting and trade, depletes populations and undermines their ability to recover. Additionally, indiscriminate collection methods and a lack of regulation contribute to the decline of targeted species, pushing them towards extinction. Balancing the socioeconomic needs of local communities with conservation efforts is essential to effectively addressing this challenge. Despite recognition of the importance of conserving MAPs, existing conservation measures in Uttarakhand are often inadequate. Limited resources, institutional capacity, and technical expertise hinder the implementation of comprehensive conservation strategies. Moreover, the lack of community involvement and stakeholder engagement undermines the sustainability of conservation initiatives. Strengthening protected area management, promoting ex-situ conservation efforts, and fostering community-led conservation initiatives are crucial steps towards addressing this challenge. To meet the challenges driven by overexploitation and unsustainable harvesting practices, cultivation of these plants might be a solution for their existence. The cultivation of such MAPs presents its own set of challenges. The harsh environmental conditions, including low temperatures, high altitude, and rocky terrain, pose constraints to successful cultivation efforts. Furthermore, limited scientific knowledge about the optimal cultivation practices for most of the MAPs hampers their domestication and commercial cultivation. Developing suitable agro-techniques, identifying appropriate germplasm sources, and establishing seed banks are essential for promoting the sustainable cultivation of such MAPs in Uttarakhand.

References

Aggarwal P, Panwar P (2020) Enhancing livelihood through medicinal and aromatic plants: a supply chain perspective. J Rural Dev 39(1):45–56

Anonymous (2019) National Medicinal Plants Board. Government of India. www.nmpb.nic.in

Attri DC, Nautiyal MC, Bisht VK, Nautiyal BP, Bahukhandi A, Trivedi VL, Sati J (2021) HPTLC fingerprinting of cultivated *Picrorhiza kurrooa* accessions in Uttarakhand, Western Himalaya. Proc Natl Acad Sci India Sect B Biol Sci 91:821–828

Bhatt VP (1999) Ethnobiology of high-altitude Himalayan communities in district Chamoli: a conservation perspective. Zoos Print 14(11):137–146

Chauhan RS, Nautiyal MC (2005) Commercial viability of cultivation of an endangered medicinal herb *Nardostachys jatamansi* at three different agroclimatic zones. Curr Sci 89(9):1481–1488

Chauhan RS, Nautiyal MC (2007) Seed germination and seed storage behaviour of Nardostachys jatamansi DC, an endangered medicinal herb of high-altitude Himalaya, Current Science 92(11):1620–1624

Dhyani S, Dhyani D (2021) Traditional knowledge and legal framework: synergistic approach for conservation of medicinal plants. Asian J Environ Ecol 15(1):15–23

Dhyani A, Nautiyal BP, Nautiyal MC (2022) Seedling performance, phenotypic traits and growing media for domestication of *Lilium polyphyllum*. Plant People Planet 4(6):685–695

Gautam RS, Mishra M, Prasad R (2020) Demand scenario of herbal products: local and global in India and China. In: Development and trade of medicinal and aromatic plants (MAPs): learnings from comparative analysis of MAPs export of India and China. Rajiv Gandhi Institute For Contemporary Studies, pp 13–18

Goraya GS, Ved DK (2017) Medicinal plants in India: an assessment of their demand and supply. National Medicinal Plants Board, Ministry of AYUSH, Government of India, New Delhi and Indian Council of Forestry Research & Education, Dehradun

Joshi R, Kumar V (2013) Alpine Flora of the Indian Himalaya: diversity, ecology, and conservation. Int J Ecol Environ Sci 39(4):221–231

Kuniyal CP, Negi BS (2018) Cultivation of the Himalayan seasoning Allium in a remote village of Uttarakhand, India. J Threat Taxa 10(11):12614–12617

Nainwal P, Singh N (2020) Sustainable use of high altitude medicinal and aromatic plant for socio-economic development in Uttarakhand: a review. Int J Pharm Sci Res 11(9):4238–4243

Nautiyal MC, Nautiyal BP (2004a) Agrotechniques for high altitude medicinal plants. Bishen Singh Mahendra Pal Singh, Dehradun, p 202

Nautiyal MC, Nautiyal BP (2004b) Collaboration between farmers, research institutions and industry: experiences of *Picrorhiza kurrooa* cultivation at Gheshe village in Chamoli district,

Uttaranchal. In: Ghayur A, Belt J (eds) Searching synergy: stakeholder view on developing a sustainable medicinal plant chain in Uttaranchal-India, Bulletin 359. KIT Publishers, Amsterdam, pp 63–71

Nautiyal BP, Prakash V, Chauhan RS, Purohit H, Nautiyal MC (2001) Assessment of germinability, productivity and cost benefit analysis of *Picrorhiza kurrooa* cultivated at lower altitudes. Curr Sci 81(5):579–585

Nautiyal BP, Prakash V, Bahuguna R, Nautiyal MC (2002) Key factors of agrotechnology for the cultivation of high altitude medicinal and aromatic plants. Ann For 10(1):85–98

Nautiyal AR, Maikhuri RK, Rao KS (2003a) Cultivation of medicinal plants in Garhwal Himalaya: a case study of *Nardostachys jatamansi* in Uttaranchal State of India. Int J Sustain Dev World Ecol 10(3):257–262

Nautiyal BP, Prakash V, Chauhan RS, Purohit H, Nautiyal MC (2003b) Germinability, productivity and economic viability of *Rheum emodi* cultivated at lower altitude. Curr Sci 84(2):143–148

Nautiyal BP, Purohit H, Parkash V, Chauhan RS, Nautiyal MC (2006) Cultivation of aconitum species. J Trop Med Plant 6(2):193–201

Negi CS, Maikhuri RK, Semwal RL (2002) Ecological and social dimensions of conservation and use of medicinal plants in a protected area of the Indian Himalayas. Conserv Biol 16(1):101–111

Pandey A, Pandey R, Negi KS, Radhamani J (2008) Realizing value of genetic resources of Allium in India. Genetic Resources and Crop Evolution 55(7):985–994

Pandey AK, Tripathi YC (2017) Role of community-based conservation in managing medicinal plants in the Indian Himalaya. Ethnobot Res Appl 14:1–10

Pangriya R (2015) Study of aromatic and medicated plants in Uttarakhand, India: with focus on role in employment generation and supply chain management. Int J Soc Sci Manag 2(2):148–156

Pant S, Samant SS, Bhandari S (2020) Climate change and medicinal plants: adaptation and mitigation measures. In: Climate change and Indian agriculture. Springer, pp 279–296

Purohit AN (1997) Medicinal plants-need for upgrading technology for trading the traditions. In: Nautiyal AR, Nautiyal MC, Purohit AN (eds) Proceedings of a seminar on harvesting herbs-2000-medicinal & aromatic plants—an action plan for Uttarakhand. Bishen Singh Mahendra Pal Singh, Dehradun, pp 49–75

Rao TM, Smitha GR (2018) Biodiversity of medicinal plants in India. In: Trainers training on conservation and cultivation of medicinal plants. ICAR—Indian Institute of Horticultural Research, pp 12–16

Sharma S, Samant SS (2014) Diversity, distribution, and indigenous uses of medicinal and aromatic plants in Parbati Valley of Kullu District, Himachal Pradesh. Indian J Tradit Knowl 13(4):665–683

Singh RK, Sureja AK (2008) Community-based conservation and management of medicinal plants in the Indian Himalaya. J Med Plant Res 2(12):345–355

Tewari S, Joshi B, Dobhal R (2018) Restoration ecology and management of degraded forests in Indian Himalaya. J For Res 23(5):299–308

Upreti K, Pant S, Tewari L (2018) Ecological niche modelling for conservation prioritization of high-altitude medicinal plants in Indian Himalayas. BMC Ecol 18(1):42

Vashistha R, Nautiyal BP, Nautiyal MC (2007) Economic viability of cultivation of the Himalayan herb *Angelica glauca* Edgew. at two different agro climatic zones. Curr Sci 93(8):1141–1145

Vashistha R, Nautiyal BP, Nautiyal MC (2008) Cultivation of *Angelica archangelica* Linn. Evaluation for economical viability at two different climatic conditions. Int J Biol Chem Sci 2(4):563–572

Ved DK, Goraya GS (2007) Demand and supply of medicinal plants in India. NMPB, New Delhi & FRLHT, Bangalore, India 18(85):210–252

Verma H, Negi MS, Joshi A, Belal B, Shukla A, Mahapatara BS, Paul J (2018) Growth attributes of Kalmegh (*Andrographis paniculata* (Burm. f.) Wall Nees) as influenced by integrated nutrient management under Tarai conditions of Uttarakhand. Int J Chem Stud 6(5):2947–2949

Yadav P (2019) Trade in medicinal and aromatic plants of India. In: An overview, Newsletter on wildlife trade in India: special issue on medicinal plants, vol 31. Traffic Newsletter, pp 7–26

Conservation and Propagation of *Podophyllum hexandrum* Royle: A Rare and Endangered Alpine Medicinal Herb

13

Hemant Lata and Suman Chandra

Abstract

Podophyllum hexandrum Royle, synonym *P. emodi* Wall, also known as Himalayan mayapple, is a high-altitude medicinal herb belonging to the family Berberidaceae. This species is a rich source of pharmacologically active compounds, most notably podophyllotoxin, a powerful lignan used as a precursor for anticancer drugs. In addition to podophyllotoxin, the plant contains other bioactive constituents such as flavonoids, lignans, and phenolic acids, contributing to its therapeutic potential. Due to unscientific overharvesting from its natural habitat in the Himalayan region and its inherently slow growth and regeneration rate, *P. hexandrum* is now considered a rare and endangered species. The increasing pharmaceutical demand, coupled with ecological pressures, underscores the urgent need for conservation. In situ and ex situ conservation strategies, along with cultivation protocols for mass propagation, are critical to prevent extinction and ensure a sustainable supply. Techniques such as micropropagation, rhizome division, and cultivation in agro-climatic conditions mimicking its native habitat have shown promise. This chapter explores the medicinal properties of *Podophyllum hexandrum,* along with its botanical characteristics, phytochemical composition, rarity, and conservation through biotechnological approaches.

H. Lata (✉) · S. Chandra (✉)
School of Pharmacy, The University of Mississippi, University, MS, USA
e-mail: hlata@olemiss.edu; suman@olemiss.edu

S. Chandra, H. Lata (eds.), *High Altitude Medicinal Plants*,
https://doi.org/10.1007/978-3-031-99406-7_13

13.1 Introduction

The Himalayas are widely recognized for their remarkable ecological and geographical diversity, supporting a vast array of flora and fauna. The majority of the Himalayan range is mountainous, encompassing a range of vegetation zones—from tropical and subtropical to temperate, subalpine, and alpine—each characterized by unique plant communities. Like other major mountain systems, the Himalayan vegetation exhibits distinct altitudinal zonation. A key feature of this zonation is the timberline, which marks the upper boundary of subalpine forests and the onset of alpine vegetation. In Himalayas, this transition typically occurs between 2900 and 3300 m above sea level (Bhadula and Purohit 1994). Although subalpine forests experience colder conditions than the temperate zones below, they host a diverse array of plant life, including both evergreen and deciduous trees, as well as a rich variety of shrubs and herbaceous species.

Alpine regions are characterized by multiple environmental stressors, including short and cool growing seasons, low temperatures, reduced partial pressure of metabolic gases, intense light during clear skies, elevated levels of UV radiation, and water stress caused by soil freezing and permafrost at higher elevations (Chandra and Lata 2022). Species inhabiting these regions possess a range of morphological, ecophysiological, and biochemical adaptations that enable them to survive under such harsh conditions. Compared to lowland species, alpine plants generally exhibit greater resistance to environmental stress. Although some adaptive traits are commonly shared, individual species often display unique strategies for coping with these challenges (Bhadula and Purohit 1994).

One of the characteristic features of alpine vegetation is the small stature of plants and the predominance of herbaceous species. A reduction in plant height is a typical adaptation observed in mountainous regions (Billings and Mooney 1968; Bhadula and Purohit 1994). Many alpine plants are perennials that survive harsh conditions through various vegetative structures such as roots, rootstocks, rhizomes, bulbs, runners, suckers, and bulbils. Studies have indicated that vegetative reproduction becomes increasingly important under greater environmental stress and shorter growing seasons (Billings 1974). In dicotyledonous perennials, underground organs function as carbohydrate storage structures. A high root-to-shoot ratio is essential to accumulate sufficient energy reserves for early spring growth.

The Himalayan alpines are home to many ecologically and economically valuable species, including medicinal herbs and aromatic plants that have been traditionally used for centuries. However, excessive grazing, unregulated harvesting, and over-exploitation of commercially significant plants—combined with natural threats such as landslides, soil erosion, and habitat fragmentation—have placed many of these species at risk. Their ongoing exploitation surpasses their ability to regenerate naturally, posing a serious threat to their survival (Nandi et al. 2016). *Podophyllum hexandrum* belonging to the family Berberidaceae is one of them. The plant produces podophyllotoxin, a key precursor in the development of powerful anticancer drugs. However, its natural populations are rapidly declining due to excessive harvesting, habitat loss, slow growth rates, and limited natural regeneration.

Consequently, *P. hexandrum* is now classified as a rare and endangered species, raising serious concerns about its future availability. The growing global demand for its bioactive compounds has intensified pressure on wild populations. To safeguard this valuable resource, immediate conservation efforts and the implementation of sustainable cultivation practices are urgently needed.

13.2 The Plant: *Podophyllum hexandrum* Royle

The genus podophyllum is comprised of twenty-two species distributed in different regions of the world. *Podophyllum hexandrum* (Himalayan mayapple) and *Podophyllum peltatum* (American mayapple) are two closely related medicinal plants belonging to the family Berberidaceae, both known for producing podophyllotoxin, a valuable precursor in the synthesis of anticancer agents such as etoposide and teniposide. While *P. hexandrum* (Fig. 13.1) is found in the alpine and subalpine regions of the Himalayas, *P. peltatum* is native to North America (Lata et al. 2002, Fig. 13.2). Except for this species, all other are distributed in the Himalayan range including Afghanistan, India, Nepal, Bhutan, China, and Japan. Among the three most commercially exploited species (Table 13.3), *Podophyllum hexandrum* has the highest podophyllotoxin content (Jackson and Dewick 1984).

Podophyllum hexandrum belongs to the family Berberidaceae. The species name *hexandrum* is derived from the Greek words "hexa" meaning six, and "andrum," meaning male, referring to the plant's characteristic of having six stamens in its flowers. The genus name *Podophyllum* is derived from the Greek "podon" (foot) and "phyllon" (leaf), referring to the shape of the plant's broad, foot-like leaves. The

Fig. 13.1 *Podophyllum hexandrum* (Himalayan Mayapple), native to the Himalayas. (**a**, **b**) *P. hexandrum* plants in its natural habitat at the alpines of Himalayas; Fruiting *P. hexandrum* plants with immature (green) berries; (**c**, **d**, **e**, **f**) Mature *P. hexandrum* plants with fully ripened (red) berries; (**g**) Leaves; (**h**) flower; and (**i**) berry of *P. hexandrum* plant grown in earthen pots at High Altitude Plant Physiology Research Center at Srinagar Garhwal (550 m amsl), Himalayas

Fig. 13.2 *Podophyllum peltatum* (American Mayapple), another species of *Podophyllum*, native to North America. (**a**, **d**) *P. peltatum* plants in their natural habitat, at the University of Mississippi; (**b**) Full-grown *P. peltatum* plants with roots, shoots, and leaves, and (**c**) *P. peltatum* grown in pots

scientific classification of *Podophyllum hexandrum* is shown in Table 13.1; however the plant is also known by several local names (Table 13.2).

Podophyllum hexandrum is a herbaceous with distinct morphological features. It has mostly large, glossy, palmate leaves that are deeply lobed, typically with 5–7 lobes, and a long petiole that can reach up to 20–30 cm in length. However, morphological variants with one, two, three and four leaves may rarely be found in nature (Purohit et al. 1998). The plant produces a single, solitary flower, usually white or pale pink, which blooms in spring. The flower is typically situated between the leaf stalks, and after pollination, and it gives rise to a berry-like fruit that turns red when ripe (Figs. 13.1 and 13.3). The plant is native to the high-altitude regions of the Himalayas and thrives in shady, moist environments. Its morphological adaptations help it survive in the cold, mountainous climate. In nature, *Podophyllum hexandrum* grows either from seeds or rhizome, which allows it to spread and regenerate. The underground part is a creeping, tortuous and knotty structure with a tuft of massive roots (Fig. 13.3a–d). The above ground portion of the plant is annual whereas the rhizome is perennial. Under the native condition seed germination and seedling establishment is very poor (Bhadula et al. 2002). The species also spreads through rhizomes which grows larger and produce more shoots as they mature. However, the underground parts are taken by traders, often before the seed setting, to meet the demand for raw material, which is the major contributing factor for its decreasing frequency in nature. An illustration of the plant podophyllum (leaves, shoots, roots, rhizomes and berries) is shown in Figs. 13.1 and 13.9.

Podophyllum hexandrum is predominantly found in the higher elevations of the Himalayan range, thriving at altitudes between 2000 and 4000 m. Its distribution

Table 13.1 Botanical nomenclature of *Podophyllum hexandrum* Royle

Rank	Scientific name
Kingdom:	Plantae
Division:	Angiosperms
Class:	Eudicots
Order:	Ranunculales
Family:	Berberidaceae
Genus:	*Podophyllum*
Species:	*Podophyllum hexandrum* Royle

Table 13.2 Local names of *Podophyllum hexandrum*

Area	Local names
Pashto:	Kakora
Urdu:	Bankakri
Nepali:	Laghupatra, Ban bhanta, Papra, Ban kakri, Golapani, Hatkaudo
Sherpa:	Haamung
Dolpali:	Meme gudhruk
Khyber Pakhtunkhwa:	Sheringut
Sanskrit:	Laghu patra
Hindi:	Papra, Papri, Banbaigan, Bakrachimaka, Bhananbakra, Ban kakdi, Aindri
Jammu:	Ban-kakdi
Ladakh:	Tandik, Drenmokukshu, Demokusu, Papra, Ol-mose, Radani, Draimukushu, Denumookusoo
Kashmir:	Wanwangun, Chhmadeh, Bankakru, Rhodadari
Sikkim:	Laghu patra
Garhwal:	Bankakhri
Ayurveda:	Bantrapushi, Giriparpat
Japanese:	Himalayan hakkakuren
Chinese:	Tao er qi, Tao er qi shu
English:	Himalayan mayapple, Indian mayapple, Podophyllum, Indian podophyllum

spans in the interior ranges of Himalayas, across northern India, Nepal, Bhutan, and parts of Pakistan. This plant prefers moist, shaded environments, often growing in temperate forests, on riverbanks, and in clearings within the dense mountainous terrain. The species is particularly abundant in regions with cool, wet conditions, making it well-suited to the climate of the Himalayas (Table 13.3).

13.3 Medicinal Uses of Podophyllum

In the last few decades, *Podophyllum hexandrum* has gained significant attention due to its medicinal properties specially in the treatment of varieties of cancers. Esteemed in various traditional medical systems—including Ayurveda, traditional

Fig. 13.3 Morphology of *Podophyllum hexandrum*. (**a**, **b**, **c**) Mature *P. hexandrum* plants collected from their natural habitat at the alpines of Himalayas; (**d**) Plant with fruit/berry, (**e**) Flowering *P. hexandrum*; (**f**, **g**) Berries of *P. hexandrum* collected from nature for the future conservation

Table 13.3 Distribution of commercially important species of *Podophyllum*

Species	Distribution range	Reference source*
Podophyllum hexandrum Royle syn *P. emodi* Wall. (Himalayan Mayapple), The most commercially exploited species.	Afghanistan, India, Nepal, Bhutan, China and Formosa	IKS (Vol. 10); Fujii (1991) and Jackson and Dewick (1984)
Podophyllum peltatum Linn. (American Mayapple), Commercially exploited species in America.	USA, Southern Canada	IKS (Vol. 12); Fujii (1991)
Podophyllum pleianthum Hance (syn *Dysosma peliantha*)	China	IK (Vol. 2); Fujii (1991)

Reference/Sources: Anonymous (1983–1970) and Anonymous (1989)
IK- Index Kewensis, *IKS*- Index Kewensis Supplementum

Chinese, Tibetan and Japanese medicine, and Eastern folk remedies—it is often regarded as a sacred remedy (Wong et al. 2000). It is believed that about 400–600 years ago, the native people of the Himalayas and America independently discovered the cathartic properties of podophyllin, the aqueous extract of podophyllum roots (Fujii 1991). Podophyllin was included in the United States Pharmacopoeia (USP) in 1820 toll 1942 as a cathartic and cholagogue. Due to its severe toxicity, it was removed from USP in 1942 (Fujii 1991).

Further, the rhizome extract of the plant exhibits a wide array of medicinal properties, such as strong antioxidant, anti-inflammatory, antiviral, and antineoplastic activities (Giri and Narasu 2000; Ganie et al. 2010). It also serves as a hepatic stimulant, cholagogue, purgative, and bitter tonic (Chauhan 1999; GoN 2016). Research suggests that it possesses anti-cancer properties by inhibiting microtubule formation, making it effective against several types of malignancies (Chauhan 1999; Chaudhari and Yamin Bibi 2021).

The plant has been traditionally used to manage numerous health conditions, including monocytic leukemia, Hodgkin's lymphoma, viral and bacterial infections, venereal warts, and rheumatoid arthralgia. It is also applied in the treatment of AIDS-associated Kaposi's sarcoma and cancers of the brain, lungs, and bladder (Cobb 1990; Beutner and Von Krogh 1990; Blasko and Cordell 1988). Topical applications of rhizome paste are employed for gynecological infections, menstrual irregularities, worm infestations, and postpartum care (Kunwar and Adhikari 2005; Kunwar et al. 2006). Moreover, the rhizome powder is used as an antiseptic, aids in healing small tumors and wounds, and is considered beneficial for skin conditions, fever, kidney issues, and childbirth support (Mahmood et al. 2011).

In Nepal and India, leaves and roots are used to address fever and gastric problems, while warmed leaves are applied to boils (Singh 2009; Kumar et al. 2022). The fruit has applications in treating menstrual disorders, colds, and coughs and acts as a mild laxative (Joshi and Joshi 2007). Despite the risk of liver toxicity, it is still used as a hepatoprotective agent in various regions such as Sikkim, Pakistan, and the Western Himalayas, where it treats ailments ranging from diabetes and infertility to wounds and digestive issues (Kunwar and Duwadee 2003; Kunwar et al. 2010; Gairola et al. 2014).

Previous studies have also reported that the aqueous-ethanolic extract of high-altitude *Podophyllum hexandrum* provides radioprotective effects against lethal gamma radiation in in-vitro models. Moreover, this extract has shown metal-chelating and DNA-protecting properties. It has also been observed that *Podophyllum hexandrum* offers radioprotection by modulating free radical flux (Chawla et al. 2006). The resin in the rhizome contains podophyllotoxin, which is utilized in the synthesis of anticancer drugs like etoposide (Issell 1982; Shrestha and Shrestha 2004) and Teniposide (Issell 1982). Traditional uses of the plant also include treatment for constipation, skin allergies, mental disorders, rheumatism, and other inflammatory conditions (Kayani et al. 2015).

13.4 Bioactive Constituents in *Podophyllum hexandrum* Royle

Although the natural distribution of this species is limited to the Himalayan region, its medicinal significance, particularly as a source of a potent anticancer drug, has gained international recognition. As a result, the species is now cultivated worldwide. For instance, it has been grown in Scotland for a considerable time (Jackson and Dewick 1984), as well as by various American organizations and universities, including the University of Connecticut and Cornell University. Additionally, the Agricultural Genome Information System (AGIS) and the Germplasm Resource

Information Network (GRIN) have developed computerized databases dedicated to this species.

The underground parts of *Podophyllum* species, particularly *P. hexandrum* (native to India) and *P. peltatum* (native to North America), are rich sources of lignans, a major class of naturally occurring polyphenolic compounds. Among these, podophyllotoxin stands out as a key bioactive compound, serving as the precursor in the chemical synthesis of anticancer drugs such as etoposide and teniposide (Lata et al. 2009; Paul et al. 2013; Newton 2023). These drugs are widely employed in the treatment of lung cancer, various forms of leukemia, and other solid tumors (Van Uden et al. 1989). Notably, the Indian species *Podophyllum hexandrum* Royle contains approximately three times more podophyllotoxin than its American counterpart, making it a more potent and valuable source for pharmaceutical applications. These lignans are primarily found in the rhizomes and roots of the plant and exhibit a wide range of pharmacological properties including antitumor, antiviral, and anti-inflammatory activities. Podophyllotoxin itself acts by inhibiting microtubule assembly, thereby preventing cell division and making it particularly effective in cancer therapy. Chemical structure of Podophyllotoxin is shown in Fig. 13.4.

13.4.1 Etoposide

Etoposide (trade name brand name Vepesid) is a semisynthetic derivative of podophyllotoxin, a natural compound extracted from the rhizome of the mayapple plant (*Podophyllum hexandrum*), also known as Himalayan mandrake.

Etoposide is a chemotherapy drug commonly used to treat various cancers, such as testicular cancer, lung cancer, lymphoma, leukemia, neuroblastoma, and ovarian cancer (Kluska and Woźniak 2021). The medication can be administered either

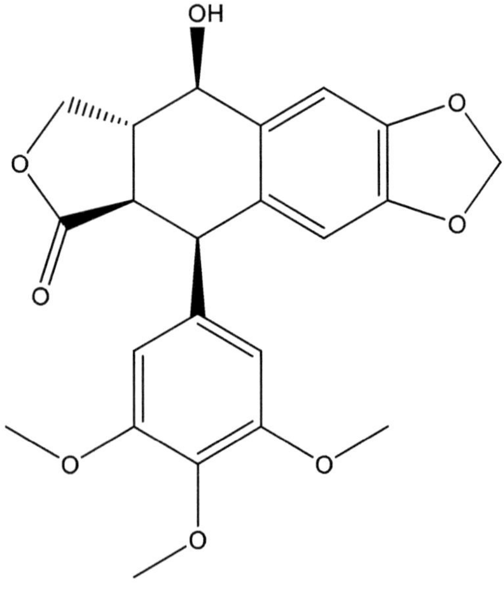

Fig. 13.4 Chemical structure of podophyllotoxin, the most bioactive lignan of podophyllum

orally or through intravenous injection. Structurally, etoposide is a glycoside of podophyllotoxin, meaning it includes a sugar molecule (a derivative of D-glucose) attached to the main compound. It is closely related to teniposide, another anticancer agent, with the primary chemical difference being a methyl group in etoposide where teniposide has a thienyl group.

In its pure form, etoposide is a white to yellow-brown crystalline powder and is soluble in organic solvents. For clinical use, it is commonly administered as etoposide phosphate, a water-soluble salt form that improves drug delivery. Molecular structure of etoposide and etoposide phosphate is shown in Figs. 13.5 and 13.6, respectively.

13.4.2 Teniposide

Teniposide is a semisynthetic analog of podophyllotoxin, a naturally occurring compound derived from the rhizome of the wild mandrake (*Podophyllum peltatum*). It is classified as a glycoside, containing a modified D-glucose moiety attached to the podophyllotoxin core (McCowage et al. 1995). Chemically, teniposide shares a close resemblance to the anticancer drug etoposide, with the key structural difference being a thienyl group in teniposide, replacing the methyl group present in

Fig. 13.5 Chemical structure of Etoposide

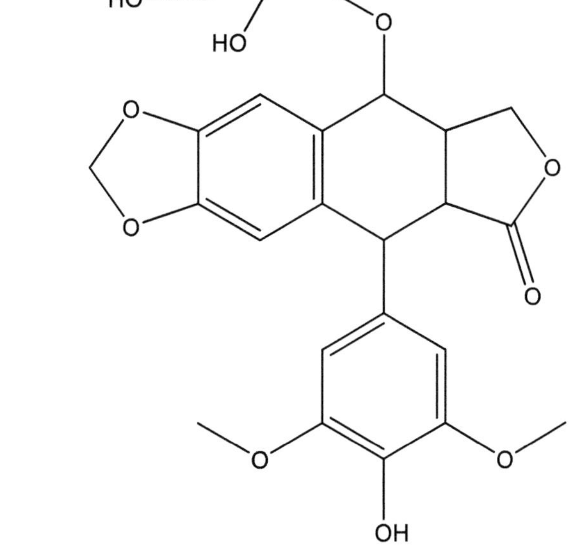

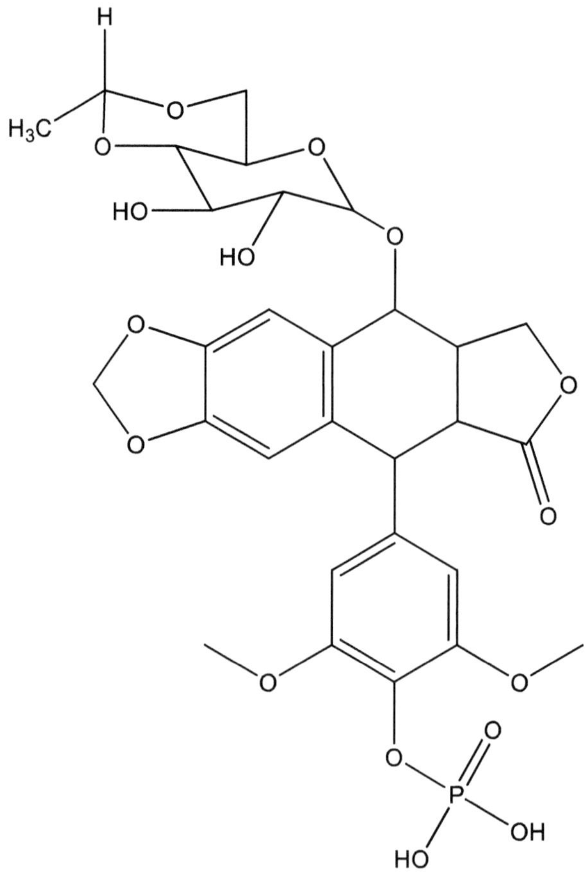

Fig. 13.6 Chemical structure of Etoposide Phosphate

etoposide. Both compounds were developed as less toxic alternatives to podophyllotoxin for use in cancer treatment. On the pharmaceutical market, teniposide is marketed under the trade name Vumon (Wang et al. 2009). Its chemical structure is illustrated in Fig. 13.7.

Since podophyllotoxin is extracted from the roots and rhizomes of *Podophyllum hexandrum*, the species has been overharvested from the nature, leading to its endangered status and highlighting the urgent need for its conservation. Moreover, the existing population size of *P. hexandrum* in Himalayas is small and if its exploitation continues at the present rate, it may eventually face the threats of genetic drift, population bottlenecks, and inbreeding depression. Therefore, there is an urgent need to enhance conservation efforts of this species.

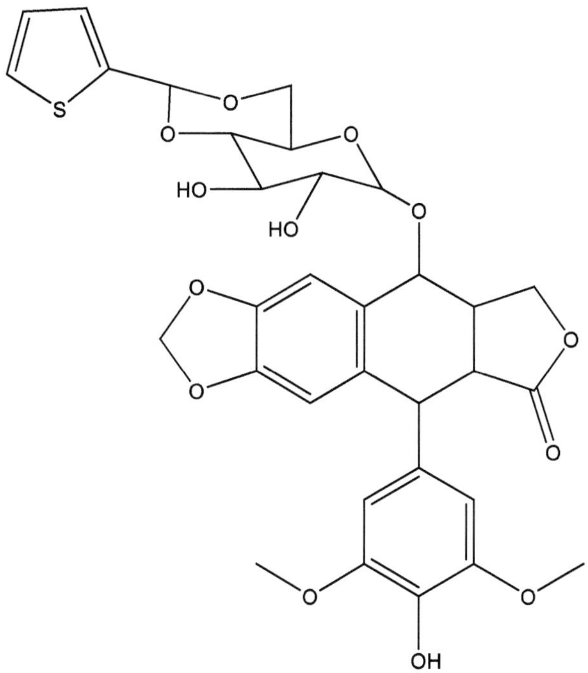

Fig. 13.7 Chemical structure of Teniposide

13.5 Need for Conservation

As described earlier, the species of *Podophyllum* yield several lignans including podophyllotoxin, which is an important anticancer compound extracted most exclusively from the rhizomes. Indian *Podophyllum (P. hexandrum)* was reported to contain three times more podophyllotoxin as compared to the American species, *P. peltatum* (Fay and Ziegler 1985). Jackson and Dewick (1984), however, reported a much higher level of podophyllotoxin in *P. hexandrum* (17 times higher) as compared to *P. peltatum*. The extraction of podophyllotoxin from *P. hexandrum* is easier as compared to other species due to the presence of higher amounts of toxin and this is also one of the reasons for its over exploitation and a world-wide demand. Therefore, serious efforts are needed to stop overexploitation, and to promote the conservation and mass multiplication of this species in order to reduce pressure on its natural habitats and to meet market demand.

13.5.1 Conservation Through Traditional Practices

Conservation of *P. hexandrum* through cultivation of its high podophyllotoxin yielding elite germplasm at its natural habitat and at comparatively lower altitudes is a sustainable alternative to wild collection. Thus, ensuring both the survival of the species and continued availability for pharmaceutical use, particularly for its potent lignan, podophyllotoxin. Cultivation initiatives can be integrated into agroforestry systems or marginal lands, offering livelihood opportunities to local communities while reducing pressure on wild populations. Limited efforts have been made in the large-scale commercial cultivation of *Podophyllum hexandrum*. Troup (1915) and Krishanamurthy et al. (1965) studied the possibility of cultivation of *Podophyllum hexandrum* at lower altitude, Chakrata, Uttar Pradesh (now Uttarakhand), India. Badhwar and Sharma (1963) and Nautiyal et al. (1987) published a short note on *Podophyllum* seeds. Purohit and Nautiyal (1988) studied inhibitory role of cotyledons on plumule development in *Podophyllum* seeds. Kamil and Dewick (1986) and Jackson and Dewick (1984) in their biosynthetic studies with seedlings in vivo found yate in a satisfactory precursor of podophyllotoxin. In the recent studies on seeds, the seed germination in *P. hexandrum* is reported to be enhanced by pre-treatment with GA3 (Nadeem et al. 2000; Zuhra et al. 2021).

The natural scarcity of *P. hexandrum*, along with its prolonged juvenile phase, low fruit and seed set, and limited propagation through conventional methods, has led to a shortage of its root and rhizome biomass makes it an expensive starting material for synthesizing its derivatives (Chuang and Chang 1987a, b). To meet the industrial demand, biotechnological production of podophyllotoxin using plant cell cultures derived plants of *P. hexandrum* has emerged as a promising alternative.

13.5.2 Conservation Through Micropropagation

The first report on tissue culture of *Podophyllum hexandrum* was given by Van Uden et al. (1989). They obtained callus cultures from roots of *P. hexandrum* Royle in B5 medium (Gamborg et al. 1968) supplemented with 2% coconut milk, 4.0 mg/l NAA and 4% sucrose. Fuji (1991) reported callus induction, suspension culture plant regeneration from callus and protoplast cultures of three *Podophyllum* species (*P. pleianthum, P. peltatum* and *P. hexandrum*) were examined under various culture conditions. 2,4-D or NAA in presence of BA promoted callus induction and callus growth. Kadkade (1981) reported that rhizome segment of *Podophyllum peltatum* formed callus on MS medium and the eight-week-old callus produced ab-peltatin and podophyllotoxin. Later, Kadkade (1982) studied the effect of growth regulators, natural factors, carbon source and monochromatic light irradiation on the growth and podophyllotoxin production of *P. peltatum* callus tissues. Berlin et al. (1986, 1988) reported that 5-methoxypodophyllotoxin and its glucoside formation by root culture of *Linum flavum.* On dry mass basis 0.7–1.3% podophyllotoxin was extracted from root culture. Van Uden et al. (1995) have recently shown that cell cultures of *Linum flavum* and *Podophyllum hexandrum* are able to convert deoxypodophyllotoxin into several lignans. Chuang and Chang (1987a) reported the embryoids production and their development into plantlets from calli of somatic tissues of *P. pleianthum* and in (1987b) reported somatic embryogenesis and planlet regeneration in callus cultures derived from immature seeds, mature zygotic embryos of *P. pleianthum.* Arumugam and Bhojwani (1990) reported somatic embryogenesis in tissue cultures of *P. hexandrum*. Callus derived from zygotic embryos on MS medium supplemented with 2.0 μM BA and 0.5 μM IAA differentiated globular embryos. These embryos develop to plantlets on higher levels of sucrose supplemented with 1–10 μM NAA. In the recent years, considerable efforts have been made to micro propagate *Podophyllum hexandrum* using various explants, including cotyledons (Rajesh et al. 2014; Khan et al. 2024), leaf (Lata et al. 2010, Zuhra et al. 2021; Khan et al. 2024), radicles (Khan et al. 2024), rhizomes (Chakraborty et al. 2010; Khan et al. 2024), roots (Zuhra et al. 2021), and seeds and seedlings (Sultan et al. 2006; Kharkwal et al. 2008; Rizwan et al. 2021; Sharma et al. 2022) (Table 13.4).

Our group, at the High Altitude Plant Physiology Research Center, have developed in-vitro propagation protocols for a screened and selected high podophyllotoxin yielding *P. hexandrum* germplasm from its different populations in Himalayas (Fig. 13.8).

Table 13.4 Advances in micropropagation of *Podophyllum hexandrum* in the last two decades

Year	Authors	Explant	Media and PGR	Results
2024	Khan et al.	Leaf, stem, rhizome, radicle, and cotyledon	Murashige and Skoog (MS) medium Shoot proliferation: benzyladenine (3.5 mg/L) and naphthalene acetic acid (0.5 mg/L) with activated charcoal Rooting: NAA from 0.5 to 1.5 mg/L	Callogenesis and Organogenesis
2023	Parasher et al.	Zygotic embryos	Vitrification and V-cryoplates using PVS2	Plantlet formation Cryopreservation
2022	Sharma et al.	Seeds, seedling, in vitro leaves and roots	Seed germination: MS medium containing BA (2.0 mg/ L), NAA (0.2 mg/L), and calcium chloride (0.7 mg/L) Callus induction: B5 medium supplemented with BA (0.1 mg/L) + NAA (1.0 mg/L) and BA (0.5 mg/L) + 2,4-D (0.5 mg/L) under dark incubation.	Germination and Callogenesis
2021	Rizwan et al.	Seedlings	MS medium, NAA 0.5–5 μM and BAP 2.5 μM	Callus induction
2021	Zuhra et al.	Leaf, root	Murashige and Skoog media (MS) media Callus induction from leaf: 6.75 μM NAA and 11.37 μM BAP Callus induction from root: IAA 1 μM and 5.0 μM BAP	Callogenesis and Organogenesis

(continued)

Table 13.4 (continued)

Year	Authors	Explant	Media and PGR	Results
2020	Srivastava and Sood	Zygotic embryos	Murashige and Skoog media (MS) media NAA and PVP, ABA and PVP, GA3 and PVP (3 g/L)	Somatic embryogenesis, Hairy root culture
2014	Rajesh et al.	Cotyledons	MS medium supplemented with 2,4-D (1.5 mg/L) under dark conditions	Callogenesis, Somatic embryogenesis, Organogenesis
2013	Guo et al.	Seedlings	Woody Plant media (WPM) IAA 1.5 mg/L and NAA 0.5 mg/L under dark conditions	Root induction
2010	Lata et al.	Leaf, zygotic embryo	Murashige and Skoog (MS) medium Callus induction: NAA (0.5–5.0 μM) and TDZ (0.5–5.0 μM) Embryogenesis: IAA 0.5 μM and 2.0 μM BA IAA 0.5 μM and 2.0 μM BA	Callogenesis Embryogenesis Organogenesis
2010	Chakraborty et al.	Rhizome	Murashige and Skoog (MS) medium Shoot proliferation: IAA 11.42 μM Rooting: IBA 100 μM	Organogenesis
2009	Li et al.	Embryo	Murashige and Skoog (MS) medium Seed germination: GA3 0.5 mg/L Root induction: IBA, GA3, Hy (0.1 ~ 10.0 mg/L) and AC (0 ~ 10.0 mg/L)	Root induction
2009	Sultan et al.	Embryo	Murashige and Skoog (MS) medium In vitro embryogenesis: BA (1.0–4.0 mg/l) and IAA (0.5–2.0 mg/l) Plantlet formation: BA (0.5 mg/l + IAA 1.0 mg/l) supplemented with activated charcoal (0.5–1.0%)	Embryogenesis Organogenesis
2008	Kharkwal et al.	Seedling, rhizome	Murashige and Skoog (MS) medium Seed germination: GA_3 (50–200 ppm) Root induction: IAA, IBA, NAA, KN, BAP (20–400 ppm)	Plantlet formation
2007	Ahmad et al.	Embryo	Murashige and Skoog (MS) medium BAP (0.5–2.5 μM), NAA (0.5-5 μM) and GA3 (0.5 μM–1 μM)	Callogenesis

(continued)

Table 13.4 (continued)

Year	Authors	Explant	Media and PGR	Results
2006	Sultan et al.	Seedlings	Gamborgs B5 media Callus induction: 2,4-D (0.5–1.5 mg/L) and BAP (0.2–1.0 mg/L) Shoot proliferation: MS supplemented with BAP and IAA (0.5–5.0 mg/L)	Callogenesis, Organogenesis
2002	Pandey et al.	Zygotic embryo	Murashige and Skoog (MS) medium Seed germination: BAP 1.0 μM Shoot proliferation: BAP 1.0 μM+ IAA1.0 μM Root induction: IAA 0.5 μM Embryogenesis: NAA 5.0 μM + BAP 0.5 μM	Somatic embryogenesis, Organogenesis
2000	Nadeem et al.	Zygotic embryo	Murashige and Skoog (MS) medium Seed germination: 250.0 mM GA3 and a combination of GA3 and BAP 250.0 mM each Embryogenesis: (IAA 1.0 ± 4.0 mM) and BA (1.0 mM); 5.0 mM NAA and 0.5 mM BAP.	Multiple shoot, plantlets, somatic embryogenesis

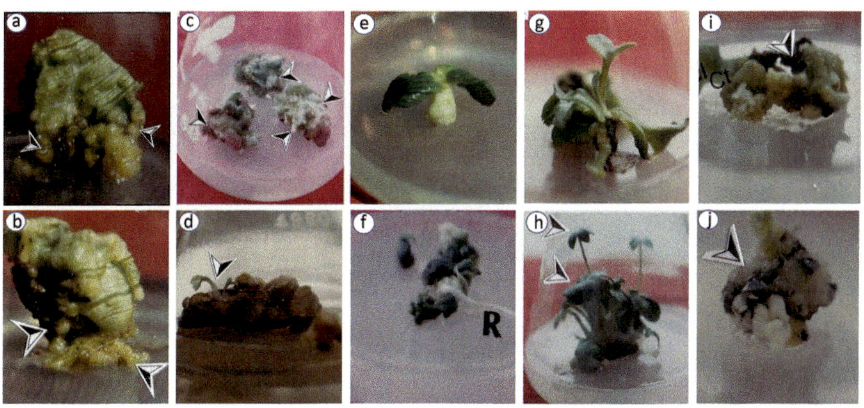

Fig. 13.8 (**a**, **b**) Callus induction from leaf in MS + NAA + TDZ (0.1 mg/L each), arrows indicate growth callus; (**c**) Compact callus and globular structures developing from leaf discs on MS + NAA + TDZ (1 mg/L each), two-week-old callus; (**d**) Shoot differentiation from callus on MS medium, 5 weeks old cultures, arrows showing shoot growth. (**e**) Enlargement of cotyledons and callus initiation from the cotyledonary tube, MS + 0.1 μM NAA + 1.0 μM BAP; (**f**) Effect of IAA on rooting of excised zygotic embryos, MS + 1 μM IAA, R indicate the roots; (**g**) Response of excised zygotic embryos after 3 weeks of growth on MS + 7.5 μM BAP; (**h**) Differentiation of multiple shoots from excised zygotic embryos, MS + 2 μM BAP, Cultures after 6 weeks of growth showing multiple shoots (arrows); (**i**) Development of shoot buds, somatic embryos and root initiation from zygotic embryo cultures, MS + 1.0 μM IAA + 0.5 μM BAP, Arrows indicate globular embryos, *cl* cotyledonary lobes; (**j**) shoot buds, somatic embryos and root initiation from zygotic embryo cultures, MS + 1.0 μM IAA + 2.0 μM BAP, arrows indicate globular embryos

13.5.2.1 Plant Material

Plant material for the study were collected from various temperate, subalpine, timberline and alpine areas of Garhwal Himalaya, namely, Tungnath (3800 m amsl), Kedarnath (3650 m amsl), Dayara (3600 m amsl), Valley of flowers (3400 m amsl), Ghangaria (3000 m amsl), Chopta (1800 m amsl) and seeds of one population were also collected from Rudranath (2600 m amsl). Seeds and rhizomes collected were planted in earthen pots under net house conditions at High Altitude, Plant Physiology Research Center, Srinagar (550 m amsl). Vegetative cuttings of rhizomes, seeds, leave disk, endosperm, and zygotic embryos of different podophyllum hexandrum populations were used for the micro propagation (Fig. 13.9).

13.5.2.2 Medium, Conditions, and Maintenance of Cultures

The surface-sterilized rhizomes, roots, and leaves were cut into 4 mm long segments and planted individually on culture medium. In the case of shoot buds, 2 or 3 outer sheathing leaves were removed before incubation. Decoated seeds were planted individually while excised embryos were planted individually or in pairs. Germinated embryos were cut into pieces consisting of leaves and cotyledons and then planted on MS medium (Murashige and Skoog 1962) supplemented with

Fig. 13.9 (**a**, **c**) Well-developed podophyllum plants, ready to be planted in nature; (**b**) A well-rooted podophyllum plant; (**d**) (1–7) Different growth stages of second-generation podophyllum plants. 1, 2, and 5: juvenile plants; 3, 4, and 6: plants with fully developed leaves; and 7: a fully developed podophyllum plant with a berry

various concentrations of BAP (N-Benzylaminopurine), Kn (Kinetin), 2, 4-D (2,4 Dichlorophenoxyacetic acid), IAA (Indole-3-acetic acid), NAA (1-Napthalene acetic acid) and TDZ (Thidiazuron), separately or in different combinations. The cultures were maintained in a culture room at 25 °C under 16 h photoperiod or continuous dark as mentioned in the text. After callus induction, the callus cultures were either maintained through regular sub-culturing every 15 days on the same medium or transferred to different medium for differentiation. For all the experiments, 10–15 cultures (replicates) were raised each time. The cultures were observed periodically and the morphological changes were recorded.

Of all the explants (derived from different organs) used in this study, only the leaf sections and zygotic embryos produced viable cultures. The leaf explants of all the populations gave viable callus cultures. Since the alpine populations of Kedarnath showed the highest podophylloresin content, greater efforts were made to propagate this population. It was interesting to note that in addition to its desirable character (high resin content), the population of Kedarnath also responded much better than other populations in the culture mediums.

13.5.2.3 Leaf Callus Culture

Experiments for the induction of callus cultures using explants derived from young leaves of all the populations were successful. Out of various concentrations of NAA (0.01, 0.2, 0.5 and 1.0 mg/L) and TDZ (0.1, 0.5 and 1.0 g/L) on MS medium tried the best callus formation was achieved with MS medium containing 0.1 mg/L NAA and 0.1 mg/L TDZ. However, under any of the combinations of various growth regulators (various combinations of IAA, NAA, 2,4-D, Kinetin, BAP and TDZ), these cultures could not be rooted.

13.5.2.4 Zygotic Embryo Culture

Excised embryos of *P. hexandrum* were cultured on MS medium supplemented with various concentrations and combination of different growth regulators including NAA, BAP, IAA and 2, 4-D. Among all the growth regulators tested, two combinations, 0.5 μM IAA + 2 μM BAP and 1 μM IAA + 0.5 μM BAP, were found to be the most effective for inducing multiple shoot and root formation. The plants raised in tissue culture using the above combination of growth regulators were transplanted in earthen pots and after hardening in soil, were transferred to the field conditions at 2000 m altitude in Himalayas.

13.6 Future Conservation Strategies

A major problem in the conservation of Himalayan medicinal species is their uncontrolled and illegal exploitation. Moreover, the rate of exploitation far exceeds the rate of their natural regeneration. Seed germination under natural conditions appears to be low and seedling establishment is poor due to harsh environmental conditions including low and fluctuating temperatures, short growing period and slow tissue maturation which leads to frost damage of young seedlings. Under natural conditions, thus, regeneration through underground parts is more common. When the roots and rhizomes are continuously taken from the natural habitat for commercial purpose, it is quite unlikely that their natural regeneration can cope up with their exploitation. Therefore, in situ conservation methods should be supplemented by ex situ conservation strategies and used for the re-introduction of laboratory raised plants in the natural habitat. Additional survey work is also needed to discover and study more populations in other localities and massive conservation measures are needed to preserve this germplasm for the future.

Under its natural habitat, insects such as *Macrima armata* were found to damage the young tender leaves of *P. hexandrum*. As the populations of this species are polymorphic, it may be possible to select elite types from nature, especially those with multiple desirable characters, e.g., wider adaptability range, insect and disease resistance and higher podophyllotoxin content. If such clones can be selected from naturally growing populations or isolated using somaclonal variation in tissue cultures, a cultivation package can be developed so the farmers in the upper Himalayan mountain ranges can start cultivating it as a cash crop. This will reduce the demand of raw material from the natural habitats and will be helpful for its in situ conservation which otherwise does not seem to be possible due to high cash value of rhizomes resulting in over exploitation by traders and pharmaceutical companies. It is also important to start a periodical monitoring of this important species under its natural habitat conditions to determine the rate of decline in its population size so that appropriate conservation measures can be initiated. In an attempt to start its conservation, laboratory tested seeds and the rhizome cuttings of one population of *P. hexandrum* have been reintroduced in its natural habitat (Chopta range, 2709 meters amsl). Also, initial attempts for the acclimatization of tissue culture raised plants under field conditions have been successful.

13.7 Conclusion

The Himalaya are well known for their rich and diverse floral, faunal, and geographical diversity. It is one of the largest mega-hotspots of critical, rare and endangered medicinal plants. One of the most important questions that needs to be answered about a given endangered species is "why is the species endangered?" There are no simple answers to this question. In some cases, the answer may be that the species may represent the last individuals of a primitive species which is shrinking due to normal course of evolution. It is, however, not a general rule that rare species are primitive. In many cases, it's quite possible that such species may be new in a particular habitat and may still be in a process of establishing their populations. On the other hand, for many species, that were known to be abundant in the past but have become rare or endangered due to habitat destruction, over-exploitation and other anthropogenic factors, the causes of rarity are simply, man-made. *Podophyllum hexandrum* faces a similar threat due to the over extraction of its roots and rhizomes from the wild, often before the plant has completed its reproductive cycle. This premature harvesting prevents berry and seed formation, and therefore results pushed the species toward rare and endangered status. Studies referred in this chapter, including our own, show that in addition to the conventional methods of propagation such as using seed and vegetative propagation, this species can also be successfully mass-propagated using tissue culture techniques. However, further studies are needed to improve the frequency of plant regeneration, mass-multiplication, and reestablishment of regenerated plants in their natural habitats in alpines.

References

Ahmad R, Sharma VK, Rai AK, Shivananda RD, Shivananda BG (2007) Production of lignans in callus culture of *Podophyllum hexandrum*. Trop J Pharm Res 6(4):803–808

Arumugam N, Bhojwani SS (1990) Somatic embryogenesis in tissue cultures of *Podophyllum hexandrum*. Can J Bot 68(3):487–491

Badhwar RL, Sharma BK (1963) A note on germination of *Podophyllum* seeds. Indian For 89:445–447

Berlin J, Wray V, Mollenschott C, Sasse F (1986) Formation of β-peltatin-A methyl ether and coniferin by root cultures of *Linum flavum*. J Nat Prod 49(3):435–439

Berlin J, Bedorf N, Mollenschott C, Wray V, Sasse F, Höfle G (1988) On the podophyllotoxins of root cultures of *Linum flavum*. Planta Med 54(03):204–206

Beutner KR, Von Krogh G (1990) Current status of podophyllotoxin for the treatment of genital warts. Semin Dermatol 9(2):148–151

Bhadula SK, Purohit AN (1994) Adaptational strategies of plants at high altitudes and future prospects for the conservation of biodiversity. Adv Plant Sci Res 1:1–24

Bhadula SK, Singh A, Lata H, Kuniyal CP, Purohit AN (2002) Distribution pattern, population diversity and propagation of some high altitude medicinal herbs from Garhwal Himalaya: problems and prospects for conservation. Gyanodaya Prakashan, Nainital 2:389–413

Billings WD (1974) Adaptations and origins of alpine plants. Arct Alp Res 6(2):129–142

Billings WD, Mooney HA (1968) The ecology of arctic and alpine plants. Biol Rev 43(4):481–529

Blasko G, Cordell GA (1988) Recent developments in the chemistry of plant derived anticancer agents. Econ Med Plant Res 2:119–182

Chakraborty A, Bhattacharya D, Ghanta S, Chattopadhyay S (2010) An efficient protocol for in vitro regeneration of *Podophyllum hexandrum*, a critically endangered medicinal plant. Indian J Biotechnol 9:217–220

Chandra S, Lata H (2022) Effect of altitude on biochemical and photosynthetic characteristics of *Aconitum balfourii* and *Podophyllum hexandrum*: high value endangered medicinal herbs from Himalayas. Int J Trop Agric 40(3–4):195–203

Chaudhari SK, Yamin Bibi MA (2021) *Podophyllum hexandrum*: an endangered medicinal plant from Pakistan. Pure Appl Biol 3(1):19–24

Chauhan NS (1999) Medicinal and aromatic plants of Himachal Pradesh. Indus Publishing

Chawla R, Arora R, Singh S, Sagar RK, Sharma RK, Kumar R et al (2006) *Podophyllum hexandrum* offers radioprotection by modulating free radical flux: role of aryl-tetralin lignans. Evid Based Complement Alternat Med 3(4):503–511

Chuang MJ, Chang WC (1987a) Embyoid formation and plant regeneration in callus culture derived from vegetative tissue of *Dysoma pleiantha* (Hance) Woodson. J Plant Physiol 128:279–283

Chuang MJ, Chang WC (1987b) Somatic embryogenesis and plant regeneration in callus culture derived from immature seeds and mature zygotic embryos of *Dysosma pleiantha* (Hance) Woodson. J. P. Plant Cell Rep 6:484–485

Cobb MW (1990) Human papillomavirus infection. J Am Acad Dermatol 22(4):547–566

Fay DA, Ziegler HW (1985) Botanical source differentiation of podophyllum resin by HPLC. J Liq Chromatogr 8(8):1501–1505

Fuji (1991) *Podophyllum* Spp. *In vitro* regeneration and production of podophyllotoxin. Biotechnology in agriculture and forestry. In: Bajaj YPS (ed) Medicinal and aromatic plants III, vol 15. Springer, Berlin/Heidelberg

Gairola S, Sharma J, Bedi YS (2014) A cross-cultural analysis of Jammu, Kashmir and Ladakh (India) medicinal plant use. J Ethnopharmacol 155(2):925–986

Gamborg OL, Miller RA, Ojima V (1968) Nutrient requirements of suspension cultures of Soyabean root cells. Exp Cell Res 50:151–158

Ganie SA, Haq E, Masood A, Zargar MA (2010) Amelioration of carbon tetrachloride induced oxidative stress in kidney and lung tissues by ethanolic rhizome extract of *Podophyllum hexandrum* in Wistar rats. J Med Plant Res 4(16):1673–1677

Giri A, Lakshmi Narasu M (2000) Production of podophyllotoxin from *Podophyllum hexandrum*: a potential natural product for clinically useful anticancer drugs. Cytotechnology 34:17–26

GoN. (2016) Medicinal plants of Nepal, 2nd edn. Ministry of Forests and Soil Conservation, Department of Plant Resources, Thapathali

Guo Q, Zhou J, Wang Z, Yang H (2013) *In vitro* rooting of *Podophyllum hexandrum* and transplanting technique. Engineering 4(10):142–145

Issell BF (1982) The podophyllotoxin derivatives VP16-213 and VM26. Cancer Chemother Pharmacol 7:73–80

Jackson DE, Dewick PM (1984) Biosynthesis of *Podophyllum* lignans-I. Cinnamic acid precursors of podophyllotoxin in *Podophyllum hexandrum*. Phytochemistry 23:1029–1035

Joshi AR, Joshi K (2007) Ethnomedicinal plants used against skin diseases in some villages of Kali Gandaki, Bagmati and Tadi Likhu watersheds of Nepal. Ethnobot Leafl 2007(1):27

Kadkade PG (1981) Formation of podophyllotoxin by *Podophyllum peltatum* tissue cultures. Naturwissenschaften 68:481–482

Kadkade PG (1982) Growth and Podophyllotoxin production in callus tissues of *P. peltatum*. Plant Sci Lett 25:107–115

Kamil WM, Dewick PM (1986) Biosynthetic relationship of aryltetralin lactone lignans to dibenzylbutyrolactone lignans. Phytochemistry 25:2093–2102

Kayani S, Ahmad M, Sultana S, Shinwari ZK, Zafar M, Yaseen G et al (2015) Ethnobotany of medicinal plants among the communities of Alpine and Sub-alpine regions of Pakistan. J Ethnopharmacol 164:186–202

Khan Z, Khan B, Shah ST, Iqbal J, Basit A, Khan MS et al (2024) Preserving nature's treasure: a journey into the in vitro conservation and micropropagation of the endangered medicinal marvel—*Podophyllum hexandrum* royle. Horticulturae 10(8):809

Kharkwal AC, Kushwaha R, Prakash O, Ogra RK, Bhattacharya A, Nagar PK, Ahuja PS (2008) An efficient method of propagation of *Podophyllum hexandrum*: an endangered medicinal plant of the Western Himalayas under ex situ conditions. J Nat Med 62:211–216

Kluska M, Woźniak K (2021) Natural polyphenols as modulators of etoposide anti-cancer activity. Int J Mol Sci 22(12):6602

Krishnamurthy T, Karira GV, Sharma BK, Bhatiya K (1965) Cultivation and exploitation of *Podophyllum hexandrum* Royle (Syn. *P. emodi* Wall. ex Hook. F. Thomas). Indian Forester 91:470–475

Kumar J, Sandal P, Singh A, Kumar A, Arya V, Devi R et al (2022) Conservation status, anticancer compounds and pharmacological aspects of Royle: a review *Podophyllum hexandrum*. Indian J Ecol 49(3):1096–1102

Kunwar RM, Adhikari N (2005) Ethnomedicine of Dolpa district, Nepal: the plants, their vernacular names and uses. Lyonia 8(1):43–49

Kunwar RM, Duwadee NPS (2003) Ethnobotanical notes on flora of Khaptad National Park, far western Nepal. Himal J Sci 1(1):25–30

Kunwar RM, Nepal BK, Kshhetri HB, Rai SK, Bussmann RW (2006) Ethnomedicine in Himalaya: a case study from Dolpa, Humla, Jumla and Mustang districts of Nepal. J Ethnobiol Ethnomed 2:1–6

Kunwar RM, Shrestha KP, Bussmann RW (2010) Traditional herbal medicine in Far-west Nepal: a pharmacological appraisal. J Ethnobiol Ethnomed 6:1–18

Lata H, Chandra S (2010) Biotechnological characterization of different populations of an endangered medicinal herb-*Podophyllum hexandrum* Royle. In: Medicinal plant biotechnology. CABI, Wallingford, pp 36–47

Lata H, Moraes RM, Douglas A, Scheffler BE (2002) Assessment of genetic diversity in *Podophyllum peltatum* by molecular markers. In: Trends in new crops and new uses. ASHS Press, Alexandria, pp 537–544

Lata H, Mizuno C, Moraes R (2009) The role of biotechnology in the production of the anticancer compound podophyllotox. In: Jain SM, Saxena PK (eds) Protocols for in vitro cultures and secondary metabolite analysis of aromatic and medicinal plants. Methods in molecular biology, vol 547. Humana Press, Totowa

Lata H, Moraes RM, Bertoni B, Pereira AM (2010) In vitro germplasm conservation of *Podophyllum peltatum* L. under slow growth conditions. In Vitro Cell Dev Biol Plant 46:22–27

Li W, Li MF, Yang DL, Xu R, Zhang YR (2009) Production of podophyllotoxin by root culture of *Podophyllum hexandrum* Royle. Electron J Biol 5(2):34–39

Mahmood A, Malik RN, Shinwari ZK, Mahmood AQEEL (2011) Ethnobotanical survey of plants from Neelum, Azad Jammu and Kashmir, Pakistan. Pak J Bot 43(1):105–110

McCowage GB, Vowels MR, Shaw PJ, Lockwood L, Mameghan H (1995) Autologous bone marrow transplantation for advanced neuroblastoma using teniposide, doxorubicin, melphalan, cisplatin, and total-body irradiation. J Clin Oncol 13(11):2789–2795

Murashige T, Skoog F (1962) A revised medium for rapid growth and bioassays with tobacco tissue culture. Physiol Plant 15:473–497

Nadeem M, Palni LMS, Purohit AN, Pandey H, Nandi SK (2000) Propagation and conservation of *Podophyllum hexandrum* Royle: an important medicinal herb. Biol Conserv 92(1):121–129

Nandi SK, Palni LMS, Pandey H, Chandra B, Nadeem M (2016) Selection of elites and in vitro propagation of selected high-value Himalayan medicinal herbs for sustainable utilization and conservation. In: Plant tissue culture: propagation, conservation and crop improvement, pp 15–44. https://doi.org/10.1007/978-981-10-1917-3_2

Nautiyal MC, Rawat AS, Bhadula SK, Purohit AN (1987) Seed germination in *Podophyllum hexandrum*. Seed Res 15:206–209

Newton J (2023) *Podophyllum peltatum*, *Podophyllum hexandrum*: the source of podophyllotoxin, etoposide and teniposide. In: Modern medicines from plants, pp 265–272. https://doi.org/10.1201/9781003413004-36

Pandey H, Chandra B, Nadeem M, Kumar A, Nandi SK, Palni LMS (2002) In vitro propagation of some alpine medicinal herbs of the Himalayan region. In: Role of plant tissue culture in biodiversity conservation and economic development. Gyanodaya Prakashan, Nainital, pp 297–323

Parasher K, Sharma S, Mukherjee P, Qazi PH (2023) Cryopreservation of zygotic embryos of *Podophyllum hexandrum* Royle, an endangered medicinal plant, by vitrification and V-cryoplate techniques. CryoLetters 44(4):219–228

Paul S, Nandi SK, Palni LMS (2013) Assessment of genetic diversity and interspecific relationships among three species of Podophyllum using AFLP markers and podophyllotoxin content. Plant Syst Evol 299:1879–1887

Purohit AN, Nautiyal MC (1988) Inhibitory effect of cotyledons on plumule development in two alpine rosettes. Can J Bot 66(1):205–206

Purohit AN, Lata H, Nautiyal S, Purohit MC (1998) Some characteristics of four morphological variants of *Podophyllum hexandrum* Royle. Plant Genetic Resources Newsletter (IPGRI/FAO) Bulletin des Ressources Phytogenetiques (IPGRI/FAO) Noticiario de Recursos Fitogeneticos (IPGRI/FAO), p 114

Rajesh M, Sivanandhan G, Jeyaraj M, Chackravarthy R, Manickavasagam M, Selvaraj N, Ganapathi A (2014) An efficient *in vitro* system for somatic embryogenesis and podophyllotoxin production in *Podophyllum hexandrum* Royle. Protoplasma 251:1231–1243

Rizwan A, Saurabh S, Vishal K, Chhavi V (2021) *In vitro* propagation of *Podophyllum hexandrum*. Int J Pharm Res Appl 6:169–172

Sharma N, Thakur M, Sharma P, Dutt B, Sharma YP (2022) *In vitro* propagation from seeds and enhanced synthesis of podophyllotoxin from root callus of *SinoPodophyllum hexandrum* Royle TS Ying (Himalayan Mayapple)—an endangered medicinal plant. Ind Crop Prod 186:115300

Shrestha UB, Shrestha S (2004) Major non-timber forest products of Nepal. Bhundipuran Parkashan, Kathmandu, 411p

Singh V (2009) Ethnobotany and medicinal plants of India and Nepal, vol 3. Scientific Publishers

Srivastava U, Sood H (2020) Propagation of *Podophyllum hexandrum* Royale to enhance production of podophyllotoxin. In: Endangered plants. IntechOpen, pp 202–219

Sultan P, Shawl AS, Ramteke PW, Jan A, Chisti N, Jabeen N, Shabir S (2006) In vitro propagation for mass multiplication of *Podophyllum hexandrum*: a high value medicinal herb. Asian J Plant Sci 5(2):179–184

Sultan P, Shawl AS, Fayaz AS, Ramteke PW (2009) *In vitro* embryogenesis and marker guided analysis of *Podophyllum hexandrum*: an endangered medicinal plant. Int J Biodivers Conserv 1(7):215–222

Troup RS (1915) A note of the cultivation of *Podoophyllum emodi*. Indian Forester 41:361–365

Van Uden W, Pras N, Battman S, Visser JF, Malingre TM (1989) Detection and identification of podophyllotoxin produced by cell culture derived from *P. hexandrum* Royle. Plant Cell Tissue Organ Cult 20:81–87

Van Uden W, Woerdenbag HJ, Pras N (1995) Cyclodextrins as a useful tool for bioconversions in plant cell biotechnology. In: Primary and secondary metabolism of plants and cell cultures III: Proceedings of the workshop held in Leiden, The Netherlands, 4–7 April 1993. Springer Netherlands, pp 103–113

Wang J, Cui Y, Tang X (2009) Chemical stability of teniposide in aqueous and parenteral lipid emulsions. Drug Dev Ind Pharm 35(4):508–513

Wong SK, Tsui SK, Kwan SY, Su XL, Lin RC (2000) Identification and characterization of *Podophyllum emodi* by API-LC/MS/MS. J Mass Spectrom 35(11):1246–1251

Zuhra Z, Saleem D, Akhtar W, Mahmood T (2021) Tissue culture optimization of *Podophyllum hexandrum* L., an endangered medicinal plant. J Anim Plant Sci 31(2):0238

Harnessing Biotechnology for the Conservation of High-Altitude Medicinal Plants

14

Meera Rai, Anshika Chauhan, Manu Pant, and Vishal Tripathi

Abstract

Medicinal plants are the treasure trove of the Himalayan ecosystems potentially contributing to the medical system and local economy of the people. However, in recent decades several species have been extinct or on the verge of extinction. Due to unsustainable harvesting, gathering, grazing, and climate warming, habitat fragmentation and unaware locals/tourists are some of the reasons for the rapid decline of the species from the Himalaya. Therefore, reliable strategies are required for the sustainable conservation of these valuable resources from extinction. Several conventions were established by the government for safeguarding the flora such as the Convention on Biological Diversity was established, the Convention on International Trade in Endangered Species of Wild Flora and Fauna, the International Treaty on Plant Genetic Resources for Food and Agriculture Convention, and the International Plant Protection Convention. Biotechnological intervention is an emerging tool of the twenty-first century for the conservation of natural resources like medicinal plants. Therefore, plant tissue culture, cryopreservation, microbial inoculants, and integration of traditional knowledge through capacity building are some of the potential strategies for the long-term utilization and conservation of medicinal plants. The biotechnological approach not only safeguards the native wild species from extinction but also supports promoting the economy of the rural people of the Himalaya.

M. Rai · A. Chauhan · M. Pant (✉) · V. Tripathi
Department of Biotechnology, Graphic Era (Deemed to Be University), Dehradun, Uttarakhand, India

S. Chandra, H. Lata (eds.), *High Altitude Medicinal Plants*,
https://doi.org/10.1007/978-3-031-99406-7_14

14.1 Introduction

The high-altitude Himalayan ecosystem is a unique repository of indigenous flora and fauna (Semwal et al. 2023). The region encompasses diverse forest ecosystems, with various endemic and high-value threatened medicinal plants (Rawat et al. 2022; Kumar et al. 2021). The 80% of the global population of developing countries depends on Medicinal and Aromatic Plants (MAPs) and some of them are also used for the development of allopathic drugs. A larger percentage of non-timber forest products (NTFPs) in the Himalayas consists of MAPs. However, the MAPs found in the alpine and subalpine regions are rare, eventually making them a Himalayan region treasure trove (Semwal et al. 2023). The World Health Organization (WHO) stated, that India annually accounts for one billion dollars in the world herbal medicine market and is known for its traditional healthcare systems. Consequently, this might increase the pressure on the market leading to uprooting of the MAPs from their natural habitat. The foremost serious threats while extirpating medicinal plants are habitat degradation, habitat loss, and over-harvesting. Therefore, the conservation of medicinal plants in their natural habitat signifies the conservation of biodiversity and accentuates the conservation of cultural diversity (Dhar et al. 2000).

To attenuate the pressure on wild species of medicinal plants and to fulfil the requirement of sustainable sources for the local population, who rely on these MAPs for their livelihood. The IUCN Red List suggests domesticating and cultivating the threatened medicinal plants near their natural habitat for the people who use them (Dhyani et al. 2022). To alleviate the unsustainable harvesting of the wild population, it is necessary to identify several habitats, climatic zones, and growing media that are beneficial for the domestication of rare and threatened medicinal plants (Dhyani et al. 2022). To ensure the enduring conservation of medicinal plants, pragmatic conservation strategies must be adopted to safeguard these valuable species. According to article 8 of the Convention on Biological Diversity (CBD) emphasized the prerequisite of in-situ conservation of natural habitats and ecosystems. Throughout the world, protected areas (PAs) are practically approachable and acceptable for the conservation of biodiversity. There are two conventional methods of conservation of medicinal plants: ex situ conservation and in situ conservation, but both techniques are time-consuming and natural. Therefore, biotechnological interventions are one of the potential tools for amplification and conservation of the rare genetic diversity of medicinal plants (Rai 2010). However, the rapid decline in the wild population of the plants has made this biotechnological method to be widely putative for conserving this invaluable treasure of the Himalayas.

Advancement in biotechnology proposes new methods for the conservation of these rare high-altitude medicinal plant through in vitro culture, micropropagation, DNA banks, genetic transformation, mycorrhization, and microbial inoculants (Rai 2010). Therefore, these techniques play a vital role in the conservation and long-term sustainability of the threatened medicinal plants of alpine regions. Manipulation of these biotechnological tools is beneficial in increasing the plant tolerance capacity for abiotic stresses such as water deficiency and extreme temperature by enhancing plant fitness, nutrient availability, and growth. Which eventually contributes to plant production and protection, especially in harsh habitats (Kushwaha et al. 2020). For instance, *Rauwofia sepentina* is an endangered, subalpine medicinal plant, used for epilepsy and sleeping disorder, regeneration through seed is quiet poor. Therefore Bhatt et al. (2008), developed a protocol for its mass cultivation through micropropagation. Similarly, *Dioscrea bulbifera*, found in the alpine and subalpine region, is effective in malaria, diarrhoea, diabetes etc., it was cryopreserved through encapsulation dehydration process in DNA bank (Rai 2010).

The rapidly increasing population, particularly in developing countries; therefore, efforts to maintain and ensure good health of this rising population is a topic of concern. On the other hand, prolonged consumption of synthetic medicines has caused health issues in both developed and developing countries. The invaluable bioactive compound from these medicinal plants has promising herbs that provide good and continuous health, in addition to their commercial worth. For instance, *Berginia ciliata* is used to treat kidney stones, *Rubia manjith* and *Berberis asiatica* to treat diabetes for skin disorders (Shinwari et al. 2024). Additionally, *Azadirachta indica* seeds contain limonoids and insecticidal triterpenoids, *Ajuga bracteosa*, has antifungal, antibacterial, antiinflamatory, and anthelminthes activities and Artemisia annua has sesquiterpene and antemisinin bioactive compounds (Shinwari et al. 2024). Harnessing biotechnological tools for the conservation of MAPs plays a crucial role in the modern era of industrialization, urbanization, and inadequate use of fertilizers. These beneficial strategies not only benefit plants but for the whole ecosystem's functioning. The implementation of conservation efforts through biotechnological interventions will help in maintaining the integrity and sustainability of the biodiversity as well as high throughput augmentation of the rare high-altitude medicinal plants in faster ways.

14.2 Importance of High-Altitude Medicinal Plants

The Indian Himalayan region (IHR) is the shelter of significant plant diversity developed from steep gradients of both precipitation and elevation. High-altitude medicinal plants are of eminent concern throughout the Himalayan region, due to their application in the Indian traditional medical system, rarity, and unsustainable harvesting for trade. Some of the medicinal plant species of higher altitudes are species of *Taxus, Fritillaria, Aconitum, Ephedra, Polygonatum, Dactylorhiza, Podophyllum, Nardostachys, Picrorrhiza* etc. Wild edible medicinal plants are one of the important sources of food supplements and might contribute to addressing world food insecurity. They are collected in form of flowers, fruits, shoots, leaves, twigs, roots, stems, and tubers (Negi et al. 2018). Unveiling the significance of high-altitude medicinal plants, the National Medicinal Plants Board (NMPB), Government of India, has established a task force on High Altitude Medicinal Plants (Singh et al. 2017; Azaizeh et al. 2003). The Himalayas have significantly contributed to several ecosystem services such as water storage, carbon sequestration, biodiversity maintenance, and food security for pastoral inhabitants for ages. For instance, the Sikkim Himalaya contains 60% and 10% of the world's global alpine plant family and genera (Ingty 2021). The occurrence of such distinct plant communities enables improved utilization of resources with amplified use of biomass, as well as encourages the establishment and existence of other co-occurring species (Kattel 2022).

Shifts in several ecological zones, particularly habitat loss, degradation, and fragmentation, are the major emerging threats to the region's plant diversity. However, in a region with higher anthropogenic activities, most of the native habitat has already been destroyed and several important medicinal plants are on the verge of extinction (Kumar et al. 2011). Medicinal plants are one of the potential key components for curing many health issues like bronchitis, dysentery, diarrhea, pulmonary, menstrual disorder, dysentery, pulmonary infections, leprosy, cancer, etc. Therefore, the extinction of these highly valuable medicinal plants impacts the associated biodiversity and treatment of many diseases lost. Therefore, attention has been required to the sustainable conservation of medicinal plants to ensure the long-term availability of the plants in their native habitat.

Conservation of higher altitude medicinal plants through biotechnological tools is not only beneficial for the native wild species conservation but also helpful for the rural people of the region for their day-to-day livelihood. Himalaya medicinal plants are a treasure trove both medicinally as well as economically. Bioprospecting products from medicinal plants hold the potential for rising income for the locals in the Himalayan region. As unemployment is one of the major concerns and locals migrate to different cities to earn, therefore, the in-vitro technique might be helpful for the economy of the Himalayas as well as promoting sustainable harvesting practices by conserving wild native species. For example, one of the bio-prospected products of the Himalayas is *Paeonia emodi*, useful in diabetes, currently least concern, the market price is approximately Rupees 300/kg (Joshi et al. 2018).

14.3 Biotechnological Tools for Conservation

Due to the high medicinal properties and increasing trade demand, biotechnological interventions are needed to address the oversupply of medicinal plants. There are several methods for the sustainable conservation of medicinal plants through biotechnological tools such as harnessing plant-microbe interactions, plant tissue culture techniques, and cryopreservation (Fig. 14.1). Plant tissue culture is a pioneering technology that has been used for decades to conserve plants using different methods like callus culture, hairy root culture, embryo culture, and somatic hybridization. Since conventional breeding techniques could not fulfill the required demand for crops, tissue culture came around as a grand leap in breeding practices. It makes use of parts of a plant to generate multiple copies of the plant in a very short duration. Tissue culture techniques enable mass multiplication and regeneration of these plants, reducing the pressure on wild populations and allowing for the production of high-quality planting material (Sati et al. 2023).

The technique exploits the property of totipotency of plant cells which means that any cell from any part of the plant can be used to generate a whole new plant. Tissue culture is used to develop thousands of genetically identical plants from one

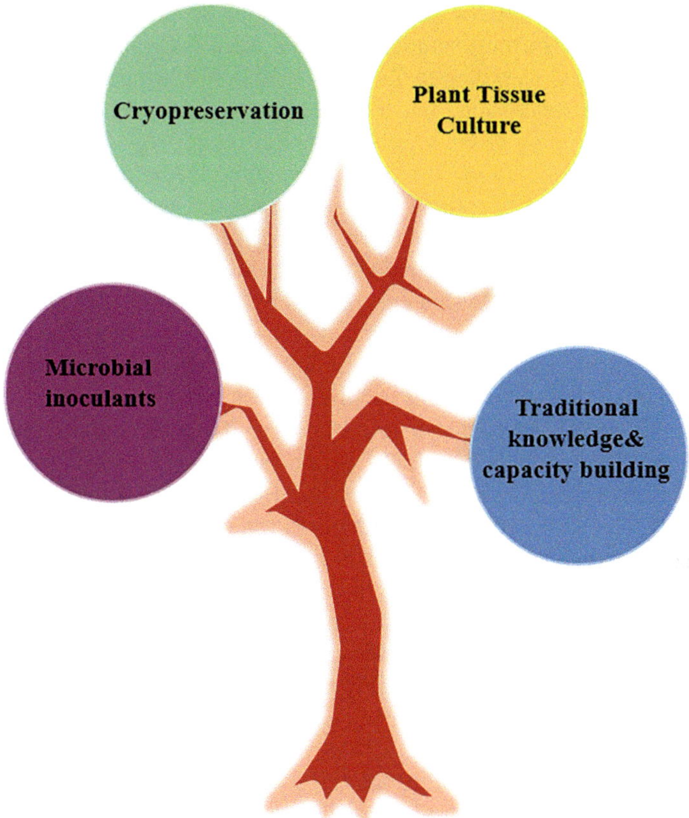

Fig. 14.1 Medicinal plants conservation techniques

single parent plant known as somaclones, and this process is known as micropropagation. The method offers an advantage over other methods as it can be used to develop disease-free plants from disease-rode plants by using their meristems (apical and axillary) as explants. Micropropagation is one of the subparts of the plant tissue culture in which the artificially producing plants vegetatively through tissue culture or cell culture techniques.

For instance, *Angelica glauca* Edgew, a higher-altitude medicinal plant with immense medicinal uses, is threatened in the Himalayas. The species is conserved through invitro propagation using buds of rhizomes in MS (Murashige and Skoog) medium supplemented with growth regulators showing a 60% survival rate in polyhouse for 1 month (Negi et al. 2024). Further, in *Geranium wallichianum* an Indigenous threatened high-value medicinal plant, in vitro propagation through callus and root was induced for plantlets growth and the study shows the successful acclimatization of plantlets while transferring in small pots containing peat moss (Masood et al. 2023). Similarly, *Rauwolfia serpentina*, micropropagation using nodal, leaf explants show multiple shoots and plantlet formation (Khan et al. 2018);

Catharanthus roseus, source of explants was shooting tip and nodal segments, has a positive response by forming multiple root and plant (Begum and Mathur 2014) (Table 14.1). Further, according to Tao et al. (2024), in-vitro propagation is a successful method for establishing the endangered high-value medicinal plant of the alpine regions, involving in vitro bulblet regeneration through MS medium callus regeneration. *Picrorhiza kurrooa is* another essential medicinal plant endemic to Himalaya useful in treating asthma, anemia, and hepatic and gastric disorders, and has been highly exploited unsustainably rendering it endangered. Hairy root culture of *Picrorhiza kurroa* has been observed as a promising approach for in vitro secondary metabolite production (Mishra et al. 2011a, b). Interestingly the phytochemical analysis of in vitro cultured plants produced high secondary metabolite in comparison to mother plant (Rawat et al. 2013a, b). Such approach reduces the pressure on endangered plants for unsustainable harvesting from its natural habitat.

Cryopreservation is another beneficial biotechnological technique widely used to conserve medicinal plants. The plants' germplasm is used in cryobanks at −196 degrees Celsius in liquid nitrogen for the sustainable utilization of the resources (Table 14.1). The ice formation inside the cells can break the cell membrane. This can be prevented by regulating the freezing rate and carefully choosing the freezing medium. Cryopreservation methods, including vitrification and encapsulation-dehydration, facilitate long-term conservation of plant genetic materials, providing a safeguard against extinction (El Merzougui et al. 2023). Genetic engineering approaches, such as RNAi and CRISPR-Cas9, can enhance stress tolerance and disease resistance in high-altitude medicinal plants, improving their adaptability and productivity (Jamwal et al. 2023). There are many benefits of the cryopreservation technique. These include fertility treatments, minimal space, and labor required, safety from genetic contamination, safeguards genetic integrity of valuable stains, safeguards the germplasm of endangered species, biological samples can be preserved for a longer period, protects the samples from disease and microbial contamination, prevents genetic drift by cryopreservation of gametes, embryos, etc. Molecular breeding, utilizing marker-assisted selection and genomic selection, enables the development of high-yielding and resilient varieties, while maintaining genetic diversity (Song and Li 2021). Next-generation sequencing and genomics-based approaches aid in understanding the genetic diversity and evolutionary relationships of these plants, informing conservation strategies and breeding programs (Haq et al. 2023). This method has proven efficient in encountering several species within the realms of horticulture, forestry, agriculture, and medicinal plant germplasm. For instance, *Delbergia odorifera* T.chen, *Alpinia oxyphylla* Miq, *Myristica fragrans* Houtt, and *Prunella Vulgaris* L are a few examples of medicinal plants that have successfully preserved in ultra-low-temperature preservation techniques by using seeds of these plants. The freezing method is quite viable for the long-term preservation of the plants for suitable applications in the future and also saves them from extinction (Zeng et al. 2024; Nits et al. 2024).

Plant-microbe interaction is one of the potential emerging methods that play a vital role in the conservation and long-term sustainability of the threatened medicinal plants of alpine regions. Microbes are beneficial in increasing the plant tolerance

Table 14.1 Potential biotechnological tools for the conservation of medicinal plants

S. no.	Plant species	Technique	Result	References
1.	*Nardostachys jatamansi*	Micropropagation	Successful regeneration of plants	Pant et al. (2021)
2.	*Podophyllum hexandrum*	Cryopreservation	Long-term conservation of genetic material	Parasher et al. (2023)
3.	*Taxus wallichiana*	Genetic engineering	Production of taxol	Patel et al. (2022)
4.	*Rhodiola rosea*	Micropropagation	Rapid multiplication of elite clones	Matvieieva et al. (2023)
5.	*Gentiana kurroo*	Somatic embryogenesis	Regeneration of plants from somatic embryos	Dutta and Sood (2022)
6.	*Valeriana wallichii*	Tissue culture	Successful regeneration of plants	Sharma et al. (2020a, b)
7.	*Bergenia cilata*	In vitro mutagenesis	Induction of desirable traits	Rafi et al. (2021)
8.	*Swertia chirata*	Cryopreservation	Long-term conservation of genetic material	Gupta and Sood (2022)
9.	*Picrohiza kurroa*	Tissue culture	Successful regeneration of plants	Pratap et al. (2022)
10.	*Hoodia gordonii*	Genetic engineering	Production of hoodia glyocides	Parveen et al. (2024)
11.	*Sausurea lappa*	Somatic embryogenesis	Regeneration of plants from somatic embryos	Groach et al. (2022)
12.	*Aconitum heterophyllum*	Tissue culture	Successful regeneration of plants	Wani et al. (2022)
13.	*Bacopa monnieri*	Cryopreservation	Long-term conservation of genetic material	Gowthami et al. (2023)
14.	*Corydalis cashmeriana*	In vitro mutagenesis	Induction of desirable traits	Pattanayak (2023)
15.	*Dioscorea deltoidei*	Micropropagation	Rapid multiplication of elite clones	Hassanein (2022)
16.	*Taxus baccata*	Genetic engineering	Production of Taxol	Sahai and Sinha (2022)
17.	*Sausurea costus*	Micropropagation	High-yield cultivation	Khan et al. (2019)
18.	*Rheum emodi*	Cryopreservation	Long-term conservation	Sharma et al. (2020a, b)
19.	*Panax pseudoginseng*	Cryopreservation	Genetic material conservation	Zeng et al. (2024)
20.	*Inula racemosa*	Micropropagation	High-yield cultivation	Moraes et al. (2021)
21.	*Hypericum perforatum*	Genetic engineering	Hypericin production	Rizzo et al. (2020)

(continued)

Table 14.1 (continued)

S. no.	Plant species	Technique	Result	References
22.	*Meconopsis aculeata*	Somatic embryogenesis	Plant regeneration	Chandra et al. (2021)
23.	*Neopicrorhiza scrophulariiflora*	Plant tissue culture	Successful regeneration of the plant	Bantawa et al. (2024)
24.	*Podophyllum emodi*	Cryopreservation	Germplasm conservation	Shukla et al. (2021)
25.	*Papaver wallichii*	Micropropagation	Rapid multiplication	Devi et al. (2022)
26.	*Angelica glauca*	Micropropagation	Direct organogenesis	Rawat et al. (2018)
27.	*Paris polyphylla*	Callus suspension culture	Optimized cultivation technique	Rawat et al. (2023)

capacity for abiotic stresses such as water deficiency and extreme temperature by enhancing plant fitness, nutrient availability, and growth. All the synergistic interactions significantly through the production of siderophore for iron chelation, potassium, phosphate, and zinc solubilization for K^+, PO_4^{2-} and Zn^{2+} ions, N_2 fixation, plant hormones, for the production of several oxidizing and reducing enzymes which eventually contributes to plant production, and protection, alleviates abiotic stresses (Kushwaha et al. 2020), and also reduces the use of chemical fertilizers. All these characteristics of using microbial inoculants might be a sustainable and effective approach that is not only beneficial in protecting plants but also helpful in maintaining environmental integrity. However, to date the application of plant growth-promoting microorganisms is more on crops and other foods as compared to higher altitude medicinal plants. For example, in China, *Rheum palmatum* Chinese medical herb is experimented with PGPM (Plant growth promoting microorganism) in a field experiment with *Bacillus amyloliquefaciens* EZZ9 + Sucrose supplement. The study shows an improvement in the quality and growth of *R. Palmatum* as well as biocontrol potential (Tian et al. 2022). A similar study, on *Astragalus mongholicus* Chinese medicine, uses several bacterial strains for consortia such as phosphate solubilizing, potassium from feldspar IAA (Indole3 acetic acid production), and siderophore-producing isolates. The finding shows that a combination of bacterial isolates improves drought tolerance capacity in *A. mongholicus*. Therefore, the application of microbial inoculants not only promotes the quality and quantity of medicinal plants but also focuses on agroecological practices as well as climate resilience (Lin et al. 2022).

14.4 Integrating Biotechnology with Traditional Knowledge

High-altitude medicinal plants, growing in unique and fragile ecosystems, are invaluable resources for traditional medicine, providing essential compounds for human health. However, their populations are threatened by habitat destruction, over-harvesting, and climate change, leading to a loss of biodiversity and genetic resources (Shivali and Namrata 2022). To address these challenges, biotechnological tools with the integration of traditional knowledge with capacity building offer innovative solutions for the conservation and sustainable utilization of high-altitude medicinal plants. The Sustainable Development Goals demand new methods to "safeguard sustainable production and consumption" and further protect, restore, and encourage the sustainable use of terrestrial ecosystems. Dietary modification, policy reforms, a rise in investments, biotechnology, and several other anticipated solutions hold the potential for understanding changes in local traditional knowledge and how Indigenous communities expedite ecosystem services, further biomanipulation of farmers' skills for biodiversity conservation (Kumar et al. 2011) (Fig. 14.1). Several efforts have been made at the national and international level for the conservation of higher altitude herbs by different agencies, and NGOs, but are also setting up slowly. However, immediate serious efforts and attention are required to conserve these therapeutic agents for the sustainability of the Himalayan ecosystem and future use (Raina and Gautam 2020).

The local people are unaware of the significance of the higher altitude medicinal plants at the global level. They gather plants unsustainably on a large scale from the wild and in most scenarios uprooted the whole plant. Local populations of the particular region are not very aware of the preservation and conservation techniques of the plants which ultimately leads to complete depletion of the plant resources. However, rare species are not only significant at the local or national level but also from the point of view of global biodiversity. Natural resources, especially medicinal plants of the particular region, are finite and precious resources that require efficient, mindful, and sustainable management with pragmatic conservation measures. Hence, immediate conservation strategies are urgently needed to protect the taxa from the verge of extinction. Several conventions were held internationally that focus on problems related to biodiversity: in (1993) the Convention on Biological Diversity was established, (1975) Convention on International Trade in Endangered Species of Wild Flora and Fauna, (2004) International Treaty on Plant Genetic Resources for Food and Agriculture Convention, and (1952) International Plant Protection Convention. All these plants biodiversity-related convention efforts to implement action at the regional, national, and international levels to attain targets of conservation and sustainable utilization. In meeting their targets, conventions have generated several complementary tactics such as site, genetic resources, species, and ecosystem-based. However, all these efforts are not able to sustain the plant biodiversity especially higher altitude medicinal plants due to negligence. Therefore, by integrating these biotechnological tools with traditional conservation approaches, we can ensure the long-term survival and sustainable utilization of high-altitude medicinal plant species. These tools offer a powerful platform for the conservation and improvement of these valuable resources, ultimately contributing to the development of sustainable healthcare solutions and long-term utilization.

14.5 Conclusion

The high-altitude ecosystems of the Himalayas are under threat due to climate change, unsustainable harvesting, and habitat degradation, requiring urgent interventions for biodiversity conservation. In this context, biotechnology offers a sustainable solution for the conservation of high-altitude medicinal plants in the Himalayas. New and emerging biotechnological approaches, such as tissue culture, molecular marker-based population mapping, and cryopreservation, can aid in conserving the genetic diversity of these plants. Furthermore, advanced propagation techniques like molecular breeding and micropropagation can facilitate large-scale cultivation for industrial purposes, reducing the pressure of unsustainable harvesting on wild populations. Integrating biotechnological interventions with conventional conservation methods can not only support the survival of threatened medicinal plants but also maintain ecological balance and the livelihoods of local communities. However, consistent research and policy support are urgently needed to leverage biotechnology for the protection of plant resources in high-altitude ecosystems.

References

Azaizeh H, Fulder S, Khalil K, Said O (2003) Ethnobotanical knowledge of local Arab practitioners in the Middle Eastern region. Fitoterapia 74(1–2):98–108

Bantawa P, Rai R, Mondal TK (2024) Enhanced somatic embryogenesis and viable synthetic seeds in *Picrorhiza scrophulariiflora* Pannell: a conservation milestone. Environ Ecol 42(3B):1287–1295

Begum T, Mathur M (2014) In vitro regeneration of *Catharanthus roseus* and *Bacopa monnieri* and their survey around Jaipur District. Int J Pure Appl Biosci 2:210–221

Bhatt R, Arif M, Gaur AK, Rao PB (2008) Rauwolfia serpentina: protocol optimization for in vitro propagation. Afr J Biotechnol 7:4265–4268

Chandra N, Singh G, Lingwal S, Rai ID, Tewari LM (2021) Alpine medicinal and aromatic plants in the Western Himalaya, India: an ecological review. Indian J Ecol 48(2):319–331

Devi S, Sinha R, Sibi G (2022) Northeast India: a treasury of medicinal plants. Emergent Life Sci Res 8:5–21

Dhar U, Rawal RS, Upreti J (2000) Setting priorities for conservation of medicinal plants—a case study in the Indian Himalaya. Biol Conserv 95(1):57–65

Dhyani A, Nautiyal BP, Nautiyal MC (2022) Seedling performance, phenotypic traits, and growing media for domestication of *Lilium polyphyllum*. Plant People Planet 4(6):685–695

Dutta I, Sood H (2022) Screening of phytochemicals in shoot cultures of endangered herb *Gentiana kurroo* Royle. Biodivers Conserv 8:48–57

El Merzougui S, Benelli C, El Boullani R, Serghini MA (2023) The cryopreservation of medicinal and ornamental geophytes: application and challenges. Plan Theory 12(11):2143. https://doi.org/10.3390/plants12112143

Gowthami R, Sharma N, Chandra R, Kurian JS, Malhotra EV, Agrawal A (2023) Development of an improved and simple shoot tip cryoconservation protocol for cryobanking of *Bacopa monnieri* (L.) Wettst. germplasm. In Vitro Cell Dev Biol Plant 59(6):744–756

Groach R, Yadav K, Singh N (2022) Actions for ex-situ conservation of *Saussurea lappa* (Decne.) Clarke. Ann Oradea Univ Biol Fasc 29(2):173–178

Gupta R, Sood H (2022) Artificial seed production and cryopreservation by encapsulation dehydration for medicinal herb of Himalayan region, *Swertia chirayita*. CryoLetters 43(5):295–302

Haq SM, Waheed M, Khoja AA, Amjad MS, Bussmann RW, Ali K (2023) A cross-cultural study of high-altitude botanical resources among diverse ethnic groups in Kashmir Himalaya, India. J Ethnobiol Ethnomed 19(1):12. https://doi.org/10.1186/s13002-023-00608-5

Hassanein AM (2022) Application of tissue culture techniques to improve the productivity of medicinal secondary products from medicinal plants, vol 7. IntechOpen, pp 113–136

Ingty T (2021) Pastoralism in the highest peaks: role of the traditional grazing systems in maintaining biodiversity and ecosystem function in the Alpine Himalaya. PLoS One 16(1):e0245221

Jamwal M, Ghora B, Kumar S, Gupta AP (2023) Applications of CRISPR/Cas9 in the synthesis of bioactive compounds from medicinal plants. In: Biosynthesis of bioactive compounds in medicinal and aromatic plants: manipulation by conventional and biotechnological approaches. Springer, Cham, pp 81–97. https://doi.org/10.1007/978-3-031-17994-9_5

Joshi P, Prakash P, Purohit VK, Joshi K (2018) Bioprospecting of *Paeonia emodi* for livelihood enhancement in Western Himalaya, India. Environ Bull Himal Ecol 26:67–70

Kattel GR (2022) Climate warming in the Himalayas threatens biodiversity, ecosystem functioning and ecosystem services in the 21st century: is there a better solution? Biodivers Conserv 31(8):2017–2044

Khan S, Banu TA, Akter S, Goswami B, Islam M, Hani U et al (2018) In vitro regeneration protocol of *Rauvolfia serpentina* L. Bangladesh J Sci Ind Res 53:133–138

Khan A, Shah AH, Ali N (2019) In-vitro propagation and phytochemical profiling of a highly medicinal and endemic plant species of the Himalayan region (*Saussurea costus*). Sci Rep 11(1):23575

Kumar M, Sheikh MA, Bussmann RW (2011) Ethnomedicinal and ecological status of plants in Garhwal Himalaya, India. J Ethnobiol Ethnomed 7:1–13

Kumar A, Sathyakumar S, Goraya GS, Gupta AK, Adhikari BS, Rawat GS (2021) Assessment of medicinal and aromatic plant species on their collection, usage, demand, markets, price trends, and life cycle in Lahaul and Pangi landscape, Himachal Pradesh. A report (TR No./2021/14) submitted to the Himachal Pradesh Forest Department and United Nations Development Programme, 161 pp

Kushwaha RK, Rodrigues V, Kumar V, Patel H, Raina M, Kumar D (2020) Soil microbes–medicinal plants interactions: ecological diversity and future prospect. In: Plant microbe symbiosis. Springer, pp 263–286

Lin Y, Zhang H, Li P, Jin J, Li Z (2022) The bacterial consortia promote plant growth and secondary metabolite accumulation in *Astragalus mongholicus* under drought stress. BMC Plant Biol 22(1):475

Masood A, Ashraf Z, Khan SA, Akhtar G, Hassan A, Shah MM et al (2023) In vitro propagation of *Geranium wallichianum*: an indigenously threatened medicinal plant in Pakistan. Kuwait J Sci 50(2):5–11

Matvieieva N, Belokurova V, Ratushniak Y, Shcherbak N, Kuchuk M (2023) In vitro direct shoot regeneration from *Rhodiola rosea* L. leaf explants. Biotechnol Acta 16(3):45–50

Mishra J, Bhandari H, Singh M, Rawat S, Agnihotri RK, Mishra S, Purohit S (2011a) Hairy root culture of *Picrorhiza kurroa* Royle ex Benth.: a promising approach for the production of picrotin and picrotoxinin. Acta Physiol Plant 33:1841–1846. https://doi.org/10.1007/s11738-011-0724-x

Mishra J, Singh M, Palni LM, Nandi SK (2011b) Assessment of genetic fidelity of encapsulated microshoots of *Picrorhiza kurrooa*. Plant Cell Tissue Organ Cult 104:181–186. https://doi.org/10.1007/s11240-010-9816-3

Moraes RM, Cerdeira AL, Lourenço MV (2021) Using micropropagation to develop medicinal plants into crops. Molecules 26(6):1752

Negi VS, Maikhuri RK, Maletha A, Phondani PC (2018) Ethnobotanical knowledge and population density of threatened medicinal plants of Nanda Devi Biosphere Reserve, Western Himalaya, India. Iran J Sci Technol Trans A: Sci. https://doi.org/10.1007/s40995-018-0545-5

Negi D, Thakur M, Dutt B, Sharma R (2024) In vitro propagation and secondary metabolites production of *Angelica glauca* Edgew: a threatened medicinal and aromatic herb of the Himalayas. Plant Cell Tissue Organ Cult 158(2):26

Nits OS, Sementsova MV, Osipova ES, Tereshonok DV, Gladkov EA (2024) IPPRAS Cryobank for the conservation of orthodox seeds of rare, endangered, medicinal, and ornamental plant species—current research. Plan Theory 13:1354

Pant HC, Pant HV, Kumar A, Tomar H, Dev M, Gaurav N (2021) In vitro clonal propagation of *Nardostachys jatamansi*: a traditional Himalayan medicinal plant. J Mt Res 16(3):87–98

Parasher K, Sharma S, Mukherjee P, Qazi PH (2023) Cryopreservation of zygotic embryos of *Podophyllum hexandrum* Royle, an endangered medicinal plant, by vitrification and V cryo-plate techniques. CryoLetters 44(4):219–228

Partap M, Chhimwal J, Kumar P, Kumar D, Padwad Y, Warghat AR (2022) Growth dynamics and differential accumulation of picrosides and its precursor metabolites in callus cell lines of *Picrorhiza kurroa* with distinct anti-steatotic potential. Process Biochem 120:85–100

Parveen I, Wang M, Lee J, Zhao J, Zhu Y, Chittiboyina AG et al (2024) Identification and functional characterization of oxidosqualene cyclases from medicinal plant *Hoodia gordonii*. Plan Theory 13(2):231

Patel P, Patel V, Modi A, Kumar S, Shukla YM (2022) Phyto-factories of anti-cancer compounds: a tissue culture perspective. Beni Suef Univ J Basic Appl Sci 11(1):43

Pattanayak S (2023) Anti-cancer plants and their therapeutic use as succulent biomedicine capsules. Explor Anim Med Res 13:59–67. https://doi.org/10.1111/exp12345

Rafi S, Kamili AN, Ganai BA, Parray JA, Jan S (2021) Variation in root morphology, enhancement in anti-oxidative enzyme responses and improved arbutin and bergenin levels in *Bergenia ciliata* (Haw.) Sternb. raised in vitro via EMS and gamma irradiations. Plant Cell Tissue Organ Cult 145:43–57

Rai MK (2010) Biotechnological strategies for conservation of rare and endangered medicinal plants. Biodiversitas 11(3):157–166

Raina R, Gautam K (2020) Conservation and utilization of high-altitude threatened medicinal plants. In: Conservation and utilization of threatened medicinal plants. Springer, pp 369–387. https://doi.org/10.1007/978-3-319-98812-2_20

Rawat B, Rawat JM, Mishra S, Mishra SN (2013a) *Picrorhiza kurrooa*: current status and tissue culture mediated biotechnological interventions. Acta Physiol Plant 35:1–12. https://doi.org/10.1007/s11738-012-1069-9

Rawat JM, Rawat B, Mehrotra S, Chandra A, Nautiyal S (2013b) ISSR and RAPD based evaluation of genetic fidelity and active ingredient analysis of regenerated plants of *Picrorhiza kurroa*. Acta Physiol Plant 35:1797–1805. https://doi.org/10.1007/s11738-013-1217-x

Rawat JM, Bhandari A, Mishra S, Rawat B, Dhakad AK, Thakur A, Chandra A (2018) Genetic stability and phytochemical profiling of the in vitro regenerated plants of *Angelica glauca* Edgew.: an endangered medicinal plant of Himalaya. Plant Cell Tissue Organ Cult 135:111–118. https://doi.org/10.1007/s11240-018-1448-z

Rawat B, Rawat JM, Purohit S, Singh G, Sharma PK, Chandra A, Shabaaz Begum JP, Venugopal D, Jaremko M, Qureshi KA (2022) A comprehensive review of *Quercus semecarpifolia* Sm.: an ecologically and commercially important Himalayan tree. Front Ecol Evol 10:961345. https://doi.org/10.3389/fevo.2022.961345

Rawat JM, Pandey S, Rawat B, Purohit S, Anand J, Negi AS, Thakur A, Mahmoud MH, El-Gazzar AM, El-Saber Batiha G (2023) In vitro production of steroidal saponin, total phenols and antioxidant activity in callus suspension culture of *Paris polyphylla* Smith: an important Himalayan medicinal plant. Front Plant Sci 14:1225612. https://doi.org/10.3389/fpls.2023.1225612

Rizzo P, Altschmied L, Ravindran BM, Rutten T, D'Auria JC (2020) The biochemical and genetic basis for the biosynthesis of bioactive compounds in *Hypericum perforatum* L., one of the largest medicinal crops in Europe. Gene 11(10):1210

Sahai P, Sinha VB (2022) Development of hairy root culture in *Taxus baccata* subsp. *wallichiana* as an alternative for increased taxol production. Mater Today Proc 49:3443–3448

Sati P, Chauhan M, Trivedi VL, Nautiyal MC, Semwal P (2023) Challenges and prospects for the in-vitro conservation of plants having anticarcinogenic potential in the Western Himalaya, India. Plant Cell Tissue Organ Cult (PCTOC) 152(2):237–252. https://doi.org/10.1007/s11240-023-02257-7

Semwal P, Rauf A, Simal-Gandara J (2023) Editorial: high altitude medicinal plants and their bioactive compounds for the prevention of oxidative stress-induced diseases and disorders. Front Pharmacol 14:1257018. https://doi.org/10.3389/fphar.2023.1257018

Sharma N, Pandey R, Gowthami R (2020a) In vitro conservation and cryopreservation of threatened medicinal plants of India. In: Conservation and utilization of threatened medicinal plants. Springer, pp 181–228

Sharma N, Rawat R, Lalremruati F, Singh A (2020b) Complete plant regeneration of *Valeriana wallichii* DC. on auxin enriched medium and phytochemical analysis. Plant Sci Today 7(4):542–550

Shinwari AK, Ahmad M, Alsahli AA, Israr A, Shinwari ZK, Shinwari S, Zhang L, Gafforov Y, Yilmaz ME (2024) Himalayan medicinal plants wealth, threats and conservation for securing the future of biodiversity. Pak J Bot, 56(6):2327–2350

Shivali V, Namrata S (2022) Threat status of three important medicinal Himalayan plant species and conservation implications. Nat Conserv Res 7(1):27–41. https://doi.org/10.24189/ncr.2022.011

Shukla S, Thomas S, Bhutani R (2021) Plant tissue culture, plant-based products and prospects of commercialization: a drive from nature towards nature. Int J Biotechnol Rec Adv 3(1):68–73

Singh M, Pandey N, Agnihotri V, Singh KK, Pandey A (2017) Antioxidant, antimicrobial activity and bioactive compounds of *Bergenia ciliata* Sternb.: a valuable medicinal herb of Sikkim Himalaya. J Tradit Complement Med 7(2):152–157. https://doi.org/10.1016/j.jtcme.2016.04.002

Song Z, Li X (2021) Recent advances in molecular marker-assisted breeding for quality improvement of traditional Chinese medicine. Curr Pharm Biotechnol 22(6):867–875. https://doi.org/10.2174/1389201022206210320135710

Tao Q, Han G, Ma B, Jia H, Zhao C, Li W, Yan Z (2024) In vitro propagation and artificial seed production of *Fritillaria cirrhosa* D. Don, an endangered medicinal plant. Phyton 93(6):1297–1310

Tian Y, Liu Y, Yue L, Uwaremwe C, Zhao X, Zhou Q et al (2022) Bacterial inoculant and sucrose amendments improve the growth of *Rheum palmatum* L. by reprograming its metabolite composition and altering its soil microbial community. Int J Mol Sci 23(3):1694

Wani TA, Kaloo ZA, Dangroo NA (2022) *Aconitum heterophyllum* Wall. ex Royle: a critically endangered medicinal herb with rich potential for use in medicine. J Integr Med 20(2):104–113

Zeng L, Sun Z, Fu L, Gu Y, Li R, He M, Wei J (2024) Cryopreservation of medicinal plant seeds: strategies for genetic diversity conservation and sustainability. Plan Theory 13(18):2577

Asteraceae Plants from Eastern Himalayan Region of India: An Overview on Potential Ethnomedicinal Application and In Vitro Conservation

15

Krishna Kant Pachauri and Rakhi Chaturvedi

Abstract

Eastern Himalayan region of India is one of the significant ecological regions across the globe and recognized as the 12th biodiversity hotspot by IUCN. The area is exceptionally rich in medicinal plant biodiversity and contains 500 species in Arunachal Pradesh, 140 species in Darjeeling Hills, and 36 species in Sikkim. Interestingly, almost all medicinal plants from this region are used by indigenous people for healing purposes. Asteraceae family plants are predominantly used by local communities like Apatani (19 species, 11 genera), Adi (8 species, 8 genera), Monpa (8 species, 5 genera), and Lohit (6 species, 5 genera) tribes of Arunachal Pradesh. The *Ageratum conyzoides, Artemisia nilagirica, A. maritima, Spilanthes acmella, S. paniculata, Blumea balsamifera, B. lacera, Vernonia cinerea, Eclipta prostrata,* and *Chrysanthemum indicum* are a few of the most commonly used medicinal plants from this family. The extracts of these plants have shown antimicrobial, antioxidant, antihelminthic, analgesic, anticancer, antimalarial, and anti-inflammatory activities. Various secondary metabolites including alkaloids, terpenoides, sesquiterpene lactones, N-alkylamides and phenolics have been reported as active ingredients from these plants. Extensive medicinal utilization and overexploitation of these plants lead to habitat depletion and pose an inevitable extinction threat. In vitro conservation strategies, such as tissue culture, micropropagation, synthetic seed technology and cryopreservation, are necessary to enhance the long-term stability of these plants. Considering the importance of Asteraceae plants, the current chapter highlighted the in vitro conservation strategies for sustainable utilization of these plants and their potential ethnomedicinal applications.

K. K. Pachauri · R. Chaturvedi (✉)
Department of Biosciences and Bioengineering, Indian Institute of Technology Guwahati, Guwahati, Assam, India
e-mail: rakhi_chaturvedi@iitg.ac.in

S. Chandra, H. Lata (eds.), *High Altitude Medicinal Plants*,
https://doi.org/10.1007/978-3-031-99406-7_15

15.1 Introduction

Medicinal plants have been the source of herbal remedies for local communities since antiquity and are even considered responsible for the origin of modern therapeutic system (Salmerón-Manzano et al. 2020). According to the WHO report (2002), 80% of the world's population depends on medicinal plants to cure various ailments. Approximately 80% of commercially available synthetic drugs are derived from herbal products (Bauer and Brönstrup 2014). The minimal side effects and affordability are prime reasons for the upsurge in traditional health care systems (THCS). In developing countries, like India, a large section of the rural population (80%) depends on this healthcare system (Jeelani et al. 2018).

India is one of the 12 megadiverse countries, representing 11% of the world's flora and sharing approximately 2.4% of the global land area. The country has a forest cover of about 22.5% of the total geographical area flourishing more than 17,000 angiosperm plant. Eastern Himalayan, Western Ghats, Indo-Burma, and Sundaland are four major biodiversity hotspots in India. Eastern Himalayan region is located at the juncture of the Indo-Malayan, Palearctic, and Sino-Japanese realms (Kandel et al. 2019). The high altitudes of eastern Himalaya lie in the eastern Nepal, Bhutan, and a few Indian territories like Darjeeling hills, Sikkim, and Arunachal Pradesh. Of these, Arunachal Pradesh (83,743 sq. km) and Sikkim (7096 sq. km) occupies a large area of this biodiversity hotspot (Tangjang et al. 2011). The area is exceptionally rich in biodiversity containing 500 species of medicinal plants in Arunachal Pradesh, 140 species in Darjeeling hills and 36 species in Sikkim (Banik et al. 2020). The Eastern Himalayan region also supports one of the world's richest alpine flora. Around one-third of alpine species are endemic to this region. The region's variable climate and changing altitudes create a diversity of habitats, from the cold deserts at very high altitudes to the evergreen tropical and temperate forests at low altitudes.

15.2 Asteraceae Family Plants

The Asteraceae family is one of the most diverse plant groups, comprising 1600–1700 genera and 24,000–30,000 plant species across the globe except Antarctica (Panda and Luyten 2018). It is India's fourth most prominent family of angiosperms after Poaceae, Orchidaceae and Leguminosae, representing more than 1050 plant species from 177 genera. The plants of this family are annual and perennial, vines, succulents, shrubs, and trees in nature. Northeast and Himalayan regions are the primary centres of these plants, containing 72.67% of total Asteraceae in India (Mitra and Mukherjee 2017; Abraham and Thomas 2016). As a global biodiversity hotspot, the region is extremely rich in medicinal herbs, and Asteraceae is the most dominant family in this area (Tangjang et al. 2011). *Ageratum conyzoides, Artemisia nilagirica, A. maritima, Spilanthes acmella, S. paniculata, Blumea balsamifera, B. lacera, Vernonia cinerea, Eclipta prostrata,* and *Chrysanthemum indicum* are some of the most commonly used local medicinal plants from this family (Mylliemngap and Arya 2021).

15.2.1 Traditional Medicinal Uses

Using medicinal plants to treat various ailments is an essential part of the health care system in India. Ayurvedic, Siddha, and Unani are three different traditional therapeutic systems of our country. The people from high altitudes of Arunachal Pradesh and Sikkim have a strong hold on herbal remedies. The local communities living there for thousands of years have built a vast knowledge base about the use of medicinal plants. These plants have been integrated into the culture of this region. The indigenous people of different tribes largely depend on medicinal plants due to their remoteness and inaccessibility to the urban areas. Asteraceae family plants are predominantly used by various tribes including Apatani (19 species, 11 genera), Adi (8 species, 8 genera), Monpa (8 species, 5 genera) and Lohit (6 species, 5 genera) from Arunachal Pradesh and Lepcha, Bhotia (4 species, 4 genera) from Sikkim (Kala 2005; Tangjang et al. 2011; Jeyaprakash et al. 2017; Chakraborty et al. 2017; Namsa et al. 2009; Tamang et al. 2023) (Fig. 15.1).

Various phytochemicals including alkaloids, terpenoids, sesquiterpene lactones, N-alkylamides, and phenolics have been reported as active ingredients from these plants (Fig. 15.2). The extracts have shown antimicrobial, antioxidant, antihelminthic, analgesic, anticancer, antimalarial, and anti-inflammatory activities (Fig. 15.1). Various medicinal plants from this family have recently been thoroughly studied to find bioactive components against multiple ailments. The popularly available genera from the Asteraceae family in the eastern Himalayan region are *Ageratum*, *Artemisia*, *Spilanthes*, *Blumea*, *Vernonia*, *Eclipta*, and *Chrysanthemum.* Table 15.1 summarizes the bioactive components and the medicinal uses of the aforementioned plants.

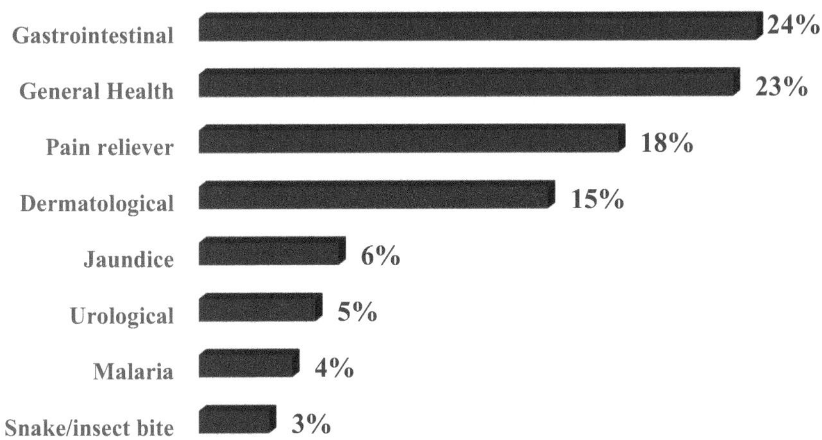

Fig. 15.1 Percentage of Asteraceae family plants used for treating various ailments (Mylliemngap and Arya 2021)

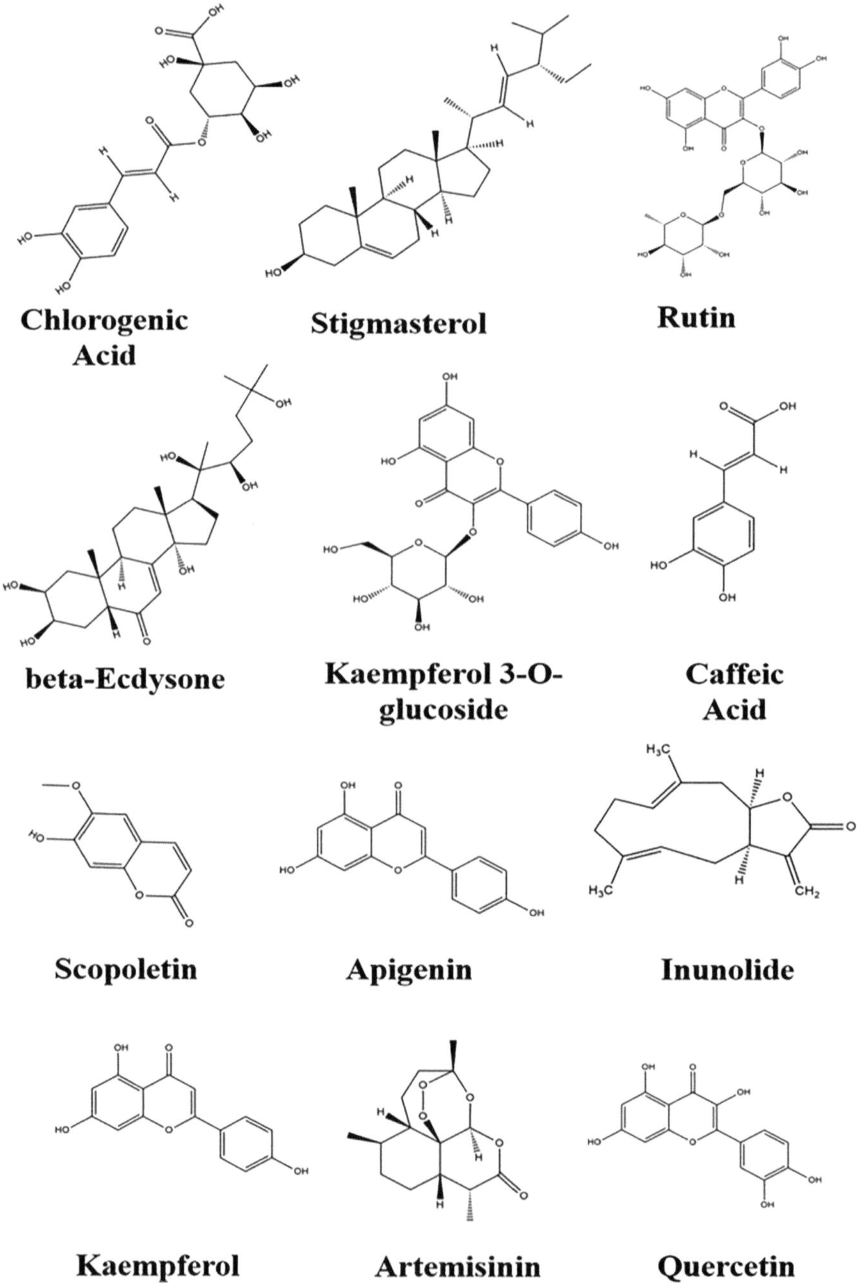

Fig. 15.2 Important bioactive metabolites from different plants of Asteraceae family from eastern Himalaya

Table 15.1 Ethnomedicinal plants belonging to Asteraceae family from Arunachal Pradesh, their important bioactive compounds and associated traditional medicinal applications

Sr. no.	Plant species	Plant part utilized	Extract	Important bioactive constituent	Medicinal uses	References
1.	*Artemisia maritima*	Whole plant	Methanolic	Epigallocatechin gallate, morin, ellagic acid, catechin hydrate, rutin, pyrogallol	Antioxidant, antimicrobial, antidiabetic, cytotoxic	Zaman et al. (2022)
2.	*Artemisia indica*	Aerial part	Aqueous	Artemisia ketone, germacrene B, borneol, chrysanthenyl acetate, *p*-cymene, α-thujone and β-pinene	Anti-tumor, antibacterial, antifungal, antioxidant, analgesic, anti-inflammatory, insecticidal and antiviral	Rashid et al. (2013)
3.	*Artemisia nilagirica*	Leaves	Ethanolic	Artemisinin, ferulic acid, luteolin, caffeic acid, quercetin, apigenin	Antifungal, anticancer, antioxidant	Albaqami et al. (2022)
4.	*Artemisia annua*	NO Data	Ethanol (80%)	Rutin, cynaroside, isorhamnetin, chrysosplenol D and casticin, scopolin and scopoletin, arteannuin B, artemisinin, dihydroartemisinic acid and artemisinic acid	Antimalarial, anticancer, hemorrhoids and bronchitis	Fu et al. (2020)
5.	*Artemisia vulgaris*	Whole plant	Ethanolic	Apigenin, chrysoeriol, diosmetin, eriodictyol, eupafolin, kaempferol, kaempferol-3-glucoside, homoeriodictyol, kaempferol 3-rhamnosid, rutin, vitexin, luteolin, quercetin, tricine	Antimalarial, anti-inflammatory, anti-hypertensive, antioxidant, anti-tumoral, immunomodulatory, hepatoprotective, antispasmodic, antiseptic	Abiri et al. (2018)
6.	*Ageratum conyzoides L*	Ariel parts	Aqueous petroleum ether	Kaempferol, stigmasterol, β-sitosterol, polymethoxy flavones, precocene I and precocene II	Antimicrobial, anti-inflammatory, analgesic, antioxidant, anticancer, antiprotozoal, antidiabetic, spasmolytic, allelopathy	Xu et al. (2023), Kamboj and Saluja (2011) and Yadav et al. (2019)
7.	*Spilanthes acmella Murr.*	Leaf and flower	Methanolic	Spilanthol, dodeca-2(E), 4(E)-dienoic acid isobutylamide and scopoletin	Antimalarial, anti-inflammatory, anaesthetic, anti-toothache	Singh and Chaturvedi (2010, 2012)

(continued)

Table 15.1 (continued)

Sr. no.	Plant species	Plant part utilized	Extract	Important bioactive constituent	Medicinal uses	References
8.	*Spilanthes paniculata*	Leaf and flower	Methanolic	Deca-2E,6Z,8E-trienoic acid isobutylamide (Spilanthol), UDA	Antimalarial, anti-inflammatory, anaesthetic, anti-toothache	Rajendran et al. (2017)
9.	*Blumea balsamifera*	Leaves	Hexane	Borneol L, veridiflorol	Anticancer	Rawati et al. (2023)
		Leaves	Petroleum ether, chloroform, and methanol	Tamarixetin, rhamnetin, luteolin, quercetin, blumeatin	Free radical-scavenging activity	Nessa et al. (2004)
10.	*Blumea lacera*	Whole plant	Methanolic	Linolenic acid, oleic acid, phytol	Antioxidant, cytotoxic, antimicrobial, and antidiarrhea	Ashrafi et al. (2022)
11.	*Vernonia cinerea*	Leaves	Methanol (90%)	Gallic acid, rutin, quercetin, caffeic acid, ferulic acid	Antidiarrhea, stomachache, cough and bronchitis	Rajamurugan et al. (2011)
12.	*Eclipta prostrata*	Whole plant	Methanolic	Ecliptal, eclalbatin, stigmasterol, wedelolactone, eclalbasaponins VII–X, eclalbasaponins I–VI, protocatechuic acid	Hepatoprotective, antimicrobial, antioxidant, anti-inflammatory, antiviral, anti-plasmodial, anti- pyretic, antihepatitis B, cytotoxic, hypotensive, hypo-lipidemic, antivenin, immunomodulatory, and analgesic	Chung et al. (2017)
13.	*Chrysanthemum indicum*	Flower	Methanol (60%)	Chlorogenic acid, 3,4-di-O-caffeoylquinic acid, 3,5-di-O-caffeoylquinic acid, 4,5-di-O-caffeoylquinic acid, linarin, luteolin-7-O-β-D-glucoside	Antibacterial, antiphlogistic, antimutagenic, antioxidant	He et al. (2016)

14.	*Chromolaena odorata* (Syn. *Eupatorium odoratum L.*)	Leaves	Ethanolic	Tamarixetin, trihydroxymonomethoxyflavanone, pentamethoxyflavanone, pentamethoxyflavanone, eupatilin, p-Coumaric acid, p-Hydroxybenzoic acid, protocatechuic acid (3,4-dihydroxyl benzoic acid)	Anti-inflammatory, antioxidant, antibacterial, anthelminthic, antifungal, anticonvulsant, antiprotozoal, cytotoxic, antipyretic, analgesic and antispasmodic	Phan et al. (2001) and Omokhua et al. (2016)
		Leaves	Methanolic	4-*O*-Methylsakuranetin, odoratin, ombuin, rutin, rhamnetin methyl caffeate, isovanillic acid, nonacosan-1-ol, β-sitosterol		Giang et al. (2024)
		Leaves	Aqueous	α-pinene, cadinene, (+)-camphor, limonene, β-caryophyllene, cadinol isomer		Inya-agha et al. (1987)
15.	*Bidens pilosa*	Whole plant	70% ethanol	Rutin, hyperoside, 4,5-O-dicaffeoylquinic acid	Anti-malarial, anti-diabetic, anti-tumoral, antimicrobial, hepatoprotective, antioxidant	Cortés-Rojas et al. (2013)
		Roots	Chloroform	1-phenyl-1,3-diyn-5-en-7-ol-acetate, antimalarial		Brandão et al. (1997)
16.	*Inula racemosa*	Roots	Ethanolic	Alantolactones, isoalantolactone, alantolides, sitosterol, daucosterol, inunolide, dihydroisoalantolactone, d-mannitol	Precordial chest pain, cough and dyspnoea, cardioprotective	Rathore et al. (2022) and Kalachaveedu et al. (2018)
17.	*Gynura bicolor*	Leaves	Ethyl acetate and water	5-p-trans-coumaroylquinic acid, 4-hydroxybenzoic acid, rutin, kampferol-3-*O*-rutinoside, 3,5-dicaffeoylquinic acid, kampferol-3-*O*-glucoside guanosine, chlorogenic acid	Anticancer, anti-hyperglycemic, antioxidant and anti-inflammatory	Teoh et al. (2016)

(continued)

Table 15.1 (continued)

Sr. no.	Plant species	Plant part utilized	Extract	Important bioactive constituent	Medicinal uses	References
18.	*Eupatorium cannabinum L.*	Whole plant	Chloroform and ethanol	β-Ecdysone, eupatorin, eupatilin, quercetin, rutin, caffeic acid	Antitumor, anti-inflammatory, hepatoprotective, immunomodulatory	Grigore et al. (2018)
19.	*Sonchus arvensis*	Leaves	Chloroform	Ascorbic acid, gallic acid, catechin, quercetin, myrecetin, kaempferol	Analgesic, sedative antibiotic, and antipyretic,	Seal (2016)
20.	*Sonchus oleraceus*	Leaves	Methanol (70%)	Caftaric acid, chicoric acid, chlorogenic acid	Antioxidant	Ou et al. (2013)

15.2.2 Geography and Climate of High Altitudes of Eastern Himalaya

The Indian portion of the Eastern Himalayas covers an area of 524,190 sq. km contributing more than 52% of the total eastern Himalayan region (Panda 2022). The high altitudes of eastern Himalaya are in Arunachal Pradesh (83,743 sq. km), Sikkim (7096 sq. km), and Darjeeling district (3149 sq. km) of West Bengal. The region owes a great altitudinal variation from foothills (<100 m) through timberline (4000 m) up to Mt. Khangchendzonga (8548 m). The annual temperature lies between less than 0 and 31 °C with 2000–8000 mm of annual rainfall. The region is thinly populated with lowest population density with 17 people per sq. km. in Arunachal Pradesh region (ISFR 2019).

15.3 Biodiversity Conservation

Eastern Himalaya is one of the significant biodiversity hotspots among 34 globally recognized biodiversity hotspots. A large number of medicinal plants are present in the region due to the unique topography and high degree of precipitation. According to an ethnobotanical survey conducted by Tangjang et al. (2011), medicinal plants have become an integral part of the traditional healthcare system of the indigenous people of Arunachal Pradesh. Over 500 medicinal plants have been identified and used by different tribes in the region for treating various ailments. Asteraceae family plants are highly used among the available medicinal plants. Recently, many of these herbs have become endangered and included in the Red-listed plants in this region. At present, some plants like *Inula racemosa* Hook F. (Rathore et al. 2022), *Spilanthes acmella*, (Sharma and Shahzad 2013), and *Artemisia vulgaris* (Govindaraj et al. 2008) are threatened and critically endangered perennial herbs throughout the Himalayan region. The overexploitation and unskilled harvesting of these plants may lead to their natural habitat depletion and pose an inevitable extinction threat. The conservation of these rich medicinal resources is necessary to stop their disappearance.

The primary objective of conserving these medicinal plants is to maintain the ecosystem stability and sustainable utilization of the phytochemicals for product development. The plant species can be conserved by two means: In situ and ex situ. In situ conservation methods, such as creating national parks, sanctuaries, and biosphere reserves can only preserve natural breeding. Ex situ conservation includes collecting, preserving, and maintaining plants in botanical gardens, gene banks, in vitro banks, seed banks and cryopreservation, keeping their characteristics and genetic information identical to the wild population (Priyanka et al. 2021) (Fig. 15.3). These methods are beneficial in case of heavy loss or sharp decline in the plant species due to habitat destruction or climate change in the region. Ex situ conservations, like gene banks, can provide excellent planting materials and authentic raw drugs. Still, these methods can never be sufficient for the sustainable utilization of medicinal plants. The conservation of these high-value medicinal plants through

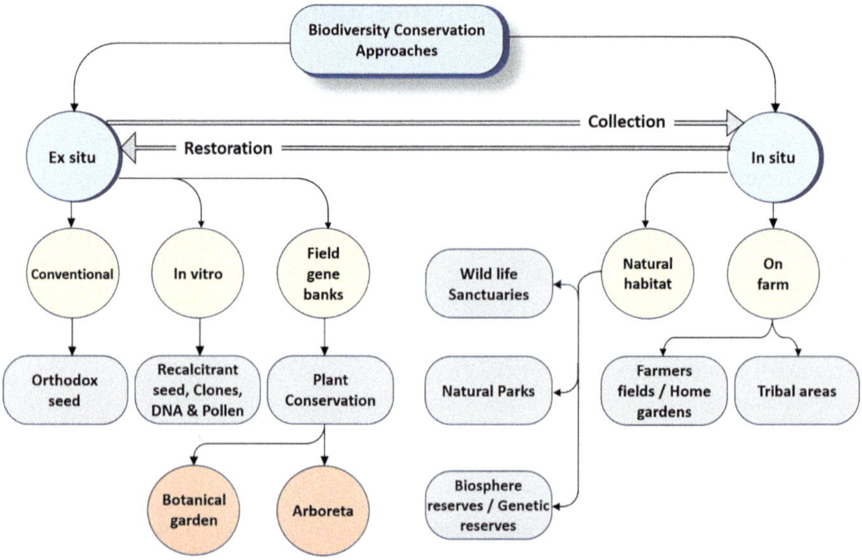

Fig. 15.3 Different methods of biodiversity conservation

modern biotechnological tools is necessary to safeguard them from extinction threats. In vitro conservation strategies pertaining to tissue culture, micropropagation, bioreactor mediated metabolite production, cryopreservation, and synthetic seed technology are most relevant methodologies to enhance the long-term stability of these plants. Micropropagation could be an excellent technique for in situ conservation, as it quickly generates many good-quality plants. This technique can improve the plant population and restore natural habitats after destructive harvesting or reproductive barriers (Krishnan et al. 2011).

15.3.1 In Vitro Methodologies for the Conservation of Plants

Ex situ conservations through in vitro methods are implemented when in situ conservations are insufficient, and sometimes, these are the only methods to save the species. It acts as a backup to in situ conservation for certain threatened plant species that might be lost if not conserved through "off-site conservation." Although these techniques have been explored for a long time, only a few practices, such as seed banks and in vitro methodologies, have been successfully implemented (Li and Pritchard 2009).

Plant biotechnology advancements related to in vitro culture and molecular biology have produced practical tools to improve and assist plant diversity conservation and management. In vitro conservation is essential, particularly for plants that do not produce viable seeds as in case of Banana and *Stevia* (Cruz-Cruz et al. 2013). In vitro methods of ex situ conservation, like micropropagation, slow growth maintenance, bioreactor mediated biomass and metabolite productions, cryopreservation and synthetic seeds, are excellent tools for providing good quality pathogen-free raw material and reducing the burden over wild flora (Kulak et al. 2022).

15.3.1.1 Micropropagation

Micropropagation (syn.: in vitro propagation) is the most frequently used term for aseptically producing true-to-type or clonal plants using a variety of tissues, cells, or cell organs. It includes inoculating small section of tissue or organ (explant) in a closed sterile vessel on a defined artificial medium under controlled physical and chemical conditions. The idea of micropropagation was first visualized by an Austrian botanist Gottlieb Haberlandt (1854–1945) in 1902 by successfully culturing the tomato roots, tobacco and carrot calluses. Unlike most animal cells, the totipotent nature of plant cells provides a potential to generate whole plant from an individual cell or group of cells expressing the entire genome (Loberant and Altman 2010). The controlled conditions include the specific growth medium with or without plant growth regulators (PGRs), photoperiod, temperature, and humidity of the environment. Various PGRs including auxin, cytokinin, and gibberellin are supplemented in the medium to control the growth and morphogenesis of the explants. A small section of meristematic (shoot-tip, apical, axillary or lateral buds) and non-meristematic tissues (roots, internodal section, leaf, petiole) are used as an explant (Deepa and Thomas 2020).

Apart from large-scale clonal propagation, the application of micropropagation has been expanded to several areas, like germplasm conservation, generation of virus free plants, genetically engineered plants, somatic hybridization, somatic embryogenesis, large biomass production through bioreactor technology and drug production through callus and root suspension cultures. Several medicinal plants in the eastern Himalayan region are under rare, endangered or threatened categories. Some of them are *Inula racemosa* (Kaur et al. 2010), *Spilanthes acmella* (Singh and Chaturvedi 2010), *Ageratum conyzoides* L. (Sharma 2008), *Artemisia annua* (Gulati et al. 1996), and *Vernonia cinerea* (Maheshwari and Kumar 2006). Unlike conventional vegetative propagation methods, viz. stem cutting and grafting, micropropagation provides the required number of true-to-type or clonal plants in less time with minimal labor requirements, irrespective of seasonal and regional variations. It is a reliable conservation method for important medicinal plants, and some of the conservation protocols are summarized in Table 15.2.

15.3.1.2 Cryopreservation

The in vitro strategies of ex-situ plant conservation can be used for 1–5 years. However, the cryopreservation method can help to store the plants or their tissues for >5 years. The in vitro generated plant tissues can be used as a starting material for cryopreservation. In this method, the plant tissues are stored at very low temperature (Approx. 165–196 °C) after treating them with a suitable cryoprotectant. Liquid nitrogen or nitrogen vapors are used to maintain these low temperatures for a long time (Kaviani and Kulus 2022). In liquid nitrogen, the metabolic processes like enzymatic activities, respiration etc. are seized and cell division process is also

Table 15.2 Summary of the in vitro conservation methods of Asteraceae plants from Arunachal Pradesh

Sr. no	Plant species	Explant	Medium composition	In vitro methodology	References
1	*Artemisia maritima*	Stem and leaf	MS + 2,4-D (11.31 μM) + BAP (6.66 μM)	Callus induction	Nabi et al. (2022)
		Nodal segment and shoot-tip	MS + BAP (6.66 μM)	Shoot regeneration	
		Green callus from stem	MS + TDZ (4.54 μM) + IBA (12.30 μM)	Somatic embryogenesis	
2	*Artemisia nilagirica*	Nodal segment	MS + IAA (14.27 μM)	Callus induction	Shinde et al. (2016)
		Callus induced from nodal segment	MS + BAP (2.50 μM) + 2iP (7.50 μM)	Adventitious shoot regeneration	
3	*Artemisia annua*	Shoot-tip and nodal explant	MS + BAP (4.44 μM) + IBA (0.49 μM)	Micropropagation	Hailu et al. (2013)
		Leaves	MS + TDZ (2.72 μM) + IBA (0.49 μM)	Somatic embryogenesis	Alejos-Gonzalez et al. (2013)
4	*Artemisia vulgaris*	Nodal explants	MS + 2iP (4.9 μM)	Micropropagation	Sujatha and Kumari (2008)
		Shoot-tip	MS + BAP (4.44 μM)	Micropropagation	Govindaraj et al. (2008)
		Shoot-tip and nodal segment	MS + BAP (4.44 μM) + kinetin (2.32 μM)	Micropropagation	Sujatha and Kumari (2007)
5	*Ageratum conyzoides*	Nodal segment	MS + BAP (13.32 μM) + IAA (17.13 μM)	Micropropagation	Sharma (2008)
6	*Spilanthes acmella*	Nodal segment	MS + BAP (5.0 μM)	Micropropagation	Singh and Chaturvedi (2010)
		Leaf	MS + BAP (10.0 μM) + NAA (1.0 μM)	Callus induction	Pandey and Agrawal (2009)
		Callus	MS + BAP (10 μM)	Shoot generation	

7	*Spilanthes paniculata*	Flowerheads	MS + 2,4-D (1 µM) + NAA (1 µM) + BAP (5 µM)	Callus induction	Rajendran et al. (2017)
8	*Blumea balsamifera*	Shoot tips	MS + BA (4.44 µM)	Micropropagation	Soriano and Cangao (2009)
9	*Blumea lacera*	Nodal segment and leaf	MS + BAP (13.32 µM) + kinetin (4.65 µM)	Micropropagation and adventitious shoot regeneration	Swaraz et al. (2020)
10	*Vernonia cinerea*	Leaf and nodal explants	MS + BAP (8.88 µM) + NAA (8.06 µM)	Shoot regeneration	Seetharam et al. (2007)
11	*Chrysanthemum indicum*	Apical meristems	MS + BAP (4.44 µM) + IAA (0.57 µM)	Micropropagation	Zafarullah et al. (2013)
12	*Chromolaena odorata*	Nodal explants	MS + BAP (2.22 µM) + IAA (0.57 µM)	Micropropagation	Mani et al. (2020)
13	*Bidens pilosa*	Stem and leaf section	MS + BAP (4.44 µM) + 2,4-D (2.04 µM)	Callus induction	Ramabulana et al. (2021)
14	*Inula racemosa*	Seeds	MS + BAP (4.44 µM)	Micropropagation	Kaur et al. (2010)
15	*Gynura bicolor*	Leaf	MS + BAP (2.22 µM) + 2,4-D (9.05 µM)	Micropropagation	Liu et al. (2011)
16	*Sonchus arvensis*	Leaf and petiole	MS + BAP (2.22 µM) + 2,4-D (4.52 µM)	Callus induction	Wahyuni et al. (2020)
17	*Sonchus oleraceus*	Stem	MS + NNA (7.42 µM)	Callus induction	Zhao et al. (2012)

BAP 6-Benzylaminopurine, *2,4-D* 2,4-Dichlorophenoxyacetic acid, *TDZ* Thidiazuron, *IBA* Indole-3-butyric acid, *IAA* 3-Indoleacetic acid, *2iP* N6-(2-Isopentenyl)Adenine, *Kinetin* 6-Furfurylaminopurine, *NAA* 1-Naphthaleneacetic acid, *NNA* Nitro-L-arginine

effectively arrested. These physiological and biochemical changes in the plant tissue under low temperatures make long-term storage possible. Cryoprotectants, such as dimethylsulphoxide (DMSO), sorbitol and mannitol, are used before freezing the tissue. The principal role of cryoprotectants is to avoid any physical and chemical injury during storage. The cryoprotecting agents lower the freezing point by increasing the viscosity of the cytosol. These compounds increase the plasma membrane stability and modulate water distribution inside and outside the cell (Kaviani and Kulus 2022).

Cryopreservation is highly evolved from the conventional two-step cooling to modern one-step methods, such as Desiccation, Pregrowth or Preculture, Pregrowth-Desiccation, Encapsulation-Dehydration, Vitrification, Encapsulation-Vitrification, Droplet-Vitrification, Cryo-Plates and Cryo-Mesh (Deepa and Thomas 2020) (Table 15.3). The development of any cryopreservation protocol is highly dependent on the stress tolerance capacity of the plants. Almost every cryopreservation method follows the six most common steps (Nagel et al. 2023).

1. *Pretreatment*—The pretreatment of in vitro generated tissues is necessary to increase their resistance against low temperatures.
2. *Treatment with cryoprotectant*—Treatment with penetrating (Glycerol, DMSO, ethylene glycol (EG) and propylene glycol (PG)) or non-penetrating cryoprotectants (sorbitol, mannitol, sucrose, dextran) is necessary to enable the cells to withstand freezing temperatures.
3. *Cooling*—Slow or fast cooling is necessary before the tissues are stored.
4. *Storage*—After cooling, tissues must be stored in a liquid nitrogen chamber.
5. *Rewarming*—It is a crucial step before recovery of the plant where the samples are rapidly transferred to a water bath at 35–40 for 2–3 min.
6. *Recovery*—In this step, the solutions or materials have been removed, and the viability of the stored tissues has been evaluated.

Inula racemosa is an important medicinal plant from the Asteraceae family. The plant is critically endangered and widely distributed throughout the eastern Himalayan region of India. Although the plant has been reported to have several medicinal properties, including antispasmodic, hypotensive, anti-inflammatory, digestive, cardiotonic, alexipharmic, aphrodisiac and antiseptic, roots of this plant are primarily used for cough treatment by the indigenous people. The plant has become critically endangered due to unskilled, illegal and unscientific harvesting to fulfil the increasing pharmaceutical and stakeholder demands (Rathore et al. 2022). In another report, Kaur et al. (2010) have developed a strategy to conserve this plant through in vitro propagation and cryopreservation. They obtained a high multiplication in MS medium supplemented by BAP (4.44 μM). The in vitro generated shoot tips were cryopreserved through vitrification, and a 64% survival rate was obtained after cryopreservation.

Two separate studies have also been conducted on the cryopreservation of transformed roots and callus of *Artemisia annua*. According to Teoh et al. (1996), the root tips of *A. annua* were conserved in a cryoprotecting mixture containing 8%

Table 15.3 Modern one-step methods of cryopreservation

Sr. no.	Cryopreservation method	Application procedure	References
1	Desiccation	In this process, tissue sample is dehydrated using sterile air (laminar airflow) and kept in a liquid nitrogen container.	Engelmann (2004)
2	Pregrowth or preculture	In this technique, the samples are cultivated in the presence of cryoprotectants like sugars (fructose, glucose, sucrose) or sugar alcohols (mannitol, sorbitol) and frozen directly into liquid nitrogen.	Engelmann (2004)
3	Pregrowth-desiccation	The rice embryos were precultured in a cryoprotectant (sucrose, sorbitol) and then air dried up to 10% water content. After drying, the samples were plunged into the liquid nitrogen	Zhang et al. (2001)
4	Encapsulation-dehydration	This technique is based on the artificial seed synthesis method. Plant material is encapsulated in alginate beads and precultured in sucrose rich medium. The sample is partially dried in up to 20–30% (fresh weight basis) and then frozen rapidly in liquid nitrogen.	Gonzalez-Arnao and Engelmann (2006)
5	Vitrification	The explant is precultured in a medium with high sucrose and salt concentrations. After preculture, sample is treated with loading solution (LS) followed by the application of PVS[a] (plant vitrification solution) for dehydration of the sample in the next step.	Roque-borda et al. (2021)
6	Encapsulation-vitrification	This method is a combination of encapsulation, dehydration, and vitrification. The samples are first encapsulated using alginate beads, and then further treated LS and PVS	Engelmann (2004)
7	Droplet-vitrification	Initial steps like preculture, LS, and PVS treatment are similar to those in the vitrification process. The cooling and rewarming are different in this method. The explant is covered in a drop of PVS on an aluminium sheet and immersed in liquid nitrogen. During rewarming, the explant is covered in aluminium foil and kept in a sucrose solution at room temperature for 20–30 min.	Roque-borda et al. (2021)
8	Cryo-plates and cryo-mesh	Aluminium cryo-plates containing 10–12 oval shaped tiny wells are used in this method. There are two types of cryo-plates: vitrification cryo-plates (V type) or dehydration cryo-plates (D type). The explant is encapsulated in alginate beads and dehydrated using PVS (V type) or air drying in laminar air flow (D type). The protocol for both methods has similar steps except the dehydration step.	Vujović et al. (2024) and Roque-borda et al. (2021)

[a]*PVS* a mixture of different cryoprotectants at different ratios. PVS1, PVS2, PVS3, PVS4 are available. LS 0.6 M sucrose, 2 M glycerol

(v/v) DMSO and 20% (w/v) sucrose at 25 °C for 1 h, followed by cooling to 4 °C at a rate of 0.09 °C/min then further cooling to −35 °C at 0.72 °C/min. After this cooling, the samples were plunged into the liquid nitrogen. An average of 65% of root tips showed regrowth after rewarming and recovery. In another study conducted by Chenshu et al. (2003), the leaf-disc generated callus of *A. annua* was conserved through cryopreservation. The in vitro generated callus was precultured in MS medium supplemented with 5% (v/v) DMSO. After preculturing a mixture of cryoprotectants containing 15% (v/v) EG, 15% (v/v) DMSO, 30% (v/v) glycerol and 13.6% (w/v) sucrose were applied before plunging the callus into the liquid nitrogen container. A high survival rate of 87% was achieved by rewarming the callus in a water bath at 25 °C and reloading in a 34% (w/v) sucrose solution. These cryopreservation methods of conserving plants like *I. racemose* and *A. annua* are practical techniques with high recovery rates. They can be applied to the other plants of the Asteraceae family and other families as well.

15.3.1.3 Synthetic Seed Technology

The idea of synthetic seed technology was first presented by Murashige (1977) in a symposium on tissue culture at Gent, Belgium. Initially, he proposed the encapsulation of in vitro generated somatic embryos. A successful synthesis of artificial seeds was reported by applying the water-soluble resin and polyoxyethylene glycol over carrot somatic embryos (Kitto and Janick 1982). Later, synthetic seeds of alfalfa were also synthesized by encapsulating the somatic embryos in alginate hydrogel (Redenbaugh et al. 1984). These seeds are produced by encapsulating the in vitro or in vivo generated explants like somatic embryos, calli, nodal segments, adventitious shoots, shoot tips, Protocorm-like bodies (orchids), hairy roots and adventitious roots. This technology is helpful for plants with non-viable seeds or limited seed supply (Gantait et al. 2015). Apart from the explant selection, the encapsulating agent is equally important in synthetic seed technology. Different gels, like agar, gelrite, alginate, carrageenan, carboxy methyl cellulose and sodium pectate, have been used for encapsulation; alginate has proved to be the best encapsulating agent. Sodium alginate is the most frequently used gelling agent due to its low cost and non-toxic nature. The encapsulated seeds are further exposed to $CaCl_2{\cdot}H_2O$ solution to harden the hydrogels (Gantait et al. 2015).

Sudarshana (2013) reported the successful encapsulation of somatic embryos of *Artemisia vulgaris* L. in 2% alginate. In this process, $CaCl_2{\cdot}H_2O$ (75 mM) was used to harden the hydrogel covering over the somatic embryos, and these embryos can be stored for 4 months at 4 °C temperature. These encapsulated embryos were cultured again on MS medium supplemented with GA_3 (4.33 μM) + IAA (2.85 μM) + Ascorbic acid (227.12 μM) with 90% regeneration frequency.

Sometimes, the generation of somatic embryos is not easy to produce. Therefore, other explants, like nodal segments, calli, adventitious shoots, shoot tips, roots, etc., are the best alternatives for encapsulation. In a study conducted by Shahzad et al. (2009), in vitro raised nodal segments of *Spilanthes acmella* were encapsulated into the 4% sodium alginate and hardened in $CaCl_2{\cdot}H_2O$ (100 mM). Encapsulated nodal segments were stored at 4 °C temperature for 1–8 weeks. These nodal segments were regenerated on MS medium with BAP (1.0 μM) and NAA (0.5 μM) with frequency of 87.8 ± 1.15%. The plants were generated from the alginate beads encapsulated nodal segments and successfully hardened with a 90% success rate. In another study, the axillary buds of *S. mauritiana* were encapsulated in 3% (w/v) sodium alginate and stored at 4 °C temperature. Approximately all synseeds retained their viability and showed maximum sprouting on ½ MS medium (Sharma et al. 2009).

15.4 *Spilanthes paniculata* Wall. ex DC.

Spilanthes paniculata wall ex. DC. is an indigenous plant of Arunachal Pradesh belonging to the Asteraceae family. It is a rich source of various bioactive N-alkylamides and is widely used for food and medicinal purposes by indigenous tribes in the region. The flowerheads of this plant are used mainly for immediate toothache relief, therefore also called an anti-toothache plant. Apart from the anti-toothache activity, the plant also possesses several other biological activities like immunomodulatory, bio-insecticidal, larvicidal, anaesthetic, and anthelminthic (Rajendran and Chaturvedi 2017). The spilanthol (Deca-2E,6Z,8E-trienoic acid isobutylamide) and dodeca-2E,4E,8Z,10E/Z-tetraenoic acid isobutylamide are the most essential therapeutic compounds available in hexane extract of flower and roots respectively. These metabolites are primary ingredients in beauty care, anti-ageing, and cosmetic application, and they reduce muscle tension and decrease visible wrinkles (Tiwari et al. 2011).

The rising global interest in *Spilanthes* and its therapeutic and cosmetic products increases the demand for plant raw materials. However, the natural habitats of this plant are diminishing at an alarming rate due to overharvesting and uncontrolled cattle grazing. Although *S. paniculata* is predominantly propagated through seeds, the geographical and seasonal restrictions can not be ignored. Obtaining pathogen free plant is also one of the major limitations of seed cultivation in the natural habitat. To overcome these limitations and fulfil continuously increasing raw material demand, micropropagation is a valuable strategy for large scale production of plants and secondary metabolites with consistent quality and quantity of the desired products.

15.4.1 Axillary Bud Proliferation and Callus Generation in *S. paniculata*

It is the most popular approach of micropropagation, where true-to-type plants are produced with uniform characteristics to the source plant (Fig. 15.4). Researchers have reported the axillary bud proliferation in different species of *Spilanthes*. *S. acmella* is a well-studied plant species in the context of micropropagation and tissue culture studies (Singh and Chaturvedi 2010). Only a few reports are available related to the micropropagation of *S. paniculata*. Mahendran et al. (2006)

Fig. 15.4 Nodal segment culture of Spilanthes paniculata,. (a) Mother plant,. (b) Nodal segments collected from the mother plant. (c) Nodal segment inoculated on MS medium, (d) and. (e) Nodal segment after 1 and 3 weeks of culture. (f) Hardening of the rooted plant

established an efficient protocol for high frequency shoot generation in *S. paniculata* using nodal segment as an explant. In this study, Murashige and Skoog (MS) medium added with BAP (13.32 μM) showed a high number of shoot multiplication. On average, 35 shoots were obtained from one nodal segment inoculated as an explant. The generated shoots showed the best rooting response in half-strength MS medium (only major salts reduced to half) supplemented with IAA (0.57 μM) and NAA (0.54 μM).

Majority of reports related to the tissue culture studies on *S. paniculata* are based on callus generation or callus mediated shoot multiplication. Varsha and Ghulam (2015) reported that leaf, nodal and internodal segments were used for callus generation in MS medium supplemented with different plant growth regulators. The maximum callus proliferation response (84.62%) was obtained from the internodal segment in a medium supplemented with BAP (7.77 μM). In another report by Rajendran and Chaturvedi (2017), callus from the leaf-disc as an explant was generated as a method of conservation and sustainable biomass generation and utilization (Fig. 15.5). The half-strength MS medium supplemented with 5% sucrose, BAP (4.82 μM) and 2,4-D (1.8 μM) was the most suitable medium for biomass and metabolite production. The callus mediated shoot proliferation of *S. paniculata* has also been investigated. The MS medium added with NAA (1 μM) and BAP (5.0 μM) shows shoot differentiation from the leaf-disc generated callus. An average of 6.5 ± 2.6 shoots per explant were obtained from each explant responded. Further increasing the BAP concentration showed a drastic decrease in the shoot proliferation per explant (Pandey et al. 2014).

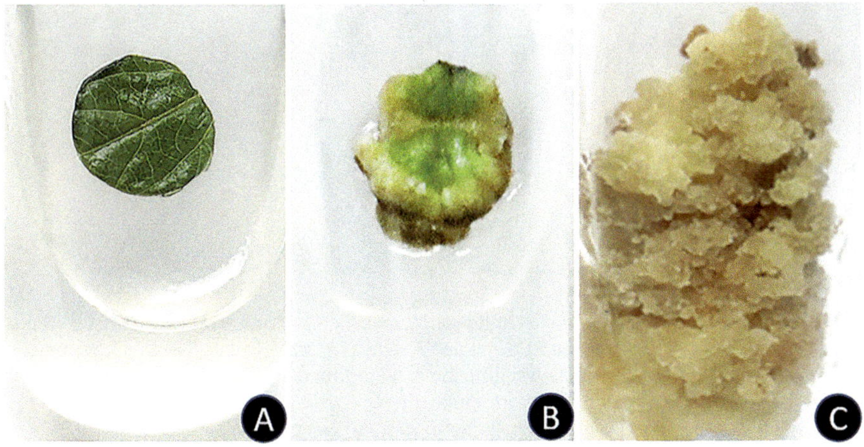

Fig. 15.5 Callus proliferation using leaf-disc as an explant. (a) Leaf-disc inoculated on MS medium supplemented with optimized growth regulators concentration. (b) Leaf-disc culture after 2 weeks of culture. (c) Callus proliferation after 5 weeks of culture

15.4.2 Adventitious Root Proliferation in *S. paniculata*

The pharmaceutical and cosmetic industry requires a large amount of biomass to extract essential metabolites like spilanthol from the different plants of *Spilanthes* sp. These metabolites can be directly used in various formulations or to generate pharmaceutically active derivatives. Regularly increasing applications of these metabolites increase the demand for raw plant materials. Adventitious root culture is an excellent alternative to produce natural plant biomass without posing any excess burden over natural flora. The roots originated from any other part of the plant except the radicle region, called adventitious roots. These roots are developed through complex molecular mechanisms based on exogenous and endogenous physiological factors. These roots are genetically more stable and have high biosynthetic capability. Adventitious roots are generated via direct organogenesis from the explant or indirectly through callus formation. These roots can be induced from the plant's petiole, leaf, internode and nodal segment by providing a suitable growth medium supplemented with or without auxin. Providing the auxin externally into the medium changes the endogenous auxin level of the explant, leading to the adventitious root formation (Khanam et al. 2022).

A study was conducted on adventitious root formation from leaf-disc and nodal segment in our laboratory. The leaf-disc (10 mm dia.) and nodal segments (2–3 cm) were inoculated on an optimized MS medium supplemented with suitable concentrations of auxins, like IAA and IBA (Fig. 15.6). After inoculation, both the explants were kept in a photoperiod of 16/8 h light/dark at 25 °C temperature. After 7–10 days of incubation, the roots were induced from both the explants. After 4 weeks of culture in the semi-solid medium, roots were inoculated in the 250 ml Erlenmeyer flasks containing 100 ml of a similar composition suspension medium. The shake

Fig. 15.6 Figure shows stages of adventitious root induction and proliferation from the leaf-disc tissue culture of *Spilanthes paniculata.* Subfigure A depicts a leaf-disc inoculated on an optimized semi-solid medium. Subfigure B shows inititiation and priliferation of adventitious roots from the leaf-disc explant after 4 weeks of culture. Subfigure C showing adventitious roots isolated and washed from semi-solid culture medium. Subfigure D displays adventitious roots inoculated in suspension medium of the similar composition

flasks were kept on an incubator shaker at 120 rpm, 25 °C and 16/8 h light/dark. After 3 weeks of culture in the liquid medium, the root biomass was harvested and further subjected to the analysis of the bioactive metabolites through chromatographic and spectrometric analysis.

15.4.3 Bioreactor Cultivation of Adventitious Roots of *S. paniculata*

The primary concern in the commercial production of phytochemicals is to convert the laboratory scale production to pilot or commercial scale. The in vitro production of plant biomass in the bioreactors is an attractive tool with better scaling-up possibilities. Unlike the traditional tissue culture system, biomass cultivation in the bioreactors also provides a real time monitoring and control of crucial process parameters like medium pH and temperature, dissolved oxygen and carbon di-oxide concentration inside the reactor vessel (Baque et al. 2012). The adventitious root cultivation has checked in different bioreactor forms, but every configuration has several advantages and disadvantages. The production of these cultures in the bioreactor is a significant challenge due to the difference between the nature of plant and microbial cells. Various factors affecting biomass and secondary metabolites production in bioreactors are impeller design, volumetric mass transfer coefficient (k_La), mixing time, agitation speed, liquid medium rheology and cell density. A suitable impeller design (Marine type, Rushton turbine, helical-ribbon type) is necessary in a stirred tank reactor for proper nutrient mixing and gaseous exchange. Modifications to the glass vessel, sparger and impeller design can be made to enhance the overall plant biomass productivity. Various explants such as

Fig. 15.7 Bioreactor cultivation of adventitious roots of Spilanthes paniculata. (a) A 3 litre capacity photobioreactor setup with controlling unit. (**b**) **A suspension root culture in bioreactor after 1-week of inoculation.** (**c**) Harvested adventitious root biomass after 3 weeks of inoculation

undifferentiated cells (callus) and differentiated tissues (root, shoot and embryos) can be used for biomass proliferation in bioreactors (Ruffoni et al. 2010).

An in vitro root culture study was performed on *S. paniculata* to produce the medicinally important N-alkylamides viz. (2E,6Z,8E)-N-(2-Methylpropyl) deca-2,6,8-trienamide (Spilanthol) and N-Isobutyl-(2E,4E,8Z,10E/Z)-dodecatetraenamide (DTAI). The roots were induced from the leaf-disc and nodal segment of the plant and cultivated into stirred tank bioreactor (STR) configurations (3 L capacity). Aeration, agitation, and inoculum size were optimized for bioreactor cultivation. Total biomass and metabolite production was higher in the bioreactor than in shake-flask cultures. The production of N-alkylamides was also observed as growth associated through growth kinetics studies in the bioreactor. The study results showed approximately 10× biomass growth, and this process can be scaled up further to a pilot plant scale or commercial scale (Fig. 15.7). Cell suspension culture in a bioreactor will pave the way for scaling up the biomass and metabolite production of medicinal plants in the future.

15.5 Conclusions

The indigenous communities of the eastern Himalayan region of India are majorly dependent over medicinal plants to cure most of their health issues. Asteraceae is one of the most widely used plant family for medicinal purposes in this region. Traditional practitioners and villagers use these species to treat cancer, snake/insect bites, malaria, kidney, jaundice, dermatological, pain relief, and gastrointestinal issues. These plants reportedly contain numerous bioactive metabolites like artemisinin, quercetin, rutin, spilanthol, apigenin etc. Although most of the plants of this family have been reported for at least one or more medicinal properties, *Artemisia*, *Spilanthes*, *Eclipta* etc. are a few important and frequently used genera. Climate change, alien species invasion, and extensive use of these plants for food and medicinal purposes destroy their natural habitats and may lead to extinction. At present, some Asteraceae herbs like *Inula racemosa* Hook F. (Rathore et al. 2022), *Spilanthes acmella*, (Sharma and Shahzad 2013), *Artemisia vulgaris* (Govindaraj et al. 2008) come under threatened and critically endangered categories throughout the Himalayan region. In vitro conservation of these plants through micropropagation (nodal segment and leaf-disc culture), bioreactor cultivation, cryopreservation, and synthetic seeds technology are excellent strategies to preserve the species.

The culture medium, plant growth regulators (PGRs), explant, and physical conditions are essential parameters that must be optimized for in vitro propagation. Bioreactor cultivation of plant biomass is also an emerging technology where large-scale production of plant biomass can be performed to meet food and pharmaceutical requirements. Finding the optimum bioreactor configuration for cultivating roots, embryos, calli, etc. is a significant challenge in this technology. The primary advantages of in vitro conservation strategies are genetic stability and an extended storage period. The explants can be stored for 3–5 years through micropropagation and more than 5 years through cryopreservation using liquid nitrogen. Synthetic seed technology is also another vital method of plant conservation. In this strategy, the somatic embryos, nodal segments, and calli are encapsulated through alginate as an explant. The current chapter includes detailed information on the medicinal aspects of Asteraceae plants from the eastern Himalayan region and their in vitro conservation methodologies. However, continuous and deepened research is required to conserve these plants of high medicinal importance. The information given in this book chapter here will help the researchers conserve medicinal plants from the Asteraceae and other plant families.

Acknowledgement We acknowledge. Ms Chetna Sharma for her kind assistance during the preparation of phytochemical structures.

References

Abiri R, Silva ALM, de Mesquita LSS, de Mesquita JWC, Atabaki N, de Almeida EB, Shaharuddin NA, Malik S (2018) Towards a better understanding of *Artemisia vulgaris*: botany, phytochemistry, pharmacological and biotechnological potential. Food Res Int 109:403–415. https://doi.org/10.1016/j.foodres.2018.03.072

Abraham J, Thomas TD (2016) Recent advances in Asteraceae tissue culture. In: Anis M, Ahmad N (eds) Plant tissue culture: propagation, conservation and crop improvement. Springer, Singapore, pp 161–195

Albaqami JJ, Benny TP, Hamdi H, Altemimi AB, Kuttithodi AM, Job JT, Sasidharan A, Narayanankutty A (2022) Phytochemical composition and in vitro antioxidant, anti-inflammatory, anticancer, and enzyme-inhibitory activities of *Artemisia nilagirica* (C.B. Clarke) pamp. Molecules 27(20):1–12. https://doi.org/10.3390/molecules27207119

Alejos-Gonzalez F, Perkins K, Winston MI, Xie D-Y (2013) Efficient somatic embryogenesis and organogenesis of self-pollination *Artemisia annua* Progeny and artemisinin formation in regenerated plants. Am J Plant Sci 04(11):2206–2217. https://doi.org/10.4236/ajps.2013.411274

Ashrafi S, Alam S, Islam A, Emon NU, Islam QS, Ahsan M (2022) Chemico-biological profiling of *Blumea lacera* (Burm. f.) DC. (Family: Asteraceae) provides new insights as a potential source of antioxidant, cytotoxic, antimicrobial, and antidiarrheal agents. Evidence-Based Complement Altern Med 2022:1–14. https://doi.org/10.1155/2022/2293415

Banik B, Das S, Das MK (2020) Medicinal plants with potent anti-inflammatory and anti-arthritic properties found in eastern parts of the Himalaya: an ethnomedicinal review. Pharmacogn Rev 14(28):121–137. https://doi.org/10.5530/phrev.2020.14.16

Baque MA, Moh SH, Lee EJ, Zhong JJ, Paek KY (2012) Production of biomass and useful compounds from adventitious roots of high-value added medicinal plants using bioreactor. Biotechnol Adv 30(6):1255–1267. https://doi.org/10.1016/j.biotechadv.2011.11.004

Bauer A, Brönstrup M (2014) Industrial natural product chemistry for drug discovery and development. Nat Prod Rep 31(1):35–60. https://doi.org/10.1039/c3np70058e

Brandão MGL, Krettli AU, Soares LSR, Nery CGC, Marinuzzi HC (1997) Antimalarial activity of extracts and fractions from *Bidens pilosa* and other *Bidens* species (Asteraceae) correlated with the presence of acetylene and flavonoid compounds. J Ethnopharmacol 57(2):131–138. https://doi.org/10.1016/S0378-8741(97)00060-3

Chakraborty T, Saha S, Bisht NS (2017) First report on the ethnopharmacological uses of medicinal plants by Monpa tribe from the Zemith ang region of Arunachal Pradesh, Eastern Himalayas, India. Plants 6(1):A1–A3. https://doi.org/10.3390/plants6010013

Chenshu A, Wang X, Yuan X, Zhao B, Wang Y (2003) Optimization of cryopreservation of *Artemisia annua* L. callus. Biotechnol Lett 25(1):35–38. https://doi.org/10.1023/A:1021717828907

Chung IM, Rajakumar G, Lee JH, Kim SH, Thiruvengadam M (2017) Ethnopharmacological uses, phytochemistry, biological activities, and biotechnological applications of *Eclipta prostrata*. Appl Microbiol Biotechnol 101(13):5247–5257. https://doi.org/10.1007/s00253-017-8363-9

Cortés-Rojas DF, Chagas-Paula DA, Da Costa FB, Souza CRF, Oliveira WP (2013) Bioactive compounds in *Bidens pilosa* L. populations: a key step in the standardization of phytopharmaceutical preparations. Rev Bras 23(1):28–35. https://doi.org/10.1590/S0102-695X2012005000100

Cruz-Cruz CA, González-Arnao MT, Engelmann F (2013) Biotechnology and conservation of plant biodiversity. Resources 2(2):73–95. https://doi.org/10.3390/resources2020073

Deepa AV, Thomas TD (2020) In vitro strategies for the conservation of Indian medicinal climbers. In Vitro Cell Dev Biol Plant 56(6):784–802. https://doi.org/10.1007/s11627-020-10084-x

Engelmann F (2004) Plant cryopreservation: progress and prospects. In Vitro Cell Dev Biol Plant 40(5):427–433. https://doi.org/10.1079/IVP2004541

Fu C, Yu P, Wang M, Qiu F (2020) Phytochemical analysis and geographic assessment of flavonoids, coumarins and sesquiterpenes in *Artemisia annua* L. based on HPLC-DAD quantification and LC-ESI-QTOF-MS/MS confirmation. Food Chem 312:1–10. https://doi.org/10.1016/j.foodchem.2019.126070

Gantait S, Kundu S, Ali N, Sahu NC (2015) Synthetic seed production of medicinal plants: a review on influence of explants, encapsulation agent and matrix. Acta Physiol Plant 37(5):1–12. https://doi.org/10.1007/s11738-015-1847-2

Giang PM, Huong DTV, Thao VM, Thuy TTT, Trang VM (2024) Flavonoids from the leaves of *Chromolaena odorata* and their α-glucosidase inhibitory activity. Pharm Chem J 57:1621–1626

Gonzalez-Arnao MT, Engelmann F (2006) Cryopreservation of plant germplasm using the encapsulation-dehydration technique: review and case study on sugarcane. Cryo-Letters 27(3):155–168

Govindaraj S, Kumari BDR, Cioni PL, Flamini G (2008) Mass propagation and essential oil analysis of *Artemisia vulgaris*. J Biosci Bioeng 105(3):176–183. https://doi.org/10.1263/jbb.105.176

Grigore A, Neagu G, Dobre N, Albulescu A, Ionita L, Ionita C, Albulescu R (2018) Evaluation of antiproliferative and protective effects of *Eupatorium cannabinum* L. extracts. Turk J Biol 42(4):334–344. https://doi.org/10.3906/biy-1803-72

Gulati A, Bharel S, Jain SK, Abdin MZ, Srivastava PS (1996) In vitro micropropagation and flowering in *Artemisia annua*. J Plant Biochem Biotechnol 5(1):31–35. https://doi.org/10.1007/BF03262976

Hailu T, Abera B, Mariam EG (2013) In vitro mass propagation of Artemisia (*Artemisia annua* L.) cv: Anamed. Plant Tissue Cult Biotechnol 23(2):165–176. https://doi.org/10.3329/ptcb.v23i2.17518

He J, Wu X, Kuang Y, Wang T, Bi K, Li Q (2016) Quality assessment of *Chrysanthemum indicum* Flower by simultaneous quantification of six major ingredients using a single reference standard combined with HPLC fingerprint analysis. Asian J Pharm Sci 11(2):265–272. https://doi.org/10.1016/j.ajps.2015.08.010

Inya-agha SI, Oguntimein BO, Sofowora A, Benjamin TV (1987) Phytochemical and antibacterial studies on the essential oil of *Eupatorium odoratum*. Int J Crude Drug Res 25(1):49–52. https://doi.org/10.3109/13880208709060911

ISFR (2019) India State of Forest Report 2019 by Forest Survey of India. Ministry of Environment and Forests & Climate Change, Government of India, Dehradun, India. Available online https://fsi.nic.in/forest-report-2019. Accessed on 20 Jan 2024

Jeelani SM, Rather GA, Sharma A, Lattoo SK (2018) In perspective: potential medicinal plant resources of Kashmir Himalayas, their domestication and cultivation for commercial exploitation. J Appl Res Med Aromat Plants 8:10–25. https://doi.org/10.1016/j.jarmap.2017.11.001

Jeyaprakash K, Lego YJ, Payum T, Rathinavel S, Jayakumar K (2017) Diversity of medicinal plants used by Adi community in and around area of D'Ering Wildlife. World Sci News 65:135–159

Kala CP (2005) Ethnomedicinal botany of the Apatani in the Eastern Himalayan region of India. J Ethnobiol Ethnomed 1:1–8. https://doi.org/10.1186/1746-4269-1-11

Kalachaveedu M, Raghavan D, Telapolu S, Kuruvilla S, Kedike B (2018) Phytoestrogenic effect of *Inula racemosa* Hook f—a cardioprotective root drug in traditional medicine. J Ethnopharmacol 210:408–416. https://doi.org/10.1016/j.jep.2017.09.001

Kamboj A, Saluja AK (2011) Isolation of stigmasterol and γ-sitosterol from petroleum ether extract of woody stem of *Abelmoschus manihot*. Int J Pharm Pharm Sci 3:94–96

Kandel P, Chettri N, Chaudhary RP, Badola HK, Gaira KS, Wangchuk S, Bidha N, Uprety Y, Sharma E (2019) Plant diversity of the Kangchenjunga Landscape, Eastern Himalayas. Plant Divers 41(3):153–165. https://doi.org/10.1016/j.pld.2019.04.006

Kaur R, Kashyap A, Majeed S, Chauhan NS, Bhardwaj SB (2010) In vitro propagation and conservation of *Inula racemosa* Hook. F. an endangered medicinal plant of temperate origin. J Adv Lab Res Biol 1(1):67–70

Kaviani B, Kulus D (2022) Cryopreservation of endangered ornamental plants and fruit crops from tropical and subtropical regions. Biology (Basel) 11(6):1–23. https://doi.org/10.3390/biology11060847

Khanam MN, Anis M, Bin JS, Mottaghipisheh J, Csupor D (2022) Adventitious root culture—an alternative strategy for secondary metabolite production: a review. Agronomy 12(5):1–18. https://doi.org/10.3390/agronomy12051178

Kitto SL, Janick J (1982) Polyox as an artificial seed coat for asexual embryos. Hortic Sci 17(3):488–488

Krishnan PN, Decruse SW, Radha RK (2011) Conservation of medicinal plants of Western Ghats, India and its sustainable utilization through in vitro technology. In Vitro Cell Dev Biol Plant 47(1):110–122. https://doi.org/10.1007/s11627-011-9344-9

Kulak V, Longboat S, Brunet ND, Shukla M, Saxena P (2022) In vitro technology in plant conservation: relevance to biocultural diversity. Plants 11(4):1–20. https://doi.org/10.3390/plants11040503

Li DZ, Pritchard HW (2009) The science and economics of ex situ plant conservation. Trends Plant Sci 14(11):614–621. https://doi.org/10.1016/j.tplants.2009.09.005

Liu LS, Li R, Zhao Y, Wen CL, Ren S, Guo YD (2011) High efficiency regeneration and genetic stability analysis of somatic clones of *Gynura bicolor* DC. Afr J Biotechnol 10(51):10380–10386. https://doi.org/10.5897/ajb10.1790

Loberant B, Altman A (2010) Micropropagation of plants. In: Encyclopedia of industrial biotechnology: bioprocess, bioseparation, and cell technology. Wiley, pp 329–346

Mahendran G, Lavanya Devi K, Narmatha Bai V (2006) Micropropagation of *Spilanthes paniculata* wall ex. DC. Plant Cell Biotechnol Mol Biol 7(1–2):85–88

Maheshwari P, Kumar A (2006) Organogenesis, shoot regeneration, and flowering response of *Vernonia cinerea* to different auxin/cytokinin combinations. In Vitro Cell Dev Biol Plant 42(6):589–595. https://doi.org/10.1079/IVP2006825

Mani M, Kannan N, Priyadarshini S, Shekhawat MS (2020) Effects of growth regulators and gelling agents on ex vitro rooting of Raspberry. In: Proceedings of the national conference on innovations in biological sciences (NCIBS), pp 78–97. https://doi.org/10.2139/ssrn.3554240

Mitra S, Mitra S, Mukherjee SK (2017) Asteraceae of India: its diversity and phytogeographical affinity. CABI, pp 36–70. https://doi.org/10.1079/9781780646947.0036

Murashige T (1977) Plant cell and organ cultures as horticultural practices. In: Acta horticulturae. International Society for Horticultural Science (ISHS), Leuven, pp 17–30

Mylliemngap W, Arya OP (2021) pharmacological potential of ethnomedicinal plants of Asteraceae family from Arunachal Pradesh Northeast India. In: Singh B (ed) Plants for novel drug molecules ethnobotany to ethnopharmacology. New India Publishing Agency, New Delhi, pp 141–168

Nabi N, Saffeullah P, Singh S (2022) Micropropagation using direct and indirect organogenesis in *Artemisia maritima* L.: scanning electron microscopy of somatic embryos and genome size analysis by flow cytometry. In Vitro Cell Dev Biol Plant 58(6):1012–1024. https://doi.org/10.1007/s11627-022-10291-8

Nagel M, Pence V, Ballesteros D, Lambardi M, Popova E, Panis B (2023) Plant cryopreservation: principles, applications, and challenges of banking plant diversity at ultralow temperatures. Annu Rev Plant Biol 16(8):7.1–7.28. https://doi.org/10.1097/00010694-198711000-00013

Namsa ND, Tag H, Mandal M, Kalita P, Das AK (2009) An ethnobotanical study of traditional anti-inflammatory plants used by the Lohit community of Arunachal Pradesh, India. J Ethnopharmacol 125(2):234–245. https://doi.org/10.1016/j.jep.2009.07.004

Nessa F, Ismail Z, Mohamed N, Haris MRHM (2004) Free radical-scavenging activity of organic extracts and of pure flavonoids of *Blumea balsamifera* DC leaves. Food Chem 88(2):243–252. https://doi.org/10.1016/j.foodchem.2004.01.041

Omokhua AG, McGaw LJ, Finnie JF, Van Staden J (2016) *Chromolaena odorata* (L.) R.M. King & H. Rob. (Asteraceae) in sub-Saharan Africa: a synthesis and review of its medicinal potential. J Ethnopharmacol 183:112–122. https://doi.org/10.1016/j.jep.2015.04.057

Ou ZQ, Schmierer DM, Rades T, Larsen L, McDowell A (2013) Application of an online post-column derivatization HPLC-DPPH assay to detect compounds responsible for antioxidant activity in *Sonchus oleraceus* L. leaf extracts. J Pharm Pharmacol 65(2):271–279. https://doi.org/10.1111/j.2042-7158.2012.01591.x

Panda RM (2022) Ecology of Eastern Himalaya. In: Panda RM (ed) Plant ecology of Indian Himalaya. Springer, Cham, pp 55–84

Panda SK, Luyten W (2018) Antiparasitic activity in Asteraceae with special attention to ethnobotanical use by the tribes of Odisha, India. Parasite 25(10):1–25. https://doi.org/10.1051/parasite/2018008

Pandey V, Agrawal V (2009) Efficient micropropagation protocol of *Spilanthes acmella* L. possessing strong antimalarial activity. In Vitro Cell Dev Biol Plant 45(4):491–499. https://doi.org/10.1007/s11627-008-9184-4

Pandey V, Sharma G, Shankar V, Agrawal V (2014) Biodiversity and in vitro conservation of three medicinally important herbs: *Spilanthes acmella* L. var. oleraceae Clarke, *S. calva* L., and *S. paniculata* Wall. ex DC. Int J Geogr Inf Syst 20(3):295–318. https://doi.org/10.1080/10496475.2013.869520

Phan TT, Wang L, See P, Grayer RJ, Chan SY, Lee ST (2001) Phenolic compounds of *Chromolaena adorata* protect cultured skin cells from oxidative damage: implication for cutaneous wound healing. Biol Pharm Bull 24(12):1373–1379. https://doi.org/10.1248/bpb.24.1373

Priyanka V, Kumar R, Dhaliwal I, Kaushik P (2021) Germplasm conservation: instrumental in agricultural biodiversity—a review. Sustainability (Switzerland) 13(12):1–18. https://doi.org/10.3390/su13126743

Rajamurugan R, Selvaganabathy N, Kumaravel S, Ramamurthy C, Sujatha V, Suresh Kumar M, Thirunavukkarasu C (2011) Identification, quantification of bioactive constituents, evaluation of antioxidant and in vivo acute toxicity property from the methanol extract of *Vernonia cinerea* leaf extract. Pharm Biol 49(12):1311–1320. https://doi.org/10.3109/13880209.2011.604334

Rajendran R, Chaturvedi R (2017) Screening and optimizing media constituents for enhanced production of medicinal N-alkylamide Deca-2E,6Z,8E-trienoic acid isobutylamide from dedifferentiated in vitro cell lines of *Spilanthes paniculata*. Biocatal Agric Biotechnol 9:95–102. https://doi.org/10.1016/j.bcab.2016.12.002

Rajendran R, Narashimman BS, Trivedi V, Chaturvedi R (2017) Isolation and quantification of antimalarial N-alkylamides from flower-head derived in vitro callus cultures of *Spilanthes paniculata*. J Biosci Bioeng 124(1):99–107. https://doi.org/10.1016/j.jbiosc.2017.02.001

Ramabulana AT, Steenkamp PA, Madala NE, Dubery IA (2021) Application of plant growth regulators modulates the profile of chlorogenic acids in cultured *Bidens pilosa* cells. Plants 10(3):1–13. https://doi.org/10.3390/plants10030437

Rashid S, Rather MA, Shah WA, Bhat BA (2013) Chemical composition, antimicrobial, cytotoxic and antioxidant activities of the essential oil of *Artemisia indica* Willd. Food Chem 138(1):693–700. https://doi.org/10.1016/j.foodchem.2012.10.102

Rathore S, Raj Y, Debnath P, Kumar M, Kumar R (2022) Ethnopharmacology, phytochemistry, agrotechnology, and conservation of *Inula racemosa* Hook f.—a critically endangered medicinal plant of the western Himalaya. J Ethnopharmacol 283:1–21. https://doi.org/10.1016/j.jep.2021.114613

Rawati S, Ginting B, Maulana I, Yahya M, Saidi N, Hasballah K (2023) Phytochemical, antioxidant, and cytotoxicity screenings of n-hexane extract from *Blumea balsamifera* L leaves. Res J Pharm Technol 16(4):1664–1668

Redenbaugh K, Nichol J, Kossler ME, Paasch B (1984) Encapsulation of somatic embryos for artificial seed production. In Vitro 20(2):256–257

Roque-borda CA, Kulus D, de Souza AV, Kaviani B, Vicente EF (2021) Cryopreservation of agronomic plant germplasm using vitrification-based methods: an overview of selected case studies. Int J Mol Sci 22(11):1–33. https://doi.org/10.3390/ijms22116157

Ruffoni B, Pistelli L, Bertoli A, Pistelli L (2010) Plant cell cultures: bioreactors for industrial production. In: Advances in experimental medicine and biology: bio-farms for nutraceuticals functional food and safety control by biosensors. Springer, pp 203–221

Salmerón-Manzano E, Garrido-Cardenas JA, Manzano-Agugliaro F (2020) Worldwide research trends on medicinal plants. Int J Environ Res Public Health 17(10):1–20. https://doi.org/10.3390/ijerph17103376

Seal T (2016) Quantitative HPLC analysis of phenolic acids, flavonoids and ascorbic acid in four different solvent extracts of two wild edible leaves, *Sonchus arvensis* and Oenanthe linearis of North-Eastern region in India. J Appl Pharm Sci 6(2):157–166. https://doi.org/10.7324/JAPS.2016.60225

Seetharam YN, Rajanna LN, Jyothishwaran G, Aravind B, Sharanabasappa G, Mallikharjun PB (2007) In vitro shoot regeneration from leaf and nodal explants of *Vernonia cinerea* (L.) Less. Indian J Biotechnol 6(3):418–420

Shahzad A, Sahai A, Sharma S (2009) Artificial seeds for propagation and preservation of *Spilanthes acmella* (L.) Murr. a pesticidal plant species Thidiazuron: from urea derivative to plant growth regulator view project artificial seeds for propagation and preservation of *Spilanthes acmella*. Int J Plant Dev Biol 3(1):62–65

Sharma L (2008) Micropropagation of *Ageratum conyzoides* L.—an important medicinal plant. J Phytol Res 21(1):47–52

Sharma S, Shahzad A (2013) Efficient micropropagation of *Spilanthes acmella* (L.) Murr.: a threatened medicinal herb. Br Biotechnol J 3(3):405–415. https://doi.org/10.9734/bbj/2013/3556

Sharma S, Shahzad A, Jan N, Sahai A (2009) In vitro studies on shoot regeneration through various explants and alginate-encapsulated nodal segments of *Spilanthes mauritiana* DC., an endangered medicinal herb. Int J Plant Dev Biol 3(1):56–61

Shinde S, Sebastian JK, Jain JR, Hanamanthagouda MS, Murthy HN (2016) Efficient in vitro propagation of *Artemisia nilagirica* var. nilagirica (Indian wormwood) and assessment of genetic fidelity of micropropagated plants. Physiol Mol Biol Plants 22(4):595–603. https://doi.org/10.1007/s12298-016-0379-6

Singh M, Chaturvedi R (2010) Improved clonal propagation of *Spilanthes acmella* Murr. for production of scopoletin. Plant Cell Tissue Organ Cult 103(2):243–253. https://doi.org/10.1007/s11240-010-9774-9

Singh M, Chaturvedi R (2012) Screening and quantification of an antiseptic alkylamide, spilanthol from in vitro cell and tissue cultures of *Spilanthes acmella* Murr. Ind Crop Prod 36:321–328

Soriano TL, Cangao EC (2009) In vitro regeneration of sambong (*Blumea balsamifera* Linn.). Philipp Agric Sci 92(1):19–24

Sudarshana MS (2013) In vitro regeneration of multiple shoots from encapsulated somatic embryos of *Artemisia vulgaris* L. IOSR J Pharm Biol Sci 6(6):11–15. https://doi.org/10.9790/3008-0661115

Sujatha G, Kumari BDR (2007) Effect of phytohormones on micropropagation of *Artemisia vulgaris* L. Acta Physiol Plant 29(3):189–195. https://doi.org/10.1007/s11738-006-0023-0

Sujatha G, Kumari BDR (2008) Micropropagation, encapsulation and growth of *Artemisia vulgaris* node explants for germplasm preservation. S Afr J Bot 74(1):93–100. https://doi.org/10.1016/j.sajb.2007.09.002

Swaraz AM, Khan Sumi S, Sultana F, Hasan M, Islam MM, Bari MW, Islam MA, Satter MA, Ahmed KS, Hossain MH (2020) Bioactive compound and bioactivity fidelitous micropropagation method of *Blumea lacera* (Burm. f.) DC.: a large scale production potential. Ind Crop Prod 151:1–17. https://doi.org/10.1016/j.indcrop.2020.112370

Tamang S, Singh A, Bussmann RW, Shukla V, Nautiyal MC (2023) Ethno-medicinal plants of tribal people: a case study in Pakyong subdivision of East Sikkim, India. Acta Ecol Sin 43(1):34–46. https://doi.org/10.1016/j.chnaes.2021.08.013

Tangjang S, Namsa ND, Aran C, Litin A (2011) An ethnobotanical survey of medicinal plants in the Eastern Himalayan zone of Arunachal Pradesh, India. J Ethnopharmacol 134(1):18–25. https://doi.org/10.1016/j.jep.2010.11.053

Teoh KH, Weathers PJ, Cheetham RD, Walcerz DB (1996) Cryopreservation of transformed (hairy) roots of *Atemisia annua*. Cryobiology 33(1):106–117. https://doi.org/10.1006/cryo.1996.0011

Teoh WY, Tan HP, Ling SK, Abdul Wahab N, Sim KS (2016) Phytochemical investigation of *Gynura bicolor* leaves and cytotoxicity evaluation of the chemical constituents against HCT 116 cells. Nat Prod Res 30(4):448–451. https://doi.org/10.1080/14786419.2015.1017726

Tiwari KL, Jadhav SK, Joshi V (2011) An updated review on medicinal herb genus *Spilanthes*. J Chin Integr Med 9(11):1170–1178. https://doi.org/10.3736/jcim20111103

Varsha NN, Ghulam MY (2015) Callus culture of *Spilanthes paniculata* (DC) Jansen—an aromatic medicinal plant. Int J Sci Res (IJSR) 4(4):3013–3015

Vujović T, Anđelić T, Marković Z, Gajdošová A, Hunková J (2024) Cryopreservation of highbush blueberry, strawberry, and saskatoon using V and D cryo-plate methods and monitoring of multiplication ability of regenerated shoots. In Vitr Cell Dev Biol Plant 60(1):85–97. https://doi.org/10.1007/s11627-023-10399-5

Wahyuni DK, Purnobasuki H, Kuncoro EP, Ekasari W (2020) Callus induction of *Sonchus arvensis* l. and its antiplasmodial activity. Afr J Infect Dis 14(1):1–7. https://doi.org/10.21010/ajid.v14i1.1

World Health Organization (2002) The world health report 2002: reducing risks, promoting healthy life. World Health Organization

Xu Z, Ji R, Zha X, Zhao H, Zhou S (2023) The aqueous extracts of *Ageratum conyzoides* inhibit inflammation by suppressing NLRP3 inflammasome activation. J Ethnopharmacol 309:1–9. https://doi.org/10.1016/j.jep.2023.116353

Yadav N, Ganie SA, Singh B, Chhillar AK, Yadav SS (2019) Phytochemical constituents and ethnopharmacological properties of *Ageratum conyzoides* L. Phyther Res 33(9):2163–2178. https://doi.org/10.1002/ptr.6405

Zafarullah A, Ilyas S, Naz S, Aslam F, Manzoor F (2013) Effect of culture media and growth regulators on in vitro propagation of *Crysanthemum indicum* L. Pak J Sci 65(4):462–466. https://doi.org/10.1007/978-94-011-3176-6_33

Zaman S, Barkatulllah, Zahoor M, Wadood Ali Shah S, Ullah Z, Ullah R, Alotaibi A (2022) Pharmacognostic evaluation of *Artemisia maritima* L. a highly medicinal specie of genus *Artemisia*. Saudi J Biol Sci 29(10):1–9. https://doi.org/10.1016/j.sjbs.2022.103419

Zhang YX, Wang JH, Bian HW, Zhu MY (2001) Pregrowth-desiccation: a simple and efficient procedure for the cryopreservation of rice (*Oryza sativa* L.) embryogenic suspension cells. Cryo-Letters 22(4):221–228

Zhao M, Zhao Y, Xing F, Bai L (2012) Components from callus cultures of *Sonchus oleraceus* L. and their cytotoxicity in vitro. In: Zhao M, Zhao Y, Xing F, Bai L (eds) Information technology and agricultural engineering. Springer, pp 107–112

Propagation and Conservation of *Taxus baccata* in Indian Himalaya 16

Mohammad Nadeem, Lok Man Singh Palni, and Fahad Al-Qurainy

Abstract

The Himalayan yew, or *Taxus baccata* L. subsp. *wallichiana*, is a native gymnosperm of the Indian Himalayas that is endangered. This species, which is well known for producing the vital anticancer drug paclitaxel (Taxol®), is severely threatened by overexploitation, poor seed germination, and slow growth rates. Although paclitaxel can be synthesized using alternative methods and derived from a variety of *Taxus* species, *T. baccata* is under extreme environmental pressure due to the high demand for natural resources.

With an emphasis on tissue culture, vegetative propagation, and seed germination, this study investigates several approaches for *T. baccata* conservation and propagation. Techniques like mechanical scarification, hormonal treatments, and stratification can help reduce seed dormancy, a significant barrier to regeneration. A practical method for mass multiplication is vegetative propagation through cuttings, which is facilitated by auxins and seasonal optimization. Commercial paclitaxel production has been made possible by tissue culture methods, such as callus and cell suspension cultures. Recent developments in methyl jasmonate-based elicitation have greatly increased paclitaxel yields, indicating the possibility of sustainable production.

Understanding *T. baccata's* growth patterns, reproduction, and resource allocation through phenological research is essential for maximizing propagation

M. Nadeem (✉) · F. Al-Qurainy
Department of Botany and Microbiology, College of Science, King Saud University, Riyadh, Saudi Arabia
e-mail: mnadeem@KSU.EDU.SA

L. M. S. Palni
G. B. Pant National Institute of Himalayan Environment, Almora, Uttarakhand, India

Nainital, Uttarakhand, India

S. Chandra, H. Lata (eds.), *High Altitude Medicinal Plants*,
https://doi.org/10.1007/978-3-031-99406-7_16

efforts and comprehending its ecological adaptations. To protect this species and guarantee its long-term use in ecological and pharmaceutical applications, integrated conservation strategies that combine conventional propagation with contemporary biotechnological techniques are essential. To reduce the threats to *T. baccata* and maintain its ecological and economic significance, this study emphasizes the urgent need for conservation action.

16.1 Introduction

Taxus baccata L. subsp. *wallichiana* (Zucc.) Pilger (the only *Taxus* species in India), an evergreen gymnosperm (Family-Taxaceae, English-Common Yew, Hindi-*Thuner*, *Tallispatra*), also called the "tree of death," has come into prominence as a source of the anticancer drug Taxol® (paclitaxel), a diterpenoid which was first isolated from the stem bark of Pacific Yew, *Taxus brevifolia* Nutt. (Wani et al. 1971). All parts of the plant except aril (a sweet, fleshy cup-like structure surrounding the 'seeds') are known to contain taxol and 10-deacetylbaccatin III, a precursor for taxol synthesis. Extracts of *T. baccata* are a well-known source of the drug "Zarnab," which is prescribed in the ancient Unani system of Indian medicine (Anonymous 1976). Poor seed germination in nature (seeds remain dormant for just under 2 years in the soil after seed fall), slow growth rate, and excessive harvesting of *T. baccata* from the forests all along the Indian Himalaya for the extraction of taxol (Shukla et al. 1994; Nandi et al. 1994; Rikhari et al. 1998) have contributed to the current environmental concern regarding *T. baccata.* The creation of toxoids, one of the main components of any anticancer medication, is still this species' most significant use (Nighat Fatima et al. 2016; Kundu et al. 2017). When the locals realized how important this plant was, they began using it extensively as a source of income, which exacerbated the plant's condition and led to its inclusion as an endangered species in the Red Data Book (Sahai and Sinha 2020).

Taxol has also been isolated from other *Taxus* species including *Taxus cuspidata, Taxus canadensis, Taxus wallichiana* and *T. x media* and also from different plant parts, e.g., pollen, seeds, needles, young stems, woody stems, bark, and roots (Wani et al. 1971; Miller et al. 1981; Witherup et al. 1990; Vidensek et al. 1990; Neto and DiCosmo 1992; Elsohly et al. 1995). Besides the natural plant source, alternative sources of taxol are being examined, and considerable efforts are being put in this direction worldwide. The chemical synthesis of taxol has continued for several years; hemisynthesis by the attachment of the side chain to biogenetic precursor of taxol such as baccatin III or 10-deacetylbaccatin have been reported earlier (Denis et al. 1988; Lavelle et al. 1993). A significant breakthrough had been its total synthesis reported recently (Holton et al. 1994; Nicolaou et al. 1994). The isolation of a fungus *Taxomyces andreanae* producing taxol has also been reported (Stierle et al. 1993). In addition, plant cell culture provides an excellent opportunity to manipulate the production of taxol in sufficient quantities (Roberts and Shuler 1997); for example, production of taxol or taxoids for semisynthesis of its derivative such as docetaxel (Taxotere®) would be highly desirable. Since the latter compound shows better water solubility and, hence, increased bioavailability than taxol, taxotere may prove to be superior to taxol for chemotherapeutical purposes. The production and recovery of the widely used anticancer drug Taxol® (paclitaxel) from culturable vascular stem cells (VSCs) of *Taxus baccata* is currently used as a successful cell line for paclitaxel production. An in situ product recovery (ISPR) technique was employed, which involved combining three commercial macro-porous resin beads (HP-20, XAD7HP and HP-2MG) with batch and semi-continuous cultivations of the *T. baccata* VSCs after adding methyl jasmonate (Me-JA) as an elicitor. The optimal resin combination resulted in 234 ± 23 mg of paclitaxel per kg of fresh-weight cells, indicating a 13-fold improved yield compared to the control (with no resins) in batch cultivation (Santoyo-Garcia et al. 2023).

16.2 *Taxus baccata*

Phenophases such as leafing, flowering, fruiting, and leaf drop provide basic knowledge about the functional rhythms of plants and plant communities. Phenological studies are also important to understand the plant response as affected by competition, e.g., light or pollinators. Knowledge of seasonality and growth of plant parts such as leaves, twigs, seeds, and other reproductive parts is a must to understand the production potential of plants, particularly economically important species. The leaf along with its axillary buds is the basic construction unit in higher plants (Harper and White 1974). The leaf is the chief photosynthetic organ in woody species, and plant/community productivity is mainly determined by its pattern of production, fall, and longevity (Newhouse and Madgwick 1968). This information is also useful to understand the adaptation of plant communities under variable weather conditions and biotic factors (Negi and Singh 1992).

Ralhan et al. (1985) reported that most of the Central Himalayan species are characterized by concentrated leaf drop and simultaneous leafing followed by flowering during summer (the warm and dry period of the year). Formation of reproductive parts just prior to favorable warm and wet monsoon season (mid-June to mid-September) facilitates pollination and fruit formation subsequently. This strategy also staggers the demand of resources (e.g., moisture and nutrients) within the forest ecosystem (Negi and Singh 1992).

In the Himalayan region, some phenological studies of trees have been carried out (Shukla et al. 1994; Ralhan et al. 1985; Rawal et al. 1991; Negi and Singh 1992). However, information on phenological studies on conifers is almost negligible. Therefore, a detailed study of leaf and shoot growth (for *T. baccata*, an economically important Himalayan species) is required in relation to its taxol production potential.

16.3 Seed Germination

Dormancy is an important stage in the life history of a plant during which growth and development are largely affected, and respiration gets greatly reduced. It ensures survival of seeds by delaying mechanism of germination under harsh conditions, which is unsuitable for seedling establishment (Fenner 2012). There are three types of seed dormancy, which are innate, enforced, and induced. The innate dormancy is inability of newly ripened seeds to germinate under optimal conditions, and it may be broken, using approaches like cold stratification (Baskin and Baskin 1995), fluctuations in light (Black and Wareing 1955), and temperature (Thompson and Grime 1983). The most effective and widely used temperature treatment to break the embryo dormancy is chilling of seeds in which seeds are layered in moist sand or soil (a process called stratification). A temperature of 5 °C is mostly applied for chilling treatment, but other temperature conditions and fluctuations in temperature have also been reported to be successful in some species, e.g., *Cyclopia* (Whitehead and Sutcliffe 1995) and *Cercis sliquastrum* L.

Seed dormancy may also be broken by applying mechanical injury to the seed coat, seed scarification (Zhang and Maun 1989), and exposure of seeds to gibberellic acid (Young and Evans 1972). Enforced dormancy occurs when the seeds are simply deprived of required conditions for germination. Induced dormancy of seeds may be found due to unsuitable condition for germination.

Post harvest ripening has also been observed in several plant species, e.g., *Podophyllum hexandrum* (Nautiyal et al. 1987), *Chaenorrhinum minus* (Arnold et al. 1996), etc. The changes that occur during storage of seeds are termed as "after ripening" (Bewley and Black 2013). Such maturation is often achieved either by applying temperature treatment or by chemicals, including natural and synthetic plant growth regulators (PGRs). In addition, a number of other compounds such as respiratory inhibitors (e.g., cyanide and azide) and nitrogenous compounds, e.g., nitrate, thiourea and even ethanol, hypochlorite, and methylene blue, may also be tried out to break the dormancy (Bewley and Black 2013).

Evidence available in the literature indicates that gibberellins play an important role in releasing seed dormancy. This becomes apparent by many reports of breaking the seed dormancy by exogenous application of most commonly used gibberellins like gibberellic acid (GA_3) or GA_4, GA_7 separately or in a mixture. These PGRs have been found to be effective in breaking the dormancy of seeds which otherwise require either chilling (Bewley and Black 2013) or exposure to light in *Potentilla* and *Picrorhiza* (Semwal and Purohit 1980; Semwal et al. 1983). Breaking of seed dormancy in many species may also be achieved using treatment of cytokinin such as kinetin and benzyladenine (BA) along with a chilling requirement (Pinfield and Stobart 1972; Van Staden et al. 1987). Cytokinins may also interact with other germination promoting hormones such as gibberellins (GA_3) and ethylene by increasing the sensitivity of seeds to GAs (Pinfield and Davies 1978; Webb et al. 1973). A recent work with *Arabidopsis thalina* indicates that the seed germination enhancement is most likely due to chilling effect in association with light in contrary to consequence of increase in the GA content (Derkx et al. 1994). It is reported that

GAs do not promote germination of primary dormant seeds, but it occurs as a result of an increased sensitivity to GAs at low temperature. However, in order to establish this fact more studies with other species are still required (Firn 1986; Arnold et al. 1996).

There is general paucity of information on seed germination of *Taxus* spp. (Le Page-Degivry 1970; Le Page-Degivry and Garello 1973), which showed that the seed germination could be achieved in *Taxus* when endogenous abscisic acid (ABA) present in the embryo was leached out into a liquid medium. The excised embryos were able to germinate on the modified MS medium, and seedlings were obtained when intact seeds were leached under running tap water for 7 days (Zhiri et al. 1994). Chang and Yang (1996) also reported in vitro embryo germination of *Taxus* species. Germination of intact seeds of *Taxus mairei* has been obtained following alternative cold and warm stratification and GA treatment (Chien et al. 1998).

16.4 Vegetative Propagation

Plant propagation is the multiplication of plants by both sexual and asexual means. Successful propagation requires knowledge of plant growth and structure. Horticulturists and foresters have been interested in rooting because it is essential for multiplication of many economically important plants. Most ornamental shrubs, commercial greenhouse crops, and numerous fruit and forest tree crops are propagated through cuttings. In addition, propagation by cuttings is the only practical means of preserving unique characteristics of some plants. Adventitious root formation in cuttings is affected by several external and internal factors, hormones being prominent among them. The role of auxins, cytokinins, gibberellins, ethylene, abscisic acid, and polyamines in rooting has been well investigated (Jarvis 1986).

Auxins are well documented for their role in inducing adventitious rooting. Exogenous application of auxins (naturally occurring and synthetic) has been shown to promote rooting in cuttings. The softwood and hardwood cuttings of deodar, silver fir, and spruce showed differential effect of auxins (Shamet and Khosla 1996). Juvenile cuttings are easy to root (Davies Jr 1984), while hardwood cuttings are difficult to root. Ease of adventitious root formation generally declines with increasing age of parent stock, which is an important problem because desirable characteristics are frequently not expressed until the plant has reached maturity (Davies Jr 1984).

Various compounds in combination with plant growth regulators or alone have been used to induce adventitious rooting (Shamet and Khosla 1996; Nandi et al. 1996, 1997). Among them dichlone (fungicide) with auxin had helped in root induction in stem cuttings of *Prunus persica.* Nandi et al. (1996, 1997) reported a systemic fungicide “carbendazim” induced root formation in *Taxus baccata* softwood cutting of current year’s growth at high frequency. Cuttings from male and female plants responded differently to season and auxin treatment. Variability in the rooting ability of cuttings from five clones of *Populus balsamifera* and *Salix planifolia* was reported (Houle and Babeux 1993) and in *Aisandra butyracea* (Tewari and Dhar

1997). Adventitious rooting in leafy cuttings has also been investigated in great detail. Leaf area and the number of leaves attached to stem cuttings play important role in inducing roots as in case of *Shorea leprosula*, cuttings having maximum leaf area could not root properly and cuttings with less leaf area rooted easily (Aminah et al. 1997). This indicates that during rooting, leaf area should be less, which minimizes excessive water loss. Photosynthetic rates and stomatal conductance were positively correlated with rooting of cuttings (Aminah et al. 1997). Non-leafy cuttings of *Irvingia gabonensis* could not be rooted on different planting material and with IBA treatments whereas in *Shorea leprosula* single leaf cuttings treated with IBA rooted better as also in case of *Leucaena leucocephala* single node leafy cuttings rooted better than non-leafy cuttings. There was a clear pattern in rooting ability of cuttings collected sequentially down the main stem. Cuttings taken from apical portion (node five to thirteen) rooted best and all cuttings had a leaf attached to it (Dick et al. 1998).

Seasonal variations play a vital role in rooting success. Some plants root throughout the year, others root during a particular season (Nanda et al. 1967; Nanda 1970, 1975). Palanisamy (Palanisamy and Pramod Kumar 1996, 1997). Generally spring or season when the fresh flush starts is considered to be the best period for rooting. Rooting in Poplar was found to be profuse if cuttings were made from March to July (Nanda 1970). However, no rooting was observed in winter (Nanda 1970). In case of *Ficus pumila*, seasonal changes influenced rooting (Davies Jr 1984). In *Prosopis chilensis*, highest percent rooting occurred during dry season, i.e., corresponding to maximum growth period (Arce and Balboa 1991). In *Quercus leucotrichophora*, maximum root regeneration was in rainy season (Bhardwaj et al. 1996; Sushma Tamta et al. 2000) and in *Azadirachta indica*, the month of February was found to be the best for rooting (Palanisamy and Pramod Kumar 1996, 1997). This indicates that season is an important factor to be considered carefully for the collection of cuttings for rooting success.

16.5 Tissue Culture

As early as in the 1950s, tissue culture of *Taxus* spp. and subsequently reports on *Taxus* callus proliferation appeared in the 1970s (Rohr 1973; David 1974, 1977; David and Plastira 1976). Earlier studies reported callus cultures that were initiated from mature stem explants with an aim to study the requirement of plant growth regulators and minerals in culture medium for improved callus proliferation; the accumulation of tanins in the culture medium was also observed. Christen et al. (1989) were the first to report the production of paclitaxel by *Taxus* cell cultures, and since then several reports have appeared (Jaziri et al. 1996). Three cell lines producing paclitaxel from *T. brevifolia* were selected by Ketchum et al. (1995) who have assessed the effect of various ingredients of B5 medium (Gamborg et al. 1968) plant growth regulators, vitamins, and sugars. Consequently, they developed a new medium for improved growth with a callus-doubling rate of 3.5–5.6 days. Srinivasan et al. (1995) have studied the kinetics of biomass accumulation and paclitaxel production in cell suspension cultures of *T. baccata* and observed that paclitaxel accumulated at a rate of 1.5 mg/l during the second phase of growth. The same amount of paclitaxel was recovered from cell suspension culture of *T. brevifolia* in a medium supplemented with 6% fructose after 10 days of incubation (Kim et al. 1995). It was reported that inclusion of carbohydrates (sucrose and fructose) in the middle part of active growth increased the rate of paclitaxel accumulation up to 14.8 mg/l (Ketchum and Gibson 1996; Mirjalili and Linden 1995). At present, reports of taxol content of 10–22 mg/l are quite common from academic laboratories (Roberts and Shuler 1997; Ketchum and Gibson 1996).

Elicitation is generally used to induce the expression of genes associated with enzymes responsible for the synthesis of secondary metabolites. It has been demonstrated that jasmonic acid and its methyl esters are signal transducers in a wide range of plant cell cultures (Gundlach et al. 1992). In the past few years, jasmonic acid and methyl jasmonate have been shown to be inexpensive elicitors of secondary metabolite production in many other systems, including *Taxus* (Roberts and Shuler 1997). Addition of 100 μM methyl jasmonate resulted in marked increase in taxol production (110 mg/l) in 2 weeks in *T. x media* cultures (Yukimune et al. 1996), this being the highest productivity reported till date. However, the highest concentration of taxol reported was 153 mg/l. Taxol production increased 19-fold when *T. cuspidata* cell cultures were exposed to 5-ppm ethylene and 10.0 μM methyl jasmonate (Mirjalili and Linden 1995). This type of induction effect points to the need to consider interactions of multiple signals to the cell. These studies have led to commercial production of taxol in the United States.

16.5.1 Micropropagation

Taxus is a very slow growing tree and "seeds" possess a long dormancy period of about 1–2 years (Steinfield 1992). Nevertheless an early germination of zygotic embryos of *Taxus baccata* was possible if embryos were leached in liquid culture medium (Le Page-Degivry 1970). Germination studies carried out with immature embryos of *T. brevifolia* and *T. x media* indicated that about 30% of precociously germinated embryos were converted into seedlings (Flores and Sgrignoli 1991); similar results were also obtained in other species of *Taxus* (Chee 1994). Germination rate was markedly improved and nearly 100% success achieved when excised mature embryos were cultured on modified MS, or Heller's (1953) medium following leaching of intact seeds under running tap water for 7 days (Zhiri et al. 1994). These results are significant in terms of obtaining enhanced seed germination.

Shoot buds were induced from mature zygotic embryos of *T. brevifolia* when cultured on ½ B5 medium supplemented with 10.0 μM BA within 14 days. The buds further elongated upon transfer to woody plant (WP) medium containing 1% activated charcoal; these shoots were then rooted on the same medium containing ABT (thiamine + NAA + IBA) rooting powder. About 58% of shoots formed roots, and plantlets with good growth were obtained within 3 months. It has been claimed that the method may also be suitable for *T. baccata* and *T. cuspidata* (Chee 1996).

16.5.2 Somatic Embryogenesis

Somatic embryogenesis has been reported in a few species of *Taxus*. Calli were reported to form when immature embryos of *T. baccata, T. brevifolia,* and *T. cuspidata* were cultured on a suitable medium (Gupta and Durzan 1987) supplemented with Kn, BA and 2,4-D (Ewald et al. 1995). Following repeated subculture of callus on the same medium supplemented with Kn and IBA, globular structures were produced (Ewald et al. 1995). More recently Chee (1996) reported somatic embryogenesis in *T. brevifolia* using immature zygotic embryos as explants. Culture of immature zygotic embryos on WP medium supplemented with a high concentration of 2-4,D (160.0 μM), 5.0 μM NAA, and 5.0 μM BA resulted in the formation of embryogenic callus within 4 weeks. Later putative embryoids were formed following transfer to a medium supplemented with 4.0 μM BA, 5.0 μM Kn, and 1.0 μM NAA in 6–8 weeks. These somatic embryos were successfully converted into plantlets (5%) after desiccation on WP medium containing 40 μM ABA and 1% activated charcoal.

All these studies indicate that plantlet production is possible on a laboratory scale; however, transfer of these plants to the field has not been reported.

16.6 Strategies for Conservation

16.6.1 Conservation In Situ

The goal of *T. baccata* in situ conservation is to maintain the species in its native environment in order to safeguard ecological relationships and genetic diversity.

- Protected Areas: National parks, nature reserves, and protected forests are home to numerous *T. baccata* populations. These places offer chances for natural regeneration with little interference from humans. To stop additional habitat fragmentation, management measures such as invasive species control, reforestation, and habitat restoration are essential.
- Methods of Forest Management: Due to its shade tolerance, yew growth may be adversely affected by excessive canopy opening, which can be avoided with selective logging techniques and reduced-impact logging.

16.6.2 Conservation In Ex Situ

A complementary strategy is ex situ conservation, which uses clone repositories, botanical gardens, and seed banks to further secure *T. baccata* genetic material.

- Seed Banks: These establishments keep *T. baccata* seeds in regulated environments while tracking their viability over time. In the event that natural populations decline, these collections enable restoration efforts and aid in genetic preservation.

- Clonal repositories and botanical gardens: *T. baccata* genotypes are protected by living collections in botanical gardens, which serve as resources for genetic research and reintroductions. Yews are preserved in botanical gardens as living genetic archives that also raise awareness of conservation and education.

16.7 Future Directions

Targeted genomic studies on *Taxus baccata* can help conservation efforts by identifying genotypes that are best suited to specific climate conditions. Research on disease resistance genetic markers will aid in the development of resilient yew populations, ensuring long-term survival. Improving micropropagation and tissue culture methods can make large-scale conservation more cost-effective and efficient. Somatic embryogenesis is a technique that allows for rapid clonal replication, making yew propagation more practical for restoration projects. Increasing legal protections and raising awareness of *T. baccata's* ecological significance are critical. Educational programs in schools, botanical gardens, and conservation centers can increase public awareness and support for yew conservation.

16.8 Conclusion

To protect *Taxus baccata* from increasing environmental pressures, an integrated approach of efficient propagation, conservation, and sustainable management is required. Future research should prioritize genetic resilience, responsible harvesting, and stricter policy enforcement to ensure the species' long-term survival.

References

Aminah H, Dick JM, Grace J (1997) Rooting of *Shorea leprosula* stem cuttings decreases with increasing leaf area. For Ecol Manag 91:247–254

Anonymous (1976) C.S.I.R. Publications and Information Directorate. Wealth of India, vol X, 591 p

Arce P, Balboa O (1991) Seasonality in rooting of *Prosopis chilensis* cuttings and in-vitro micropropagation. For Ecol Manag 40:163–173

Arnold RM, Arnold JA, Arnold TH (1996) Germination of *Chaenorrhinum minus* seeds in response to gibberellin treatments. J Plant Physiol 148:677–683

Baskin CC, Baskin JM (1995) Warm plus cold stratification requirement for dormancy break in seeds of the woodland herb *Cardamine concatenata* (Brassicaceae), and evolutionary implications. Can J Bot 73:608–612

Bewley JD, Black M (2013) Seeds: physiology of development and germination. Springer

Bhardwaj D, Mishra V, Shamet G (1996) Rooting response of *Quercus leucotrichophora* Linn. cuttings to chemical treatments and physiochemical status. J Tree Sci 15:49–51

Black M, Wareing P (1955) Growth studies in woody species VII. Photoperiodic control of germination in *Betula pubescens* Ehrh. Physiol Plant 8:300–316

Chang S, Yang J (1996) Enhancement of plant formation from embryo cultures of *Taxus mairei* using suitable culture medium and PVP. Bot Bull Acad Sin 37:35–40

Chee PP (1994) In vitro culture of zygotic embryos of Taxus species. HortScience 29:695–697

Chee PP (1996) Plant regeneration from somatic embryos of *Taxus brevifolia*. Plant Cell Rep 16:184–187

Chien C-T, Kuo-Huang L-L, Lin T-P (1998) Changes in ultrastructure and abscisic acid level, and response to applied gibberellins in *Taxus mairei* seeds treated with warm and cold stratification. Ann Bot 81:41–47

Christen A, Bland J, Gibson D (1989) Cell cultures as a means to produce taxol. Proc Am Assoc Cancer Res 30:566

David A (1974) Histophysiologie vegetale. Realisation de cultures de tissue d'If (Taxus baccata L.) a partir de fragments de pousses ages d'un an; obtention d'une culture indefinie. CR Acad Sci Paris 279:1757–1759

David A (1977) La culture des tissus de deux gymnospermes Pinus pinaster Sol. et *Taxus baccata* L. In: La Culture des Tissus et des Cellules des V6g6taux. Masson, Paris, p 9

David A, Plastira V (1976) Comportement En Culture In Vitro De Cellules Isolees De Deux Gymnospermes: *Taxus Baccata* L. Et Pinus Pinaster Sol. CR Acad Sci Paris 110:5917–5919

Davies F Jr (1984) Shoot RNA, cambrial activity and indolebutyric acid effectivity in seasonal rooting of juvenile and mature *Ficus pumila* cuttings. Physiol Plant 62:571–575

Denis JN, Greene AE, Guenard D, Gueritte-Voegelein F, Mangatal L, Potier P (1988) Highly efficient, practical approach to natural taxol. J Am Chem Soc 110:5917–5919

Derkx M, Vermeer E, Karssen C (1994) Gibberellins in seeds of *Arabidopsis thaliana*: biological activities, identification and effects of light and chilling on endogenous levels. Plant Growth Regul 15:223–234

Dick J, Magingo F, Smith R, Mcbeath C (1998) Rooting ability of *Leucaena leucocephala* stem cuttings. Agrofor Syst 42:149–157

Elsohly HN, Croom EM Jr, Kopycki WJ, Joshi AS, Elsohly MA, Mcchesney JD (1995) Concentrations of taxol and related taxanes in the needles of different Taxus cultivars. Phytochem Anal 6:149–156

Ewald D, Weckwerth W, Naujoks G, Zocher R (1995) Formation of embryo-like structures in tissue cultures of different yew species. J Plant Physiol 147:139–143

Fenner M (2012) Seed ecology. Springer

Firn RD (1986) Growth substance sensitivity: the need for clearer ideas, precise terms and purposeful experiments. Physiol Plant 67:267–272

Flores HE, Sgrignoli PJ (1991) In vitro culture and precocious germination of Taxus embryos. In Vitro Cell Dev Biol Plant 27:139–142

Gamborg OLC, Miller RA, Ojima K (1968) Nutrient requirements of suspension cultures of soybean root cells. Exp Cell Res 50:151–158

Gundlach H, Müller MJ, Kutchan TM, Zenk MH (1992) Jasmonic acid is a signal transducer in elicitor-induced plant cell cultures. Proc Natl Acad Sci 89:2389–2393

Gupta PK, Durzan DJ (1987) Biotechnology of somatic polyembryogenesis and plantlet regeneration in loblolly pine. Bio/Technology 5:147–151

Harper J, White J (1974) The demography of plants. Annu Rev Ecol Syst 5:419–463

Heller R (1953) Recherches sur la nutrition minerales des tissues vegetaux cultives, in vitro. Ann Sci Nat Bot Biol Veg 14:223

Holton RA, Kim HB, Somoza C, Liang F, Biediger RJ, Boatman PD, Shindo M, Smith CC, Kim S (1994) First total synthesis of taxol. 2. Completion of the C and D rings. J Am Chem Soc 116:1599–1600

Houle G, Babeux P (1993) Temporal variations in the rooting ability of cuttings of *Populus balsamifera* and *Salix planifolia* from natural clones–populations of subarctic Quebec. Can J For Res 23:2603–2608

Jarvis B (1986) Endogenous control of adventitious rooting in non-woody cuttings. In: New root formation in plants and cuttings. Springer

Jaziri M, Zhiri A, Guo Y-W, Dupont J-P, Shimomura K, Hamada H, Vanhaelen M, Homès J (1996) Taxus sp. cell, tissue and organ cultures as alternative sources for taxoids production: a literature survey. Plant Cell Tissue Organ Cult 46:59–75

Ketchum R, Gibson D (1996) Paclitaxel production in suspension cell cultures of Taxus. Plant Cell Tissue Organ Cult 46:9–16

Ketchum R, Gibson D, Gallo LG (1995) Media optimization for maximum biomass production in cell cultures of pacific yew. Plant Cell Tissue Organ Cult 42:185–193

Kim J-H, Yun J-H, Hwang Y-S, Byun SY, Kim D-I (1995) Production of taxol and related taxanes in *Taxus brevifolia* cell cultures: effect of sugar. Biotechnol Lett 17:101–106

Kundu S, Jha S, Ghosh B (2017) Metabolic engineering for improving production of taxol. Springer, Cham

Lavelle F, Gueritte-Voegelein F, Guenard D (1993) Le taxotère: des aiguilles d'if à la clinique. Bull Cancer 80:326–338

Le Page-Degivry MT (1970) Acide abscissique et dormance chez les embryons de *Taxus baccata* L. CR Acad Sci Paris 271:482–484

Le Page-Degivry MT, Garello G (1973) La dormance embryonnaire chez *Taxus baccata*: Influence de la composition du milieu liquide sur I'induction de la germination Par. Physiol Plant 29:204–207

Nautiyal MC, Rawat AS, Bhadula SK, Purohit AN (1987) Seed germination in *Podophyllum hexandrum*. Seed Res 15:206–209

Miller RW, Powell RG, Smith CR Jr, Arnold E, Clardy J (1981) Antileukemic alkaloids from *Taxus wallichiana* Zucc. J Organomet Chem 46:1469–1474

Mirjalili N, Linden JC (1995) Gas phase composition effects on suspension cultures of *Taxus cuspidata*. Biotechnol Bioeng 48:123–132

Nanda K (1970) Investigations on the use of auxins in vegetable reproduction of forest plants, Final report of PL 480. Research Project A 7FS-11 (FG In 255), pp 1–215

Nanda K (1975) Physiology of adventitious root formation. Indian J Plant Physiol 18:80–89

Nanda K, Purohit AN, Bala A, Anad VK (1967) Seasonal rooting response of stem cuttings of some forest tree species to auxins. Indian Forester 94:154–162

Nandi S, Pandey A, Palni L (1994) Cancer cure from Himalayan yew. Cent Himalayan Environ Assoc Bull 7:1–7

Nandi S, Rikhari H, Nadeem M, Palni L (1997) Clonal propagation of *Taxus baccata* L.—a Himalayan asset under threat. Physiol Mol Biol Plants 3:15–24

Nandi SK, Palni LMS, Rikkari H (1996) Chemical induction of adventitious root formation in *Taxus baccata* cuttings. Plant Growth Regul 19:117–122

Negi G, Singh S (1992) Leaf growth pattern in evergreen and deciduous species of the Central Himalaya, India. Int J Biometeorol 36:233–242

Neto AGF, Dicosmo F (1992) Distribution and amounts of taxol in different shoot parts of *Taxus cuspidata*. Planta Med 58:464–466

Newhouse M, Madgwick H (1968) Comparative seedling growth of four hardwood species. For Sci 14:27–30

Nicolaou K, Yang Z, Liu J, Ueno H, Nantermet P, Guy R, Claiborne C, Renaud J, Couladouros E, Paulvannan K (1994) Total synthesis of taxol. Nature 367:630–634

Nighat Fatima NF, Kondratyuk T, Park Eunjung PE, Marler L, Muniba Jadoon MJ, Qazi M, Mirza H, Ibrar Khan IK, Naima Atiq NA, Chang Lengchee CL (2016) Endophytic fungi associated with Taxus fuana (West Himalayan Yew) of Pakistan: potential bio-resources for cancer chemopreventive agents. Pharm Biol 54(11):2547–2554

Palanisamy K, Kumar P (1997) Effect of position, size of cuttings and environmental factors on adventitious rooting in neem (*Azadirachta indica* A. Juss). For Ecol Manag 98:277–280

Palanisamy K, Pramod Kumar PK (1996) Seasonal effect on induction of adventitious rooting in stem cuttings of neem (*Azadirachta indica* A. Juss). Indian J For 19(2):183–186

Pinfield N, Davies H (1978) Hormonal changes during after-ripening of *Acer platanoides* L. seeds. Z Pflanzenphysiol 90:171–181

Pinfield N, Stobart A (1972) Hormonal regulation of germination and early seedling development in *Acer pseudoplatanus* (L.). Planta 104:134–145

Ralhan P, Khanna R, Singh S, Singh J (1985) Certain phenological characters of the shrub layer of Kumaun Himalayan forests. Vegetatio 63:113–119

Rawal R, Bankoti N, Samant S, Pangtey Y (1991) Phenology of tree layer species from the timber line around Kumaun in Central Himalaya, India. Vegetatio 93:108–118

Rikhari H, Palni L, Sharma S, Nandi S (1998) Himalayan yew: stand structure, canopy damage, regeneration and conservation strategy. Environ Conserv 25:334–341

Roberts SC, Shuler ML (1997) Large-scale plant cell culture. Curr Opin Biotechnol 8:154–159

Rohr R (1973) Production de cals par les gamétophytes mâles de *Taxus baccata* L. cultivés sur un milieu artificiel. Etude en microscopie photonique et électronique. Caryologia 25:177–189

Sahai P, Sinha VB (2020) Collection of Taxus from Tirthan valley, Himachal Pradesh, India for conservation in non natural environment. Plant Arch 20:2080–2083

Santoyo-Garcia JH, Valdivia-Cabrera M, Ochoa-Villarreal M, Casasola-Zamora S, Ripoll M, Escrich A, Moyano E, Betancor L, Halliday KJ, Loake GJ (2023) Increased paclitaxel recovery from *Taxus baccata* vascular stem cells using novel in situ product recovery approaches. Bioresour Bioprocess 10:68

Semwal J, Purohit A (1980) Germination of Himalayan alpine and temperate Potentilla. Proc Plant Sci 89:61–65

Semwal J, Purohit A, Gaur R (1983) Seed germination in some Himalayan alpine plants. Seed Res 11(1):42–46

Shamet G, Khosla P (1996) Vegetative propagation of deodar (*Cedrus deodara*) in north-west Himalaya. Ann For 4(1):21–24

Shukla G, Rao K, Haridasan K (1994) *Taxus baccata* in Arunachal Pradesh. Arunachal For News 12:1–7

Srinivasan V, Pestchanker L, Moser S, Hirasuna TJ, Taticek RA, Shuler ML (1995) Taxol production in bioreactors: kinetics of biomass accumulation, nutrient uptake, and taxol production by cell suspensions of *Taxus baccata*. Biotechnol Bioeng 47:666–676

Steinfield D (1992) Early lessons from propagating Pacific yew. Rocky Mountain Forest and Range Experiment Station Technical Research Report, p 221

Stierle A, Strobel G, Stierle D (1993) Taxol and taxane production by *Taxomyces andreanae*, an endophytic fungus of Pacific yew. Science 260:214–216

Sushma Tamta ST, Purohit V, Nandi S, Palni L (2000) Chemical induction of root formation in *Quercus leucotrichophora* L. stem cuttings. Indian J For 23(2):135–138

Tewari A, Dhar U (1997) Studies on the vegetative propagation of the Indian butter tree (*Aisandra butyracea* (Roxb.) Baehni). J Hortic Sci 72:11–17

Thompson K, Grime J (1983) A comparative study of germination responses to diurnally-fluctuating temperatures. J Appl Ecol 20:141–156

Van Staden J, Manning J, Dickens C (1987) A phylogenetic analysis of the role of plant hormones in the development and germination of legume seeds. In: Advances in legume systematics. Royal Botanic Gardens, Kew, pp 387–442

Vidensek N, Lim P, Campbell A, Carlson C (1990) Taxol content in bark, wood, root, leaf, twig, and seedling from several Taxus species. J Nat Prod 53:1609–1610

Wani, M. C., Taylor, H. L., Wall, M. E., Coggon, P. And Mcphail, A. T. 1971. Plant antitumour agents VI. The isolation and structure of taxol, a novel antileukemic and antitumour agents from *Taxus brevifolia*. J Am Chem Soc, 93, 2325–2327

Webb D, Staden JV, Wareing P (1973) Seed dormancy in Acer: changes in endogenous cytokinins, gibberellins and germination inhibitors during the breaking of dormancy in *Acer saccharum* Marsh. J Exp Bot 24:105–116

Whitehead C, Sutcliffe M (1995) Effect of low temperatures and different growth regulators on seed germination in Cyclopia spp. J Plant Physiol 147:107–112

Witherup KM, Look SA, Stasko MW, Ghiorzi TJ, Muschik GM, Cragg GM (1990) Taxus spp. needles contain amounts of taxol comparable to the bark of *Taxus brevifolia*: analysis and isolation. J Nat Prod 53:1249–1255

Young JA, Evans RA (1972) Germination and persistence of achenes of Scotch thistle. Weed Sci 20:98–101

Yukimune Y, Tabata H, Higashi Y, Hara Y (1996) Methyl jasmonate-induced overproduction of paclitaxel and baccatin III in Taxus cell suspension cultures. Nat Biotechnol 14:1129–1132

Zhang J, Maun M (1989) Seed dormancy of *Panicum virgatum* L. on the shoreline sand dunes of Lake Erie. Am Midl Nat 122:77–87

Zhiri A, Jaziri M, Homes J, Vanhaelen M, Shimomura K (1994) Factors affecting the in vitro rapid germination of *Taxus embryos* and the evaluation of taxol content in the plantlets. Plant Cell Tissue Organ Cult 39:261–263

Conservation and Cultivation of Medicinal and Aromatic Plants in the High Altitudes of Uttarakhand, India

17

Vijay Kant Purohit, Jaidev Chauhan, Pradeep Dobhal, Sudeep Chandra, Babita Patni, Vijay Laxmi Trivedi, P. C. Nautiyal, and M. C. Nautiyal

Abstract

The Indian Himalayan Region (IHR) is home to a rich variety of medicinal and aromatic plants (MAPs). However, their biodiversity is declining due to overharvesting, rising global demand, low agricultural productivity, migration, habitat loss, and climate change. MAPs are essential to traditional medicine and have great economic value, yet many species, such as *Aconitum heterophyllum, Picrorhiza kurroa*, and *Nardostachys jatamansi*, are now endangered or vulnerable. Despite efforts to cultivate MAPs, only 12–15 hectares of land in Uttarakhand are under cultivation. Cultivation efforts and government initiatives aim to conserve these endemic species and support mountain communities, yet face challenges like technical difficulties, strict regulations, and financial barriers for farmers. The study utilized both primary and secondary data and reported the urgent need to promote MAP cultivation and conservation strategies to ensure sustainable resource management and secure local livelihoods. It identifies priority MAP species for conservation and reports research gaps in understanding their distribution and ecological status. This chapter also points out the lack of information on MAP supply chains, value addition, and adulteration. Addressing these gaps is important for the development of the policies and effective conservation management of Himalayan bioresources.

V. K. Purohit · J. Chauhan · P. Dobhal · S. Chandra (✉) · B. Patni · V. L. Trivedi · M. C. Nautiyal
High Altitude Plant Physiology Research Centre, HNB Garhwal University (A Central University),
Srinagar, Garhwal, Uttarakhand, India

P. C. Nautiyal
Department of Seed Science, HNB Garhwal University, Srinagar, Garhwal, Uttarakhand, India

S. Chandra, H. Lata (eds.), *High Altitude Medicinal Plants*,
https://doi.org/10.1007/978-3-031-99406-7_17

17.1 Introduction

The Indian Himalayan Region (IHR) is a vast mountain area stretching across five regions: Trans-Himalaya, North-West Himalaya, West Himalaya, Central Himalaya, and East Himalaya (Pal and Samant 2005). The IHR covering an area of 70,074.54 km^2 is a repository of biodiversity and is home for a wide variety of unique flora and fauna. Half of all flowering plant species found in India are harboured within the IHR. This region is rich in plant diversity, including edible, sacred, and economically important species such as those yielding oils, medicines, and fragrances. The unique combination of high species richness and endemism in the IHR has made it a focus of extensive scientific research (Bargali et al. 2022).

Uttarakhand, located in the western Himalayan region, is endowed with a great wealth of flora that includes a number of medicinal and aromatic plants, and so is called the herbal state of India. The state shares a border with Tibet and China in the north, Himachal Pradesh in the northwest, Uttar Pradesh to the south, and Nepal to the east. 24% of the topographical area of the state is under alpine zone and is extended between 28° 53′ to 31° 27′ latitudes and 77° 34′ to 81° 02′ longitudes (Pandey et al. 2021). Uttarakhand is home to 1642 medicinal plants belonging to 177 families and 869 genera (Khajuria et al. 2021); nearly 350–400 species are reported from alpine regions of the state (Chandra et al. 2021).

The people and village communities living in far-flung high-altitude regions of Himalayan states, including Uttarakhand, often depend on natural resources, since other sources of livelihood are either scarce or in some cases not available. Remoteness, low agricultural production, uneven terrain, and climatic hazards (irregular precipitation, extreme heat, etc.) further exacerbate their problems, consequently leading to higher migration rate, abandonment of agricultural land in mountain villages, and illicit exploitation of natural wealth for income generation, especially plants of medicinal and aromatic importance, which is ultimately causing the depletion of biodiversity in the high-altitude regions.

In high-altitude regions of Uttarakhand, various MAPs are used in treatments of various ailments (Table 17.1). Due to the presence of active ingredients or secondary metabolites in these plants, they have huge market demands among the pharmaceutical industries. Species like *Aconitum heterophyllum*, *Aconitum balfourii*, *Picrorhiza kurrooa*, *Nardostachys jatamansi*, and *Allium stracheyi* are some of the species in high trade from mountain regions of Uttarakhand. MAPs of the mountain region of western Himalaya are declining and facing various anthropogenic (habitat destruction by over exploitation, grazing, and unregulated tourism) and natural threats, i.e. climate change, genetic drifts, etc. (Chandra et al. 2021).

However, the government and different organizations had introduced cultivation practices for conservation of these endemic species and supplemented livelihood generation of mountain communities and are actively engaged in providing technical assistance, quality planting material, field demonstration, training, and marketing linkages to farmers and other stakeholders for livelihood upliftment. But these efforts are not enough, and there is still requirement of providing better incentives and developing supply chain management, on-site marketing, value addition, etc.

Table 17.1 Some highly traded Himalayan MAPs, their active contents, and uses

Plant	Active contents	Ailment/use	References
Aconitum balfourii	Pseudoaconitine, 8-Omethylveratroylpseudaconine, veratroylbikhaconine, veratroylpseudaconine, indaconitine, ludaconitine, 8-deacetylyunaconitine, bikhaconitine, neoline, and chasmanine	Used in Ayurvedic and Unani medicines. Anti-inflammatory, antinociceptive, hypotensive, bradycardic, analgetic, and cardiotonic activities. Diaphoretic, diuretic, febrifuge, anti-inflammatory, anti-rheumatic, antipyretic, and vermifuge. In large doses, it acts as powerful sedative, narcotic, and poison. Aconitum species possess pharmacological activities such as antifungal, antibacterial, and insecticidal properties	Khetwal (2007), Sharma and Gaur (2012), Ameri (1998) and Anwar et al. (2003)
Aconitum heterophyllum	Aconitine, N-diethyl-N-formyllaconitine, N-succinoylanthranilate, anthorine, 12-secohetisan-2-ol, and atesinol-6-benzoylheterastine	Antidiarrheal, expectorant, diuretic, hepatoprotective, antipyretic and analgesic, antioxidant, alexipharmic, anodyne, anti-atrabilious, anti-flatulent, anti-periodic, anti-phlegmatic, and carminative properties	Ali et al. (2021), Zhou et al. (2015), Verma et al. (2010) and Ukani et al. (1996)
Angelica glauca	Myrcene, a-phellandrene, sabinene, p-cymene and a-pinene. b-phellandrene, (Z)-ligustilide, (Z)-butylidene phthalide	Stimulant, cordial, appetizer, dyspepsia, cardioactive, carminative, expectorant, diaphoretic and also in stomach troubles and urinary disorders, constipation and gastric problems, bilious complaints, rheumatism, bronchitis, menorrhiza, for treating rinderpest	Purohit et al. (2015) and Vashistha et al. (2010)
Coptis teeta	Berberine, coptisine, palmatine, jatrorrhizine, columbamine, phenylpropanoids, epiberberine lignans, flavonoids, phenolic acids, steroids, and saccharides	Fevers, nausea, dysentery, jaundice, flatulence, haemorrhages, visceral obstruction, conjunctivitis, eye troubles, nervous diseases, appetiser, indigestion, constipation, dyspepsia, and cough	Bajpay et al. (2019)
Dactylorhiza hatagirea	Dactylorhin A, B, C, D, and E and dactyloses A-B	Dysentery, chronic fever, diarrhoea, cough, stomach ache, cuts, wounds, burns, and fractures	Wani et al. (2020)

(continued)

Table 17.1 (continued)

Plant	Active contents	Ailment/use	References
Hedychium species	β-Pinene, terpinolene, terpineol, linalyl acetate, cadinene, 1,8-cineole, α-pinene, β-terpinene, camphene, β-myrecene, limonene, ɣ-terpinene, α-pinene oxide, α-Farnesene, β-cubebene, β-cardinene, 2-heptadecanone, and tetradecanal	Halitosis, nausea, diminished appetite, vomiting, hiccups, and inflammation	Singh et al. (2023)
Inula racemosa	Sesquiterpene lactones: alantolactone and isoalantolactone, dihydro-iso alantolactone, dihydroinunolide, isoinunolide, isoallolantolactone, allolantolactone, isoinunal, inunal, isoalantodiene and alantodiene, e B-sitosterol, daucosterol, D-mannitol, aplotaxene, and phenlyacetonitirile	Hypoglycemic, hypocholesterolemic, and hypolipidemic properties, anti-cancer activity	Mohan and Gupta (2017) and Chaturvedi et al. (1995)
Nardostachys species	Jatamansone, α-cadinol, β-gurjunene, maaliol, patchouli alcohol,9-aristolan-12-ol	Nervous, cardiovascular, digestive, epilepsy, mental disorders, insomnia, migraine, heart palpitations, melasma, and hyperactivity	Gottumukkala et al. (2011)
Paris polyphylla	Flavonoid glycosides, sterols, triterpenoid, polysaccharides, polyphyllin I, II, V, VI, and VII, gracillin, L-arabinose, and D-galactose	Snake bites, narcotic effects, insect bites, wounds, food poisoning, and fever	Rawat et al. (2023)
Picrorhiza kurroa	Picroside I, II, d-mannitol, kutkiol, kutki sterol, and apocynin	Immunomodulatory, anti-allergic, anti-anaphylactic and anti-neoplastic activities. Used to treat respiratory diseases, allergies, inflammatory conditions, fever, diarrhoea, chronic scorpion stings, asthma, and liver-related diseases	Upadhyay et al. (2013), Bhandari et al. (2008, 2010) and Dwivedi et al. (1992)

Podophyllum hexandrum	Podophyllin, podophyllotoxin, epipodophyllotoxin, podophyllotoxone, quercetin, podophyllotoxin glucoside, quercetin-3-glycoside 4-demethylpodophyllotoxin, 4- kaempferol, kaempferol-3-glucoside, dimethyl podophyllotoxin glucoside, deoxypodophyllotoxin, isopicropodophyllone, picropodophyllotoxin, 4-methyl deoxypodophyllotoxin, S-peltatin, and α-peltatin	Diabetes, joint pain, jaundice, heart problem, skin disease, wound, cancer and as blood purifier, cough, hepatic stimulant, purgative, diarrhoea, and liver ailment	Anand et al. (2022)
Polygonatum verticillatum Linn.	α-bulnesene, eicosadienoic, linalyl acetate, pentacosane, docasane, piperitone, diosgenin, calarene, and santonin	Cardiotonic, demulcent, antiperiodic, diuretic, aphrodisiac, energizer, sedative, hypoglycemic, pain reliever, general tonic, antitumor, and nervine tonic	Saboon et al. (2016)
Rheum species	Emodin, aloe-emodin, emodin, rhein, chrysophanol, physcion, resveratrol, and piceatannol	Diabetes, heart problem, skin disease, wound, cancer, cough, hepatic stimulant, purgative, diarrhoea, and liver ailment	Tabin et al. (2022)
Saussurea species	Linoleic acid, palmitic acid, doconexent, methyl palmitate, stearic acid, 1-docosanol, gondoic acid, henicosanoic acid, methyl oleate, pentacosanoic acid, dehydrocostus lactone, α-terpineol, γ-curcumene, piperine, bupleuronol	Used against disorders including paralysis, cerebral ischemia, wounds, and cardiac and mental disorders; some people also use it as antiseptic and in healing cuts	Semwal et al. (2020) and Semwal and Painuli (2019)
Selinum vaginatum	Monoterpene hydrocarbons, 6E-farnesol, oxygenated sesquiterpenes 2E, β-pinene, γ-terpinene, oxygenated sesquiterpenes, monoterpenes, and elemol, β-phellandrene, selinum vaginatum, β-phellandrene, and β-phellandrene	Menstrual problems, epilepsy, hysteria, skin disease, dysmenorrhea, asthma, cough, hysteria, hypotensive, sedative, and analgesic	Joshi et al. (2018) and Saraswat et al. (2020)
Trillium govanianum	Diosgenin, govanosides, borassoside, and pennogenin	Used against pain and inflammation. Anti-cancerous activity, also used in GI disorders, sexual disorders, backache, kidney problems and as vermicide	Rahman et al. (2015, 2016) and Chauhan (1999)
Valeriana species	Valerenic acid, acetoxyvaleranone, valernyl acetate, bornyl acetate, valerenol, bornyl acetate, valerenic acid, (Z)-valernyl acetate, and acetoxyvaleranone	Eye troubles, liver, spleen, wounds, kidney ulcers, cardiac debility, asthma, dry cough, and chronic fever	Safaralie et al. (2008)

The rising demands of MAPs in international market require continuous assessment of status, supply, demand, and associated challenges. Documentation of information is important for effective conservation and management strategies. However, data deficit affects the accurate evaluation and decision-making for biodiversity conservation initiatives (Chandra et al. 2021). The compilation of primary and secondary data is therefore a significant effort to help implicate policymaking decisions. This study is compiled using primary data (collected by the authors) and secondary data (published reports and articles) and is important as it:

A. Elaborates the importance of MAPs cultivation and conservation especially with reference to local livelihood and natural resource sustainability
B. Highlights the priority species of MAPs that are under cultivation and are in urgent need of conservation due to their high demand and illicit exploitation
C. Enumerates the type and nature of studies that are required to be undertaken to understand the distribution and status of MAPs in high-altitude regions at ground level

Provides critical insights into the status of high-altitude MAP supply and demand chain, value addition, substitution and adulteration in MAPs, etc.

17.2 Distribution and Status of Medicinal and Aromatic Plants in Uttarakhand

Approximately 1748 species of medicinal and aromatic plants (MAPs) reported from the Indian Himalayan Region (IHR), nearly 40% occur in Uttarakhand (Samant and Pal 2003) of which 28 species are recognized as Globally Significant Medicinal Plants (GSMP) (Ved and Goraya 2008; Kuniyal and Sundriyal 2013). Utilization of MAPs for traditional uses may be sustainable, but commercial collection of MAPs is supposed to be unsustainable (van Odenhoven and de Groot 2011) with a majority of share of the plant species used commercially today being extracted from the wild especially from the sub-alpine and alpine zones of the Himalayan regions. Excluding some innovative attempts and recent facilitation-driven developments, the trade of Himalayan medicinal plants generally relies on harvesting from the natural habitats (Olsen 1998; Kuniyal et al. 2005).

Some of the species, which are in great demand from various pharmaceutical companies, include *Picrorhiza kurrooa, Podophyllum hexandrum, Nardostachys grandiflora, Dactylorhiza hatagirea, Aconitum heterophyllum, and Saussurea costus* (Uniyal et al. 2002). Of the reported MAPs, some species namely *Aconitum balfourii*, *Aconitum heterophyllum*, *Aconitum laeve*, *Allium stracheyi*, *Fritillaria roylei*, *Nardostachys jatamansi*, *Angelica archangelica*, *Angelica glauca*, *Meconopsis aculeata*, *Paris polyphyll*a, *Rheum australe*, *Rhododendron anthopogon*, *Saussurea gossypiphora*, *Saussurea obvallata*, and *Trillium govaninaum* are under different conservation threat categories at global level (IUCN). Also, about 27 species are also under various threat categories at regional level (Ved et al. 2003).

There are only a few studies focusing on the status and distribution of high-altitude MAPs in Uttarakhand. The most extensively studied top ten species in meadows or high-altitude areas in Uttarakhand include *Angelica glauca, Aconitum heterophyllum, Picorrhiza kurroa, Malaxis muscifera, Dactylorhiza hatagirea, Arnebia benthamii, Fritillaria roylei, Aconitum ferox, Aconitum balfourii, and Rheum australe* (Bisht et al. 2006; Bhatt et al. 2014). Although in the recent years most of these species have come under the Rare, Endangered and Threatened (RET) categories, but this was not always the case. Earlier reports suggest that *N. jatamansi* had a high density in Pindari area, *Malaxis* species was reported to have a high density in Dronagiri area, *Aconitum* species in Panwali Kantha, *Podophyllum hexandrum* around Tungnath (Airi et al. 2000) but after the standardization of medicinal plant estimation techniques, population of several high-altitude species started to decline (Rawat et al. 2012).

Uniyal et al. (2002) in a study, estimated population status and biomass availability of 14 threatened MAPs extracted from higher altitudes of Kumaon Himalaya for trade purpose. A stratified random sampling done in the mid 2001 by them in distinct landscapes revealed that the 14 target species were distributed within nine types of habitats, i.e. species like *Rheum* and *Nardostachys* were confined to very high alpine zones (3900 masl), whereas *Dactylorhiza hatagirea, Aconitum heterophyllum, Jurinea macrocephala* and *Arnebia benthamii* had a broad distribution range occurring in even sub-alpine zones (3000 masl). They also found that altitude influences the distribution pattern of MAPs, and other micrographic features viz., water availability, canopy cover, and slope also play a significant role in the distribution of the studied MAPs.

A report by Rawat et al. (2012) provides a detailed account of various priority MAP species and areas for conservation and development distributed in a total of 15 forest divisions and 81 ranges of Uttarakhand. The forest division (FD) of Badrinath (>45 species) followed by Tehri FD (45 species) and Uttarkashi FD (41 species) were found to be the richest in MAP species including the common ones like *Aconitum ferox, A. heterophyllum, Allium humile, A. stracheyi, Angelica glauca, A. archangelica, Dactylorhiza hatagirea, Hedychium spicatum, Jurinea* species, *Lillium polyphyllum, Megacarpea polyandra, Malaxis muscifera, M. acuminata, Meconopsis aculeata, Nardostachys grandiflora, Picrorhiza kurrooa, Podophyllum hexandrum, Polygonatum cirrhifolium, P. verticillatum, Rheum australe, Selinum wallichianum, Taxus wallichiana, Trillium govanianum, Valeriana jatamansi,* etc.

A study carried out by Chandra et al. (2021) to determine the availability, quality, sufficiency, and directionality of past and ongoing research on MAPs in context to the excessive resource utilization, changing climatic conditions, and the increasing developmental and tourism activities suggests that of the 350–400 MAPs that are reported to be found in the high-altitude areas of Uttarakhand, 30–35 species are in high demand for trade and commerce. Their results also reveal that of the 400 Km^2 area of alpine meadows of Uttarakhand, most of the studies on alpine plants are site specific (45%) with abundance of 31 MAPs assessed in a few meadows having only 1–2 individuals/m^2. The study also shed light on the lack of studies involving information of geospatial distribution of MAPs, demand and supply ratio, impacts of climate change and management, and monitoring strategies for MAPs in Uttarakhand (Table 17.2).

Table 17.2 Threat status of important MAPs in Uttarakhand

Scientific name	Family	Local name	Habit	Life cycle	Habitat	Conservation status
Aconitum balfourii Stapf	Ranunculaceae	Meetha	Herb	Perennial	The plant is found in the temperate alpine ravines; 2800 m and 4200 m	V
Aconitum heterophyllum Wall ex Royle	Ranunculaceae	Atees	Herb	Biennial	Sub-alpine and alpine zones; 2000–5000 m	En
Angelica glauca Edgew.	Apiaceae	Choru	Herb	Perennial	Moist places in the mountains and in mountain valleys; 2600–3700 m	En
Arisaema propinquum Schott.	Araceae	Naagphani	Herb	Perennial	Near rocks and moist places; 2500–3600 m	T
Dactylorhiza hatagirea (D. Don)	Orchidaceae	Salampanja, Hatajari	Herb	Perennial	Grassland slopes; 2800–3900 m	En
Hedychium spicatum Sm.	Zingiberaceae	Van haldi	Herb	Perennial	Subtropical areas of Himalayan region; 1000–3000 m	V
Inula racemosa Hook. f.	Asteraceae	Pushkarmool	Herb	Perennial	Temperate alpine Himalaya; 1500–4200 m	CE
Nardostachys jatamasi (D. Don) DC.	Caprifoliaceae	Balchhar, Jatamansi	Herb	Perennial	Alpine meadows, Juniper scrub, Rhododendron forests; 3200–5000 m	CE
Paris polyphylla Sm.	Melanthiaceae	Satuwa	Herb	Perennial	Moist habitat of Rhododendron, Oak, and Laurel forests; 1800–3300 m	V

Picrorhiza kurroa Royle ex Benth	Scrophulariacea	Kutki; Kadwi	Herb	Perennial	Alpine meadows and rocky slopes; 2700–4000 m	En
Podophyllum hexandrum Royle	Berberidaceae	Van Kakri	Herb	Perennial	Open mountain slopes and shady coniferous forests; 2000–3500 m	En
Polygonatum verticillatum Linn.	Asparagaceace	Meda	Herb	Perennial	Temperate forests: 1800–3900 m	V
Rheum australe D. don (syn. R. emodi)	Polygonaceae	Rhubarb; Archa	Herb	Perennial	Steep rocky slopes; 3400–3800 m	En
Rheum moorcroftianum Royle	Polygonaceae	Rhubarb; Dolu	Herb	Perennial	Rocky slopes; 3500–5000 m	T
Saussurea obvallata (DC.) Edgew.	Asteraceae	Brahma Kamal	Herb	Perennial	Near glacial zone, amongst boulders; 3000–5100 m	En
Saussurea simpsoniana (Fielding & Gardner) Lipsch.	Asteraceae	Phen Kamal	Herb	Perennial	Glacial zone; 4300–5000 m	En
Selinum vaginatum (Edgew.) C. B. Clarke.	Apiaceae	Bhutkeshi	Herb	Perennial	Sloping meadow; 1800–3800 m	T
Trillium govanianum Wall. ex D. Don	Melanthiaceae	Nagchatri, Tipatti	Herb	Perennial	Moist and shaded places; 2600–3400 m	En
Valeriana jatamansi Jones. (syn. V. wallichii)	Caprifoliaceae	Mushkbala, Tagar	Herb	Perennial	Temperate forests; 1800–3000 m	CE (FRLHT)

T threatened, *En* endangered, *CE* critically endangered, *V* vulnerable

17.3 Cultivation and Conservation Status of MAPs

MAPs are core ingredients in many traditional medicinal systems in rural areas. They also play an important role in the livelihood and income generation. In the western Himalaya, 30% of the total trade of MAPs alone is operated from alpine and high-altitude regions. Such high demand and low cultivation cause illegal exploitation from wild sources, consequently the biodiversity and habitats are declining. The cultivation of these plants on a larger scale is very important for sustainable management of demand and supply ratio (Tripathi et al. 2019).

The cultivation of medicinal plants holds significant importance for preserving biodiversity and enhancing rural livelihoods in Uttarakhand. Cultivated medicinal plants are expected to provide an alternative supply source for the market, thereby lifting off the pressure on wild populations to some extent. Additionally, the cultivation of MAPs can offer a supplementary income for the state's marginal farmers. This is particularly crucial as 80% of Uttarakhand's working population relies on agriculture; however, half of the families own less than 2 acres land, only 5% have more than 5 acres of land, such small landholdings often lead to low income (Maikhuri et al. 2001).

The government is also working to protect and grow valuable MAPs in the Himalaya. They are implementing programs like, creating gardens, development of nurseries to provide high-quality seeds, and encouraging research on cultivation of rare and important MAPs. They are also protecting natural habitat of these species (Phondani et al. 2016; Maikhuri et al. 2017). Despite these efforts, many valuable MAPs are still at risk of rarity. To address these challenges, comprehensive research and deeper understanding of the medicinal and aromatic plant sector will facilitate the development of effective conservation and cultivation strategies (Negi et al. 2018).

In Uttarakhand, roughly 12–15 hectares of land is under cultivation of high-altitude MAPs. In Garhwal region, Chamoli, Rudraprayag, Tehri Garhwal, and Uttarkashi districts and in Kumaon region, Bageshwar and Pithoragarh districts are engaged in the cultivation of high-altitude MAPs. The district Chamoli is alone cultivating 71% of the total area under MAPs followed by Uttarkashi (10%). Other districts have less than 10% area under MAPs cultivation (Fig. 17.1).

The farmers or stakeholders are mainly engaged in cultivation and trade of MAPs, i.e. *Sassurea costus*, *Picrorhiza kurrooa*, *Aconitum heterophyllum*, *A. balfourii*, *Rheum emodi*, *Swertia chirayita*, *Valleraina jatamansi*, *Bunium persicum, Angelica glauca, Allium stracheyi, Hedychium spicatum and Nardostachys grandiflora.* Species *P. kurrooa* and *S. costus* have highest trade from high-altitude regions of Uttarakhand (Fig. 17.2).

Farmers are utilizing the cultivation packages developed by Nautiyal and Nautiyal (2004), including seed/planting material propagation rate, density/spacing, intercultural practices, and yield production of MAP species of high-altitude regions (Table 17.3). The commercial cultivation of most of the species has been conducted by means of vegetative propagation. The livelihood of marginal farmers engaged in cultivation of these species has improved; however, due to lack of infrastructure,

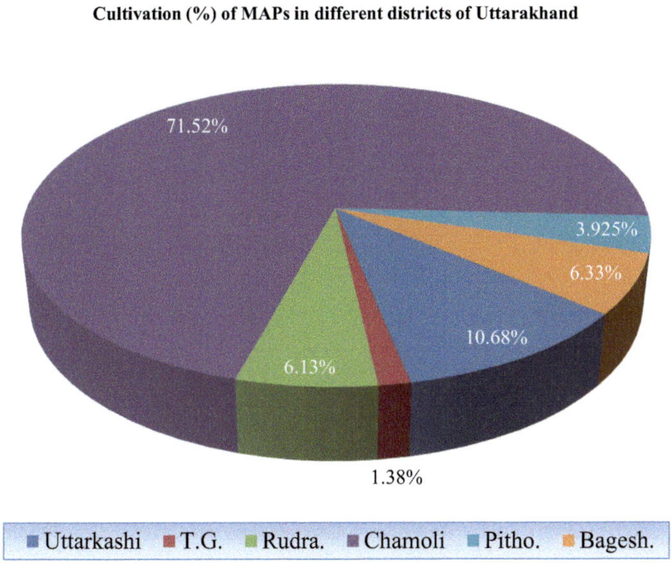

Fig. 17.1 Present cultivation (%) of medicinal and aromatic plants in different districts of Uttarakhand. *Abbreviations: TG* Tehri Garhwal, *Rudra* Rudraprayag, *Pitho* Pithoragarh, *Bagesh* Bageshwar. (Courtesy: Purohit et al. 2021)

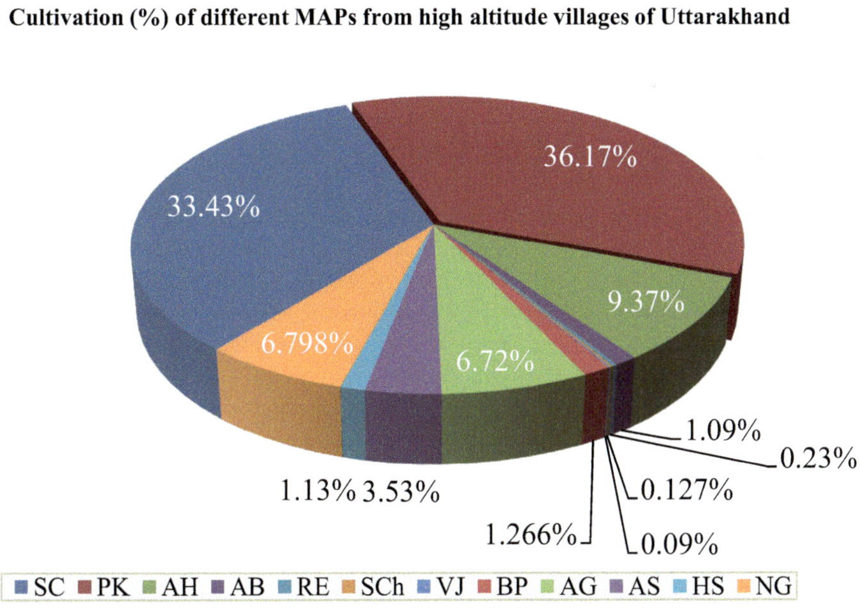

Fig. 17.2 Cultivation % of MAPs in trade from high-altitude regions of western Himalaya. *Abbreviations: SC Saussurea costus, PK Picrorhiza kurrooa, AH Aconitum heterophylum, AB Aconitum baulfourii, RE Rheum emodi, SCh Swertia chirayita, VJ Valleriana jatamansi, BP Bunium persicum, AG Angelica glauca, AS Allium stracheyi, HS Hedychium spicatum, NG Nardostachys grandiflora.* (Courtesy: Purohit et al. 2021)

Table 17.3 Standardized cultivation methods for important high-altitude MAP species

S. N.	Plant species	Seed rate/ hectare	Plant spacing	Seedlings rate/ hectare	Harvesting time	Yield (Kg)/ hectare
1.	*Picrorhiza kurrooa*	1.5 kg	30 × 30 cm	110,000	3 years	650–700
2.	*Saussurea costus*	3.75 kg	30 × 30 cm	37,000	3 years	4000–5000
3.	*Aconitum heterophyllum*	695 gm	30 × 30 cm	110,000	3 years	84–110
4.	*Aconitum balfourii*	695 gm	30 × 30 cm	110,000	3 years	273–340
5.	*Hedychium spicatum*	1000–1500 kg	30 × 45 cm	65,000	3–4 years	10,000
6.	*Nardostachys grandiflora*	650 gm	30 × 30 cm	110,000	3 years	1100–1150
7.	*Angelica glauca*	6.25 kg	45 × 45 cm	80,000	3 years	1000
8.	*Rheum emodi*	5.75 kg	75 × 75 cm	15,000	3 years	9850
9.	*Valeriana jatamansi*	2.5 kg	20 × 40 cm	80,000	2 years	7000
10.	*Allium stracheyi*	1.5 kg	30 × 30 cm	180,000	Thrice a year	79–102

transportation, and technologies required for cultivation of quality raw or planting material, the stakeholders are holding back the production of high-quality seeds and plants needed to meet market demand.

17.4 Supply Chain and Marketing

The cultivation of medicinal plants on a large, organized scale is important for sustainability of biodiversity and establishing a reliable supply chain for the pharmaceutical sector. The large-scale cultivation of MAPs can effectively reduce the pressure on wild populations, protecting endangered species and ultimately preserving their natural ecosystem. The supply of MAPs mainly relies on market demands; this is where the engagement of local communities becomes paramount. Conservation initiatives are most effective when the people who directly depend on these resources are actively involved in their management. Through participatory decision making and shared governance processes communities can play key role in ensuring the long-term sustainability of both conservation and cultivation efforts. This collaborative approach not only conserves the medicinal plants but also empowers rural communities, creating a symbiotic association between nature and people (Joshi and Joshi 2014). In Uttarakhand, MAPs are traded in both raw and processed form; the rural farmers have been selling raw MAPs. Lack of direct market access and unorganized and unregulated marketing cause high wastage and low returns to farmers. The supply chain for medicinal plant (MAP) products involves

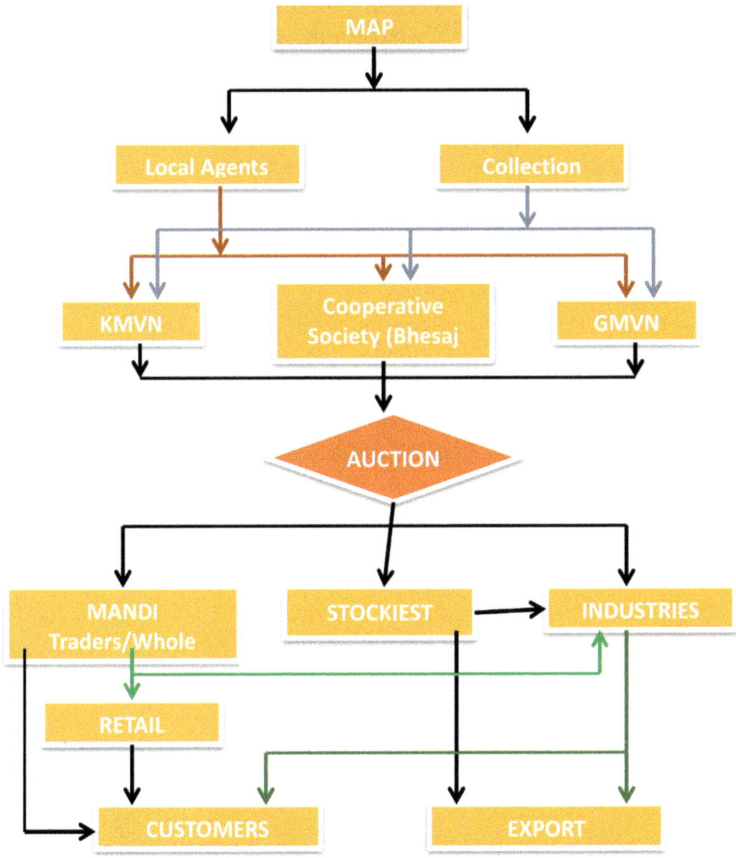

Fig. 17.3 The supply chain and marketing of medicinal and aromatic plants in Uttarakhand. (Courtesy: Pangriya 2015)

several key players. These include farmers who cultivate the plants, forest collectors, village-level collection centres, local traders, cooperative societies (Bheshaj Sangh), regional development corporations (Kumaon Mandal Vikas Nigam and Garhwal Mandal Vikas Nigam), wholesalers, stockists, manufacturing companies, and finally, the end consumers (Pangriya 2015, Fig. 17.3).

17.5 Challenges in MAP Production

The government, NGOs and other different organizations have been working to enhance the MAP production through (a) promoting the cultivation of medicinal plants among farmers, (b) technical support and training to the farmers, (c) establishing nurseries to propagate and supply planting material, and (d) streamlining the registration process for farmers as medicinal plant growers.

Despite these efforts, medicinal plant production in Uttarakhand is still low. Alam and Peppelenbos (2009) reported the following reasons behind limiting MAP production in Uttarakhand:

(i) *Technical Challenges*: Growing high-altitude medicinal plants is difficult as these plants require specific soil and environmental conditions. Research institutes are trying to find better ways to grow these plants, but limited financial and technical resources are major factors affecting their capabilities. Unorganized linkage between farmers and industries is affecting supply chain management in Uttarakhand.

(ii) *Regulative Challenges*: To prevent wild plants from being sold as cultivated, the state government requires growers and traders to obtain a permit for cultivation or trading protected species. In the past, getting the permit was difficult because many government agencies were involved, causing significant delays. However, now the things have been better; the procedures have been streamlined, resulting in a large number of farmers now registered as growers of medicinal plants.

(iii) *Economic Challenges*: The cultivation of medicinal plants faces significant economic challenges, including long gestation periods, low yields, and low market prices. Many marginal farmers are hesitant to invest in unfamiliar crops due to the associated risks. Additionally, several medicinal plants require 3 years or more to be ready for harvest, especially in high-altitude regions. For impoverished farmers, this lengthy wait for returns represents a substantial sacrifice. Consequently, farmers often either convert only a small area of their land or use distant, frequently neglected plots for growing medicinal plants. Typically, these plots range in size from 0.05 to 0.10 acre. Given the limited benefits of cultivating medicinal plants on such small areas, farmers often lack the motivation to maintain the plantations effectively.

(iv) *Challenges in Supply Chain Management*: The medicinal plant sector is characterized by a highly informal supply chain, posing significant challenges for cultivators. A complex supply chain involving numerous formal and informal intermediaries (wholesalers and traders) prioritizes low-cost sourcing over traceability and sustainability. Competition from wild-harvested and cultivated planting materials drives market prices down, resulting in insufficient economic incentives for farmers.

17.6 Conclusion

Uttarakhand is blessed with a number of endemic flora. Its treasure trove of medicinal and aromatic plants provides livelihoods and contributes to the economic development of mountain communities. The rich diversity of MAPs is in high demand both domestically and internationally. However, unsustainable extraction from natural habitats and challenges faced by local communities necessitates a comprehensive conservation strategy. Strengthening legal frameworks, research and development, supply chain sustainability, and capacity building programs are essential areas for government policy. Additionally, strategies focused on crop diversification, value addition, quality control, infrastructure development, climate change adaptation, and ecotourism are crucial for the long-term sustainability of the medicinal and aromatic plant sector.

Acknowledgement The data presented in the chapter was collected during research and development projects funded by the National Mission on Himalayan Studies, Ministry of Environment, Forestry and Climate Change (Grant No. GBPNI/NMHS-PMU-2017/MG03/9.18e36), and the Department of Biotechnology (Grant No. BT/PR 45245/NER/95/1907/2022), Government of India, New Delhi, India.

References

Airi S, Rawal RS, Dhar U, Purohit AN (2000) Assessment of availability and habitat preference of *Jatamansi*—a critically endangered medicinal plant of west Himalaya. Curr Sci 79(10):1467–1471

Alam G, Peppelenbos L (2009) Cultivation of medicinal plants in Uttarakhand. Econ Polit Wkly 44(10):99–104

Ali S, Chouhan R, Sultan P, Hassan QP, Gandhi SG (2021) A comprehensive review of phytochemistry, pharmacology and toxicology of the genus *Aconitum L.* Adv Tradit Med 23(5):1–22

Ameri A (1998) The effect of *Aconitum* alkaloids on the central nervous system. Prog Neurobiol 56:211–235

Anand U, Biswas P, Kumar V, Ray D, Ray P, Loake VI et al (2022) *Podophyllum hexandrum* and its active constituents: novel radioprotectants. Biomed Pharmacother 146:112555

Anwar S, Ahmad B, Sultan M, Gul W, Islam N (2003) Biological and pharmacological properties of *Aconitum chasmanthum*. J Biol Sci 3(11):989–993

Bajpay A, Nainwal RC, Singh D (2019) *Coptis teeta*: a potential endemic and endangered medicinal plant of Eastern Himalayas. J Pharmacogn Phytochem 8(4):245–248

Bargali H, Kumar A, Singh P (2022) Plant studies in Uttarakhand, Western Himalaya—a comprehensive review. Trees For People 8:100203. https://doi.org/10.1016/j.tfp.2022.100203

Bhandari P, Kumar N, Singh B, Kaul VK (2008) Simultaneous determination of sugars and picrosides in *Picrorhiza* species using ultrasonic extraction and high-performance liquid chromatography with evaporative light scattering detection. J Chromatogr A 1194(2):257–261

Bhandari P, Kumar N, Singh B, Ahuja PS (2010) Online HPLC-DPPH method for antioxidant activity of *Picrorhiza kurroa* Royle ex Benth. and characterization of kutkoside by Ultra-Performance LC-electrospray ionization quadrupole time-of-flight mass spectrometry. Indian J Exp Biol 48:323–328

Bhatt D, Joshi GC, Kumar R, Tewari LM (2014) Phytosociological features and threat categorization of *A. heterophyllum* Wall. ex Royle and A. ferox Wall. ex Ser. in Kumaun Himalaya. J Ecol Nat Environ 6(3):11–118

Bisht AK, Bhatt A, Rawal, Dhar U (2006) Assessment, extraction pattern and restoration through propagation of *Angelica glauca*, a critically endangered Himalayan medicinal herb. Int J Biodivers Sci Ecosyst Serv Manage 2(2):105–113

Chandra N, Singh G, Lingwal S, Rai ID, Tewari LM (2021) Alpine medicinal and aromatic plants in the Western Himalaya, India: an ecological review. Indian J Ecol 48(2):319–331

Chaturvedi P, Shukla S, Tripathi P, Chaurasia S, Singh SK, Tripathi YB (1995) Comparative study of *Inula racemosa* and *Saussurea Lappa* on the glucose level in Albino rats. Anc Sci Life 15(1):62–70

Chauhan NS (1999) Medicinal and aromatic plants of Himachal Pradesh. Indus Publishing Company, New Delhi

Purohit VK, Mengwal BS, Dobhal P, Chauhan J (2021) Cultivation of High Altitude Medicinal and Aromatic Plants. Bioresources and sustainable livelihood of rural India. Deb CR, Paul A (Eds.), Mittal Publications, New Delhi, India. pp. 978-93.

Dwivedi Y, Rastogi R, Garg NK, Dhawan BN (1992) Picroliv and its components kutkoside and picroside I protect liver against galactosamine-induced damage in rats. Pharmacol Toxicol 71:383–387

Gottumukkala VR, Annamalai T, Mukhopadhyay T (2011) Phytochemical investigation and hair growth studies on the rhizomes of *Nardostachys jatamansi* DC. Pharmacogn Mag 7(26):146

Joshi BC, Joshi RK (2014) The role of medicinal plants in livelihood improvement in Uttarakhand. Int J Herb Med 1(6):55–58

Joshi D, Melkani AB, Nailwal M, Bisht L, Prasad R (2018) *Selinum vaginatum* cb clarke: terpenoid composition and antibacterial activity of whole aerial parts and root essential oil. J Essent Oil-Bear Plants 21(5):1176–1185

Khajuria AK, Manhas RK, Kumar H, Bisht NS (2021) Ethnobotanical study of traditionally used medicinal plants of Pauri district of Uttarakhand, India. J Ethnopharmacol 276:114204

Khetwal KS (2007) Constituents of high altitude Himalayan herbs. Part XX. A C-19 diterpenoid alkaloid from *Aconitum balfourii*. Indian J Chem 46B:1364–1366

Kuniyal CP, Sundriyal RC (2013) Conservation salvage of Cordyceps sinensis collection in the Himalayan mountains is neglected. Ecosyst Serv 3:e40–e43. https://doi.org/10.1016/j.ecoser.2012.12.004

Kuniyal CP, Rawat YS, Oinam SS, Kuniyal JC, Vishvakarma SCR (2005) Kuth (*Saussurea lappa*) cultivation in the cold desert environment of Lahaul valley: arising threats and need to revive socio-economic values. Biodivers Conserv 14:1035–1045

Maikhuri RK, Rao KS, Semwal RL (2001) Changing scenario of Himalayan agroecosystems: loss of agrobiodiversity, an indicator of environmental change in Central Himalaya, India. Environmentalist 21:23–39

Maikhuri RK, Negi VS, Rawat LS, Pharswan DS (2017) Bioprospecting of Medicinal Plants in Nanda Devi Biosphere Reserve:Linking Conservation with Livelihood. Current Science 113(04) 571–577. https://doi.org/10.18520/cs/v113/i04/571-577

Mohan S, Gupta D (2017) Phytochemical analysis and differential in vitro cytotoxicity assessment of root extracts of *Inula racemosa*. Biomed Pharmacother 89:781–795. https://doi.org/10.1016/j.biopha.2017.02.053

Nautiyal MC, Nautiyal BP (2004) Agro-techniques for high altitude medicinal and aromatic plants. Bishen Singh Mahendra Pal Singh, International Book Distributors, Dehradun, pp 1–202

Negi VS, Kewlani P, Pathak R, Bhatt D, Bhatt ID, Rawal RS, Sundriyal RC, Nandi SK (2018) Criteria and indicators for promoting cultivation and conservation of medicinal and aromatic plants in Western Himalaya, India. Ecol Indic 93:434–446

Olsen CS (1998) The trade in medicinal and aromatic plants from Central Nepal to Northern India. Ecol Bot 52:279–292

Pal M, Samant SS (2005) Diversity, distribution and conservation of economically important medicinal and aromatic plants of the Indian Himalayan region. Blood Purif 1500:2500

Pandey A, Singh G, Palni S, Chandra N, Rawat JS, Singh AP (2021) Application of remote sensing in alpine grasslands cover mapping of western Himalaya, Uttarakhand, India. Environ Monit Assess 193(4):166

Pangriya R (2015) Study of Aromatic and Medicated Plants in Uttrakhand India: With Focus on Role in Employment Generation and Supply Chain Management. International Journal of Social Sciences and Management 2(2):148–156. https://doi.org/10.3126/ijssm.v2i2.12396

Phondani PC, Bhatt ID, Negi VS, Kothyari BP, Bhatt A, Maikhuri RK (2016) Promoting medicinal plants cultivation as a tool for biodiversity conservation and livelihood enhancement in Indian Himalaya. Journal of Asia-Pacific Biodiversity 9(1): 39–46. https://doi.org/10.1016/j.japb.2015.12.001

Purohit VK, Andola HC, Haider SZ et al (2015) Essential oil constituents of *Angelica glauca* Edgew. Roots: an endangered species from Uttarakhand Himalaya (India). Natl Acad Sci Lett 38:445–447. https://doi.org/10.1007/s40009-015-0395-z

Rahman SU, Ismail M, Khurram M, Haq IU (2015) Pharmacognostic and ethnomedicinal studies on *Trillium govanianum*. Pak J Bot 47:187–192

Rahman SU, Adhikari A, Ismail M, Shah MR, Khurram M, Shahid M, Ali F, Haseeb A, Akbar F, Iriti M (2016) Beneficial effects of *Trillium govanianum* rhizomes in pain and inflammation. Molecules 21:1095

Rawat GS, Adhikari BS, Tiwari UL, Raut N, Chandola S (2012) Status and distribution of medicinal plants in various forest division of Garhwal region, Uttarakhand. Wildlife Institute of India, Dehradun, p 392

Rawat JM, Pandey S, Rawat B, Rai N, Preeti P, Thakur A, Butola JS, Bachheti RK (2023) Traditional Uses Active Ingredients and Biological Activities of Paris polyphylla Smith: A Comprehensive Review of an Important Himalayan Medicinal Plant. Journal of Chemistry 2023(1): 7947224. https://doi.org/10.1155/2023/7947224

Saboon YB, Arshad M, Sabir S, Amjad MS, Ahmed E, Chaudhari SK (2016) Pharmacology and biochemistry of *Polygonatum verticillatum*: a review. J Coast Life Med 4(5):406–415

Safaralie A, Fatemi S, Sefidkon F (2008) Essential oil composition of Valeriana officinalis L. roots cultivated in Iran: comparative analysis between supercritical CO_2 extraction and hydrodistillation. J Chromatogr A 1180(1–2):159–164

Samant SS, Pal M (2003) Diversity and conservation status of medicinal plants in Uttaranchal state. Indian Forester. In: Negi SS, Srivastava RK, Bisht NS (eds) Medicinal and aromatic plants (taxonomy, conservation, cultivation, economics and marketing). Dehradun, pp 275–291

Saraswat N, Sachan N, Chandra P (2020) A detailed review on the rarely found Himalayan herb *Selinum vaginatum*: its active constituents, pharmacological uses, traditional and potential benefit. Pharmacophore 11(2):40–52

Semwal P, Painuli S (2019) Antioxidant, antimicrobial, and GC-MS profiling of *Saussurea obvallata* (Brahma Kamal) from Uttarakhand Himalaya. Clin Phytoscience 5(1):12. https://doi.org/10.1186/s40816-019-0105-3

Semwal P, Painuli S, Tewari D, Bussmann RW, Palni LMS, Thapliyal A (2020) Brahma Kamal (*Saussurea obvallata* (DC.)) Edgew. An important Himalayan medicinal plant. Ethnobot Res Appl 19(40):1–15

Sharma E, Gaur AK (2012) *Aconitum balfourii* Stapf: a rare medicinal herb from Himalayan Alpine. J Med Plant Res 6(22):3810–3817

Singh M, Kumar R, Sharma S et al (2023) *Hedychium spicatum*: a comprehensive insight into its ethnobotany, phytochemistry, pharmacological and therapeutic attributes. S Afr J Bot 161:638–647

Tabin S, Gupta RC, Kamili AN, Parray JA (2022) Medical and medicinal importance of *Rheum* spp. collected from different altitudes of the Kashmir Himalayan range. Cell Mol Biomed Rep 2(3):187–201

Tripathi S, Kumar A, Kohli K, Kumar J (2019) Organic cultivation of medicinal plants in Uttarakhand: an overview. Progress Hortic 51(1):45–49

Ukani MD, Mehta NK, Nanavati DD (1996) *Aconitum heterophyllum* (ativisha) in ayurveda. Anc Sci Life 16(2):166–171

Uniyal SK, Awasthi A, Rawat GS (2002) Current status and distribution of commercially exploited medicinal and aromatic plants in upper Gori valley, Kumaon Himalaya, Uttaranchal. Curr Sci 82(10):1246–1252

Upadhyay D, Dash RP, Anandjiwala S, Nivsarkar M (2013) Comparative pharmacokinetic profiles of picrosides I and II from kutkin, *Picrorhiza kurroa* extract and its formulation in rats. Fitoterapia 85:76–83

van Oudenhoven APE, de Groot RS (2011) Analysing and monitoring human impacts on biodiversity and ecosystem services. Int J Biodivers Sci Ecosyst Serv Manag 7(4):245–246. https://doi.org/10.1080/21513732.667243.2011.667243

Vashistha RK, Butola JS, Nautiyal BP, Nautiyal MC (2010) Phenological attributes of *Angelica glauca* and *A. archangelica* expressed at two different climatic zones in Western Himalaya. Open Access J Med Aromat Plants 1:7–12

Ved DK, Goraya GS (2008) Demand and supply of medicinal plants India. National Medicinal Plants Board, New Delhi, p 216

Ved DK, Kinhal GA, Ravikumar K, Prabhakaran V, Ghate U, Vijaya SR, Indresha JH (2003) Conservation assessment and management prioritization for the medicinal plants of Jammu & Kashmir, Himachal Pradesh & Uttaranchal. Foundation for Revitalisation of Local Health Traditions, Bangalore

Verma S, Ojha S, Raish M (2010) Anti-inflammatory activity of *Aconitum heterophyllum* on cotton pellet-induced granuloma in rats. J Med Plant Res 4(15):1566–1569

Wani IA, Kumar V, Verma S, Tasleem Jan A, Rather IA (2020) *Dactylorhiza hatagirea* (D. Don) Soo: a critically endangered perennial orchid from the North-West Himalayas. Plan Theory 9(12):644

Zhou G, Tang L, Zhou X, Wang T, Kou Z, Wang Z (2015) A review on phytochemistry and pharmacological activities of the processed lateral root of *Aconitum carmichaelii* Debeaux. J Ethnopharmacol 160:173–193

18 Phytochemistry, Pharmacology, and Nanotechnology of *Nepeta leucophylla Benth.*: A Himalayan Aromatic Medicinal Plant

Garima Bhardwaj, Ravjit Kaur, Anita Devi, and Sakshi Thakur

Abstract

Mother Nature is a source of numerous invaluable bioactive secondary metabolites and lead molecules, which form the backbone of numerous ancient medicinal systems all around the world. *Nepeta leucophylla Benth.* is an aromatic medicinal plant found in the western *Himalayas* at elevation of >2000 m. It is one of the less explored species of the genus *Nepeta* found in the western *Himalayas*. Traditionally, the plant is used as a remedy for malarial fever and to prepare herbal tea. Plants are also used for the cure of different ailments in livestock by local tribes in Himalayas, but there is no documented proof for the same. Apart from these, *N. leucophylla* is also known to show antibacterial, antifungal, antiviral, antioxidant, and anticancer potential. Flavonoids, phenolic acid, terpenoids, steroids, and hydrocarbons are the key classes of secondary metabolites isolated from the different parts of the plant. This chapter represents the status of species with respect to its distribution, taxonomy, traditional uses with more emphasis on phytochemical composition, pharmacology, and future prospective.

G. Bhardwaj (✉)
Department of Chemistry, Sant Longowal Institute of Engineering and Technology, Sangrur, Punjab, India

R. Kaur
SBMDSOE School Longowal, Sangrur, Punjab, India

A. Devi
Dr Shanti Swarup Bhatnagar University Institute of Chemical Engineering and Technology, Panjab University, Chandigarh, India

S. Thakur
Department of Chemistry, Chandigarh University, Mohali, Punjab, India

S. Chandra, H. Lata (eds.), *High Altitude Medicinal Plants*,
https://doi.org/10.1007/978-3-031-99406-7_18

18.1 Introduction

The Himalayan biodiversity is the richest reservoirs of biologically potent secondary metabolites (SMs) and lead molecules. The Himalayan region is represented by a total of 10,503 species, which belongs to 2322 genera and 240 families (Rana and Rawat 2017). The literature revealed that the species found at higher altitude have potentially high medicinal value and concentration of SMs (Kumar et al. 2011). Plant species found in the Himalayas region have been well known for their traditional uses but so for many species are still unexplored with respect to their pharmacological potential and phytochemical composition. This is mainly attributed to geographical position of Himalayas, tough environment conditions, and inaccessibility to the region due to dangerous means of transport (Singh 2006; Ballabh and Chaurasia 2009). SMs have played an important role in drug discovery from ancient times and are still adding substantial contributions to drug development and discovery. The use of SMs for the cure of various kinds of ailments has been the oldest and cheapest type of healthcare known to human beings. On account of which scientists from all over the world have been making considerable efforts to solve mysteries of nature. Nowadays, scrutinizing the treasure of nature has become the key objective of various researchers from all over the world in order to isolate new bioactive SMs and to develop new and more efficient strategies for their isolation as compared to traditional ones (Cragg and Newman 2013; Atanasov et al. 2015).

Genus *Nepeta* of Lamiaceae family is represented by 300 species. These are native to Europe, North Africa, and Asia. In Asia, the species mainly grows in the mountainous region of Turkey, Iran, Pakistan, and India. In India, genus *Nepeta* is composed of around 41 species and are mainly found in the Western Himalayan region Uttarakhand, Himachal Pradesh, and Jammu & Kashmir (Sharma and Cannoo 2013, 2016a; Patial et al. 2019; Sharma et al. 2021a). *Nepeta* species are known to possess variety of biological properties such as sedative, diuretic, antiseptic, antitussive, emmenagogue, diaphoretic, etc., and these are extensively used in different traditional medicine systems (Formisano et al. 2011; Sharma and Cannoo 2013; Suntar et al. 2017; Sharma et al. 2019, 2021a, b, 2023). Further, many species of genus *Nepeta* are also known to have innumerable biological properties such as antimicrobial, antiviral, antifungal, antioxidant, cytotoxic, anti-inflammatory, antinociceptive, nematocidal, insecticidal, anthelmintic, trypanocidal, bronchodilatory, and spasmolytic activities (Formisano et al. 2011; Sharma and Cannoo 2013; Suntar et al. 2017; Sharma et al. 2019, 2021a, b, 2023).

N. leucophylla is an aromatic medicinal plant, widely used as source of ethnomedicine in the western Himalayan region by local tribes (Bisht et al. 2010; Sharma and Cannoo 2016a, 2017; Sharma et al. 2021a, 2023). The scientific studies revealed that the essential oil and extracts isolated from different parts of *N. leucophylla* have been well recognized for their antibacterial, antiviral, antifungal, anticancer, and antioxidant properties (Bisht et al. 2010; Sharma and Cannoo 2016a, b, 2017; Sharma et al. 2021a, 2023) (Fig. 18.1). The headspace GC-MS and GC-MS analysis of essential oil obtained from the different plant parts revealed the presence of various bioactive SMs viz. caryophyllene, germacrene D, δ-cadinene, β-chamigrene,

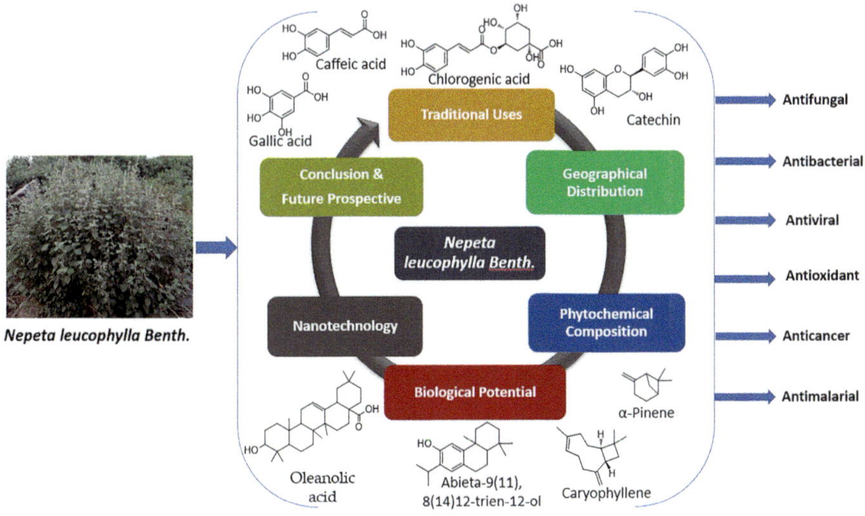

Fig. 18.1 Phytochemistry and pharmacology of *N. leucophylla*

α-pinene, viridiflorol, α-phellandren, germacrene B, α-selinene, iridodial dienol diacetate, coleon U 12-methyl ether, iridodial β-monoenol acetate, and dihydroiridodial diacetate, etc. (Mathela et al. 1991; Bottini et al. 1992; Saxena and Mathela 1996; Sharma et al. 2021a). Further, chromatographic analysis of various extracts of *N. leucophylla* revealed the presence of various classes of bioactive SMs such as flavonoids (catechin hydrate, rutin trihydrate, and myricetin), phenolic acids (chlorogenic acid, caffeic acid, ρ-coumaric acid, syringic acid, vanillic acid, and gallic acid), diterpenoids (totarol and abieta-9(11),8(14),12-trien-12-ol), triterpenoids (oleanolic acid, ursolic acid, and squalene), steroids (stigmasterol, stigmasta-5,20(22)-dien-3-ol, stigmast-5-en-3-ol, stigmast-4-en-3-one), fatty acids and their esters (pentadecenoic acid, methyl ester of linolenic acid, linoleic acid, and linolenic acid) (Sharma and Cannoo 2016a, b, 2017; Sharma et al. 2021a, 2023) (Fig. 18.1). Till date, no article has been published on the complete phytochemical composition, pharmacological potential, and nanotechnology of different parts of *N. leucophylla*. So, the aim of the present book chapter is to compile all the research data of *N. leucophylla* under single article along with mandate for future research.

18.2 Taxonomy

N. leucophylla is a wild aromatic herb with white flowers. The botanical classification of the plant is presented in Fig. 18.2 (Sharma et al. 2021a, 2023) (https://plants.jstor.org/stable/10.5555/al.ap.specimen.k000911181; https://iiim.res.in/herbarium/lamiaceae/nepeta_leucophylla.htm).

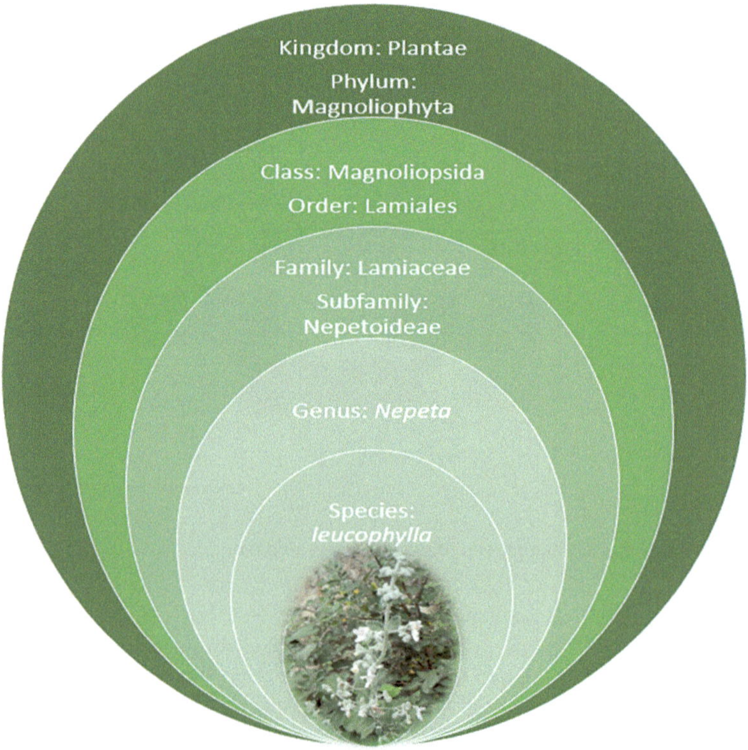

Fig. 18.2 Botanical classification of *N. leucophylla*

18.3 Geographical Distribution

N. leucophylla usually grows on the open slopes of mountain in the dry and wet regions of western Himalayas at an altitude of >2000 m particularly in Uttarakhand, Himachal Pradesh (Chamba and Lahul), Ladhakh, Jammu & Kashmir (Bisht et al. 2010; Kumar et al. 2014; Sharma and Cannoo 2016a, 2017; Sharma et al. 2021a; 2023).

18.4 Phytochemical Composition

18.4.1 Isolation Techniques and Solvent

The percentage extractive yield of various extracts prepared from different plant parts was considerably affected by the type of extraction solvents/techniques used and their extraction superiority. The extraction superiority subsequently relates to the isolation costs, viz., extraction time, energy consumed, solvent volume used, etc., effects on living beings and their environment (Azmir et al. 2013; Karabegovic et al. 2014). Consequently, it is very significant to select an appropriate extraction solvent/technique for the isolation of bioactive NPs from plant resources, which is environmentally friendly, cost effective, and most efficient in nature.

Sharma and Cannoo (2016a, 2016b, 2017) investigated three different extraction techniques {traditional (MM—maceration method and SEM—Soxhlet extraction method) and modern (UAEM—ultrasonication-assisted extraction method)} solvents of different polarity (hexane, chloroform, and methanol) to isolate various extracts from different shade dried parts {aerial parts (leaves and flowers), stems, and roots) of *N. leucophylla*. The results revealed that the extraction solvent/method and the plant part used significantly affected the percentage extractive yield (PEY) of different extracts that ranged between 3.66 and 17.66, 0.99 to 12.16, and 0.9 to 8.07% for aerial parts, stem and roots, respectively (Table 18.1). The highest PEY was observed in the extracts obtained from aerial parts followed by the extracts isolated from stems and roots, respectively. On the other hand, irrespective of the extraction method and plant part used, the methanol extracts showed the highest PEY followed by chloroform and hexane extracts, which may be ascribed to the high dielectric constant and polar protic nature of methanol (as compared to chloroform and hexane extracts) that assisted the isolation of some nonpolar (long change hydrocarbons) and medium polar (fatty acid, long chain alcohols, esters, etc.) to polar compounds (phenolic acids, flavonoids, glycosides, etc.) consistent with the fact that "like dissolves like" (Kalia et al. 2008). The results revealed that there should be a direct relation between polarity of solvent and the percentage yield, where polar solvents showed the highest PEY as compared to medium polar and non-polar ones. Among the extraction methods, the results of PEY follow the order SEM > UAEM > MM in the case of methanol/chloroform, methanol/hexane, and chloroform extracts obtained from aerial parts, stems, and roots, respectively. Further, the order of PEY in the case of methanol extract of roots and chloroform extract obtained from stems was SEM > MM > UAEM.

Table 18.1 The percentage yield of various extracts isolated from the different parts of *N. leucophylla* using different extraction methods (Sharma and Cannoo 2016a, b, 2017)

		Percentage yield (%) (w/w)		
Extraction method used	Solvent used	Aerial parts (Leaves + flowers)	Stems	Roots
SEM	*Methanol*	17.66	12.16	8.07
	Chloroform	13.92	3.08	1.24
	Hexane	3.7	1.16	1.03
UAEM	*Methanol*	17.57	9.69	5.15
	Chloroform	13.79	2.32	1.14
	Hexane	4.18	1.10	1.15
MM	*Methanol*	10.86	7.99	5.43
	Chloroform	9.34	2.47	1.13
	Hexane	3.66	0.99	0.90

w/w weight/weight (w/w—yield of dry extract in g/total powdered dry material taken in g × 100), *SEM* Soxhlet extraction method, *UAEM* ultrasonic assisted extraction method, *MM* maceration method

The high PEY in various extracts isolated by SEM was attributed to the complete extraction of the plant material by frequent washings with new and warm solvent (Haizhou et al. 2004; Karabegovic et al. 2014), while in the case of UAEM, high PEY in very less time as compared to MM may be ascribed to the phenomenon of acoustic cavitations that facilitated the penetration of solvent into the plant cell by rupturing the cell wall, hence increasing the isolation of SMs by mass transfer into the solvent (Vinatoru 2001; Karabegovic et al. 2014). Further, the high temperature in the case of SEM and UAEM as compared to MM could be another reason that contributed towards higher PEY in the extracts obtained by SEM and UAEM. The high temperature increased the diffusion coefficients and solubility of SMs and reduced the solvent viscosity, which in turn improved the mass transfer of SMs from plant material into the solvent (Wang et al. 2008; Rababah et al. 2010).

18.4.2 Qualitative Analysis of Phytochemicals

The qualitative phytochemical investigations revealed that reducing sugars, glycosides, polyphenols, coumarins, flavonoids, steroids, diterpenoids, and triterpenoids were the key classes of SMs present in various extracts obtained from different parts of *N. leucophylla.* The overall best results of qualitative phytochemical analysis were reported in the case of aerial parts (among the plant parts used), methanol (among solvents used), and SEM (among the extraction method used) (Table 18.2) (Sharma and Cannoo 2016a, b, 2017). The reason for supremacy of methanol over chloroform and hexane as extraction solvent was its high dielectric constant and polar protic nature, which assisted the isolation of polar classes of SMs equally with medium polar and less polar ones. Previously, Seladji et al. (2014), Fareed et al. (2013) and Edewor and Usman (2011) also reported alike results for phytochemical composition in the case of *N. nepetella* (aerial parts), *N. praetervisa* (aerial parts), and *N. cataria* (leaves), respectively.

18.4.3 Quantitative Analysis

Total Polyphenolic Content (TPC) The values of TPC of different extracts with regard to aerial parts, stems, and roots ranged between 0.98 ± 0.47 to 141.9 ± 3.86, 0.34 ± 0.17 to 101.31 ± 0.62 and 2.25 ± 0.42 to 87.04 ± 1.22 mg GAE/g DPE (Table 18.2), respectively. Independent of the extraction methods/solvents used, the various extracts obtained from the aerial parts showed the highest values of TPC followed by the extracts isolated from stems and roots, except for methanol extract obtained by MM where the order of TPC was stem>roots>aerial parts. With regard to solvents used, methanol extracts showed the highest TPC followed by chloroform and hexane extracts. In the case of extraction method used, the highest TPC values were observed in the extracts obtained from SEM followed by UAEM and MM with respect to methanol/chloroform and chloroform/hexane extracts isolated from aerial parts and stems, respectively (Sharma and Cannoo 2016a, b, 2017).

Table 18.2 TPC results of various extracts isolated from the different parts of *N. leucophylla* using different extraction methods (Sharma and Cannoo 2016a, b, 2017)

		TPC (mg GAE/g DPE)		
Extraction method used	Solvent used	Aerial parts (Leaves + flowers)	Stems	Roots
SEM	*Methanol*	141.9 ± 3.86	133.7 ± 1.23	94.02 ± 2.09
	Chloroform	35.96 ± 1.08	11.13 ± 0.5	10.12 ± 0.76
	Hexane	1.66 ± 0.46	1.3 ± 0.07	–
UAEM	*Methanol*	83.90 ± 1.63	79.21 ± 0.63	85.10 ± 2.9
	Chloroform	9.46 ± 2.22	6.53 ± 0.21	2.25 ± 0.42
	Hexane	7.11 ± 1.84	0.74 ± 0.27	–
MM	*Methanol*	77.16 ± 0.58	101.31 ± 0.62	87.04 ± 1.22
	Chloroform	8.37 ± 1.23	5.38 ± 0.16	2.56 ± 0.42
	Hexane	0.98 ± 0.47	0.34 ± 0.17	–

mg GAE/g DPE mg of gallic acid equivalent per g of dry plant extract

The higher value of TPC in the extracts isolated by SEM over UAEM and MM was attributed to the exhaustive extraction of the SMs by continuous and repeated washing of the plant material with new solvent in the case of former. Further, higher/comparable TPC values in different extracts isolated by UAEM over MM in shorter time span may be ascribed to the acoustic cavitations phenomenon of ultrasound waves and high temperature as discussed previously in the case of PEY (Vinatoru 2001; Karabegovic et al. 2014).

Total Flavonoid Content Among the aerial parts, stems, and roots, the values of TFC varied from 6.21 ± 0.49 to 394.48 ± 15.45, 20.21 ± 3.01 to 282 ± 74 ± 0.86, and 5.73 ± 1.31 to 228.75 ± 1.98 mg RE/g DPE, respectively (Table 18.3). Among the different plant parts used, the TFC values followed the order stems > roots > aerial parts with respect to methanol, chloroform, and hexane extract obtained by MM; stems > aerial parts > roots in the case of chloroform/hexane and hexane extracts isolated by UAEM and SEM, respectively; aerial parts > stem > roots among the methanol/chloroform extracts of SEM and roots > stems > aerial parts with respect to methanol extract of UAEM. The flavonoids are usually present deep in the plant cell mainly in the cell vacuoles and chloroplasts (Agati et al. 2012). Further, the cell wall is stronger in root and stem cells as compared to leaf cells because the former have harder tissue than the latter. So, the high TFC values in the extracts obtained from stems and roots by MM and UAEM as compared to the extracts obtained from aerial parts by SEM may be due to larger extraction time in the case of MM and the effect of ultrasound waves (acoustic cavitation phenomenon) in the case of UAEM. The higher extraction time creates larger osmotic pressure due to enhanced flow of solvent into the cell vacuoles and results in the rupture of cell wall of the latter, while in the case of ultrasound waves, the phenomenon of acoustic cavitations help to rupture the cell wall. Due to these, the rate of diffusion of flavonoids into the solvent increased (Vinatoru 2001; Karabegovic et al. 2014; Sharma and Cannoo 2016a, b, 2017).

In the case of various solvents used, methanol extracts showed the highest TFC values followed by chloroform and hexane extracts irrespective of the extraction method and the plant part used. The reason behind this was same (high dielectric constant and polar protic nature of methanol) as discussed earlier in the case of PEY and TPC. Among the extraction methods used, the results of TFC follow the order SEM > UAEM > MM in the case of methanol/chloroform/hexane and hexane extracts isolated from aerial parts and roots, while in the case of methanol and chloroform extracts obtained from stems and roots, the order was SEM > MM > UAEM. Further, the order UAEM > SEM > MM for TFC values was noted in the methanol and hexane extracts of roots and stems, respectively (Sharma and Cannoo 2016a, b, 2017).

RP-HPLC–DAD Analysis The results of RP-HPLC–DAD analysis showed the presence of total seven (three flavonoids and four phenolic acids), eight (three flavonoids and five phenolic acids), and five (two flavonoids and three phenolic acids)

Table 18.3 TFC results of various extracts isolated from the different parts of *N. leucophylla* using different extraction methods (Sharma and Cannoo 2016a, b, 2017)

Extraction method used	Solvent used	TFC (mg RE/g DPE)		
		Aerial parts (Leaves + flowers)	Stems	Roots
SEM	*Methanol*	394.48 ± 15.45	282.74 ± 0.86	207.45 ± 3.37
	Chloroform	90.92 ± 14.69	24.79 ± 6.67	28.35 ± 2.27
	Hexane	17.38 ± 2.62	20.78 ± 0.86	10.56 ± 3.87
UAEM	*Methanol*	165.1 ± 3.48	192.76 ± 18.47	228.75 ± 1.98
	Chloroform	37.7 ± 11.23	40.25 ± 12.18	7.45 ± 0.99
	Hexane	15.69 ± 2.63	22.79 ± 2.62	5.73 ± 1.31
MM	*Methanol*	136.76 ± 5.24	228.85 ± 6.7	190.1 ± 9.18
	Chloroform	18.21 ± 0.86	31.95 ± 10.45	21.17 ± 4.9
	Hexane	6.21 ± 0.49	20.21 ± 3.01	–

mg RE/g DPE mg of rutin tryhydrate equivalent per g of dry plant extract

polyphenolics out of fourteen polyphenolic standards (four flavonoids and ten polyphenolic acids) analyzed in the different methanol extracts obtained from aerial parts, stems, and roots of *N. leucophylla*, respectively (Table 18.4), while none of polyphenolic compound was detected in chloroform and hexane extracts obtained from different plant parts. Further, myricetin had the highest concentration among all detected polyphenolic compounds, followed by chlorogenic acid, rutin trihydrate, caffeic acid, ρ-coumaric acid, catechin hydrate, syringic acid/vanillic acid, and gallic acid, respectively. Furthermore, the overall best results were noted in the case of methanol extracts obtained by SEM irrespective of the plant part used (Sharma and Cannoo 2016a, b, 2017). Polyphenolics such as vanillic acid, rutin, syringic acid, chlorogenic acid, and caffeic acid were also identified previously in many other *Nepeta* species (Koksal et al. 2017; Misic et al. 2015; Micelia et al. 2005).

GC-MS Analysis of Extracts The GC-MS analysis of methanol, chloroform, and hexane extracts (most bioactive) obtained from aerial parts, stems, and roots using SEM revealed the presence of 27, 18, and 24; 19, 25, and 20; and 14, 29, and 21 compounds, respectively. Analysis of these extracts showed the presence of long chain hydrocarbons, terpenoids, long chain alcohols, phenolics, steroids, fatty acids, and their esters as the major classes of volatile SMs. The results also revealed that the hexane and chloroform extracts had more content of long chain hydrocarbons as compared to methanol extracts, whereas steroids, fatty acids, and their ester were present in higher concentration in methanol extracts as compared to hexane and chloroform extracts. The results revealed that as the polarity of extraction solvent decreases from methanol to hexane (polar to nonpolar), accordingly the content of long chain hydrocarbons also increases in extraction solvents. Thus, the polarity of extraction solvents used significantly affected the composition of volatile SMs in these extracts (Sharma and Cannoo 2016a, b, 2017).

Apart from above compounds oleanolic acid, ursolic acid, squalene, abieta-9(11), 8(14), 12-trien-12-ol (Sharma et al. 2023), coleon U 12-methyl ether and (R)-2,3,4,4a-tetrahydro-5,8,10-trihydroxy-6-methoxy-1,1,4a-trimethyI-7-(1-methylethyl)-9(1-H) phenanthrenone) (Mathela et al. 1991) are key compounds isolated from different parts of *N. leucophylla* (Fig. 18.3).

18.4.4 Essential Oil Composition

Headspace GC-MS Analysis The headspace GC-MS analysis of roots, stems, leaves, and inflorescence revealed the presence of aromatic compounds, ketones, organic acids, oxygenated sesquiterpenes hydrocarbons (OSTH), oxygenated monoterpene hydrocarbons (OMTH), sesquiterpene hydrocarbons (STH), and monoterpene hydrocarbons (MTH) as the key classes of SMs. Germacrene B, germacrene D, α-phellandrene, α-pinene, viridiflorol, and caryophyllene are major SMs reported in the inflorescence; α-phellandren, α-pinene, acetic acid, δ-cadinene,

Table 18.4 RP-HPLC-DAD analysis results of polyphenolic content (μg/g of DPE) of methanol extracts obtained from different parts of *N. leucophylla* using different extraction methods (Sharma and Cannoo 2016a, b, 2017)

	Polyphenolic content (μg/g of DPE)								
	Aerial parts (Leaves + flowers)			Stems			Roots		
Standards	SEM	UAEM	MM	SEM	UAEM	MM	SEM	UAEM	MM
CH	107.11 ±1.47	89.6 ±0.64	88.94 ±1.35	103.04 ±1.45	98.10 ±0.22	96.47 ±0.68	92.36 ±0.08	93.59 ±1.77	99.05 ±0.42
CLA	324.82 ±1.44	310.88 ±0.43	310.26 ±0.07	313.04 ±0.20	312.0 ±0.22	310.46 ±0.47	–	–	–
CA	226.38 ±0.17	228.03 ±0.31	228.25 ±1.17	227.80 ±0.74	242.49 ±1.44	233.06 ±0.49	248.37 ±0.35	248.3 ±0.75	242.37 ±0.64
ρ-CA	–	–	–	221.17 ±0.03	220.99 ±0.03	221.05 ±0.02	–	–	–
SA	65.14 ±1.36	61.68 ±0.66	66.41 ±1.99	45.10 ±0.07	48.89 ±0.95	45.94 ±0.06	–	–	–
VA	63.35 ±1.70	61.64 ±0.1	58.13 ±0.1	55.37 ±2.47	55.95 ±0.07	54.09 ±0.09	54.02 ±0.33	54.58 ±0.26	54.51 ±0.11
RT	292.98 ±1.65	309.30 ±1.97	289.97 ±0.87	256.92 ±0.14	261.47 ±0.37	259.72 ±0.26	–	–	–
MY	667.75 ±4.05	666.22 ±5.81	555.52 ±4.06	639.8 ±21.14	559.70 ±20.46	396.96 ±17.69	614.2 ±4.04	521.5 ±4.9	546.37 ±4.34
GA	–	–	–	–	–	–	1.69 ±0.31	1.72 ±0.11	3.0 ±0.43

CH catechin hydrate, *CLA* chlorogenic acid, *CA* caffeic acid, *ρ-CA* ρ-coumaric acid, *SA* syringic acid, *VA* vanillic acid, *RT* rutin trihydrate, *MY* myricetin, *GA* gallic acid

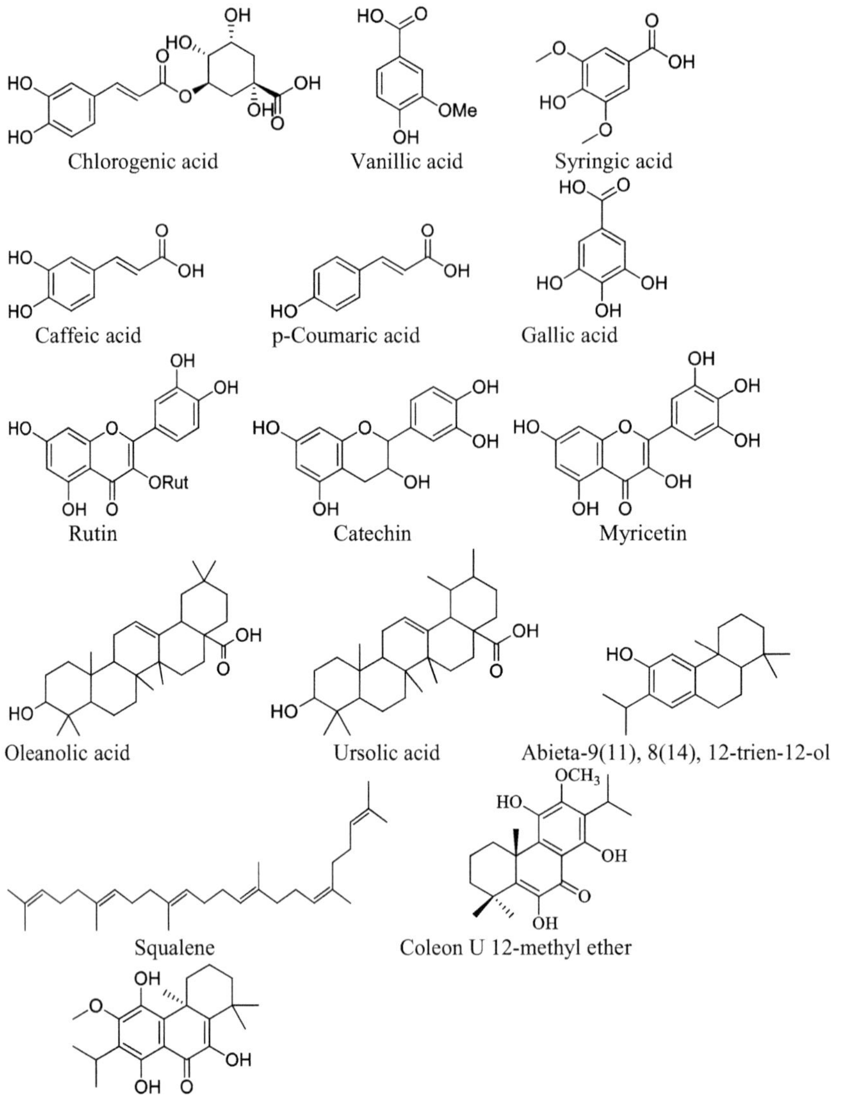

(R)-2,3,4,4a-tetrahydro-5,8,10-trihydroxy-6-methoxy-1,1,4a-trimethyl-7-(1-methylethyl)-9(1H)-phenanthrenone

Fig. 18.3 Structure of various key compounds reported in different extracts *N. leucophylla*

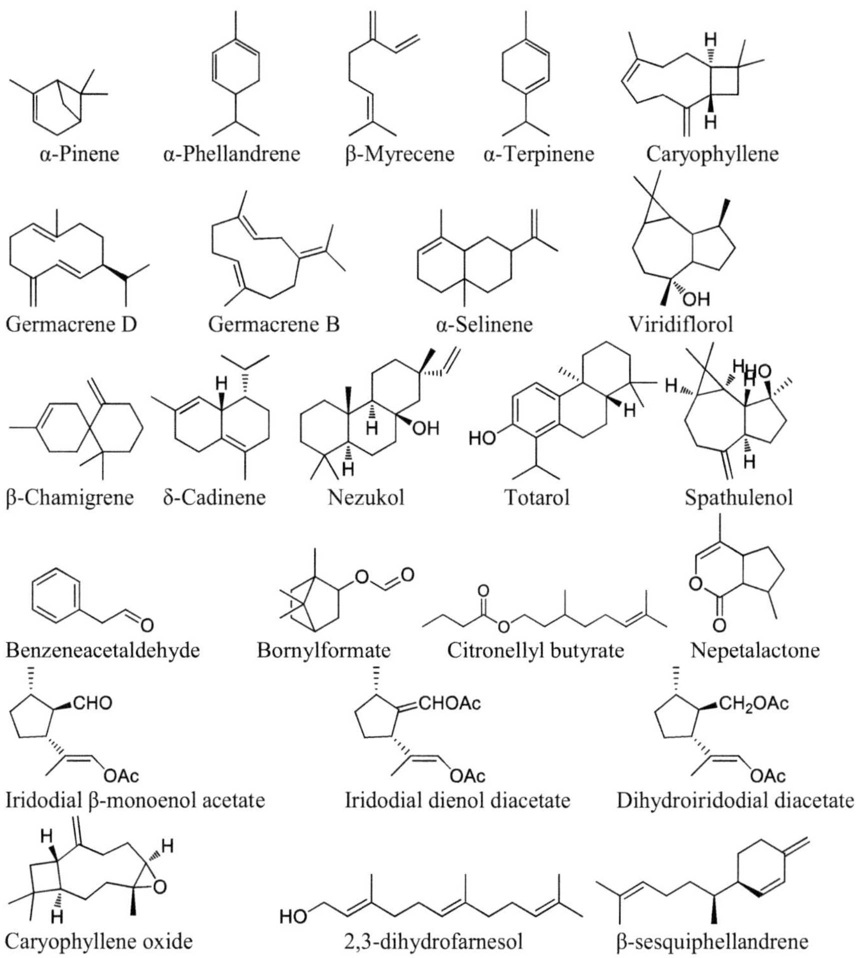

Fig. 18.4 Structure of the major volatile phytoconstituents identified in different parts of *N. leucophylla*

germacrene D, α-selinene, and caryophyllene are the key SMs of the leaves; caryophyllene and acetic acids are the principal SMs observed in stems, while α-pinene, caryophyllene, and viridiflorol are the major SMs found in roots (Fig. 18.4, Table 18.5) (Sharma et al. 2021a).

GC-MS Analysis of Essential Oils (EOs) The GC-MS analysis of EOs isolated from the roots, stems, and aerial parts revealed the presence of 31, 30, and 30 compounds, respectively. The isolated EOs were a complex mixture of esters, organic acids, epoxide, phenols, ketones, alcohols, aromatic compounds, phenolic diterpene hydrocarbons, oxygenated diterpene hydrocarbons (ODTH), OSTH, MTH, and

Table 18.5 Results of headspace analysis (volatile phytochemical composition) of different parts of *N. leucophylla* (Sharma et al. 2021a)

Name of compound	Root (%/SI)	Stem (%/SI)	Leaves (%/SI)	Inflorescence (%/SI)
Carbon dioxide (flu, leaves and stem)/1-Alanine ethylamide (roots)	21.10/90	42.68/97	15.08/96	26.34/92
Acetic acid	3.49/98	35.03/98	3.70/97	0.26/96
α-Pinene	6.00/97	1.13/97	3.13/97	9.83/98
β-Myrcen	0.92/95	–	2.13/95	3.47/96
α-Phellandrene	0.85/95	0.96/96	3.03/95	5.14/96
Caryophyllene	9.87/97	4.28/97	23.98/97	16.20/98
Germacrene D	0.81/96	–	8.68/98	4.83/98
Germacrene B	–	–	–	3.94/91
α-Selinene	–	–	10.98/91	–
δ-Cadinene	0.97/89	1.06/95	3.74/91	2.27/95
Viridiflorol	36.05/96	–		8.77/96

SI similarity index, *UI* unidentified; only the compounds having concentration more than 0.25% are given in the table

STH. Spathulenol, α-phellandren, δ-cadinene, abieta-9(11),8(14),12-trien-12-ol, β-chamigrene, germacrene D, and caryophyllene were the major compounds present in the EOs obtained from aerial parts; totarol, undecane, α-terpine, benzeneacetaldehyde, diisobutyl phthalate, hydroxy-4-methyl-2,6-di-tert-butylbenzene, caryophyllene, abieta-9(11),8(14),12-trien-12-ol and bis (6-methylheptyl) phthalate were the key constituents found in the EOs isolated from stems, while α-terpine, bornyl formate, and totarol were the major compounds identified in the EOs of the root (Table 18.6) (Sharma et al. 2021a).

Gupta et al. (1971 and Gupta (1973) isolated EO from the aerial parts of *N. leucophylla* and found that sesquiterpene alcohol nepetol, citronellyl butyrate, and nepetalactone are the key component present in the EO. Later, some other groups reported that iridodial β-monoenol acetate, dihydroiridodial diacetate, and iridodial dineol diacetate were the key component present in EOs isolated from aerial parts of *N. leucophylla* (Bottini et al. 1992; Saxena and Mathela 1996; Bisht et al. 2010). Further, Kumar et al. (2014) reported that iridodial β-monoenol acetate and caryophyllene oxide were the major compounds present in the EOs obtained from leaves of *N. leucophylla*. Furthermore, Kumar et al. (2019) reported that caryophyllene oxide, 2,3-dihydrofarnesol, iridodial-β-monoenol acetate, and β-sesquiphellandrene were major components present in the EOs of *N. leucophylla*. These results were considerably different both quantitatively and qualitatively compared to the results obtained by Sharma et al. (2021a) from EOs (dominated by β-chamigrene, germacrene D, and caryophyllene) isolated from the different parts of *N. leucophylla* collected from Himachal Pradesh area. These types of compositional differences in the EOs may be ascribed to different harvesting periods, agro-climatic and geographical conditions (Rather and Hassan 2011; Dhifi et al. 2016).

Table 18.6 Results of GC-MS analysis of essential oils isolated from different parts of *N. leucophylla* (Sharma et al. 2021)

Name of compound	Root (%/SI)	Stem (%/SI)	Leaves (%/SI)
α-Phellandrene	–	–	3.67/96
α-Terpinene	3.43/98	4.33/98	2.30/98
Benzeneacetaldehyde	1.59/97	4.73/98	1.66/84
Undecane	–	3.84/94	–
Caryophyllene	1.16/98	8.25/98	25.49/98
Germacrene D	–	–	11.17/98
β-Chamigrene	–	–	8.27/90
1-Hydroxy-4-methyl-2,6-di-tert-butylbenzene	–	7.21/96	–
Bornyl formate	3.89/89	–	–
δ-Cadinene	–	1.63/92	4.49/92
Spathulenol	–	–	3.18/89
Diisobutyl phthalate	–	7.02/96	–
Phytol	–	3.02/96	–
Nezukol	–	3.01/81	1.63/82
Totarol	30.55/93	3.14/91	–
Abieta-9(11),8(14),12-trien-12-Ol	2.64/88	9.01/91	4.99/91
Bis (6-methylheptyl) phthalate	–	16.60/93	–

SI similarity index, *UI* unidentified

18.5 Pharmacological Potential

18.5.1 Antifungal Potential

Saxena and Mathela (1996) evaluated iridodial b-monoenol acetate obtained from the *Nepeta leucophylla* Benth. EOs for its antifungal potential against four mycotoxin producing fungus viz. *Penicillium viridicatum, Penicillium citrinum, Aspergillus ochraceus and Aspergillus flavus*, and two soybean pathogens *Macrophomina phaseolina* and *Sclerotium rolfsii* using zones of inhibition method. The results revealed that the iridodial b-monoenol acetate completely inhibited the growth of all fungi at concentration of 0.012 g per flask. Bisht et al. (2010) screened EOs obtained from six *Nepeta* species against two fungal strains, viz. *Trichophyton rubrum* and *Candida albicans.* The results showed that EO (rich in iridodial *β*-monoenol acetate, iridodial dienol diacetate, and dihydroiridodial diacetate) obtained from *N. leucophylla* exhibited highest antifungal potential against both fungal stains as compared to all other EOs. Further, Kumar et al. (2014) screened EOs obtained from three *Nepeta* species (*N. leucophylla*, *N. clarkei*, and *N. ciliaris*) for their in vitro antifungal potential against soil born fungi viz. *Helminthosporium maydis*, *Curvularia lunata*, *Fusarium oxysporum*, *Rhizoctonia solani*, *Alternaria solani*, *Sclerotonia sclerotiorum*, and *Albugo candida*. The *N. leucophylla* EO showed highest antifungal potential against *H. maydis* (IC_{50}: 43.6 ug/mL) and for rest of the strain, it showed moderate activity. *N. leucophylla* EO showed significant inhibition potential (IC_{50} values 43.6–557.4 ug/mL; 44.0–83.9%) against *S. sclerotiorum, F. oxysporum, R. solani, H. maydis, and A. solani.*

18.5.2 Antibacterial

EOs isolated from six Himalayan *Nepeta* species, viz. *Nepeta clarkei* Hook. f., *Nepeta leucophylla* Benth., *Nepeta govaniana* Benth., *Nepeta erecta* Benth., *Nepeta elliptica* Royle ex Benth. and *Nepeta discolor* Royle ex Benth., were screened for their antibacterial potential against five gram negative bacteria (*Pasteurella multocida, Pseudomonas aeruginosa, Serratia marcescens, Escherichia coli*, and *Proteus vulgaris*) and one gram positive bacterium (*Staphylococcus aureus*) using disc diffusion method. Out of all the examined bacteria, *N. leucophylla* EO (rich in iridodial β-monoenol acetate, iridodial dienol diacetate, and dihydroiridodial diacetate) showed significant antibacterial potential against *P. aeruginosa, P. vulgaris*, and *S. aureus* with zone inhibition diameter of 20–27.4 mm (Bisht et al. 2010).

Sharma et al. (2023) evaluated the oleanolic acid, squalene, abieta-9(11),8(14),12-trien-12-ol and methyl ester of linolenic acid obtained from *N. leucophylla* extract for their antibacterial potential against *Staphylococcus aureus, Mycobactrium tuberculosis*, and *E. coli* by in silico method. The result revealed that all the isolated compounds reported to have a better Moldock score compared to ethambutol and near to ceftriaxone a reference drug with each target.

18.5.3 Antiviral

Sharma et al. (2023) evaluated *in-silico* antiviral potential of the oleanolic acid, squalene, abieta-9(11),8(14),12-trien-12-ol and methyl ester of linolenic acid obtained from *N. leucophylla* extract against H1N1 (Influenza virus), SARS-CoV-2 (Acute Respiratory Syndrome Coronavirus) and HIV (Human immunodeficiency virus). The results showed that oleanolic acid derivative (3-Phthaloyl oleanolic acid) provided best results against the target proteins reverse transcriptase (HIV) and hemagglutinin (H1N1), while oleanolic acid showed the best results against SARS-CoV-2 (RNA dependent RNA polymerase). Rest of all the molecules showed poor antiviral properties.

18.5.4 Antioxidant

The DPPH radical scavenging potential of *N. leucophylla* ranged between 8.93 ± 2.09 to 92.57 ± 0.11, 14.93 ± 1.50 to 90.43 ± 2.74, and 9.3 ± 0.59 to 91.4 ± 0.143% in the case of various extracts (methanol, chloroform, and hexane) obtained from aerial parts, stem, and roots, respectively. However, the results of NO percentage inhibition of various extracts obtained from aerial parts, stems, and roots ranged between 6.13 ± 0.41 to 64.03 ± 2.29, 1.29 ± 0.50 to 35.56 ± 0.18, and 2.58 ± 0.66 to 51.6 ± 0.05, respectively. Similarly, the results of FRAP values varied from 4.64 ± 0.71 to 196.38 ± 0.31, 4.41 ± 0.56 to 148.65 ± 0.38, and 2.54 ± 0.43 to 130.75 ± 0.09 mg Fe (II) E/g DPE in the case of different extracts obtained from aerial parts, stems, and roots of *N. leucophylla*, respectively, while, the results TAC assay in the various extracts isolated from aerial parts, stems, and roots ranged between 10.98 ± 1.36 to 74.72 ± 1.1, 0.11 ± 0.10 to 40.74 ± 1.97, and 2.81 ± 0.18 to 40.8 ± 0.34 mg AAE/g DPE, respectively (Sharma and Cannoo 2016a, b, 2017). High correlation was observed among the results of different antioxidant assays and polyphenolic content (TPC, TFC) of various extracts isolated via different extraction techniques from the aerial parts, stems, and roots of *N. leucophylla* confirmed that the polyphenolics were the foremost constituents responsible for the antioxidant activity of these extracts. Further, high concentration of polyphenolic compounds (TFC and TPC) in methanol extracts over chloroform and hexane extracts was the key reason behind the higher antioxidant activity of the former compared to the latter. Similarly, the higher antioxidant activity of different extracts isolated by SEM over the extracts obtained by MM and UAEM was also attributed to the higher TFC and TPC values in the extracts isolated by former method (Sharma and Cannoo 2016a, b, 2017; Sharma et al. 2020).

Singh and Dhaliwal (2019) screened *N. leucophylla* root extract and silver metal nanoparticles synthesized from root extract for it antioxidant potential. The results revealed that synthesized silver nanoparticles showed better DPPH scavenging potential (79.41% at 250 μg/ml concentration) as compared to the root extract of *N. leucophylla* (68.29% at 250 μg/ml concentration).

Kumar et al. (2019) evaluated the antioxidant potential of EOs isolated from five Lamiaceae and six Asteraceae species. Out of five Lamiaceae species, four belongs to genus *Nepeta* (*N. leucophylla, N. hindostana, N. ciliaris,* and *N. clarkei*). The antioxidant potential was evaluated by using five different methods viz. radical scavenging method (DPPH and superoxide radical scavenging assay), metal chelating, ferric-reducing antioxidant power (FRAP), lipid peroxidation inhibition (LPI), and total antioxidant capacity (TAC) assay. Among *Nepeta* species, *N. ciliaris* (IC_{50}: 0.9 mg/mL and IC_{50}: 1.0 mg/mL) and *N. leucophylla* (IC_{50}: 1.2 mg/mL and IC_{50}: 1.4 mg/mL) showed better DPPH and super oxide radical scavenging potential. Same trend was reported in the case of metal chelating assay. In FRAP assay, *N. leucophylla* showed lowest antioxidant potential. Further, in the case of LPI and TAC, *N. leucophylla* showed the highest inhibition of linoleic acid oxidation (38.9%) and TAC value (6.6 mg AAE/g) among all the Nepeta species.

The antioxidant potential of various EOs isolated from different parts of *N. leucophylla* ranged between 12.21 ± 0.37 to 14.33 ± 0.14% (DPPH assay), 21.27 ± 0.14 to 38.19 ± 0.39 mg AAE/g DPE (TAC assays) and18.66 ± 0.12 to 20.31 ± 0.07 mg Fe (II) E /g EO (FRAP assay) (Sharma et al. 2021a).

18.6 Nanotechnology

Previous studies revealed that the different extracts obtained from the various parts of *N. leucophylla* are the rich source of polyphenolic compounds (Sharma and Cannoo 2016a, b, 2017). These polyphenolic compounds are well known for their reduction potential and extensively used from the greener synthesis of metal nanoparticles. The synthesized metal nanoparticles showed various application in different fields, i.e., pharmaceutical, cosmetic, agriculture, and food. These metal nanoparticles also showed various pharmacological properties loke antimicrobial, anticancer, antifungal, and antioxidant. Table 18.7 represents the comparative analysis of various metal nanoparticles synthesized from different extracts of *N. leucophylla* based on structural, morphological, and functional properties (Singh and Dhaliwal 2019; Singh et al. 2020; Kathuria et al. 2022). Figure 18.5 represents the different steps involved in synthesis of metal nanoparticles of different *N. leucophylla* extracts, while Fig. 18.6 presents the mechanism involved in the synthesis of metal nanoparticles from polyphenol-rich extracts and stabilization of synthesis metal nanoparticles by polyphenols.

Table 18.7 Comparative analysis of metal nanoparticles obtained from different extracts of *N. leucophylla*

References	Part	NPs	Particle size	Shape	Image	Antioxidant property
Singh and Dhaliwal (2019)	Root	Ag	20 nm	Spherical	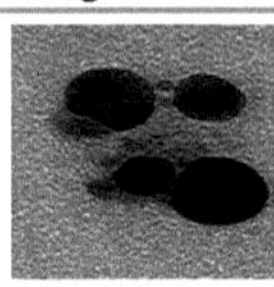	119.28 ± 0.011 μg/ml
Singh and Dhaliwal (2020)	Stem	Ag	15–25 nm	Spherical	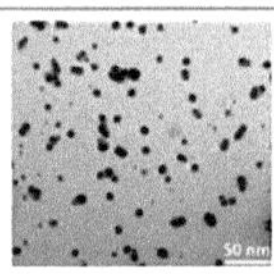	–
Singh et al. (2020)	Stem	Au	75 ± 3 nm	Spherical	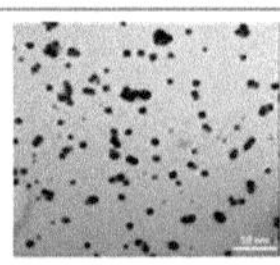	68.57%
Kathuria et al. (2022)	Stem	Au	15–20 nm	Spherical, octagonal, hexagonal	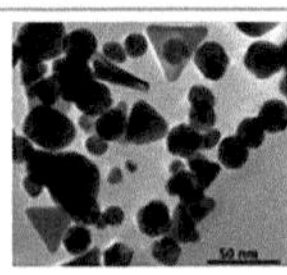	54.47 ± 1.47%

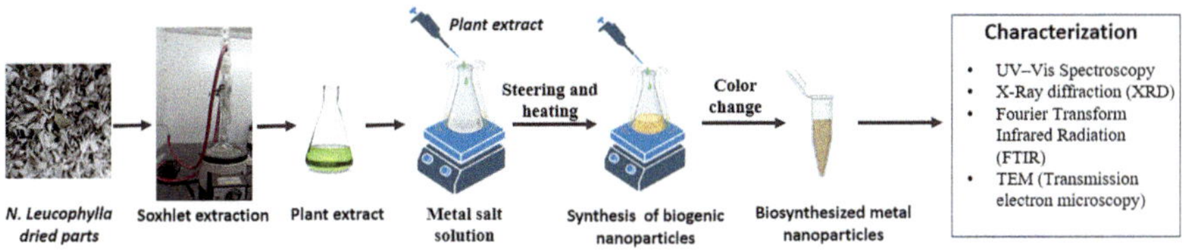

Fig. 18.5 Process of synthesis of metal nanoparticles of *N. leucophylla* extracts. (Kathuria et al. 2022)

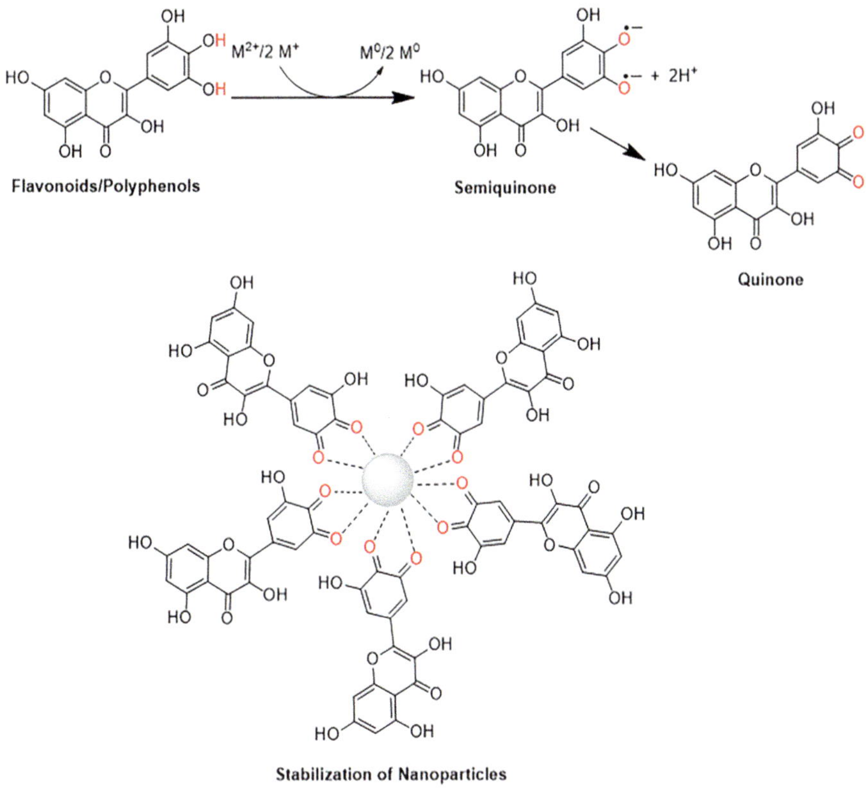

Fig. 18.6 Mechanism involved in the synthesis of metal nanoparticles from polyphenol-rich extracts and stabilization of synthesized nanoparticles. (Kathuria et al. 2022)

18.7 Future Perspectives

Figure 18.7 presents the various potent areas in which various extracts of *N. leucophylla* can be explored in future for their potential applications.

Nanotechnology *N. leucophylla* is a rich source of various classes of secondary metabolites viz. flavonoids, phenolic acids, terpenoids, and steroids. The biomole-

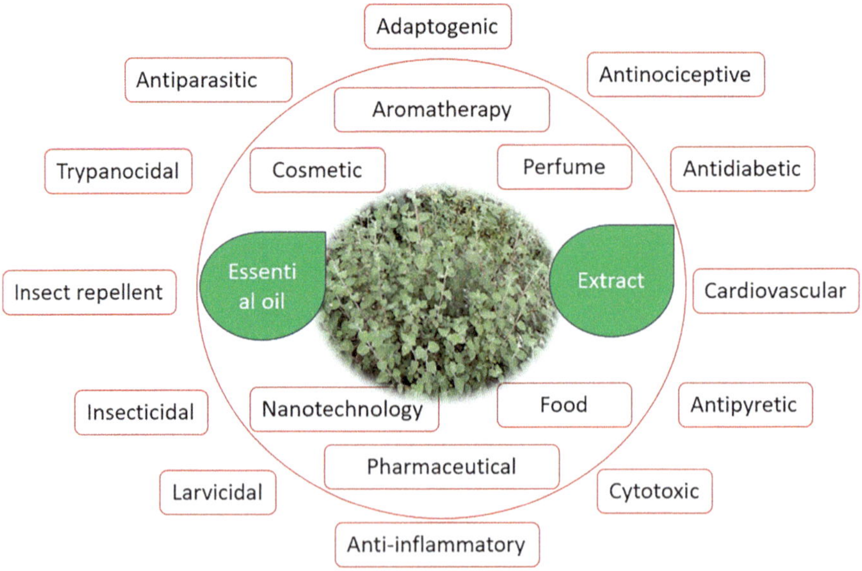

Fig. 18.7 Potential areas for future exploration of various extracts of *N. leucophylla* in terms of their possible applications

cules such as flavonoids and phenolic acids are well known as metal reduction, stabilizers, and capping agents. Therefore, the different alcoholic and hydroalcoholic extracts obtained from different parts of *N. leucophylla* can be explored for their potential in the field of nanotechnology for the greener synthesis of metal nanoparticles, nano emulations, and carbon nano dots. Further these synthesized nanoparticles and nano dots can be explored for their applications in pharmaceuticals, nutraceuticals, food, and agriculture industries.

Food and Nutraceuticals Owing to the high concentration of polyphenolic compounds various extracts of *N. leucophylla* can be evaluated for their potential as a source of natural antioxidant and food preservative.

Aromatherapy and Insect Repellent Different parts of *N. leucophylla* are rich source of aromatic essential oils (rich in mono, sesqui, and diterpenes), which can be explored for their potential in the aromatherapy and as insect repellent.

Cosmetic Industry The various extracts of *N. leucophylla* are rich in bioactive SMs such as polyphenols, terpenoids, and steroids; therefore, these extracts can act as a rich source of natural antioxidant, antibacterial, antifungal, antiviral, and anti-inflammatory agents for various cosmetic formulations.

Environmental Remediation The application of various extracts of *N. leucophylla* and synthesized nanoparticles can be explored in the field of green catalysis, heavy metal removal, and dye degradation by various sensing techniques.

Ethnopharmacological Prospective The different extracts obtained from *N. leucophylla* have been reported to have prominent antibacterial, antifungal, antiviral, antioxidant potential, and are rich source of bioactive SMs. The species of genus *Nepeta* are well known for their various biological properties such as cytotoxic, anticancer, anti-inflammatory, anti-nociceptive, nematocidal, insecticidal, anthelmintic, trypanocidal, spasmolytic, bronchodilator, vasorelaxant, platelet aggregation, and anti-atherosclerotic phytotoxic activity. All these species have comparable chemical composition to *N. leucophylla*. Therefore, there has been noteworthy scope for exploring various extracts of *N. leucophylla* for their diverse pharmacological potential.

Toxicological and Clinical Studies Further, there are some specific research gaps like lack of toxicological data, standardized and effective dosing strategies, and lack of clinical trials. These gaps need to be filled to ensure long-term usage of various *N. leucophylla* extracts with safety. Preclinical studies highlight the pharmacological potential of *N. leucophylla*, but still there is a notable lack of clinical trials to validate its efficacy and safety.

18.8 Conclusion

N. leucophylla has been one of the representative species of genus *Nepeta* in the western Himalayas. The phytochemical diversity of *N. leucophylla* has mainly been characterized by flavonoids, phenolic acids, terpenoids, and steroids. These phytochemicals have been mainly responsible for its different pharmacological properties such as antibacterial, antifungal, antiviral, and antioxidant. The current review revealed that the different extracts and EOs isolated from different parts of *N. leucophylla* have significant potential in the field of nanotechnology, pharmaceutical, cosmetics, aromatherapy, environmental remediation, food, and nutraceuticals. So for, many bioactive SMs have been reported in the different parts of *N. leucophylla*, yet the nature likely holds many more in her basket. Further, *N. leucophylla* has not been extensively investigated for its complete pharmacological potential. Therefore, the different extracts and EOs obtained from different parts of *N. leucophylla* can be further investigated in vitro and in vivo for their pharmacological potential against different pathogenic viruses, fungus, bacteria, and various degenerative diseases. Thus, the present review can play a significant role in progressing research focused at identifying and characterizing new bioactive SMs for the treatment of several degenerative diseases, while also providing important insights into elite germplasm for future research and advancements.

References

Agati G, Azzarello E, Pollastri S, Tattini M (2012) Flavonoids as antioxidants in plants: location and functional significance. Plant Sci 196:67–76

Atanasov AG, Waltenberger B, Pferschy-Wenzig EM, Linder T, Wawrosch C, Uhrin P, Temml V, Wang L, Schwaiger S, Heiss EH, Rollinger JM, Schuster D, Breuss JM, Bochkov V, Mihovilovic MD, Kopp B, Bauer R, Dirsch VM, Stuppner H (2015) Discovery and resupply of pharmacologically active plant-derived natural products: a review. Biotechnol Adv 33:1582–1614

Azmir J, Zaidul IS, Rahman MM, Sharif KM, Mohamed A, Sahena F, Jahurul MH, Ghafoor K, Norulaini NA, Omar AK (2013) Techniques for extraction of bioactive compounds from plant materials: a review. J Food Eng 117:426–436

Ballabh B, Chaurasia OP (2009) Medicinal plants of cold desert Ladakh used in the treatment of stomach disorders. Indian J Tradit Knowl 8:185–190

Bisht DS, Mathela CS, Lal P, Singh L, Pande V, Padalia RC (2010) Constituents and antimicrobial activity of the essential oils of six Himalayan *Nepeta* species. J Serb Chem Soc 75:739–747

Bottini AT, Dev V, Shah GC, Mathela CS, Melkani AB, Nerio AT, Sturm NS (1992) Cyclopentanomonoterpene enol acetates from *Nepeta leucophylla*. Phytochemistry 31:1653–1657

Cragg GM, Newman DJ (2013) Natural products: a continuing source of novel drug leads. Biochim Biophys Acta 1830:3670–3695

Dhifi W, Bellili S, Jazi S, Bahloul N, Mnif W (2016) Essential oils' chemical characterization and investigation of some biological activities: a critical review. Medicine 3(4):25

Edewor TI, Usman LA (2011) Phytochemical and antibacterial activities of leaf extracts of *Nepeta cataria*. Afr J Pure Appl Chem 5:503–506

Fareed G, Afza N, Malik A, Mughal UR, Tareen RB, Varsiani MA, Fareed M, Lateef M (2013) Phytochemical screening, total phenolic contents and biological evaluation of aerial parts of *Nepeta praetervisa*. J Chem Soc Pak 35:1366–1370

Formisano C, Rigano D, Senatore F (2011) Chemical constituents and biological activities of *Nepeta* species. Chem Biodivers 8:1783–1818

Gupta YN (1973) Indian Perfum 17(pt. 2):28

Gupta YN, Baslas KK, Chauhan RNS (1971) Chemical examination of essential oil from the plant *Nepeta leucophylla*. Indian Oil Soap J 36:257

Haizhou L, Pordesimo L, Weiss J (2004) High intensity ultrasound-assisted extraction of oil from soybeans. Food Res Int 37:731–738

https://iiim.res.in/herbarium/lamiaceae/nepeta_leucophylla.htm

https://plants.jstor.org/stable/10.5555/al.ap.specimen.k000911181

Kalia K, Sharma K, Singh HP, Singh B (2008) Effects of extraction methods on phenolic contents and antioxidant activity in aerial parts of *Potentilla atrosanguinea* lodd. and quantification of its phenolic constituents by RP-HPLC. J Agric Food Chem 56:10129–10134

Karabegovic IT, Stojicevic SS, Velickovic DT, Todorovic ZB, Nikolic NC, Lazic ML (2014) The effect of different extraction techniques on the composition and antioxidant activity of Cherry Laurel (*Prunus laurocerasus*) leaf and fruit extracts. Ind Crop Prod 54:142–148

Kathuria D, Bhattu M, Sharma A, Sareen S, Verma M, Kumar S (2022) Catalytic reduction of water contaminants using green gold nanoparticles mediated by stem extract of *Nepeta leucophylla*. Top Catal 65(19):1899–1909

Koksal E, Tohma H, Kılıc M, Alan Y, Aras A, Gulcin I, Bursal E (2017) Assessment of antimicrobial and antioxidant activities of *Nepeta trachonitica*: analysis of its phenolic compounds using HPLC-MS/MS. Sci Pharm 85:24–37

Kumar M, Sheikh MA, Bussmann RW (2011) Ethnomedicinal and ecological status of plants in Garhwal Himalaya, India. J Ethnobiol Ethnomed 7:32–44

Kumar V, Mathela CS, Tewari AK, Bisht KS (2014) In vitro inhibition activity of essential oils from some Lamiaceae species against phytopathogenic fungi. Pestic Biochem Physiol 114:67–71

Kumar V, Mathela CS, Kumar M, Tewari G (2019) Antioxidant potential of essential oils from some Himalayan Asteraceae and Lamiaceae species. Med Drug Discov 1:100004

Mathela CS, Gupta A, Upreti P, Pant AK, Olmstead MM, Hope H, Bottini AT (1991) Coleon U 12-methyl ether from *Nepeta leucophylla*. J Nat Prod 54:910–912

Micelia N, Taviano MF, Giuffrida D, Trovato A, Tzakou O, Galati EM (2005) Anti-inflammatory activity of extract and fractions from *Nepeta sibthorpii* Bentham. J Ethnopharmacol 97:261–266

Misic D, Siler B, Gasic UM, Avramov S, Zivkovic S, Zivkovic JN (2015) Simultaneous UHPLC/DAD/(+/−)HESI-MS/MS analysis of phenolic acids and nepetalactones in methanol extracts of *Nepeta* species: a possible application in chemotaxonomic studies. Phytochem Anal 26:72–85

Patial PK, Sharma A, Cannoo DS (2019) Correlation study among the extraction techniques, phytochemicals and antioxidant activity of *Nepeta spicata* aerial parts. Biocatal Agric Biotechnol 20:1–14. https://doi.org/10.1016/j.bcab.2019.101275

Rababah TM, Banat F, Rababah A, Ereifej K, Yang W (2010) Optimization of extraction conditions of total phenolics, antioxidant activities and anthocyanin of oregano, thyme, terebinth and pomegranate. J Food Sci 75:C626–C632. https://doi.org/10.1111/j.1750-3841.2010.01756.x

Rana SK, Rawat GS (2017) Database of Himalayan plants based on published floras during a century. Data 2(4):36

Rather MA, Hassan T (2011) Analysis of the diterpene rich essential oil of *Nepeta clarkei* Hooke from Kashmir Himalayas by capillary GC-MS. Int J ChemTech Res 3(2):959–962

Saxena J, Mathela CS (1996) Antifungal activity of new compounds from *Nepeta leucophylla* and *Nepeta clarkei*. Appl Environ Microbiol 62:702–704

Seladji M, Bekhechi C, Beddou F, Dib H, Bendimerad N (2014) Antioxidant activity and phytochemical screening of *Nepeta nepetella* aqueous and methanolic extracts from Algeria. J Appl Pharm Sci 4:12–16

Sharma A, Cannoo DS (2013) Phytochemical composition of essential oils isolated from different species of genus Nepeta of Labiatae family: a review. Pharmacophore 4(6):181–211

Sharma A, Cannoo DS (2016a) Effect of extraction solvents/techniques on polyphenolic contents and antioxidant potential of the aerial parts of *Nepeta leucophylla* and the analysis of their phytoconstituents using RP-HPLC-DAD and GC-MS. RSC Adv 6(81):78151–78160

Sharma A, Cannoo DS (2016b) Comparative evaluation of extraction solvents/techniques for antioxidant potential and phytochemical composition from roots of *Nepeta leucophylla* and quantification of polyphenolic constituents by RP-HPLC-DAD. J Food Meas Charact 10(3):658–669

Sharma A, Cannoo DS (2017) A comparative study of effects of extraction solvents/techniques on percentage yield, polyhenolic composition, and antioxidant potential of various extracts obtained from stems of *Nepeta leucophylla*: RP-HPLC-DAD assessment of its polyhenolic constituents. J Food Biochem 41(2):e12337

Sharma A, Nayik GA, Cannoo DS (2019) Pharmacology and toxicology of *Nepeta cataria* (catmint) species of genus nepeta: a review. In: Plant and human health, vol 3. Springer, Cham, pp 285–299

Sharma A, Bhardwaj G, Gaba J, Cannoo DS (2020) Natural antioxidants: assays and extraction methods/solvents used for their isolation. In: Antioxidants in fruits: properties and health benefits. Springer, Singapore, pp 1–33

Sharma A, Bhardwaj G, Cannoo DS (2021a) Antioxidant potential, GC/MS and headspace GC/MS analysis of essential oils isolated from the roots, stems and aerial parts of *Nepeta leucophylla*. Biocatal Agric Biotechnol 32:101950

Sharma A, Cooper R, Bhardwaj G, Cannoo DS (2021b) The genus *Nepeta*: traditional uses, phytochemicals and pharmacological properties. J Ethnopharmacol 268:113679. https://doi.org/10.1016/j.jep.2020.113679

Sharma A, Kathuria D, Kolita B, Gohain A, Das AK, Bhardwaj G, Simal-Gandara J (2023) Greener approach for the isolation of oleanolic acid from *Nepeta leucophylla* Benth. Its derivatization and their molecular docking as antibacterial and antiviral agents. Heliyon 9(8):e18639

Singh JS (2006) Sustainable development of the Indian Himalayan region: linking ecological and economic concerns. Curr Sci 90:784–788

Singh J, Dhaliwal AS (2019) Novel green synthesis and characterization of the antioxidant activity of silver nanoparticles prepared from *Nepeta leucophylla* root extract. Anal Lett 52:213–230. https://doi.org/10.1080/00032719.2018.1454936

Singh J, Dhaliwal AS (2020) Plasmon-induced photocatalytic degradation of methylene blue dye using biosynthesized silver nanoparticles as photocatalyst. Environ Technol 41(12):1520–1534

Singh J, Kumar S, Dhaliwal AS (2020) Controlled release of amoxicillin and antioxidant potential of gold nanoparticles-xanthan gum/poly (Acrylic acid) biodegradable nanocomposite. J Drug Deliv Sci Technol 55:101384

Suntar I, Nabavi SM, Barreca D, Fischer N, Efferth T (2017) Pharmacological and chemical features of *Nepeta* L. genus: its importance as a therapeutic agent. Phytother Res 32:185–198

Vinatoru M (2001) An overview of the ultrasonically assisted extraction of bioactiveprinciples from herbs. Ultrason Sonochem 8:303–313

Wang J, Sun BG, Cao Y, Tian Y, Li XH (2008) Optimization of ultrasound-assisted extraction of phenolic compounds from wheat bran. Food Chem 106:804–810

Phytoconstituents, Pharmacology, Production Technology and In Vitro Micropropagation Studies of High-Value-Trans-Himalayan Medicinal Plant *Rhodiola imbricata Murr*

19

Shradulya Shukla, Puspinder Bhardwaj, Nitish Kumar, and Rajni Sharma

Abstract

The trans-Himalayan cold desert region is a great repository of natural botanical resources having diverse utility towards development of phytofoods, health promoting natural products and other medicinal and therapeutic formulations. These plants contain a plethora of various classes of bioactive compounds like flavonoids, glycosides, phenylpropanoids, etc., which have a very high value towards the human health. Among them, *Rhodiola imbricata* is a high-value medicinal plant of trans-Himalayan mountain passes. This plant is highly sought after in national and international herbal product markets due to its unique phytochemical composition and resultant medicinal properties. The present chapter aims at providing comprehensive and current information regarding geographical distribution, phyto-fingerprinting, pharmacological potential, followed by production technology and in vitro micropropagation studies. The present chapter critically analyses the findings of scientific studies conducted on *R. imbricata* to date.

S. Shukla · N. Kumar
Defence Institute of High Altitude Research (DIHAR), Defence Research and Development Organization (DRDO), Ministry of Defence, Leh-Ladakh, Jammu & Kashmir, India

P. Bhardwaj · R. Sharma (✉)
Natural Products Medicinal Chemistry Divisions, Indian Institute of Integrative Medicine (CSIR), Jammu, Jammu & Kashmir, India

S. Chandra, H. Lata (eds.), *High Altitude Medicinal Plants*,
https://doi.org/10.1007/978-3-031-99406-7_19

19.1 Introduction

Owing to its extreme environmental conditions, the cold desert is known for its ethano-botanical wealth, especially unique high-value medicinal plants. The trans-Himalayan region is a land of valleys and high mountain passes varying in altitudes, thus encompassing unique microclimatic conditions that uniquely affect the phytochemical contents of native medicinal plants (Ballabh et al. 2007; Dhar et al. 2019). Among these high-value medicinal plants, Rhodiola is one of the perennial herbs that belongs to the Campanulaceae family and also resembles the sebum, which is known as stonecrops, a genus within the Rhodiola family, comprising approximately 90 species found in the alpine regions of the Northern Hemisphere. This genus is characterized by its fleshy rhizomes, variably thick or thin, with basal or scaly leaves. Among them, *Rhodiola imbricata* is an important medicinal herb also known as Himalayan stone crop or *golden arctic root*, commonly known by different names such as rose root, arctic root, golden root, Himalayan stone crop, shrolo marpo and lalphool, belonging to Crassulaceae family (stonecrops) with a plethora of established medicinal properties. Most of the *Rhodiola* species are dispersed in cold areas of northern hemisphere, which includes Asia and Europe, whereas distribution of *Rhodiola imbricata* is only restricted to trans-Himalayan region of Leh-Ladakh in India. Extracts of *R. imbricata* have been studied for many pharmacological activities like neurogenerative disorder, antioxidant, radioprotective, immunomodulatory and antiproliferative due to the presence of bioactive constituents like phenylpropanoids and phenylethanoids (Gupta et al. 2007; Tayade et al. 2013a, b; Rattana et al. 2020; Bhardwaj et al. 2018a, b). Since wild resources are the major source of raw material for *Rhodiola*-based products, selection of quality raw material is imperative for optimum efficacy of the finished product. Traditionally, it has been used by indigenous communities for enhancing physical and mental endurance in extreme environmental conditions, making it a significant subject for modern pharmacological research. However, despite such extensive folk use, there is still a need to explore this indigenous plant, although Pundir et al. (2021) have already compiled the data till 2021 for *R. imbricata*. So here we just want to highlight the development more and studies related to cultivation as well as phytoconstituents.

19.2 Geographical Distribution and Botanical Description

19.2.1 Geographical Distribution

Rhodiola imbricata Edgew is an herbaceous perennial plant of the family Crassulaceae (Bhardwaj et al. 2018a), which grows in the cold regions of the northern hemisphere and mountainous areas of Asia, Canada, Mongolia and Europe (Khanna et al. 2017; Grech-Baran et al. 2015; Kaur et al. 2023). The genus *Rhodiola* contains more than 90 different species (Wal et al. 2019) out of which 6 species of Rhodiola specifically *R. imbricata, R. quadrifida, R. heterodonta, R. tibetica, R.*

sinuate and R. wallichiana are found in India (Chaurasia and Gurmet 2003). In India, its distribution is restricted to the trans-Himalayan Leh-Ladakh region of Chang La (5360 m) and Khardung La (5359 m) (Gupta et al. 2012). It is also known as Himalayan rose root, golden root and Himalayan stone crop (Chaurasia et al. 2003). According to the data reported on the Indian portal of Biodiversity, some parts of vegetation are reported in Chang La, Pensi La, and Kumaon;, Rohtang Pass, Himachal Pradesh; Khardung La Pass, Jammu Kashmir; and Birje Ganj pass, Uttrakhand (Pundir et al. 2021).

19.2.2 Botanical Description

Rhodiola is a dioecious perennial herb that grows to a height of 70–75 cm (Wal et al. 2019). Its leaves, which are densely arranged, measure 1.2–3.5 cm in length and range from oblong to narrowly elliptic. The stem is 8–18 cm long and smooth. The flowers, which can be four or five, parted, are either dioecious or hermaphroditic. The angular-oblanceolate petals are accompanied by stamens that are notably longer, with purplish-red anthers measuring 2.6–5.2 mm and filaments 4.8–8.2 mm long. The plant bears —three to six fruits, each containing multiple seeds, with a subcylindrical rhizome measuring 8–40 cm in length (Chaurasia et al. 2007; Gupta et al. 2012; Bhardwaj et al. 2018a) (Fig. 19.1).

19.3 Phytochemistry

As already been discussed, *R. imbricata* is naturally decreasing due to its overexploitation, high altitude region and traditional usage in the Amchi system of medicine (Rattan et al. 2020). Very little has been done to explore the secondary metabolities of this plant, so if we talk about isolation part, most of the work has been based on qualitative as well as quantitative analysis. There are many herbal formulations based on Rhodiola where salidroside and rosavins have been found as the most potent ingredients. Several secondary metabolites such as phenylethanol derivatives, phenylpropanoids, phenolic acids and flavonoids were reported from *R. imbricata* (Singamaneni et al. 2020). Senthilkumar et al. (2014) have done the high-performance liquid chromatography (HPLC) profiling of rhizomes, which showed the presence of some important phenolic compounds like gallic acid and rutin in a good amount. Most of the study has been done to explore the secondary metabolites via a GC-MS analysis by Tayade et al. (2013a, b) on different extracts of *R. imbricata* roots, which have been discussed below. Very little work has been done on the isolation part; only 19 compounds have been isolated till now (Fig. 19.5) from ethyl acetate and n-butanol extracts through preparative HPLC (Choudhary et al. 2015).

Fig. 19.1 Botanical description of *Rhodiola imbricate*

19.3.1 Qualitative and Quantitative Analysis: Dietary Amino Acids, Fatty Acids, Minerals and Vitamins

Tayade et al. 2017 have done the qualitative and quantitative analysis of amino acids, fatty acids and dietary minerals in *R. imbricata* root using pre-column phenylisothiocyanate (PITC) derivatization following reverse phase-high-performance liquid chromatography (RP-HPLC), fatty acid methyl esters (FAMEs) derivatization with gas chromatography coupled with flame ionization detector (GC-FID) and inductively coupled plasma optical emission spectrometry (ICP-OES) analysis. Result has shown the presence of 9 essentials and 12 non-essential amino acids range from 91.33 ± 7.77 μg/g to 1640.67 ± 11.85 μg/g. Gly (1640.67 lg/g) was found to exhibit the highest content followed by His (1434.33 lg/g), Lys (1329.33 lg/g), Pro (1263.67 lg/g), Ala (1142.33 lg/g), Cys HCL (1136.33 lg/g) and

Nor Leu (1038.67 lg/g). His and Lys were found to be the dominant contributors of essential amino acids, while the major non-essential amino acids were Gly, Pro, Ala, Cys HCL and Leu. L-2-amino-nbutyric acid, Trp and Orn were not detected. Val was found to be in the below detection limit (BDL) range and hence its content was not quantified (Hemant soo et al. 2021). Analysis of fatty acids in R. imbricata root showed the occurrence of 10 fatty acids contributing 99.99% of the total lipid (Table 19.3). The total lipid was found to be a rich source of SFAs viz. capric acid (19.91%), caproic acid (10.87%), palmitic acid (9.42%), lignoceric acid (6.16%) and behenic acid (5.71%), which together constituted 52% of the total lipid (Fig. 19.2) Among the Unsaturated Fatty Acids (UFA), linoleic acid (15.06%), oleic acid (12.38%), arachidonic acid (8.38%), linolelaidic acid (6.11%) and docosadienoic acid (5.99%) were prominent. MUFA and PUFA were 35.54% and 12.38% of the total lipid content respectively in the root sample whereas found to contain thirteen important mineral elements required for human nutrition. Calcium (11034.17 mg/kg) was quantified in highest amount tracking down to potassium (2143.25 mg/kg), iron (1441.17 mg/kg), magnesium (581.99 mg/kg), phosphorus (376.72 mg/kg), sodium (109.75 mg/kg), manganese (75.78 mg/kg), zinc (16.27 mg/kg), chromium (7.27 mg/ kg), nickel (4.89 mg/kg), copper (3.49 mg/kg), cobalt (2.98 mg/kg) and molybdenum (2.65 mg/kg). These micronutrients could be beneficial for normal growth and development for human and other animals. Rapid resolution liquid chromatography/tandem mass spectrometry (RRLC–MS/MS) with electrospray ionization (ESI) source operating in multiple reactions monitoring (MRM) mode was optimized for the sequential analysis of nine water-soluble vitamins (B1, B2, two B3 vitamins, B5, B6, B7, B9 and B12) and six fat-soluble vitamins (A, E, D2, D3, K1 and K2) in *R. imbricata* roots (Fig. 19.2).

19.3.2 GC-MS-Based Chemoprofiling of Different Extracts of *R. imbricata* Roots

Earlier reports of GC-MS chemo profiling of different extract (n-hexane, ethyl acetate, chloroform, methanol, ethanol and aqueous) of *R. imbricata* roots have revealed the presence of different phytochemicals like fatty ester, flavonoids, terpenoids, phytosterols, alkanes, alkylaldheydes, *etc* (Tayade et al. 2013a, b; Hemant et al. 2021; Singamaneni et al. 2020) (Tables 19.1 and 19.2, and Fig. 19.4). A recent study by Sharma et al. (2022) analyzed the chemometric profiles of *Rhodiola imbricata* root extracts (50:50 hydromethanolic) collected from three distinct locations in Ladakh—Chang La, Khardung La and Shashi La—using GC-MS (Figs. 19.3 and 19.4). The study is significant for being the first to report the presence of apocynin, a potent ROS down-regulating compound, in the *Rhodiola* genus. The results revealed notable variations in the volatile phytochemical compositions across the regions, underscoring the influence of geographic origin on phytoconstituent profiles and the need for location-based quality markers in product development. Chang La samples were dominated by phenols (96.78%), primarily piceol (94%) and apocynin (2.78%); Khardung La samples also showed high piceol (73%) and apocynin (3.25%) content with other minor constituents, while Shashi La samples had a

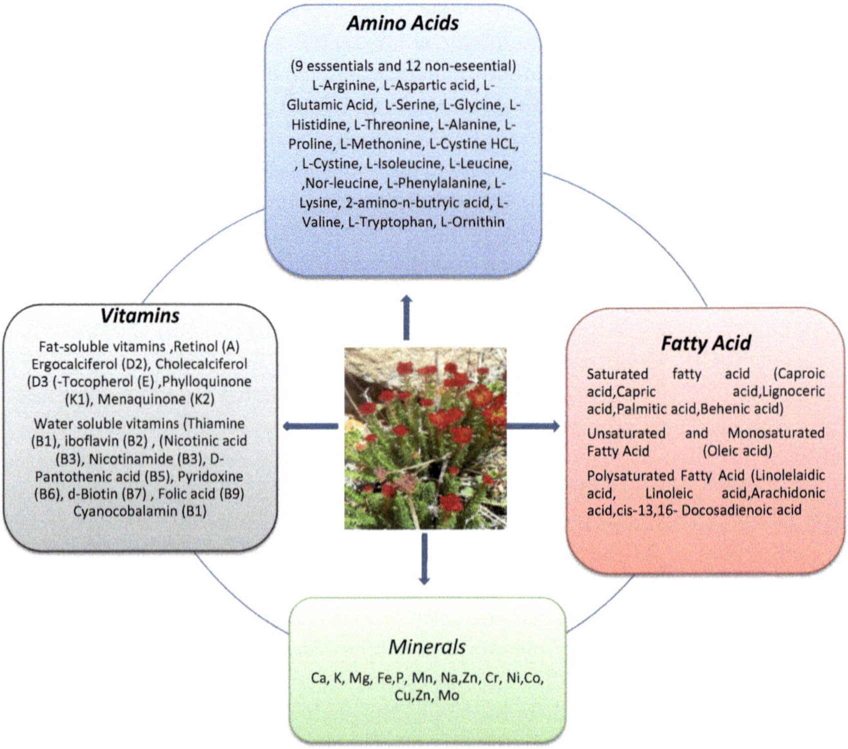

Fig. 19.2 Primary metabolites reported from *Rhodiola imbricate*

different profile dominated by alkanes and fatty acids. A total of 13 bioactive compounds (with similarity index ≥80%) were identified based on molecular weight, retention time and structural match with the NIST library. Significant statistical differences ($p \leq 0.001$) in peak areas of major compounds (piceol and apocynin) were observed, particularly between samples from Chang La and Khardung La regions (Table 19.3).

19.3.3 Isolation and Characterization of Compounds by Column Chromatography

Fifteen compounds, including four new compounds, were isolated from the ethyl acetate extract and four compounds from n-butanol extract 3,5-dihydroxy-benzyl, 3-methoxy-5-hydroxybenzyl alcohol, orcinol, O-methylorcinol, Tyrosol, p-hydroxybenzylalcohol, p-hydroxybenzaldehyde, p-hydroxyacetophenone, 4-methoxyphenethyl alcohol, 3-hydroxy-5-methylphenylβ-D-glucopyranoside, 3-methyl-5-methoxyphenyl-β-D-glucopyranoside,2-hydroxymethyl-6-methoxyphenyl-β-D-glucopyranoside, phenyl-β-D-glucopyranoside, 3,5-dimethoxyphenyl-β-D-glucopyranoside and 3,4,5-trimethoxyphenyl-β-D-glucopyranoside (Choudhary et al. 2015) (Fig. 19.5).

Table 19.1 Class of Compounds reported via a GC-MS in different extracts of *R. imbricata*

Hexane extract	Peak area (%)	DCM:$CHCl_3$	Peak area (%)
1-pentacosanol	28.2	Stigmasterol	24.30
17-pentatriacontene	7.1	Methyl tri-butyl NH4Cl	14.64
Stigmast-5-en3-ol, (3β,24S)	13.4	Bis (2-ethylhexyl) phthalate	11.50
1-tetracosanol	9.3	7,8-dimethylbenzocyclooctene	7.97
1-henteracontanol	8.5	Ethyl-linoleate	4.75
13-tetradecen-1-ol acetate	6.4	3-methoxy-5-methylphenol	4.16
1-heptacosane	3.5	Hexadecanoic acid	4.13
1-hentriacontane	3.6	Campesterol	3.94
1-tericosanol	2.5	Benzene methanol, 3-hydroxy-5-methoxy	2.62
α-tocopherol-β-D-mannoside	0.7	17-pentatriacontene	3.38
13-docosan-1-ol, (Z)	2.2	Benzene sulfonic acid, 4-amino-3-nitro	3.21
Eicosen-1-ol, cis-9	1.9	Orcinol	2.93
Stigmast-4-en-3-one	1.3	1-tetracosanol	1.86
Bis (2-ethylhexyl) phthalate	1.2	Stigmast-4-en-3-one	1.82
Hexadecanoic acid	1.2	α-Tocopherol	1.31
1-tetrateracontane	0.9	Eicosen-1-ol, cis-9	1.13
Campesterol	0.9	Eucalyptol	1.95
Stigmastanol	0.7	Thujone	4.73
3-methoxy-5-methylphenol	0.5	Camphor	0.85
		Borneol	1.22
		b-fenchyl alcohol	0.91
		Benzenethiol, 4-chloro	1.02

19.4 Pharmacological Properties

In past years, an extensive study has been conducted to decipher the effect of *Rhodiola imbricata* on various biological ailments. Studies have shown that the plant is known to have antioxidant, adaptogenic, antifatigue, antiaging, antimicrobial, antidepressant, radioprotective and anticancer properties. For these activities shown by the flora is majorly associated with its key secondary metabolites reported in the plant.

Table 19.2 Class of Compounds reported via a GC-MS in different extracts of *R. imbricata*

EtOH	Peak area (%)	Methanol	Peak area (%)
1,3-dimethoxybenzene	27.6	3-methoxy-5-methylphenol	0.43
1,3-benzenediol, 5-pentadecyl	16.9	Ethanone, 1-(4-hydroxyphenyl)	11.07
1,3-benzenediol, 5-methyl	8.4	1-dodecane	0.66
3-methoxy-5-methylphenol	10.1	1-dotriacontane	2.21
Benzenemethanol, 3-hydroxy, 5-methoxy	5.7	Octadecane, 1-chloro	17.01
Cholest-4-ene-3,6-dione	5.7	Hexadecanoic acid, methyl ester	1.73
Dodecanoic acid, 3-hydroxy	4.5	Hexadecanoic acid	3.67
7,8-dimethylbenzocyclooctene	3.6	9,12-octadecadienoic acid (Z,Z)	3.99
3,5-dimethoxyphenyl acetate	3.4	2-hydroxy-1-(hydroxymethyl)ethyl ester	
1-dodecanol, 3,7,11-trimethyl	0.6	9,12,15-octadecatrienoic acid, 2,3-dihydroxypropyl ester (Z,Z,Z)	1.74
Stigmasterol (2.2)	0.7	Ethyl linoleate	3.35
Hexadecanoic acid	1.8	1,3-dimethoxybenzene	3.57
Oleic acid	1.4	Ascaridole	5.92
Stigmast-4-en-3-one	1.3	d-tocopherol	0.56
Bis (2-ethylhexyl) phthalate	1.2	a-tocopherol	8.42
Hexadecanoic acid	1.2	Campesterol	4.98
1-tetrateracontane	0.9	Stigmast-5-en-3-ol, (3b,24S)	21.91
Campesterol	0.9		
Stigmastanol	0.7	Stigmast-4-en-3-one	1.55
3-methoxy-5-methylphenol	0.5		
Dotriacontane	5.7		
Heptadecane, 9-hexyl	5.4		
Hexadecanoic acid methyl ester	2.3		
Bis (2-ethylhexyl) phthalate	3.6		

19.4.1 Antioxidant Properties

Authors have also reported the antioxidant properties of *Rhodiola imbricata* using various assays i.e. 1,1-diphenyl-2-picrylhydrazyl radical (DPPH) radical-scavenging activity, 2,2′-azinobis-(3-ethylbenzothiazoline-6-sulfonic acid) diammonium salt (ABTS) radical-scavenging activity, nitric oxide radical-scavenging activity, hydroxyl radical-scavenging activity, total antioxidant capacity and reducing power assay (Kanupriya et al. 2005; Raj et al. 2010; Chen et al. 2012; Senthilkumar et al. 2014; Tayade et al. 2013a, b; Choudhary et al. 2015; Kapoor et al. 2018, 2022). Secondary metabolite i.e. salidroside has been reported to protect human erythrocytes from H_2O_2-mediated erythroptosis and also reduces the overall production of reactive oxygen species (ROS) (Qian et al. 2012). Qian et al. (2011) have reported to improved increase in thioredoxin-1 (Trx1) and glutathione peroxidase-1 (GPx1) after treatment with salidroside promoting erythropoiesis and protecting erythroblasts against oxidative stress.

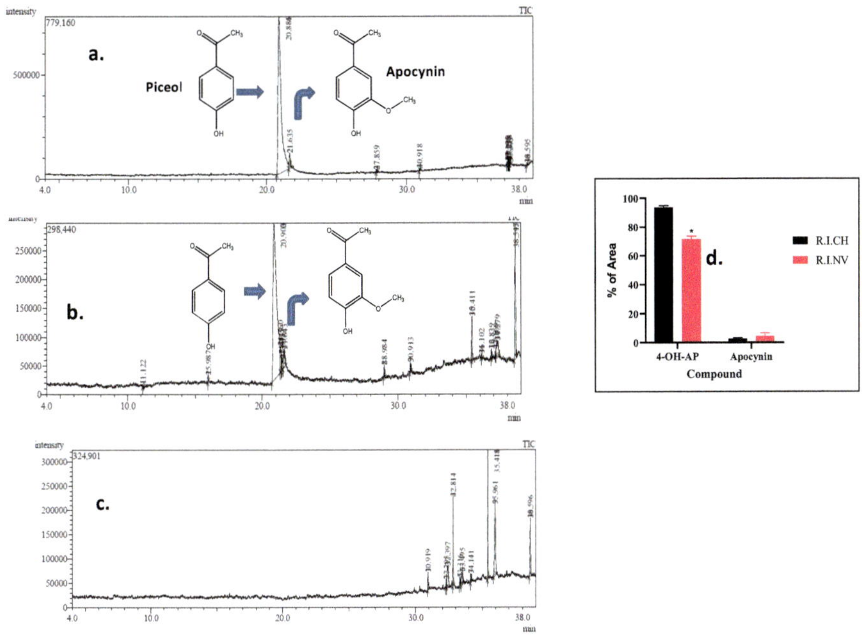

Fig. 19.3 (**a**) GC-MS profiling of *R. imbricata* Chang La sample (R.I.C) R., (**b**) GC-MS profiling of R. imbricata Khardung La sample (R.I.K), (**c**) GC-MS profiling of *R. imbricata* Shashi La sample (R.I.S), (**d**) Comparative profile of 4-hydroxyacetophenone and apocynin peak areas in R.I.C and R.I.K based on statistical analysis

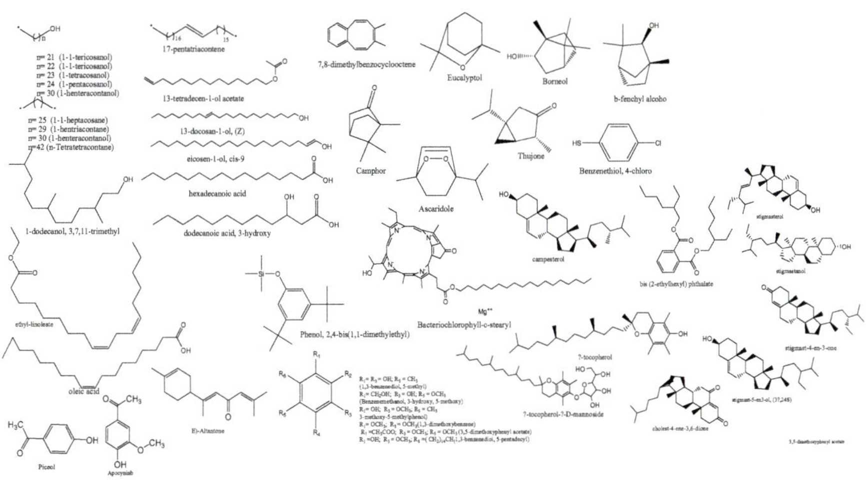

Fig. 19.4 Phytochemicals reported via a different analysis (LC-MS, GC-MS etc.)

Table 19.3 Comparative Chemical composition of the volatile compounds from R.I.C, R.I.K and R.I.S

R.I.C	Peak area (%)	R.I.K	Peak area (%)	R.I.S	Peak area (%)
Piceol[a]	94.0	Piceol[a]	73.80	N-Hexadecanoic acid	3.2
Apocynin[a]	2.78	Apocynin[a]	3.25	Eicosyl acetate[a]	10.9
(E)-Altantone[a]	0.29	Phytone[b]	0.62	Heneicosane[a]	12.5
Eicosanoic acid[a]	0.25	N-Hexadecanoic acid	0.61	2-(9,12-octadecadienyloxy-Ethanol)[a]	1.84
		Heneicosane[a]	2.37	Tetracosanal[a]	5.1
		Tridecanol,2-ethyl-2-methyl[a]	0.64	Hexacosane[a]	1.4
				Octacosane[a]	30.4

R.I.C= *R. imbricata* Chang La extract, *R.I.K* =*R. imbricata* Khardung La extract, *R.I.S* =*R. imbricata* Shashila extract

[a]Reported for the first time

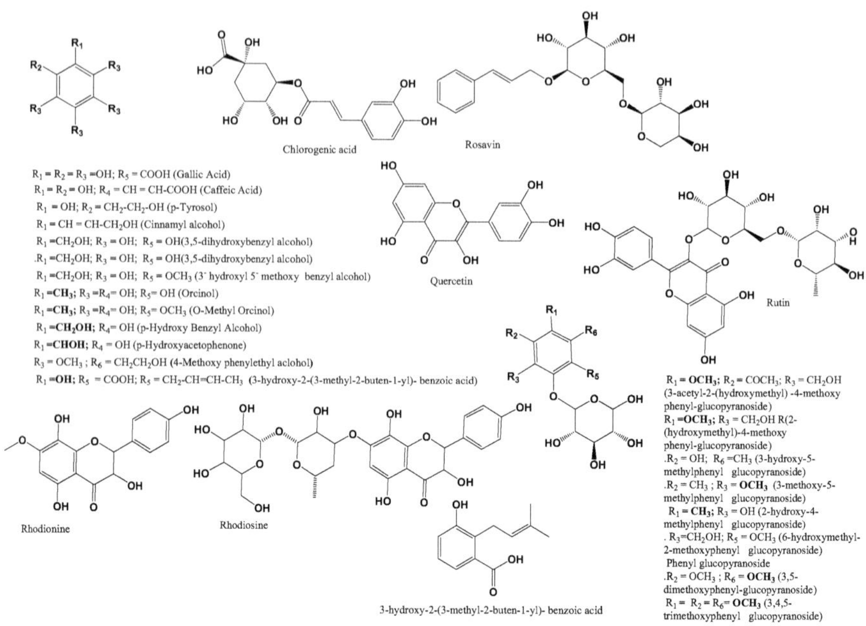

Fig. 19.5 List of phytocomponents reported via an isolation as well as HPLC

19.4.2 Anticancer Properties

Choudhary et al. (2015) have reported the anti-cancer properties of compounds isolated from *Rhodiola imbricata* against A-549 and MCF-7 cell line. Mishra et al. (2008, 2012) have reported *Rhodiola imbricata rhizome* anti-cancer properties.

19.4.3 Immunomodulatory Properties

Modulation of immune response to alleviate diseases has long since been of interest. Various studies by authors have mentioned the immunomodulatory effect. Studies have shown to up regulate TLR-4, interleukin-6 (IL-6) and tumour necrosis factor-α (TNF-α) in human peripheral blood mononuclear cells (PBMCs) and mouse macrophage cell line RAW 264.7 (Mishra et al. 2006, 2012). Authors have reported to reduce the parasite load of visceral leishmaniasis (VL) and increase the delayed type of hypersensitivity, IL-12, IFN-γ and TNF-α cytokines and $CD4^+$, $CD8^+$ T cells (Chauhan et al. 2024). Mishra et al. (2009, 2010) have reported immunological stimulatory effect of *Rhodiola imbricata* response equivalent to Freund's adjuvant (CFA) against strong antigen tetanus toxoid (TT) and to stimulate innate immune response pathway.

19.4.4 Antiaging and Adaptogenic Properties

Different species of *Rhodiola* are known for the various biological effects in the beneficial frame of humans. A key secondary metabolite salidroside is known for its antiaging properties. Mao et al. (2010) have reported to reverse the senescence like phenotype effect i.e. morphology, cell cycle, DNA Damage and the major molecular markers p53, p21 and p16. Jafari et al. (2007) have reported an increase in the lifespan of *Drosophila melanogaster* on supplementation of *Rhodiola* in comparison to control groups. Authors have also reported the adaptogenic properties of *Rhodiola imbricata* under hypoxia conditions with no acute or sub-acute toxicity (Gupta et al. 2008).

19.4.5 Radioprotective Properties

A bunch of studies has also reported the flora radioprotective properties. Arora et al. (2005) have reported the radioprotective activity of *R. imbricata* on U87 cell line and mice in addition to the antioxidant activities. Chawla et al. (2010) has reported *Rhodiola imbricata* as a portent prophylactic agent that can be efficiently utilized in nuclear and radiological emergencies. A pre-irradiation administration of *R. imbricata* with aqueous or aqua-alcoholic extract has significantly increased the survival rate by up to 90% against whole-body lethal gamma irradiation induced mortality in mice (Goel et al. 2006).

19.5 Production Technology and Micropropagation or In Vitro Propagation

19.5.1 Climatic and Soil

It can be cultivated in cool, sufficiently moist climates with evenly distributed precipitation (Grech-Baran et al. 2015) at altitudes ranging from 9000 to 18,500 feet above sea level (Chaurasia et al. 2003). The soil must be deep and easily penetrable to accommodate its deep root system. Well-drained loam or sandy loam soils with a pH of 6–8 are ideal for its cultivation (Galambosi 2006, 2014).

19.5.2 Propagation

Multiplication of plant by sexual and asexual methods is called propagation (Galambosi 2014). *Rhodiola* plant can be propagated by different ways, which are given below:

19.5.2.1 By Seed

Rhodiola seeds are very small, measuring about 1.0–2.5 mm in length and 0.2–0.5 mm in width, with a thousand-seed weight of 200 mg. The seeds are sown in October to November on the surface of a sand/peat mixture and left outdoors under the snow throughout the winter (Galambosi 2006). In April to May, the seed boxes are moved to a glasshouse with a temperature of 18–22 °C, where germination begins within 3–7 days. Once the seedlings develop one to three pairs of leaves, they are transplanted into the main field (Ampong-Nyarko 2014).

Dried *Rhodiola* seeds have a low germination rate of 3–35%, but after 30 days of stratification at −4.5 °C, the germination rate increases to 55–80%. Revina et al. (1976) reported that soaking the dry seeds in water overnight, then mixing them with moist sand and storing them in a suitable container at 2–4 °C for 1 month, can raise the germination rate to about 75%. Another experiment by Aiello and Fusani (2003) showed that treating the seeds with growth hormones, specifically GA_3 (100 ppm), resulted in an 87% germination rate.

19.5.2.2 By Vegetative (Root Cutting)

Rhodiola is commercially propagated through root cuttings. The rhizome-like underground roots of the mother plant are cut into sections 2–10 cm long, ensuring that each piece has at least one bud and some root hairs (Galambosi 2006). Root division can be carried out in early spring or late summer, typically on plants that are 3 years old. The cuttings should be potted and kept outdoors or in light shade, either in a cold frame or greenhouse, until they establish roots. Within 3 weeks, the plantlets should have rooted and begun to grow (Galambosi 2006, 2014).

19.5.3 Micro Propagation or In Vitro Propagation

The commercial cultivation of *Rhodiola* species faces challenges such as low seed viability and germination rates, delayed maturation and the plants' inability to thrive outside their native habitat (Bhardwaj et al. 2018a). In this regard, micropropagation could be highly beneficial for rapid multiplication of plantlets or cloning desired genotypes and conservation of these plants (Kaur et al. 2023). Furthermore, the plant tissue culture or in vitro propagation substrate offers an alternate strategy for mass cultivation and manufacturing of medicinally important compounds (Rattan et al. 2022). The different approaches explored for its plant tissue culture and developmental work done in *Rhodiola* spp. (Fig. 19.6).

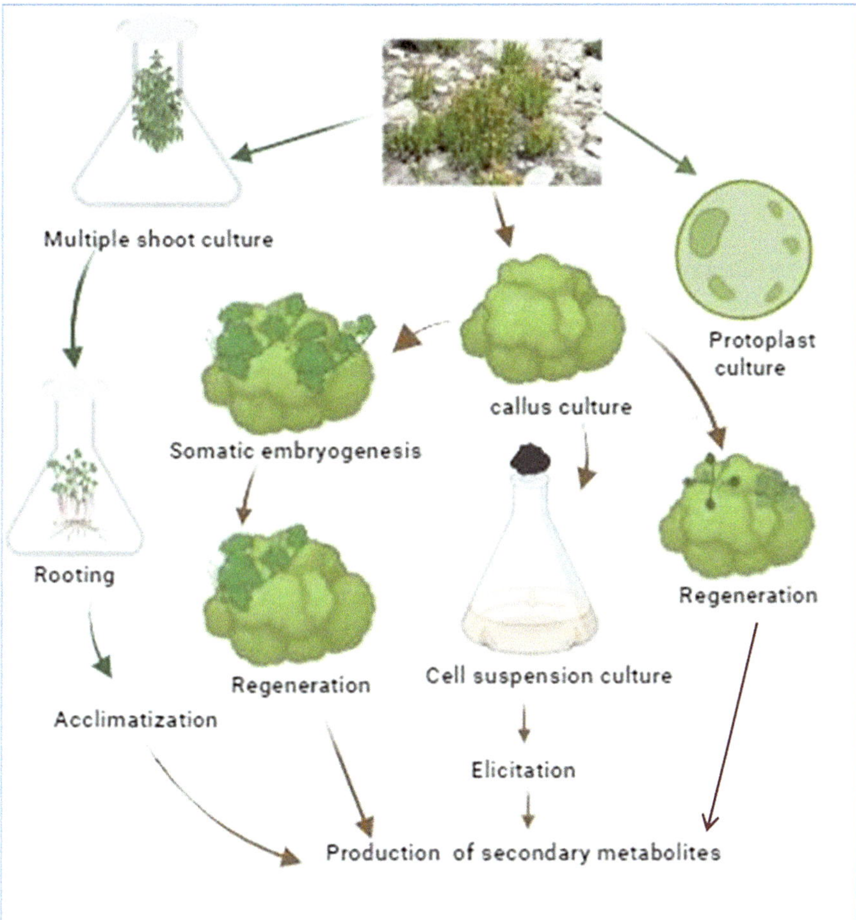

Fig. 19.6 Diagram shows the research area of in vitro culture of *Rhodiola* spp.

Table 19.4 Different explants used in *Rhodiola* spp. for in vitro culture

Explant	Response	References
Leaf	Multiple shoot and root development	Bhardwaj et al. (2018b)
	Callus development	Jianfeng et al. (2007), Jianfeng et al. (2009), Kapoor et al. (2019) and Rattan et al. (2021)
	Multiple shoot development	Wu et al. (2003)
Leaf discs	Multiple shoot development	Wu et al. (2003)
Axenic leaves and roots tissues	Callus development	Rattan et al. (2020)
Stem	Bud development	Haijun et al. (2006)
	Multiple shoot development	Wu et al. (2003)
Shoot nodes	Multiple shoot development	Wu et al. (2003)
Seed	Callus induction, shoot and root	Bhardwaj et al. (2018a)
	Shoot and root proliferation	Dolker et al. (2024)

19.5.3.1 Explant Used in Plant Tissue Culture of *Rhodiola* Species

Selecting the appropriate explant is crucial for any in vitro culture study. Factors such as the developmental stage, regenerative potential and disease-free explants are considered during selection (Kaur et al. 2023; Shukla et al. 2020). Various explants, including leaf, leaf discs, axenic leaves, shoot nodes, seeds and root tissue, have been utilized by researchers for both direct and indirect regeneration in *Rhodiola* spp. (Table 19.4).

19.5.3.2 Cleaning and Disinfecting of Explants

After selecting the explant, surface disinfection is a critical step to minimize contamination. The concentration and duration of the disinfection process depend on the type and developmental stage of the explant, as improper conditions can negatively affect its growth and development. For *Rhodiola* spp., the typical protocol involves treating the explant with two to three drops of Tween-20 for 5–10 min, followed by a 70% ethanol treatment for 2–3 min, and then thoroughly washing it with sterile water (Bhardwaj et al. 2018b). Another experiment conducted by Rattan et al. (2020) revealed that seeds of *Rhodiola* spp. treated with 70% ethanol for 1 min and 2% sodium hypochlorite for 7 min significantly reduced the contamination.

19.5.3.3 Nutrient Medium for In Vitro Culture of *Rhodiola* Spp.

The basal medium provides essential macronutrients and micronutrients to the explant in in vitro conditions (Bhardwaj et al. 2018a), along with vitamins and other organic components necessary for plant growth and development. In studies on *Rhodiola* spp., almost all researchers have utilized Murashige and Skoog (MS) medium for various experiments (Bhardwaj et al. 2018b; Kapoor et al. 2019; Rattan et al. 2021, 2022; Dolker et al. 2024). Sucrose was commonly used as the carbon source in the nutrient medium, though its concentration varied. Generally, 3% (w/v) sucrose was used by most researchers (Bhardwaj et al. 2018b; Kapoor et al. 2019; Rattan et al. 2021, 2022; Dolker et al. 2024).

19.5.3.4 Plant Growth Regulator for In Vitro Culture and Production of Secondary Metabolites of *Rhodiola* Spp.

Callus, shoot and root regeneration can be achieved by supplementing the culture medium with cytokinin and auxin. Typically, a higher cytokinin concentration promotes shoot development, whereas a higher auxin concentration encourages rooting and callus induction. Various cytokinins, such as N6-benzylaminopurine (BAP) and kinetin (Kn), have been used for shoot regeneration, with BAP being the most commonly used and yielding promising results when applied alone or in combination with auxin (Kaur et al. 2023). Several studies have reported successful callus formation, shoot and root development and the production of secondary metabolites using MS medium supplemented with auxins and cytokinins, as shown in Table 19.5.

19.5.3.5 Acclimatization of In Vitro Plant of *Rhodiloa* Spp.

Acclimatization involves transferring in vitro plantlets to soil and assessing the survival rate of micropropagated plants, with the primary goal of evaluating their performance under natural conditions. After successful in vitro rooting, the shoots were washed with tap water to remove any remaining media and then planted in

Table 19.5 Different explants used in *Rhodiola* spp. for in vitro culture

In vitro development	Growth hormones used	Secondary metabolites production	References
Callus development	MS + 2,4,D + NAA + BA + kinetin	Salidroside	Sheng et al. (2005) and Li et al. (2005)
Callus development	MS + BA + IAA	–	Jianfeng et al. (2007) and Hong et al. (2008)
Shoot and leaf proliferation	MS + BAP and Thidiazuron (TDZ)	Caffeic acid and p-coumaric acid	Bhardwaj et al. (2018a)
Callus, shoot and root proliferation	MS + NAA and BAP	Cinnamyl alchol glyco-sides	Bhardwaj et al. (2018b)
Shoot and root proliferation and leaf	Liquid MS+ *meta*-Topolin+TDZ and IBA	Tyrosol, salidroside, and rosavin	Dolker et al. (2024)
Callus and shoot proliferation	MS + NAA, BAP and Jasmonic acid	Salidroside	Kapoor et al. (2019)
Callus, shoot and root proliferation	MS, TDZ and NAA	Salidroside, rosavin and rosavins	Rattan et al. (2020, 2021)

hardening boxes or small pots containing a sterile mixture of soil, perlite and vermiculite (Bhardwaj et al. 2018a). Some studies have explored acclimatization of *Rhodiola* using various combinations of these components. Successful acclimatization was achieved with a mixture of autoclaved sand, perlite and vermiculite in a 1:1:1 ratio (Bhardwaj et al. 2018b). The plants were kept in a culture room for 1 week to facilitate the transition from in vitro to outdoor conditions. Subsequently, the plantlets were transferred to a greenhouse and planted in larger pots containing a soil, sand and vermiculite mix in a 2:1:1 ratio (Dolker et al. 2024).

19.5.3.6 Transplanting

Rhodiola grows slowly and its growth and root yield are significantly affected by the age of the seedlings. Seedlings that are 1–2 years old can be transplanted into the field. In an experiment conducted in Finland, Galambosi (2006) found that 12-month-old seedlings produced higher root yields compared to younger ones, confirming earlier Russian guidelines (Nukhimovskii et al. 1987).

19.5.3.7 Spacing and Weeding

The optimal spacing for Rhodiola seedlings is 40 × 40 cm. In plain fields, seedlings can be transplanted at a spacing of 60 × 20 cm, while raised beds or plastic mulch systems allow for a spacing of 30 × 30 cm (Galambosi 2014; Ampong-Nyarko 2014).

For weed management, mechanical harrowing between rows 2–4 times and hand weeding between plants during the summer months is recommended. Both organic and inorganic mulch can be used to control weeds in Rhodiola fields. An experiment conducted in Canada by Kwesi et al. (2005) showed that using plastic mulch significantly suppressed weed growth and increased yield.

19.5.3.8 Manure and Fertilizer

Information on fertilizer application for Rhodiola is limited due to its natural habitat, where it grows on rocks with a very thin soil layer. However, an organic manure experiment conducted in Finland from 1997 to 2000 (Galambosi 2006; Ampong-Nyarko 2014) tested different doses of compost (farmyard manure) ranging from 100 to 250 q/ha, which were incorporated into the soil before seed sowing or transplanting. No additional manure or fertilizer was applied during the 4 years of plant growth, and the results showed that increasing compost doses improved the growth of vegetative shoots and the fresh weight of the roots. These findings were confirmed by Russian scientists, who reported the highest root weight using a fertilizer mix of nitrogen (50 kg/ha), phosphorus (50 kg/ha) and potassium (70 kg/ha), along with 500 q/ha of additional manure in northern Russia (Elsakov and Gorelova 1999).

19.5.3.9 After Care

Occasionally, Rhodiola plants may be affected by insect pests and diseases. Aphids, for example, can damage the leaves and shoots by sucking sap from the soft parts of the plant. They can be controlled by spraying insecticidal soaps or using neem or canola oil. A disease caused by the saprophytic fungus *Armillaria mellea* can harm the roots or rhizomes of the plant (Grech-Baran et al. 2015; Galambosi 2014).

19.5.3.10 Harvesting and Yield

The yield of fresh and dry roots is influenced by plant density and age. Roots are harvested manually for smaller-scale cultivation, while larger-scale operations utilize mechanical harvesting. In the third and fourth years of growth, fresh root yields range from 2.50 to 4.85 kg/m^2, and dried root yields vary from 0.5 to 1.3 kg/m^2. Older plants tend to produce higher yields; for example, after 6 years, each plant can yield more than 2 kg of fresh root, resulting in a calculated fresh yield of 9.0 kg/m^2 (Galambosi 2006, 2014; Ampong-Nyarko 2014).

Rhodiola is a valuable medicinal plant, but its cultivation is limited due to specific climatic requirements and a lack of knowledge regarding good agricultural practices. This section covers good agricultural practices and discusses large-scale production of planting material through micropropagation. Additionally, it addresses the production of secondary metabolites in in vitro grown plants.

19.6 Conservation Strategies

Due to overharvesting and habitat destruction, *Rhodiola imbricata* is facing the threat of extinction in its native habitat. Conservation strategies include:

- *Ex Situ Conservation:* Micropropagation, tissue culture and cryopreservation techniques are crucial for maintaining a genetic reservoir of *Rhodiola imbricata* outside its natural habitat.
- *In Situ Conservation:* Efforts are underway to establish protected areas and cultivate the plant in its natural environment while maintaining the ecological balance.

19.7 Products Available in Market

Several companies produce *Rhodiola imbricata* supplements, primarily in capsule form, as it is valued for its adaptogenic, anti-stress and antioxidant properties. Most of the *Rhodiola* supplements on the market are derived from *Rhodiola rosea*, which is the most researched and widely used species. Due to its well-documented adaptogenic effects, such as reducing fatigue, improving mental clarity and combating stress, *Rhodiola rosea* is favoured by manufacturers for adaptogenic formulations. In contrast, *Rhodiola imbricata* is less studied and less commonly available, though it is gaining attention for its unique properties, particularly in high-altitude applications. Some companies now combine both species in their products to enhance their overall effectiveness, especially for individuals seeking to improve stamina, manage stress, or alleviate altitude sickness (Table 19.6).

Table 19.6 Available information for *Rhodiola* products from various companies

No.	Product common name	Company name	Available	Use	Reference/website link
1	Rhodiola Root Extract	NOW Foods	Capsules	Reduce stress, improve energy, enhance cognitive function	https://www.nowfoods.com/products/supplements/rhodiola-500-mg-veg-capsules
2	Rhodiola Complex	Nature's Way	Capsules	Reduce stress, boost mood, improve physical performance	https://naturesway.com/products/rhodiola
3	Rhodiola rosea	Gaia Herbs	Liquid extract	Reduce stress, improve mood, enhance cognitive function	https://www.gaiaherbs.com/products/rhodiola-rosea
4	Solaray Rhodiola	Solaray	Capsules	Reduce stress, improve energy, enhance cognitive function	https://www.amazon.com/Solaray-Rhodiola-Extract-Adaptogen-Guaranteed/dp/B000I4DEVU
5	Rhodiola rosea Capsules	Humming Herbs	Capsules	Stress and anxiety relief, sleep restoration, improvement in energy, and focus and increasing stamina	https://hummingherbs.com/product/humming-herbs-rhodiola-rosea/?gad_source=4&gclid=Cj0KCQjwgL-3BhDnARIsAL6KZ68%2D%2DbKgpPkQSTqXMj9niSGkn1Tx4SMyQSvQMuwfzialKrxJIJobwmQaAoBSEALw_wcB
6	Rhodiola rosea Capsules	NutraChamps	Capsules	Soothe Tension & Support Stress Relief Boost Smooth Energy & Stamina Enhance Focus, Clarity & Memory Support Overall Wellness & Vitality Immune System Support & Protection	https://www.nutrachamps.com/products/rhodiola-rosea?_pos=1&_sid=b99d18501&_ss=r
7	Rhodiola	Thorne	Capsules	A stress-relieving botanical to support neurotransmitters that enhance mood, sleep, and mental focus	https://www.thorne.com/products/dp/rhodiola
8	Rhodiola Extract	Life Extension	Capsules	Promotes ATP production and cellular energy metabolism	https://www.lifeextension.com/search#q=Rhodiola&t=coveo4A2453FD
9	Dual Action Rhodiola™	Paradise	Capsules	Supporting concentration, memory, increased physical endurance and sexual vitality.	https://paradiseherbs.com/products/rhodiola/

10	Rhodiola Extract	California Gold Nutrition	Capsules	Help support energy production, stamina, and potentially reduce fatigue.	https://www.californiagoldnutrition.com/products/california-gold-nutrition-rhodiola-extract-500-mg-60-veggie-capsules-106155?variant=43790079623353
11	Rhodiola, Alcohol-Free, 4 fl oz	Herb Pharm		Athletic and physical endurance, Uplifting energy to help make life's daily stress more manageable, Energizing nootropic and adaptogen for mind and body.	https://www.herb-pharm.com/products/rhodiola
12	Rhodiola, Vegetarian Capsules	Nature's Answer	Capsules	Elps the body adapt to stress	https://www.naturesanswer.com/product/rhodiola-root-60-standardized-veggie-capsules/
13	SFP Rhodiola Root Extract Vegetable Capsules	Solgar	Capsules		https://www.solgar.com/products/sfp-rhodiola-root-extract-vegetable-capsules/
14	Rhodiola Rosea Powder	Nutricost	Powder		https://nutricost.com/products/nutricost-rhodiola-rosea-powder?_pos=2&_sid=b624c731a&_ss=r#tabe93fe9d7-86fe-4f4c-8a25-586b3b13ddba
15	Rhodiola-Plus, 75 Liquid Soft-Gels	Irwin Naturals	Liquid soft-gels		https://in.iherb.com/pr/irwin-naturals-rhodiola-plus-75-liquid-soft-gels/116158
16	Rhodiola Extract 60 Capsules	Nutramagik	Capsules	Supports brain health, reduces stress and anxiety	https://www.mystore.in/en/product/nutramagik-rhodiola-extract-60-capsules-supports-brain-health-reduces-stress-and-anxiety?srsltid=AfmBOoqNPH2FE_6uY156mkPbu_vat0vqUdnv38T31-wJUTSrIFbJnA8J9YE
17	Rhodiola 500 mg with Piperine Absorption Complex Veg Capsule	Trexgenics		Useful in fighting stress and anxiety. Used to enhance cognitive functioning. Boosts memory and focus	https://www.1mg.com/otc/trexgenics-rhodiola-500mg-with-piperine-absorption-complex-veg-capsule-otc580010?srsltid=AfmBOooAiZfjFjBrScZ3tusaZX4Vb9pQmU6i12XXWqBJimOhtGrfrVf3TUE&wpsrc=Google+Organic+Search
18	Rhodiola Capsules	Biotrex nutraceuticals	Capsules	Anti-fatigue, stress relief, Supports mental and physical stamina	https://www.biotrexnutritions.com/products/biotrex-rhodiola-500mg-60-capsules

19.8 Conclusion

We have critically reviewed the bioactive components with their respective structures, biological activities and in vitro micropropagation techniques relative to future use from *R. imbricata*. The research findings showed that tissue culture techniques can play an important role in clonal propagation of elite genotypes, which have diverse medicinal applications, and eventually due to over exploitation and irregular concern, this plant is facing local extinction. In view of the medicinal properties and increased demand of this plant in the pharmaceutical industry, the outlined procedure offers a potential system for improvement and conservation.

Acknowledgements This work was supported in part by the Defence Research Development Organization and Council of Industrial Research Institute India.

References

Aiello N, Fusani P (2003) Effetti della prerefrigeazione e dell'acido gibberellico sulla germinazione del seme di rodiola rosea. Sementi Elette, Edagricole, Bologna

Ampong-Nyarko K (2014) Rhodiola rosea cultivation in Canada and Alaska. Rhodiola rosea 14:125

Arora R, Chawla R, Sagar R, Prasad J, Singh S, Kumar R, Sharma RK (2005) Evaluation of radioprotective activities of Rhodiola imbricata Edgew–A high altitude plant. Molecular and cellular biochemistry, 273(1):209–223

Ballabh B, Chaurasia OP, Ahmed Z (2007) Herbal products from high altitude plants of Ladakh Himalaya. Curr Sci 92:1664–1665

Bhardwaj AK, Singh B, Kaur K et al (2018a) In vitro propagation, clonal fidelity and phytochemical analysis of *Rhodiola imbricata* Edgew: a rare trans-Himalayan medicinal plant. Plant Cell Tissue Organ Cult 135:499–513. https://doi.org/10.1007/s11240-018-1482-x

Bhardwaj AK, Naryal A, Bhardwaj P, Warghat AR, Arora B, Dhiman S, Chaurasia OP (2018b) High efficiency in vitro plant regeneration and secondary metabolite quantification from leaf explants of *Rhodiola imbricata*. Pharm J 10(3):470–475

Chawla R, Jaiswal S, Kumar R, Arora R, Sharma RK (2010) Himalayan bioresource Rhodiola imbricata as a promising radioprotector for nuclear and radiological emergencies. J of Pharmacy And Bioallied Sciences, 2(3):213–219

Chauhan K, Kaur G, Anand N, Kaur S (2024) Medicinal *Rhodiola imbricata* induces protective immune responses against susceptible and resistant strains of *Leishmania donovani* in murine model. Biologia (Bratisl) 79:1725–1747. https://doi.org/10.1007/S11756-024-01656-Y/FIGURES/16

Chaurasia OP, Basant Ballabh BB, Brahma Singh BS (2003) Himalayan rose root, *Rhodiola* spp.-potential health drink and drug plant. Nat Prod Rad 2:71–92

Chen TS, Liou SY, Chang YL (2012) Antioxidant evaluation of three adaptogen extracts. Am J Chin Med 36:1209–1217. https://doi.org/10.1142/S0192415X08006533

Choudhary A, Kumar R, Srivastava RB, Surapaneni SK, Tikoo K, Singh IP (2015) Isolation and characterization of phenolic compounds from *Rhodiola imbricata*, a Trans-Himalayan food crop having antioxidant and anticancer potential. J Funct Foods 16:183–193

Dhar P, Das KS, Barhwal K, Hota KS, Mishra PK, Singh BS (2019) Trans-Himalayan phytococktail confers protection against hypobaric hypoxia-induced hippocampal neurodegeneration and memory impairment in male sprague dawley rats. High Alt Med Biol 20(3):279–292. https://doi.org/10.1089/ham.2019.0011

Dolker D, Behera S, Justine AK, Kumari V, Pati PK (2024) Production of large-scale genetically identical and phytochemically stable in vitro plants of *Rhodiola imbricata* using meta-Topolin and liquid culture system. Plant Cell Tissue Organ Cult 156(1):18

Elsakov GV, Gorelova AP (1999) Fertilizer effects on the yield and biochemical composition of rose-root stonecrop in North Kola region. Agrokhimiya 10:58–61

Galambosi B (2006) Demand and availability of *Rhodiola rosea* L. raw material. In: Medicinal and aromatic plants. Springer, Dordrecht, pp 223–236

Galambosi B (2014) Cultivation of *Rhodiola rosea* in Europe. In: Cuerrier A, Ampong-Nyarko K (eds) *Rhodiola rosea*. CRC Press, Boca Raton, pp 87–124

Goel HC, Bala M, Prasad J et al (2006) Radioprotection by *Rhodiola imbricata* in mice against whole-body lethal irradiation. J Med Food 9:154–160. https://doi.org/10.1089/JMF.2006.9.154

Grech-Baran M, Syklowska-Baranek K, Pietrosiuk A (2015) *Approaches of Rhodiola kirilowii* and *Rhodiola rosea* field cultivation in Poland and their potential health benefits. Ann Agric Environ Med 22(2):281–285

Gupta A, Kumar R, Upadhyay NK, Pal K, Kumar R, Sawhney RC (2007) Effects of *Rhodiola imbricata* on dermal wound healing. Planta Med 73(8):774–777

Gupta V, Saggu S, Tulsawani RK et al (2008) A dose dependent adaptogenic and safety evaluation of *Rhodiola imbricata* Edgew, a high altitude rhizome. Food Chem Toxicol 46:1645–1652. https://doi.org/10.1016/J.FCT.2007.12.027

Gupta S, Bhoyar MS, Kumar J, Warghat AR, Bajpai PK, Rasool M et al (2012) Genetic diversity among natural populations of *Rhodiola imbricata* Edgew. from trans-Himalayan cold arid desert using random amplified polymorphic DNA (RAPD) and inter simple sequence repeat (ISSR) markers. J Med Plant Res 6(3):405–415

Haijun L, Bin G, Qiong Y, Yujun L, Chunchao L (2006) Tissue cultures of four Rhodiola species. Acta Botan Boreali-Occiden Sin 26(10):2023–2027

Jafari M, Felgner JS, Bussel II, Hutchili T, Khodayari B, Rose MR, Vince-Cruz C (2007) Rhodiola: a promising anti-aging Chinese herb. Rejuvenation Res 10:587–602. https://doi.org/10.1089/rej.2007.0560

Jianfeng L, Xiufeng Y, Xia L, Yunqing C (2007) Callus induction and plant regeneration from leaf explants of *Rhodiola sachalinensis* A. Bor. J North East For Univ 15:33–34

Jian-Feng L, Yun-Qing C, Zhi-Wen C (2009) Protoplast isolation and plant regeneration from leaves of *Rhodiola sachalinesis*. Chin Tradit Herb Drug 7:2010–2014

Kanupriya, Prasad D, Ram MS et al (2005) Cytoprotective and antioxidant activity of *Rhodiola imbricata* against tert-butyl hydroperoxide induced oxidative injury in U-937 human macrophages. Mol Cell Biochem 275:1–6. https://doi.org/10.1007/S11010-005-7637-1/METRICS

Kapoor S, Raghuvanshi R, Bhardwaj P, Sood H, Saxena S, Chaurasia OP (2018). Influence of light quality on growth, secondary metabolites production and antioxidant activity in callus culture of Rhodiola imbricata Edgew. Journal of Photochemistry and Photobiology B: Biology, 183:258–265

Kapoor S, Sharma A, Bhardwaj P, Sood H, Saxena S, Chaurasia OP (2019) Enhanced Production of Phenolic Compounds in Compact Callus Aggregate Suspension Cultures of Rhodiola imbricata Edgew. Applied Biochemistry and Biotechnology, 187(3):817–837

Kapoor S, Sood H, Saxena S, Chaurasia OP (2022) Green synthesis of silver nanoparticles using Rhodiola imbricata and Withania somnifera root extract and their potential catalytic, antioxidant, cytotoxic and growth-promoting activities. Bioprocess and Biosystems Engineering, 45(2):365–380

Kaur G, Kaur K, Singh V, Panag NK, Saluja P (2023) Species of the genus Rhodiola found in the Western Himalayas. In: Immunity boosting medicinal plants of the Western Himalayas. Springer, pp 393–431

Khanna K, Mishra KP, Ganju L, Singh SB (2017) Golden root: A wholesome treat of immunity. Biomedicine and Pharmacotherapy, 87:496–502

Kwesi AN, Sowiak V, Brown J et al (2005) Guidelines for the cultivation of *Rhodiola rosea* in Alberta, Manuscript, Crop Diversification Centre North, Edmonton

Li Wei LW, Du GuiSeng DG, Huang QinNi HQ (2005) Salidroside contents and related enzymatic activities in *Rhodiola kirilowii* callus. Acta Botan Boreali-Occiden Sin 25(8):1645–1648

Mao GX, Wang Y, Qiu Q, Deng HB, Yuan LG, Li RG, Wang Z (2010) Salidroside protects human fibroblast cells from premature senescence induced by H2O2 partly through modulating oxidative status. Mechanisms of ageing and development, 131(11–12):723-731

Mishra KP, Padwad YS, Jain M et al (2006) Aqueous extract of *Rhodiola imbricata* rhizome stimulates proinflammatory mediators via phosphorylated IκB and transcription factor nuclear factor-κB. Immunopharmacol Immunotoxicol 28:201–212. https://doi.org/10.1080/08923970600815139

Mishra KP, Padwad YS, Dutta A et al (2008) Aqueous extract of *Rhodiola imbricata* rhizome inhibits proliferation of an erythroleukemic cell line K-562 by inducing apoptosis and cell cycle arrest at G2/M phase. Immunobiology 213:125–131. https://doi.org/10.1016/J.IMBIO.2007.07.003

Mishra KP, Ganju L, Chanda S, Karan D, Sawhney RC (2009) Aqueous extract of Rhodiola imbricata rhizome stimulates Toll-like receptor 4, granzyme-B and Th1 cytokines in vitro. Immunobiology, 214(1):27–31

Mishra KP, Chanda S, Shukla K, Ganju L (2010) Adjuvant effect of aqueous extract of Rhodiola imbricata rhizome on the immune responses to tetanus toxoid and ovalbumin in rats. Immunopharmacology and immunotoxicology, 32(1):141–146

Mishra KP, Ganju L, Singh SB (2012) Anti-cellular and immunomodulatory potential of aqueous extract of *Rhodiola imbricata* rhizome. Immunopharmacol Immunotoxicol 34:513–518. https://doi.org/10.3109/08923973.2011.638307

Pundir A, Kad A, Sood H (2021) *Rhodiola imbricata*. In: Himalayan medicinal plants. Academic, pp 151–171. https://doi.org/10.1016/B978-0-12-823151-7.00014-3

Qian EW, Ge DT, Kong SK (2011) Salidroside promotes erythropoiesis and protects erythroblasts against oxidative stress by up-regulating glutathione peroxidase and thioredoxin. J Ethnopharmacol 133:308–314. https://doi.org/10.1016/J.JEP.2010.09.025

Qian EW, Ge DT, Kong SK (2012) Salidroside protects human erythrocytes against hydrogen peroxide-induced apoptosis. J Nat Prod 75:531–537. https://doi.org/10.1021/NP200555S/ASSET/IMAGES/LARGE/NP-2011-00555S_0005.JPEG

Raj K, Amol T, Chaurasia OP et al (2010) Evaluation of anti-oxidant activities and total phenol and flavonoid content of the hydro-alcoholic extracts of *Rhodiola* sp. Pharm J 2:431–435. https://doi.org/10.1016/S0975-3575(10)80027-6

Rattan S, Sood A, Kumar P, Kumar A, Kumar D, Warghat AR (2020) Phenylethanoids, phenylpropanoids, and phenolic acids quantification vis-à-vis gene expression profiling in leaf and root derived callus lines of *Rhodiola imbricata* (Edgew.). Ind Crop Prod 154:112708

Rattan S, Kumar D, Warghat AR (2021) Growth kinetics, metabolite yield, and expression analysis of biosynthetic pathway genes in friable callus cell lines of *Rhodiola imbricata* (Edgew). Plant Cell Tissue Organ Cult 146(1):149–160

Rattan S, Kumar P, Kaur E, Sood A, Acharya V, Warghat AR (2022) Comparative transcriptome and tissue-specific expression analysis of genes reveal tissue-cultured plants as an alternative source for phenylethanoids and phenylpropanoids in *Rhodiola imbricata* (Edgew.). Gene 836:146672

Rattana S, Sood A, Kumar P, Kumar A, Kumar D, Warghat RA (2020) Phenylethanoids, phenylpropanoids, and phenolic acids quantification vis-àvis gene expression profiling in leaf and root derived callus lines of Rhodiola imbricata (Edgew.). Ind Crop Prod 154:112708. https://doi.org/10.1016/j.indcrop.2020.112708

Revina TA, Krasnov EA, Sviridova TP et al (1976) Biological characteristics and chemical composition of *Rhodiola rosea* grown in Tomsk. Rastit Resur 12(3):355–360

Senthilkumar R, Chandran R, Parimelazhagan T (2014) Hepatoprotective effect of *Rhodiola imbricata* rhizome against paracetamol-induced liver toxicity in rats. Saudi J Biol Sci 21:409–416. https://doi.org/10.1016/j.sjbs.2014.04.001

Sharma R, Kumar R, Thakur M, Patel M, Chaurasia O, Saxena S (2022) GC MS based comparative phytochemical profiling of *Rhodiola imbricata* roots collected from different high mountain passes of Ladakh India and a first report of apocynin from genus *Rhodiola*. Def Life Sci J 7(2):71–78. https://doi.org/10.14429/dlsj.7.17313

Sheng CZ, Hu TQ, Bi H, Yuan YJ, Jiang Y (2005) Effects of plant growth substances on induction and culture of callus from *Rhodiola quadrifida*. Zhongguo Zhong yao za zhi/China J Chin Mater Med 30(16):1237–1240

Shukla S, Rajbhar Y, Sharma A, Bhardwaj P, Tiwari VK, Tiwari H et al (2020) Effect of auxins supplementation on callus induction and root regeneration in banana *(Musa paradisiaca* L.) cv. Udhayam under in vitro cultures. J Exp Biol Agric Sci 8(1):30–34

Singamaneni V, Sharma U, Lone B, Gupta P (2020) Phytochemistry and pharmacological activities of *Rhodiola imbricata* Edgew., a high value medicinal herb of Cold Desert Himalaya. In: Singh B (ed) Botanical leads for drug discovery. Springer, Singapore. https://doi.org/10.1007/978-981-15-5917-4_8

Tayade AB, Dhar P, Sharma M, Chauhan RS, Chaurasia OP, Srivastava RB (2013a) Antioxidant capacities, phenolic contents, and GC/MS analysis of *Rhodiola imbricata* Edgew. root extracts from trans-Himalaya. J Food Sci 78(1-2-3):C402–C410. https://doi.org/10.1111/1750-3841.12054

Tayade AB, Dhar P, Sharma KJ, Chauhan M, Rajinder S, Chaurasia PO, Srivastava BR (2013b) Chemometric profile of root extracts of *Rhodiola imbricata* Edgew. with hyphenated gas chromatographymass spectrometric technique. PLoS One 8(1):e52797. https://doi.org/10.1371/journal.pone.0052797

Tayade AB, Dhar P, Kumar J, Sharma M, Chaurasia OP, Srivastava RB (2017) Trans-Himalayan *Rhodiola imbricata* Edgew. root: a novel source of dietary amino acids, fatty acids and minerals. J Food Sci Technol 54(2):359–367. https://doi.org/10.1007/s13197-016-2469-4

Wal A, Wal P, Rai AK, Tiwari R, Prajapati SK (2019) Adaptogens with a special emphasis on *Withania somnifera* and *Rhodiola rosea*. In: Nutrition and enhanced sports performance. Academic, pp 407–418

Wu S, Zu Y, Wu M (2003) High yield production of salidroside in the suspension culture of *Rhodiola sachalinensis*. J Biotechnol 106(1):33–43

Endophytes of *Taxus wallichiana* Zucc. (Himalayan Yew): Host-Microbe Interaction Benefits

20

Priyanka Adhikari and Anita Pandey

Abstract

Taxus wallichiana Zucc. (Himalayan yew) is a well-known medicinal plant of the Taxaceae family. The anticancer properties of *Taxus* species are well-known across the medical community. It was mentioned in the IUCN Red List a few years ago; therefore, it is crucial to protect and propagate the species. In this perspective, an understanding of the root microbiome with special reference to plant growth-promoting (PGP) activities is a prerequisite for developing rapid propagation approaches for this slow-growing plant species. In this context, *T. wallichiana* roots have recently been used to isolate beneficial bacteria and fungi. The isolated bacteria and fungi had genotypic and phenotypic traits of *Burkholderia* and *Enterobacter*, and *Aspergillus* and *Penicillium* species, respectively. Phosphate solubilization was demonstrated by all endophytes, as evidenced by the generation of phosphatases in a medium containing calcium, iron, and aluminum phosphate and phytases in a medium containing calcium and sodium phytate. Additionally, yellow pigment with antagonistic or antibacterial activity was generated by *Burkholderia* sp. Apart from their ability to promote plant development in vitro, the endophytic bacteria *Burkholderia* and *Enterobacter* had favorable effects on the growth of test plants under net house conditions. Additionally, endophytic fungi were capable of producing taxol and other highly significant secondary metabolites, i.e., phenols, fatty acids, aromatic dicarboxylic acid, and alkaloids in ethyl acetate extracts. Furthermore, both bacterial and fungal crude extracts and rich fractions showed antibacterial and anti-

P. Adhikari
Central Ayurveda Research Institute, Bidhannagar-700091, Kolkata, West Bengal, India

A. Pandey (✉)
Department of Biotechnology, Graphic Era (Deemed to be University), Clement Town-248002, Dehradun, Uttarakhand, India
e-mail: anita.pandey@geu.ac.in

S. Chandra, H. Lata (eds.), *High Altitude Medicinal Plants*,
https://doi.org/10.1007/978-3-031-99406-7_20

fungal activity, respectively. The capacity of these endophytic fungi to generate a range of secondary metabolites and their antimicrobial qualities are likely to have potential applications in biotechnological, medicinal, and agricultural domains.

20.1 Introduction

The term "plant microbiome" refers to the variety of endophytic microorganisms that are present in plants in natural habitats, such as bacteria, fungi, and archaea (Quiza et al. 2015). Of all the plant-associated microbial communities, endophytes the microbiota that lives inside the plant endosphere without causing any symptoms have the strongest explicit and intimate relationship with their host plants (Mishra and Sharma 2022). All ecosystems depend heavily on microorganisms, which perform a variety of bioactivities to maintain the ecosystem's viability (Schloter et al. 2018). In plant-microbiome research, bioprospection of these endophytes is a developing field with special attention to their ecological, biotechnological, therapeutic, and evolutionary aspects (Sharma et al. 2020).

Taxus (common name yew; family: Taxaceae) is a genus of evergreen tree, native to western, central, and southern Europe, northwest Africa, northern Iran, and southwest Asia (Behman et al. 2016). In the Indian subcontinent, the species grows in the northern hemisphere, with its distribution in the hills of northern Jammu & Kashmir, Himachal Pradesh, Uttarakhand, and the states in northeast namely Meghalaya, Nagaland, and Manipur, at an altitude range of 1800–3300 m (Adhikari et al. 2022a). The species has received considerable attention on account of its existing exploitation for the extraction of the drug (taxol) from its bark (Wani et al. 1971). Further, it is also known for its various ethnomedicinal uses, e.g., the leaf paste is used in the treatment of asthma and bronchial disorders. Tea, made out of the stem bark of Himalayan yew, has been popular in Himalayan tribal communities for curing colds, coughs, and hypertension. The species is also known as a source of antioxidants (Juyal et al. 2014).

Taxus species have received considerable attention on account of the existing exploitation of their bark. However, the taxol content in *Taxus* species is very low (0.01–0.03% dry weight) for the extraction of the drug taxol, meaning that the bark of countless plants is required for procurement of taxol, and lots of yews are devastated in the procedure. As a result, *Taxus* species have been placed on the IUCN Red List of Threatened Species (Thomas and Farjon 2011). Based on the literature review, it was noticed that research on taxol production by endophytes was prioritized at first; other important biological activities of endophytes were recognized at a later stage. Therefore, this book chapter summarizes the research done so far on the diversity of *Taxus wallichiana*-associated microbial endophytes with particular reference to taxol production along with the other reported ecological and biotechnological applications. We classified the present chapter into the following research disciplines: diversity of endophytes, taxol-producing endophytes, biological activities of the endophytes, and production of secondary metabolites from the

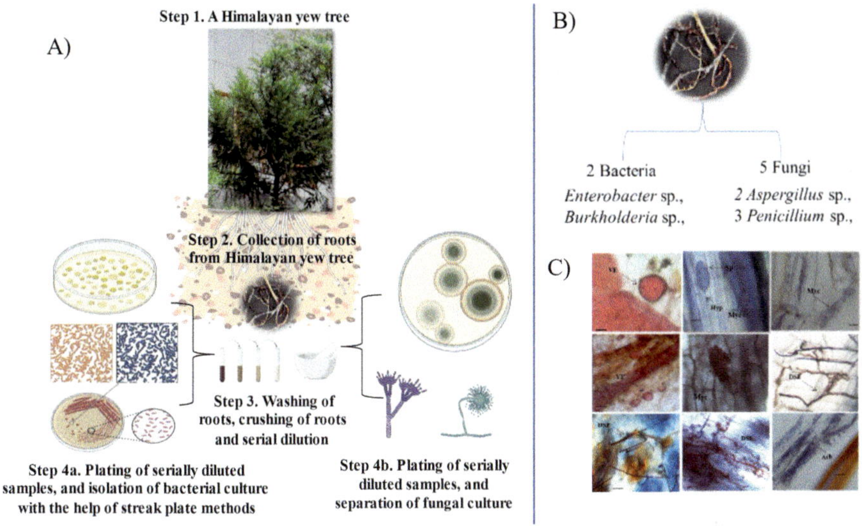

Fig. 20.1 (**a**) Isolation of endophytic bacteria and fungi. (**b**) Endophytes from *T. wallichiana*. (**c**) Endophytic fungal colonization of *T. wallichiana* roots. *DSE* dark septate endophytes, *Sp* spore, *Hyp* hypha, *Myc* mycosclerosia, *FM* fungal mycelium, *Arb* arbuscules, *My* mycelium. Bar = 2 μm

endophytes and their host competitiveness (Fig. 20.1a–c). Secondary metabolite identification in all reported studies has involved the isolation of the endophyte followed by cultivation in the lab and then extraction from mycelial cultures. Here, in the present chapter, we have focused on the microbial endophytes that have been reported from the internal tissues (endosphere) of different plant parts.

20.2 Ecological and Biotechnological Applications

Endophytes in the root microbiome are essential to the host plant's ability to operate. These microbes can be screened, characterized, selected based on PGP traits, and employed in the propagation process to generate healthy plants. As we indicated at the outset, the Himalayan yew is a priority candidate for conservation because it is included in the IUCN Red List of Threatened Species. Under the microscope, the roots of *T. wallichiana* clearly show extensive mycorrhizal and other endophyte colonization (Fig. 20.1c). A long-term study demonstrated the use of root-associated microbiome to inoculate the stem cuttings of *T. wallichiana* to raise healthy plantlets for field transfer as an environmentally benign nursery technique. Observations recorded on the interactions between plants and microbes emphasized the significance of root microbiomes in the establishment of host plant species in a particular ecological niche (Pandey et al. 2020). A schematic diagram of the different steps involved in the stem-cutting methods is given in Fig. 20.2.

Arbuscular mycorrhizal colonization in Himalayan yew has been documented in some forest areas in the Indian Himalayan region with the species *Glomus*

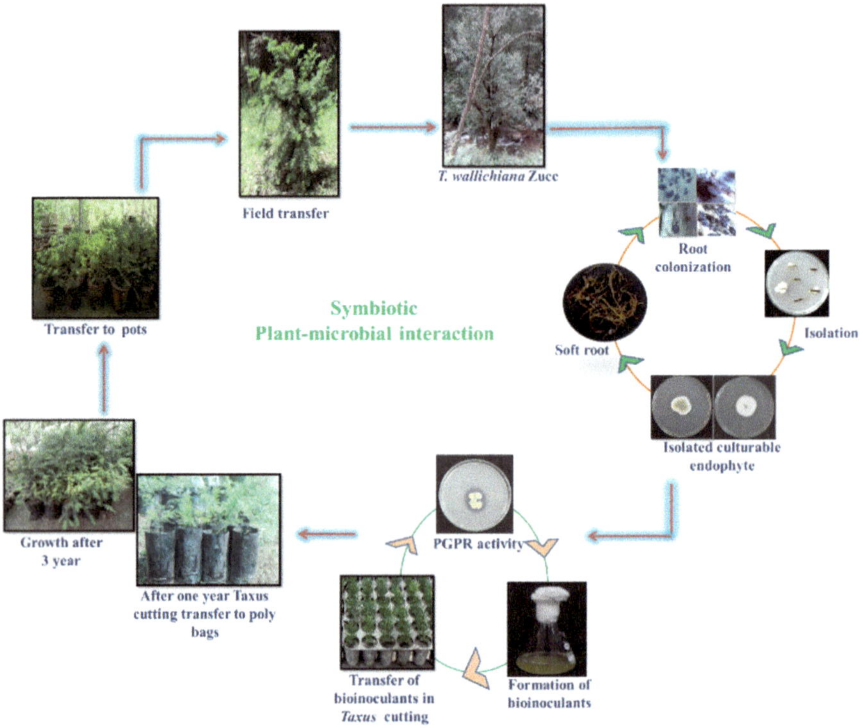

Fig. 20.2 Cultivation of *T. wallichiana* through stem cuttings using microbial inoculation

exhibiting a dominant role (Chaurasia et al. 2004). These documents demonstrate how these partners have co-evolved with the host plant (*T. wallichiana*) in a symbiotic way that is mutually advantageous. From an ecological standpoint, plant-endophyte symbiosis is becoming a significant area of study (Wani et al. 2015). Particularly important for plant growth is the root microbiome, which helps the host plant absorb nutrients and provides defense against biotic and abiotic stressors. When it comes to tree species that coevolve with their microbial partners, this idea probably works better.

Bioprospection of endophytes is an emerging topic in plant-microbe interaction research with particular reference to their ecological, biotechnological, medicinal, evolutionary, and environmental sustainability considerations (Sharma et al. 2020). Endophytic microorganisms play an important role in plant growth by developing symbiotic associations with plants. Colonization of the host by endophytic microbes while primarily contributing to plant growth also provides resistance to the host against a range of biotic and abiotic stresses. In plant-microbe association, the bacterial endophytes that colonize the internal tissues of plants are being recognized for their benefits through direct as well as indirect mechanisms. The direct mechanisms involve nitrogen fixation, phosphate solubilization, and IAA production, while in the case of indirect mechanisms, the endophytes provide defense against pathogen attacks by producing diffusible and volatile antimicrobials, siderophores, ammonia, HCN, and the lytic enzymes (Shen et al. 2019; Yadav 2018).

Literature on plant-microbe interactions emphasizes beneficial relationships between endophytic microbes and medicinal plants (Andreozzi et al. 2019; Shen et al. 2019). In addition to forming a symbiotic relationship with their host plant, endophytic microorganisms are being researched for their biotechnological and industrial significance because of their capacity to secrete secondary metabolites, which can be used as immunosuppressants, biocontrol agents, antimicrobials, and antitumor agents. Furthermore, their capacity to release antiviral substances and produce naturally occurring antioxidants, antidiabetic medicines, antibiotics, and insecticidal chemicals has been documented (Gouda et al. 2016; Yadav 2018).

20.3 Bacterial Endophytes

In recent years, endophytic bacteria, isolated from plant species that flourish in low-temperature mountain ecosystems, have drawn interest due to their role in plant growth and potential for biocontrol (Adhikari and Pandey 2020; Pandey and Yarzabal 2019). Bacterial species, namely *Burkholderia* sp. and *Enterobacter* sp., have recently been reported from the roots of *T. wallichiana* (Adhikari and Pandey 2020, 2021). These endophytic bacteria have huge potential in plant growth promotion and development and also have industrial applications, as they are the source of many biologically active secondary metabolites. Applications of the endophytes isolated from the roots of *T. wallichiana* are shown in Fig. 20.3.

20.4 PGP Traits

T. wallichiana endophytic bacteria, *Burkholderia* sp., and *Enterobacter* sp. could thrive across a wide range of temperatures (5–40 °C, opt = 25 °C) and pH (1.5–11.0, opt = 6–7), and they could withstand salt concentrations up to 12%. In addition to producing siderophore, HCN, ammonia, and salicylic acid, both bacterial endophytes also showed the ability to produce IAA and ACC deaminase. *Burkholderia* sp. using tricalcium, iron, and aluminum phosphate as substrates, the bacteria was found to be a promising phosphate solubilizer at a broad temperature range (5–35 °C). Furthermore, under both acidic and alkaline conditions, the bacterial isolates produced the enzymes phosphatase and phytase. Overall, *Burkholderia* sp. outperformed *Enterobacter* sp. in terms of PGP characteristics, despite *Enterobacter* sp. producing more salicylic acid (Adhikari and Pandey 2020).

In net house-based bioassays, the rice (*Oryza sativa*) and soybean (*Glycine max*) test crops showed the beneficial effects of the injection with formulations of *Burkholderia* sp. and *Enterobacter* sp. on physicochemical and plant growth parameters. When compared to the control plants, the treated crops' seedlings (black soybean and rice) showed an increase in length and total biomass. The biomass of rice roots and shoots grew by 13% and 24%, respectively. Under net house

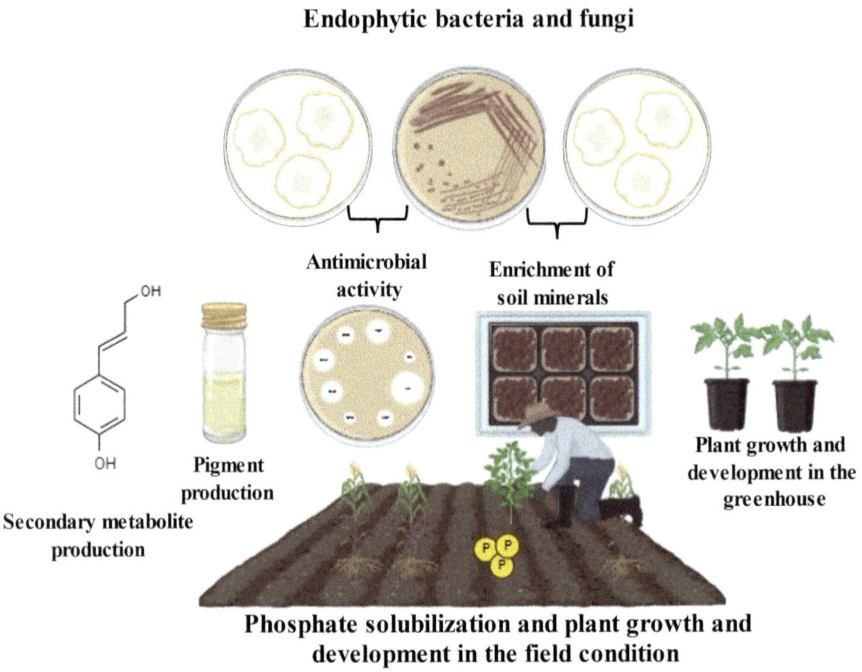

Fig. 20.3 Applications of endophytic bacteria and fungi isolated from roots of *T. wallichiana*

circumstances, the biomass of soybean roots, shoots, leaves, and nodules rose by 11%, 37%, 24%, and 9%, respectively, in comparison to the control plants. The injection of both test crops with *Enterobacter* sp. and *Burkholderia* sp. produced statistically significant positive outcomes. Because of environmental sustainability, the findings will have ramifications for creating formulations, particularly for low-temperature conditions (Adhikari and Pandey 2020).

20.5 Pigment Production

A variety of bioactive metabolites are produced by endophytic microbes, which help plants grow and remain competitive. *Burkholderia* sp. is an endophytic, antagonistic, and pigment-producing bacterium that has been shown to produce diffusible and volatile antimicrobials in plate bioassays against the phytopathogens *Alternaria alternata*, *Fusarium oxysporum*, and *F. solani*. To increase the pigment yield, the effects of physicochemical parameters such as temperature, pH, carbon, and nitrogen supply were also investigated. It was shown that bacterial pigment synthesis increased at pH 6 and 25 °C. Furthermore, maltose and ammonium chloride, as sources of carbon and nitrogen, respectively, improved the synthesis of pigments. Three categories of microorganisms (bacteria, actinobacteria, and fungi) were targets of potential antimicrobial activity in its rich fractions and crude organic extracts.

Additionally, the results of the mass spectra indicated the presence of antimicrobial secondary metabolites. These included saturated hydrocarbons (Heneicosane, Nonacosan-14-one), cycloalkyls (cyclo dodecane), aromatic dicarboxylic acids (benzene dicarboxylic acid, phthalic acid), and fatty acids (hexadecanoic acid, eicosanoic acid, octadecenoic acid, and docosanoic acid) (Adhikari and Pandey 2021).

20.6 Fungal Endophytes and PGP Traits

Endophytic fungi reside inside the host plant and provide a variety of biological functions without impairing the host plant. One of the important roles that endophytic microorganisms can play is the solubilization of inorganic phosphate, which is one of the foremost mechanisms behind plant development. Adhikari and Pandey (2019) emphasized the potential of two *Aspergillus* and three *Penicillium* species for their ability to produce phosphatases, phytases, and organic acids to solubilize insoluble phosphates in the presence of tricalcium, aluminum, and iron phosphate at varying temperatures. Moreover, they have additional potential to solubilize phosphate and phytate under acidic and alkaline conditions. Organic acids such as lactic acid, succinic acid, oxalic acid, and malic acid are present in greater concentrations during the phosphate solubilization process.

20.7 Secondary Metabolite Production from the Endophytic Fungi-Taxol Production

The most promising chemotherapeutic drug, Taxol® (generic name: Paclitaxel), was extracted from the bark of many *Taxus* species. Since the Himalayan *Taxus* species are at risk of going extinct, it is essential to create sustainable, alternative approaches for commercializing and increasing the production of paclitaxel. In this regard, physical and chemical properties are significant and useful factors, especially when employing endophytic microorganisms to produce active compounds. Adhikari and Pandey (2021) used biochemical and molecular techniques to test five endophytic fungi for the synthesis of paclitaxel. The impact of physico-chemical factors, such as temperature, pH, length of incubation, and medium components, such as salt concentration, and carbon and nitrogen sources, on the synthesis of paclitaxel was then examined. It was found that two of the isolates of *Penicillium* sp. and *Aspergillus* sp. produced paclitaxel. The amplicons of C-13 phenylpropanoid side chain-CoA acyltransferase (BAPT) and 10-deacetylbaccatin III-10-O-acetyl transferase (DBAT) were present in both the endophytes. Further, other species like *Annulohypoxylon* sp., *Pestalotiopsis microspore, Seimatoantlerium nepalense,* and *Sporormia minima* have also been reported from the *T. wallichiana* for taxol production (Adhikari et al. 2022b; Gauchan et al. 2021; Strobel and Hess 1997; Zaiyou et al. 2017). A detailed description of taxol production by the endophytes from different *Taxus* species worldwide is given in the recently published review by Adhikari et al. (2023).

20.8 Secondary Metabolite Analysis and In Vitro Biological Activities

Endophytic fungi coexist with the host plant in a mutually dependent relationship and have the capacity to produce the same bioactive secondary metabolites that the host plant is known for. In this line, Adhikari and Pandey (2024) have highlighted the secondary metabolite production by endophytic fungi, i.e., *Penicillium* sp., *P. daleae*, *Aspergillus versicolor*, and *Aspergillus* sp. isolated from *T. wallichiana* roots. Furthermore, all the isolates exhibited strong biological activities, specifically antibacterial and antioxidant. The presence of phenols, flavonols, flavonoids, tannins, and glycosides were demonstrated in all the endophytes. Seven significant organic acids, namely ascorbic acid, citric acid, lactic acid, malic acid, oxalic acid, pyruvic acid, and succinic acid, were produced by all of the endophytes. Furthermore, using ABTS and DPPH, all of the endophytes demonstrated antioxidant activity. Additionally, against *Bacillus subtilis*, *B. megaterium*, *Escherichia coli*, *Serratia marcescens*, *Pseudomonas chlororaphis*, and *P. palleroniana*, all the endophytes exhibited antibacterial activity. The ethyl acetate extracts included phenols, fatty acids, aromatic dicarboxylic acid, and alkaloids, as determined by GC-MS analysis. These endophytic fungi are being studied for their ability to produce a variety of secondary metabolites and for their antibiotic activity which could be beneficial to the biotechnological, pharmaceutical, and agricultural sectors.

20.9 Host Competitiveness

Taxus-associated endophytes are gaining importance because of their potential ecological responsibilities in addition to their potential medicinal significance. Additionally, dark septate endophyte colonization was noted in *T. wallichiana* roots (Adhikari et al. 2023). Bundles of mycelial filaments that are dark in appearance due to a higher quantity of melanin-like chemicals are what make up this category of endophytes; the color does not transfer when stained. Due to reports of these endophytes colonizing high-altitude plants flourishing in low-temperature environments, they have been identified as climatic bioindicators (Adhikari and Pandey 2018; Pandey 2019). Under laboratory conditions, the endophytic fungus of *T. wallichiana* demonstrated a broad range of growth temperatures and pH values along with a high degree of tolerance to salt. These ecological characteristics of the host-associated endophytes are probably going to help the host-symbiont connection survive and remain competitive. The microbial endophytes not only promote plant growth but also protect the host from a variety of biotic and abiotic stressors (Adhikari and Pandey 2018).

The role of microbial endophytes in helping host plants adapt and survive in stressful conditions is becoming more widely acknowledged. It was discovered that a variety of endophytic microorganisms colonized the roots of *T. wallichiana* with the most common ones being vesicular-arbuscular fungus, dark septate endophytes, and fungal mycelium. Fungal mycelium showed the highest rate of colonization

(84%) followed by vesicular-arbuscular mycorrhizae (52%) and dark septate endophytes (34%). Observations on the endophytic structures were hampered by the melanin concentration and other debris in the roots. The solution to this was to soak the cleaned roots in distilled water 2–3 weeks. As a result, when the fungi were examined under a microscope, their size and shape were more apparent (Adhikari and Pandey, 2018).

20.10 Concluding Remarks and Future Prospects

Certain kinds of medicinal plants, including *Taxus*, are well known for growing slowly and having a difficult time regrowing in the wild. For instance, it has been noted that *Taxus wallichiana* (Himalayan yew), which naturally regenerates by seeds, has extremely poor seed germination, taking over a year and a half to germinate (Pandey et al. 2002). For pharmaceutical and nutraceutical applications, endophytic microbes offer a different and sustainable supply of physiologically active metabolites. Given that *Taxus* species are known worldwide for producing the anticancer drug taxol, extracting the drug from its endophyte partners is undoubtedly a viable alternative strategy. This strategy is desperately needed for the *Taxus* species conservation.

Furthermore, to get insight into the biochemical pathways of taxol production related to the microbial partner as well as the plant host, molecular experiments are necessary. Upscaling the production of taxol and related substances will also be aided by the design and manufacture of the biosynthetic genes involved in taxol production and their transfer to bacterial or yeast cell factories. Such research and its results, when combined with microbial fermentation technologies, will definitely aid in the global production of taxol since they provide biologically appropriate and economically viable taxol in its purest form at a reduced cost.

Only a few endophytic microorganisms have been identified and thoroughly investigated, despite their enormous richness. Two key obstacles in endophyte-related investigations are the isolation of microbial endophytes from woody plants and the optimization of media for their growth and the required biological compound(s). The targeted species will probably influence the methodology utilized in the isolation of endophytes, particularly from the plant portions of woody species. The surface cleaning and sterilization of plant components, especially the roots, becomes challenging due to the accumulation of chemicals such as melanins and phenolics. Furthermore, a great deal of endophytes requires special growth conditions and specialized medium to be culturable.

Researchers are increasingly focusing on the role that microbial endophytes play in numerous ecological processes, including plant development, biodegradation, and other activities. Of these, one of the main areas of focus for research is the advantages of the relationships between endophytic microbes and medicinal plants. Gaining an understanding of the rhizosphere and plant microbiome will enable discoveries of the ecological and biotechnological roles played by microbial endophytes. Furthermore, because of their importance to industry and ecology, endophytes such as dark septate endophytes should be included. Worldwide *Taxus* species offer a wealth of opportunities for additional structural and functional research. This will additionally bolster the findings about the biology and ecological facets of host-microsymbiont relationships in alpine environments.

Acknowledgments The authors thank the National Mission on Himalayan Studies (Ministry of Environment, Forest & Climate Change, Govt. of India, New Delhi) for financial support. Dr. Kamal Ghanshala, President of the Graphic Era Group of Institutions, is acknowledged for extending the facilities to AP.

References

Adhikari P, Pandey A (2018) Diversity of endophytic fungi associated with Himalayan yew (*Taxus wallichiana* Zucc.) roots. Proc Himalayan Res Consortium 1:63–72

Adhikari P, Pandey A (2019) Phosphate solubilization potential of endophytic fungi isolated from Taxus wallichiana Zucc. roots. Rhizosphere. 9:2–9. https://doi.org/10.1016/j.rhisph.2018.11.002

Adhikari P, Pandey A (2020) Bioprospecting plant growth promoting activities of endophytic bacteria isolated from Himalayan yew (*Taxus wallichiana* Zucc.). Microbiol Res 239(2):126536

Adhikari P, Pandey A (2021) *Burkholderia* sp. (GBPI_TWL), an endophytic bacterium from *Taxus wallichiana* Zucc., produces pigment with antimicrobial metabolites. Biologia 76:3567–3578

Adhikari P, Pandey A (2024) Phytochemicals, and antioxidants and antibacterial activities of *Taxus wallichiana* Zucc. root-associated endophytic fungi. Biologia 79:1895–1907

Adhikari P, Agnihotri V, Suman SK, Pandey A (2022a) Deciphering the antimicrobial potential of *Taxus wallichiana* Zucc: identification and characterization using bioassay-guided fractionation. Chem Biodivers 20(1):e202200572

Adhikari P, Singh M, Pandey A (2022b) Taxol production by endophytic fungi isolated from the roots of *Taxus wallichiana* Zucc (Himalayan Yew). J Graphic Era Univ 10(2):195–216

Adhikari P, Joshi K, Pandey A (2023) *Taxus-associated* fungal endophytes: anticancerous to other biological activities. Fungal Biol Rev 45:100308

Andreozzi A, Prieto P, Mercado-Blanco J, Monaco S, Zampieri E, Romano S, Vale G, Defez R, Bianco C (2019) Efficient colonization of the endophytes *Herbaspirillum* hut tense RCA24 and *Enterobacter cloacae* RCA25 influences the physiological parameters of *Oryza sativa* L. cv. Baldo rice. Environ Microbiol 21(9):3489–3504

Benham SE, Houston-Durrant T, Caudullo G, de Rigo D (2016) In: San-Miguel-Ayanz J, de Rigo D, Caudullo G, Houston Durrant T, Mauri A (eds) *Taxus baccata* in Europe: distribution, habitat, usage and threats. European Atlas of forest tree species, The Publications Office of the European Union, Luxembourge, p 015921

Chaurasia B, Pandey A, Palni LMS (2004) Occurrence of arbuscular mycorrhizae in the rhizosphere of Himalayan Yew (*Taxus baccata* L. subsp. *wallichiana* (Zucc.) Pilger)—a case study. In: Podila GK, Varma AK (eds) Basic research and applications of mycorrhizae. IK International Pvt. Ltd, New Delhi, pp 26–35

Gauchan DP, Velez H, Acharya A, Ostman JR, Karl L (2021) *Annulohypoxylon* sp. strain MUS1, an endophyte isolated from *Taxus wallichiana* Zucc. produces taxol and other bioactive metabolites. 3 Biotech 11:152

Gouda S, Das G, Sen SK, Shin HS, Patra JK (2016) Endophytes: a treasure house of bioactive compounds of medicinal importance. Front Microbiol 7:1538

Juyal D, Thawani V, Thaledi S, Joshi M (2014) Ethnomedical properties of *Taxus wallichiana* Zucc (Himalayan Yew). J Tradit Complement Med 4(3):159–161

Mishra S, Sharma P (2022) Metabolomics insights into endophyte-derived bioactive compounds. Front Microbiol 13:835931

Pandey A (2019) Are dark septate endophytes bioindicators of climate in mountain ecosystems? Rhizosphere 29:110e111

Pandey A, Yarzabal LA (2019) Bioprospecting cold-adapted plant growth promoting microorganisms from mountain environments. Appl Microbiol Biotechnol 103(2):643–657

Pandey A, Nadeem M, Palni LMS (2002) Improvement in seed germination of Himalayan yew through simple soil treatments. Indian J For 25(2):109–113

Pandey A, Adhikari P, Dhyani A, Palni LMS (2020) Use of root microbiome in the conservation of Himalayan yew (*Taxus wallichiana* Zucc.): an ecofriendly approach. Natl Acad Sci Let 44(3):249–251

Quiza L, St-Arnaud M, Yergeau E (2015) Harnessing phytomicrobiome signaling for rhizosphere microbiome engineering. Front Plant Sci 6:507

Schloter M, Nannipieri P, Sorensen SJ, Elsas JDV (2018) Microbial indicator of soil quality. Biol Fertil Soils 54(1):1e10

Sharma S, Rani V, Saini R, Verma ML (2020) Bioprospecting and biotechnological applications of microbial endophytes. In: Arora PK (ed) Microbial technology for health and environment: microorganisms for sustainability. Springer, pp 191–228

Shen FT, Yen JH, Liao CS, Chen WC, Chao YT (2019) Screening of rice endophytic biofertilizers with fungicide tolerance and plant growth promoting characteristics. Sustainability 11(4):1133

Strobel GA, Hess WM (1997) Glucosylation of the peptide leucinostatin A, produced by an endophytic fungus of European yew, may protect the host from leucinostatin toxicity. Chem Biol 4(7):529–536

Thomas P, Farjon A (2011) *Taxus wallichiana*. The IUCN Red List of Threatened Species. T46171879A9730085

Wani MC, Taylor HL, Wall ME, Coggon O, McPhail AT (1971) Plant antitumor agents. VI. The isolation and structure of taxol, a novel antileukemic and antitumor agent from *Taxus brevifolia*. J Am Chem Soc 93(9):2325e2327

Wani ZA, Ashraf N, Mohiuddin T, Riyaz-Ul-Hassan S (2015) Plant-endophyte symbiosis, an ecological perspective. Appl Microbiol Biotechnol 99(7):2955–2965

Yadav AN (2018) Biodiversity and biotechnological applications of host-specific endophytic fungi for sustainable agriculture and allied sectors. Acta Sci Microbiol 1(5):1–5

Zaiyou JMD, Li MBS, Xiqiao HMS (2017) An endophytic fungus efficiently producing paclitaxel isolated from *Taxus wallichiana* var. mairei. Medicine 96(27):e7406

Elevated Remedies: The Botany, Conservation, and Cultivation of Saffron and Native New Mexican Herbs

21

Saeid Zehtab Salmasi, Connie M. Maxwell, Kevin Lombard, and Ivette Guzman

Abstract

New Mexico's high-altitude region supports several medicinal plants adapted to grow in challenging environments. This chapter explores the botany, conservation, and cultivation of prominent herbs that exemplify major categories of medicinal herbs and highlights their significant medicinal, ecological, and economic value. These include saffron (*Crocus sativus*), a non-native that shows promise in New Mexico, and native New Mexican herbs, osha (*Ligusticum porteri*), a high elevation forest perennial, ephedra (*Ephedra viridis*), a shrub found in high aridlands, cota (*Thelesperma megapotamicum*), a perennial gathered in lower elevation aridlands, and yerba mansa (*Anemopsis californica*), a perennial found and can be further developed within traditional irrigated valleys. Saffron, known for its vibrant color and health benefits, thrives in well-drained soil and

The original version of the chapter has been revised. A correction to this chapter can be found at https://doi.org/10.1007/978-3-031-99406-7_26

S. Zehtab Salmasi (✉)
Sustainable Agriculture Science Center, Department of Plant and Environmental Sciences, New Mexico State University, Alcalde, NM, USA
e-mail: saeidzs@nmsu.edu

C. M. Maxwell
New Mexico Water Resources Research Institute, New Mexico State University, Las Cruces, NM, USA

K. Lombard
Agricultural Science Center at Farmington, Department of Plant and Environmental Sciences, New Mexico State University, Farmington, NM, USA

I. Guzman
Department of Plant and Environmental Sciences, New Mexico State University, Las Cruces, NM, USA

S. Chandra, H. Lata (eds.), *High Altitude Medicinal Plants*,
https://doi.org/10.1007/978-3-031-99406-7_21

cool temperatures. The native herbs such as osha, ephedra, cota, and yerba mansa are integral to Indigenous healing practices and continue to be vital for local New Mexico cultures. The ecological conservation of these important medicinal plants is becoming increasingly endangered by impacts from climate change, habitat destruction, overexploitation (harvesting), the introduction of invasive species, and pollution. Community participation in conservation has long been essential for saving these species, allowing communities to use traditional ecological knowledge, cooperate on a communal basis, and participate in education and awareness so that several local communities are interested in the conservation aspect of such efforts. Eco-tourism and sustainability-based commerce provide supplementary economic incentives to conservation objectives. The cultivation of saffron and native herbs requires local traditional knowledge, modern agricultural practices, habitat restoration, and sustainable farming. Sustainable farming practices such as crop rotation, polyculture, permaculture, and water-saving irrigation methods are indispensable for reducing environmental degradation and promoting long-term agricultural abundance. Saffron and native herbs may offer substantial agricultural benefits in the state of New Mexico, providing an economic boon for local growers. Local farmers could benefit from this potential economic boom while preserving these amazing plants by promoting sustainable harvesting. This provides a basic structure for the sustainable utilization of high-altitude medicinal plants and their various operations in terms of botany, conservation, and cultivation practices. Since the arid regions that host saffron and these native New Mexican herbs are fragile, understanding and respecting their ecosystem will help preserve them as medicine and cultural heritage identity for future generations.

21.1 Introduction

The cooler temperatures, along with lower oxygen levels and higher UV radiation in high-altitude regions, have led to the development of a unique profile of medicinal plants. Plants are found to grow only at high elevations, for example, 1500 m (5000 feet) above sea level. However, other plants can thrive at even higher altitudes (up to 14,000 feet or more). These plants have not been conditioned to grow under normal circumstances; therefore, their medicinal power has evolved alongside these adverse growth conditions (Zhang et al. 2023). These adaptations are important for both energy farmers, for example, in New Mexico, who face a variety of severe challenges due to climate change—among them, the loss of significant amounts of surface water, well-depth drops, increased temperatures, and heightened storm intensity leading to frequent floods. There are increases in flooding around the state, with high sediment transport coming from lower lands typically supported by upland watersheds (Maxwell et al. 2022). Among the botanical treasures found in these regions are saffron (*Crocus sativus*) and various native herbs of New Mexico. These plants have garnered attention not only for their medicinal benefits but also for their economic and cultural significance. Herbs in these categories are also a key part of the holistic management of agricultural valley watershed ecosystems. If farming communities add a focus on the conservation of these herbs to their management approaches, significant multifunctional benefits are possible.

Saffron, often referred to as "red gold," is renowned for its vibrant color, distinctive aroma, and myriad health benefits. Cultivated primarily in high-altitude regions, 1200–1800 m above sea level, this precious spice has a storied history dating back to ancient civilizations. Its cultivation requires meticulous care and specific climatic conditions, making it one of the most expensive spices in the world. New Mexico, with its unique high-altitude terrain, presents an ideal environment for saffron cultivation, promising both ecological and economic benefits (Siddiqui et al. 2022). Alongside saffron, New Mexico's high-altitude ecosystems are home to a variety of native herbs with significant medicinal value (Allred and Jercinovic 2020a; Zehtab Salmasi et al. 2023). These plants have been integral to the traditional healing practices of Indigenous communities and continue to be a vital part of local culture and heritage. The unique adaptations of these herbs to high-altitude conditions not only enhance their therapeutic properties but also underscore the importance of their conservation. However, the survival of these valuable plants is threatened by factors such as climate change, habitat destruction, and overharvesting—challenges that New Mexico farmers also face, with numerous resulting effects. Effective conservation strategies are essential to protect these species and ensure their sustainable use. Local communities play a crucial role in these efforts, bringing traditional knowledge and practices to the forefront of conservation initiatives. New Mexico has significant sources of traditional knowledge, as it is home to 23 tribes and pueblos and over 700 acequias, which are systems of mutually managed irrigation channels based upon traditional practices (Arellano 2014; NMAA 2019). The researchers on this team, in collaboration with local communities, are developing pilot projects where young farmers would implement watershed restoration practices to expand habitat to conserve and harvest these species, resulting in stewardship of overall watershed health to address community impacts from climate change.

Cultivating high-altitude medicinal plants like saffron and native herbs requires specialized knowledge and sustainable practices. Farmers and horticulturists must navigate the challenges posed by altitude, climate, and soil conditions to successfully grow these plants. Emphasizing sustainable farming techniques and habitat restoration can help mitigate environmental impacts and promote long-term viability. This chapter explores the fascinating world of high-altitude medicinal plants, focusing on the botany, conservation, and cultivation of saffron and native herbs of New Mexico. Through an in-depth examination, we aim to highlight the importance of these plants and the efforts necessary to preserve their legacy for future generations.

21.2 Botany of High-Altitude Medicinal Plants

This section focuses botany few well-known high-altitude medicinal plants: saffron (*Crocus sativus*) and various native herbs found in New Mexico.

21.2.1 Characteristics of Saffron

Saffron, scientifically known as *Crocus sativus*, is a perennial plant renowned for its vivid crimson stigmas. These stigmas are harvested and dried to produce the highly valued spice used in culinary and medicinal applications. Botanically, saffron thrives in high-altitude climates that offer a combination of well-drained soil, abundant sunlight, and cool temperatures. The saffron plant typically blooms in autumn, producing purple flowers with three red stigmas. Each flower has a corm, a bulb-like structure that stores nutrients and facilitates the plant's survival during harsh conditions. The leaves are slender and grass-like, emerging in clusters from the base of the plant. Saffron's ability to withstand the rigors of high-altitude environments is attributed to its corm, which can endure cold temperatures and periods of dormancy (Kothari et al. 2021; Ibrahim et al. 2022).

21.2.1.1 Historical Context and Uses

Saffron (Fig. 21.1) has been cultivated for over 3000 years, with its origins tracing back to ancient Persia. Historical records indicate that saffron was used by Egyptian pharaohs for its aromatic and coloring properties. It was also prized in ancient Greece and Rome for its medicinal properties, believed to treat a variety of ailments, including gastrointestinal issues, insomnia, and menstrual discomfort (Kumar et al. 2022). Today, saffron is used not only in culinary dishes but also in traditional

Fig. 21.1 A photograph taken by Saeid Zehtab Salmasi showcases a saffron flower cultivated at the Sustainable Agriculture Science Center of New Mexico State University in Alcalde. The center's altitude reaches 5680 feet (1731 m)

medicine systems such as Depression, Alzheimer's, Ayurveda and Traditional Chinese Medicine (TCM) (Caser et al. 2019).

21.2.2 Native Herbs of New Mexico

New Mexico's diverse high-altitude landscapes are home to a variety of native herbs with significant medicinal properties. Among these, the following species are particularly noteworthy:

1. Osha (*Ligusticum porteri*): Often referred to as "bear root," osha is a member of the carrot (Apiaceae/Umbelliferae) family. It can be found in the forest and meadow landscapes at elevations of around 2000–3000 m. This beautiful plant has large, split leaves and clusters of small white flowers. Osha root is also very aromatic and has long been an herbal remedy for respiratory ailments (Zehtab Salmasi et al. 2023).

   Traditional and Modern Uses: Osha root has been used by Native Americans to alleviate cold, flu, and bronchitis symptoms. Just as traditional herbalists have done for centuries, modern-day practitioners still use osha in antiviral and antibacterial formulas.
2. Ephedra (*Ephedra* spp.): There are 69 species of identified Ephedra species in Asia and North American arid regions, including 6 species found in New Mexico (*Ephedra apera*, *Ephedra coryi*, *Ephedra cutleri*, *Ephedra torreyana*, *Ephedra trifurca*, and *E. viridis*) (Gonzalez-Juarez et al. 2020; Allred and Jercinovic 2020b). The New Mexican species are commonly known as Mormon tea, this shrub thrives in arid, high-altitude regions with less than 200 mm rainfall. Jointed green stems and small scale-like leaves Ephedra (may contain ephedrine-like alkaloids) were once used for bronchodilation and stimulation (Meyer 2008).

   Traditional and Modern Uses: Historically, ephedra from China, commonly known as "ma huang," is the largest producer of the stimulant ephedrine and has been regulated in many countries due to its effects on the cardiovascular system (Tang et al. 2023). Ephedrine has been used to treat asthma, nasal congestion, and as a stimulant. Even though the New Mexican Mormon tea may not contain the highly regulated ephedrine compound, it contains ephedrine-like compounds with the potential and promising effects to treat asthma and nasal congestion, and its use as a mild stimulant, the American Mormon tea has been used in traditional herbal remedies (Caveney et al. 2001; Moore 2003).
3. Cota (*Thelesperma* spp.): Cota, also known as Navajo Tea, Hopi Tea, or Green Thread, is a perennial herb that grows in the arid regions of New Mexico. In New Mexico, the most widespread species is *Thelesperma megapotamicum*; however, there are five more species (*Thelesperma ambiguum, Thelesperma filifolium, Thelesperma longipes, Thelesperma simplicifolium,* and *Thelesperma subnudum*) that can also be used for tea (Allred and Jercinovic 2020b). Cota is characterized by its slender, thread-like leaves and bright yellow, daisy-like flowers.

Cota is traditionally used by Indigenous communities to make a medicinal tea known for its anti-inflammatory and antioxidant properties (Siben et.al. 2021).

Traditional and Modern Uses: Traditionally, Cota has been used by numerous Native American tribes as a tool against myriad health challenges, such as digestive issues and inflammation. The tea that brews from its leaves and flowers is not only medicinal but also valued in cultural ways. By modern herbalists, Cota is widely known for its possible health effects and for cleansing any herbs used for alternative remedies.

4. Yerba Mansa (*Anemopsis californica*): Yerba mansa may be related to black cohosh more than to white, although it still grows in the same moist, high-altitude environments, such as riverbanks and wetlands. Yerba mansa has large, heart-shaped leaves and spike-like white flowers. For its anti-inflammatory and antimicrobial properties, it is used in the form of roots (35%) and leaves (65%) (Zehtab Salmasi et al. 2023).

   Traditional and Modern Uses: Yerba mansa has also been used by Native American tribes to treat wounds, inflammation, and as a digestive aid. Nowadays, it is also used in herbal tinctures and teas as a relaxant.

21.3 Adaptations to High Altitude

New Mexico's high-altitude medicinal plants have a number of adaptations that help them grow and thrive in some extremely difficult environments:

1. *Morphological Adaptations:* The plants that grow at high altitudes, facing various difficulties in the realms of feeding, have developed several morphological traits. For example, saffron has a compact growth form that minimizes water loss, and native herbs typically have dense root systems to optimize nutrient uptake. The importance of these adaptations lies in water conservation and nutrient deprivation, which are fundamental for survival in habitats lacking life-sustaining, changing environments (Enebe and Erasmus 2023).
2. *Physiological Differences:* Other examples include producing secondary metabolites such as flavonoids and alkaloids that can protect against UV damage or herbivory. This not only protects the plant but is also considered responsible for its medicinal properties (Yeshi et al. 2022).
3. *Reproduction:* For high-altitude plants, a unique reproduction strategy may involve nakedness. For example, saffron reproduces vegetatively by means of corms, which are storage organs that allow for slow dispersal and colonization, even in unfavorable years or when seed germination is inhibited. This method of reproduction is crucial for the plant to survive and persist in adverse environmental conditions (Peng et al. 2012).

The high-altitude medicinal plants of botany are, therefore, proof that nature is a creator above all else. The metabolic particularities and adaptations of saffron, as well as native herbs from New Mexico, can only be fully understood when their cultivation in a conservation context is investigated. Apart from their medicinal properties, these plants play a role in maintaining biodiversity and supporting the ecological stability of cold desert ecosystems.

21.4 Conservation of Medicinal Plants

The preservation of high-altitude herbs, such as saffron and native herbal plants from New Mexico, is essential for biodiversity preservation, traditional medicine, and sustainable agriculture. The different threats to these plants are also covered in this chapter, along with successful conservation approaches and the essential involvement of local communities.

21.4.1 Threats to Saffron and Native Herbs

Species of saffron and native herbs in high-altitude ecosystems are under increasing anthropogenic pressure due to various reasons:

1. *Habitat Loss:* Urbanization, deforestation, and agricultural expansion have resulted in extensive loss of natural habitats. Habitat loss from land development projects, for example, frequently disturbs the fragile ecosystems in which such saffron and native New Mexican herbs grow. Habitat fragmentation may result in populations of a plant becoming isolated, which increases their risk of extinction (Liu et al. 2016).
2. *Climate Change:* The problem of changing weather patterns, temperature levels in the atmosphere, and erratic precipitation can alter growth cycles and may reduce the survival rates of these plants. These high-altitude regions are especially sensitive to climate change, making the plants that live in them uniquely vulnerable. A change in snowmelt patterns and more extreme weather events can disrupt the delicate balance these plants rely on (Aguirrebengoa et al. 2020).
3. *Overharvesting:* The demand for medicinal plants has become so high that certain areas are now being overharvested. Overharvesting practices are also responsible for reducing the overall population of plants, which is not a good sign as it makes them harder to regrow. This is a particular worry for plants such as osha, where the root is accumulated typically with fatal consequences for the plant (Groner et al. 2022).
4. *Invasive Species:* Non-native species that are moved to the highlands can colonize and displace indigenous medicinal plants, compounding threats from habitat loss. By affecting soil composition, water availability, and light penetration, invasive species can create a hostile environment for native plants (Tobin 2018).
5. *Pollution:* Dirt or wickedness can also acutely affect the crop series, where medicinal flora undergoes decelerated growth and discoloration syndrome due to particulate matter deposition in particle flow. It can also change the natural

chemical composition of soil, which influences plant nutrition and health (O'Donnell and Sharrock 2017).

21.4.2 Conservation Strategies

In this chapter, we propose conservation strategies to mitigate these threats and secure the future of high-altitude medicinal plants:

1. *Restoration Projects:* Launching habitat restoration projects to support the recovery of conditions in degraded areas ensures they are once again conducive to high-altitude medicinal plants. Supporting vegetation in arid lands often involves water harvesting practices that spread and slow flows, reducing erosive flow energies (Norman et al. 2022), including soil restoration measures, tree planting, as well as removing invasive species that alter plant communities of lower population densities, such as deforestation and thinning overgrown juniper stands. This type of activity in high-altitude, dryland, and semi-continent areas has a significant impact on their energy management during monsoonal growing seasons (i.e., similar to New Mexico's climates) because it influences the powerhouse drivers or feedback favoring some resilience options, as well as yielding multiple functional properties under standardized conditions (Maxwell et al. 2021). The expansion of flood flows produces extended inundated zones, increased amounts and rates of soil infiltration, and larger reservoirs for moisture storage in the landscape, which supports revegetation and reduces both erosion and flooding. In addition, restoration projects can include the reintroduction of native plants and culturally valued species to restore ecological balance (Munzel et al. 2023). The harvest of these species produces a durable focus on the restoration site and yields self-sustaining labor required to perpetuate correctly performed practices.
2. *Protected Areas:* Conservation of the natural habitats where these plants naturally grow can be facilitated by establishing protected areas and conservation zones. These include national parks, reserves, and community-conserved areas where human activity is controlled. As refugia for plant species, protected areas provide an opportunity to assess the recovery of these populations in the absence of habitat destruction and overharvesting (Hoffmann 2022).
3. *Sustainable Harvesting:* The development and promotion of sustainable harvesting guidelines can ensure that the demand for medicinal plants remains within their regenerative capacity. These methods include engaging in the practice of rotational harvesting, setting harvest quotas, and collecting without destroying plant populations. When only part of the entire plant is harvested or when a method is used that does not kill off the plant and allows for regeneration, such as renting, it would be considered sustainable (Manohar 2012).
4. *Ex Situ Conservation:* Botanical Gardens, Seed Banks, and Tissue Culture Labs are mechanisms responsible for ex situ conservation; they play a vital role in conserving plant genetic material. In a pinch, these facilities could provide res-

ervoirs for replanting indigenous plant species back into their natural habitats. In this way, ex situ conservation helps preserve plants from becoming extinct due to unexpected environmental adaptations (Atkinson et al. 2022).

5. *Research and Monitoring:* Continuous research and monitoring are needed to understand the ecological requirements of these plants, as well as the effectiveness of conservation interventions. Adaptive management practices can also utilize this data. Long-term monitoring can also better detect future threats and enable timely management interventions (Sparrow et al. 2020).

21.4.3 Role of Local Communities

High-altitude medicinal plants and traditional agricultural practices are often absorbed locally by human populations, and local communities (e.g., Indigenous tribes and pueblos in New Mexico) hold crucial importance for their conservation. The inclusion of these communities plays a critical role in the success of conservation strategies because they have a comprehensive understanding of the local environment and traditional practices. Their role includes the following:

1. *Traditional Knowledge:*

   Intimate Understanding: New Mexico Indigenous tribes and pueblos have vast plant knowledge in recognition of these high-altitude flora, including medicinal applications and growth habits, as well as ecological requirements—from osha to yerba mansa to ephedra—indelibly tied to the land.

   Sustainable Practices: Traditionally, the way that we have grown and harvested these resources naturally (sustainably) is often inherently rooted in hundreds of years of tried methods to make them ultimately renewable. For instance, collecting osha roots in a sustainable manner helps the population regenerate and thrive.

   Integration with Modern Science: However, integration needs to occur with modern science, where traditional knowledge can sit alongside current scientific research for more effective conservation. The integration of this information can be used to assist in the understanding of effective iterated adaptive management practices (Falk et al. 1999; Kleitz et al. 2003, 2008; Otieno and Analo 2012; Lombard et al. 2013).

   For example, Charles Martin, a horticulturist from New Mexico State University, conducted multiple trials on medicinal plant cultivation in New Mexico and worked across the state (Martin et al. 2010; Lombard et al. 2013; Zehtab Salmasi et al. 2023). He found it challenging to transplant osha from its natural surroundings, bringing the intact plant and as much soil around and on the roots as possible.

   Dr. Mary O'Connell, PhD in Physiochemistry, has published high-impact journal articles characterizing the medicinal and ecological dynamics of high-altitude plants, such as osha and yerba mansa. She identifies bioactive compounds in situ within these plants with potent anti-inflammatory, antimicrobial, and antiviral activities, confirming their traditional use as folk medicine and possible applications in modern pharmacology. She also points to the stabilizing function of these plants in buffer zones and their contribution to moisture retention. Dr. O'Connell supports the combination of traditional ecological knowledge and scientific research and stresses the essential role that community harvesting will play in perpetuating these resources—stocks for sustainable conservation strategies of plant resources (Medina-Holguín et al. 2008).

2. *Community-Based Conservation:*

   Community-Based Decision-Making: Engaging native tribes and acequias in the decision-making process will ensure that conservation endeavors are respectful of culture and are more likely to be embraced by those on whom it will have an impact.

   Equitable Benefit-Sharing: For conservation benefits (income from eco-tourism and sustainable trade) to be socially appropriate motivational drivers for the community, they must benefit all or most stakeholders living among critical large species so that a sense of ownership and obligation is created.

   Community-Managed Protected Areas: When tribes and acequias manage their own protected areas, those lands result in higher conservation efficiency. These areas serve as safe havens for medicinal plants, allowing them to grow wild without habitat loss or overharvesting pressure (Armitage et al. 2020).

3. *Education and Awareness:*

   Promoting Education: Teaching communities why medicinal plants need to be conserved helps create a sense of ownership and accountability among them. Arguably, these programs could effectively raise awareness of the ecological, cultural, and economic value dimensions that are associated with such plants. Mobilizing Youth: Programs directed at engaging youth in traditional knowledge and conservation practices can guarantee that valuable information will not be lost because this is a learning process where school community groups can be very beneficial. Workshops and Training: Such workshops can focus on sustainable harvesting techniques, organic farming, and conservation practices, allowing locals to manage their local environment efficiently (Lombard et al. 2013; Chen et al. 2016).

4. *Economic Incentives:*

   Eco-tourism: Eco-tourism projects may, on the one hand, be beneficial for helping to establish an eco-friendly economy while, at the same time, acting as a potent driving force in maintaining and conserving medicinal plants. They can be used to educate tourists about the significance of these plants and why they need to be conserved.

   Sustainable Trade: Promoting the sustainable trade of medicinal plants, which safeguards effective harvesting (over-harvesting leads to depletion of stocks), certification schemes, and fair-trade practices can help ensure that money benefits the local people.

   Payments for Ecosystem Services (PES): Organizations could also include more on-the-ground local development with PES schemes, providing financial incentives to communities in return for ensuring that the destruction of nature does not happen. These payments compensate for ecosystems that deliver and/or safeguard ecosystem services (e.g., water purification, soil fertility, biodiversity preservation) (Samal and Dash 2023).

5. *Case Studies and Examples*

   Example 1:

   Pueblo Tribes and Acequias in NM: Super important herbs for protecting the land; traditional medicine is everyone's resource, says this plan when it grows back and people are all over. Thus, the plan tribal considers trespassing constituents: osha (aka bear root), yerba mansa, and ephedra. Criminals court/seeds perfect prosecute? Bringing these tribes into conservation can also call on the knowledge of their traditional methods to develop harvest guidelines and protected area management (Shelley 2006).

   Example 2:

   Community forest management in the Jemez Mountains: The community resource approach to managing forests has conserved higher altitude medicinal plants. These botanicals are sustainably managed by local communities for the sustainable use of forest resources (Armitage et al. 2020).

6. *Challenges and Solutions*

   Balancing Conservation with Livelihoods: One of the biggest challenges is managing to ensure that conservation does not sting local community livelihoods. Solutions: Diversification of income (ecotourism, trade). This is a challenge in and of itself, as many conservationists find it difficult to sustain long-term engagement from stakeholders involved. Long-lasting participation requires ongoing investment in education, the fair distribution of benefits, and regular community consultations. Problem: Traditional and research knowledge integration—merging traditional understanding with established data analysis may present challenges. This integration may be facilitated by collective projects and collaborative research.

We close by stressing that respect and dialogue are key for the involvement of Native tribes, pueblos, and acequias in the conservation of high-altitude medicinal plants from New Mexico. Any effective and sustainable conservation strategy will have to build on the traditional knowledge of local peoples, integrating them into decision-making processes as far as possible; raising awareness about why there is scarcity even though finfish are highly abundant (but not down by fish), so that they better understand what their own individual incentive might be for complying with rules or enabling others who wish it upon themselves through undercover support of perpetrators/fakers-NRA. This helps preserve these precious plants not just for medicinal purposes but for the culture and high-altitude biodiversity associated with Native tribes, pueblos, and acequias.

21.5 Cultivation Practices

The practices for growing saffron and native herbs in New Mexico at its altitude require a blend of traditional knowledge and modern agricultural techniques, particularly those focused on green agriculture suited to their specific environment. The following pages (when discovered) will explore the specific methods and considerations involved in cultivating precious saffron alongside native herbs of New Mexico, as well as the sustainable agricultural practices needed to preserve these delicate wild occurrences from becoming mere myths.

21.5.1 Growing Saffron

One of the most expensive and highly prized spices in the world, saffron is made from the flower of Crocus sativus. An ideal altitude for growing saffron is 1200–1800 m above sea level. It is then planted in earthenware pots at higher altitudes, where they are allowed to mature and multiply. Furthermore, the mountainous terrain reduces rainfall and pests at these elevations, resulting in more delicate saffron flowers and a high-grade harvest. These conditions of elevation and climate provide the optimal atmosphere for growing saffron, resulting in high-quality strands containing a characteristic taste, scent, and rich dyes. Growing saffron at high elevations requires careful attention to soil standards, weather, and harvesting techniques.

1. *Soil Preparation:* Saffron grows well in sandy loam soil, which is loose and has a pH of 6–8. The soil should be deeply tilled prior to sowing and enriched with organic material for improved fertility. Good drainage is a necessary pre-step to avoid bulb rot in saffron cultivation. To improve the soil structure and nutrient content, one may add organic compost or well-rotted manure (Kothari et al. 2021). Cultivate or plow deep enough to break up compacted layers and incorporate organic matter for fertility. Compost or well-rotted manure works organic matter into the soil to improve structure, increase microbial activity, and make nutrients more available. Applying a layer of organic compost or well-rotted manure will enhance the structure and fertility of the soil. Soil amendments can also be used to balance the pH and Cation Exchange Capacity (CEC) in the soil, both of which are important for nutrient retention and release (Kothari et al. 2021).
2. *Planting:* Saffron corms have to be planted in late summer so that they will settle before winter. Corms should be planted 15–20 cm apart and at a depth of 10–15 cm. This way, there is enough space between plants to avoid air stagnation and problems associated with it, such as fungal diseases. They drain very well, alleviating issues like waterlogging, which was a problem with this soil (Samal and Dash 2023).
3. *Climate:* Saffron is a winter-hardy perennial that requires cold winters and hot, dry summers. The plants need a frost-free period during flowering and grow best when temperatures are between 15 and 20 °C during the growing season.

Successful cultivation of these plants requires protection from adverse weather conditions such as heavy frost or severe rainfall (Kothari et al. 2021, Kumar et al. 2022).

4. *Irrigation and Maintenance:* Saffron crops require very little water. Waterlogging causes fungal infections and root rot. Instead, they do not need a lot of watering; when planted, they should only be misted occasionally during dry spells. Weeding is also very important to avoid nutrient competition with weeds. Mulching is beneficial for retaining soil moisture and suppressing weeds (Kothari et al. 2021; Caser et al. 2019).
5. *Harvesting:* Saffron blooms are harvested soon after blooming in the early morning when the flowers are fully opened. The crimson threads are picked by hand, and the drying process is an exacting procedure that takes great care to retain the commercially important spice properties. Drying in a cool, well-ventilated place away from direct sunlight preserves their original color and fragrance (Koocheki et al. 2019; Kothari et al. 2021).

21.5.2 Cultivating Native Herbs

The native herbs of New Mexico, such as osha (*Ligusticum porteri*), ephedra *(Ephedra viridis), cota (Thelesperma megapotamicum), and* yerba mansa *(Anemopsis californica)*, have long been used for their medicinal properties. Cultivating these herbs involves understanding their specific growing conditions and ecological roles.

Osha (Ligusticum porteri)
Osha, a member of the parsley family and also called "bear root," is traditionally used for respiratory illnesses.

1. *Soil and Location:* Osha does best in well-drained, organic-rich loam. It is most often found in shaded areas and typically inhabits forested regions, generally at an elevation of 7000–10,000 feet (2100–3000 m). Choose partial shade sites, too, imitating its natural position in the understory. Add organic compost to improve soil fertility and drainage.
2. *Reproduction:* Seeds or root division can be used to propagate osha. The process of seed germination is difficult due to dormancy problems. It requires stratification (exposure of the seeds to cold and moist conditions) over a few months in order for their dormancy to be broken. This is a winter and spring process to help seeds germinate more efficiently. If you are aiming to plant in a new area, it is best to do so in early spring or late fall. Healthy, mature plants are the ideal source of divisions to maintain robust growth (Zehtab Salmasi et al. 2023).
3. *Watering and Care:* Osha needs regular watering during its first growing season. Though once rooted, it is quite drought-tolerant; it appreciates some deep watering at times. Composting and mulching help retain soil moisture while also sup-

pressing weeds and regulating the temperature of the ground. Leaf litter or straw would be perfect organic mulch.

4. *Pest and Disease Management:* Osha does not appear to suffer from excessive pest problems or diseases. You can take this to a whole new level with organic methods by using neem oil or releasing beneficial insects in your garden, which will naturally maintain ecological balance.

Ephedra (Ephedra viridis)

Mormon tea (Ephedra) is an arid-loving, high-elevation plant that offers bronchodilator and stimulant effects.

1. *Soil and Site:* Ephedra requires well-drained, sandy, or rocky soils. Naturally, this plant does well in full sun and is often found growing under desert-like conditions. Mimic natural habitats: Choose sites where organic matter has nothing to hold on to and where it gets a full dose of sunlight. Add sand or gravel to the soil for better drainage.
2. *Propagation:* Ephedra can be grown from seeds or cuttings. Sow seeds in a well-drained seedbed and keep them just moist until germination. Even waiting for scarification (lightly scratching the seed coat) will enhance germination rates. Propagate by semi-hardwood cuttings in late spring or early summer; root them in a well-drained medium (Hammoudi et al. 2022).
3. *Watering and Maintenance:* Water ephedra regularly until it is established. After establishment, it is very drought-resistant and needs to be watered infrequently in order to avoid root rot.

   Mulching: Mulch is very important since Ephedra does not require it in nature; though gravel mulch will help the soil maintain necessary moisture and suppress weed growth.

   Pest and Disease Management: Ephedra is fairly resistant to pests and diseases. Nonetheless, continuous monitoring is required for the early discovery of potential issues. Natural pest control: Employ beneficial insects or organic sprays to maintain plant health without pesticides.

Cota (Thelesperma megapotamicum)

To grow Cota, you need to understand its natural lifestyle and environment, and then reproduce them for easy and sustainable germination.

1. *How to Prepare the Soil:* Cota likes sandy or loamy soil and does well with well-drained, relatively poor soil. Careful soil preparation, including tilling and organic amendments to add structure and help with proper drainage.
2. *Propagation:* Cota plants are grown or propagated through seeds and root divisions. For planting, sow seeds in the spring after your last frost. These seeds should be covered with a light layer of soil and kept well-watered until they germinate. Plant root divisions in early spring or fall.

3. *Climate Considerations:* Cota loves full sun, excels in high heat and drought, and is well-adapted for dry, low annual rainfall areas and semiarid regions.
4. *Irrigation and Maintenance:* After it is established, Cota seldom needs watering. Avoid over-irrigation, which can cause root rot. Additionally, mulching is a technique used to retain soil moisture and suppress weeds. In some cases, Cota is suitable for cultivation under drip irrigation for production purposes. For example, in Farmington, NM, Cota is intercropped with grapes in a vineyard under drip irrigation where sandy, well-drained soils are present. In this environment, it acts as a weed suppressant (Fig. 21.2).
5. *Harvesting:* Stems and flowers are best harvested in the summer when miner's lettuce is in full bloom. These are then set out in the sun to dry and later stored for use as tea or various other remedies.
6. *Pest and Disease Management:* Cota is not highly susceptible to pests or diseases. Routine checks of the plants and the application of organic pest deterrents like neem oil or beneficial insects can help keep your plant healthy.

Fig. 21.2 *Thelesperma megapotamicum* (Cota) as a dual crop with grapes under drip irrigation at the Agricultural Science Center at Farmington

Yerba Mansa (Anemopsis californica)

Yerba mansa grows in the wet, high-elevation areas such as riverbanks and marshes. It is considered to have anti-inflammatory and antimicrobial properties associated with its phytochemical profiles that are influenced by the soil and location in which it is grown (Medina-Holguin et al. 2008).

1. *Soil and Location:* The plant is best grown in well-drained, moist soils that are high in organic matter. It thrives in full sun to part shade and is often found near water sources (Zehtab Salmasi et al. 2023). Choose a sunny site rich in organic matter with consistent moisture for optimal growth.
2. *Propagation:* Yerba mansa is easily propagated via seed, cutting, or division. Cultivation: Good germination can be achieved by planting the seeds in moist seedbeds, where the plants require constant moisture until they sprout. This is best done in early spring. Start plants by rooting cuttings from healthy stock in a soilless medium that is well-drained.
3. *Growing and Maintenance:* Yerba mansa will need regular watering, especially during lengthy periods of drought. It requires even moisture to do well. For moisture retention, weed suppression, and protecting soil temperature, use soft plant cover (organic mulch) like compost and grass clippings.
4. *Pest and Disease Management:* Yerba mansa generally does not suffer from pests or diseases. Frequent surveillance with prompt intervention is important. Nonorganic methods: Implement natural pest control methods or sprays for plants to stop the pests in their tracks and keep your plants healthy.

   Habitat restoration methods: In the habitats of these endemic seed species, which should be restored by integrating watershed principles with water harvesting before adequate rainfall sets in, reseed them just at the onset of the monsoon. These methods date back to dryland farming in New Mexico and arid America, where Indigenous cultures seeded floodplains before they were covered by flood flows (Bryant and Holloway 1983). Reaping every area likely to be flooded and covering it with dirt is the only sure strategy. Although seed balls protect against such predation, a strategy commonly used in the Mediterranean (but which has also largely failed to take off in this region, likely due to the hardening of aggregates held together by the harsh sun). Since Yerba mansa is not an example of a low water-use crop, it may be integrated into agro-ecological landscaping by being added along irrigation canals to tap into continuous and frequent flows with little or no manual/slickwater application.

21.6 Sustainable Farming Practices

High-altitude medicinal plants need to be conserved using sustainable farming techniques. Such methods save the environment and promote plant health while increasing crop yield. A study examined the successful cultivation of medicinal plants at high altitudes in New Mexico. It was found that, by using organic practices, some medicinal plants prefer seeding over transplanting (Kleitz et al. 2008).

1. *Crop Rotation and Polyculture:* Plant species engage in a wide range of ecosystem services, such as contributing to soil fertility, regulating insect pests, and inhibiting weeds synergistically through crop rotation or polyculture. For instance, saffron can be intercropped with legumes to fix nitrogen in the soil for use by the saffron plants. Polyculture also increases biodiversity, which leads to a more stable agricultural system (Petrie and Bates 2017).
2. *Organic Farming:* Not using synthetic fertilizers and pesticides preserves soil health, preventing chemical runoff. The use of organic compost and natural pest deterrents helps in making it a climatically friendly production. Organic farming, above all, the use and implementation of simple green manure practices, can reduce some environmental impact, improve soil structure, and increase biodiversity (Panday et al. 2024).
3. *Water Conservation:* Effective methods of irrigation, such as drip or sprinkler irrigation, can be effective in saving water and reducing wastage. Rainwater harvesting is yet another helpful way of saving water. In the water-restrictive environments of high altitudes, which are often arid and devoid of root-watered vegetation like alpine tundra ecosystems, efficient water conservation practices are of key importance (Petrie and Bates 2017).
4. *Community Engagement:* Involving local communities in sustainable farming initiatives helps develop a sense of ownership and responsibility. Wider acceptance and long-term continuous success depend on training farmers regarding sustainable approaches to be followed. Community engagement could also allow for the transmission of indigenous knowledge and initiatives, which are key for sustainable agriculture (Evans and Sadler 2008; Tay et al. 2024).

Through practices like these, growers can successfully produce altitude-specific medical plants such as saffron and native New Mexican herbs while also helping to conserve incredibly valuable species in their native habitats.

21.7 Economic Importance for Local Farmers

21.7.1 Saffron (*Crocus sativus*)

1. *Saffron Conserves a High Market Value:* Saffron is often referred to as red gold, and it remains one of the most expensive spices in the world due to its labor-intensive form of cultivation. Saffron can be a cash crop specially for small farms and water deficit conditions of northern New Mexico. According to Kothari et al. saffron fetches thousands of dollars per kilo (Kothari et al. 2021).
2. *Employment Opportunities:* Growing, harvesting, and processing saffron is a job-intensive occupation that can create livelihoods in rural areas. Kumar et al. found that this result is consistent with their report, which explains how saffron farming provides seasonal employment and therefore contributes significantly to local economies (Kumar et al. 2022).

3. *Export Potential:* With a huge international demand for saffron (for culinary and medicinal uses), this offers local farmers avenues of sale in which they can participate. The unique, high-altitude environment of New Mexico can produce some very advanced saffron, which really appeals to markets globally.
4. *Value-Added Products:* Farmers can broaden their revenue streams by manufacturing saffron-based value products like flavored oils, teas, and cosmetics to reach more markets with the promise of improved profit margins. In addition, it is estimated that value-added products can result in substantial economic returns from saffron cultivation (Caser et al. 2019).

21.7.2 Native Herbs of New Mexico

Osha (Ligusticum porteri)

1. *Traditional Medicine:* As an excellent remedy for respiratory afflictions, osha is highly prized in traditional medicine and therefore always has a local market. In New Mexico, this manifests as regional cultural values that vary between the sovereign tribes and pueblos, including Indo-Hispano communities, as well as demographic groups of relative newcomers who have migrated into the region. Sacred plants of high elevation and many high-mountain herbal remedies are considered important. Higher elevations can be sacred within the communities. Before getting started on any kind of community engagement, it is a great idea first to reach out and have discussions with keepers or formal associations known in your area.
2. *Wildcrafted Teas and Herbal Supplements:* The rise of natural/organic products, particularly herbal supplements, makes it possible for farmers to sell their osha roots to companies that process them into proprietary blends marketed as dietary supplements or medicinal teas. Ibrahim et al. (2022) back up this trend with a discussion of their results, attributing this value of natural remedies to the increase in consumer interest.

Ephedra (Ephedra spp.)

1. *Bronchodilator and Stimulant:* Despite regulatory restrictions, ephedra remains in demand for its traditional uses, providing niche market opportunities. Meyer (2008) discusses the historical and modern applications of ephedra, underscoring its economic potential.
2. *Drought-Tolerant Crop:* Healthy plants that happily resist drought, ensuring success for farmers in desiccated landscapes and a plethora of options. This adds a density role in water-stress regions.

Cota (Thelesperma megapotamicum)

1. *Traditional Medicine and Herbal Tea:* Cota, commonly known as Navajo tea, is traditional medicine. The plant is then picked, dried, and brewed into a tea most famous for its anti-inflammatory properties, along with added diuretic and digestive benefits. Since Cota has traditional usage in Ayurveda, there is always a demand for it, which ultimately benefits these farmers as they never have to search the market for their harvest. Zehtab Salmasi et al. (2023) reported that Cota is still widely used in Indigenous healing practices; this represents a continued demand for it and an opportunity to cultivate it rationally.
2. *Health and Wellness Market:* The consumer bandwagon on natural and herbal products gives farmers a chance to sell Cota due to its high demand among health and wellness product manufacturers. The medicinal qualities of Cota have led to its use in herbal supplements and home remedies; as a result, its geographic access has been extended. Ibrahim et al. (2022) confirmed this trend.
3. *Value-Added Products:* Cota-based value-added products can be produced, which will help farmers improve their income. These products are often sold as herbal teas or tinctures and can be marketed by retailers with health claims. These value-added products bring additional income but also expand the market for Cota. Caser et al. (2019) conclude that producing value-added products can markedly improve the economic returns from medicinal plant production.
4. *Cultural Heritage Tourism:* Cultivating and selling Cota can serve to promote cultural heritage tourism by drawing tourists who are interested in Indigenous healing practices as well as traditional uses of native plants. This model of ecotourism can bring extra income opportunities to specific villages, as tourists are willing to pay more for an authentic experience and locally oriented goods. Local economies and the conservation aspects of medicinal plants can be promoted by cultural heritage tourism (Samal and Dash 2023).
5. *Drought-Tolerant Crop:* Its broader adaptive range to arid and semiarid areas makes Cota an ideal crop for cultivation in drought-prone environments. With this kind of resilience, it guarantees repeated cash flow for farmers, especially in drought-prone regions. Due to the deep taproot system of Cota, its consistency in obtaining water from deeper soil layers makes it less dependent on regular irrigation and more resistant to climate change. In New Mexico, access to water is a major consideration for farmers, and this trait can be extremely beneficial.

   Investing in a potentially lucrative Cota industry offers local farmers an opportunity to help conserve and sustainably utilize this economically valuable native plant. The cultivation of Cota is for economic reasons as well as for the preservation and conservation of traditional knowledge and cultural heritage.

Yerba Mansa (Anemopsis californica)

1. *Anti-inflammatory and Antimicrobial Products:* Yerba mansa is used in numerous products as a common natural health product, including tinctures and teas for their anti-inflammatory and antimicrobial properties. Medina-Holguín et al.

(2008) illuminated the bioactive compounds present in this plant with therapeutic effects.
2. *Cultural Heritage:* Yerba mansa is cultivated or foraged on lands belonging to several Indigenous Nations in the southwest USA, bringing increased tourism from cultural tourists interested in local medicinal practices, which can create more streams of revenue for local communities.

21.8 Conclusion

In addition to their aromatic qualities, high-altitude herbal medicine plants are well-respected for their powerful healing properties and cultural significance (e.g., saffron from the Asiatic region) of the state. This is an adaptation to their fantastic, detailed botanical features that allow them to survive in some of the most difficult environments on our planet. Such is the keystone function they provide in preserving cultural and ecological agricultural valley watersheds. The vivid crimson stigmas of saffron are a perfect example of the creative powers that humans have been able to harness while simultaneously showcasing nature's intricate dance with agriculture. For instance, their native herbs also boast a wide range of medicinal properties that have been valued and employed by generations steeped in Indigenous roots. It is also an excellent entry opportunity for working on multiple conservation benefits desperately needed by New Mexico rural agricultural communities. Yet climate change, the loss of habitats, and overexploitation are threatening these unique species at an alarming rate. We must employ strong conservation strategies NOW to protect these species for the next generation. This involves work to protect their natural habitats, efforts to promote sustainable harvesting, and engagement with the local community in conservation plans. Indigenous knowledge and practices are central to these measures as they provide a valuable source of tools for the effective implementation of nature-based solution projects that will safeguard socio-cultural integrity.

Both the cultivation of saffron and native herbs demand particular practices, so both are under the Royal Protection Act to be conducted appropriately. The fact that they are native herbs requires certain practices to mimic their natural high-altitude environmental conditions, while the cultivation of saffron requires as much attention, from soil preparation to meticulous harvesting techniques. To reduce these adverse environmental effects and ensure the sustainable future of these plants, a tool system should be designed with the whole plant life cycle in mind. This is where sustainability practices like organic farming or permaculture come into play. In short, botany meets conservation and cultivation for the sustainable utilization of medicinal high-altitude plants. Through an understanding and respect for their ecosystems, we can enjoy saffron as well as native herbs forever in New Mexico. These plants provide medicine and a way to connect with the natural world, both on a cultural level and through personal experience. We are guardians of our environment, and the obligation exists to protect these jewels for all time.

Funding This research was supported by the intramural research program of the U.S. Department of Agriculture, National Institute of Food and Agriculture, Organic Transitions, Award No 2022-51106-38025. The findings and conclusions in this book chapter have not been formally disseminated by the U.S. Department of Agriculture and should not be construed to represent any agency determination or policy.

References

Aguirrebengoa M, Menéndez R, Müller C, González-Megías A (2020) Altered rainfall patterns reduce plant fitness and disrupt interactions between below- and aboveground insect herbivores Abstract Ecosphere 11(5). https://doi.org/10.1002/ecs2.v11.5 https://doi.org/10.1002/ecs2.3127

Allred KW, Jercinovic EM (2020a) Flora Neomexicana III: an illustrated identification manual, part 1, 2nd edition. Identification keys and descriptions of families, genera, and species, notes on habitat and ecology, and distribution maps are provided for the 766 species (and 828 total taxa) of spore plants, gymnosperms, and monocotyledonous plants found in New Mexico. Many of the species are illustrated. https://floraneomexicana.org/wp-content/uploads/2021/07/fnm-iii-part-1-2020.pdf

Allred KW, Jercinovic EM (2020b) Flora Neomexicana III: an illustrated identification manual, part 2, 2nd edition. https://floraneomexicana.org/wp-content/uploads/2021/07/fnm-iii-part-2-2020.pdf

Arellano JE (2014) Enduring acequias: wisdom of the land, knowledge of the water. UNM Press

Armitage D, Mbatha P, Muhl E-K, Rice W, Sowman M (2020) Governance principles for community-centered conservation in the post-2020 global biodiversity framework. Conserv Sci Pract 2:e160. https://doi.org/10.1111/csp2.160

Atkinson J, Brudvig LA, Mallen-Cooper M, Nakagawa S, Moles AT, Bonser SP (2022) Terrestrial ecosystem restoration increases biodiversity and reduces its variability, but not to reference levels: a global meta-analysis. Ecol Lett 25(7):1725–1737. https://doi.org/10.1111/ele.14025

Bryant VM, Holloway RG (1983) The role of floodplain agriculture in prehistoric New Mexico. J Arid Environ 6(1):53–66. https://doi.org/10.1016/S0140-1963(18)31325-4

Caser M, Victorino ÍMM, Demasi S, Berruti A, Donno D, Lumini E, Bianciotto V, Scariot V (2019) Saffron cultivation in marginal alpine environments: how AMF inoculation modulates yield and bioactive compounds. Agronomy 9:12. https://doi.org/10.3390/agronomy9010012

Caveney S, Charlet DA, Freitag H, Maier-Stolte M, Starratt AN (2001) New observations on the secondary chemistry of world Ephedra (Ephedraceae). Am J Bot 88:1199–1208. https://doi.org/10.2307/3558330

Chen SL, Yu H, Luo HM, Wu Q, Li CF, Steinmetz A (2016) Conservation and sustainable use of medicinal plants: problems, progress, and prospects. Chin Med 30(11):37. https://doi.org/10.1186/s13020-016-0108-7

Enebe MC, Erasmus M (2023) Symbiosis—a perspective on the effects of host traits and environmental parameters in arbuscular mycorrhizal fungal richness, colonization and ecological functions. Agriculture 13(10):1899. https://doi.org/10.3390/agriculture13101899

Evans RG, Sadler EJ (2008) Methods and technologies to improve efficiency of water use. Water Resour Res 44:W00E04. https://doi.org/10.1029/2007WR006200

Falk CL, van Voorthuizen H, Wall MM, Kleitz KM, Guldan SJ, Martin CA (1999) Costs and returns of growing selected medicinal herbs in New Mexico indicate positive return to land and risk likely. HortTechnology 9(4):681–686. https://doi.org/10.21273/HORTTECH.9.4.681

González-Juárez DE, Escobedo-Moratilla A, Flores J, Hidalgo-Figueroa S, Martínez-Tagüeña N, Morales-Jiménez J, Muñiz-Ramírez A, Pastor-Palacios G, Pérez-Miranda S, Ramírez-Hernández A et al (2020) A review of the ephedra genus: distribution, ecology, ethnobotany, phytochemistry and pharmacological properties. Molecules 25:3283. https://doi.org/10.3390/molecules25143283

Groner VP, Nicholas O, Mabhaudhi T, Slotow R, Akçakaya HR, Mace GH, Pearson RG (2022) Climate change, land cover change, and overharvesting threaten a widely used medicinal plant in South Africa. Ecol Appl 32(4):e2545. https://doi.org/10.1002/eap.2545

Hammoudi HD, Krayem M, Khaled S, Younes S (2022) A focused insight into thyme: biological, chemical, and therapeutic properties of an Indigenous Mediterranean Herb. Nutrients 14(10):2104. https://doi.org/10.3390/nu14102104

Hoffmann S (2022) Challenges and opportunities of area-based conservation in reaching biodiversity and sustainability goals. Biodivers Conserv 31:325–352. https://doi.org/10.1007/s10531-021-02340-2

Ibrahim IA, Jabbour AA, Abdulmajeed AM, Elhady ME, Almaroai YA, Hashim AM (2022) Adaptive responses of four medicinal plants to high altitude oxidative stresses through the regulation of antioxidants and secondary metabolites. Agronomy 12(12):3032. https://doi.org/10.3390/agronomy12123032

Kleitz KM, Wall MM, Falk CL, Martin CA, Guldan SJ, Remmenga MD (2003) Yield potential of selected medicinal herbs grown at three plant spacings in New Mexico. HortTechnology 13(4):631–636. https://doi.org/10.21273/HORTTECH.13.4.0631

Kleitz KM, Wall MM, Falk CL, Martin CA, Remmenga MD, Guldan SJ (2008) Stand establishment and yield potential of organically grown seeded and transplanted medicinal herbs. HortTechnology 18(1):116–121. https://doi.org/10.21273/HORTTECH.18.1.116

Koocheki A, Rezvani Moghaddam P, Aghhavani-Shajari M, Fallahi HR (2019) Corm weight or number per unit of land: which one is more effective when planting corm, based on the age of the field from which corms were selected? Ind Crop Prod 131:78–84. https://doi.org/10.1016/j.indcrop.2019.01.026

Kothari D, Thakur R, Kumar R (2021) Saffron (*Crocus sativus* L.): gold of the spices—a comprehensive review. Hortic Environ Biotechnol 62:661–677. https://doi.org/10.1007/s13580-021-00349-8

Kumar A, Devi M, Kumar R, Kumar S (2022) Introduction of high-value *Crocus sativus* (saffron) cultivation in non-traditional regions of India through ecological modelling. Sci Rep 12:11925. https://doi.org/10.1038/s41598-022-15907-y

Liu Z, He C, Wu J (2016) The relationship between habitat loss and fragmentation during urbanization: an empirical evaluation from 16 world cities. PLoS One 11(4):e0154613. https://doi.org/10.1371/journal.pone.0154613

Lombard KA, Martin C, Greenfield J (2013) Risk management education in southwest medicinal herb production and marketing in New Mexico: assessing grower and potential grower outcomes after a two-day workshop. Int J Geogr Inf Syst 20(1):31–44. https://doi.org/10.1080/10496475.2013.819477

Manohar PR (2012) Sustainable harvesting of medicinal plants: some thoughts in search for solutions. Anc Sci Life 32(1):1–2. https://doi.org/10.4103/0257-7941.113789

Martin CA, Rodriguez J, O'Connell M, Steiner R, Yao S (2010) Lavender cultivar trial results for North-Central New Mexico, 2003 through 2005, Research report 700. NMSU

Maxwell CM, Fernald AG, Cadol D, Faist AM, King JP (2021) Managing flood flow connectivity to landscapes to build buffering capacity to disturbances: an ecohydrologic modeling framework for drylands. J Environ Manag 278:111486

Maxwell CM, Sheely M, Conrad L, Perez K, Hanson A, Fernald S, Langarudi S. (2022) Community conversations for the New Mexico 50-year water plan: managing water for future resilience. New Mexico Water Resources Research Institute Miscellaneous Report No. 34. Retrieved from Las Cruces, NM https://nmwrri.nmsu.edu/miscellaneous-reports-2/m34/

Medina-Holguín AL, Omar Holguín F, Micheletto S, Goehle S, Simon JA, O'Connell MA (2008) Chemotypic variation of essential oils in the medicinal plant, *Anemopsis californica*. Phytochemistry 69(4):919–927. https://doi.org/10.1016/j.phytochem.2007.11.006

Meyer SE (2008) Ephedra L.: ephedra or Mormon-tea. In: Bonner FT, Karrfalt RP (eds) The woody plant seed manual, Agric handbook no. 727. U.S. Department of Agriculture, Forest Service, Washington, DC, pp 492–494

Moore M (2003) Medicinal plants of the mountain west, 2nd edn. Museum of New Mexico Press

Münzel T, Hahad O, Daiber A, Landrigan PJ (2023) Soil and water pollution and human health: what should cardiologists worry about? Cardiovasc Res 119(2):440–449. https://doi.org/10.1093/cvr/cvac082

NMAA (2019) About the mission New Mexico Acequia Association. Retrieved from https://lasacequias.org/about-us/

Norman LM, Lal R, Wohl E, Fairfax E, Gellis AC, Pollock MM (2022) Natural infrastructure in dryland streams (NIDS) can establish regenerative wetland sinks that reverse desertification and strengthen climate resilience. Sci Total Environ 849:157738. https://doi.org/10.1016/j.scitotenv.2022.157738

O'Donnell K, Sharrock S (2017) The contribution of botanic gardens to ex situ conservation through seed banking. Plant Divers 39(6):373–378. https://doi.org/10.1016/j.pld.2017.11.005

Otieno NE, Analo C (2012) Local indigenous knowledge about some medicinal plants in and around Kakamega forest in Western Kenya. F1000Res 1:40. https://doi.org/10.12688/f1000research.1-40.v2

Panday D, Bhusal N, Das S, Ghalehgolabbehbahani A (2024) Rooted in nature: the rise, challenges, and potential of organic farming and fertilizers in agroecosystems. Sustainability 16(4):1530. https://doi.org/10.3390/su16041530

Peng D, Zhang Z, Niu Y, Yang Y, Song B, Sun H, Li Z (2012) Advances in the studies of reproductive strategies of alpine plants. Biodivers Sci 20(3):286–299

Petrie CA, Bates J (2017) 'Multi-cropping', intercropping and adaptation to variable environments in Indus South Asia. J World Prehist 30(2):81–130. https://doi.org/10.1007/s10963-017-9101-z

Samal R, Dash M (2023) Ecotourism, biodiversity conservation and livelihoods: understanding the convergence and divergence. Int J Geoheritage Parks 11(1):1–20. https://doi.org/10.1016/j.ijgeop.2022.11.001

Shelley B (2006) Integrative medicine research in New Mexico: lessons from the published literature. Complement Health Pract Rev 11(2):107–119. https://doi.org/10.1177/1533210106291988

Siben BA, Castro MJ, Faraoni MB (2021) Polyphenols from Thelesperma megapotamicum and Their Antioxidant and Neuroprotective Activities. Chemistry Proceedings 3(1):122. https://doi.org/10.3390/ecsoc-24-08393

Siddiqui SA, Ali Redha A, Snoeck ER, Singh S, Simal-Gandara J, Ibrahim SA, Jafari SM (2022) Anti-depressant properties of crocin molecules in saffron. Molecules 27(7):2076. https://doi.org/10.3390/molecules27072076

Sparrow BD, Edwards W, Munroe SEM, Wardle GM, Guerin GR, Bastin JF, Morris B, Christensen R, Phinn S, Lowe AJ (2020) Effective ecosystem monitoring requires a multi-scaled approach. Biol Rev Camb Philos Soc 95(6):1706–1719. https://doi.org/10.1111/brv.12636

Tang S, Ren J, Kong L, Yan G, Liu C, Han Y, Sun H, Wang X-J (2023) Ephedrae Herba: a review of its phytochemistry, pharmacology, clinical application, and alkaloid toxicity. Molecules 28(2):663. https://doi.org/10.3390/molecules28020663

Tay M, Ng T, Lim Y (2024) Fostering sustainable agriculture: an exploration of localised food systems through community supported agriculture. Environ Sustain Indic 22:100385. https://doi.org/10.1016/j.indic.2024.100385

Tobin PC (2018) Managing invasive species. F1000Res 7:F1000 Faculty Rev-1686. https://doi.org/10.12688/f1000research.15414.1

Yeshi K, Crayn D, Ritmejerytė E, Wangchuk P (2022) Plant secondary metabolites produced in response to abiotic stresses has potential application in pharmaceutical product development. Molecules 27(1):313. https://doi.org/10.3390/molecules27010313. PMID: 35011546; PMCID: PMC8746929

Zehtab Salmasi S, Heyduck R, Martin C (2023) Medicinal and high value native plants suitable for small farms and water deficit conditions of northern New Mexico. J Med Plants Stud 11(5):26–28. https://doi.org/10.22271/plants.2023.v11.i5a.1582

Zhang X, Kuang T, Dong W, Qian Z, Zhang H, Landis JB, Feng T, Li L, Sun Y, Huang J (2023) Genomic convergence underlying high-altitude adaptation in alpine plants. J Integr Plant Biol 65:1620–1635. https://doi.org/10.1111/jipb.13485

Cryobanking: The New Noah's Ark for Diverse Medicinal and Aromatic Plants

22

Esther E. Uchendu, Olubunmi Akomolede, and Michael Abberton

Abstract

Plants with medicinal and aromatic values have been in use for centuries across the cultures and traditions of the world to treat illnesses, prevent ill health and produce fragrances. The World Health Organization (WHO) reported that 80% of people in developing regions depend on plant-based medicines for treatment of several recalcitrant diseases; thus there is a need to document effective conservation techniques to secure the future of propagules with short lifespans in order to reduce loss of the plants' biodiversity. This compilation provides information on the available conservation methods, particularly in vitro long-term conservation protocols through cryopreservation for several medicinal plant types, and covers diverse explants. Medicinal and/or aromatic plants can be preserved using medium-term storage (MTS) or long-term storage (LTS). Both in vitro methods of conservation are crucial for safeguarding the plant species biodiversity, especially those of unique, rare, endangered and at-risk status. The in vitro methods rely on the principles of cellular totipotency during micropropagation. Ensuring that the germplasm is in good physiological status (optimal) prior to or during in vitro conservation is important toward the maintenance of quality and therapeutic properties. This review provides vital information toward the long-term conservation of medicinal plants while also highlighting the critical factors that affect successful cryopreservation.

E. E. Uchendu (✉)
Genetic Resources Center (GRC), International Institute of Tropical Agriculture (IITA), Ibadan, Nigeria

Department of Crop and Horticultural Sciences, University of Ibadan, Ibadan, Nigeria
e-mail: E.Uchendu@cgiar.org

O. Akomolede · M. Abberton
Genetic Resources Center (GRC), International Institute of Tropical Agriculture (IITA), Ibadan, Nigeria

S. Chandra, H. Lata (eds.), *High Altitude Medicinal Plants*,
https://doi.org/10.1007/978-3-031-99406-7_22

22.1 Introduction

The depletion of medicinal plant biodiversity and increased numbers of endangered species continue to be major concerns, owing to climate change, human activities and high human population density (Mir et al. 2021). In order to mitigate these challenges, several improved protocols for micropropagation and conservation were developed. Many of these methods have shortcomings and challenges, including poor growth, limited protocols for long-term conservation and plant regeneration of diverse propagules, maintenance of vigor following several years of in vitro culture, hyper-sensitivity to desiccation and inability to tolerate freezing conditions (Kundu et al. 2018).

Broad types of propagules or tissues are suitable for cryopreservation, including shoots, axillary buds, hairy and adventitious roots, protocorms or protocorm-like bodies, somatic embryos, embryogenic and non-embryogenic cell cultures, adventitious buds, rhizome sections, embryos or embryonic axes and seeds (Gowthami et al. 2024; Zeng et al. 2024).

The steps required for in vitro plant conservation include (1) establishing germplasm in the field or screen house, (2) in vitro culture (tissue culture), with phytosanitary cleaning, and (3) cryobanking (Panis et al. 2020). For those that are in need of long-term conservation via cryopreservation, ranges of protocols are available and have been tested widely on medicinal plants (Tables 22.1, 22.2 and 22.3). However, quite a number of plant genotypes are freeze intolerant due to high sensitivity to dehydration (Pence et al. 2020; Lee et al. 2021). Prior to undergoing cryopreservation regime, the tissues or organs to be cryopreserved, need to be preconditioned in order to obtain a high recovery following liquid nitrogen exposure. Long conditioning is commonly referred to as preculture while short treatments just prior to cryopreservation are pre-treatments.

22.2 Preculture

The in vitro plants are grown under conditions that do not automatically adapt them to dehydration or cold stresses. Thus, preculture is critical before the cryopreservation protocol. Preculture treatments mimic natural environmental conditions that plants encounter in nature. These may include cold, dehydration or osmotic conditioning that last up to or more than a week. This may vary depending on the type of plant and the protocol to be used. Some examples include a preculture on medium with 0.7 M sucrose at 5 °C which produced high levels of shoot formation in apple (Sakai and Engelmann 2007) and tea (Sakai et al. 2008). Preculture conditions for improving plant growth after cryopreservation have been reported. For example, the preculture of Gentian and Wasabi on 0.3 M and 0.5 M sucrose respectively for 16 h at 25 °C significantly increased shoot formation (58% and 47% respectively) (Sakai and Engelmann 2007). Tobacco suspension cell cultures were successfully cryopreserved following preculture on 0.3 M mannitol for 1 day, prior to vitrification with

Table 22.1 Summary of effective protocols for cryopreservation of low-altitude medicinal plants

S/N	Common names	Scientific names	Medicinal uses	Protocols	References
1	Syrian rue	*Peganum harmala* L.	Treatment of depression, inflammation, and parasitic infection	Vitrification	Pavlov et al. (2019)
2	Tansy	*Tanacetum ulutavicum* T.	Digestive disorders, fever, and anti-inflammatory	Vitrification	Pavlov et al. (2019)
3	Sawwort	*Serratula dissecta* L.	Anti-inflammatory and antioxidant	Vitrification	Pavlov et al. (2019)
4	Spike savory	*Satureja spicigera*	Cramps, muscle aches, and nausea	Vitrification	Ghaffarzadeh-Namazi et al. (2017)
5	Common bulge	*Ajuga reptans* var. *atropurpurea*	Wound healing and anti-inflammatory	Encapsulation dehydration	Hirata et al. (2002)
6	Horseradish	*Armoracia rusticana*	Respiratory health, and antibacterial and digestive aids	Encapsulation dehydration	Hirata et al. (1998)
7	Rugula	*Eruca sativa*	Antioxidant, anti-inflammatory, and digestive health	Encapsulation vitrification	Xue et al. (2008)
8	Wild banana	*Ensete glaucum*	Digestive health and wound healing	Simple air dehydration	Singh et al. (2021)

PVS2 (Reinhoud 1996). This initial step is necessary because many plants are not naturally adapted to cold and thus intolerant to low temperature conditions.

22.3 Cryopreservation Protocols Effective for Medicinal Plants

22.3.1 Vitrification

This procedure involves the transition of water (following rapid immersion of the samples into liquid nitrogen) from liquid to a glassy state or amorphous state without ice crystallization. Glass formation causes metabolic inactivity because it stops metabolic reactions requiring molecular diffusion (Benson 2008). This glassy state lacks organized structure and displays physical and mechanical properties like matter in solid state. For vitrification to occur successfully, the solutes in the cytosol have to be concentrated or viscous. Increased viscosity prevents water from coming together to form ice crystals (Benson 2008). Therefore, the cryoprotectants used for vitrification are exceedingly viscous. For example, the highly effective cryoprotectant, Plant Vitrification Solution #2 (PVS2) contains a mixture of 30% glycerol

Table 22.2 Summary of effective protocols for cryopreservation of high-altitude medicinal plants

S/N	Common names	Scientific names	Medicinal uses	Protocols	References
1	Mountain tea	*Sideritis leucantha*	Anti-inflammatory, antiseptic, and antioxidant	Vitrification	Juan-Vicedo et al. (2021)
2	Indian gentian	*Gentiana kurroo* Royle	Treatment of jaundice and digestive disorders	Droplet vitrification	Sharma et al. (2021a)
3	Muskroot	*Valeriana jatamansi* J.	Treatment of insomnia and anxiety	Vitrification	Sharma et al. (2021b)
4	Triangular yam	*Dioscorea deltoidea*	Treatment of skin disorders and type 2 diabetes	Vitrification and droplet vitrification	Sharma et al. (2022)
5	Kutki	*Picrorhiza kurroa*	Respiratory issues and liver problems; aids digestion	Vitrification	Sharma and Sharma (2003)
6	Annual wormwood	*Artemisia annua*	Antimalarial, anticancer, and antiviral	Slow freezing	Toeh et al. (1996)
7	Chinese gentian	*Gentiana macrophylla*	Fever, jaundice, and digestion	Encapsulation vitrification	Xue et al. (2008)
8	Chinese yew	*Taxus chinensis*	Chemotherapy treatment	Slow freezing	Kim et al. (2001)
9	Gentians	*Gentiana* spp.	Appetite and digestion stimulation	Vitrification and encapsulation vitrification	Tanaka et al. (2004)
10	Slender savory	*Ziziphora tenuior* L.	Digestive issues and respiratory issues	Encapsulation-dehydration and encapsulation vitrification	Al-Baba et al. (2015)
11	Bachtiari savory	*Satureja bachtiarica*	Blood sugar regulation, antibacterial, and antifungal	Vitrification	Ghaffarzadeh-Namazi et al. (2015)

(w/v), 15% ethylene glycol (w/v), 15% Dimethylsulfoxide (w/v DMSO) and 0.4 M sucrose (Sakai et al. 1990, 1991a, b) in liquid Murashige and Skoog (MS) basal salts and vitamins medium (Murashige and Skoog 1962). It supercools to below −100 °C and turns into a metastable glass at ~115 °C without undergoing crystallization (Sakai et al. 2008). Being metastable means, it can readily revert (a process called de-verification) and crystalize if adequate precaution is not taken at each step.

The vitrification protocol is applicable to a range of plant types ranging from temperate, tropical or sub-tropical species and explant types including shoot tips, meristems, somatic embryos etc. So far, many medicinal plants have been cryopreserved using the vitrification protocol with over 50% recovery rate (Towill 1990;

Table 22.3 Summary of effective protocols for cryopreservation of low-moderate altitude medicinal plants

S/N	Common names	Scientific names	Medicinal uses	Protocols	References
1	True ginseng	*Panax ginseng*	Immune support, energy, and stamina	Slow freezing	Butenko et al. (1984)
2	Milkvetch root	*Astragalus membranaceus*	Fatigue, respiratory infections, and heart disease	Encapsulation vitrification	Xue et al. (2008)
3	Belladonna	*Atropa belladonna*	Muscle spasms and motion sickness	Vitrification	Touno et al. (2006)
4	Beet	*Beta vulgaris*	Lower blood pressure and improved blood flow	Slow cooling	Benson and Hamill (1991)
5	Black henbane	*Hyoscyamus niger L.*	Pain relief and sedative	Vitrification	Jung et al. (2001)
6	African false olive	*Maesa lanceolata*	Wound healing, cough, and cold relief	Vitrification	Lambert et al. (2009)
7	Barrel medic	*Medicago truncatula*	Digestive aid and anti-inflammatory	Vitrification	Lambert et al. (2009)
8	Wild tobacco	*Nicotiana rustica*	Respiratory and pain relief	Slow cooling	Benson and Hamill (1991)
9	Periwinkle	*Vinca minor L.*	Blood sugar regulation and wound healing	Encapsulation dehydration	Hirata et al. (2002)
10	Sweet wormwood	*Artemisia annua*	Malaria treatment and antiviral	Two-step freezing	An et al. (2003)
11	Madagascar periwinkle	*Catharanthus roseus*	Diabetes management and blood pressure regulation	Slow freezing	Chen et al. (1984)
12	Foxglove	*Digitalis* spp.	Heart failures and atrial fibrillation	Slow freezing	Moran et al. (1999)
13	Woolly foxglove	*Digitalis lanata*	Atrial fibrillation and heart failure	Slow freezing and ultrarapid freezing	Dietrich et al. (1967)
14	Ringed vine	*Holostemma annulare*	Anti-inflammatory, analgesic, and antipyretic	Encapsulation dehydration	Decruse et al. (1999)

(continued)

Table 22.3 (continued)

S/N	Common names	Scientific names	Medicinal uses	Protocols	References
15	Hop	*Humulus* spp.	Sedative	Slow freezing	Reed et al. (2003)
16	Zhi Mu	*Anemarrhena asphodeloides* Bunge	Fever, dry mouth, kidney, and liver health	Vitrification	Sen-Rong and Ming-Hua (2012)
17	Aromatic ginger	*Kaempferia galanga* L.	Digestive issues, respiratory problems, and skin conditions	Vitrification	Preetha et al. (2013)
18	Chinese white atractylodes	*Atractylodes macrocephala*	Promotes urination and immune system	Droplet vitrification	Zhang et al. (2015)
19	Muruci	*Byrsonima Intermedia* A.	Gastrointestinal issues and skin ailments	Droplet vitrification	Silva et al. (2013)
20	Redwood	*Sequoia sempervirens*	Respiratory ailments	Vitrification, encapsulation-vitrification, and droplet vitrification	Ozudogru et al. (2011)
21	Varuna	*Crateva nurvala* Buch	Treatment of kidney, liver, eczema, and boils	Vitrification	Sanayaima et al. (2006)
22	Jivanti	*Holostemma annulare*	Diabetes, arthritis, indigestion, and ulcers	Encapsulation-vitrification	Decruse et al. (1999)
23	Spider flower	*Cleome spinosa* C.	Anti-inflammatory and antimicrobial	Vitrification	Vilardo et al. (2019)
24	Tobacco	*Nicotiana tabacum*	Fatigue, painkiller, and respiratory treatment	Vitrification	Uchendu et al. (2013)
25	Sweet flag	*Acorus calamus* Linn.	Anxiety, appetite, arthritis, and asthma	Vitrification	Parasher et al. (2024)

Towill and Bonnart 2003; Volk et al. 2004, Uchendu et al. 2011). Cryopreservation using the vitrification protocol involves firstly, exposing the samples to viscous solution commonly known as a loading solution (LS) (2 M glycerol + 0.4 M sucrose + MS solution), followed by exposure to cryoprotectant, in most cases the PVS2. The optimal duration of exposure to these solutions varies depending on the genotype and explant type (Sakai and Engelmann 2007). *Thymus moroderi* shoot tips had 71.4% post-cryopreservation survival rate following 60 min exposure to PVS2 dehydration and vitrification (Marco-Medina et al. 2010). The following diagram illustrates the vitrification procedure (Fig. 22.1).

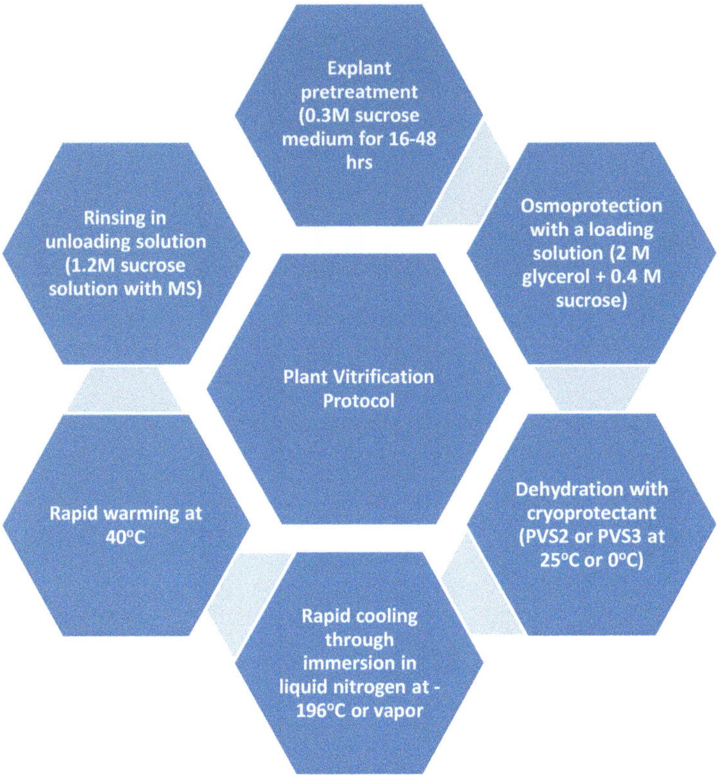

Fig. 22.1 A generalized plant vitrification procedure

22.3.2 Droplet Vitrification (DV)

This vitrification-based protocol was developed using recalcitrant shoot tips of *Musa* and Ensete (Panis et al. 2005). The different steps in developing this protocol have been reported (Panis 2024). Several vitrification solutions were developed but PVS2 and PVS3 are currently used the most (Sakai et al. 1991a, b). In the Sakai protocol, the explants are treated with loading solution (2 M glycerol + 0.4 M sucrose), and then transferred to individual micro-droplets (5–10 μl) of the cryoprotectant, PVS2 on aluminum foil strips (Fig. 22.2a–c), followed by rapid immersion in liquid nitrogen (Sakai and Engelmann 2007). During the DV process, samples are in direct contact with liquid nitrogen, which guarantees quicker cooling and often raises the percentage of post-cryopreservation recovery and regeneration rates. A specific workflow procedure for the DV is illustrated in Fig. 22.3. Briefly, meristems are excised from in vitro stock plants and either precultured for 1 day or not on solidified medium with MS + 0.3 M sucrose. The meristems are transferred to loading solution for 30 min, removed and then to PVS2 in aluminum strips for 15 min before a rapid cooling in liquid nitrogen. The cryopreserved meristems are rapidly

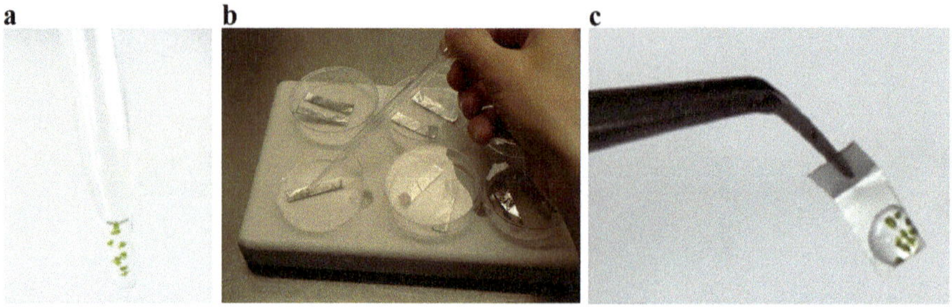

Fig. 22.2 (**a**) Shoot tips in plant vitrification solution #2 (PVS2), (b) Shoot tips transferred to aluminum foil, (**c**) Shoot tips in micro droplets (5–10 μl) of PVS2, on aluminum foil strips

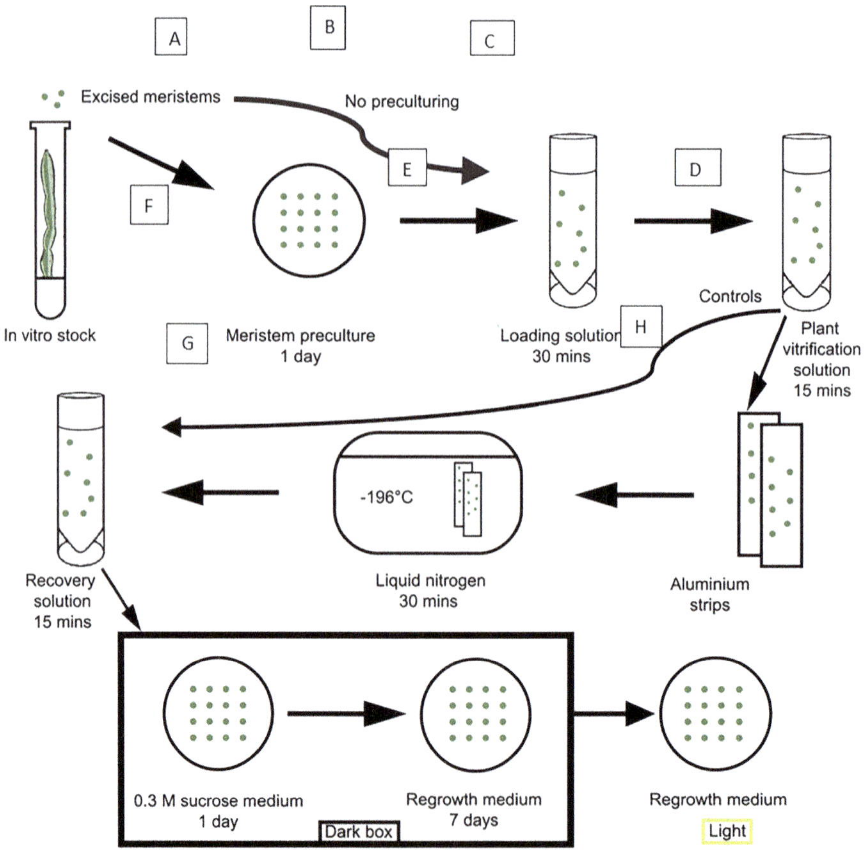

Fig. 22.3 A generalized workflow diagram for cryopreservation by droplet vitrification (DV)

rewarmed (with vitrification there is no freezing or thawing which involve ice) and rinsed in a recovery solution containing MS with 1.2 M sucrose before plating on recovery medium with 0.3 M sucrose for 1 day under dark conditions, followed by transfer to regrowth medium for additional 7 days under dark conditions. The meristems are transferred to light conditions until full regeneration.

The DV technique has been used on a vast number of plant tissues from temperate, tropical or sub-tropical plant types where they are routinely applied for germplasm conservation. It makes a big difference on plant recovery after cryopreservation, thus is useful toward the protection of endangered medicinal plant species.

22.3.3 Encapsulation Vitrification

Explants are encapsulated in alginate beads prior to performing the vitrification procedure, followed by rapid immersion into liquid nitrogen. A full review of this methodology was provided (Sakai and Engelmann 2007). This protocol works well and is used to conserve garlic germplasm by three major institutions (Leibniz Institute of Plant Genetics and Crop Plant Research (IPK), Gatersleben, Germany; National Agrobiodiversity Center (NAAS), RDA, Suwon, South Korea; and Crop Research Institute, Prague, Czech Republic). The Genetic Resource Center, International Institute of Tropical Agriculture (IITA) also uses this technique for cryopreservation of *Dioscorea* spp.

22.3.4 Encapsulation Dehydration

This is a valuable procedure for cryopreserving different germplasm. However, it takes a much longer time than the vitrification procedure because the samples to be cryopreserved have to be handled singly. Shoot tips of *Thymus moroderi* had 50% survival following cryopreservation by encapsulation-dehydration when the source plants were cold-hardened at 10 °C for 5 weeks, precultured on MS medium with 0.08 M sucrose for 48 h, encapsulated and pretreated with 0.75 M sucrose for 19 h (22% moisture content, FW) prior to freezing in liquid nitrogen (Marco-Medina et al. 2010). Popova et al. (2023) has reviewed the critical role of microenvironment on plant recovery from cryopreservation.

22.4 Importance of Medicinal Plants Preservation via In Vitro Methods

In vitro conservation techniques have several benefits and applications. The major benefits include that they are useful for crop improvement. It makes materials available for the production of secondary metabolites from the medicinal or aromatic plants. Other benefits include the ability to store a large amount of plant germplasm in the genebank (Lusty et al. 2021). The in vitro conserved medicinal plants can be distributed and reintroduced into their natural habitats where they may be used for research, breeding or commercial purposes to ensure the sustainability of the plant's genetic resources.

22.5 Few Limitations of the In Vitro Methods

Many plants produce exudates or secondary metabolites, some of which breakdown upon contact with light. Examples include phenolic compounds which breakdown to form toxic compounds that affect plant's growth and development (Uchendu et al. 2010). There could also be the problem of somaclonal variation (genetic variation among somatic cells or tissues) that promotes the production of off-types (Singh and Singh 2021). Progeny plants may be susceptible or vulnerable to diseases as they are vegetative propagated, which in the end may affect their availability.

22.6 In Vitro Conservation Workflow Process

The workflow process for in vitro germplasm conservation involves different critical stages. This starts with the selection of conserved plants from the field bank (in-situ conservation), followed by introduction into ex-situ conservation via tissue culture, and health check, then cryobanking (Fig. 22.4). It is best practice to select

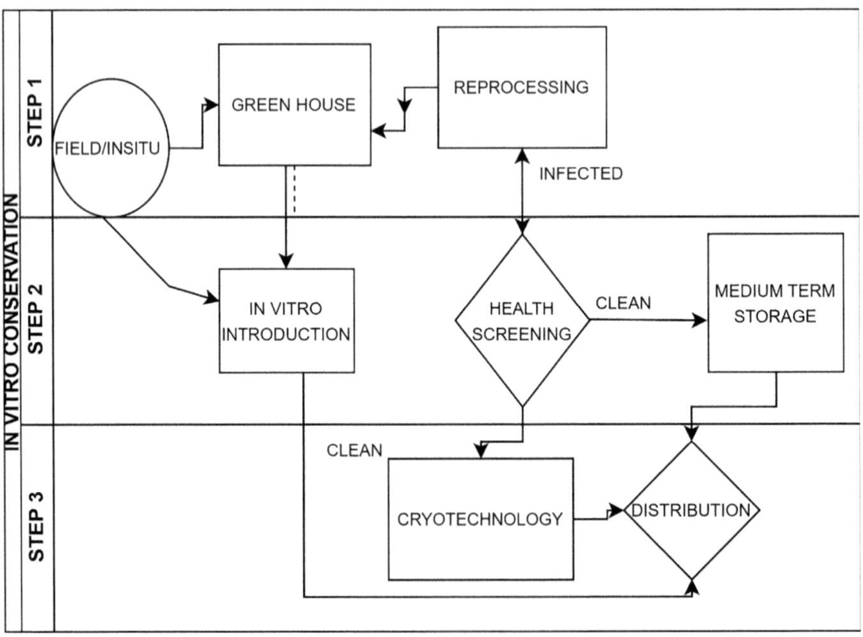

Fig. 22.4 Quality management chart demonstrating steps from in situ to ex situ conservation

plants during the wet or rainy season to reduce plant stress due to dehydration. Minimizing stress on growing plants, increases explant chances of establishment in vitro. Specifically, the collection of plants early in the morning or during the evening when the weather is more favorable reduces the stress level of tissues. Alternatively, the collected explants are first established in the green house (under controlled growth conditions) to reduce contamination on the surface area of plants before in vitro processing. After completing a cycle of processing, if the plants remain infected, the entire procedure is repeated until a clean plant is available for conservation via medium term storage or cryotechnology. All germplasm in conservation can be recovered for distribution upon request (Fig. 22.4).

Following the selection of plants from the field bank (in situ) or the green house, the explants are surface disinfected or sterilized (Fig. 22.5) and then planted in the culture medium. Once introduced and established, health check is conducted including DNA fingerprinting to ascertain the genetic fidelity status. If the indexing result is negative including the endophyte tests, the germplasm is migrated to medium term storage (MTS) with periodic or routine checks in order to ensure that quality germplasm is conserved.

With a clean health status, the explants are processed for cryopreservation in liquid nitrogen (Fig. 22.6). Plants processed for cryobanking are expected to meet certain criteria including unique, true to type and of clean health status. Conserved germplasm program for future distribution can be sourced from either MTS or long-term storage (LTS).

Fig. 22.5 Explants undergoing surface sterilization before introduction to in vitro conditions

Fig. 22.6 Plant cryopreservation in liquid nitrogen (−196 °C)

22.7 Medium-Term Storage (MTS)

The protocol for MTS starts from collection of plant germplasm either from the field, greenhouse or wild (Fig. 22.1). Priority is often given to species that are endangered, rare or at risk of extinction for establishment of the explants (shoots, meristem, leaf tissue, seeds etc.) into clean in vitro cultures. The explants are disinfected of any surface contaminants that may risk the successful establishment and growth of the propagules in vitro. After which they are planted in a sterile culture medium (composition of ingredients depends on the plant species and explant type), under controlled environmental conditions [temperature (25–27 °C), photoperiod (16 h light/8 h darkness), and humidity (80–98%)] for growth and development of the explants to occur. Following growth, multiplication and nutrient depletion, the explants are further subcultured or transferred into fresh culture media to reduce overcrowding in the culture vessels and prolong the viability of the cultures. Soon after the establishment of cultures, the health status of each genotype is checked. With a virus negative result, the cultures are migrated into MTS and assigned permanent location and accession numbers (Fig. 22.7). Regular monitoring of cultures is critical and helps detect contamination or other issues.

22.8 Long-Term Storage (LTS)

In order to stop all metabolic activities and maintain genetic stability over an indefinite period, the explants are subjected to a cryopreservation procedure. Plant cryopreservation technique involves the maintenance of cells, tissue or organ in an ultralow temperature, usually that of liquid nitrogen (LN, −196 °C) or its vapor

Fig. 22.7 In vitro cultures stored in medium term storage (MTS), under controlled environmental conditions (25–27 °C, 16 h light/8 h darkness, and 80–98% relative humidity)

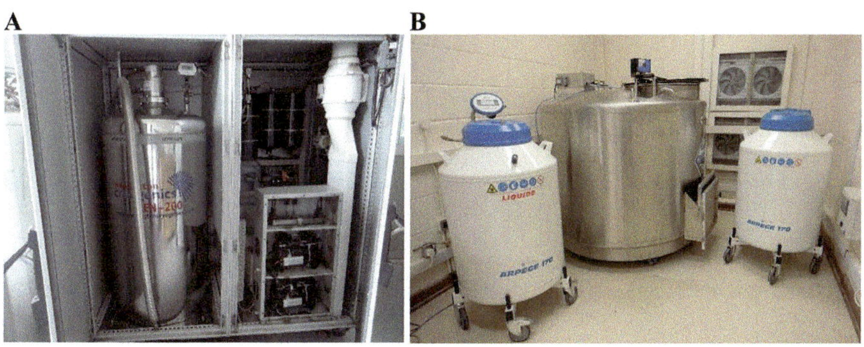

Fig. 22.8 (**a**) Liquid nitrogen generator (**b**) Dewars for indefinite cryobanking of samples at −196 °C

(−160 °C), in a manner such that they remain viable after thawing or rewarming. It is the final stage of in vitro conservation of any germplasm. Once the process for cryopreservation is completed, the cryopreserved samples are booked into the database system and transferred into dewars filled with liquid nitrogen from a generator (Fig. 22.8a) where they are cryobanked for indefinite period (Fig. 22.8b). Inventories can be maintained in a database system.

22.9 Conclusion

The ex situ conservation of plants with medicinal or aromatic qualities guarantees their availability and protection in a changing environment. This study reviews the application of in vitro methods to medicinal plants conservation and compiles information on the effective protocols for medicinal plants of low, moderate, and high altitudes, showing the most commonly used cryopreservation protocol(s) for these groups of medicinal plants. The vitrification or vitrification-based protocols is common for several medicinal plants with or without slight medication. The medium-term storage and long-term storage serve different purposes and are both unique strategies suitable for conservation of medicinal plants if critical factors are maintained.

References

Al-Baba H, Shibli RA, Akash M, Al-Qudah TS, Tahtamouni RW, Al-Ruwaiei H (2015) Cryopreservation and genetic stability assessment of threatened medicinal plant (*Ziziphora tenuior* L.) grown wild in Jordan. Jor J Biol Sci 8(4):247–256

An C, Wang X, Yuan X, Zhao B, Wang Y (2003) Optimization of cryopreservation of *Artemisia ama* L. callus. Biotechnol Lett 25:35–35

Benson EE (2008) Cryopreservation theory. In: Reed BM (ed) Plant cryopreservation: a practical guide. Springer, New York, pp 15–32. https://doi.org/10.1007/978-0-387-72276-4_2

Benson EE, Hamill JD (1991) Cryopreservation and post-freeze molecular and biosynthetic stability in transformed roots of *Beta vulgaris* and *Nicotiana rustica*. Plant Cell Tissue Org Cult 24:163–172

Butenko RG, Popov AS, Volkova LA, Chernyak ND, Nosov AM (1984) Recovery of cell cultures and their biosynthetic capacity after storage of *Dioscorea deltoidea* and *Panax ginseng* cells in liquid nitrogen. Plant Sci Lett 33:235–292

Chen TH, Kartha KK, Leung NL, Kurz WG, Chatson KB, Constabel F (1984) Cryopreservation of alkaloid-producing cell cultures of Periwinkle (*Catharanthus roseus*). Plant Physiol 75(3):726–731. https://doi.org/10.1104/pp.75.3.726. PMID: 16663695; PMCID: PMC1066984

Decruse SW, Seeni S, Pushpangadan P (1999) Cryopreservation of alginate-coated shoot tips of *in vitro* grown *Holostemma annulare* (Roxb.) *K. Schum,* an endangered medicinal plant: influence of preculture and DMSO treatment on survival and regeneration. CryoLetters 20:243–250

Dietrich B, Wolf T, Bormann A, Popoy A, Butenko RG, Luckner M (1967) Cryopreservation of *Digitalis lanata* shoot tips. Planta Med 53:359–363

Ghaffarzadeh-Namazi L, Babaeian N, Ghamari-zare A, Nematzadeh G (2015) Cryopreservation the seeds of the medicinal plant *Satureja bachtiarica* Bunge. Int J Biosci 6:24–29

Ghaffarzadeh-Namazi L, Keller ERJ, Senula A, Babaeian N (2017) Investigations on various methods for cryopreservation of callus of the medicinal plant *Satureja spicigera*. J Appl Res Med Aromat Plants 5:10–15

Gowthami R, Martinez-Montero ME, Rajasekharan PE (2024) Cryoconservation of antiallergic medicinal plants: techniques and applications. In: Gantait S, Majumder J, Sharangi AB (eds) Biotechnology of medicinal plants with antiallergy properties. Springer, Singapore. https://doi.org/10.1007/978-981-97-1467-4_21

Hirata K, Goda S, Phunchindawan M, Du D, Ishio M, Sakai A, Miyamoto K (1998) Cryopreservation of horseradish hairy root cultures by encapsulation-dehydration. J Ferment Bioeng 86:418–420

Hirata K, Mukai M, Goda S, Ishio-Kinugasa M, Yoshida K, Sakai A, Miyamoto K (2002) Cryopreservation of hairy root cultures of *Vinca minor* (L.) by encapsulation-dehydration. Biotechnol Lett 24:371–376

Juan-Vicedo J, Ramírez-Luna JE, Piqueras A, Casas JL (2021) Micropropagation and cryopreservation by vitrification of the Spanish endemic medicinal plant *Sideritis leucantha Cav. subsp. leucantha (Lamiaceae).* In Vitro Cell Dev Biol Plant 57:1057–1065

Jung DW, Sung CK, Touno K, Yoshimatsu K, Shimomura K (2001) Cryopreservation of *Hyoscyamus niger* adventitious roots by vitrification. J Plant Physiol 158:801–805

Kim SI, Choi HK, Son JS, Yun JH, Jang MS, Kim HR, Song JY, Kim JH, Choi HJ, Hong SS (2001) Cryopreservation of *Taxus chinensis* suspension cell cultures. CryoLetters 22(1):43–50. PMID: 11788843

Kundu S, Salma U, Gantait S (2018) Cryopreservation of medicinal herbs: major breakthroughs, hurdles and future. In: Kumar N (ed) Biotechnological approaches for medicinal and aromatic plants. Springer, Singapore. https://doi.org/10.1007/978-981-13-0535-1_16

Lambert E, Goossens A, Panis B, Van Labeke MC, Geelen D (2009) Cryopreservation of hairy root cultures of *Maesa lanceolata* and *Medicago ammontila.* Plant Cell Tissue Org Cult 96:289–296

Lee H-E, Popova E, Park H-N, Park S-U, Kim H-H (2021) Optimization of a cryopreservation method for the endangered Korean species *Pogostemon yatabeanus* using a systematic approach: the key role of ammonium and growth regulators. Plants 10(10):2018

Lusty C, van Beem J, Hay FR (2021) A performance management system for long-term germplasm conservation in CGIAR genebanks: aiming for quality, efficiency and improvement. Plants 10(12):2627. https://doi.org/10.3390/plants10122627

Marco-Medina A, Casas JL, Gonzalez-Benito ME (2010) Comparison of vitrification and encapsulation-dehydration for cryopreservation of *Thymus moroderi* shoot tips. CryoLetters 31(4):301–309. PMID: 20818458

Mir TA, Jan M, Khare RK, Bhat MH (2021) Medicinal plant resources: threat to its biodiversity and conservation strategies. In: Aftab T, Hakeem KR (eds) Medicinal and aromatic plants. Springer, Cham. https://doi.org/10.1007/978-3-030-58975-2_28

Morán M, Cacho M, Fernández-Tárrago J, Corchete P (1999) A protocol for the cryopreservation of *Digitalis haps* L cell cultures. CryoLetters 20:193–194

Murashige T, Skoog F (1962) A revised medium for rapid growth and bio assays with tobacco tissue cultures. Plant Physiol 15:473–497. https://doi.org/10.1111/j.1399-3054.1962.tb08052.x

Ozudogru EA, Kirdok E, Kaya E, Capuana M, Benelli C, Engelmann F (2011) Cryopreservation of redwood (*Sequoia sempervirens* (D. Don.) Endl.) *in vitro* buds using vitrification-based techniques. CryoLetters 32(2):99–110

Panis B (2024) Different steps in developing a droplet vitrification protocol. Presented for Rural Development Administration (RDA) on 24 June 2024 at Suwon (Korea). 32 sl. https://hdl.handle.net/10568/158437

Panis B, Piette B, Swennen R (2005) Droplet vitrification of apical meristems: a cryopreservation protocol applicable to all Musaceae. Plant Sci 168:45–55

Panis B, Nagel M, Van den Houwe I (2020) Challenges and prospects for the conservation of crop genetic resources in field genebanks, in *in vitro* collections and/or in liquid nitrogen. Plants 9(12):1634. https://doi.org/10.3390/plants9121634

Parasher K, Sharma C, Sharma S, Mukherjee S, Mukherjee P (2024) Cryopreservation and assessment of genetic fidelity *Acorus calamus* Linn., an endangered medicinal plant. CryoLetters 45(2):122–133

Pavlov AV, Verzhuk VG, Orlova SY, Radchenko OY, Yerastenkova MV, Gavrilkova YA, Murashev SV (2019) Cryopreservation as a method to preserve some fruit and berry crops and wild medicinal plants. Probl Cryobiol Cryomed 29(1):44–57

Pence VC, Ballesteros D, Walters C, Reed BM, Philpott M, Dixon KW, Pritchard HW, Culley TM, Vanhove AC (2020) Cryobiotechnologies: tools for expanding long-term ex situ conservation to all plant species. Biol Conserv 250:108736

Popova E, Kulichenko I, Kim HH (2023) Critical role of regrowth conditions in post-cryopreservation of *in vitro* plant germplasm. Biology 12:542

Preetha TS, HemanthaKumar AS, Krishnan PN (2013) Biotechnology and Bioinformatics Division, Jawaharlal Nehru Tropical Botanic Garden and Research. Shoot tip cryopreservation by vitrification in *Kaempferia galanga* L. An endangered, overexploited medicinal plant in Tropical Asia. IOSR J Pharm Biol Sci 8(3):19–23

Reed BM, Okut N, Achino JD, Narver L, DeNoma J (2003) Cold storage and cryopreservation of hops (*Humulus* L.) shoot cultures through application of standard protocols. CryoLetters 24:389–396

Reinhoud PJ (1996) Cryopreservation of tobacco suspension cells by vitrification. Doctoral Paper, Rijks University, Leiden, The Netherlands

Sakai A, Engelmann F (2007) Vitrification, encapsulation-vitrification and droplet-vitrification: a review. CryoLetters 28(3):151–172

Sakai A, Kobayashi S, Oiyama I (1990) Cryopreservation of nucellar cells of navel orange (*Citrus sinensis* Osb. var. *brasiliensis* Tanaka) by vitrification. Plant Sci 9:30–33

Sakai A, Kobayashi S, Oiyama I (1991a) Cryopreservation of nucellar cells of navel orange (*Citrus sinensis* Osb.) by simple freezing method. Plant Sci 74:243–248

Sakai A, Kobayashi S, Oiyama I (1991b) Survival by vitrification of nucellar cells of navel orange (*Citrus sinensis* Osb. var. *brasiliensis* Tanaka) cooled to -196°C. J Plant Physiol 137:465–470

Sakai A, Hirai D, Niino T (2008) Development of PVS-based vitrification and encapsulation–vitrification protocols. In: Reed B (ed) Plant cryopreservation: a practical guide. Springer, New York, pp 33–57

Sanayaima RK, Kaur A, Agrawal A, Babbar SB (2006) Cryopreservation of *in vitro*-grown shoot tips of *Crateva nurvala* Buch. Ham, an important medicinal tree. CryoLetters 27(6):375–386

Sen-Rong H, Ming-Hua Y (2012) A simple and efficient protocol for cryopreservation of embryogenic calli of the medicinal plant *Anemarrhena asphodeloides* Bunge by vitrification. Plant Cell Tissue Org Cult 109:287–296

Sharma N, Sharma B (2003) Cryopreservation of shoot tips of *Picoriza kurra* Royle ex Benth, an indigenous endangered medicinal plant through vitrification. CryoLetters 24:131–190

Sharma N, Gowthami R, Devi SV et al (2021a) Cryopreservation of shoot tips of *Gentiana kurroo* Royle—a critically endangered medicinal plant of India. Plant Cell Tissue Organ Cult 144:67–72

Sharma S, Parasher K, Mukherjee P, Sharma YP (2021b) Cryopreservation of a threatened medicinal plant, *Valeriana jatamansi* Jones, using vitrification and assessment of biosynthetic stability of regenerants. CryoLetters 42(5):300–308

Sharma N, Malhotra EV, Chandra R et al (2022) Cryopreservation and genetic stability assessment of regenerants of the critically endangered medicinal plant *Dioscorea deltoidea* Wall. ex Griseb. for cryobanking of germplasm. In Vitro Cell Dev Biol Plant 58:521–529

Silva LC, Paiva R, Swennen R, Andre E, Panis B (2013) Shoot-tip cryopreservation by droplet vitrification of *Byrsonima intermedia* A. Juss.: a woody tropical and medicinal plant species from Brazilian cerrado. CryoLetters 34(4):338–348. PMID: 23995401

Singh PR, Singh LJ (2021) *In vitro* propagation for improvement of medicinal plants: a review. J Pharmacogn Phytochem 10(1):1484–1489. E-ISSN: 2278-4136 P-ISSN: 2349-8234

Singh S, Thangjam R, Harish GD et al (2021) Conservation protocols for *Ensete glaucum,* a crop wild relative of banana, using plant tissue culture and cryopreservation techniques on seeds and zygotic embryos. Plant Cell Tissue Org Cult 144:195–209

Tanaka D, Niino T, Isuzugawa K, Hikage T, Uemura M (2004) Cryopreservation of shoot apices of *in-vitro* grown gentian plants: comparison of vitrification and encapsulation-vitrification protocols. CryoLetters 25:167–176

Teoh KH, Weathers PJ, Cheetham RD, Walcerz DB (1996) Cryopreservation of transformed (hairy) roots of *Artemisia annua*. Cryobiology 33(1):106–118

Touno K, Yostimatsa K, Shimomura K (2006) Characteristics of *Atropa belladonna* hairy roots cryopreserved by the vitrification method. CryoLetters 27:63–72

Towill LE (1990) Cryopreservation of isolated mint shoot tips by vitrification. Plant Cell Rep 9:178–180

Towill LE, Bonnart R (2003) Cracking in a vitrification solution during cooling or warming does not affect growth of cryopreserved mint shoot tips. CryoLetters 24:341–346

Uchendu EE, Leonard SW, Traber MG, Reed BM (2010) Vitamin E and C reduces lipid peroxidation and improves regrowth of blackberry shoot tips following cryopreservation. Plant Cell Rep 29(1):25–35

Uchendu EE, Brown DCW, Saxena PK (2011) Cryopreservation of shoot tips and cotyledons of the North American ginseng (*Panax quinquefolius* L.). CryoLetters 32(6):463–472

Uchendu EE, Shukla MR, Reed BM, Saxena PK (2013) An efficient method for cryopreservation of St John's wort and tobacco: role of melatonin. In: II international symposium on plant cryopreservation, vol 1039, pp 233–241

Vilardo AFRM, Mendonça TF, Engelmann F, Cordeiro LS, Albarello N, Simões-Gurgel C (2019) Cryopreservation of *in vitro*-grown shoot tips of the medicinal species *Cleome spinosa* (Cleomaceae) applying vitrification-based techniques. CryoLetters 40(4):237–246

Volk GM, Maness N, Rotindo K (2004) Cryopreservation of garlic (*Allium sativum* L.) using plant vitrification solution 2. CryoLetters 25:219–226

Xue SH, Luo XJ, Wu ZH, Zhang HL, Wang XY (2008) Cold storage and cryopreservation of hairy root cultures of medicinal plant *Eruca sativa* Mill., *Astragalus membranceus,* and *Gentiana macrophylla* Pall. Plant Cell Tissue Org Cult 92:251–260

Zeng L, Sun Z, Fu L, Gu Y, Li R, He M, Wei J (2024) Cryopreservation of medicinal plant seeds: strategies for genetic diversity conservation and sustainability. Plan Theory 13(18):2577. https://doi.org/10.3390/plants13182577

Zhang JM, Huang B, Lu XX, Volk GM, Xin X, Yin GK et al (2015) Cryopreservation of *in vitro* grown shoot tips of Chinese medicinal plant *Atractylodes macrocephala* Koidz using a droplet-vitrification method. CryoLetters 36(3):195–204

23 Illicit Trade and Genetic Erosion: The Plight of *Trillium govanianum* in the Himalayan Region

Praveen Dhyani, Harsh Kumar Chauhan, Dharam Chand Attri, and Ram Kumar Sharma

Abstract

Trillium govanianum, a perennial and endangered high-altitude herb, is valued for its high content of steroidal saponins. Traditionally, the species has been part of many ethnobotanical applications with the local communities for various health and spiritual benefits. As the species has found its use in modern medicine owing to its anti-inflammatory, antioxidant and anticancer activity, it has been harvested/extracted illegally and unsustainably from its natural habitat. This harvesting and subsequent trade have left the species in a dwindling state, with habitat fragmentation, degradation and a loss of genetic diversity. To highlight the urgent need for the ecological preservation of *Trillium govanianum*, this chapter highlights the species' different characteristics, such as habitat and distribution, biological attributes, and traditional and modern medicinal uses. The chapter then examines the conservation challenges and subsequently discusses the multidimensional conservation strategy, which is to balance the economic needs of local communities with ecological preservation, thereby affecting the genetic resilience of *T. govanianum* in its natural habitat.

P. Dhyani (✉)
Institute for Integrated Natural Sciences, University of Koblenz, Koblenz, Germany

H. K. Chauhan
Department of Botany, D.S.B. Campus, Kumaun University, Nainital, India

D. C. Attri
Department of Environment Sustainability and Climate Change, Islamic University of Science and Technology, Pulwama, Jammu & Kashmir, India

R. K. Sharma
Department of Biotechnology, CSIR-Institute of Himalayan Bioresource Technology, Palampur, Himachal Pradesh, India

S. Chandra, H. Lata (eds.), *High Altitude Medicinal Plants*,
https://doi.org/10.1007/978-3-031-99406-7_23

23.1 Introduction

Trillium govanianum is a perennial herbaceous plant species that belongs to the family Melanthiaceae (Table 23.1). Characterized by its tri-lobed leaves and solitary, often inconspicuous flower, the growth of this plant species is restricted to the high-altitude environment (Rashid et al. 2023). *Trillium govanianum* is well-recognized for its therapeutic properties (Ur Rahman et al. 2015a), which have been used in traditional remedies for centuries. The species encompasses several alkaloids that are of pharmacological interest primarily for their potential anti-inflammatory and anti-cancer properties, making the plant essential for curing deadly diseases (Chauhan and Bisht 2020; Mishra et al. 2023).

Trillium govanianum, also known as the Himalayan Trillium, is ecologically unique due to its specialized adaptations to the high-altitude Himalayan ecosystem. Being an indigenous species, it plays a vital role in maintaining soil health, promoting biodiversity, and cultivating symbiotic connections with other creatures. Its existence indicates a healthy ecosystem as it thrives in certain environmental circumstances and supports the forest understory by giving insects and microbes a place to live (Kumar et al. 2024; Sharma et al. 2018a; Ur Rahman et al. 2017b).

23.2 Habitat and Geographic Distribution

It is primarily found in temperate and sub-alpine zones, in an altitude range of 2400–3500 m above mean sea level (amsl). However, there are rare reports of the species occurring at elevations as high as 4000 m (Chauhan and Bisht 2020; Polunin and Stainton 1984). The species prefer growing under shady conditions below the rocks or dense tree canopies in thick humus, slowly decomposing litter, and well-moistured soil. Nevertheless, it competes with other understory plants for light, nutrients, and space (Chandola et al. 2019; Kumar et al. 2021b).

In the temperate Himalayan forests, the species is reported to be found in mixed forests (Fig. 23.1) (*Abies pindrow, Quercus* spp., *Cedrus deodara, Juglans regia, Juniperus indica, Betula* spp., *Rhododendron* spp., and *Salix* spp.) while in

Table 23.1 Botanical classification of *Trillium govanianum*

Kingdom	Plantae
Subkingdom	Tracheobionta
Super division	Spermatophyta
Division	Magnoliophyta
Class	Liliopsida
Subclass	Liliidae
Order	Liliales
Family	Melanthiaceae
Genus	*Trillium* L.
Species	*T. govanianum*

Fig. 23.1 *Trillium govanianum* habitat in sub-alpine forests

sub-alpine areas it is found under the canopy of *Rhododendron,* predominantly *R. campanulatum*. In both types, it forms an efficient symbiotic association with mycorrhizal fungi, which help absorb soil nutrients (Kanaujia et al. 2023). Insects, particularly bees, are primary pollinators of *T. govanianum*.

T. govanianum is well recognized for its wide distribution and spans several countries along the Himalayan range (Fig. 23.2) (Kumar et al. 2021b). The species is spread across Afghanistan, China, Tibet, India, Nepal, and Pakistan (Roskov et al. 2013). Each region presents slightly varying climatic and ecological conditions, contributing to the species' adaptability. It is widely distributed across the Himalayan states in India, mainly in Himachal Pradesh, Jammu and Kashmir, Sikkim, and Uttarakhand. In Himachal Pradesh, the species has been reported from areas such as Kullu, Shimla, Kinnaur, and Lahaul Spiti, characterized by their cold, alpine environments (Chauhan et al. 2018; Ur Rahman et al. 2015a). Similarly, in Jammu and Kashmir, plant populations occurred in various districts, including Doda, Kishtwar, and Bandipora, as well as in the famous meadows of Gulmarg and Sonamarg (Ur Rahman et al. 2015a). The plant is also found in Sikkim, which claims one of the most diverse collections of flora in the Indian Himalayas. In Uttarakhand Himalaya, populations have been reported in regions such as Gangotri, Harkidun, Kedarnath, Valley of Flower, and parts of the Bhagirathi Valley (Chandola et al. 2019; Chauhan and Bisht 2020; Chauhan et al. 2018; Roskov et al. 2013).

In Pakistan, species are mainly found in the northern regions, particularly in the Khyber Pakhtunkhwa province (Ur Rahman et al. 2015a). The climatic conditions in these areas are similar to those of the Indian Himalayas, with cold winters and relatively mild summers, which provide a suitable habitat for the species.

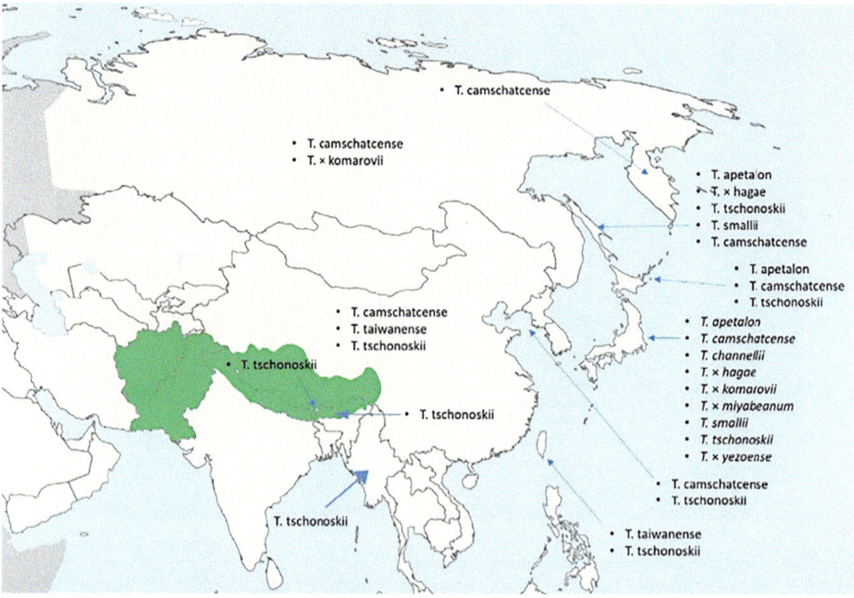

Fig. 23.2 Geographical distribution of *Trillium govanianum* (in green color) and other trillium species. *Map for representation only and not to scale. (*Reprinted from Dhyani and Sharma (2022). Copyright (2022), with permission from Elsevier)

In China, *T. govanianum* is primarily reported from the Tibetan region of Southern Xizang, which lies within the broader Himalayan range (Liang and Soukup 2000). The Tibetan Plateau, often called the "Roof of the World," provides a unique environment for the species. The presence of species in Tibet highlights its ability to thrive in some of the harshest climates, where temperatures fluctuate intensely, and the growing season is short.

The species is also found in the Himalayan regions of Nepal and Afghanistan (Roskov et al. 2013). In Nepal, the species is likely to be found in high-altitude areas such as the Annapurna and Langtang regions. Conversely, Afghanistan presents a more challenging situation to record existing distribution due to the ongoing conflict and lack of detailed botanical surveys (Rashid et al. 2023; Thapa et al. 2021).

23.3 Botanical Characteristics

23.3.1 Morphological Features

T. govanianum is a rhizomatous herbaceous plant, recognized as a notable species of the Himalayan region due to its unique biological characteristics and adaptability to high-altitude environments (Chauhan et al. 2019; Rathore et al. 2020). The plants bloom from May to June, and experience senesce before snowfall. The plant

Fig. 23.3 Physical characteristics of *Trillium govanianum*

typically grows to 10–20 cm in height and exhibits a characteristic tri-lobed arrangement of its leaves (Fig. 23.3). At maturity, these leaves appear in a whorl of three, each broadly ovate and acute, borne on short stalks, especially its reproductive organs, and are equally distinctive (Chauhan et al. 2020a; Kumar et al. 2023).

Trillium govanianum has a unique and complex reproductive process, characterized by sexual reproduction through flowering, seed dispersal, and asexual reproduction. However, asexual reproduction occurs sometimes in natural habitats, but it is uncommon and mainly observed in older plants with more extensive rhizomes (Chandola et al. 2023; Chauhan and Bisht 2020; Ohara 1989). During the reproductive phase, it produces a solitary flower that develops at the shoot apex, measuring about 2–3 cm in diameter, and features six bright yellow stamens, which are basifixed, meaning that the filaments of the stamens are attached at their base. The flower also displays a whorl of sepals and petals, which protect and support the reproductive structures (Fig. 23.2) (Chandola et al. 2023; Chauhan et al. 2020a). At the center lies a three-celled ovary that is purplish brown and eventually gives rise to multiple seeds. These seeds are dispersed once the ovary matures into a berry (Chauhan et al. 2019; Rathore et al. 2020).

23.3.2 Life Cycle and Longevity

The life cycle of *T. govanianum* is characterized by its perennial nature and long life span; overall, the species has a life expectancy of over 30 years (Chauhan et al. 2018). The species' life cycle can be divided into three distinct phases: the single-leaf vegetative stage, the three-leaf vegetative stage, and the three-leaf reproductive stage (Chauhan et al. 2018, 2019; Ohara 1989). The transition from one stage of growth and reproduction is controlled by the species rhizome in coordination with the external environment. During the initial phase, the plant stays in the single-leaf phase for several years, allowing the rhizome to accumulate resources and increase its biomass rather than producing flowers (Chauhan et al. 2020b). Once a sufficient biomass threshold is reached in the rhizome, the plant transitions to the reproductive phase. Nevertheless, this transition can take several years as the plant slowly builds up the energy required to support flowering and seed production. However, this phenomenon is common in perennial plants, where the balance between vegetative growth and reproduction is tightly regulated by environmental conditions and the availability of internal reserves (Chauhan et al. 2019; Kumar et al. 2024).

The blossom stage of this species' life cycle represents a major change, although a brief one—the plant blooms once during the reproductive phase after not blossoming for years. If the rhizome cannot sustain the necessary nutrients due to environmental stresses, it may revert to the previous vegetative stage in the next growing season. The ability of the plant to reverse phases is most likely an adaptive technique that allows it to survive under less-than-ideal conditions by delaying the use of energy for reproduction until favorable conditions are met (Chauhan et al. 2019; Kumar et al. 2023).

The climate of a species' habitat directly impacts its seasonal cycle. The plant usually appears in April when the winter snow melts in the Himalayas, where it is widely distributed. It must capitalize on the little growth season before the next winter arrives; therefore, the time it emerges is crucial. Typically, flowering occurs in May and June, when there is plenty of sunshine for photosynthesis and mild temperatures. The berries of plants mature in the later summer months, with the exact timing depending on the altitude at which it is growing. At lower altitudes, berry maturity and leaf senescence occur in July and August, while at higher altitudes, this process is delayed until September or October (Chandola et al. 2023; Chauhan et al. 2019). Once the plant has produced seeds, it enters a period of dormancy.

In the Himalayan area, dormancy is essential for plant populations to survive harsh winters. The rhizome stays alive underground during dormancy while the plant's aboveground portions wither away. The plant may grow again when the following year's conditions are more favorable because the rhizome stores nutrients and energy. This perennial life cycle, with alternating periods of growth, reproduction, and dormancy, is a common survival strategy among plants in higher-elevation environments, where the growth season is shorter, and the winters are more prolonged and more severe (Chandola et al. 2023; Chauhan et al. 2019; Kanaujia et al. 2023).

The species' growth and reproductive habits are greatly influenced by the prevailing climatic conditions such as temperature, light, snow and moisture in the higher altitudes. For instance, the snowmelt time can have a decisive effect on the plant dormancy. Nevertheless, the snowmelt should be on time, as if the snow melts too fast, the plants may emerge too soon, making them susceptible to late spring frost. However, if the snowmelt is delayed, the whole process of plant development may be delayed, thus affecting crucial seed germination and flowering (Chauhan et al. 2018, 2019; Ohara 1989). Moreover, the snowmelt time is also a crucial factor closely related to snowmelt time if the plant rhizome does not get proper amounts of water and nutrients, it cannot endure dry periods and would go for wilting (Chauhan et al. 2019, 2020b). Likewise, Light availability also has a decisive effect on the growth of *Trillium govanianum*. It has been found that plants in shader environments tend to grow slower. Therefore, delayed flowering is observed in these plants. In contrast, the sunnier side plants can grow faster and reach reproductive maturity sooner (Chauhan et al. 2018, 2019; Ohara 1989).

The reproductive strategy of *T. govanianum* is also noteworthy for its dependence on insect pollinators. Like other flowering plants, the species depends on insects to transfer pollen from one flower to another, facilitating fertilization and seed production. Although the specific pollinators have not been studied for the species, the distinctive aroma of flowers with bright yellow stamen can be an adaptive feature to attract pollinators (Chauhan et al. 2020a). Notably, the plant's need for exclusive pollinators can render the plant populations more vulnerable to the changing population forces of insects at higher altitudes, where they are already limited to small numbers due to climate (Chauhan et al. 2019). Seed dispersal is another aspect of its reproductive behavior. Trillium, being myrmecochorous, suffers a limited dispersal ability, potentially leading to the dense and patchy distribution of the species. This limited seed dispersal ability may also contribute to its vulnerability as the denser and closer populations may compete for limited resources, while too sparse populations may struggle with the fragmentation of resources, ultimately leading to reproductive struggle (Chauhan et al. 2019, 2021; Dhyani et al. 2020).

23.3.3 Ethnobotanical and Cultural Significance

Indigenous populations in the Himalayan region have a long history of using *T. govanianum* for medical purposes. Locally, the plant is known as "Nagchatri" in India and "Trishul" in Tibetan regions (Sharma et al. 2018a). In some places, the roots are crushed and combined with other medicines to treat different ailments, while in some areas, the leaves are brewed into tea (Kumar et al. 2021b).

In line with the holistic approach to well-being that prioritizes harmony between the body and environment, *T. govanianum* is incorporated into different formulations in Tibetan medicine (Chauhan et al. 2021; Hussain et al. 2023). *T. govanianum* is used in Ayurveda to balance the doshas (biological energies), especially "kapha," and relieve joint pain, respiratory problems, skin infections, and stomach discomfort. When used as a topical poultice, it fights bacteria and lessens phlegm, which helps treat indigestion, bronchitis, and asthma and promotes wound healing by reducing surrounding inflammation through its anti-inflammatory, analgesic, and antibacterial qualities (Ur Rahman et al. 2016; Chauhan et al. 2019).

Ayurveda and Himalayan traditional medicine also acknowledge *T. govanianum* as an aphrodisiac, with its steroidal saponins said to enhance male energy and reproduction and in the treatment of hemorrhoids (Chauhan et al. 2019). It is also utilized as a tonic in Uttarakhand and Himachal Pradesh to increase overall vigor and vitality. *T. govanianum* is also used in Unani medicine to treat uterine problems and irregular menstruation through its steroidal chemicals that resemble estrogen (Chauhan et al. 2019).

The cultural importance of *T. govanianum* extends beyond its medicinal value to its role in spiritual practices and rituals. The plant's versatility and importance in sustaining the spiritual well-being of the locals can be understood from the fact that for many Himalayan Indigenous communities, *T. govanianum* is revered for its symbolic associations with health, healing, and protection from hostile forces

(Sharma and Parashar 2017; Sharma 2017). These practices are deeply rooted in the belief that the plant's innate power can balance the body's energies and reinstate health (Payyappallimana 2013; Ranganathan 2018; Reddy 2023). According to folklore, the plant is a kind spirits that guards the population and provides assistance and healing when needed (Rashid et al. 2023; Thapa et al. 2021), capable of curing the sick and protecting the vulnerable. For example, In Buddhism, *T. govanianum* is used as a Talisman or amulet for its spiritual energy to establish harmony with the nature god (Chandola et al. 2019; Kanaujia et al. 2023; Rathore et al. 2020). These Cultural chronicles are widely spread in many generations of the local communities, as Oral traditions have enacted a critical part in sustaining the knowledge of *T. govanianum* and its uses (Radha and Pundir 2019; Rathore et al. 2020; Ur Rahman et al. 2016). and 5-hydroxy-sitosteroid, have been isolated from *T. govanianum* (Ismail et al. 2015; Ur Rahman et al. 2015a). Govanoside B, a novel steroidal saponin.

23.3.4 Medicinal Properties and Applications

Trillium govanianum's varied chemical composition and bioactive chemicals have been uncovered by contemporary pharmacological research, strengthening the plant's therapeutic qualities. Several phytochemical studies have identified various compounds from the Trillium genus, including saponins, steroids, flavonoids, terpenoids, phenolics, and fatty acid esters, with steroidal saponins emerging as the primary active constituents (Huang and Zou 2011; Ur Rahman et al. 2017a; Yokosuka and Mimaki 2008). Over 30 steroidal saponins have been isolated from Trillium species, demonstrating significant anticancer, anti-inflammatory, antifungal and antioxidant properties (Ur Rahman et al. 2017a). These findings suggest that steroidal saponins could serve as marker compounds for the quality assessment of the genus Trillium. However, a limited number of steroidal compounds (Fig. 23.4), such as diosgenin, pennogenin, govanoside A, borassoside E, sitosteroid, were also isolated from the rhizomes of *T. govanianum* (Singh et al. 2021) The crude extracts, fractions, and steroidal saponins of *T. govanianum* exhibited notable biological activities, including anticancer, anti-leishmanial, antibacterial, anti-inflammatory, antifungal and antioxidant properties (Khan et al. 2024; Ur Rahman et al. 2015a; Rathore et al. 2020). Given its widespread therapeutic applications, complex phytochemical profile and global use, it is necessary to establish a comprehensive quality control framework to ensure the safety and efficacy of this precious species.

The species possesses potent anti-inflammatory, antibacterial, and antioxidant qualities because to its saponins, glycosides, and flavonoids (Verma et al. 2021). The results validate traditional uses for inflammation, intestinal issues, and respiratory infections (Sharma et al. 2018a; Ur Rahman et al. 2016). Moreover, *T. govanianum* extracts have shown promise in preventing the growth of particular cancer cells, indicating a potential use in cancer treatment. Because of its antioxidant qualities, it is also an attractive choice for creating natural therapies for illnesses linked to oxidative stress (Chauhan et al. 2018; Ur Rahman et al. 2015a, 2017b). Moreover, the plant's strong analgesic and anti-inflammatory properties have been validated by

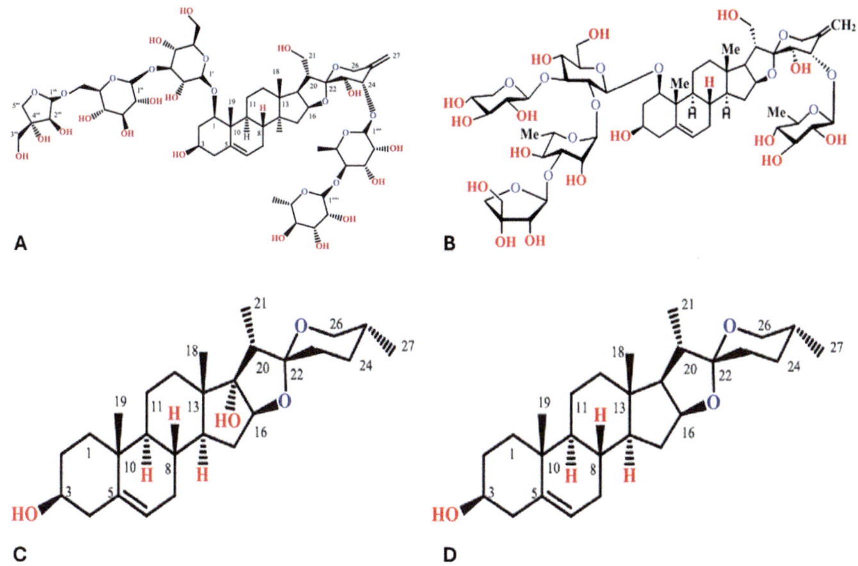

Fig. 23.4 Chemical structures of steroidal compounds of *Trillium govanianum*. (**a**) govanoside A; (**b**) govanoside B; (**c**) pennogenin; (**c**) diosgenin

both in vitro and in vivo research (Ur Rahman et al. 2016), supporting its potential as a successful natural remedy.

Detailed studies of these properties are described here.

23.4 Antioxidant Activity

According to studies, *T. govanianum* exhibits intense scavenging activity with hexane and chloroform fractions. However, the extract's antioxidant potential may be lower than ascorbic acid and butylated hydroxytoluene (BHT) standards. The presence of more extensive fat contents is responsible for the reduced scavenging ability, and it has become a prominent contender for its antioxidant qualities (Kumar et al. 2021a; Kumar and Kumari 2023; Kundra et al. 2020; Ur Rahman et al. 2015b).

Additionally, the effectiveness of the *T. govanianum* extracts in neutralizing free radicals has been reliably confirmed through DPPH and ABTS assays (Chandola et al. 2022; Chaudhary et al. 2024). Furthermore, their capability to prevent lipid peroxidation, a detrimental process/condition to cells, exhibited *T. govanianum's* defensive function against damage induced by oxidative stress. Validation of precise bioactive compounds, such as flavonoids and phenolic acids, highlights their collective role in augmenting the antioxidant ability of species. However, to assess the effects of *T. govanianum* antioxidant traits on human well-being, more in vitro research insights involving clinical trials and animal models are required (Kundra et al. 2020). Ethanolic extract prepared from rhizomes of *T. govanianum* exhibited significant antioxidant action and scavenged free radicals block the oxidation of LDL cholesterol (Kumar and Kumari 2023). Another in vitro research revealed the maximum amount of DPPH, ABTS, and FRAP activity in methanol extracts of rhizome, followed by leaf and stem (Kundra et al. 2020).

23.4.1 Anticancer Activity

The extract from *T. govanianum* exhibited cytotoxic solid and antiproliferative effects on prostate and cervical cancer cells, outperforming the well-known anticancer medication doxorubicin (Ur Rahman et al. 2015a). The anticancer activity of *T. govanianum* was further confirmed by Khan et al. According to their study, *T. govanianum* roots were cytotoxic to cancer cells of the liver (HepG2), lungs (A549), breast (MCF7), and bladder (EJ138) (Khan et al. 2016). Additionally, Sharma et al. described the insignificant toxicity of *T. govanianum* extract for healthy cells, with still remarkable antiproliferative effects against cancer cells, reducing their viability (Sharma et al. 2016). In a subsequent study, leaf and root extracts demonstrated angiogenesis inhibition and caused programmed cell death, essential for stopping cancer spread. While flavonoids, alkaloids, and phenolic chemicals are considered factors in the study, specific bioactive substances are still being investigated. Nevertheless, extensive preclinical and clinical trials are necessary for in vitro and clinical applications (Sharma et al. 2018b).

23.4.2 Antimicrobial Activity

Studies indicated that the leaves, stem, and rhizome of *T. govanianum* encompass nine genera of fungi, including *Alternaria, Aspergillus, Fusarium, Mucor, Pythium, Phoma, Rhizopus, Trichoderma* and *Stachybotrys* as endophytes (Sagar et al. 2017). In addition to that, its potential on diverse microorganisms, especially *Bacillus subtilis, Escherichia coli, Klebsiella pneumoniae, Pseudomonas aeruginosa, Staphylococcus aureus, Yersinia pestis, Xanthomonas campestris* and *Salmonella typhimurium* (Kanaujia et al. 2023). The in vitro study assessed the antibacterial ability of rhizome extract of *T. govanianum* using diverse solvent systems, including methanol, ethanol, acetone, and distilled water (Sagar et al. 2017). Applying the agar well diffusion method, these extracts were investigated against Y. pestis, S. aureus, and E. coli—three human pathogenic microorganisms. Furthermore, the study revealed that methanolic extract of *T. govanianum* rhizomes suppressing *S. aureus* growth effectively (Kanaujia et al. 2023). Similarly, an investigation was also carried out into *S. aureus, E. coli, K. pneumoniae, S. typhimurium and P. aeruginosa* (Verma et al. 2021).

23.4.3 Antidiarrheal and Antispasmodic Activities

T. govanianum is a prominent species used to treat the disorders associated with the digestive system. In their 2022 study, Muhammad et al. investigated the antispasmodic potential of *T. govanianum* on isolated rabbit jejunum, with acetylcholine as a tissue stabilizer and verapamil as a calcium channel blocker. The study found that by boosting stool latency time and dramatically lowering fecal weight and quantity, *T. govanianum* extracts alleviated diarrhea. Additionally, it completely stopped the isolated rabbit jejunum's spontaneous contractions caused by elevated potassium, similar to what verapamil does. It generated a right shift in the calcium concentration-response curve, proving the calcium channel-blocking effect. These results recommend the use of *T. govanianum* as a favorable treatment for diarrhea and stomach spasms, although a more in-depth study is warranted (Kanaujia et al. 2023; Kundra et al. 2020; Rauf et al. 2021).

23.4.4 Analgesic and Anti-inflammatory Activity

Pharmaceuticals derived from plants have been used for centuries and are frequently linked to fewer adverse effects. Numerous herbal remedies have been used for many years with few side effects, making them a desirable substitute for synthetic medications (Dina et al. 2010). One such traditional medicinal plant is *T. govanianum*, which is used in various cultures to treat pain and inflammation. The pharmacological potential of this herb has gained attention in recent years as researchers continue to investigate its efficacy in alleviating pain and inflammation. A study conducted by Ur Rahman et al. examined the crude methanolic (MeOH) extract of *T. govanianum* and its solvent sections to evaluate their analgesic and anti-inflammatory properties (Ur Rahman et al. 2016). The results were promising; the extracts demonstrated significant pain-relieving and anti-inflammatory effects. Specifically, the extracts were shown to mitigate paw oedema, a standard indicator of inflammation, and alleviate both visceral chemical and acute thermal nociception in experimental models. These findings suggest this plant species possesses notable pharmacological activities that could provide a natural alternative for pain and inflammation management (Ur Rahman et al. 2016). Further supporting these observations, Suresh et al. (2021) investigated the anti-inflammatory effects of *T. govanianum* using lipopolysaccharide (LPS)-stimulated RAW 264.7 macrophages, a commonly used model for studying inflammation. Macrophages stimulated by LPS typically release inflammatory mediators, such as nitric oxide and pro-inflammatory cytokines, including tumor necrosis factor-alpha (TNF-α), interleukin-1 beta (IL-1β), and interleukin-6 (IL-6). The study found that the extract, along with its solvent fractions and bioactive compounds, substantially lowered the levels of these inflammatory mediators in a concentration-independent manner. These findings recommend that *T. govanianum* contains bioactive components capable of modulating inflammatory pathways and suppressing the production of pro-inflammatory molecules (Sofi et al. 2022; Suresh et al. 2021).

The term “oxidative burst” describes the immune cells’ quick generation of reactive oxygen species (ROS), essential to the body’s inflammatory response. In a study, the crude methanolic extract and its solvent fractions exhibited vigorous oxidative burst inhibitory activity and analgesic and anti-inflammatory properties (Ur Rahman et al. 2015b). When compared to ibuprofen, a common anti-inflammatory medication, the crude methanolic extract and the n-butanol soluble fraction of *T. govanianum* showed considerable inhibitory effects in the oxidative burst experiment, followed by the chloroform and hexane soluble fractions (Ur Rahman et al. 2016).

23.4.5 Antifungal Activity

A novel spirostane steroidal saponin, named govanoside A, was isolated from the rhizomes of *T. govanianum*, along with three previously known compounds: borassoside E, pennogenin, and diosgenin, in a study by Ur Rahman et al. (2015b) Govanoside A and borassoside E showed promising antifungal action. These compounds were tested against several pathogenic fungal strains, including *Aspergillus niger* ATCC 16888, *Aspergillus flavus* ATCC 9643, *Candida albicans* ATCC 18804, and *Candida glabrata* ATCC 90030, where they exhibited notable inhibitory effects (Ismail et al. 2015; Singh et al. 2020, 2021).

23.5 The Economic Value and Trade of *Trillium govanianum*

23.5.1 Market Demand and Trade Networks

The global demand for medicinal plants is increasing rapidly. Because *T. govanianum* contains diosgenin, a corticosteroid hormone that can be extracted from its roots and utilized to synthesize sex hormones, the market for the plant has increased. Apropos, Prices have unexpectedly climbed, even felt by the local collectors in the Dhauladhar mountain range of Himachal Pradesh (Uniyal and Datta 2012) from Rs 800 per kg to between Rs. 2500 and Rs. 3000 (Dhyani et al. 2020; Rathore et al. 2020; Sharma et al. 2018a).

23.5.2 Domestic and International Trade Dynamics

The domestic and international trade dynamics of *T. govanianum* have gained attention due to its application in the pharmaceutical and herbal industries. This increment is evident because the species was not listed in a 2007 compendium of medicinal plants traded from India (Ved and Goraya 2007) but appeared in the 2017 update due to its significant contribution (Goraya et al. 2017). The rhizomes of the species are highly sought after for their bioactive compounds, including steroidal saponins. Its rhizomes possess trillarin, which hydrolyses to produce diosgenin, an anti-aging and anti-cancer compound used to make steroidal medications (Chaudhary et al. 2015). Moreover, the Diosgenin content in the species is much higher than that of other medicinal plants, making it highly profitable for pharmaceutical industries (Singh et al. 2017). In recent years, the domestic need for *T. govanianum* has surged due to its inclusion in various ayurvedic formulations (Chauhan et al. 2019; Rathore et al. 2020). *T. govanianum* is sold as a raw product or processed into powders, extracts, or oils. These products are widely used in traditional healing practices by local practitioners, known as vaidyas, hakims, or herbalists. Local healers often source *T. govanianum* from community members who harvest the plant, creating a mutually beneficial economic relationship in the region (Chandola et al. 2019; Chauhan et al. 2019; Chauhan 2011; Rathore et al. 2020).

The international trade of species is also growing, with the plant being exported to countries such as Europe, and North America (Kanaujia et al. 2023; Kumar et al. 2021a; Shivali and Namrata 2022; Ur Rahman et al. 2016). Eventually, it is now among southern Asia's most traded medicinal plants, particularly from the Himalayas (Khan et al. 2016). This has, however, opened the door for smuggling of the plant rhizome from countries like India, Nepal and Pakistan, as the rhizomes are sold at significantly higher prices in international markets than in local markets (Kala 2005). Resultingly, the total trade from Pakistan's Swat Valley was estimated to be 8 tons in 2012 (Sher et al. 2014), while in India, it ranged from 200 to 500 tons in 2014–2015 (Goraya et al. 2017). In many cases, the plant is harvested in violation of local and national conservation and trade laws (Kala et al. 2006; Vidyarthi et al. 2013). While some countries have begun to implement stricter controls on the import of endangered plant species, *T. govanianum* has not yet been widely recognized as a species in need of such protection on a global scale. As a result, it continues to be traded internationally, often without adequate oversight, which leads to tremendous conservation challenges.

23.6 Conservation Challenges

The conservation of *Trillium govanianum* is facing significant challenges due to increasing human activity and the species' biological limitations, which together have contributed to its current "Endangered" status (https://www.iucnredlist.org/species/175804005/176257695) (Sofi et al. 2022). In the past, local populations and traditional healers harvested *T. govanianum* for ethnobotanical purposes. Following seed distribution, these groups harvested only mature plants using sustainable approaches. Natural regeneration and long-term population levels were made possible by such techniques. But since 2010, these collecting methods have changed significantly because of an increase in commercial demand (Chauhan et al. 2020b). Nowadays, commercial collectors, frequently inexperienced and driven by financial gain, engage in destructive harvesting, uprooting whole plants regardless of age or reproductive state. This approach threatens population viability, as it removes both mature and immature plants from the wild (Fig. 23.5) (Ahmed and Akhtar 2016).

Fig. 23.5 *Trillium govanianum* extraction by collectors in the Indian Himalayan Region (**a**) species gathering; (**b**) gathered rhizomes; (**c**, **d**) processed rhizomes; (**e**) gatherers with the produce; (**f**) camping sites of gatherers. (*Reprinted from Chauhan et al. (2018), Copyright (2018), with permission from Elsevier)

The market value of *T. govanianum* is tied to the plant's age and maturity, with older three-leaf plants having greater biomass and economic value than younger, one-leaf plants (Chauhan et al. 2018). Due to Plants phenology and the Himalayas' typical subalpine/alpine climatic conditions, the limited collection window (April to June) encourages premature harvesting, affecting the plant's reproductive cycle and reducing yield. Harvesters must collect more plants to meet market demands, uprooting approximately 1100 individuals to produce 1 kg of rhizomes (Chauhan et al. 2018)—an unsustainable ratio given the species' slow growth rate and prolonged reproductive cycle. Therefore, Overexploitation and related habitat degradation by human activity are the top significant threats to natural resources. For instance, a study by Hougue and Breon, 2022, studied the tremendous threats to species identified by the International Union for Conservation of Nature (IUCN) as "threatened with extinction." The study finds that of 20,784 species, 88.3% were impacted by habitat destruction and 26.6% by overexploitation, testifying that habitat destruction and overexploitation threaten more species than all other categories combined (Hogue and Breon 2022).

The combined effects of destructive harvesting, a short harvesting season, and the species' unique life cycle create an environment where *T. govanianum* populations struggle to regenerate. Beyond human impacts, the species' slow growth and distinctive reproductive strategy (Chauhan et al. 2020b) also hinder its ability to rebound. Nonetheless, climate change poses an increasing threat; increases in temperature and precipitation may result in fewer viable habitats for the species making climate change a more severe concern (Sofi et al. 2022). Efforts to conserve *T. govanianum* are further complicated by both in-situ and ex-situ limitations. Seed banking, a commonly used ex-situ conservation technique, is also challenging as the trillium seeds resist the conventional storage techniques available. The other ex-situ technique of artificial propagation also faces a challenge in the plant's behavior of slow growth and specialized environment requirements. It is noteworthy that even if artificial propagation succeeds, restoring healthy plants to natural environment remains problematic.

23.7 Overharvesting Impacts on Population Genetics of *Trillium govanianum*

Genetic diversity allows a species to evolve in response to changing environmental conditions and developing illnesses. *Trillium govanianum* cannot successfully adapt to selection pressures such as soil degradation, climate change, and disease without a diverse set of genetic variations. Restricted adaptation, in turn, makes the species more vulnerable to extinction in the wild. *T. govanianum's* genetic diversity is significantly threatened by overharvesting and habitat fragmentation.

The population's capacity to reproduce and recover is weakened by the ongoing removal of healthy people due to overharvesting. Small, poorly scattered populations with fewer reproductive individuals in the wild result from this ongoing decline. Reduced genetic diversity is closely correlated with smaller population numbers (Ellstrand and Elam 1993; Sun 1996). In these populations, genetic drift—a random process where specific alleles become more or less standard—causes a further reduction in genetic variation. In *Trillium govanianum*, a low diversity and allelic diversity has been confirmed by the studies (Dhyani et al. 2020; Islam et al. 2023). The studies establish that the small populations left in the Indian Himalayan Region frequently lose unique alleles because of the random fluctuations in the allele frequency, ultimately reducing genetic diversity. This decrease can result in a population bottleneck in which only a few individuals survive long-term harvesting pressures, severely limiting genetic diversity. This steady decline in genetic diversity causes genetic erosion, permanently limiting the range and combination of local alleles in *T. govanianum* populations (Maxted and Guarino 2006).

Genetic erosion brings a variety of negative consequences, including a loss of valuable traits, which reduces the species' short-term resilience, evolutionary adaptability, and long-term survival prospects in changing and challenging environments (Huenneke 1991; Jump et al. 2009). While genetic erosion impacts many species, its effects are particularly severe in rare, endangered, and threatened (RET) species like *T. govanianum*, where habitat loss and overharvesting significantly heighten extinction risk. When Individuals are overharvested from the populations in specific areas, this creates a gap within and among the populations. The gap created thus increases the physical and genetic distance between the remnants of these populations, leading to isolated populations scattered into smaller patches. Nevertheless,

Glossary

Alleles: One of two or more alternative versions of a gene.

Genetic diversity: Total number of genetic characteristics in the genetic makeup of a species.

Gene flow: the introduction of genetic material (by interbreeding) from one population of a species to another, thereby changing the composition of the gene pool of the receiving population.

Inbreeding: The mating of individuals or organisms that are closely related through common ancestry,

Genetic structure: any pattern in the genetic makeup of individuals within a population.

Gene pool: Sum of a population's genetic material at a given time.

Genetic drift: a change in the gene pool of a small population that takes place strictly by chance.

Population bottleneck: a sharp reduction in the size of a population due to environmental events.

Isolation by Distance: term used to refer to the accrual of local genetic variation under geographically limited dispersal.

Definitions adopted from www.britannica.com and www.wikipedia.org

such fragmentation can have lasting consequences on the overall fitness of the species to changing climate (Frank et al. 2021).

In *T. govanianum*, habitat loss and the resulting small population sizes also damage its genetic composition by fostering genetic differentiation, which refers to variations between different populations. Studies by Dhyani et al. and Islam et al. have found significant genetic differentiation among *T. govanianum* populations, likely due to habitat fragmentation in the Himalayan region (Dhyani et al. 2020; Islam et al. 2023).

As fragmentation separates populations, gene flow—the interchange of genetic material—decreases, causing different populations to evolve separately in response to their unique environmental stresses. More compact, isolated communities have fewer mating options, increasing the risk of inbreeding and further limiting gene flow, ultimately leading to genetic drift. This random process diminishes genetic diversity over generations. This fragmentation also intensifies Isolation by Distance (IBD), creating "genetic islands" where populations are physically and genetically separated. *T. govanianum* populations in the Himalayas, for example, show high IBD (Dhyani et al. 2020), highlighting how populations are becoming increasingly distinct due to fragmentation. This genetic isolation is compounded by the plant's limited seed dispersal capability, often relying on ants (myrmecochory), and by anthropogenic activities, which prevent populations from exchanging genetic material effectively.

The weak genetic structure of Trillium species worldwide makes *T. govanianum* susceptible to these pressures. Factors such as limited dispersal ability, small and fragmented populations, a long lifespan with slow reproduction, self-incompatibility, endemic distribution, and environmental sensitivity all contribute to the vulnerability of its genetic structure. Furthermore, *T. govanianum* matures slowly, requiring years to attain reproductive age. This causes a reproductive lag, in which the population cannot generate sufficient seeds to maintain itself if key reproductive individuals are lost. The absence of genetic variety and reproductive capacity hastens genetic degradation, which may exacerbate as populations decline. Sustainable management and focused conservation methods are thus critical for saving *T. govanianum* from extinction and protecting its genetic integrity in the face of overfishing and habitat fragmentation.

23.8 Conservation Solutions

23.8.1 Community-Based Conservation Initiatives

Community-driven conservation initiatives have emerged as practical tools for promoting sustainable management of resources in the Himalayas. For example, the cultivation of *Picrohiza kurroa* is successfully being carried out with public-private partnerships in some villages of Garhwal, Indian Himalayas (Chauhan 2021b). In Uttarakhand, community-led projects have introduced alternative income-generating activities, such as eco-tourism and the cultivation of medicinal plants in home gardens, reducing pressure on wild populations.

In Nepal, several projects have been carried out involving local stakeholders, villagers, traditional healers, and local governing bodies to conserve and manage medicinal plant resources. For example, The Darwin Initiative of the Uttarakhand government has supported a project to create a suitable market model for the sustainable and lawful commerce of *Nardostachys jatamansi* (Chauhan 2021a). To our knowledge, no reports on the community-based conservation of *T. govanianum are available.*

23.8.2 Sustainable Harvesting Guidelines

The practical and sustainable conservation and management of *T. govanianum* can be achieved through changes in the method of harvesting. Selective and rotational harvesting can immediately affect the status of *T. govanianum*. Extracting only reproductive adult plants, leaving younger plants to develop and reproduce and implementing rotational harvesting can help rejuvenate species. Harvesting may also be limited by the season, for example, during the early growing season (March to June). The plants will be able to bloom and yield more seeds. Above all, sustaining *T. govanianum* populations requires long-term conservation by teaching locals and collectors sustainable gathering practices.

23.8.3 Certification and Fair-Trade Models

In recent years, eco-labelling and fair trade have emerged as promising tools for the sustainable utilization of medical plants as these strategies primarily focus on ecosystem-social-economic equilibrium. Eco-labelling is a certification process that identifies and certifies products/services produced/managed/sourced in an environmentally friendly and sustainable manner. For example, the Forest Stewardship Council (FSC) is a known eco-labelling agency that certifies several non-timber forest products (NTFPs), including chicle latex, maple syrup, Brazil nuts, and palm hearts forest products (Schmitt et al. 2008; Shanley et al. 2015). FSC-NTFP standards benefit producers and support biodiversity conservation (Singh and Chatterjee 2022). For Himalayan plants, eco-labelling ensures that the plants harvested for trade follow sustainable guidelines without threatening survival. The Forest Stewardship Council (FSC) has authorized the harvest of *Nardostachys jatamansi*, a critically endangered medicinal plant, in some Nepali forest regions, including the Bajhan district, to promote its sustainable use (Chauhan 2021a).

Fair-trade models can ensure local communities receive fair compensation for efforts in conserving the species. The traceable supply chain can help control illegal trade and ensure the benefits of sustainable harvest to all stakeholders. A classic example is the case of *N. jatamansi* in Nepal, which is being harvested and legally traded following Fairwild Standard principles. The initiative is focused on delivering a locally appropriate market model for sustainable and legal trade to protect biodiversity and sustain local livelihoods. No agency currently carries out the certification and fair-trade model of *T. govanianum,* to the best of our knowledge, based on an extensive literature search. However, it presents an opportunity to promote sustainable trade practices for *T. govanianum.*

Ecological, economic, and social dimensions related to the sustainable utilization of *T. govanianum.*

The sustainable use of *T. govanianum* should involve a multifaceted approach that considers ecological preservation, economic development, and social inclusion. The *T. govanianum* is associated with 57 plant species, 864 bacterial and 1209 fungal operational taxonomic units, providing ecological niches and maintaining the community structure (Rashid et al. 2023). The species is a prominent component of the local ecosystem, and it can provide essential ecosystem services, such as soil stabilization and Nutrient cycling. Therefore, Sustainable utilization of the species can provide economic benefits to local communities. Sustainable markets for diosgenin-rich rhizomes involving fair trade practices can promote economic growth in the region. Outmigration is a severe issue having socio-economic dimensions in the high-altitude areas of the Indian Himalayas (Kumar and Sati 2023). Resources like *T. govanianum* could deter outmigration from the region. Timmermann and Smith-Hall reported that migrants in Nepal were returning in the medicinal plant-gathering period to take a stake in the precious harvest (Timmermann and Smith-Hall 2019).

23.8.4 Policy and Regulatory Framework

India has several laws to control the exploitation of biodiversity, such as the Indian Forest Act (1927), the Wildlife Protection Act (1972), the Environment Protection Act (1986), and the Biological Diversity Act (2002). Likewise, in Nepal, biodiversity is protected under The Forest Act of 1993 and the National Parks and Wildlife Conservation Act of 1973. However, these laws often lack specific provisions for conserving rare and endemic species like *Trillium govanianum*. While the transportation of its rhizomes requires a permit in Himachal Pradesh, India, the regulatory framework remains inconsistent, and enforcement is weak (Uniyal and Datta 2012). Policies governing the permits appear in flux, and an INR 8000 fee per quintal is required, as outlined in the Himachal Pradesh Forest Produce Transit (Land Routes) Rules of 2013. Globally, *T. govanianum* should be protected under international agreements, such as the Convention on International Trade in Endangered Species of Wild Fauna and Flora (CITES).

23.8.5 Enforcement and Compliance Challenges

The enforcement of laws related to the trade of *Trillium govanianum* has several challenges. The Himalayan ranges where *T. govanianum* is found—Afghanistan, Bhutan, China (Tibet), India, Nepal, and Pakistan (Fukuda 2001; Roskov et al. 2013)—Various laws and regulations on biodiversity commercialization make regulation challenging. Enforcement operations are made more difficult by ineffective collaboration across nations (Chauhan and Gallacher 2021). The legal frameworks and regulations controlling the species' trade and collection are frequently unknown to the local populations. Even when laws are put in place, it might be difficult to enforce them. A lack of infrastructure, resources, and conflicts frequently hampers law enforcement in transboundary areas. Therefore, governments, non-governmental organizations, and international organizations may play essential roles in the conservation and sustainable use of *Trillium govanianum* by strengthening the legal framework, promoting community involvement and collaboration, and in situ and ex-situ species conservation.

23.8.6 Strengthening Legal Frameworks

23.8.6.1 Government

By offering the resources, coordination, and legal and regulatory framework required to safeguard this species and its delicate environment, governance can play a critical role in *T. govanianum* conservation. Local communities can be granted access to the species' natural environment by putting rules governing spatiotemporal harvesting into effect. Using community-based management strategies that allow local stakeholders to be involved in decision-making is crucial. Implementing regulations controlling harvesting quotas, trading permits, and penalties for illegal trade can reduce the overexploitation of the species.

23.8.6.2 Promoting Community Engagement

Local communities can perform an essential role in the long-term protection of *T. govanianum*. NGOs can sensitize and facilitate community participation by training on the methods of sustainable harvest, cultivation, and long-term benefits of legal trade.

23.8.6.3 International Collaboration

Collaboration with neighbouring countries to address transboundary conservation challenges will be essential to address illegal trade issues. For example, the International Centre for Integrated Mountain Development (ICIMOD) in Nepal and other organizations can enhance this collaboration through knowledge exchange, capacity building, and coordinated efforts.

23.8.6.4 Enhancing In Situ and Ex Situ Conservation Efforts

Several populations of *T. govanianum* thrive in conservation reserves, protected areas, and community-managed forests. Cultivation of the species in these areas through participatory conservation programs involving benefit-sharing mechanisms should be promoted. Ex-situ conservation involving cultivating *Trillium govanianum* outside its natural habitat, such as in botanical gardens and controlled environments, is equally essential for conserving *T. govanianum*. Only a few reports are available on the species' ex-situ propagation methods (Chauhan et al. 2020b). There is a need for refinement in these methods for mass propagation of the species. Promoting the cultivation of *Trillium govanianum* as a high-value medicinal plant in agricultural settings can also help meet market demand with a conclusion for out-compromising wild populations.

23.9 Conclusion

Trillium govanianum is a prized natural resource for its high steroidal content and medicinal value to many known diseases. The species not only holds a medicinal but also promises a potent economic value. However, the species faces complex survival challenges due to increasing demand and illegal trade in the Region. The associated unsustainable harvesting worsens the critical situation of the species, such as weak genetic structure and complex reproduction. Effective conservation strategies are required to the best of their capacity to save this species. Among many, the strategy may require a multipronged approach, which includes strict lawful enforcement, community-driven conservation, and, most importantly, accompanying sustainable trade models. Like always in any conservation effort, empowering local stakeholders and spreading conservation messages among the masses, we can work towards a balanced framework to secure the survival of *Trillium govanianum* in nature.

References

Ahmed MJ, Akhtar T (2016) Indigenous knowledge of the use of medicinal plants in Bheri, Muzaffarabad, Azad Kashmir, Pakistan. Eur J Integr Med 8(4):560–569

Chandola V, Nautiyal A, Chandra S, Ranjan R (2019) A review on prioritizing conservation of Himalayan medicinal plant species: case of *Trillium govanianum* (Wall. ex. D. Don) Kunth. J Med Plants Stud 7(5):23–27

Chandola V, Chandra S, Nautiyal A, Concenço G (2022) Antioxidant potential and impact of different extraction solvents on the free, esterified and insoluble-bound phenolics, flavonoid and tannin content of *Trillium govanianum* Wall ex D. Don, a rare Himalayan herb. Vegetos 35(4):953–960

Chandola V, Chandra S, Kumar RR, Nautiyal A (2023) Reproductive behavior of an endangered Himalayan medicinal herb *Trillium govanianum* Wall. ex D. Don. Vegetos 36(4):1437–1445

Chaudhary S, Chikara SK, Sharma MC, Chaudhary A, Alam Syed B, Chaudhary PS et al (2015) Elicitation of diosgenin production in *Trigonella foenum-graecum* (fenugreek) seedlings by methyl jasmonate. Int J Mol Sci 16(12):29889–29899

Chaudhary D, Kumar V, Sharma R, Devi S, Sharma P, Kumar P (2024) Isolation, characterization and management of endophyte (s) prevailing during the in vitro establishment of endangered medicinal herb Himalayan *Trillium* (*Trillium govanianum* Wall. ex D. Don). Plant Cell Tissue Organ Cult 156(1):16

Chauhan NS (2011) Economic, medicinal and aromatic plants: their distribution, trade and utilization in the Cold Desert areas of Himachal Pradesh. Himalayan Ecol 19:17

Chauhan H (2021a) *Nardostachys jatamansi*. The IUCN Red List of Threatened Species 2021. e. T50126627A88304158. https://doi.org/10.2305/IUCN.UK.2021-3.RLTS.T50126627A88304158.en

Chauhan H (2021b) *Picrorhiza kurroa*. The IUCN Red List of Threatened Species 2021: e.T88304131A88304135. https://doi.org/10.2305/IUCN.UK.2021-3.RLTS.T88304131A88304135.en

Chauhan HK, Bisht AK (2020) *Trillium govanianum*. The IUCN Red List of Threatened Species 2020: e. T175804005A176257695. https://dx.doi.org/10.2305/IUCN.UK.2020-3.RLTS.T175804005A176257695.en

Chauhan HK, Gallacher D (2021) Can citizen reporting apps plug the data gap in the Himalayan wildlife trade? Trees For People 6:100150

Chauhan HK, Bisht AK, Bhatt ID, Bhatt A, Gallacher D, Santo A (2018) Population change of *Trillium govanianum* (Melanthiaceae) amid altered indigenous harvesting practices in the Indian Himalayas. J Ethnopharmacol 213:302–310

Chauhan HK, Bisht AK, Bhatt ID, Bhatt A, Gallacher D (2019) Trillium–toward sustainable utilization of a biologically distinct genus valued for traditional medicine. Bot Rev 85:252–272

Chauhan HK, Bhatt ID, Bisht AK (2020a) Biology, uses and conservation of *Trillium govanianum*. In: Socio-economic and eco-biological dimensions in resource use and conservation: strategies for sustainability. Springer, pp 235–247

Chauhan HK, Bisht AK, Bhatt ID, Bhatt A (2020b) Protocol for vegetative propagation of *Trillium govanianum* Wall ex D. Don. J Appl Res Med Aromat Plants 16:100233

Chauhan HK, Gallacher D, Bisht AK, Bhatt ID, Bhatt A, Dhyani P, Kewlani P (2021) Variations in phytochemistry, morphology, and population structure in *Trillium govanianum* (Melanthiaceae). Botany 99(10):651–664

Dhyani P, Sharma RK (2022) Plant genetic diversity: implications for conservation of Himalayan *Trillium*. In: DellaSala DA, Goldstein MI (eds) Imperiled: the encyclopedia of conservation. Elsevier

Dhyani P, Sharma B, Singh P, Masand M, Seth R, Sharma RK (2020) Genome-wide discovery of microsatellite markers and, population genetic diversity inferences revealed high anthropogenic pressure on endemic populations of *Trillium govanianum*. Ind Crop Prod 154:112698

Dina TA, Taslima MA, Ahmed NU, Uddin MN (2010) Analgesic and anti-inflammatory properties of *Argyreia argentea* methanol extract in animal model. J Taibah Univ Sci 3(1):1–7

Ellstrand NC, Elam DR (1993) Population genetic consequences of small population size: implications for plant conservation. Annu Rev Ecol Evol Syst 24:217–242. https://doi.org/10.1146/annurev.es.24.110193.001245

Frank SC, Pelletier F, Kopatz A, Bourret A, Garant D, Swenson JE et al (2021) Harvest is associated with the disruption of social and fine-scale genetic structure among matrilines of a solitary large carnivore. Evol Appl 14(4):1023–1035

Fukuda I (2001) The origin and evolution in *Trillium* 1. The origin of the Himalayan *Trillium govanianum*. Cytologia 66(1):105–111

Goraya G, Ved D, Ravikumar K, Rawat R (2017) Domestic trade of herbal raw drugs. Medicinal plants in India: an assessment of their demand and supply. National Medicinal Plants Board, Ministry of AYUSH, Government of India, New Delhi and Indian Council of Forestry Research & Education, Dehradun

Hogue AS, Breon K (2022) The greatest threats to species. Conserv Sci Pract 4(5):e12670. https://doi.org/10.1111/csp2.12670

Huang W, Zou K (2011) Cytotoxicity of a plant steroidal saponin on human lung cancer cells. Asian Pac J Cancer Prev 12(2):513–517

Huenneke LF (1991) Ecological implications of genetic variation in plant populations. In: Falk DA, Holsinger KE (eds) Genetics and conservation of rare plants. University Press, Oxford, pp 31–44

Hussain K, Awan MS, Nasir M, Sabir S, Mehmood A, Iqbal T et al (2023) Phytochemical and antioxidant screening of *Trillium govanianum* Wall. ex. D. Don., from Western Himalayas Azad Jammu & Kashmir, Pakistan. KEPES 21(3):372–378

Islam SU, Mangral ZA, Tariq L, Bhat BA, Tantray WW, Ahmad R et al (2023) Conservation genetics of endangered *Trillium govanianum* Wall. ex D. Don—a pharmaceutically prized medicinal plant from the Himalaya and implications for species recovery. Gene 888:147748

Ismail M, Shah MR, Adhikari A, Anis I, Ahmad MS, Khurram M (2015) Govanoside A, a new steroidal saponin from rhizomes of *Trillium govanianum*. Steroids 104:270–275

Jump AS, Marchant R, Peñuelas J (2009) Environmental change and the option value of genetic diversity. Trends Plant Sci 14(1):51–58. https://doi.org/10.1016/j.tplants.2008.10.002

Kala CP (2005) Indigenous uses, population density, and conservation of threatened medicinal plants in protected areas of the Indian Himalayas. Conserv Biol 19(2):368–378

Kala CP, Dhyani PP, Sajwan BS (2006) Developing the medicinal plants sector in northern India: challenges and opportunities. J Ethnobiol Ethnomed 2:1–15

Kanaujia KA, Maurya N, Arya DK (2023) Exploring the medicinal potential of trillium govanianum (a threatened plant): current insights, challenges, and future prospects. World J Pharm Pharm Sci 12(11):433–463

Khan KM, Nahar L, Al-Groshi A, Zavoianu AG, Evans A, Dempster NM et al (2016) Cytotoxicity of the roots of *Trillium govanianum* against breast (MCF7), liver (HepG2), lung (A549) and urinary bladder (EJ138) carcinoma cells. Phytother Res 30(10):1716–1720

Khan KM, Nadeem MF, Mannan A, Chohan TA, Islam M, Ansari SA et al (2024) Biochemical, toxicological, and in silico aspects of *Trillium govanianum* Wall. ex D. Don (Trilliaceae): a rich source of natural bioactive compounds. Chem Biodivers 21(1):e202301375

Kumar D, Kumari V (2023) Metabolite profiling, antidiabetic, and antioxidant potential of different tissues of *Trillium govanianum* Wall. ex D. Don. S Afr J Bot 153:102–108

Kumar S, Sati VP (2023) Patterns and determinants of rural–urban migration in the Garhwal Himalaya. GeoJournal 88(4):3679–3698

Kumar D, Joshi R, Sharma A, Nadda G (2021a) A comprehensive search of the primary and secondary metabolites and radical scavenging potential of *Trillium govanianum* Wall. ex D. Don. Chem Biodivers 18(10):e2100300

Kumar V, Singh P, Singh PK, Rahim MS, Sharma V, Roy J, Sharma H (2021b) *Trillium govanianum*. In: Himalayan medicinal plants. Elsevier, pp 243–257

Kumar P, Singh K, Lone JF, Bhushan A, Gupta P, Gairola S (2023) Morpho-anatomical, molecular, and chemical standardization of *Trillium govanianum* Wall. ex D. Don: an endangered medicinal herb native to the Himalayas. Pharmacogn Mag 19(1):128–143

Kumar V, Garima, Sharma R, Thakur A, Sharma P, Sharma YP et al (2024) Conserving industrially important endangered medicinal herb *Trillium govanianum* (Himalayan Trillium) through biotechnological based interventions. Plant Cell Tissue Organ Cult 159(1):3

Kundra R, Samant SS, Sharma RK (2020) Assessment of antioxidant potential of *Trillium govanianum* Wall. ex D. Don, a critically endangered medicinal plant of Northwestern Indian Himalaya. Proc Natl Acad Sci India Sect B Biol Sci 90:95–101

Liang SY, Soukup VG (2000) Paris L. In: Flora of China, vol 24. Science Press, Beijing, pp 88–95

Maxted N, Guarino L (2006) Genetic erosion and genetic pollution of crop wild relatives. In: Genetic erosion and pollution assessment methodologies. Proceedings of PGR Forum workshop. CAB International

Mishra AP, Kumar A, Yadav SN (2023) Ecology and conservation of threatened medicinal plants in the Trans-Himalayan region of Nanda Devi Biosphere Reserve, Western Himalaya. Trees For People 14:100451

Ohara M (1989) Life history evolution in the genus *Trillium*. Plant Species Biol 4(1):1–28

Payyappallimana U (2013) Health and well-being in Indian local health traditions. In: An integrated view of health and well-being: bridging Indian and western knowledge. Springer, pp 99–112

Polunin O, Stainton A (1984) Flowers of the Himalaya. Oxford University Press, New Delhi

Radha SP, Pundir A (2019) Review on ethnomedicinal plant: *Trillium govanianum* Wall. Ex D. Don. Int J Theor Appl Sci 11:4–9

Ranganathan S (2018) Indigenous healing practices in India: shamanism, spirit possession, and healing shrines. In: Psychosocial interventions for health and well-being. Springer, pp 109–122

Rashid K, Rashid S, Ganie AH, Nawchoo IA, Tantry MA, Khuroo AA (2023) *Trillium govanianum*—a promising endemic medicinal herb of the Himalaya. In: Bioprospecting of tropical medicinal plants. Springer, pp 381–408

Rathore S, Walia S, Devi R, Kumar R (2020) Review on *Trillium govanianum* Wall. ex D. Don: a threatened medicinal plant from the Himalaya. J Herb Med 24:100395

Rauf A, Akram M, Semwal P, Mujawah AA, Muhammad N, Riaz Z et al (2021) Antispasmodic potential of medicinal plants: a comprehensive review. Oxidative Med Cell Longev 2021(1):4889719

Reddy S (2023) Revival of local health traditions from healers perspectives: urgency of recognition and state support. In: Ethnomedicine and tribal healing practices in India: challenges and possibilities of recognition and integration. Springer, pp 347–368

Roskov Y, Kunze T, Paglinawan L, Orrell T, Nicolson D, Culham A et al (2013) Species 2000 & ITIS catalogue of life, 2013 annual checklist. Species 2000, Reading

Sagar A, Thakur L, Thakur JS (2017) Studies on endophytes and anti-bacterial activity of *Trillium govanianum* Wall. ExD. Don. Int J Bot Stud 2:63–67

Schmitt J, Pokorny B, Ying L (2008) Certification of non-timber forest products in China: effects on food quality, forest conservation and rural development. For Trees Livelihoods 18(1):81–89

Shanley P, Pierce AR, Laird SA, Binnqüist CL, Guariguata MR (2015) From lifelines to livelihoods: non-timber forest products into the twenty-first century. In: Tropical forestry handbook. Springer, pp 1–50

Sharma DK (2017) Review on traditional medicinal plant *Trillium govanianum* (Nagchatri). J Med Plants Stud 5:120–122

Sharma A, Parashar B (2017) A review on *Trillium govanianum*. World J Pharm Res 6(2):500–511

Sharma S, Sharma A, Mehta V, Chauhan RS, Malairaman U, Sood H (2016) Efficient hydroalcoholic extraction for highest diosgenin content from *Trillium govanianum* (nag chhatri) and it's in vitro anticancerous activity. Asian J Pharm Clin Res 9(4):386–392

Sharma OR, Arya D, Goel S, Vyas K, Shinde P (2018a) *Trillium govanianum* Wall. ex D. Don (Nagchatri): an important ethnomedicinal plant of Himalayan region (Himachal Pradesh). J Med Plants Stud 6(1):11–13

Sharma S, Mehta V, Sharma P, Jaggi K, Udayabanu M, Sood H (2018b) Antifertility activity and contraceptive potential of the hydroalcoholic rhizome extract of *Trillium govanianum* in female wistar rats. Asian J Pharm Clin Res 11(11):329–332

Sher H, Aldosari A, Ali A, de Boer HJ (2014) Economic benefits of high value medicinal plants to Pakistani communities: an analysis of current practice and potential. J Ethnobiol Ethnomed 10:1–16

Shivali V, Namrata S (2022) Threat status of three important medicinal Himalayan plant species and conservation implications. Nat Conserv Res 7(1):27–41

Singh S, Chatterjee S (2022) Forest certification related to non-timber forest products (NTFPs) in India: study of NTFP harvest of rhododendrons in Western Himalayas for its sustainable use. Environ Sci Proc 13(1):19

Singh P, Singh G, Bhandawat A, Singh G, Parmar R, Seth R, Sharma RK (2017) Spatial transcriptome analysis provides insights of key gene (s) involved in steroidal saponin biosynthesis in medicinally important herb *Trillium govanianum*. Sci Rep 7(1):45295

Singh PP, Bora PS, Suresh PS, Bhatt V, Sharma U (2020) Qualitative and quantitative determination of steroidal saponins in *Trillium govanianum* by UHPLC-QTOF-MS/MS and UHPLC-ELSD. Phytochem Anal 31(6):861–873

Singh PP, Suresh PS, Bora PS, Bhatt V, Sharma U (2021) Govanoside B, a new steroidal saponin from rhizomes of *Trillium govanianum*. Nat Prod Res 36(1):37–45

Sofi II, Zargar SA, Ganie AH, Shah MA (2022) Distribution dynamics of *Arnebia euchroma* (Royle) IM Johnst. and associated plant communities in trans-Himalayan Ladakh region in relation to local livelihoods under climate change. Trees For People 7:100213

Sun M (1996) Effects of population size, mating system, and evolutionary origin on genetic diversity in *Spiranthes sinensis* and *S. hongkongensis*. Conserv Biol 10(3):785–795. https://doi.org/10.1046/j.1523-1739.1996.10030785.x

Suresh PS, Singh PP, Sharma A, Padwad YS, Sharma U (2021) Steroidal saponins of *Trillium govanianum*: quality control, pharmacokinetic analysis, and anti-inflammatory activity. Biocatal Agric Biotechnol 35:102071

Thapa S, Kunwar RM, Adhikari B, Paudel HR, Subedi S (2021) *Trillium govanianum* (Himalayan Trillium): production, distribution, use and conservation in Nepal. Nord J Bot 39(12):njb.03356

Timmermann L, Smith-Hall C (2019) Commercial medicinal plant collection is transforming high-altitude livelihoods in the Himalayas. Mt Res Dev 39(3):R13–R21

Uniyal SK, Datta A (2012) Nagchhatri—a plant in Peril. J Biodivers Manag For 1(1):1–2

Ur Rahman S, Ismail M, Khurram M, Haq IU (2015a) Pharmacognostic and ethnomedicinal studies on *Trillium govanianum*. Pak J Bot 47:187–192

Ur Rahman S, Ismail M, Shah MR, Iriti M, Shahid M (2015b) GC/MS analysis, free radical scavenging, anticancer and beta-glucuronidase inhibitory activities of *Trillium govanianum* rhizome. Bangladesh J Pharmacol 10(3):577–583

Ur Rahman S, Adhikari A, Ismail M, Raza Shah M, Khurram M, Shahid M et al (2016) Beneficial effects of *Trillium govanianum* rhizomes in pain and inflammation. Molecules 21(8):1095

Ur Rahman S, Adhikari A, Ismail M, Shah MR, Khurram M, Anis I, Ali F (2017a) A new trihydroxylated fatty acid and phytoecdysteroids from rhizomes of *Trillium govanianum*. Antiseptic 3(5):323–327

Ur Rahman S, Ismail M, Khurram M, Ullah I, Rabbi F, Iriti M (2017b) Bioactive steroids and saponins of the genus *Trillium*. Molecules 22(12):2156

Ved D, Goraya G (2007) Demand and supply of medicinal plants in India. NMPB, New Delhi & FRLHT, Bangalore, India 18(85):210–252

Verma R, Tapwal A, Kumar D, Puri S (2021) Antimicrobial potential and phytochemical profiling of ethnomedicinal plant *Trillium govanianum* Wall. ex D. Don in Western Himalaya. J Herb Med 29:100491

Vidyarthi S, Samant S, Sharma P (2013) Dwindling status of *Trillium govanianum* Wall. ex D. Don—a case study from Kullu district of Himachal Pradesh, India. J Med Plant Res 7(8):392–397

Yokosuka A, Mimaki Y (2008) Steroidal glycosides from the underground parts of *Trillium erectum* and their cytotoxic activity. Phytochemistry 69(15):2724–2730

Alpine Medicinal Plants in the Himalaya: Contribution to Mountain Livelihood, Socioeconomy, and Their Marketing

24

Laxman Singh, Basant Singh, Vikram S. Negi, C. P. Kala, I. D. Bhatt, and Sunil Nautiyal

Abstract

The alpine region of the Himalaya is well acknowledged as a repository of therapeutically important medicinal plants (MPs) that are extensively used in various traditional systems of medicine across history and cultures. With over 1748 species of MPs, the Indian Himalayan Region (IHR) accounts for a large proportion of native (31%) and endemic species (15.5%). Besides their great relevance to the healthcare system, these species play an important role in the socioeconomic development of rural communities. Also, for over a decade now, Himalayan herbal medicines have gained considerable recognition worldwide as complementary and alternative medicines against a broad spectrum of human ailments. These herbs are a rich source of several biologically active substances, encompassing a major role in disease prevention alone or in synergy. Despite significant advancements in modern medicine and technology, over 80% of the current medications and one-third of the population in developing countries still rely on plant products directly or indirectly for primary healthcare.

Consequently, the use of herbal medicines as dietary supplements is increasing due to the widespread perception that herbs are natural, innately safe, and cost-effective. Given the above and to meet the escalating demand (both national and international), over 70–90% of Himalayan medicinal plants under trade are harvested directly from the wild, wherein nearly 70% involves destructive harvesting. Moreover, issues are intensified by the biotic and abiotic pressures (like grazing, habitat fragmentation and loss, illegal harvesting and trade, the dominance of native/invasive species, chaotic tourism, and climate anomalies) and inherent characteristics of alpine species, including restricted distribution, high habitat specificity, pollinator deficiency, and poor germination rates. Thus, the

L. Singh · B. Singh · V. S. Negi · I. D. Bhatt (✉) · S. Nautiyal
G. B. Pant National Institute of Himalayan Environment, Almora, Uttarakhand, India

C. P. Kala
Indian Institute of Forest Management, Bhopal, Madhya Pradesh, India

S. Chandra, H. Lata (eds.), *High Altitude Medicinal Plants*,
https://doi.org/10.1007/978-3-031-99406-7_24

question arises: Are we reaching a point of no return? With this continuous depletion of Himalayan medicinal plant resources, their conservation and mainstreaming these species into cultivation has assumed considerable importance. Further, to strengthen this, the government has initiated various schemes (e.g., 30–70% financial support) and missions that have strategized self-reliance and narrowed the gap between demand and supply. Doing this can reap maximum benefits and sustain the Himalayan bioresources.

24.1 Introduction

The term alpine biome depicts the boundary layer above the tree line and below the permanent snow line, commonly called "sky islands" subtended by distinctly different low-statured and non-arboreal vegetation (Germino 2014). Several factors define this biome, including elevation, solar irradiation, wind speed, seasonal vegetation, soils, and topographic relief, among which heterogeneity in microclimate is the most determinant abiotic factor that shapes the physiognomy, species composition, and richness (Singh et al. 2021). Among the life forms, specialized perennial angiosperms that are either dwarf shrubs or herbaceous meadows, tussock grasslands, and marsh meadows dominated by sedges or mat-like cushion forms mostly prevail (Billings 1973; Kala 1998). The ecosystems here are mostly heterogeneous yet fragile and dynamic, displaying an intricate array of successional phases, from pioneer lichen-moss communities to well-established communities (Rawat 2007). Globally, the alpine zone covers approximately 2.6% of the terrestrial landform (Körner 2023), while in the Himalayas region, it covers nearly 33% (26% of the area is vegetated and 7% of the area falls under perpetual snow or barren rocks) of the geographical area (Lal et al. 1991). Notably, the global alpine flora (angiosperm) is estimated to be around 8000–10,000 species, about 4% of the total global number (Spehn and Körner 2005).

The representative diversity in the Indian Himalayan Region (IHR)—one among the 36 biodiversity hotspots, includes over 8000 species of angiosperms (40% endemic), 44 gymnosperms (16% endemic), 6000 pteridophytes (25% endemic), 1737 bryophytes (33% endemic), 1159 lichens (11% endemic), and 6900 species of fungi (27% endemic) (Singh and Hajra 1996; Rawal et al. 2013). Rich biodiversity is mainly attributed to diverse habitat types, influenced by a wide altitudinal range (300–8000 m), varied climatic conditions, temperature regimes, and complex topographical features (Samant et al. 1998). Among these, Samant et al. (1998) reported over 1748 plant species (1685—angiosperms, 12—gymnosperms, and 5—pteridophytes) of therapeutic values with a high percentage from the alpine region (Fig. 24.1). Further, these medicinal plants have been in great demand both in traditional healing practices and among pharmaceutical companies. As per the International Union for Conservation of Nature (IUCN) and the World Wildlife Fund (WWF), over 50,000–80,000 plant species are globally used for medicinal purposes. The main markets of medicinal plants include Europe, North America, and Asia, with China and India being the leading exporting countries (i.e., >70% of

Fig. 24.1 Threatened Himalayan medicinal plants in commerce. (**a**) *Aconitum heterophyllum*, *Podophyllum hexandrum*, (**c**) *Dactylorhiza hatagirea*, (**d**) *Rheum emodi*, (**e**) *Nardostachys jatamansi*, (**f**) *Paris polyphylla*, (**g**) *Picrorhiza kurroa*, and (**h**) *Trillium govanianum*

the demand for herbal medicines worldwide is produced in both countries) (Parvin et al. 2023). The global market for plant-derived drugs was valued at US ($)1.2 billion in the 1990s, and has increased to $25.6 billion in 2015 (Roosta et al. 2017). By 2050, the trade in herbal products is expected to cross US $ 5 trillion with a compound annual growth rate of ~7% (Booker et al. 2012; Negi et al. 2018a).

In India, the herbal sector has emerged as a promising economic sector, with 8.13% of the global share and 22% growth, the highest in the world (Maitra and Satya 2015). Nearly 40,000 distinct formulations of traditional herbal drugs, prepared using 1100 species of MPs, are documented in different traditional medicine systems in India (Maitra and Satya 2015). Available information shows that 1800 species are used in Classical Indian systems of medicines; Ayurveda uses 1200, Siddha 900, Unani 700, Amchi 600, and Tibetans system-450 species (Sati 2013). Considering the economic prosperity, the trade of these commodities has led to the development of vital resource enterprises, particularly in developing nations. However, the backdrop of surplus demand, limited availability of the resource (raw material), and lack of cultivation have put extra pressure on wild medicinal plant populations and consequently encouraged illegal harvesting (Table 24.1). Illicit and rampant MPs collection is recognized as one of the major threats to their wild population in the western Himalaya. As a result, the magnitude of population decline has been quite dramatic over the years (Singh 2022).

To an estimate, about 70–90% of MPs under trade are harvested from the wild (Bhattacharya et al. 2008). The abrupt harvesting practices are also in full swing in the IHR, wherein approximately 90% of MPs are harvested directly from the wild; nearly 70% involve destructive harvesting (MOEF 2014). The region (i.e., IHR), although exceptionally rich in biodiversity, is among the most fragile and vulnerable ecosystems in the world. In addition to considerable anthropogenic disturbances, environmental perturbations have seriously affected the structural integrity of this

Table 24.1 Threatened Himalayan medicinal plants in trade, their IUCN categories, source of raw material, and active constituents

Species	Trade name	Altitude range	Traded form	IUCN status	Demand (MT)	Raw material	Active constituents	References
Aconitum heterophyllum	Atis	3200–4500	Root (tuber)	EN	100–200	W	Aconitine, heteratisine, heterophyllidine	Malhotra and Sharma (2021)
Angelica glauca	Choru	2000–3800	Rhizome	EN	<10	W	(Z)-ligustilide, (Z)-butylidinenephathalie, ascorbic acid	Kumar et al. (2022)
Arnebia benthamii	Gauzaban, Ratanjot	3000–3900	Root/ whole plant	DD	100–200	W	Shikonin, β-Dimethyllacryl shikonin	Katoch et al. (2016)
Bunium persicum	Kala jeera	1800–3500	Seed	–	<10	W/C	γ-Terpinene, cuminaldehyde, α-terpinene-7-al, γ-terpinene-7-al, ρ-cymene, limonene	Shahsavari et al. (2008)
Dactylorhiza hatagirea	Salampanja	2500–4500	Tuber	EN	10–50	W	Dactylorhin (A, B, C, D, E), dactyloses A & B, and polyphenol	Singh et al. (2022a, b, 2023)
Fritillaria roylei	Kakoli, Van lahsun	2800–4000	Root/bulb	VU	<10	W	Sipeimine, peimine	Kumar et al. (2020)
Gentiana kurrooa	Trahimaan, Kadu	2000–3000	Root, whole plant	CR	<10	W	Loganic acid, swertisin, gentisin, gentioside, isogentisin, gentianine	Mubashir et al. (2017)
Habenaria intermedia	Vriddhi	2000–3300	Tuber	EN	10–50	W	Sinapic acid, gallic acid, catechin, hydroxy benzoic acid	Virk et al. (2020) and Giri et al. (2012)
Hippophae rhamnoides	Sea buckthorn	3300–4500	Fruit	LC	50–100	W	Oleanolic acid, ursolic acid, dulcioic acid, palmitic acid	Zheng et al. (2017)
Nardostachys jatamansi	Jatamansi	3300–5000	Root and rhizome	CR	500–1000	W	Nardostachone, valeranol, angelicin, nardal	Pathak and Godela (2023)

Polygonatum cirrhifolium	Meda	2500–4000	Tuber	VU	100–200	W	Dracaenoside F, dioscin, diosgenin, isosprirostanol	Luo et al. (2022)
Paris polyphylla	Satua	1800–3500	Rhizome	VU	10–50	W	Polyphyllin VI, polyphyllin VII, polyphyllin C	Shah et al. (2012)
Picrorhiza kurrooa	Kutki	3000–4000	Root	CR	1000–2000	W	Picroside I and picroside II, inidazole	Upadhyay et al. (2013)
Podophyllum hexandrum	Bankakri	2600–3800	Fruit/root	EN	10–50		Podophyllotoxin, epipodophyllotoxin, podophyllotoxone	Anand et al. (2022)
Rheum emodi	Dolu	3200–4200	Root/ rhizome	EN	100–200		Emodin, aloe emodine, revandchinone, chrysophanol glucoside	Zargar et al. (2011)
Saussurea costus	Kuth	2000–3500	Root	CR	100–200	C	Costunolide, dehydrocostus lactone & cynaropicrin	Pandey et al. (2007)
Swertia chirayita	Chiraiyata	1500–3000	Whole plant	EN	500–1000	W/C	Mangiferin, swerchirin, amarogentin, swertianin, oleanolic acid, bellidifolin, sweroside, swertiamarin	Swati et al. (2023)
Taxus wallichiana	Yew	1000–4400	Leaf	EN	100–200	W/C	Taxol, taxicin-I, taxicin-II	Jenniskens et al. (1996)
Trillium govanianum	Nagchatri	2700–4000	Root	EN	200–500		Diosgenin (1), borassoside D (2), govanoside (B)	Suresh et al. (2021)
Valeriana jatamansi	Tagar	1500–3600	Root/ rhizome		1000–2000	W/C	Valerenic acid, valeric acid, gallic acid, caffeic acid	Paniagua-Zambrana et al. (2024)

pristine ecosystem. Some of the major responsible anthropogenic activities that topple the integrity of the Himalayan ecosystem include habitat fragmentation and degradation, unscientific and over-exploitation of resources, grazing and trampling, illegal trade, and developmental projects (viz. motorway, dams, and recreation centers) (Singh et al. 2023). It has been generally found that anthropogenic activities are closely associated with species and population declines, apart from climate change, which is also beginning to play a key role (Singh et al. 2022a, b). Several Himalayan species are already or on the verge of extinction, while many are under different threats status. Therefore, it is important to understand the factors contributing to the loss of plant diversity, including the geographical distribution of these dynamics and the mechanisms driving these processes. This knowledge will be invaluable in guiding conservation efforts and planning for the future.

24.2 Diversity and Distribution of Medicinal Plants

The floral diversity of the IHR (including the alpine region) is well documented in respective floras/ checklists and bears close affinities with the Hindu-Kush Himalayan (HKH) region. In the western Himalaya, the dominant angiosperm families are Asteraceae, Poaceae, Brassicaceae, Scrophulariaceae, Ranunculaceae, Fabaceae, Rosaceae, and Ranunculaceae (Rau 1975; Rawat 2007). Very recently, considering the threatened plant diversity and realizing the importance of location-specific datasets, efforts are being made to grid-map threatened and endemic plant diversity of the IHR (Mehta et al. 2023; Tiwari et al. 2024). Of the total 456 threatened taxa (belonging to 233 genera and 87 families) from IHR (as per IUCN 2020), 85 taxa were such that fall under higher extinction risk categories (i.e., 18 critically endangered, 34 endangered, 31 vulnerable, and 02 extinct in the wild) including 41 trees, 34 herbs, 3 shrubs, 3 mosses, 2 vines, and 2 bryophytes belonging to 56 genera under 41 families (Mehta et al. 2023). Of these, Orchidaceae (12 threatened species) was the dominant family, followed by Dipterocarpaceae (7 species), Fabaceae, and Zingiberaceae (6 species each) (Mehta et al. 2023). Further, species such as *Aconitum heterophyllum*, *Aconitum chasmanthum*, *Aquilaria khasiana*, *Angelica glauca*, *Lilium polyphyllum*, *Magnolia pleiocarpa*, *Nardostachys jatamansi*, *Saussurea costus*, etc. were prioritized species for conservation action that need priority attention (Mehta et al. 2023). As per the latest IUCN report (IUCN 2024), among the Himalayan states, Arunachal Pradesh was reported to harbor the maximum number of threatened taxa with 80 species followed by Assam (55 species), Meghalaya (49 species) and Sikkim (36 species) (Fig. 24.2), whereas, previously Mehta et al. (2023) reported (as per IUCN 2020) Arunachal Pradesh to harbor 34 species followed by Meghalaya (33) and Assam (28 species).

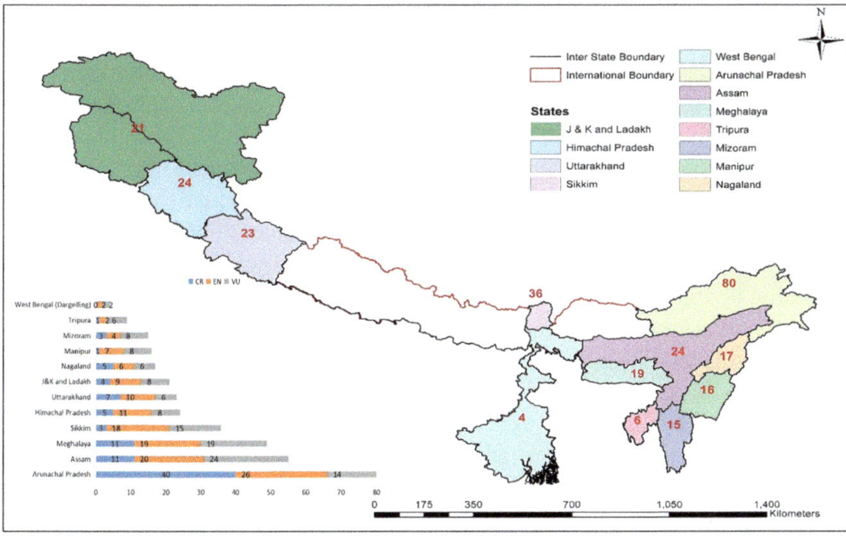

Fig. 24.2 State-wise distribution of IUCN species (IUCN 2024) in the Indian Himalayan region. *CR* critically endangered, *EN* endangered, *VU* vulnerable

24.3 Medicinal Plant Wealth in Traditional Healthcare Practices

Medicinal plants (MPs) have been the historical mainstay of traditional healthcare practices. According to the World Health Organization (WHO 2011), about 80% of the world's population uses herbs and other traditional medicines for their primary healthcare and has established three kinds of herbal medicines: raw plant material, processed plant material, and herbal products. Herbal medicines are known as mixtures of chemical products synthesized in plants and have been proven safe due to their fewer side effects. Asian medicinal plants account for about 50% of export quantity and 45% of global earnings from traditional medicines (Vasisht et al. 2016). India has strong traditional healthcare practices represented by classical medicine systems such as Ayurveda and Siddha, which use about 7000 species of flowering plants (Mukherjee 2001). Nearly 40,000 distinctive formulations of traditional herbal drugs, prepared using 1100 species of MPs, are documented in different traditional medicine systems in India (Maitra and Satya 2015). Traditional herbal medicines are rooted in indigenous knowledge systems and continue to play an important role in the healthcare system in tribal and rural areas of India (Kala 2005a, b; Negi et al. 2018a). Nearly 138 million rural households in India and thousands of folk healers are believed to consume significant quantities of diverse medicinal plants for wellness and health care (Goraya and Ved 2017). A survey of the traditional folk healers, identified by using a priori information in the sampled villages, revealed the use of 340 MPs, of which 202 species are in active trade, at an average consumption of 5.89 MT of herbal raw drugs per folk healer (Goraya and Ved 2017).

In the Himalayan region, traditional herbal medicines play a crucial role in decision-making with respect to the use of MP resources (Kala 2005a, b; Negi et al. 2018a; Phondani et al. 2010). The IHR supports over 1748 (23.4% of India) plant species of known medicinal value (Samant et al. 1998). Inhabitants of IHR states are traditionally well familiar with the healing properties of plants and therefore, dependency on MPs for primary healthcare forms an integral part of their life (Dhar et al. 2000; Hamilton 2004; Kuniyal et al. 2015). Interestingly, the majority of the old generation in high-altitude villages is less educated in this region, but they possess a deep understanding and knowledge of traditional healthcare. They have acquired this knowledge by trial and error and have contributed to the origin and evolution of many traditional herbal therapies (Dhar et al. 2000; Kala 2005a, b; Negi et al. 2018b).

Medicinal plants are used extensively to treat various ailments in rural and remote areas of the Himalayan region (Table 24.2). For example, a total of 243 herbal medical formulations prepared by Vaidyas (the traditional herbal healers) treating 73 different ailments were documented in Uttarakhand (Kala 2005a; Phondani et al. 2010). Likewise, a total of 86 plant species from the Garhwal region (Phondani et al. 2010) and 53 from the Kumaon region (Negi et al. 2018b) were identified as being used for the treatment of various ailments by Bhotiya tribal communities. The Bhotiya people use over 150 different medicinal plants to treat more than 105 combinations of diseases. Vidyarthi et al. (2013) reported 75 species from Himachal Pradesh used in traditional herbal medicine systems. About 158 MP are used for the treatment of various ailments by the Apatanis tribes of Arunachal Pradesh (Kala 2005b). Amchis use about 337 plant species for their traditional health care in the trans-Himalayan region (Kala 2003). The method for using these plants varies among age groups according to the nature of the disease and specific dose (Kala 2005b). In the majority of cases, a decoction of leaves, tuber/root, fruit, and root of MPs is used in the treatment of various diseases (Kala 2006, 2011; Negi et al. 2018a; Phondani et al. 2013, 2014). The paste of some plants can even be plastered on the skin to set dislocated or fractured bones or ease muscular pain. Often, during meditation, some food restrictions also need to be followed (Singh 2017).

Instead of making a great contribution to traditional health care, knowledge of herbal medicines is gradually being lost since this knowledge is mainly restricted to local and indigenous healers. There is a sharp decline in the transfer of knowledge across generations, particularly since younger generations are no longer interested (Maikhuri et al. 2005; Phondani et al. 2013, 2014). The knowledge of many important medicinal plants remained confined to traditional herbal healers—Vaidyas, who hold knowledge on the correct timing of collection, the parts to be used, storage techniques, and methods of preparation. Simultaneously, human-caused biodiversity losses affect MPs and associated traditional knowledge (Tali et al. 2019). Depletion and loss of MP diversity and associated traditional knowledge base may significantly impact human health and livelihood (Kala 2002; Maikhuri et al. 2000). Hence, protection of the genetic pool of this valuable wealth in nature is urgently

Table 24.2 Ethnobotanical uses of high-value medicinal plants from the Himalayan region

Species	Family	Part used	Medicinal value/use	References
Allium stracheyi	Liliaceae	Whole plant	Cold, cough, jaundice, improve digestion, relieve stress	Payal et al. (2020) and Maikhuri et al. (2017)
Arnebia benthamii	Boraginaceae	Root	Fever, wound healing, hair tonic, cardiac disorders	Negi et al. (2019), Manzoor et al. (2024) and Ganie et al. (2014)
Angelica glauca	Apiaceae	Root	Dysentery, bronchitis, rheumatism, menorrhiza, urinary disorders, constipation, gastric problems	Thakur et al. (2020) and Purohit et al. (2015)
Aconitum heterophyllum	Ranunculaceae	Roots	Fever, abdominal pain, reproductive disorders, coughs, diarrhoea, indigestion	Thakur et al. (2024), Kumar and Singhal (2019) and Barman et al. (2021)
Aconitum violaceum	Ranunculaceae	Roots	Antipyretic, abdominal pain, antidote	Sulaiman et al. (2020) and Negi et al. (2019)
Aconitum balfouri	Ranunculaceae	Roots	Diaphoretic, anti-pyretic	Sharma and Gaur (2012)
Arnebia euchroma	Boraginaceae	Root	Antiseptic, purify blood, abortifacient, hair tonic, regulate blood pressure, cure cough, cold, cuts, wounds, backache, headache	Thakur et al. (2020)
Bergenia stracheyi	Saxifragaceae	Rhizome	Cuts, wounds, kidney stones, liver diseases, lung infection	Negi et al. (2019)
Dactylorhiza hatagirea	Orchidaceae	Tubers	Aphrodiasic, health tonic, cough and cold, cuts, wounds healing, antidiabetic	Singh et al. (2021)
Fritillaria roylei	Liliaceace	Bulb	Healing wounds, bronchitis, health tonic	Poudel et al. (2020)

(continued)

Table 24.2 (continued)

Species	Family	Part used	Medicinal value/use	References
Lilium polyphyllum	Liliaceae	Tuber/ bulbs	Abdominal bloating, revitalizing night cream, and in Chyawanprash	Khoja et al. (2022) and Devi et al. (2024)
Malaxis muscifera	Asteraceae	Root	Puerperal fever, colic, and as a poultice to eruptions, aromatic, gout, rheumatism	Singh et al. (2024)
Nardostachys jatamansi	Valerianaceae	Rhizome/ root	Digestion, hysteria, syncope epilepsy, mental weakness	Pathak et al. (2023) and Sahu et al. (2016)
Picrorhiza kurrooa	Scrophulariaceae	Root/ stem	Anaemia, asthma, jaundice, skin disorders, fever, burning sensation, respiratory diseases, hepatitis, anorexia	Kumari et al. (2021) and Debnath et al. (2020)
Podophyllum hexandrum	Berberidaceae	Fruits/ rhizome	Cough, asthma	Yaqoob et al. (2022) and Kumar and Singhal (2019)
Polygonatum cirrhifolium	Asparagaceae	Bulbs	Tuberculosis, debility, leucorrhoea	Bhatt et al. (2014)
Rheum emodi	Polygonaceae	Rhizome	Fever, cough, indigestion, stomach problems, cuts, wound, muscular swelling, tonsillitis, mumps, ulcer, jaundice, swelling	Mohtashami et al. (2021)
Rheum moorcroftianum	Polygonaceae	Rhizome	Hyperlipidemia, anticancer	Gillani et al. (2024)
Swertia chirayita	Gentianaceae	Whole plant	Skin diseases, ulcers, diabetes, fever	Pradhan and Badola (2015) and Bhatt et al. (2024)
Selinum vaginatum	Apiaceae	Rhizome	Epilepsy and hysteria, mental disorders, asthma, cough, hysteria, as an analgesic and an antibacterial	Saraswat et al. (2020)
Trillium govanianum	Melanthiaceae	Rhizomes	Cancer boils, dysentery, menstrual and sexual disorders, antiseptic and wound healing	Chauhan et al. (2018)

required for long-run sustainability and making it available for future generations (Salgotra and Chauhan 2023). Further, documentation of indigenous knowledge for future use, as well as benefit-sharing mechanisms, is extremely important as sociocultural practices contribute significantly toward the conservation of these threatened plants. Considering the serious threat of complete loss of such a rich heritage of information restricted to only higher age groups, there is an urgent need for attention and progressive exposure at the government level. Traditional knowledge and practices have contributed to gene pool conservation and, therefore need to be promoted. Traditional knowledge of herbal medicine associated with various tribal communities requires exploration, analysis, popularization, and documentation with support of incentives for the protection of the system.

24.4 Pharmacological Activities

Medicinal plants from the Himalayan region are renowned for their exceptional properties, which extend to their nutraceutical, pharmaceutical, and industrial applications. Several studies have highlighted effective pharmacological applications of these species against treating different diseases, i.e., cancer, diabetes, epilepsy, inflammation, hepatoprotection, Alzheimer's, and malaria, which is attributed to the presence of numerous secondary metabolites, including phenolics, flavonoids, alkaloids, lignins, and saponins (Table 24.3). Recent advancements in the application of herbal and traditional medicines have resulted in identifying and characterizing pure and active substances for enhanced efficacy. One such compound is decursin, a coumarin-isolated from *Angelica galuca*, and has demonstrated anticancer activity against breast and prostate cancer (Jiang et al. 2007; Kumar et al. 2022). Further, decursin is reported to reduce ERα at the protein and mRNA levels and block estrogen-stimulated genes in MCF-7 breast cancer cells. Similarly, podophyllotoxin and taxol are well-known anticancer compounds isolated from *Podophyllum hexandrum* and *Taxus wallichiana*, respectively. Motyka et al. (2023) reported numerous pharmaceutical studies of podophyllotoxin against various cancer types, including lung, digestive, colorectal, liver, bladder, breast, and leukemia. Likewise, etoposide (VP-16)—a derivative of podophyllotoxin has been reported to play a significant role against several cancer therapy treatments since 1983 (Zhao et al. 2021). Etoposide has recently been reported for treating cytokine storms in COVID-19 patients during phase II (April 2020) clinical trials (Zhao et al. 2021). Likewise, paclitaxel (taxol), an isoprenoid member, has been extensively used in various cancer treatments since 1994. Paclitaxel is reported to be particularly effective in treating ovarian, breast, non-small-cell lung carcinoma (NSCLC), Kaposi's sarcoma, and leukopenia cancers (Gallego-Jara et al. 2020). Another category of the compound is shikonin reported from *Arnebia benthamii* with potential anticancer properties. Alam et al. (2021), reported that shikonin at a concentration of 2.5 mg/kg significantly reduced effusion lymphoma by reducing spleen weight and ascites volume in treated mice. Further, several in vivo studies have shown significant positive effects of shikonin on different cancer cells including leukemia, lymphoma, gastric, colon, liver, pancreatic, lung, and breast (Guo et al. 2019).

Besides, traditional medicines derived from Himalayan medicinal plants have been used against diabetes, a chronic condition that has garnered widespread

Table 24.3 Description of pharmacological activities and chemical constituents of Himalayan medicinal plants

Species	Pharmacological activities	Description	Chemical structures of principal active constituents	References
Angelica glauca	Anticancer	Decursin-coumarins found in *A. glauca* show effective cytotoxic effects against breast and prostate cancer	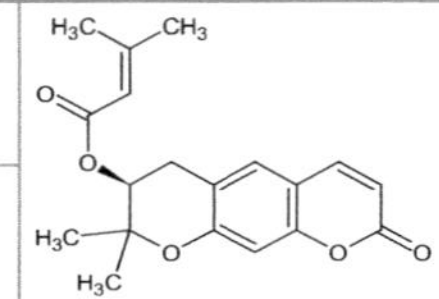 Decursin	Kumar et al. (2022) and Jiang et al. (2007)
	Broncho relaxant	Dexamethasone is a standard drug used to treat bronchial asthma. Similar results were obtained when the essential oil obtained from A. glauca was tested in Guinea pigs at 400 μL/Kg was reported to show comparable results to that of the standard compound		
Arnebia benthamii	Antioxidant assays	*Arnebia benthamii* ethyl acetate; extract showed potent DPPH inhibition (IC_{50}) potential, lipid peroxidation, and DNA damage protection activity	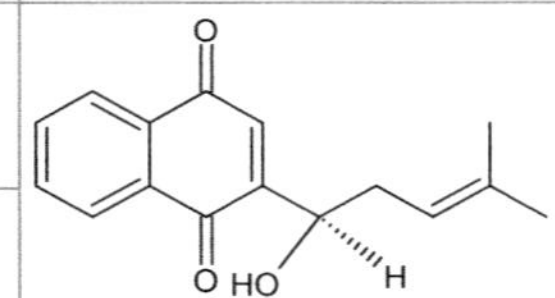 Shikonin	Ganie et al. (2013)
	Anticancer activity	*Arnebia benthamii* extract showed effective results against six human cancer cells (i.e., HOP-62, A549, PC-3, THP-1, HCT-116, MIA-Pa-Ca)		
Dactylorhiza hatagirea	Anti-inflammatory	The hydro-ethanolic *D. hatagirea* tuber extract (200 mg/kg) showed potent anti-inflammatory activity in the cotton pellet granuloma model and carrageenan-induced rat paw edema model	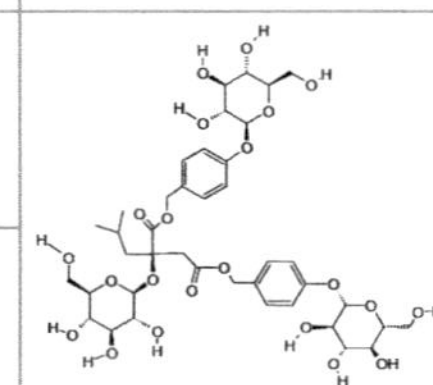 Dactylorhin A	Sirohi et al. (2019)
	Neuro-pharmacological	*D. hatagirea* extract was tested in insomnia (hypnosis) in the Swiss albino male mice model, and it was found that the 300 mg/kg concentration significantly increased the mean sleep duration time from 20.51 ± 2.12 (control) to 57.00 ± 13.24.		

Nardostachys jatamansi	Anticancer	The ethanolic extract showed a neuroprotective activity against Aβ42 neurotoxicity	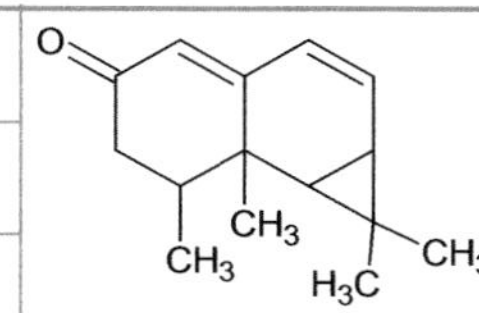 Nardostachone	Pathak and Godela (2023) and Joshi and parle (2006)
	Anti-inflammatory	Water extract of *N. jatamansi* showed potent anti-inflammatory activity against female mice		
	Anti-Alzheimer	Ethanolic and diethyl ether extract of *N. jatamansi* significantly treated dementia in elder people through memory-enhancing medication		
	Antidiabetic	Ethanolic extract of *N. jatamansi* improves hyperglycemia by enhancing sensitivity in skeletal muscle and decreasing glucose production		
Podophyllum hexandrum	Anticancer	Podophyllotoxin isolated from the *Podophyllum hexandrum* showed effective results against lung, testicular, neuroblastoma, hepatoma, and other cancers. Similarly, 4-demethyl-deoxypodophyllotoxin glucoside isolated exhibits potential anticancer activities by altering the Chk-2 signaling pathway in MCF-7 breast cancer cells	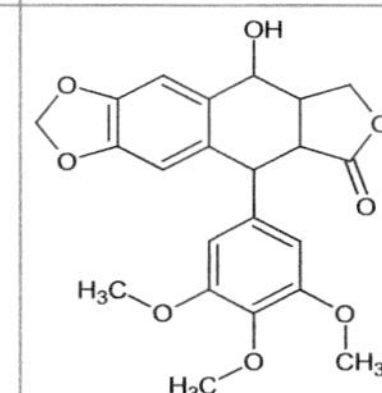 Podophyllotoxin	Giri and Lakshmi Narasu (2000) and Zilla et al. (2014)
Polygonatum cirrhifolium	Hepatoprotection	Aqueous ethanolic extract was used to determine hepatoprotection potential in Wistar rats at doses of 200 mg/kg and 400 mg/ kg p.o. per day. The results reported both doses significantly downplayed hepatotoxicity in a dose-dependent manner	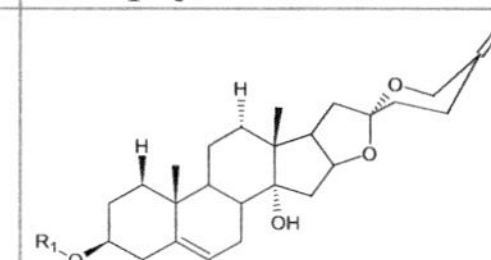 R_1: glc-(1_2)-[xyl-(1_3)]-glc-(1_4)-gal Polygodosides A	Ghaia et al. (2023)
Paris polyphylla	Immune stimulating activity	Diosgenyl saponins isolated from *P. polyphylla* have a noticeable effect on immune-stimulating activity in mouse macrophage cells and observed a significantly enhanced phagocytic activity	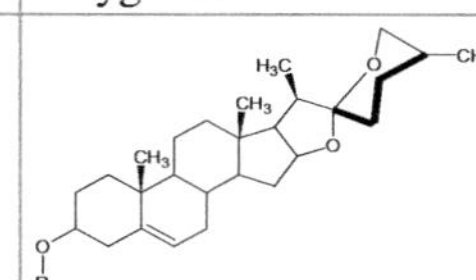 Diosgenin	Rawat et al. (2023)

(continued)

Table 24.3 (continued)

Species	Pharmacological activities	Description	Chemical structures of principal active constituents	References
Picrorhiza kurrooa	Hepatoprotective	*P. kurrooa* showed significant hepatoprotective potential against carbon tetrachloride-intoxicated rats and *Amanita* poisoning	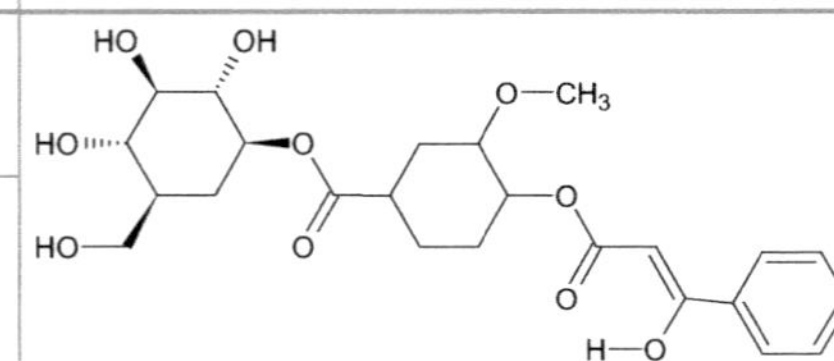 Kutkin 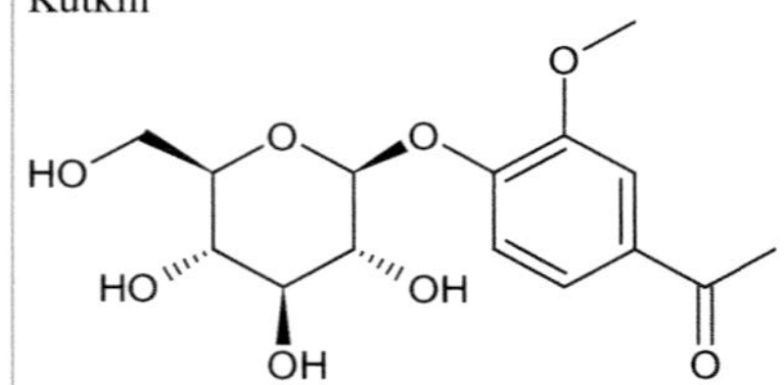 Androsin	Gaddipati et al. (1999) and Almeleebia et al. (2022)
	Nephroprotective	*P. kurrooa* extract was checked against nimesulide-induced toxicity in mice and reported mean serum urea of the nimesulide group decreased from 60 to 23 mg/dl and 25 mg/dl, respectively		
Saussurea costus	Anticancer	Dehydrocostus lactone is a thioredoxin isolated from the methanolic extract of the species that showed potent anticancer activity against cervical cancer HeLa cells	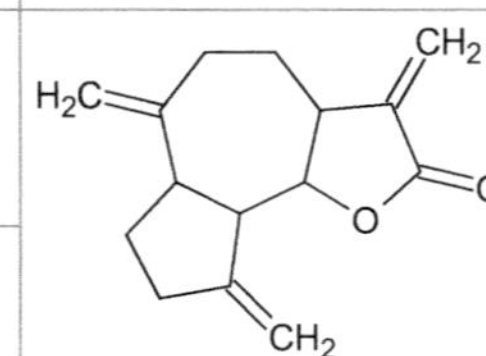 Dehydrocostus	Yang et al. (2017) and Ali and Venkatesalu (2022)
	Antiepileptic activity	Ethanolic root extract showed potent activity against seizures induced by both maximal electroshock (MES) and pentylenetetrazol (PTZ)		

Swertia chirayita	Anticancer	The anticancer activity of *S. chirayita* (ethanolic extract) was tested against Dimethylbenz(a) anthracene (DMBA)-induced skin carcinogenesis in Swiss albino male mice. The result thus obtained showed a significant reduction in lipid peroxidation and inhibition of multiplicity of DMBA-induced papillomas	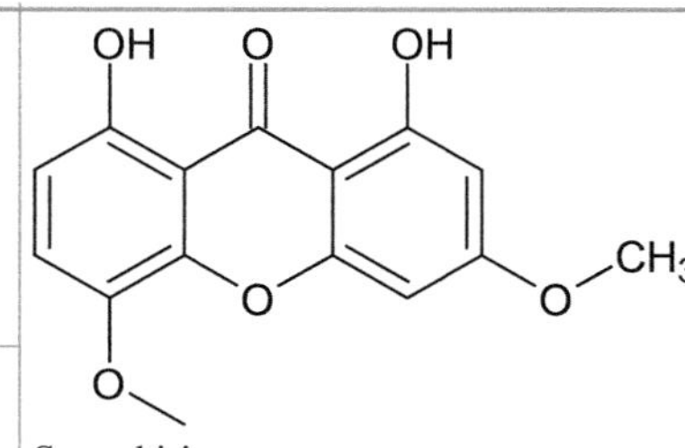 Swerchirin 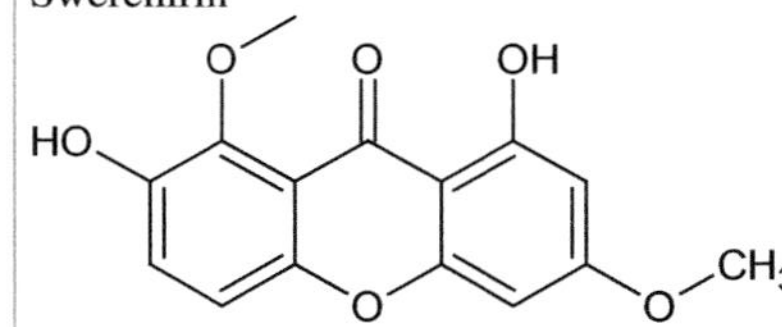 Gentiacauleine	Swati et al. (2023)
	Anti-malarial	Swerchirin or Methylbellidifolin isolated showed effective results in the antimalarial test		
Taxus wallichiana	Anticancer	Several compounds including taxol, taxiresinol 1, and 1-hydroxy-2-deacetoxy-5-decinnamoyltaxinine showed an effective result in different kinds of cancers (breast, liver, lung, blood, and gynecological cancers)	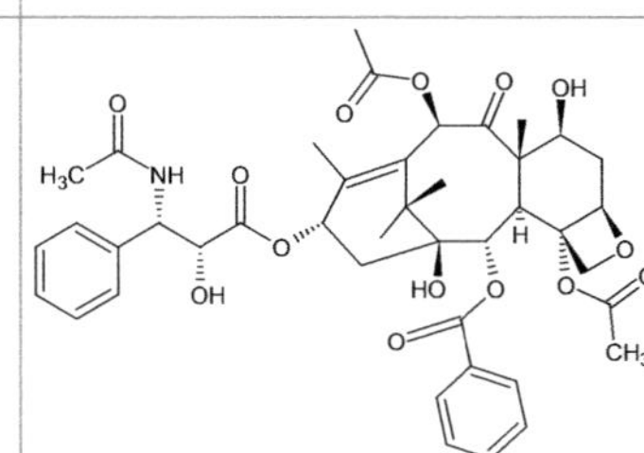 Taxol	Sharma and Garg (2015)

CR critically endangered, *EN* endangered, *VU* vulnerable, *LL* least concern, *DD* data deficit, *W* wild, *C* cultivated

attention, especially with increasing cases among children post-COVID-19 (D'Souza et al. 2023). Diabetes treatment has traditionally relied on a range of synthetic drugs, including biguanides, sulfonylureas, meglitinides, dipeptidyl peptidase IV inhibitors, thiazolidinediones, incretin agonists, and α-glucosidase inhibitors (Pandey et al. 2022). Although these medications have provided relief to a great extent, but in the longer run, significant serious side effects/complications are reported (Kerru et al. 2018). Thus, medicinal plants are being looked up on for treating diabetes. In the search for safer options, several Himalayan plants are reported to show promising antidiabetic potential. Notably, among these are *Swertia chirayita*, *Aconitum heterophyllum*, *Picrorhiza kurroa*, *Saussurea costus*, *Dactylorhiza hatagirea*, and *Nardostachys jatamansi* (Table 24.3). Research indicates that these plants can enhance insulin production or inhibit key enzymes such as α-amylase and α-glucosidase, thereby helping to regulate blood sugar levels effectively (Pandey et al. 2022). Similarly, picroside I, II, kutoside, and androsine, isolated and extracted from *Picrorhiza kurrooa*, have demonstrated multiple pharmacological activities. A standardized herbal drug, "Picroliv," combines picroside I, II, and kutoside in a 1:5:1 ratio and protects against various ailments, including respiratory, hepatic, inflammatory, and leishmanial diseases (Soni and Grover 2019). Despite the significant pharmacological activity of Himalayan medicinal plants, there is a pressing need to translate this knowledge into phyto-pharmaceutical formulations/products. One of the primary obstacles to the integration of herbal medicine into modern medical practices is the scarcity of scientific and clinical data demonstrating their efficacy and safety. It is essential to conduct clinical research on herbal drugs, develop straightforward bioassays for biological standardization, and carry out pharmacological and toxicological evaluations. Additionally, developing various animal models for toxicity and safety assessments is crucial; however, doing this will narrow the gap between laboratory discoveries and translation into effective formulations.

24.5 Significance of Medicinal Plants for Socioeconomic Transformation

MPs are a major resource base for the traditional medicine and herbal industry and provide livelihood and health security to a large segment of the region (Negi et al. 2018b; Goraya and Ved 2017). The export of value-added extracts of medicinal herbs/herbal products during 2017–2018 stood at US $ 456.12 million, recording a growth rate of 12.23% over the year. There are only 242 species that witness high-volume trade in the Himalayan region (>100 MT) annually (Goraya and Ved 2017). MPs can help farmers to strengthen their livelihoods directly through income generation from their trade and health care provision, particularly in rural and remote areas. The estimated value of MP from both cultivation and forest sources in Uttarakhand was $70 billion in 2013 (Kuniyal et al. 2015), and this is increasing at a rate of 12–15% annually. The forest or wild-sourced MP includes 701 recognized species (Sati 2013), which is around 40% of the MP found in the IHR.

The sale of MPs accounts for 17–35% of the total income of some households in a few high-altitude villages in IHR (Hamilton 2004; Chandra and Sharma 2018). For example, Phondani et al. (2016) reported that a total of 132 farmers adopted cultivation of MPs under 14 hectares of land in the Champawat district of Uttarakhand. The study reported the highest net profit to the villagers about US $1469.69/hectare/year for Rosmarinus officinalis and US $523.24/hectare/year for Matricaria chamomilla after 3 years of cultivation. Ankur NGO in Uttarakhand has worked with a group of farmers at Ramni of Ghat block, at an altitude of 2600 m for the cultivation of Picrorhiza kurrooa (kutki). With the help of NGOs, the farmers cultivated 2 million kutki plants in 2007 on an area of about 5 hectares (Alam and Belt 2009). In addition to this, about 100 farmers working with Ankur were reported to be successfully cultivating kutki in their fields; the average number of kutki plants cultivated by Ghat farmers was 4000 plants, and sell the products after 2 years for Rs. 400/kg of dry root weight. Kuniyal et al. (2013) reported that the cultivation and marketing of Saussurea lappa provided income between 1744 US$ in 2007 and 3178 US$ in 2010, and the cultivation and marketing of Picrorhiza kurrooa provided income between 5246 US$ (2007) and 7677 US$ (2010). Further, the Himalayan Research Group (HRG), Shimla, identified Swertia cordata as a potential species for commercial cultivation in Chachiot tehsil in Mandi, Himachal Pradesh. The total number of farmers enrolled for cultivation of Swertia cordata was about 1000, and the total area under cultivation was about 70 hectares, making it the largest cluster of farmers growing MP of wild origin in the IHR (Goraya and Ved 2017). Another success story of MP cultivation and livelihood enhancement is from Gheshe village in the Chamoli district of Uttarakhand, where farmers of the village have organized themselves into a society under the name "Gheshe Kishan Samiti" with the State Bhaisaj Sangh, mainly to acquire permits for commercial cultivation of kutki (Picrorhiza kurrooa). After 13 years of its initiation, kutki is now cultivated by 35 farmers over 4.5 ha of land in this village. The present annual production from this area is about 1 MT of dried roots, which costs about Rs 800 per kg.

MP sector in the IHR is widely viewed not only as an option for meeting current and future demands for large-volume production of plant-based drugs but also as support for herbal remedies used in traditional systems of healthcare. IHR states have a huge advantage in promoting the cultivation of high-altitude MPs because of the considerable area under such habitat and environmental conditions. Promoting MP cultivation can play a vital role in livelihood diversification for the marginalized population living in remote areas, as well as conserving the market-demanding species. However, the cultivation of the MPs sector in the IHR is experiencing several key challenges, on which a comprehensive understanding and timely actions are vital for sustainable livelihood and employment opportunities (Negi et al. 2018b; Chandra and Sharma 2018). These include (i) lack of availability of good quality seeds/ planting material, (ii) longer gestation period (2 or 3-year crop cycle) of many high-value MPs, (iii) small land holdings are less supportive for mass scale cultivation, (iv) lack of marketing infrastructure, (v) illicit trade practices in the region, (vi) long supply chain and distribution network, (vii) extensive involvement of the middlemen, and (viii) policy complexity and ineffectiveness.

24.6 From Farm to Industry

With a growing interest in alternative healthcare and systematic advancements in scientific validation of traditional remedies, the modern market for herbal and natural healthcare products is rapidly expanding. Therefore, the commercial use of medicinal plant resources is on the rise, creating a wide range of herbal and natural healthcare products (Table 24.4). It is projected that 80% of the world's population is dependent on herbal medicine, and approximately 60% of modern pharmaceutical drugs encompass plant-derived compounds (Chandra 2020). Similarly, the commerce in medicinal plants from the IHR is flourishing due to the unique therapeutic characteristics of Himalayan medicinal plants, which are found only in this region. Some of the prominent Himalayan medicinal plants that are actively traded in India are presented in Table 24.4.

In India, around 7500 species of medicinal plants are utilized in various traditional, folk, and herbal medicines (Dubey et al. 2004), among which 242 species are such which have an annual consumption volume of over 100 metric tons (MT)/year (Goraya and Ved 2017). In addition, 650–700 species are commercially exploited and used in phytopharmaceutical formulations (Maitra and Satya 2015). Thus, India, with an estimated trade value of US$ 140 million (Mishra et al. 2016), is recognized as a repository for the medicinal plant industry in Asia, supporting nearly 35 million workdays of employment (Traffic—the Wildlife Trade Monitoring Network 2018). The domestic demand for medicinal plants in India was estimated at 195,000 tons, and export demand was estimated at 134,500 tonnes. Over the years, the exports of medicinal plants have grown significantly from Rs. 345.80 crore in 2005–2006 to Rs. 3211 crore in 2014–2015, a ninefold increase in the last decade (www.nmpb.nic.in). Although there has been a sharp increase in demand for medicinal plants, approximately 70–90% of these species are being harvested directly from the wild, and as a result, 315 of its 6560 known medicinal plants are endangered under various threat IUCN categories (Kling 2016). Some of the prominent species with high market demand and under the IUCN categories include Aconitum heterophyllum, Angelica glauca, Saussurea costus, Dactylorhiza hatagirea, Trillium govanianum, Podophyllum hexandrum, and Nardostachys jatamansi (Table 24.4).

Considering the above, many medicinal plant conservation and cultivation programs are being organized by the respective State Medicinal Plant Board (SMPBs), National Medicinal Plant Board (NMPB), and research institutions. To strengthen the supply chain, the cultivators are registered with the statutory body for licensing. For example, in Uttarakhand alone, more than 24,000 farmers are cultivating around 4000 hectares of land (Negi et al. 2018a). Also, the State Medicinal Plant Board offers 50% financial support and subsidies to farmers for cultivating 28 (i.e., *Carum carvi*, *Cinnamomum tamala*, *Ephedra gerardiana*, *Dactylorhiza hatagirea*, *Rheum emodi*, *Saussurea lappa*, *Fritillaria roylei*, *Aconitum heterophyllum*, *Paris polyphylla*) species categorized as "Globally Significant Medicinal Plants" (GSMP) by the State Government with an objective of cultivation and conservation from socio-economic as well as sustainability perspectives (Kuniyal et al. 2015). Further, efforts

Table 24.4 Market Products derived from Himalayan medicinal plants

Species	Product name	Manufacturer	Quantity	Price (₹)
Allium stracheyi	Jamboo	Miho fresh, Kota bagh, Uttarakhand	30 gm	220/–
	Jamboo, Faran	The Pahari life, Nainital Uttarakhand	35 gm	169/–
	Jambu	Peerul Herbs, Uttarakhand	35 gm	220/–
Angelica glauca	Gandrayani	The Pahari life, Nainital Uttarakhand	40 gm	169/–
	Angelica root	Herbaveda, New Delhi	200 gm	439/–
Aconitum heterophyllum	Aconite 200CH	Homeopathic Diluted Globules	30 gm	140/–
	Atish	Indian Jadi Booti	100 gm	2220/–
	Anchery drugs	Avees Corporation	100 gm	1300/–
	Atish tablets-60	Shree Ram Ayurvedic	60 tablets	660/–
Bunium persicum	Kala-Jeera		100 gm	55/–
	Dr. Arya Kala Jeera	Arya Vastu Bhandar B-8 Transport Nagar Dehradun Uttarakhand	100 gm	299/–
Gentiana kurroo	Trahimaan	DKC Agrotech Private Limited	100 gm	499/–
	Neel kanhti powder	K.R. Impex Enterprises, Amritsar	100 gm	275/–
	Trayman essential oil	Razial Craft Dhauli squre, Bhubaneswar	30 mL	550/–
	Yogimate	K.R. Impex Enterprises, Amritsar	500 gm	1149/–
	Herbal tea	K.R. Impex Enterprises, Amritsar	250 gm	1050/–
Nardostachys jatamansi	Herb essential Jatamansi root powder	Herb Essential, Jain Lifesciences Pvt. Ltd., Telangana, India	100 gm	184/–
	Triphaladi taila	Nagarjuna	200 mL	150/–
	Jatamansi tablets	Merlion Naturals, Infinator Pvt. Ltd	120 tablets	520/–
	Powder	Indian Jadi Booti	400 gm	1920/–
	Jata mansi root	Golden nature	100 gm	349/–
	Pure Jatamansi essential oil	R V Essential	10 mL	959/–
	Capsule	Aarogyam Healthcare	60 capsules	360/–
	Fixuphealth Melation tablets	Nutralike Formulation	20 tablets	460/–
	Jatamansi (spikenard)	Devprayag	2 mL	999/–

(continued)

Table 24.4 (continued)

Species	Product name	Manufacturer	Quantity	Price (₹)
Picrorhiza kurroa	Kutki powder	Health & Herbs	100 gm	399/–
	Kutki powder	Darshini	100 gm	575/–
	Kutki powder	Sierra India Organics	500 gm	2575/–
	Kutki capsule	Saptamveda	60 capsules	339/–
	Kutki (natural liver detox pitta + Kapha Balance	Jiva Ayurveda Pharmacy Pvt. Ltd., India	60 capsules	363/–
	Kutki churn	Vyas	100 gm	325/–
	Kutki capsules	Planet Ayurveda	60 capsules	1215/–
	Liv Amrit	Smartveda	60 tablets	699/–
	Yakritas	myUpchar Ayurveda	60 capsules	488/–
	Uriral	Nuealz	60 tablets	448/–
	Punarva Makoy & kutki	Sansu Health Care	100 mL	449/–
Polygonatum cirrhifolium	Mahameda essential oil	Xetomos	30 mL	555/–
	Mahameda seeds	Padmawati distributers	3000 gm	799/–
	Chyawanprash	Patanjali foods limited, Haridwar, India	1000 gm	246/–

are being made to establish a value chain for respective species in the IHR to bridge the supply chain from farm to industry.

24.7 Issues, Challenges, and Opportunities of Medicinal Plants

A recent study reports that an average of 2.3 species have become extinct each year for the past 2.5 centuries, with the rate of ongoing extinction [18–26 extinctions per million species years (E/MSY)] is up to 500 times the background extinction rate for plants (0.05–0.35 E/MSY) (De Vos et al. 2015; Humphreys et al. 2019). Further, as per the report of the International Union for Conservation of Nature (IUCN) about 13.49% (~40,468) of the flowering plant species, a subgroup of vascular plants, are under varying degrees of threat (IUCN 2021). In the IHR, a total of 85 species representing 41 trees, 34 herbs, 3 shrubs, 3 mosses, 2 vines, and 2 bryophytes belonging to 56 genera under 41 families are presented in various IUCN threat categories (Mehta et al. 2023). Also, out of 242 species (>100 MT/year trade volume) categorized under high commercial demand, 114 (47%) are collected directly from forests, 59 species (25%) are wild and collected from landscapes outside forests, 54 species (22%) are obtained from cultivation, and 15 species (6%) are imported (Goraya and Ved 2017; Gowthami et al. 2021). With over 70–90% of the species in trade collected directly from the wild and nearly 70% involving destructive harvesting practices, thus questions arise: are we reaching a point of no return? (Rawal et al. 2013). Further, the issues are exacerbated by additional biotic and abiotic pressures such as grazing, habitat fragmentation and loss, premature/illegal harvesting and trade, over-dominance of native/invasive species, chaotic tourism, and climate aberration. Additionally, the inherent characteristics of alpine species, such as restricted distribution range, high habitat specificity, pollinator deficiency, poor germination rate (in nature), and lack of cultivation practices, have worsened the sustainability of these species in the longer term (Singh et al. 2023). All these estimates suggest that there is an urgent need to develop strategies for the conservation and mainstreaming of biological wealth. Doing this can help reap maximum benefits and equivocally sustain the Himalayan bioresources.

Sector-wise, medicinal plants are among the important bioresources in the Himalaya for livelihood diversification for marginalized communities. On the other hand, the cultivation of medicinal plants in the region is not receiving the desired attraction it should have (from the farmer's point of view). Notably, it falls short of youth aspiration and expectations. This is so because the Himalayan species have a 3–4 years gestation period before accumulating the desired amount (mg/g or lit/kg) of chemical constituents. Moreover, species under cultivation lack post-harvest processing technology, value-addition, and value-chain development. Most often, the species are cultivated by the farmers, and when it comes to monetary appreciation, they get very little out of it. Considering this and for promoting cultivation, the Government has now subsidized (30–70%) listed plants under the different subsidized categories. Also, many missions, such as the "Aroma Mission," "National Ayush Mission," and "National Mission on Himalayan Studies," are working on promoting medicinal plant cultivation. Further, the Government of India (GoI) has established the "National Medicinal Plant Board" (NMPB) under the Ministry of AYUSH, which provides funding support for research and skill-building programs on medicinal plant cultivation and postharvest-processing. In addition to NMPB, "Department of Science and Technology" (DST), "Department of Biotechnology" (DBT), "Indian Council of Forestry Research and Education" (ICFRE), Council of Scientific and Industrial Research (CSIR), and "Indian Council of Agricultural Research" (ICAR) have also supported researchers in researching on medicinal plants.

24.8 Conclusion and Recommendations

To achieve commercial success from farm to industry, it is crucial to possess a deep understanding of demand and production processes and the value chain associated with them. Despite various challenges that have been identified, mainstreaming Himalayan biodiversity for conservation and sustainable development is of utmost importance given the increasing number of threatened species year-wise. A comprehensive approach, such as institutional partnerships, integration of indigenous knowledge, sustainable harvesting practices, and educational initiatives, can significantly contribute to mainstreaming Himalayan biodiversity. Furthermore, it is important to involve local communities and stakeholders in decision-making processes and to establish appropriate legal and policy frameworks for sustainable conservation. Moving forward, it is imperative to prioritize research and development initiatives to bridge knowledge gaps and enhance conservation planning. The following are some of the action points for effective conservation and economic dimensions:

- Sustainable wild harvest management schemes supported by governments and authorities can be implemented.
- Cluster-based cultivation approaches need to be strengthened.
- Gene banks need to be developed, particularly for habitat-specific, slow-growing species with high susceptibility to being over-harvested.
- Mainstreaming into the cultivation of commercially important species needs to be promoted.
- User rights over the resource and access to it need to be clarified. This is particularly the case where MPs are considered common property.
- Intellectual property rights need to be acknowledged so that local users or other entities can be adequately compensated.

Acknowledgments The authors are thankful to the Director (NIHE) for providing the necessary facilities and encouragement.

Conflicts of Interest The authors declare no conflict of interest.

Funding Financial support from NMHS (NMHS 2023-24/SC-XI/LG13/03) funded project is gratefully acknowledged.

References

Alam G, Belt J (2009) Developing a medicinal plant value chain: lessons from an initiative to cultivate Kutki (*Picrorhiza kurrooa*) in Northern India, KIT working papers series C5. KIT, Amsterdam

Alam MM, Kariya R, Boonnate P, Kawaguchi A, Okada S (2021) Induction of apoptosis by Shikonin through ROS-mediated intrinsic and extrinsic apoptotic pathways in primary effusion lymphoma. Transl Oncol 14(3):101006. https://doi.org/10.1016/j.tranon.2020.101006

Ali SI, Venkatesalu V (2022) Botany, traditional uses, phytochemistry and pharmacological properties of *Saussurea costus*—an endangered plant from Himalaya—a review. Phytochem Lett 47:140–155. https://doi.org/10.1016/j.phytol.2021.12.008

Almeleebia TM, Alsayari A, Wahab S (2022) Pharmacological and clinical efficacy of *Picrorhiza kurroa* and its secondary metabolites: a comprehensive review. Molecules 27(23):8316. https://doi.org/10.3390/molecules27238316

Anand U, Biswas P, Kumar V, Ray D, Ray P, Loake VI, Dey A (2022) *Podophyllum hexandrum* and its active constituents: novel radioprotectants. Biomed Pharmacother 146:112555. https://doi.org/10.1016/j.biopha.2021.112555

Barman T, Samant SS, Jyoti, Dey A, Nandy S, Maitra R, Anjana (2021) Sustainable employment of folkloric botanicals and conservation practices adopted by the inhabitants of Parbati Valley of North Western Himalaya, India in healing substantial corporeal disorders. Adv Tradit Med 23:443–482. https://doi.org/10.1007/s13596-021-00605-3

Bhatt D, Kumar R, Tewari LM, Joshi GC (2014) *Polygonatum cirrhifolium* Royle and *Polygonatum verticillatum* (L.) Allioni: status assessment and medicinal uses in Uttarakhand, India. J Med Plant Res 8(5):253–259. https://doi.org/10.5897/JMPR2013.5234

Bhatt S, Kumar A, Arunachalam A, Arunachalam K (2024) Ethnomedicinal diversity and traditional knowledge system of the Jaunsari tribe in Uttarakhand, Western Himalaya. Proc Natl Acad Sci India Sect B Biol Sci 94(1):177–192. https://doi.org/10.1007/s40011-023-01514-y

Bhattacharya P, Prasad R, Bhattacharyya R, Asokan A (2008) Towards certification of wild medicinal and aromatic plants in four Indian states. Unasylva 59(230):35–44

Billings WD (1973) Arctic and alpine vegetation: similarities, differences, and susceptibility to disturbance. Bioscience 23:697–704. https://doi.org/10.2307/1296827

Booker A, Johnston D, Heinrich M (2012) Value chains of herbal medicines—research needs and key challenges in the context of ethnopharmacology. J Ethnopharmacol 140(3):624–633. https://doi.org/10.1016/j.jep.2012.01.039

Chandra P (2020) The medicinal and aromatic plants business of Uttarakhand: a mini review of challenges and directions for future research. Nat Res Forum 44(3):274–285. https://doi.org/10.1111/1477-8947.12208

Chandra P, Sharma V (2018) Strategic marketing prospects for developing sustainable medicinal and aromatic plants businesses in the Indian Himalayan region. Small-Scale For 17(4):423–441. https://doi.org/10.1007/s11842-018-9396-3

Chauhan HK, Bisht AK, Bhatt ID, Bhatt A, Gallacher D, Santo A (2018) Population change of *Trillium govanianum* (Melanthiaceae) amid altered indigenous harvesting practices in the Indian Himalayas. J Ethnopharmacol 213:302–310. https://doi.org/10.1016/j.jep.2017.11.003

D'Souza D, Empringham J, Pechlivanoglou P, Uleryk EM, Cohen E, Shulman R (2023) Incidence of diabetes in children and adolescents during the COVID-19 pandemic: a systematic review and meta-analysis. JAMA Netw Open 6(6):e2321281. https://doi.org/10.1001/jamanetworkopen.2023.21281

De Vos JM, Joppa LN, Gittleman JL, Stephens PR, Pimm SL (2015) Estimating the normal background rate of species extinction. Conserv Biol 29(2):452–462. https://doi.org/10.1111/cobi.12380

Debnath P, Rathore S, Walia S, Kumar M, Devi R, Kumar R (2020) *Picrorhiza kurroa*: a promising traditional therapeutic herb from higher altitude of western Himalayas. J Herb Med 23:100358. https://doi.org/10.1016/j.hermed.2020.100358

Devi U, Bagri A, Bajpai AB (2024) Ethno-medicinal plants used by Jadh Bhotiya Community of District Uttarakashi, Uttarakhand, India. Ecol Quest 35(2):1–23. https://doi.org/10.12775/EQ.2024.017

Dhar U, Rawal RS, Upreti J (2000) Setting priorities for conservation of medicinal plants—a case study in the Indian Himalaya. Biol Conserv 95(1):57–65. https://doi.org/10.1016/S0006-3207(00)00010-0

Dubey NK, Kumar R, Tripathi P (2004) Global promotion of herbal medicine: India's opportunity. Curr Sci 86(1):37–41. https://www.jstor.org/stable/24109515

Gaddipati JP, Madhavan S, Sidhu GS, Singh AK, Seth P, Maheshwaril RK (1999) Picroliv—a natural product protects cells and regulates the gene expression during hypoxia/reoxygenation. Mol Cell Biochem 194:271–281. https://doi.org/10.1023/A:1006982028460

Gallego-Jara J, Lozano-Terol G, Sola-Martínez RA, Cánovas-Díaz M, de Diego Puente T (2020) A compressive review about Taxol®: history and future challenges. Molecules 25(24):5986. https://doi.org/10.3390/molecules25245986

Ganie SA, Dar TA, Hamid R, Zargar O, Abeer SU, Masood A, Zargar MA (2014) In vitro antioxidant and cytotoxic activities of *Arnebia benthamii* (Wall ex. G. Don): a critically endangered medicinal plant of Kashmir valley. Oxidative Med Cell Longev 2014(1):792574. https://doi.org/10.1155/2014/792574

Germino MJ (2014) Plants in alpine environments. In: Monson R (ed) Ecology and the environment. The plant sciences, vol 8. Springer, New York. https://doi.org/10.1007/978-1-4614-7501-9_12

Ghaia R, Nagarajanb K, Groverb P, Kumara V, Goelc R, Kaura C, Chauhana R (2023) Bioassay guided hepatoprotective activity of *Polygonatum cirrhifolium* against isoniazid and rifampicin induced hepatotoxicity in rats. Indian Drugs 60(07):80–87. https://doi.org/10.53879/id.60.07.13557

Gillani SW, Ahmad M, Zafar M, Manzoor M, Shah GM, Shaheen H, Khishlatovna KK (2024) Ethnobotanical exploration of traditional medicinal plants among the rural inhabitants of District Muzaffarabad, Kashmir Himalayan Region. Plant Sci Today 11(sp1):21–33. https://doi.org/10.14719/pst.3265

Giri A, Lakshmi Narasu M (2000) Production of podophyllotoxin from *Podophyllum hexandrum*: a potential natural product for clinically useful anticancer drugs. Cytotechnology 34:17–26. https://doi.org/10.1023/A:1008138230896

Giri L, Dhyani P, Rawat S, Bhatt ID, Nandi SK, Rawal RS, Pande V (2012) In vitro production of phenolic compounds and antioxidant activity in callus suspension cultures of *Habenaria edgeworthii*: a rare Himalayan medicinal orchid. Ind Crop Prod 39:1–6. https://doi.org/10.1016/j.indcrop.2012.01.024

Goraya GS, Ved DK (2017) Medicinal plants in India: an assessment of their demand and supply. National Medicinal Plants Board, Ministry of AYUSH, Government of India, New Delhi and Indian Council of Forestry Research & Education, Dehradun, pp 1–430

Gowthami R, Sharma N, Pandey R, Agrawal A (2021) Status and consolidated list of threatened medicinal plants of India. Genet Resour Crop Evol 68(6):2235–2263. https://doi.org/10.1007/s10722-021-01199-0

Guo C, He J, Song X, Tan L, Wang M, Jiang P, Peng C (2019) Pharmacological properties and derivatives of Shikonin—a review in recent years. Pharmacol Res 149:104463. https://doi.org/10.1016/j.phrs.2019.104463

Hamilton AC (2004) Medicinal plants, conservation and livelihoods. Biodivers Conserv 13:1477–1517. https://doi.org/10.1023/B:BIOC.0000021333.23413.42

https://nmpb.nic.in/. Assessed on 20–25 June 2024

https://www.iucnredlist.org/. Assessed on 05–07 July 2024

Humphreys AM, Govaerts R, Ficinski SZ, Nic Lughadha E, Vorontsova MS (2019) Global dataset shows geography and life form predict modern plant extinction and rediscovery. Nat Ecol Evol 3(7):1043–1047. https://doi.org/10.1038/s41559-019-0906-2

IUCN (2020) The IUCN redlist of Threatened Species. https://www.iucnredlist.org

IUCN (2021) The IUCN redlist of Threatened Species. https://www.iucnredlist.org

IUCN (2024) The IUCN redlist of Threatened Species. https://www.iucnredlist.org

Jenniskens LH, van Rozendaal EL, van Beek TA, Wiegerinck PH, Scheeren HW (1996) Identification of six taxine alkaloids from *Taxus baccata* needles. J Nat Prod 59(2):117–123. https://doi.org/10.1021/np960033l

Jiang C, Guo J, Wang Z, Xiao B, Lee HJ, Lee EO, Lu J (2007) Decursin and decursinol angelate inhibit estrogen-stimulated and estrogen-independent growth and survival of breast cancer cells. Breast Cancer Res 9:1–12. https://doi.org/10.1186/bcr1790

Joshi H, Parle M (2006) *Nardostachys jatamansi* improves learning and memory in mice. J Med Food 9(1):113–118. https://doi.org/10.1089/jmf.2006.9.113

Kala CP (1998) Ecology and conservation of alpine meadows in the Valley of Flowers National Park, Garhwal Himalaya. PhD thesis, Forest Research Institute (Deemed University), Dehradun

Kala CP (2002) Indigenous knowledge of Bhotiya tribal community on wool dyeing and its present status in the Garhwal Himalaya, India. Curr Sci 83(7):814–817. https://www.jstor.org/stable/24107081

Kala CP (2003) Medicinal plants of Indian trans-Himalaya: focus on Tibetan use of medicinal resources. Bishen Singh Mahendra Pal Singh

Kala CP (2005a) Ethnomedicinal botany of the Apatani in the Eastern Himalayan region of India. J Ethnobiol Ethnomed 1:1–8. https://doi.org/10.1186/1746-4269-1-11

Kala CP (2005b) Indigenous uses, population density, and conservation of threatened medicinal plants in protected areas of the Indian Himalayas. Conserv Biol 19(2):368–378. https://doi.org/10.1111/j.1523-1739.2005.00602.x

Kala CP (2006) Medicinal plants of the high altitude cold desert in India: diversity, distribution and traditional uses. Int J Biodivers Sci Ecosyst Serv Manage 2(1):43–56. https://doi.org/10.1080/17451590609618098

Kala CP (2011) Traditional ecological knowledge, sacred groves and conservation of biodiversity in the Pachmarhi Biosphere Reserve of India. J Environ Prot 2(07):967. https://doi.org/10.4236/jep.2011.27111

Katoch P, Rana S, Kumar D, Kumar S, Bhushan S (2016) Concurrent NP-HPTLC determination of Shikonin and β, β-Dimethylacryl Shikonin in *Arnebia benthamii*. J Chromatogr Sci 54(8):1421–1427. https://doi.org/10.1093/chromsci/bmw096

Kerru N, Singh-Pillay A, Awolade P, Singh P (2018) Current antidiabetic agents and their molecular targets: a review. Eur J Med Chem 152:436–488. https://doi.org/10.1016/j.ejmech.2018.04.061

Khoja AA, Andrabi SAH, Mir RA, Bussmann RW (2022) Ethnomedicine and culture: exploitation of plant species for traditional use in the remote area of Kashmir Himalayas. Ethnobot Res Appl 24:1–22. https://doi.org/10.32859/era.24.20.1-22

Kling J (2016) Protecting medicine's wild pharmacy. Nat Plants 2:16064. https://doi.org/10.1038/nplants.2016.64

Körner C (2023) Concepts in alpine plant ecology. Plants 12(14):2666. https://doi.org/10.3390/plants12142666

Kumar R, Singhal VK (2019) Traditional knowledge and conservation status of some selected medicinal herbs from Uttarkashi district in Uttarakhand, Western Himalayas. Taiwania 64(1):52–64. https://doi.org/10.6165/tai.2019.64.52

Kumar P, Partap M, Rana D, Kumar P, Warghat AR (2020) Metabolite and expression profiling of steroidal alkaloids in wild tissues compared to bulb derived in vitro cultures of *Fritillaria roylei*—high value critically endangered Himalayan medicinal herb. Ind Crop Prod 145:111945. https://doi.org/10.1016/j.indcrop.2019.111945

Kumar P, Rana V, Singh AN (2022) *Angelica glauca* Edgew.—a comprehensive review. J Appl Res Med Aromat Plants 31:100397. https://doi.org/10.1016/j.jarmap.2022.100397

Kumari I, Kaurav H, Chaudhary G (2021) Ethnobotanical significance of *picrorhiza kurroa* (kutki), a threatened species. Int J Res Rev 8(4):363–375. https://doi.org/10.52403/ijrr.20210443

Kuniyal CP, Kuniyal PC, Butola JS, Sundriyal RC (2013) Trends in the marketing of some important medicinal plants in Uttarakhand, India. Int J Biodivers Sci Ecosyst Serv Manage 9(4):324–329. https://doi.org/10.1080/21513732.2013.819531

Kuniyal CP, Bisht VK, Negi JS, Bhatt VP, Bisht DS, Butola JS, Singh SK (2015) Progress and prospect in the integrated development of medicinal and aromatic plants (MAPs) sector in Uttarakhand, Western Himalaya. Environ Dev Sustain 17:1141–1162. https://doi.org/10.1007/s10668-014-9595-9

Lal JB, Gulati AK, Bist MS (1991) Satellite mapping of alpine pastures in the Himalayas. Int J Remote Sens 12(3):435–443. https://doi.org/10.1080/01431169108929664

Luo L, Qiu Y, Gong L, Wang W, Wen R (2022) A review of *Polygonatum* Mill. genus: its taxonomy, chemical constituents, and pharmacological effect due to processing changes. Molecules 27(15):4821. https://doi.org/10.3390/molecules27154821

Maikhuri RK, Nautiyal S, Rao KS, Semwal RL (2000) Indigenous knowledge of medicinal plants and wild edibles among three tribal sub communities of the Central Himalayas, India. J Indigenous Knowl Dev Monitor 8(2):7–13

Maikhuri RK, Rao KS, Kandari LS, Joshi R, Dhyani D (2005) Does the outreach programme make an impact? A case study of medicinal and aromatic plant cultivation in Uttaranchal. Curr Sci 88(9):1480–1486. https://www.jstor.org/stable/24110718

Maikhuri RK, Negi VS, Rawat LS, Pharswan DS (2017) Bioprospecting of medicinal plants in Nanda Devi Biosphere Reserve: linking conservation with livelihood. Curr Sci 113(4):571–577. https://www.jstor.org/stable/26293888

Maitra S, Satya P (2015) Value added products from medicinal plants. In: Sharangi A, Datta S (eds) Value addition of horticultural crops: recent trends and future directions. Springer, New Delhi. https://doi.org/10.1007/978-81-322-2262-0_7

Malhotra N, Sharma S (2021) *Aconitum heterophyllum*. In: Himalayan medicinal plants. Academic, pp 5–25. https://doi.org/10.1016/B978-0-12-823151-7.00015-5

Manzoor M, Ahmad M, Zafar M, Gillani SW, Shah GM, Shaheen H, Khishlatovna KK (2024) Exploration of traditional Ethno-gynaecological knowledge: advances to ethnobotanical studies from indigenous communities of Neelum Valley in the Himalayan Region. Plant Sci Today 11(sp1):77–86. https://doi.org/10.14719/pst.3264

Mehta P, Bisht K, Sekar KC, Tewari A (2023) Mapping biodiversity conservation priorities for threatened plants of Indian Himalayan region. Biodivers Conserv 32(7):2263–2299. https://doi.org/10.1007/s10531-023-02604-z

Mishra P, Kumar A, Nagireddy A, Mani DN, Shukla AK, Tiwari R, Sundaresan V (2016) DNA barcoding: an efficient tool to overcome authentication challenges in the herbal market. Plant biotechnologyj, 14(1):8–21

MOEF (2014) India's Fifth National Report to the Convention on Biological Diversity. Ministry of Environment and Forests and Climate Change, Government of India

Mohtashami L, Amiri MS, Ayati Z, Ramezani M, Jamialahmadi T, Emami SA, Sahebkar A (2021) Ethnobotanical uses, phytochemistry and pharmacology of different *Rheum* species (Polygonaceae): a review. In: Barreto GE, Sahebkar A (eds) Pharmacological properties of plant-derived natural products and implications for human health, Advances in experimental medicine and biology, vol 1308. Springer, Cham. https://doi.org/10.1007/978-3-030-64872-5_22

Motyka S, Jafernik K, Ekiert H, Sharifi-Rad J, Calina D, Al-Omari B, Cho WC (2023) Podophyllotoxin and its derivatives: potential anticancer agents of natural origin in cancer chemotherapy. Biomed Pharmacother 158:114145. https://doi.org/10.1016/j.biopha.2022.114145

Mubashir K, Ganai BA, Ghazanfar K, Akbar S, Rah B, Tantry M, Masood A (2017) Anti-inflammatory and immuno-modulatory studies on LC-MS characterised methanol extract of *Gentiana kurroo* Royle. BMC Complement Altern Med 17:1–10. https://doi.org/10.1186/s12906-017-1593-7

Mukherjee PK (2001) Evaluation of Indian traditional medicine. Drug Inf J 35(2):623–632. https://doi.org/10.1177/009286150103500235

Negi VS, Pathak R, Sekar KC, Rawal RS, Bhatt ID, Nandi SK, Dhyani PP (2018) Traditional knowledge and biodiversity conservation: a case study from Byans Valley in Kailash Sacred Landscape, India. J Environ Plan Manag 61(10):1722–1743. https://doi.org/10.1080/09640568.2017.1371006

Negi VS, Kewlani P, Pathak R, Bhatt D, Bhatt ID, Rawal RS, Sundriyal RC, Nandi SK (2018a) Criteria and indicators for promoting cultivation and conservation of medicinal and aromatic plants in Western Himalaya, India. Ecol Indic 93:434–446. https://doi.org/10.1016/j.ecolind.2018.03.032

Negi VS, Giri L, Sekar KC (2018b) Floristic diversity, community composition and structure in Nanda Devi National Park after prohibition of human activities, Western Himalaya, India. Curr Sci 115(6):1056–1064. https://www.jstor.org/stable/26978357

Negi VS, Maikhuri RK, Maletha A, Phondani PC (2019) Ethnobotanical knowledge and population density of threatened medicinal plants of Nanda Devi Biosphere Reserve, Western Himalaya, India. Iran J Sci Technol Trans A: Sci 43:63–73. https://doi.org/10.1007/s40995-018-0545-5

Pandey MM, Rastogi S, Rawat AKS (2007) *Saussurea costus*: botanical, chemical and pharmacological review of an ayurvedic medicinal plant. J Ethnopharmacol 110(3):379–390. https://doi.org/10.1016/j.jep.2006.12.033

Pandey V, Bhatt ID, Nandi SK (2022) Himalayan medicinal plants for treatment of diabetes: their active components and underlying mechanism. Stud Nat Prod Chem 75:119–160. https://doi.org/10.1016/B978-0-323-91250-1.00001-X

Paniagua-Zambrana NY, Bussmann RW, Kikvidze Z (2024) *Valeriana officinalis* L. Caprifoliaceae. In: Bussmann RW, Paniagua-Zambrana NY, Kikvidze Z (eds) Ethnobotany of the mountain regions of Eastern Europe. Ethnobotany of mountain regions. Springer, Cham. https://doi.org/10.1007/978-3-030-98744-2_298-1

Parvin S, Reza A, Das S, Miah MMU, Karim S (2023) Potential role and international trade of medicinal and aromatic plants in the world. Eur J Agric Food Sci 5(5):89–99. https://doi.org/10.24018/ejfood.2023.5.5.701

Pathak S, Godela R (2023) *Nardostachys jatamansi*: phytochemistry, ethnomedicinal uses, and pharmacological activities: a comprehensive review. Fitoterapia 172:105764. https://doi.org/10.1016/j.fitote.2023.105764

Payal K, Rao KS, Maikhuri RK, Kandari LS (2020) Ethnobotanical assessment of endangered *Allium stracheyi* (Baker): a lesser known species of the Nanda Devi Biosphere Reserve, Central Himalaya, India. In: Ethnomedicinal plant use and practice in traditional medicine. IGI Global, pp 1–16. https://doi.org/10.4018/978-1-7998-1320-0.ch001

Phondani PC, Maikhuri RK, Kala CP (2010) Ethnoveterinary uses of medicinal plants among traditional herbal healers in Alaknanda catchment of Uttarakhand, India. Afr J Tradit Complement Altern Med 7(3):195–206. https://doi.org/10.4314/ajtcam.v7i3.54775

Phondani PC, Maikhuri RK, Bisht NS (2013) Endorsement of ethnomedicinal knowledge towards conservation in the context of changing socio-economic and cultural values of traditional communities around Binsar Wildlife Sanctuary in Uttarakhand, India. J Agric Environ Ethics 26:573–600. https://doi.org/10.1007/s10806-012-9428-5

Phondani PC, Maikhuri RK, Saxena KG (2014) The efficacy of herbal system of medicine in the context of allopathic system in Indian Central Himalaya. J Herb Med 4(3):147–158. https://doi.org/10.1016/j.hermed.2014.05.004

Phondani PC, Bhatt ID, Negi VS, Kothyari BP, Bhatt A, Maikhuri RK (2016) Promoting medicinal plants cultivation as a tool for biodiversity conservation and livelihood enhancement in Indian Himalaya. J Asia-Pac Biodivers 9(1):39–46. https://doi.org/10.1016/j.japb.2015.12.001

Poudel HR, Kunwar RM, Bussmann RW, Paniagua-Zambrana NY (2020) *Fritillaria cirrhosa* D. Don. Liliaceae. In: Ethnobotany of the Himalayas. Springer, Cham, pp 1–8. https://doi.org/10.1007/978-3-030-45597-2_105-3

Pradhan BK, Badola HK (2015) Local knowledge on the use of *Swertia chirayita* as traditional medicine: conservation challenges in Sikkim Himalaya, India. Ethnobot Res Appl 14:345–355. https://doi.org/10.17348/era.14.0.345-355

Purohit VK, Andola HC, Haider SZ, Tiwari D, Bahuguna YM, Gairola KC, Arunachalam K (2015) Essential oil constituents of *Angelica glauca* Edgew. roots: an endangered species from Uttarakhand Himalaya (India). Nat Acad Sci Lett 38:445–447. https://doi.org/10.1007/s40009-015-0395-z

Rau MA (1975) High altitude flowering plants of West Himalaya. Botanical Survey of India, Indian Botanic Garden, Howrah

Rawal RS, Bhatt ID, Sekar KC, Nandi SK (2013) The Himalayan biodiversity: richness, representativeness, uniqueness and life-support values. GB Pant Institute of Himalayan Environment and Development (GBPIHED), Kosi-Katarmal, Almora, Uttarakhand, India, 84pp

Rawat GS (2007) Alpine vegetation of the western Himalaya: species diversity, community structure, dynamics and aspects of conservation. Doctoral dissertation, DSc Thesis, Kumaun University, Nainital, 239pp

Rawat JM, Pandey S, Rawat B, Rai N, Preeti P, Thakur A, Bachheti RK (2023) Traditional uses, active ingredients, and biological activities of *Paris polyphylla* Smith: a comprehensive review of an important Himalayan medicinal plant. J Chem 2023(1):7947224. https://doi.org/10.1155/2023/7947224

Roosta RA, Moghaddasi R, Hosseini SS (2017) Export target markets of medicinal and aromatic plants. J Appl Res Med Aromat Plants 7:84–88. https://doi.org/10.1016/j.jarmap.2017.06.003

Sahu R, Dhongade HJ, Pandey A, Sahu P, Sahu V, Patel D, Kashyap P (2016) Medicinal properties of *Nardostachys jatamansi* (a review). Orient J Chem 32(2):859–866

Salgotra RK, Chauhan BS (2023) Genetic diversity, conservation, and utilization of plant genetic resources. Gene 14(1):174. https://doi.org/10.3390/genes14010174

Samant SS, Dhar U, Palni LMS (1998) Medicinal plants of Indian Himalaya. Gyanodaya Prakashan, Nainital

Saraswat N, Sachan N, Chandra P (2020) A detailed review on the rarely found Himalayan herb *Selinum Vaginatum*: its active constituents, pharmacological uses, traditional and potential benefits. Pharmacophore 11(2-2020):40–52

Sati VP (2013) Cultivation of medicinal plants and its contribution to livelihood enhancement in the Indian Central Himalayan Region. Adv Med Plant Res 1(2):17–23

Shah SA, Mazumder PB, Choudhury MD (2012) Medicinal properties of *Paris polyphylla* Smith: a review. J Herb Med Toxicol 6(1):27–33

Shahsavari N, Barzegar M, Sahari MA, Naghdibadi H (2008) Antioxidant activity and chemical characterization of essential oil of *Bunium persicum*. Plant Foods Hum Nutr 63:183–188. https://doi.org/10.1007/s11130-008-0091-y

Sharma H, Garg M (2015) A review of traditional use, phytoconstituents and biological activities of Himalayan yew, *Taxus wallichiana*. J Integr Med 13(2):80–90. https://doi.org/10.1016/S2095-4964(15)60161-3

Sharma E, Gaur AK (2012) *Aconitum balfourii* Stapf: a rare medicinal herb from Himalayan Alpine. J Med Plant Res 6(22):3810–3817. https://doi.org/10.5897/JMPR11.1213

Singh V (2017) Medicinal plants and bone healing. Natl J Maxillofac Surg 8(1):4–11. https://doi.org/10.4103/0975-5950.208972

Singh L (2022) Source-specific variability in morphological, biochemical, and genetic attributes of *Dactylorhiza hatagirea* (D. Don) Soo in the west Himalaya, India. Doctoral thesis, Kumaun University, Nainital, India

Singh DK, Hajara PK (1996) Changing perspectives of biodiversity status in the Himalaya. British Council Division, New Delhi, pp 23–38

Singh L, Bhatt ID, Negi VS, Nandi SK, Rawal RS, Bisht AK (2021) Population status, threats, and conservation options of the orchid *Dactylorhiza hatagirea* in Indian Western Himalaya. Reg Environ Chang 21:1–16. https://doi.org/10.1007/s10113-021-01762-6

Singh L, Kanwar N, Bhatt ID, Nandi SK, Bisht AK (2022a) Predicting the potential distribution of *Dactylorhiza hatagirea* (D. Don) Soo-an important medicinal orchid in the West Himalaya, under multiple climate change scenarios. PLoS One 17(6):e0269673. https://doi.org/10.1371/journal.pone.0269673

Singh L, Singh B, Kewlani P, Belwal T, Bhatt ID, Nandi SK, Bisht AK (2022b) Process optimization and bioactive compounds quantification from *Dactylorhiza hatagirea* tuber for alleviating glycemic and oxidative stress. J Appl Res Med Aromat Plants 26:100352. https://doi.org/10.1016/j.jarmap.2021.100352

Singh L, Nandi SK, Bhatt ID, Bisht AK (2023) Inventorization of ecology, ethnobotany, and conservation status of *Dactylorhiza hatagirea*: problems, progress, and prospects. In: Jha S, Halder M (eds) Medicinal plants: biodiversity, biotechnology and conservation, Sustainable development and biodiversity, vol 33. Springer, Singapore. https://doi.org/10.1007/978-981-19-9936-9_12

Singh K, Kumar P, Kumar B, Gairola S (2024) Study on magico-religious plants in Paddari tribe of Jammu and Kashmir, India. Ethnobot Res Appl 27:1–20. https://doi.org/10.32859/era.27.10.1-20

Sirohi B, Sagar R, Kori ML (2019) Comparative antioxidant activity of hydroalcoholic extract of roots of *Dactylorhiza hatagirea* and aerial part of *Lavandula stoechas*. J Adv Sci Res 10(04 Suppl 2):311–316

Soni D, Grover A (2019) "Picrosides" from *Picrorhiza kurroa* as potential anti-carcinogenic agents. Biomed Pharmacother 109:1680–1687. https://doi.org/10.1016/j.biopha.2018.11.048

Spehn EM, Körner C (2005) A global assessment of mountain biodiversity and its function. In: Huber UM, Bugmann HKM, Reasoner MA (eds) Global change and mountain regions. Advances in global change research, vol 23. Springer, Dordrecht. https://doi.org/10.1007/1-4020-3508-X_39

Sulaiman SS, Khan S, Bussmann RW, Ali M, Hussain D, Hussain W (2020) Quantitative ethnobotanical study of Indigenous knowledge on medicinal plants used by the tribal communities of Gokand Valley, District Buner, Khyber Pakhtunkhwa, Pakistan. Plan Theory 9(8):1001. https://doi.org/10.3390/plants9081001

Suresh PS, Singh PP, Sharma A, Padwad YS, Sharma U (2021) Steroidal saponins of *Trillium govanianum*: quality control, pharmacokinetic analysis, and anti-inflammatory activity. Biocatal Agric Biotechnol 35:102071. https://doi.org/10.1016/j.bcab.2021.102071

Swati K, Bhatt V, Sendri N, Bhatt P, Bhandari P (2023) *Swertia chirayita*: a comprehensive review on traditional uses, phytochemistry, quality assessment and pharmacology. J Ethnopharmacol 300:115714. https://doi.org/10.1016/j.jep.2022.115714

Tali BA, Khuroo AA, Ganie AH, Nawchoo IA (2019) Diversity, distribution and traditional uses of medicinal plants in Jammu and Kashmir (J&K) state of Indian Himalayas. J Herb Med 17:100280. https://doi.org/10.1016/j.hermed.2019.100280

Thakur M, Sharma PK, Asrani RK, Patil RD, Gautam H (2020) Traditional therapeutic uses of some important medicinal and aromatic plants of the tribal area of Lahaul valley of Himachal Pradesh, India. Indian J Tradit Knowl 19(4):761–775. https://doi.org/10.56042/ijtk.v19i4.44520

Thakur A, Kumari R, Kumar A, Chaudhary A (2024) Quantitative ethnobotanical study of medicinal plants used by native people of selected areas of Chhota Bhangal, Himachal Pradesh. Ethnobot Res Appl 28:1–34

Tiwari H, Sekar KC, Pandey A, Tiwari A, Mehta P, Kanwal KS, Arya D (2024) Diversity, distribution and need of urgent conservation of endemic plants in Himalaya. Biodivers Conserv 33:2285–2303. https://doi.org/10.1007/s10531-024-02815-y

TRAFFIC (2018) Wild at home. Retrieved from https://www.traffic.org/site/assets/files/9241/wild-athome.pdf on 24 July 2019

Upadhyay D, Dash RP, Anandjiwala S, Nivsarkar M (2013) Comparative pharmacokinetic profiles of picrosides I and II from kutkin, *Picrorhiza kurroa* extract and its formulation in rats. Fitoterapia 85:76–83. https://doi.org/10.1016/j.fitote.2013.01.004

Vasisht K, Sharma N, Karan M (2016) Current perspective in the international trade of medicinal plants material: an update. Curr Pharm Des 22(27):4288–4336. https://doi.org/10.2174/1381612822666160607070736

Vidyarthi S, Samant SS, Sharma P (2013) Traditional and indigenous uses of medicinal plants by local residents in Himachal Pradesh, North Western Himalaya, India. Int J Biodivers Sci Ecosyst Serv Manag 9(3):185–200. https://doi.org/10.1080/21513732.2013.823113

Virk JK, Gupta V, Maithani M, Rawal RK, Kumar S, Singh R, Bansal P (2020) Isolation of sinapic acid from *Habenaria intermedia* D. Don: a new chemical marker for the identification of adulteration and substitution. Curr Tradit Med 6(4):380–387. https://doi.org/10.2174/2215083804666181030101709

WHO (2011) The world medicines situation, traditional medicines: global situation, issues and challenges. WHO Press, Geneva

Yang HJ, Kim MJ, Kang S, Moon NR, Kim DS, Lee NR, Park S (2017) Topical treatments of *Saussurea costus* root and *Thuja orientalis* L. synergistically alleviate atopic dermatitis-like skin lesions by inhibiting protease-activated receptor-2 and NF-κB signaling in HaCaT cells and Nc/Nga mice. J Ethnopharmacol 199:97–105. https://doi.org/10.2174/2215083804666181030101709

Yaqoob A, Singh DP, Yunus M, Bhat GA, Shukla G, Bhat JA, Pokhrel S (2022) Folkloric knowledge of plant species used by local communities in a protected area of Kashmir Himalayas. In: Öztürk M, Khan SM, Altay V, Efe R, Egamberdieva D, Khassanov FO (eds) Biodiversity, conservation and sustainability in Asia. Springer, Cham. https://doi.org/10.1007/978-3-030-73943-0_39

Zargar BA, Masoodi MH, Ahmed B, Ganie SA (2011) Phytoconstituents and therapeutic uses of *Rheum emodi* wall. ex Meissn. Food Chem 128(3):585–589. https://doi.org/10.1016/j.foodchem.2011.03.083

Zhao W, Cong Y, Li HM, Li S, Shen Y, Qi Q, Tang YJ (2021) Challenges and potential for improving the druggability of podophyllotoxin-derived drugs in cancer chemotherapy. Nat Prod Rep 38(3):470–488. https://doi.org/10.1039/D0NP00041H

Zheng L, Shi LK, Zhao CW, Jin QZ, Wang XG (2017) Fatty acid, phytochemical, oxidative stability and in vitro antioxidant property of sea buckthorn (*Hippophaë rhamnoides* L.) oils extracted by supercritical and subcritical technologies. LWT Food Sci Technol 86:507–513. https://doi.org/10.1016/j.lwt.2017.08.042

Zilla MK, Nayak D, Amin H, Nalli Y, Rah B, Chakraborty S, Ali A (2014) 4′-Demethyl-deoxypodophyllotoxin glucosides isolated from *Podophyllum hexandrum* exhibits potential anticancer activities by altering Chk-2 signaling pathway in MCF-7 breast cancer cells. Chem Biol Interact 224:100–107. https://doi.org/10.1016/j.cbi.2014.09.022

Commercial Harvesting and the Future of Alpine Medicinal Plants

25

Yang Niu and Hang Sun

Abstract

The commercial harvesting, or large-scale exploitation, of alpine medicinal plants can lead to a range of consequences. This includes a direct decrease in population size and even local extinction if the harvest rate surpasses an irreversible threshold. Indirect ecological impacts may involve alterations to habitats and changes in interspecific interactions, which can reduce the survival and reproductive success of other organisms within the same community. Furthermore, commercial harvesting may influence plant phenotypes and affect their evolution. We present a case study on *Fritillaria delavayi* (Liliaceae) to illustrate how human collection affects the color evolution of this alpine medicinal herb. Ultimately, we emphasize that sustainable utilization, informed by local knowledge, is essential for the conservation of alpine medicinal plants.

25.1 Commercial Harvest of Alpine Medicinal Plants

Overexploitation is one of the main threats to biodiversity. Based on a recent estimate, 4543 species, on land and at sea, are threatened by unsustainable commercial harvesting (Di Minin et al. 2019). Commercial harvesting involves the large-scale collection, hunting, or extraction of natural resources—such as animals, plants, and other organisms—primarily for sale or trade to generate economic profit. This process often imposes significant and continuous pressure on these resources, distinguishing it from random collection. Numerous examples are known from the past. African elephants and rhinos are targeted for their ivory and horns (reviewed in

Y. Niu · H. Sun (✉)
State Key Laboratory of Plant Diversity and Specialty Crops, Kunming Institute of Botany, Chinese Academy of Sciences, Kunming, Yunnan, People's Republic of China
e-mail: sunhang@mail.kib.ac.cn

S. Chandra, H. Lata (eds.), *High Altitude Medicinal Plants*,
https://doi.org/10.1007/978-3-031-99406-7_25

Allendorf and Hard 2009), while bluefin tuna and sharks are harvested for their meat and fins. Sea turtles are collected for their shells. In plants, species like African Mahogany (*Khaya* spp.) and Rosewood (*Dalbergia* spp.) are overharvested for their valuable wood, used in furniture and musical instruments. Orchid and cycad species face heavy collection for horticultural purposes.

Medicinal plants are also the main targets of commercial harvesting. The international market for medicinal plants is valued at $20–40 billion annually and is projected to reach $5 trillion by 2050 (Larsen and Olsen 2007). According to recent estimates, approximately 24,000 species are used as medicine (Pironon et al. 2024), but only a little over 3200 species are cultivated commercially. In other words, most medicinal plants in the market are harvested from the wild (Brinckmann et al. 2022).

Alpine medicinal plants face particularly severe harvest pressures for several reasons. First, these plants are challenging to cultivate due to their need for specific conditions, such as cold temperatures, significant diurnal temperature ranges, high moisture, and radiation, which are difficult to replicate on a large scale. Second, there is a belief that wild alpine plants, that experienced harsh environmental conditions, possess greater efficacy and value compared to cultivated ones. Third, underdeveloped mountain regions often lack sustainable harvesting practices. Additionally, in these regions, wild harvesting is frequently a crucial income source (Timmermann and Smith-Hall 2019). Based on a study conducted in mountainous Nepal, the authors found that income from selling wild collected medicinal plant products constituted an average of 58% of the total annual household income and 78% of cash income. Medicinal plant income increased through time. Even though medicinal plant income per collection day decreased, income at the community level doubled (Timmermann and Smith-Hall 2019).

Medicinal plants from high altitudes are typically small herbs or shrubs uprooted from their bases. In the western Himalayas, the often-reported plants include *Aconitum heterophyllum*, *Aconitum violaceum*, *Allium stracheyi*, *Angelica glauca*, *Bergenia stracheyi*, *Dactylorhiza hatagirea*, *Fritillaria roylei*, *Heracleum candicans*, *Jurinea dolomiaea*, *Polygonatum verticillatum*, *Rheum australe*, *Selinum vaginatum*, *Sinopodophyllum hexandrum*, and *Trillium govanianum* (Sharma et al. 2022). In Nepal, *Nardostachys jatamansi* and *Neopicrorhiza scrophulariiflora* are also the most common targets (Ghimire 2023). *Rhodiola crenulata*, *Fritillaria delavayi*, *Saussurea laniceps*, and other congeners can also be added to this list in western China (Fig. 25.1).

Most of these plants are directly harvested from the wild. In India, 90% of the medicinal plants traded and used are collected from the wild, with a substantial portion coming from the alpine Himalayas (Goraya and Ved 2017; Uniyal 2017). Similar challenges are observed in other alpine regions such as the Andes. In the Cajamarca market of Peru, 51% of medicinal plants are sourced from the wild. The strong demand for plant products has not led to an increase in their cultivation, and the collection of some wild plants has increased the disturbance to high-altitude ecosystems, posing a threat to the sustainability of the use and market of medicinal species (Seminario et al. 2019).

Fig. 25.1 Alpine medicinal plants from Qinghai-Xizang Plateau. (**a**) *Rhodiola renulata* (Crassulaceae), (**b**) *Neopicrorhiza scrophulariiflora* (Plantaginaceae), (**c**) *Nardostachys jatamansi* (Caprifoliaceae), (**d**) *Fritillaria cirrhosa* (Liliaceae), (**e**) *Bergenia purpurascens* (Saxifragaceae), (**f**) *Meconopsis horridula* (Papaveraceae), (**g**) *Sinopodophyllum hexandrum* (Berberidaceae)

25.2 Factors Influencing the Strength of Commercial Harvest

Whether an alpine medicinal plant is suffering heavy harvesting depends on several factors.

25.2.1 Economic Value

Species with high market demand and commanding high prices are particularly susceptible to overharvesting. Market dynamics, such as supply and demand, drive this trend. Rare plants often become more valuable, which in turn attracts more harvesters seeking to capitalize on their worth. For instance, *Ophiocordyceps sinensis* and *Fritillaria cirrhosa* bulbs, both highly prized in traditional medicine, have seen significant price increases due to their rarity and high demand. The price for *Ophiocordyceps sinensis* can reach up to USD 16 (CNY 110) per single individual, while *Fritillaria cirrhosa* bulbs can cost up to USD 715 (CNY 5000) per kilogram (https://www.zyctd.com, accessed in September 2024, Fig. 25.2). Such high prices reflect their value and contribute to the intense harvesting pressure on these species. A study reported a phenomenon that the harvesting seasons of *Fritillaria* and *Nardostachys* overlap, but because *Fritillaria* is more valuable, harvesters tend to prioritize collecting it. This reduces the harvesting pressure on *Nardostachys* to some extent (Chang 2022).

25.2.2 Resource Availability

The availability of a plant in the wild plays a crucial role in determining its level of harvesting. When the population of a medicinal plant drops below a certain threshold, the difficulty of collecting it increases. This scarcity can lead to the search for alternative or substitute species, shifting the harvesting pressure to these related plants. For example, in Northwestern Yunnan, the snow lotus *Saussurea laniceps* was once heavily harvested due to its unique appearance and medicinal properties (Fig. 25.3). However, as its population declined, collectors turned to the more

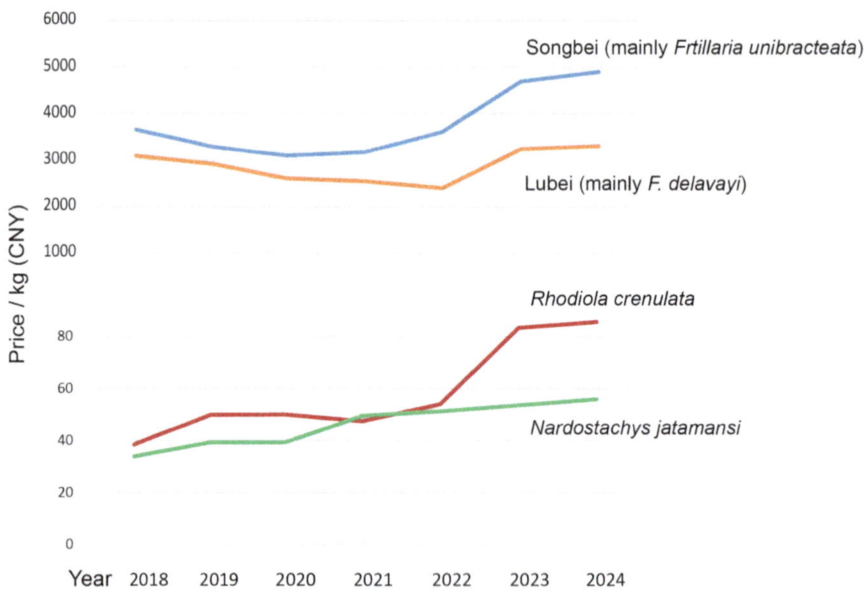

Fig. 25.2 Price change through time in four common alpine medicinal plants from Qinghai-Xizang Plateau

Fig. 25.3 Medicinal plants that are heavily harvested from Qinghai-Xizang Plateau. (**a–b**) *Saussurea laniceps* (Asteraceae), (**c**) *Fritillaria delavayi* (Liliaceae) in its habitat, (**d**) a digger left by local fritillary collectors

abundant, though less conspicuous, *Saussurea medusa*, which has now become a more common target for harvesting.

25.2.3 Accessibility

The accessibility of harvesting areas is greatly influenced by infrastructure and habitat types. Improved transportation routes, such as roads, enhance access to previously remote regions, making it easier for collectors to reach and harvest plants. For example, the expansion of road networks in mountainous areas has opened up new populations of *Bergenia ciliata* to commercial harvesting (Pyakurel et al. 2018). As a result, this species, once less accessible, has become more vulnerable to overharvesting due to enhanced collector access and increased commercial interest. Additionally, the difficulty of collection, influenced by habitat characteristics, affects collectors' willingness to harvest. For example, Li et al. (2017) examined human disturbance on the characteristics of *Fritillaria cirrhosa* bulbs in Hengduan Mountains of southwest China. Their findings revealed that the abundance of bulbs increased with distance from the main road. They also noted that bulbs located closer to the road tended to be deeper, which may indicate a difficulty-dependent collection process.

25.3 Ecological Consequences of Commercial Harvesting

25.3.1 Population Decline and Extinction

Excessive collection of medicinal plants can cause rapid population declines. If harvesting surpasses the natural regeneration rate, species may face endangerment or extinction. In Nepal, the regeneration of *Nardostachys jatamansi* can be very sensitive to harvesting. When 10% and 25% individuals in the plots were harvested, ramet density tended to increase in 2 years after harvesting. However, only at the 10% harvesting level did ramets return to pre-harvest density, after 4 years. In contrast, at 50% and 75% levels, ramet density decreased throughout the study period (Ghimire et al. 2005). This impact of harvesting differs between species. For example, according to Ghimire et al. (2005) *Neopicrorhiza scrophulariiflora* seems more tolerant than *Nardostachys jatamansi*. Negative effect on the change of ramet density was significant only at higher (>50%) harvesting levels.

Many medicinal plants are endangered due to over-exploitation. In the Western Himalaya, alpine herbs such as *Picrorhiza kurrooa*, *Aconitum heterophyllum*, *Fritillaria roylei*, etc. are commercially harvested, leading to decreased populations (Uniyal et al. 2011). The Hengduan Mountains in SW China face similar issues, with 11 plant species subject to large-scale collection; nine of these are experiencing resource depletion. The fatality rate of harvesting is 64%, rising to 78% in high-altitude areas (Ma and Zhuang 2010). It has been estimated that nearly one-third of commonly used Chinese medicinal plants are significantly declining, with 60–70% of all endangered species being medicinal (Guo 2017). On the Tibetan Plateau, about 10% of medicinal species are listed in the 'China Species Red List.' Similar situation also found in other alpine regions. For instance, *Krameria lappacea*, valued for its antioxidant and antimicrobial properties, has nearly extinct populations in Peru and Europe (Weigend and Dostert 2005). In Colombia's Andean region, local farmers are forced to climb higher for useful plants that have migrated due to climate change and overharvesting (Rodríguez et al. 2018). In Lesotho's mountainous regions, 20% of highland medicinal plants are at risk of extinction (Mugomeri et al. 2016).

25.3.2 Altered Composition and Interactions in Community

Intense harvesting significantly impacts the physical environment by damaging ground cover and altering the physical and chemical properties of the surrounding soil. Some medicinal plants play a vital role in maintaining soil health by fixing nitrogen, stabilizing soil, and contributing organic matter. The removal of these plants can lead to degraded soil quality and negatively affect the overall productivity of the ecosystem.

Commercial harvesting reduces the general biodiversity of the community. For example, Xu et al. (2013) reported that the harvesting of *Cordyceps sinensis* reduces grassland species diversity, coverage, and aboveground biomass, while increasing community dominance. Due to the harvesting activities, the Shannon-Wiener diversity index decreased by 10.3–12.5%, community coverage dropped by 10.6–19.1%, and aboveground biomass declined by 20.7–46.1%. The Berger-Parker dominance index increased by 16.7–31.8%, but no significant changes were observed in belowground biomass. These impacts may due to the interactions among plant, animal, and microbiomes, such as pollination, herbivory, and dispersal. For example, overharvesting of *Corydalis* spp. and *Rhodiola* spp. (medicinal herbs in Qinghai-Xizang Plateau) may inevitably influence the population of *Parnassius* spp. (Papilionidae), a group of alpine butterflies that feed on these plants, although such studies have not been conducted yet.

Different plant species within a community often share pollinators, leading to either dependence or competition among them. Certain species, especially generalists that provide significant rewards to various pollinators, play a vital role in maintaining the stability of the pollination network (González et al. 2010). The removal of these key species can lead to substantial changes in the network's structure, negatively impacting the reproductive success of other plants. Additionally, plants employ physical and chemical mechanisms to defend against threats such as herbivores and pathogens. The types and abundance of secondary metabolites vary among species, and at the community level, the diversity of defensive strategies is closely tied to overall species diversity. Research has shown that more diverse communities are better equipped to resist generalist herbivores (Prieur-Richard et al. 2002; Castagneyrol et al. 2014). Therefore, the loss of certain key species could weaken the ability of other community members to defend themselves against herbivores.

25.4 Evolutionary Consequences of Commercial Harvesting

25.4.1 Harvest-Driven Selection

Continuous and targeted harvesting of wild animals and plants can exert strong selection pressure on specific traits (Darimont et al. 2009). Most documented examples pertain to animals. For instance, Atlantic cod (*Gadus morhua*) populations in heavily fished areas have evolved to reach sexual maturity at younger ages and smaller sizes as a response to intense size-selective harvesting, where larger individuals are removed from the population before they can reproduce (Sharpe and Hendry 2009). Similarly, in populations of bighorn sheep (*Ovis canadensis*), individuals with smaller horns are less likely to be hunted, leading to selective pressure favoring smaller horn sizes. Over time, this has resulted in a reduction in the average size of horns in some populations (Pigeon et al. 2016).

In plants, documented examples are much rarer, potentially due to a lack of attention. A study in Western China found that over half (59.70%) of households involved in harvesting *Nardostachys jatamansi* reported a trend of decreasing quantity or smaller root sizes in recent years, attributing this mainly to human harvesting activities (Chang 2022). The snow lotus (*Saussurea* spp.), a medicinal plant native to the Qinghai-Xizang Plateau, includes the notable species *Saussurea laniceps*, recognized for its distinctive appearance and believed to treat headaches, high blood pressure, and menstrual issues. This species has been heavily harvested over time and has also been sold as a souvenir in recent decades. Law and Salick (2005) found that herbarium specimens of *S. laniceps* have gradually decreased in height over the past century. In contrast, the less conspicuous species *S. medusa* has not shown this trend. Having observed *S. laniceps* in the wild, I note that its situation may involve not only dwarfism but also increased rarity. It is now primarily found on remote cliff faces at very high altitudes. Meanwhile, *S. medusa* and other congeners are experiencing increasing harvesting pressure. If the authors' conclusions hold true, these plants may also evolve toward dwarfism in the future.

We can expect that fatal harvesting may exert stronger selection pressure than non-fatal one. Traits that direct influence survivor and reproduction may be selected more often than other traits. These include size (influence apparency and collection willingness), color (influence apparency), flowering season (influence reproduction), and seed mature duration (reproduction). For example, if plants are harvested before they have a chance to flower and set seed, their ability to reproduce and maintain population levels is compromised. Over time, this can lead to evolutionary changes where only plants that reproduce quickly.

25.4.2 Genetic Diversity

Intense harvesting can lead to a genetic bottleneck, where only a limited number of individuals contribute to the next generation. This reduction in genetic variation may result in inbreeding depression, causing offspring to be less fit and more vulnerable to diseases and environmental stresses. Such studies in alpine medicinal plants are rare, but examples in animals are common. For instance, African elephant (*Loxodonta africana*, Whitehouse and Harley 2001), Sea Otter (*Enhydra lutris*, Larson et al. 2002) and Sika deer (*Cervus nippon*, Nabata et al. 2004) have all experienced significant population size bottleneck caused by extensive hunting.

25.5 A Case Study of Commercial Harvest Induced Color Evolution in *Fritillaria delavayi*

The genus *Fritillaria* includes about 170 species (Hill 2024). Many of them are used ornamentally or medicinally. "Chuanbei" (川贝) is an important traditional medicinal material in China and neighboring regions, comprising six species, namely *F.* cirrhosa, *F. unibracteata, F. przewalskii, F. delavayi, F. taipaiensis* or *F. unibracteata* (Chinese Pharmacopoeia Commission 2010). The medicinal part used is the underground bulb, primarily for treating coughs, with active ingredients of various alkaloids. Based on species and distribution area, it is divided into three subcategories "songbei" (松贝, mainly *F. cirrhosa*) "qingbei" (青贝, mainly *F. unibracteata* and *F. prewalskii*) and "lubei" (炉贝, *F. delavayi*), with their bulbs slightly different from each other.

Fritillaria delavayi, known as "lubei," is distributed in alpine scree environments at altitudes above 3800 m, with the Hengduan Mountains being a key distribution area. At several sites we have visited, this plant exhibits significant geographical color variations. The leaves can range from green to gray to brown, while the flowers can be yellow, yellow-green, gray, or brown. The brown or gray leaves provide excellent camouflage against the rocky background.

Such camouflage has also been observed in other alpine scree plants. Our previous research indicates that the camouflage of the *Corydalis benecincta* is linked to butterfly foraging preferences. This plant is a host for the alpine Apollo butterfly (*Parnassius cephalus*), whose females use visual cues to find the host and lay eggs on nearby rocks. The larvae then feed on the food plant, often damaging the entire plant. Thus, if specific leaf colors reduce the likelihood of butterfly oviposition, these color variations may be subject to selective pressure. Niu et al. (2014) studied populations with natural leaf color dimorphism and found that camouflaged individuals had a higher survival rate than green ones, demonstrating the adaptive value of camouflage.

However, our long-term observations at various locations indicate that few herbivores consume *F. delavayi*. This plant contains toxic alkaloids, which help it resist generalist herbivores. Since *Fritillaria* is a key source of traditional medicine in the region and faces heavy harvesting (Cunningham et al. 2018), we frequently see groups of local people, equipped with basic tools, searching for and collecting *F. delavayi*. This situation leads us to consider whether commercial harvesting could impact the evolution of Fritillaria's coloration. We base this on several factors: (1) Humans are visual animals, relying on visual cues for target identification, similar to natural selection; (2) the harvesting pressure is intense: we measured samples and found an average bulb dry weight of 0.28 g, meaning over 3500 plants are needed to obtain 1 kg of bulbs, which is a much higher selection pressure than natural forces; and (3) the duration of selection is significant: Fritillaria has been used in China for over 2000 years, with large-scale harvesting spanning more than 150 years, providing sufficient time for rapid evolution under strong selection pressure. Field observations reveal that camouflage varies by location and harvesting intensity. Thus, we can investigate if the efficacy of camouflage correlates with harvesting intensity. If our hypothesis is correct, we would expect better camouflage in areas with intense harvesting.

To address this, we quantified both camouflage efficacy and harvesting intensity. Camouflage efficacy measures the color difference between the target and its background, determined by analyzing their reflected spectra through a human color vision model. A smaller color difference indicates better camouflage. Quantifying harvesting intensity is more challenging; while regional or national supply data can be obtained, it only reflects general trends. Therefore, we focused on tracking the purchasing volume of Fritillaria (locally known as "Zhimu") at the village level. Since each village manages its own area, village-level purchasing data provides a more precise measure of local harvesting. This data is usually in kilograms, but bulb size varies between regions, so we also assessed the dry weight of bulbs from each mountain to estimate annual harvest quantities. Harvesting intensity depends not just on the absolute amount of harvesting (numerator) but also on the total resource available locally (denominator). To estimate resource totals, we established plots on each mountain, counted as the individual numbers of *F. delavayi*, and estimated suitable habitat area using satellite maps. This allowed us to estimate resource totals and calculate harvesting intensity as the ratio of harvesting volume to total resources. Additionally, we also evaluated actual harvesting intensity by marking plots and measuring the proportion of harvested individuals, which ranged from 1% to 83% and aligned with the results based on the previous methods (Fig. 25.4). These results support our hypothesis that higher harvesting intensity correlates with better camouflage.

Since we can only obtain harvest data from recent years, we cannot determine if the collection intensity for each group has been historically stable. If there were a parameter reflecting collection intensity that is independent of historical data, we could further test this conclusion. We considered collection difficulty as a potential parameter. Due to significant variations in rock types across different hilltops, the

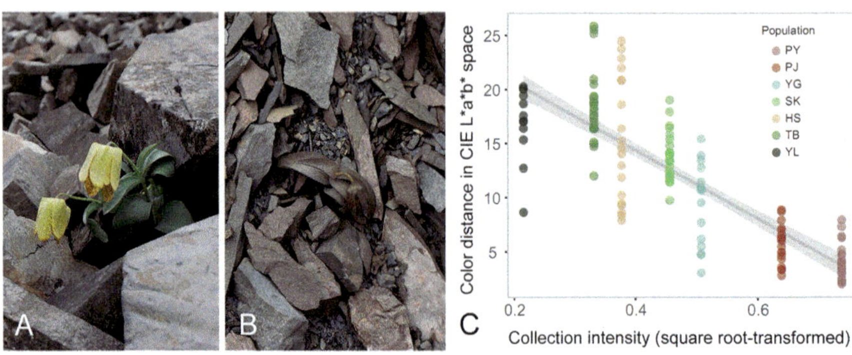

Fig. 25.4 Geographical color variation in *Fritillaria delavayi* and its correlation with collection intensity. (**a**) Individuals with green leaves and yellow flowers in NW Yunnan province, SW China. (**b**) Individuals with leaf and flower colors that match the rocks. (**c**) Correlation between camouflage efficacy (indicated by color distance between leaf and rock in L*a*b* color space) and collection intensity. (Transformed from Niu et al. 2021)

size and arrangement of rocks in scree slopes differ widely, resulting in substantial variations in excavation difficulty. Field experience indicates that this difficulty can be measured by the time required to collect a single bulb (where each specimen has only one bulb): some take only a few seconds, while others take several minutes. The results align with expectations: groups that are easier to excavate have better camouflage, while those that are harder to collect have poorer camouflage.

The results suggest that commercial harvest, as a major selective pressure, likely affects the phenotypic evolution of this medicinal plant. Thus, we further explored the specifics of this selection process. Ideally, we would follow the local collectors to see if their specific behaviors directly impact the color phenotype. However, numerous uncontrollable variables in real-world conditions, such as weather, timing, and experience, could influence the results. Moreover, studying the excavation process in the field might inadvertently encourage excavation behavior, which we want to avoid.

Based on a large number of field-captured images, we developed an online visual game called "Spot the plant" to simulate human search behavior. In the game, players are tasked with locating specific targets (*F. delavayi*) and can choose their role—either as a human (trichromatic vision) or a yak (dichromatic vision)—and provide their experience level (whether they have played the game before). The game presents players with 14 randomly chosen field images, each containing a Fritillary hidden in a rock crevice (with its prominence evaluated by the Salience value, Pike, 2018). Players must find the targets as quickly as possible, and the backend records the search time. We then analyzed the relationship between search time and the Fritillary's salience. Results showed that better camouflage increased the search time.

While color itself may not be a primary trait for excavators, it impacts the excavation process or survival probability and thus is subject to selection. Commercial harvest exerts strong selective pressure, significantly affecting the evolution of this alpine medicinal plant.

25.6 Conclusions

The habitat in alpine regions is highly fragile, and biodiversity faces significant threats. The commercial harvesting of alpine medicinal plants can lead to population declines or even species extinction. This alters habitats and disrupts species interactions, ultimately affecting the survival and reproduction of various organisms. Furthermore, commercial harvesting serves as an often-overlooked selective pressure that can influence the phenotypes of wild plants, resulting in profound evolutionary consequences.

During our literature review, we noted that while human utilization of plants is widespread, research on plant species is significantly less extensive compared to studies on animals. Additionally, although there are many alpine regions worldwide, literatures on the harvesting and utilization of alpine medicinal plants that we can get are largely confined to the Himalayas. This may partly because the language bias, i.e., studies in other alpine regions are not written in English. Nevertheless, we believe that similar challenges exist in alpine areas across Asia, Africa, and America. In these regions, rich biodiversity confronts relatively underdeveloped economical condition, posing challenges for biodiversity conservation.

Based on the findings from studies, recommendations can be made regarding the conservation and utilization. On one hand, it is advisable for government agencies to enhance their management efforts. Our previous research on *Fritillaria* (Niu et al. 2021) has garnered attention from both the public and policymakers. We are encouraged to see that all wild species of the *Fritillaria* genus have been classified as protected species in the updated national key protected wild plant list of 2021 in China. On the other hand, it is essential to recognize that the use of alpine medicinal plants is a crucial part of traditional culture and knowledge, holding potential benefits for humanity. By changing harvest model, harnessing the unique mountain climates, developing local breeding and cultivation practices, expanding utilization avenues, and boosting the incomes of local residents, we may reduce the harvesting of wild alpine medicinal plants and discover sustainable approaches to their conservation.

References

Allendorf FW, Hard JJ (2009) Human-induced evolution caused by unnatural selection through harvest of wild animals. Proc Natl Acad Sci USA 106:9987–9994

Brinckmann JA, Kathe W, Berkhoudt K, Harter DEV, Schippmann U (2022) A new global estimation of medicinal and aromatic plant species in commercial cultivation and their conservation status. Econ Bot 76:319–333

Castagneyrol B, Jactel H, Vacher C, Brockerhoff EG, Koricheva J (2014) Effects of plant phylogenetic diversity on herbivory depend on herbivore specialization. J Appl Ecol 51:134–141. https://doi.org/10.1111/1365-2664.12175

Chang W (2022) An ecological study of the plateau medicinal plant *Nardostachys Jatamansi*. Dissertation, Chongqing University

Chinese Pharmacopoeia Commission (2010) Pharmacopoeia of the People's Republic of China. China Medical Science Press

Cunningham AB, Brinckmann JA, Pei SJ, Luo P, Schippmann U, Long X, Bi YF (2018) High altitude species, high profits: can the trade in wild harvested *Fritillaria cirrhosa* (Liliaceae) be sustained? J Ethnopharmacol 223:142–151

Darimont CT, Carlson SM, Kinnison MT, Paquet PC, Reimchen TE, Wilmers CC (2009) Human predators outpace other agents of trait change in the wild. Proc Natl Acad Sci USA 106:952–954

Di Minin E, Brooks TM, Toivonen T, Butchart SHM, Heikinheimo V, Watson JEM et al (2019) Identifying global centers of unsustainable commercial harvesting of species. Sci Adv 5:eaau2879

Ghimire SK, McKey D, Aumeeruddy-Thomas Y (2005) Conservation of Himalayan medicinal plants: harvesting patterns and ecology of two threatened species, *Nardostachys grandiflora* DC. and *Neopicrorhiza scrophulariiflora* (Pennell) Hong. Biol Conserv 124:463–475

Ghimire NP, Poudeyal MR, Gurung A, Poudel MR, Bhattarai HD, Ghimire SK (2023) Traditional socio-cultural utilization and harvesting practices of an alpine medicinal herb, *Neopicrorhiza scrophulariiflora* in Khumbu Valley, Nepal. Ethnobotany Research and Applications 26:1–15

González AMM, Dalsgaard B, Olesen JM (2010) Centrality measures and the importance of generalist species in pollination networks. Ecol Complex 7:36–43

Goraya, G.S., Ved, D.K., 2017. Medicinal plants in India: an assessment of their demand and supply. National Medicinal Plants Board, Ministry of Ayush, Government of India, New Delhi and Indian Council of Forestry Research & Education, Dehradun

Guo Q (2017) Resources of medicinal plants. Higher Education Press. (in Chinese)

Hill L (2024) Fritillaria: a list of published names. version 5.7. Available at http://www.fritillariaicones.com/info/names/frit.names.pdf. Last accessed September 2024

Larsen HO, Olsen CS (2007) Unsustainable collection and unfair trade? Uncovering and assessing assumptions regarding Central Himalayan medicinal plant conservation. Biodivers Conserv 16:1679–1697

Larson S, Jameson R, Bodkin J, Staedler M, Bentzen P (2002) Microsatellite DNA and mitochondrial DNA variation in remnant and translocated sea otter (*Enhydra lutris*) populations. J Mammal 83(3):893–906

Law W, Salick J (2005) Human-induced dwarfing of Himalayan snow lotus, *Saussurea laniceps* (Asteraceae). Proc Natl Acad Sci USA 102:10218–10220

Li X, Liu L, Gu X, Xiang J (2017) Heavy collecting induces smaller and deeper Fritillariae Cirrhosae Bulbus in the wild. Plant Divers 39:208–213

Ma J, Zhuang H (2010) From valley to alpine mountain: diversity and utilization of Tibetan medicinal plants in Deqin. Acta Bot Yunnanica 32:67–73. (in Chinese, with English abstract)

Mugomeri E, Chatanga P, Raditladi T, Makara M, Tarirai C (2016) Ethnobotanical study and conservation status of local medicinal plants: towards a repository and monograph of herbal medicines in Lesotho. Afr J Tradit Complement Altern Med 13:143–156

Nabata D, Masuda R, Takahashi O (2004) Bottleneck effects on the sika deer Cervus nippon population in Hokkaido, revealed by ancient DNA analysis. Zool Sci 21:473–481

Niu Y, Chen G, Peng D-L, Song B, Yang Y, Li Z-M, Sun H (2014) Grey leaves in an alpine plant: a cryptic colouration to avoid attack? New Phytol 203:953–963

Niu Y, Stevens M, Sun H (2021) Commercial harvesting has driven the evolution of camouflage in an alpine plant. Curr Biol 31:446–449

Pigeon G, Festa-Bianchet M, Coltman DW, Pelletier F (2016) Intense selective hunting leads to artificial evolution in horn size. Evol Appl 9:521–530

Pironon S, Ondo I, Diazgranados M, Allkin R, Baquero AC, Cámara-Leret R et al (2024) The global distribution of plants used by humans. Science 383:293–297

Prieur-Richard AH, Lavorel S, Linhart YB, Dos Santos A (2002) Plant diversity, herbivory and resistance of a plant community to invasion in Mediterranean annual communities. Oecologia 130:96–104

Pyakurel D, Bhattarai Sharma I, Smith-Hall C (2018) Patterns of change: the dynamics of medicinal plant trade in far-western Nepal. J Ethnopharmacol 224:323–334

Rodríguez MA, Angueyra A, Cleef AM, Van Andel T (2018) Ethnobotany of the Sierra Nevada del Cocuy-Güicán: climate change and conservation strategies in the Colombian Andes. J Ethnobiol Ethnomed 14:1–12

Seminario J, Zumaeta SBE, Cunya AS (2019) Status of research on medicinal plants in the Cajamarca's region, Peru. In Ethnobotany (pp. 70–89). CRC Press.

Sharma R, Kaur S, Uniyal SK (2022) Population and vulnerability assessment of high value medicinal plants in the alpine regions of western Himalaya. J Appl Res Med Aroma 26:100353

Sharpe DMT, Hendry AP (2009) Life history change in commercially exploited fish stocks: an analysis of trends across studies. Evol Appl 2:260–275. https://doi.org/10.1111/j.1752-4571.2009.00080.x

Timmermann L, Smith-Hall C (2019) Commercial medicinal plant collection is transforming high-altitude livelihoods in the Himalayas. Mt Res Dev 39:R13–R21

Uniyal SK (2017) Classifying medicinal plant collectors: their approach and attitude. Curr Sci 113:2085–2089

Uniyal A, Uniyal SK, Rawat GS (2011) Commercial extraction of *Picrorhiza kurrooa* Royle ex Benth. in the Western Himalaya. Mt Res Dev 31:201–208

Weigend M, Dostert N (2005) Towards a standardization of biological sustainability: wildcrafting Rhatany (*Krameria lappacea*) in Peru. Med Plant Conserv 11:24–27

Whitehouse AM, Harley EH (2001) Post-bottleneck genetic diversity of elephant populations in South Africa, revealed using microsatellite analysis. Mol Ecol 10:2139–2149

Xu Y, Xu C, Zhai Y, Chao S, Du J, Liu C, Zhang L (2013) The impact of *Cordyceps sinensis* excavation on grassland vegetation in the Sanjiangyuan region. Res Environ Sci 26:1194–1200. (in Chinese, with English abstract)

Correction to: Elevated Remedies: The Botany, Conservation, and Cultivation of Saffron and Native New Mexican Herbs

Saeid Zehtab Salmasi, Connie M. Maxwell, Kevin Lombard, and Ivette Guzman

Correction to:
Chapter 21 in: S. Chandra, H. Lata (eds.),
High Altitude Medicinal Plants,
https://doi.org/10.1007/978-3-031-99406-7_21

The original version of Chapter 21 was inadvertently published without the funding information. The following information has now been included in the end of the chapter.

Funding: This research was supported by the intramural research program of the U.S. Department of Agriculture, National Institute of Food and Agriculture, Organic Transitions, Award No 2022-51106-38025. The findings and conclusions in this book chapter have not been formally disseminated by the U.S. Department of Agriculture and should not be construed to represent any agency determination or policy.

The updated version of this chapter can be found at
https://doi.org/10.1007/978-3-031-99406-7_21

S. Chandra, H. Lata (eds.), *High Altitude Medicinal Plants*,
https://doi.org/10.1007/978-3-031-99406-7_26